Nanoimaging - Future of Precision Medicine

Sikandar Shaikh

Nanoimaging - Future of Precision Medicine

Springer

Sikandar Shaikh
Department of Radiodiagnosis
Shadan Institute of Medical Sciences
Hyderabad, India

ISBN 978-981-97-7291-9 ISBN 978-981-97-7292-6 (eBook)
https://doi.org/10.1007/978-981-97-7292-6

This Springer imprint is published by the registered company Springer Nature Singapore Pte Ltd.
The registered company address is: 152 Beach Road, #21-01/04 Gateway East, Singapore 189721, Singapore

This book is dedicated to
Almighty God whose blessings are always there
My parents
My wife
My daughters
My mentors
My teachers
My well-wishers
My friends and colleagues
My patients because of whom I learnt the art of medicine and still learning

Foreword 1

Perelman School of Medicine
University of Pennsylvania

Radiology

Abass Alavi, M.D., MD (Hon), PhD (Hon), DSc (Hon)
Professor of Radiology and Neurology
Director of Research Education

It is my great pleasure to write this foreword for Dr. Sikandar Shaikh's book *Nanoimaging—Step Towards Precision Medicine*. Nanoimaging is growing fast as a modality of molecular imaging and will be revolutionizing this domain for the foreseeable future. This innovation has led to enhancing the concept of "Personalised Medicine" which will be very effective for early detection and management of various diseases and disorders. In recent years, many applications of nanotechnology have been developed and improved by technological innovations, and this has become a handy futuristic tool for the practice of medicine. The role of nanoimaging is not only for the management of cancer and neurological disorders, but also in cardiovascular diseases, infections, and inflammation and many more. Nanoimaging provides better sensitivity and specificity for various disease processes, leading to higher precision and accuracy. In this comprehensive book about the evolution of the nanoimaging in medicine, different imaging techniques that have been enhanced by this technology are described, such as conventional ultrasound to high-end molecular imaging modalities. This is possible due to the varied nanoparticles that are used, including dendrimers, nanotubes, nanocages and many more types. The role of the various metals is also important for the specific applications, for example, the use of gold nanoparticle in CT imaging. In this book, various aspects of nanoimaging are included in different chapters such as its safety and risks, the use of nanorobots, as well as its translation from preclinical imaging to therapeutic applications in patients for the purposes of diagnosis and theranostics.

I have known Dr. Sikandar Shaikh as a distinguished colleague and an academician par excellence in our discipline. He is currently involved as a major contributor to many international organizations. He is the Past President of the ISMRM India Chapter, Indian Ambassador of the European Society of Hybrid Imaging, Committee member of the World Molecular Imaging Society, Society of Pediatric Radiology, ISMRM International, European Society of Onco-imaging, Board of Director of the Asian Society of Magnetic Resonance in Medicine and various National Indian Radiological subspecialties. He is Associate Editor of JMRI, Advisory Editor of Clinical Radiology, and reviewer of AJR, BIR, European Radiology, JMRI, Radiology,

RadioGraphics, Radiology Cardiothoracic Imaging and many more. He has made many presentations in various international and national societies. Therefore, with his expertise and calibre, this book will be a great source of knowledge of nanoimaging for clinicians, radiologists, nuclear physicians, researchers, biomedical engineers, biotechnology scientists and residents across various specialties. For his efforts on this wide range of topics involving nanoimaging, I wish Dr. Shaikh all the best for publishing his third book and in his future endeavours.

University of Pennsylvania
Philadelphia, PA, USA
January 31, 2024

Abass Alavi

Foreword 2

Empowered by the concepts of evidence-based medicine and personalized therapy, diagnostic imaging developed into a pillar of therapy guidance in medical care. The evolution of imaging exponentially progressed from planar morphological imaging, over three-dimensional high-resolution imaging towards the non-invasive assessment of physiological tissue properties by functional imaging and the non-invasive characterization of tissues on a cellular level by molecular imaging. Nanoimaging in particular allows for the visualization of nanoscale pathophysiology of tissue, cells, and biomolecules and a direct observation of nanostructural details of cells, surface proteins, and DNA. Increasing knowledge of the complex pathophysiology of disease in oncology, neurology, cardiology, and inflammation provides the basis for individualized therapy concepts that rely on biomarkers for patient stratification and early therapy response assessment. Multimodality nanoimaging concepts include MRI, CT, and ultrasound as well as PET, SPECT, and advanced methods of microscopy and spectroscopy to characterize tissue on an ever more detailed level in vivo. Nanoimaging target characterization also contributed significantly to the development of theranostic concepts in oncology, where the same biological structure (e.g., membrane receptors) is targeted for diagnostic imaging and molecular therapy. With accelerated clinical translation of molecular therapy concepts and the rise of competing diagnostic methods, such as liquid biopsy, the role of medical imaging for personalized therapy guidance undergoes rapid transformation. While the characterization of tumor clonal heterogeneity may be dominated by liquid biopsy, precise localization of metastatic disease is highly relevant for localized therapy concepts and optimization of survival in oligometastatic disease. Nanoimaging concepts drive the understanding of pathophysiology and redefine the future role of medical imaging for individualized patient care. This third book of Dr. Sikandar Shaikh on Nanoimaging provides a comprehensive overview and proficient insights into the current status of this energetic and transformational field in medical imaging.

Ludwig-Maximilians University Munich
Munich, Germany
January 31, 2024

Clemens Cyran

Preface

Nanomedicine is a newer terminology that is being used in medicine based on the nano-enabled technique which is being used for the prevention, diagnosis, monitoring, and management of various diseases. This potential has led to having a significant role in imaging, various imaging protocols, various drug delivery systems, different engineering aspects, and various therapeutic options comprising theranostics. This has revolutionized the practice of medicine and imaging which includes several pathologies comprising of cardiovascular diseases, cancers, psychiatric and neurodegenerative diseases, various bacterial and viral infections, and diabetes. The most important step in the disease is to have a fast and accurate diagnosis with precise accuracy. The use of nanoimaging has led to a more specific non-invasive imaging technique that can use various nanomaterials that can diagnose diseases at the cellular level. Many high-precision imaging nanomaterials are being used with the technological advances of various imaging modalities for better evaluation and diagnosis of diseases. The therapeutic component theranostics is also being used for the precise management of various conditions at the cellular level.

This book is important for basic clinicians and all specialties who are involved in nanoimaging like basic sciences, chemistry, biology, engineering, imaging, nuclear medicine, various other subspecialty branches of robotic engineering, etc.

This book gives a wide spectrum of information and scientific content from the nanotechnology basics, nanomaterials, different nanomaterial applications, and modality-wise roles of ultrasound, CT, MRI, PET-CT, nuclear medicine, SPECT-CT, PET-MR, and all modalities are covered.

Along with this, the application of nanoimaging in different systems and diseases is also discussed in detail in cancer, neurology, cardiology, infection and inflammation, theranostics, and in a wide range of applications.

Other important topics that are covered are nanorobots, in vivo nanoimaging, preclinical imaging, nano spectroscopy imaging, and nano safety are covered in detail.

Each chapter is meticulously described, and the different applications are explained in detail.

After reading this book, it allows readers to understand the value of nanoimaging in various diseases like various cancers, cardiac diseases, and neurological disorders including neurodegenerative diseases, various infections, inflammations, and many more pathologies.

The most important aspect that it overviews is the role of nanoimaging in preclinical and in vivo imaging whereas the role of conventional imaging modalities is very less.

The role of robotics by the use of nanoimaging is explained in detail. This gives the perfect blend of the various aspects of the robotic role in imaging.

Safety is an important aspect in any imaging modalities which can be the toxicity or radiation aspects; all these are explained in detail.

With all this content *Nanoimaging—A Step Towards Precision Medicine* is justified and also this book will act as the review book for multiple specialties that deal with various diverse applications.

The use of ultrasound, computed tomography, nuclear medicine, magnetic resonance imaging, PET-CT, PET-MR, and various hybrid and molecular imaging modalities are well established due to ongoing biochemical and medical research. The nanoimaging technologies are developing better and more powerful contrast agents by using all the imaging modalities, as these nanomaterials have less toxicity, more and better permeability, and retention effects in various tissues. Another advantage is the size of the nanoparticles which have good biodistribution, blood circulation half-life, cellular uptake, tissue penetration, and targeting cells. Due to this, nanoimaging has become an important imaging component which is leading to the concept of personalized imaging and ultimately precision medicine.

Hyderabad, India Sikandar Shaikh

Acknowledgments

Dr. Abas Alavi, MD, MD(Hon), PHD(Hon), DSc(Hon)
University of Pennsylvania, Philadelphia, PA, USA
Prof. Clemens Cyran
Professor of Radiology
Section Chief Hybrid and Molecular Imaging and Vice Chair at the Department of Radiology, Ludwig-Maximilians University Munich, Munich, Germany
Dr. Prabhakar Reddy, MD
Vice President, Asian Musculoskeletal Society, AMS
Past President AOCR and National IRIA
Dr. Anand Abkari, MD(RD), DMR, FICR, FAAMS
Professor Emeritus
Deccan College of Medical Sciences, Hyderabad, India
Prof. S. Rammurthi, MD
Professor and HOD Radiology
Nizam's Institute of Medical Sciences, Hyderabad, India
Dr. Hemant Patel, MD
Treasurer AOSR
Immediate Past President National IRIA
International Academic Coordinator
Director, Gujarat Imaging Centre, Ahmedabad, India
Dr. Arvind Chaturvedi
Prof. and Head Radiology
Rajiv Gandhi Cancer Institute, New Delhi
Prof. T. Mandapal
HOD—Radiology
Care Hospitals, Banjara Hills, Hyderabad
Prof. C. Amarnath
Professor and HOD
Past President IRIA
Stanley Medical College, Chennai, India
Dr. Suresh Kawathekar
Retd. Professor and HOD—Radiology
MNR Medical College, Sanga Reddy, Hyderabad, India

Contents

About the Author

Sikandar Shaikh, DMRD, DNB, EDiR, MNAMS, FICR, graduated from B R Ambedkar Marathwada University, Aurangabad, Maharashtra, India. He is an adjunct professor at the Department of Biomedical Engineering, Indian Institute of Technology. He is a professor in the Department of Radiology at Shadan Institute of Medical Sciences, Hyderabad and adjunct faculty at Kasturba Medical College, Manipal, India. He has many international publications and has delivered more than 126 oral and poster presentations at various international meetings. He has authored two books *PET-CT in Infection and Inflammation* and *Advances in Imaging—A Step Towards Precision Medicine* and authored several book chapters. Dr. Sikandar was the immediate past president of ASMRM and ISMRM India Chapter. He is also the Indian ambassador of the European Society for Hybrid, Molecular and Translational Imaging (ESHI), an education committee member of the World Molecular Imaging Society (WMIS), an education committee member and MRI committee member of International Society for Pediatric Radiology (SPR), vice president of Indian Society of Neuroradiology (ISNR), vice chairman of the Indian College of Radiology and Imaging (ICRI), ethics committee member at IIT Hyderabad, associate editor of JMRI (*International Journal of MRI*), consulting editor of *Clinical Radiology*, mentor European Society of Head and Neck Imaging (ESHI), joint secretary of Society of Oncoimaging India (SOII), membership committee member of the European Society of Cardiovascular Imaging (ESCI). He is also the reviewer of the RSNA, ISMRM, ASMRM, and other meetings. He was awarded the best radiologist award in India in 2018 and 2020. He is also instrumental in conducting record daily non-stop teaching programs for ICRI.

Dr. Sikandar is the reviewer of many international radiology journals, such as *Radiology*, *Radiographics, Radiology Cardiothoracic Imaging, Radiology Cancer Imaging, European Radiology, American Journal of Roentgenology, Frontiers in Radiology, Frontiers in Nuclear Medicine,* and *Indian Journal of Radiological Imaging.*

Abbreviations

^{111}I	111 Indium
^{13}C	13 Carbon
^{18}O	18 Oxygen
^{1}H	1 Hydrogen
3D	3 Dimensional
^{64}Cu	64 Copper
^{90}Y	90 Yttrium
^{99m}Tc	99m Technetium
ACS	Acute coronary syndrome
AD	Alzheimer's disease
AFMIR	Atomic force microscopy infrared
AIDS	Acquired immunodeficiency disease
$Al2O_3$	Aluminum oxide
ALARA	As low as reasonably achievable
AMF	Alternating magnetic field
ARF	Acoustic radiation force
As	Arsenic
Au	Gold
AuNP	Gold nanoparticle
BBB	Blood–brain barrier
Bi	Bismuth
BOLD	Blood oxygen level dependent imaging
C3F8	Octafluoropropane
CA	Contrast agents
CAA	Cerebrovascular amyloid angiopathy
CAM	Cytokine-induced adhesion molecules
Cd	Cadmium
CD	Cluster of differentiation
CDs	Carbon quantum dots
CdSe	Cadmium selenic
CdTe	Telluric cadmium
CL	Cerenkov luminescence
CLIO	Cross-linked iron oxide
CMRI	Contrast MRI
CMT	Carrier-mediated transport
CNS	Central nervous system
CNT	Carbon nanotubes

Co	Cobalt
COL	Chitosan oligosaccharide lactate
COVID-19	Coronavirus disease 19
CPP	Cell-penetrating peptides
CRET	Cerenkov resonance energy transfer
CrO	Chromium oxide
CT	Computed tomography
Cu	Copper
Cucl	Copper chloride
CuO	Copper oxide
CuS	Copper sulfide
CuSe	Copper and selenium
CW	Continuous wave
Cy3	Cyanine 3
Cy5	Cyanine-5
DCE	Dynamic contrast imaging
DF	Desferrioxamine
DLS	Dynamic light scattering
DMARDs	Disease-modifying antirheumatic drugs
DMSO	Dimethyl sulfoxide
DNA	Deoxyribonucleic acid
DOTA	1,4,7,10-tetraazacyclododecane-1,4,7,10-tetraacetic acid
DOX	Doxorubicin
DPPA-TPA NP	Diketopyrrolopyrrole–triphenylamine organic nanoparticles
DPPC/DSPC	1,2-dipalmitoyl-*sn*-glycero-3-phosphocholine
DTPA	Diethylenetriaminepentaacetic acid
DWI	Diffusion-weighted imaging
EGFR	Epidermal growth factor receptor
ELIPs	Echogenic immunoliposome
EM	Electron microscopy
EPR effect	Enhanced permeability and retention effect
ERHNPs	Electro responsive hydrogel NPs
ESC	Embryonic stem cells
EV	Extracellular vehicles
F MRI	Functional MRI
FA	Folic acid
FDA	Food and Drug Administration
Fe	Ferrous
Fe_3O_4	Ferrous oxide
FeCo	Ferrous cobalt
FeO	Ferrous oxide
FIONs	Ferrimagnetic iron oxide nanoparticles
FR	Folate receptor
FRET	Fluorescence resonance energy transfer
FSGS	Focal segmental glomerulosclerosis
Ga	Gallium
Gd	Gadolinium
Ge	Germanium

GLC	Glucocorticoid
GLUT	Glucose transporters
GNPs	Gold nanoparticles
GO	Graphene oxide
GPC	Glypican
HEP3	Hepatitis genotype 3
HER	Human epidermal growth factor
HIFU	High frequency ultrasound
HIV	Human immunodeficiency virus
HMOS	Human milk oligosaccharides
HMPAO	Hexamethylpropylene amine oxime
HPG	Hyperbranched polyglycerols
HU	Hounsfield unit
I	Iodine
ICAM	Intercellular Adhesion Molecule
ICG	Indocyanine green
IGF	Insulin-like factors
IgG	Immunoglobulin
IL	Interleukin
In	Indium
IONPs	Iron oxide nanoparticles
IR s-SNOM	Infrared scattering-type scanning near-field optical microscopy
KCK	Kidney clearance kinetics
Kvp	Kilo voltage
LAT1	L-type amino acid transporter
LDL	Low-density lipoprotein
Le	Lutetium
LHRH	Luteinizing hormone-releasing hormone
LNC	Lipid nanocapsules
LNs	Lymph nodes
LSD	Light-sensitive drugs
MBs	Microbubbles
MCT	Micro-computed tomography
MDIO	Modified nanoparticulate
MDR	Multidrug resistance
MI	Mechanical index
MMPs	Matrix metalloproteinases
Mn	Manganese
MnO_2	Manganese oxide
MPA	Mercaptopropionic acid
MPO	Myeloperoxidase
MPS	Mononuclear phagocyte system
MR FM	MR force microscopy
MRI	Magnetic resonance imaging
MSOT	Multi-spectral optoacoustic tomography
MTB	Mycobacterium
MTX	Methotrexate

Na2S	Sodium sulfide
Nano CT	Nano-computed tomography
Nano MRI	Nano-Magnetic Resonance Imaging
NBs	Nanobubbles
NBs	Nonbinding surface
NFTs	Neurofibrillary tangles
NIR	Near-infrared dyes
NIR light	Near-infrared light
NIR	Near-infrared
NIRF	Near-infrared fluorescence
Nm	Nanometer
NMR	Nuclear magnetic resonance
NNMs	Nanomedicines
NO	Nitrous oxide
NOTA	N-(2-[18F]fluoropropionyl)-L-glutamate ([18F]FPGLU)
NPs	Nanoparticles
NS	Nanosystems
NSAIDs	Nonsteroidal anti-inflammatory drug
NSCs	Neural stem cells
O_2	Oxygen
p DNA	Plasmid DNA
PAI	Photoacoustic imaging
PAMAM	Polyamidoamine
PASP	Polyaspartic acid
PASPIO	Poly aspartic coated iron oxide
PAT	Photoacoustic tomography
PC	Prostate C
PCA	Principle component analysis
PDT	Photodynamic therapy
PEG	Polyethylene glycol
PEI	Photoelectric imaging
PEI	Polyethylenimine amine
PEO	Polyethylene oxide
PET-CT	Positron emission tomography and computed tomography
PET-MR	Positron emission tomography and magnetic resonance imaging
PFOB	Perfluorooctyl bromide
pH	Potential of Hydrogen
PiFM	Photon-induced force microscopy
PL	Persistence luminescence
PLGA	Poly Lactic-co-Glycolic Acid
PLNPs	Persistent luminescence particles
PLPA	Paclitaxel-loaded polymeric nanoparticles
PnP	Peak negative pressure
PS	Presenilin
PTT	Partial thromboplastin time
PTT	Photothermal therapy
PTX	Paclitaxel

RA	Rheumatoid arthritis
RBC	Red blood cells
Re	Rhenium
RECIST	Response evaluation criteria in solid tumors
RES	Reticuloendothelial cell
RF pulse	Radiofrequency pulse
RGD	Arginine-glycine-aspartic acid
RGD peptide	Arginine-glycine-aspartic acid peptide
RNA	Ribonucleic acid
RNS	Reactive nitrogen species
ROS	Reactive oxygen species
ROS	Redox oxidative stress
RT	Radiotherapy
SARS	Severe acute respiratory syndrome
SDT	Self-determination theory
Se	Selenium
SERS	Surface-enhanced Raman scattering
SERS	Surface-enhanced Raman Spectroscopy
SF6	Sulfur hexachloride
SiNPs	Silica nanoparticles
SiO_2	Silicon dioxide
siRNA	Small interfering RNA
SLN	Sentinel lymph nodes
SMP	Super magnetic
SnO_2	Stannic oxide
SiNP	Silica coated Nanoparticles
SNV	Silica Nano Vehicle
SPECT-CT	Single-photon emission computed tomography and computed tomography
SPIO	Superparamagnetic iron oxides
TALEN	Transcription activator-like effector nuclease
TAMs	Tumor-associated macrophages
TCL-SPION	Thermally cross-linked superparamagnetic iron oxide nanoparticle
TCR	T-cell receptor cells
TEM	Transmission electron microscopy
TEM	Transmission electron microscope
TENG	Triboelectric nanogenerator
TiO_2	Titanium dioxide
TNPs	Theranostic nanoparticles
TRPV 1	Temperature-sensitive transient receptor potentials
UCNPs	Upconversion nanoparticle
US	Ultrasound
UV	Ultraviolet
VCAM	Vascular cytokine induced adhesion molecules
w/cm	Weight by centimeter
WINCS	Wireless neurotransmitter syndrome
ZFN	Zinc finger nuclease

Zn	Zinc
ZnO	Zinc oxide
Zr	Zirconium

Overview of the Nanoimaging 1

1.1 Introduction

The human body comprises various cells with different functions based on unique biochemical components and structures. The amalgamation of these functions is assessed to study cells and organs. The different physiological components of the cells form various biological structures and components, requiring precise physiological functions. To better understand these different biological processes in the cells, various biological outcomes can be evaluated. Thus, nanoimaging helps in a better understanding of molecular biology, made possible by the specific structures and functions of cells. Based on these functions, various new biomarkers are being developed using nanoparticles. Nanoimaging is now more widely used compared to conventional anatomical imaging. It enables molecular-level applications like sampling, labeling, and identifying molecular derangements in cells. Consequently, nanoimaging has greater sensitivity and specificity in evaluating molecular biological structures.

Nanoscience has become a promising specialty, advancing through the use of nanomaterials that range from 1–100 nm. Based on this, various nanoparticles are prepared and used in various applications [1–3]. Different properties of nanomaterials, such as optical, electrical, and/or magnetic properties, are being evaluated. The advantage of nanoparticles is that they are very small in size but have a larger surface area to volume ratio, which enables newer applications. Quantum mechanics laws govern these particles rather than the conventional laws of physics and chemistry, facilitating their functional applications at the molecular or atomic scale [4]. Nanotechnologies are considered better, safer, cleaner, and long-lasting for various applications [2]. Nanomaterials have diverse uses, such as adding nanoparticles for better performance, like nanocrystals [5]. Non-medical applications include sunscreens, everyday cosmetics, sporting goods, and various electronic items [5]. Nanotechnology is also employed in diagnosis, imaging, and targeted drug delivery. The functional properties of nanomaterials are being optimized to reduce costs and improve healthcare [6].

These nano-enabled imaging techniques can be used for various purposes, including diagnosing disease, monitoring, and managing different conditions [7]. Newer techniques employ different imaging modalities for diagnosis, drug delivery to targeted tissues, tissue engineering, implants, and various pharmaceutical applications [8]. Nanoimaging is now applied to a range of diseases, such as neurological, cardiovascular, musculoskeletal, psychiatric, neurodegenerative diseases, and infectious and inflammatory conditions like bacterial, viral, fungal, and systemic pathologies such as diabetes [9].

S. Shaikh, *Nanoimaging - Future of Precision Medicine*,
https://doi.org/10.1007/978-981-97-7292-6_1

Nanotechnology in imaging and diagnosis offers a quicker, more accurate, non-invasive technique for various pathologies [10]. The role of biological markers is to detect disease processes as early as possible, making them valuable diagnostic tools [11]. Precision molecular imaging has also emerged as an important concept, enabling focused, controlled drug release, uniform, targeted drug distribution, and efficient monitoring of therapeutic response. This approach helps to reduce generalized drug toxicity, as drug distribution can be more precisely monitored [12].

The safety of nanotechnology is also essential, with potential health hazards needing careful consideration. Identifying, predicting, and evaluating health risks associated with different nanomaterials is a current challenge. Smaller nanoparticles can be more hazardous than larger particles [13]. Different catalytic effects can pose risks, as they are more active and mobile, potentially threatening human health and the environment [14]. Risks are categorized based on their health impact, especially concerning nanoparticles.

Thus, nanotechnology's role in imaging enables earlier and more accurate diagnoses than conventional imaging, providing broader therapeutic options that contribute to personalized medicine [15, 16]. The appropriate use of nanomaterials can prevent side effects, reduce potential hazards to the body, and improve the safety and effectiveness of nanotechnology applications [17].

References

1. Drexler KE. Nanosystems: molecular machinery, manufacturing, and computation. New York, NY: John Wiley & Sons; 1989.
2. Hu T, Mei X, Wang Y, Weng X, Liang R, Wei M. Two-dimensional nanomaterials: fascinating materials in biomedical field. Sci Bull. 2019;64:1707–27.
3. Belkin A, Hubler A, Bezryadin A. Self-assembled wiggling nano-structures and the principle of maximum entropy production. Sci Rep. 2015;5:8323.
4. Ede JD, Lobaskin V, Vogel U, Lynch I, Halappanavar S, Doak SH, et al. Translating scientific advances in the AOP framework to decision making for nanomaterials. Nanomaterials. 2020;10:1–22.
5. Allhoff F, Patrick L, Daniel M. What is nanotechnology and why does it matter? From science to ethics. John Wiley & Sons; 2010. p. 3–5.
6. Mashaghi S, Jadidi T, Koenderink G, Mashaghi A. Lipid nanotechnology. Int J Mol Sci. 2013;14:4242–82.
7. Farokhzad OC, Langer R. Nanomedicine: developing smarter therapeutic and diagnostic modalities. Adv Drug Deliv Rev. 2006;58:1456–9.
8. Filipponi L, Nicolau DV. Cell patterning. Wiley encyclopedia of biomedical engineering. John Wiley & Sons; 2006.
9. Lombardo D, Kiselev MA, Caccamo MT. Smart nanoparticles for drug delivery application: development of versatile nanocarrier platforms in biotechnology and nanomedicine. J Nanomater. 2019;12:1–26.
10. Medina C, Santos-Martinez MJ, Radomski A, Corrigan OI, Radomski MW. Nanoparticles: pharmacological and toxico-logical significance. Br J Pharmacol. 2007;150:552–8.
11. Ratner BD, Bryant SJ. Biomaterials: where we have been and where we are going. Annu Rev Biomed Eng. 2004;6:41–75.
12. Ma B, Martín C, Kurapati R, Bianco A. Degradation-by-design: how chemical functionalization enhances the biodegradability and safety of 2D materials. Chem Soc Rev. 2020;49:6224–47.
13. Zhu X, Vo C, Taylor M, Smith BR. Non-spherical micro- and nanoparticles in nanomedicine. Mater Horizons. 2019;6:1094–121.
14. Cronin JG, Jones N, Thornton CA, Jenkins GJS, Doak SH, Clift MJD. Nanomaterials and innate immunity: a perspective of the current status in nanosafety. Chem Res Toxicol. 2020;33:1061–73.
15. Hoshyar N, Gray S, Han H, Bao G. The effect of nanoparticle size on in vivo pharmacokinetics and cellular interaction. Nanomedicine (Lond). 2016;11:673–92.
16. Khafoor AA, Karim AS, Sajadi SM. Recent progress in synthesis of nano based liposomal drug delivery systems: a glance to their medicinal applications. Results Surfaces Interfaces. 2023;11:100124.
17. Li Y, Boraschi D. Endotoxin contamination: a key element in the interpretation of nanosafety studies. Nanomedicine. 2016;11:269–87.

2 Nanoimaging Principles

2.1 Introduction

Nanoscience is one of the most developing sciences that studies the properties of nanomaterials ranging between 1 and 100 nm in size. Nanotechnology has various applications in research, allowing for the identification and modification of new materials that can be tailored to different formats for specific uses. For this, the atomic scale is being used to identify and create the various nanomaterials that will be used in various applications [1–3]. Nanomaterials are based on the various optical, electrical, and/or magnetic properties that have different nanoscales, and these are also used in various non-medical applications as well. Another important property of the nanomaterials is that they provide a large surface area to volume ratio, and due to this, more and more applications can be used. These nanomaterials are being analyzed by the laws of quantum mechanics instead of the classical laws of physics and chemistry. The primary application of nanotechnology is to create functional systems at the molecular or atomic scale [4]. These nanotechnologies are commonly used as due to their structural advantages, safety, cleanliness, longer-lasting effects, and practical applications in daily life [5]. Nanomaterials are being widely used every day and are being used in two categories. The nanoparticle use can be enhanced in various ways by adding different agents or biomarkers, which will enhance the overall performance of the nanoimaging. Another way is the use of nanomaterials like nanocrystals and nanoparticles (NPs), which can be used directly due to their specific properties. With this, the use of nanomaterials will have a significant impact on medical management [6]. Nanomaterials are also being used in the form of sunscreens, cosmetics, sporting goods, tires, electronics, and several other everyday items [6]. Due to this, property nanotechnology use has changed the medical concept in the form of diagnostic methods, imaging, and drug delivery. The use of nanomaterials is significantly increased in function, which has lower costs and fewer hazards, thus resulting in a significant impact in health care as well as in the surrounding environment [7].

2.2 Nanotechnology in Medicine and Healthcare

Nanomedicine refers to the application of nanotechnology in medicine and healthcare. Here the concept is the use of nanomedicine technologies that are at the smallest nanoscale and with this the various techniques will lead to prevent, diagnose, monitor, and treat various diseases [8]. Nanotechnology has significantly impacted imaging techniques, diagnostic tools, drug deliv-

S. Shaikh, *Nanoimaging - Future of Precision Medicine*,
https://doi.org/10.1007/978-981-97-7292-6_2

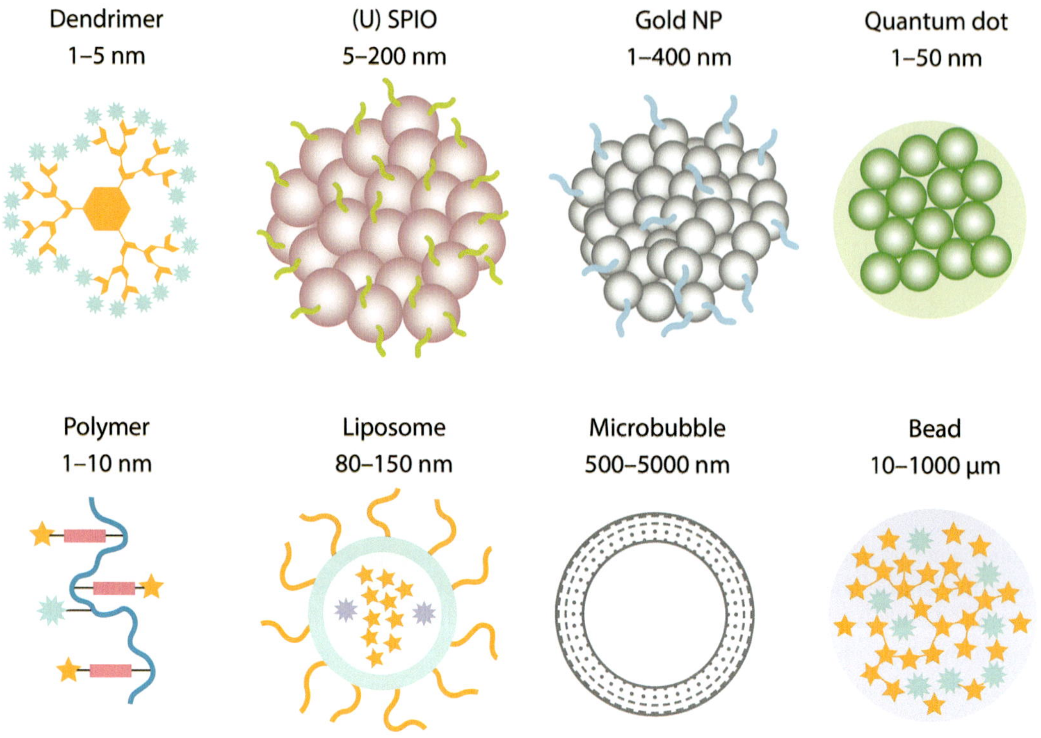

Diagram 2.1 Diagrammatic representation of different types of nanoparticles like dendrimers, USPIO, gold nanoparticles, quantum dots, polymers, liposomes and microbubbles

ery systems, tissue engineering, implants, and therapeutics [9]. These are being used in several diseases like cardiovascular diseases, cancer, various musculoskeletal conditions, psychiatric and neurodegenerative diseases, bacterial and viral infections, and inflammations, along with diabetes [10].

NPs structurally are in the form of polymers, proteins, lipids, lipoproteins, various metals, like silica, carbon nanotubes, and selenium-cadmium nanocrystals like quantum dots and microbubbles, which are being used as the diagnostic and theranostic agents (Diagram 2.1). Nanodiagnostics (1 nm to 1 µm) are in the range of their size. Theranostic nanomedicine is an emerging field where NPs are used as both diagnostic and therapeutic agents. Again, the particular particle design and size will be giving specific diagnostic information like a molecular target in the tissue and vascular permeability so that the particle can be easily transported, followed by the retention of the molecule, releasing of the drug from the particle, and ultimately the response to the therapy evaluation. Various research articles are being seen based on the use of the materials, along with their synthesis as well as functionalization [10–16]. Thus, much of the research was being done from the 1970s, 1980s, and 1990s on lipoproteins and other contrast-based agents.

2.3 Nanoparticles

Different types of NPs have been designed and evaluated recently. Different types of NPs are being used, including liposomes, polymers, proteins, dendrimers, and micelles. These agents can be combined with various chemotherapeutic agents for better drug delivery (Diagram 2.2). Some of the NPs used are Doxil (Janssen, Horsham, PA), Ambisome (Gilead, Foster City, CA), and Abraxane (Celgene, Summit, NJ), which are being called the nanomedicine formulations and are being used in various preclinical and clinical trials [17–20]. Nanocages are also important components of the NPs (Diagram 2.3). The use of NPs for diagnostic purposes is usually slower initially. Despite being identified two decades ago, these NP formulations are not yet fully established for human use. As will be outlined below, despite substantial progress in synthesizing state-of-the-art nanodiagnostics, the use of the bench is not precise [19]. Significant differences exist between nanotherapeutics and nanodiagnostics. The pharmacologic activity is able to differentiate between various physiologic or pathophysiologic effects. And when the treatment happens, some of the side effects are always encountered for sure. Low-quantity, pharmacologically inactive doses used in positron emission tomography (PET) studies often undergo rapid biodegradation or elimination, which reduces side effects. Consequently, nanodiagnostics must be excreted or metabolized relatively quickly after local, oral, or parenteral administration and should not have any pharmacologic and/or toxicologic activities. These intrinsic properties of the various NPs are based on their various pharmacokinetic and pharmacodynamic properties. Due to the small size of the nanometers, the smaller polymers and dendrimers, when added, can restrict the biodistribution in various compartments after injection. They will be there around the site of injection and will be slowly cleared via lymphatic drainage. NPs administered orally will enter the gastrointestinal tract and be excreted in the feces. Intravenously, this will be based on the size, charge, and surface properties (e.g., stealth coating); the biodistribution of NPs is confined to the vascular system and to the fenestrated endothelium, in the liver and spleen, as well as to tumors and sites of inflammation; these are due to the fenestrated endothelium and enhanced vascular leakiness. The intravenously administered nanotherapeutics have many advantages. The majority of the NPs will be in the vascular system, and the vascular endothelium will limit the side effects. The tumor will have abnormal vascularity of tumors and also in inflammation where there will be extravasation noted leading to more permeability and retention (EPR) effect, leading to better and increased therapeutic efficacy [17–19, 21–23]. And due to this, these properties will not be helpful if it is administered intravenously with the nanodiagnostics. These will cause the EPR-mediated NPs in the tumors and inflammatory lesions. The EPR will be increasing in time, and this will reach peak levels between 12 and 48 h. NPs will not be able to increase rapidly with more specific contrast enhancement, and this is needed for better disease diagnosis by the "standard" low molecular weight imaging agents, leading to the high signal-to-noise ratio. These smaller low molecular agents will be excreted easily and will be excreted faster. Some of the drawbacks are the very poor tissue-penetrating

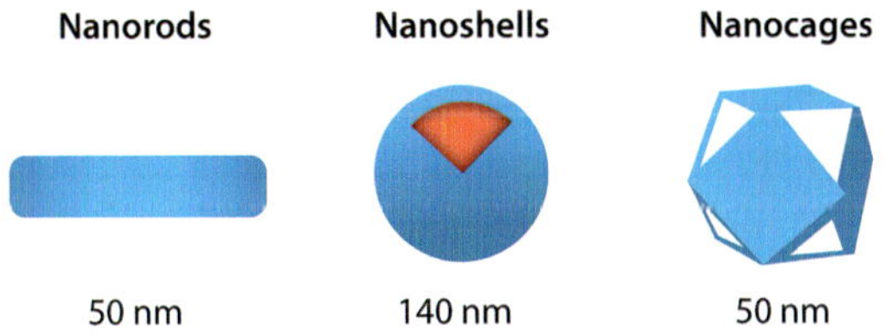

Diagram 2.2 Showing different types of nanorods, nanoshells, and nanocages

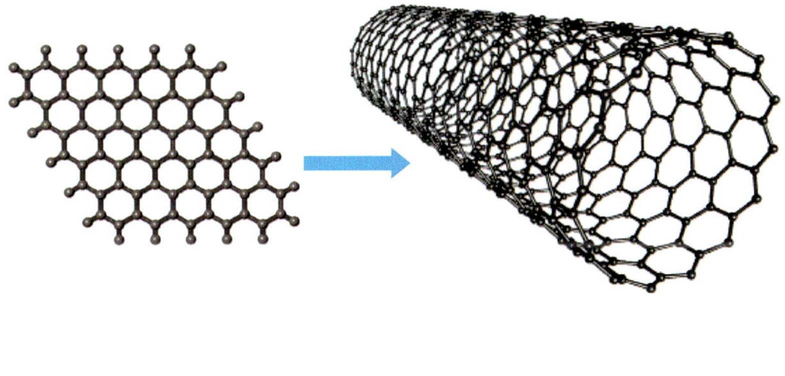

Diagram 2.3 The basic structure of the nanocage

properties due to the very large size, preventing them from accessing the target cells, which are more distantly placed from the vessels. Another disadvantage is more prolonged times in some of the vessels, like the liver and spleen, which will cause more retention in these organs and can lead to side effects. The heavy metal-based quantum dots can cause concerns of biocompatibility. The blood pool contrast agents that are being used in the imaging of the mononuclear phagocyte system (MPS) will be labeling, tracking of stem cells, monitoring the vascular targets, and helping the interventional radiologic management with the various theranostic options with the combination of the drug targeting and imaging.

2.4 Nanoparticle Properties

NP imaging uses the signal intensity for the stability in the biodistribution of the submicron diameter molecular imaging agents. The concept of using these NPs is useful in radionuclide and magnetic resonance imaging.

The synthesis, properties, and applications of NPs are seen in different forms. These properties are based on various shapes or sizes, the typical physical and chemical properties along with the high surface area and nanoscale size. Their optical properties are based on the size that is being displayed in the different colors. The specific features like reactivity, toughness, and different properties will be based on their unique size, shape, and structure. Due to these properties, they are used in various applications like catalysis, imaging, medical applications, energy-based research, and environmental applications. The various heavy metal NPs like lead, mercury, and tin are rigid and stable.

Nanotechnology has been an important research area since the last century. This "nanotechnology" concept was presented by Nobel laureate Richard P. Feynman in a 1959 lecture, which he delivered. "There's Plenty of Room at the Bottom" (Feynman, 1960), which has led to an increase in NP use. Nanotechnology produces materials of various types at the nanoscale level. NPs are a wide class of materials that include particulate substances, which have one dimension less than 100 nm at least [24]. The shape criteria will be based on differentiating these materials on the 0D, 1D, 2D, or 3D [25]. The size criteria are also an important criterion having an impact on the physiochemical properties of a substance, and this is again based on the various optical properties. The 20-nm size of the various NPs like gold (Au), platinum (Pt), silver (Ag), and palladium (Pd) NPs can be identified easily based on the various physiochemical properties. The bioimaging applications will be changing the various colors, sizes, and shapes, and this can be identified easily [26].

NPs are usually made up of three layers, which are (a) the surface layer, functioning with many small molecules, metal ions, surfactants, and polymers. (b) The shell layer is different from the core in all aspects, and (c) the core, being the central portion of the NP, is also referred to as the NP itself [27]. Due to this peculiar feature, they are interested in multidisciplinary fields and one of the examples is the electron microscopy (SEM) and transmittance electron microscope (TEM) images of mesoporous and nonporous methacrylate functionalized silica (MA-SiO2). The important feature of the mesoporousity will be imparting the special status in the use of the NPs. The NPs are being used for various aspects like drug delivery [28], chemical and biological sensing [29], gas sensing [30–32], CO_2 capturing [33, 34], and many other applications [35].

2.5 Nanoparticle Imaging Modalities

Various important existing imaging modalities that are being used are X-ray radiography, computed tomography (CT), magnetic resonance imaging (MRI), ultrasound (US), PET, single photon emission computed tomography (SPECT), and fluorescence imaging [36]. For better lesion detection, now due to technological advancements, more than one modality can be combined [37, 38]. And to add to this, the contrast agents are being used for better sensitivity in the evaluation of the anatomic and functional analysis and also for the differentiation of the normal tissue and abnormal lesions. Even with the use of the various contrast agents, the evaluation of the tumor has limitations due to the spatial resolution that will be generated by hardware, for example, the contrast-enhanced CT for the hypervascular hepatoma detection, which is as small as 3 mm [39]. Nowadays the commonly used small molecules are for the fast metabolism and have a non-specific distribution with varying toxicities [40, 41]. Nanomaterials with unique targeting properties of active and passive nature can be used as targeting agents. The NPs will be showing enhanced permeability and retention (EPR) effects in tumors and increased contrast agent in the local tumors [42]. NP size is a critical factor in determining biodistribution, blood circulation half-life, cellular uptake, tumor penetration, and targeting [43]. The average renal filtration pore is 10 nm [44], and these NPs having less than 10 nm in size will be cleared faster by the renal excretion system [45]. On the other hand, the NPs that have sizes greater than 100 nm are detected easily by the macrophages and they will accumulate in these organs with the MPS, like lymph nodes, liver, spleen, and lung [46]. Few literature studies show that the NP sizes between 10 and 60 nm show increased cellular uptake [43, 47]. Passive targeting involves labeling NPs with ligands that bind specifically to target receptors, enhancing localization and contrast in imaging [48–50]. One of the examples is that the gold NP surface, when it is attached to the prostate-specific membrane antigen RNA aptamer, has more CT density for prostate cancer cell imaging [51]. Similarly the use of the nanosized superparamagnetic iron oxide (SPIO) agents shows a higher affinity for the anti-EGFR antibody in lung tumors by MRI [52]. The use of antibodies and antibody fragments are the most common and very efficient active targeting ligands. The use of the antibodies will have more affinity for the antigens, and this can cause more NP concentration in a specific location [53]. The aptamer is another targeting aptamer, and this is also called the chemical antibody. It is a single DNA or RNA sequence that will be folding the secondary structure, which will have a high targeting affinity to the cell surface receptors. Aptamers are small in size, easy to synthesize, and can show lower immunogenicity [54].Aptamers can be easily cleared by the renal system, and all these will be degraded by nucleases, and this will prevent the circulation time for tumor localization [55, 56]. The use of peptides can be used like the additional ligand targeting moiety, and this can be for chemical stability, ease of synthesis, and also reduced immunogenicity [57]. The arginine glycine-aspartic acid (RGD) peptide is the important integrin binding domain and can bind to multiple integrin species [58], and this binding is very common for various NP applications [59]. The use of other proteins and molecules having active targeting roles is transferrin, folic acid, and biotin [60]. The use of the various stimuli will be used for the physical targeting, external sources, or fields guiding the NPs to the target site as is seen in the photothermal and magnetic hyperthermia therapy [61]. The use of various acidic pH/reduction dual-stimuli responsive nanoprobes for enhanced CT imaging of tumors is also seen [62]. The drug release is being increased as a change in pH, temperature, or a combination of both of these. The different endogenous parameters like pH, temperature, enzyme activity, and redox gradient belong to endogenous stimuli, and light, magnetic, and ultrasound belong to external stimuli [63].The important factor for the prolonged blood circulation time in NP-based contrast agents is due to the increased contrast signal. NPs can also be seen by the circulation time, which is more relevant for imaging performance. One of the common modification methods is the encapsulation of the various hydrophobic nanomaterials in a polyethylene glycol (PEG) shell [64], which causes the increased solubility and also prolongs blood circulation time. The important characteristic feature of the backbone of PEG is the NP binding to opsonin's dye, to which the macrophage recognition is reduced, and due to this NP clearance is also reduced by the reticuloendothelial system (RES) [65, 66]. Another use of dextrose and polysaccharides like chitosan, hyaluronic acid, and fucoidan also plays an important role in prolonging NP circulation [67, 68]. The zwitterionic modification will cause the various surface properties of the NPs, which will be resistant to aggregation, then binding plasma proteins and macrophage uptake, and in this way cause prolonged circulation time [69, 70]. The albumin surface modification of NPs will result in increased blood circulation time by maintaining biological activity and also decreasing immunogenicity [71].

2.6 Nanoparticle Biomechanism

Nanoimaging agents have a high surface area to volume ratio as compared to routine contrast agents and thus can have better surface labeling with specific molecules and ligands, leading to less toxicity [72, 73]. The other advantages of NP imaging are the functional visualization and monitoring of biological processes, like macrophage detection in atherosclerotic lesions with the use of CT [74] and molecular imaging of angiogenesis by MRI [75]. Again, longer plasma circulation time of NPs will lead to better biodistribution, resulting in better and greater lesion-to-background contrast [76]. The shape and size of the NPs are also adjustable for better optimizing the loading of imaging compounds, along with their intrinsic physical properties, which can also be changed to meet specific clinical needs [77, 78].

NP use in cell biology is now one of the most demanding developments in the last couple of years. Most of the studies are focused on the various interactions between NPs and cells, which will help us understand better the synergy between the organisms and nanomaterials. This also includes the various in vivo therapeutic or imaging applications when applied to the cells for using various applications. The important aspect is the biocompatibility of the various cells, which is important for the viability of the NPs. Cellular proliferation along with toxicity needs to be taken care and all these are important for better biocompatibility [79]. Depending on the target cells, different mechanisms such as micropinocytosis and phagocytosis are utilized, influenced by the NP's size, surface charge, and cellular uptake properties [80]. NP surfaces can be modified to enhance the specificity and functionality of nanoconjugates. The role of the surface modifiers is to increase the cellular uptake of nanoconjugates, the specificity of cellular uptake, and the efficiency of intracellular targeting or retention of nanoconjugates. The NP modifiers/conjugants like the various antibodies and peptides will increase the subcellular focused compartment targeting, and some of the nucleic acids will change the subcellular retention of nanoconjugates. The detection of NPs in cells can be done by the use of optical microscopy and electron microscopy. Along with this, the new concepts are four-photon microscopy [81], near-infrared surface-enhanced Raman scattering [82], X-ray fluorescence micro- and nano-probe imaging [83], and coherent X-ray diffraction imaging [84]. These can be used to significantly increase the various aspects of the NP imaging. The futuristic approach is the use of 3D imaging with a resolution as good as 5 nm 3 voxels (coherent X-ray diffraction imaging) will permit the frozen cells with the NPs to focus targeting within the cells also.

2.7 Conclusion

Despite these advances, the diagnostic applications of NPs remain limited. NP uptake in the MPS and by EPR effects shows higher signals. Another aspect is the very low penetration in tissues; the molecularly targeted micro- and nanoparticles are not completely delivered to the targeted cells, especially when the targets are outside the vasculature. The targets inside the vasculature can use the larger microparticles, which can extravasate and cause less a specific background. Other factors impacting are efficacy to target, and by these, the various diagnostic and therapeutic potentials can be identified. The other aspects are uniformity, batch-to-batch, particle size, shape, and charge, blood half-life, biodistribution, and elimination of the particles are also important [10]. The NPs used for diagnostics and theronostics are also evolving due to their very high clinical potential. Despite some limitations, the growth of nanodiagnostics and theronostics is phenomenal and will be one of the important hallmarks of the practice of molecular nanoimaging in the coming years.

References

1. Nikalje AP. Nanotechnology and its applications in medicine. Med Chem. 2015;5:81–9. https://doi.org/10.4172/2161-0444.1000247.
2. Adabi M, Naghibzadeh M, Adabi M, Zarrinfard MA, Esnaashari S, Seifalian AM, et al. Biocompatibility and nanostructured materials: applications in nanomedicine. Artif Cells Nanomed Biotechnol. 2017;45:833–42.
3. Belkin A, Hubler A, Bezryadin A. Self-assembled wiggling nano-structures and the principle of maximum entropy production. Sci Rep. 2015;5:8323.
4. Buzea C, Pacheco II, Robbie K. Nanomaterials and nanoparticles: Sources and toxicity. Biointerphases. 2007;2:MR17–71.
5. Albanese A, Tang PS, Chan WC. The effect of nanoparticle size, shape, and surface chemistry on biological systems. Annu Rev Biomed Eng. 2012;14:1–16. https://doi.org/10.1146/annurev-bioeng-071811-150124.
6. Allhoff F, Patrick L, Daniel M. What is nanotechnology and why does it matter? From science to ethics. John Wiley & Sons; 2010. p. 3–5.
7. Desai N. Challenges in development of nanoparticle-based therapeutics. AAPS J. 2012;14:282–95. https://doi.org/10.1208/s12248-012-9339.
8. Modi S, Prajapati R, Inwati GK, Deepa N, Tirth V, Yadav VK, Yadav KK, Islam S, Gupta P, Kim D-H. Recent trends in fascinating applications of nanotechnology in allied health sciences. Crystals. 2022,12:39. https://doi.org/10.3390/cryst12010039.
9. Badrunnisa S, Menghani SS. Review on health sciences applications of nanotechnology. J Coast Life Med. 2023;11:94–100.
10. Lombardo D, Kiselev MA, Caccamo MT. Nanocarrier platforms in biotechnology and nanomedicine. J Nanomater. 2019;12:1–26.
11. Dessale M, Mengistu G, Mengist HM. Nanotechnology: a promising approach for cancer diagnosis, therapeutics and theragnosis. Int J Nanomedicine. 2022;17:3735–49. https://doi.org/10.2147/IJN.S378074.
12. Jakhmola A, Anton N, Vandamme TF. Inorganic nanoparticles-based contrast agents for x-ray computed tomography. Adv Healthc Mater. 2012;1(4):413–31.
13. Wahajuddin AS, Arora S. Superparamagnetic iron oxide nanoparticles: magnetic nanoplatforms as drug carriers. Int J Nanomedicine. 2012;7:3445–71.
14. Kiessling F, Morgenstern B, Zhang C. Contrast agents and applications to assess tumor angiogenesis in vivo by magnetic resonance imaging. Curr Med Chem. 2007;14(1):77–91.
15. Liu Z, Kiessling F, Gaetjens J. Advanced nanomaterials in multimodal imaging: design, functionalization, and biomedical applications. J Nanomater. 2010;2010:894303.
16. Heneweer C, Gendy SE, Peñate-Medina O. Liposomes and inorganic nanoparticles for drug delivery and cancer imaging. Ther Deliv. 2012;3(5):645–56.
17. Duncan R, Gaspar R. Nanomedicine(s) under the microscope. Mol Pharm. 2011;8(6):2101–41.
18. Lammers T, Aime S, Hennink WE, Storm G, Kiessling F. Theranostic nanomedicine. Acc Chem Res. 2011;44(10):1029–38.
19. Rizzo LY, Theek B, Storm G, Kiessling F, Lammers T. Recent progress in nanomedicine: therapeutic, diagnostic and theranostic applications. Curr Opin Biotechnol. 2013;24(6):1159–66.
20. Grimm J, Scheinberg DA. Will nanotechnology influence targeted cancer therapy? Semin Radiat Oncol. 2011;21(2):80–7.
21. Matsumura Y, Maeda H. A new concept for macromolecular therapeutics in cancer chemotherapy: mechanism of tumoritropic accumulation of proteins and the antitumor agent smancs. Cancer Res. 1986;46(12 Pt 1):6387–92.
22. Jain RK, Stylianopoulos T. Delivering nanomedicine to solid tumours. Nat Rev Clin Oncol. 2010;7(11):653–64.
23. Lammers T, Kiessling F, Hennink WE, Storm G. Drug targeting to tumors: principles, pitfalls and (pre-) clinical progress. J Control Release. 2012;161(2):175–87.
24. Laurent S, Forge D, Port M, Roch A, Robic C, Vander Elst L, Muller RN. Magnetic iron oxide nanoparticles: synthesis, stabilization, vectorization, physicochemical characterizations, and biological applications. Chem Rev. 2010;110:2574. https://doi.org/10.1021/cr900197g.
25. Tiwari JN, Tiwari RN, Kim KS. Zero-dimensional, onedimensional, two-dimensional and three-dimensional nanostructured materials for advanced electrochemical energy devices. Prog Mater Sci. 2012;57:724–80.
26. Dreaden EC, Alkilany AM, Huang X, Murphy CJ, El-Sayed MA. The golden age: gold nanoparticles for biomedicine. Chem Soc Rev. 2012;41:2740–79.
27. Shin W-K, Cho J, Kannan AG, Lee Y-S, Kim D-W. Cross-linked composite gel polymer electrolyte using mesoporous methacrylate-functionalized SiO2 nanoparticles for lithium-ion polymer batteries. Sci Rep. 2016;6:26332.
28. Lee JE, Lee N, Kim T, Kim J, Hyeon T. Multifunctional mesoporous silica nanocomposite nanoparticles for theranostic applications. Acc Chem Res. 2011;44:893–902.
29. Barrak H, Saied T, Chevallier P, Laroche G, Adel M, Hamzaoui AH. Synthesis, characterization,

and functionalization of ZnO nanoparticles by N-(trimethoxysilylpropyl) Nanoparticles 927 ethylenediamine triacetic acid (TMSEDTA): investigation of the interactions between phloroglucinol and ZnO@TMSEDTA. Arab J Chem. 2016; https://doi.org/10.1016/J.ARABJC.2016.04.019.
30. Mansha M, Khan I, Ullah N, Qurashi A. Synthesis, characterization and visible-light-driven photoelectrochemical hydrogen evolution reaction of carbazole-containing conjugated polymers. Int J Hydrogen Energy. 2017;42:10952–61.
31. Rawal I, Kaur A. Synthesis of mesoporous polypyrrole nanowires/nanoparticles for ammonia gas sensing application. Sens Actuators A Phys. 2013;203:92–102.
32. Ullah H, Khan I, Yamani ZH, Qurashi A. Sonochemicaldriven ultrafast facile synthesis of SnO2 nanoparticles: growth mechanism structural electrical and hydrogen gas sensing properties. Ultrason Sonochem. 2017;34:484–90.
33. Ganesh M, Hemalatha P, Peng MM, Jang HT. One pot synthesized Li, Zr doped porous silica nanoparticle for low temperature CO_2 adsorption. Arab J Chem. 2017;10:S1501–5.
34. Ramacharyulu PVRK, Muhammad R, Praveen Kumar J, Prasad GK, Mohanty P. Iron phthalocyanine modified mesoporous titania nanoparticles for photocatalytic activity and CO_2 capture applications. Phys Chem Chem Phys. 2015;17:26456–62.
35. Shaalan M, Saleh M, El-Mahdy M, El-Matbouli M. Recent progress in applications of nanoparticles in fish medicine: a review. Nanomed Nanotechnol Biol Med. 2016;12:701–10.
36. Ryvolova M, Chomoucka J, Drbohlavova J, Kopel P, Babula P, Hynek D, Adam V, Eckschlager T, Hubalek J, Stiborova M, Kaiser J, Kizek R. Modern micro and nanoparticle-based imaging techniques. Sensors. 2012;12:14792–820.
37. Wang W, Pang J, Su J, Li F, Li Q, Wang X, Wang J, Ibarlucea B, Liu X, Li Y. Applications of nanogenerators for biomedical engineering and healthcare systems. InfoMat. 2022;4:e12262. https://doi.org/10.1002/inf2.12262.
38. Fan Q, Cheng K, Hu X, Ma X, Zhang R, Yang M, Lu X, Xing L, Huang W, Gambhir SS, Cheng Z. Transferring biomarker into molecular probe: melanin nanoparticle as a naturally active platform for multimodality imaging. Journal of the American Chemical Society. J Am Chem Soc. 2014;136:15185–94.
39. Cuccurullo V, Di Stasio GD, Mazzarella G, Cascini GL. Microvascular invasion in HCC: the molecular imaging perspective. Contrast Media Mol Imaging. 2018;2018:9487938.
40. Chronos NA, Goodall AH, Wilson DJ, Sigwart U, Buller NP. Profound platelet degranulation is an important side effect of some types of contrast media used in interventional cardiology. Circulation. 1993;88:2035–44.
41. Petrik M, Weigel C, Kirsch M, Hosten N. No detectable nephrotoxic side effect using a dimer, nonionic contrast media in cerebral perfusion computed tomography in case of suspected brain ischemia. Rofo. 2005;177:1242–9.
42. Oh IH, Min HS, Li L, Tran TH, Lee YK, Kwon IC, Choi K, Kim K, Huh KM. Cancer cell-specific photoactivity of pheophorbide a–glycol chitosan nanoparticles for photodynamic therapy in tumor-bearing mice. Biomaterials. 2013;34:6454–63.
43. Hoshyar N, Gray S, Han H, Bao G. The effect of nanoparticle size on in vivo pharmacokinetics and cellular interaction. Nanomedicine. 2016;11:673–92.
44. Scott RP, Quaggin SE. The cell biology of renal filtration. J Cell Biol. 2015;209:199–210.
45. Mazumder S, Biswas GR, Majee SB. Applications of nanorobots in medical techniques. IJPSR. 2020;11:3150.
46. Zhou Y, Dai Z. New strategies in the design of nanomedicines to oppose uptake by the mononuclear phagocyte system and enhance cancer therapeutic efficacy. Chem Asian J. 2018;13:3333–40.
47. Huang Y, He S, Cao W, Cai K, Liang X-J. Biomedical nanomaterials for imaging-guided cancer therapy. Nanoscale. 2012;4:6135–49.
48. Chen LQ, Xiao SJ, Hu PP, Peng L, Ma J, Luo LF, Li YF, Huang CZ. Aptamer-mediated nanoparticle-based protein labeling platform for intracellular imaging and tracking endocytosis dynamics. Anal Chem. 2012;84:3099–110.
49. Das M, Duan W, Sahoo SK. Multifunctional nanoparticle–EpCAM aptamer bioconjugates: a paradigm for targeted drug delivery and imaging in cancer therapy. Nanomedicine. 2015;11:379–89.
50. Ke R, Yang W, Xia X, Xu Y, Li Q. Tandem conjugation of enzyme and antibody on silica nanoparticle for enzyme immunoassay. Anal Biochem. 2010;406:8–13.
51. Mbunge E, Muchemwa B, Batani J. Sensors and healthcare 5.0: Transformative shift in virtual care through emerging digital health technologies. Glob Health J. 2021;5:169–77. https://doi.org/10.1016/j.glohj.2021.11.008.
52. Wang Z, Qiao R, Tang N, Lu Z, Wang H, Zhang Z, Xue X, Huang Z, Zhang S, Zhang G, Li Y. Active targeting theranostic iron oxide nanoparticles for MRI and magnetic resonance-guided focused ultrasound ablation of lung cancer. Biomaterials. 2017;127:25–35.
53. Alibakhshi A, Abarghooi Kahaki F, Ahangarzadeh S, Yaghoobi H, Yarian F, Arezumand R, Ranjbari J, Mokhtarzadeh A, de la Guardia M. Targeted cancer therapy through antibody fragments-decorated nanomedicines. J Control Release. 2017;268:323–34.
54. Anselmo AC, Mitragotri S. Nanoparticles in the clinic: an update. Bioeng Transl Med. 2019;4:e10143. https://doi.org/10.1002/btm2.10143.

55. Alshaer W, Hillaireau H, Fattal E. Aptamer-guided nanomedicines for anticancer drug delivery. Adv Drug Deliv Rev. 2018;134:122–37.
56. Zhou J, Rossi J. Aptamers as targeted therapeutics: current potential and challenges. Nat Rev Drug Discov. 2017;16:181–202. https://doi.org/10.1038/nrd.2016.199.
57. Inaba H, Matsuura K. Peptide nanomaterials designed from natural supramolecular systems. Chem Rec. 2019;19:843–58. https://doi.org/10.1002/tcr.201800149.
58. Welch EC, Powell JM, Clevinger TB, Fairman AE, Shukla A. Advances in biosensors and diagnostic technologies using nanostructures and nanomaterials. Adv Funct Mater. 2021;31:2104126. https://doi.org/10.1002/adfm.202104126.
59. Li Y, Xiao Y, Lin HP, Reichel D, Bae Y, Lee EY, Jiang Y, Huang X, Yang C, Wang Z. In vivo β-catenin attenuation by the integrin α5-targeting nano-delivery strategy suppresses triple negative breast cancer stemness and metastasis. Biomaterials. 2018;188:160–72.
60. Zhong Y, Meng F, Deng C, Zhong Z. Ligand-directed active tumor-targeting polymeric nanoparticles for cancer chemotherapy. Biomacromolecules. 2014;15:1955–69.
61. Yu X, Trase I, Ren M, Duval K, Guo X, Chen Z. Design of nanoparticle-based carriers for targeted drug delivery. J Nanomater. 2016;2016:1087250.
62. Wang A, Yin L, He L, Xia H, Chen F, Zhao M, Ding J, Shi H. An acidic pH/reduction dual-stimuli responsive nanoprobe for enhanced CT imaging of tumours in vivo. Nanoscale. 2018;10:20126–30.
63. Yang L, Sun H, Liu Y, Hou W, Yang Y, Cai R, Cui C, Zhang P, Pan X, Li X, Li L, Sumerlin B, Tan W. Self-assembled aptamer-grafted hyperbranched polymer nanocarrier for targeted and photoresponsive drug delivery. Angew Chem Int Ed Engl. 2018;57:17048–52.
64. Wang JL, Du XJ, Yang JX, Shen S, Li HJ, Luo YL, Iqbal S, Xu CF, Ye XD, Cao J, Wang J. The effect of surface poly (ethylene glycol) length on in vivo drug delivery behaviors of polymeric nanoparticles. Biomaterials. 2018;182:104–13.
65. Hu J, Sheng Y, Shi J, Yu B, Yu Z, Liao G. Long circulating polymeric nanoparticles for gene/drug delivery. Curr Drug Metab. 2018;19:723–38.
66. Mozar FS, Chowdhury EH. Pegylation of carbonate apatite nanoparticles prevents opsonin binding and enhances tumor accumulation of gemcitabine. J Pharm Sci. 2018;107:2497–508.
67. Abdollah MRA, Carter TJ, Jones C, Kalber TL, Rajkumar V, Tolner B, Gruettner C, Zaw-Thin M, Baguna Torres J, Ellis M, Robson M, Pedley RB, Mulholland P, de Rosales RTM, Chester KA. ACS Nano. 2018;12:1156–69.
68. Mustafa S, Devi VK, Pai RS. Effect of PEG and water-soluble chitosan coating on moxifloxacin-loaded PLGA long-circulating nanoparticles. Drug Deliv Transl Res. 2017;7:27–36.
69. Liu X, Li H, Chen Y, Jin Q, Ren K, Ji J. Mixed-charge nanoparticles for long circulation, low reticuloendothelial system clearance, and high tumor accumulation. Adv Healthc Mater. 2014;3:1439–47.
70. Xiao W, Lin J, Li M, Ma Y, Chen Y, Zhang C, Li D, Gu H. Prolonged in vivo circulation time by zwitterionic modification of magnetite nanoparticles for blood pool contrast agents. Contrast Media Mol Imaging. 2012;7:320–7.
71. Esfandyari-Manesh M, Mostafavi SH, Majidi RF, Koopaei MN, Ravari NS, Amini M, Darvishi B, Ostad SN, Atyabi F, Dinarvand R. Improved anticancer delivery of paclitaxel by albumin surface modification of PLGA nanoparticles. Daru. 2015;23:28.
72. Montet X, Montet-Abou K, Reynolds F, Weissleder R, Josephson L. Nanoparticle imaging of integrins on tumor cells. Neoplasia. 2006;8:214–22.
73. Montet X, Weissleder R, Josephson L. Imaging pancreatic cancer with a peptide– nanoparticle conjugate targeted to normal pancreas. Bioconjug Chem. 2006;17:905–11.
74. Qin J, Peng C, Zhao B, Ye K, Yuan F, Peng Z, Yang X, Huang L, Jiang M, Zhao Q, Tang G, Lu X. Noninvasive detection of macrophages in atherosclerotic lesions by computed tomography enhanced with PEGylated gold nanoparticles. Int J Nanomedicine. 2014;9:5575–90.
75. Sadeghi M, Sadeghi S, Naghib SM, Garshasbi HR. A comprehensive review on electrochemical nano biosensors for precise detection of blood-based oncomarkers in breast cancer. Biosensors. 2023;13:481. https://doi.org/10.3390/bios13040481.
76. Ouaras K, Lombardi G, Couëdel L, Arnas C. Hassouni K. Microarcing-enhanced tungsten nano and micro-particles formation in low pressure high-density plasma. Phys Plasmas. 2019;26:023705.
77. Nasrollahzadeh M, Sajjadi M, Sajadi SM, Issaabadi Z. Interface science and technology. In: Green nanotechnology, vol. 28. Amsterdam: Elsevier; 2019. p. 145–98.
78. Bhavane R, Badea C, Ghaghada KB, Clark D, Vela D, Moturu A, Annapragada A, Johnson GA, Willerson JT, Annapragada A. Dual-energy computed tomography imaging of atherosclerotic plaques in a mouse model using a liposomal-iodine nanoparticle contrast agent. Circ Cardiovasc Imaging. 2013;6:285–94.
79. Dirisala A, Osada K, Chen Q, Tockary TA, Machitani K, Osawa S, Liu X, Ishii T, Miyata K, Oba M, et al. Optimized rod length of polyplex micelles for maximizing transfection efficiency and their performance in systemic gene therapy against stroma-rich pancreatic tumors. Biomaterials. 2014;35:5359–68. https://doi.org/10.1016/j.biomaterials.2014.03.037.
80. Li M, Hu X, Zhao Y, Jiao N. An overview of recent progress in micro/nanorobots for biomedical applications. Adv Mater Nanotechnol. 2023;8:2201928. https://doi.org/10.1002/admt.202201928.
81. Vogt A, Combadiere B, Hadam S, Stieler KM, Lademann J, Schaefer H, Autran B, Sterry W, Blume-Peytavi U. 40 nm, but not 750 or 1,500 nm, nanoparticles enter epidermal CD1a+ cells after

transcutaneous application on human skin. J Invest Dermatol. 2006;126:6.
82. Chattopadhyay PK, Price DA, Harper TF, Betts MR, Yu J, Gostick E, Perfetto SP, Goepfert P, Koup RA, De Rosa SC, Bruchez MP, Roederer M. Quantum dot semiconductor nanocrystals for immunophenotyping by polychromatic flow cytometry. Nat Med. 2006;12(8):972–7.
83. Reddy GR, Bhojani MS, McConville P, Moody J, Moffat BA, Hall DE, Kim G, Koo YE, Woolliscroft MJ, Sugai JV, Johnson TD, Philbert MA, Kopelman R, Rehemtulla A, Ross BD. Clin Cancer Res. 2006;12:22.
84. Liu S, Tong W, Gao C, Liu Y, Li X, Zhang Y. Environmentally friendly natural materials for triboelectric nanogenerators: a review. J Materials Chem A. 2023;11:9270–99. https://doi.org/10.1039/D2TA10024J.

3 Nanoparticle Design

3.1 Introduction

The National Nanotechnology Initiative (www.nano.gov/) has defined nanotechnology being capable of measuring the dimensions between 1.0 and 100.0 nm, or between one and 100 billionths of a meter. Different manipulations can modify the various atoms or molecules to create nanoparticles [1]. Nanomaterials are similar to the molecules and cells that are affected by disease processes, and they can be manipulated and adjusted for the use of various functions. The size and physical properties are two important criteria that can form the basis for the use of nanotechnologies in medicine and other non-medical applications [2]. Nanomedicine uses the properties and physical characteristics for better diagnosis and evaluation at the cellular level [2]. Thus, nanotechnology use can be an important aspect of the more focused and targeted use in different pathologies. Nanoparticle-based contrast agents are developed for all existing modalities like single-photon emission computed tomography (SPECT) and positron emission tomography (PET), magnetic resonance (MR) imaging, ultrasonography (US), computed tomography (CT), and optical imaging. Along with the diagnostic approach and the use of nanoparticle-based therapeutics, there is a more focused and targeted drug delivery to novel molecular targets, as seen in various conditions. Targeted multifunctional nanoparticle-based therapeutics will substantially increase the efficacy, for example, of antineoplastic agents. The tumor biology and the barriers can be studied by identifying the amount of microvasculature and their environment for the targeted imaging and therapeutics. The various properties of nanotechnologies will include the newer drug-delivery options, contrast agents, and also devices for the diagnosis. The use of the nanomaterials with their large surface area–to–volume ratio will show the different properties. Few nanoparticles can be single and can allow a larger amount of contrast agent or drug molecules. Few of the NPs, like quantum dots and gold nanoparticles, are being used for novel imaging and therapeutics. The use in interventional radiology is the localized delivery of therapeutic nanoparticles in the tumor, which can reduce the toxicity levels if not administered in a focused manner. Here, ultimately, the clinical care in patients by various interventional techniques can be increased [3, 4]. The statement of the Society of Interventional Radiology Interventional Oncology Task Force's vision statement has mentioned that the improved integration of current imaging-guided oncologic technologies in combination with the different newer developments in genomics, proteomics, molecular imaging, and nanotechnology by 2015 will have a major impact on health care [5].

S. Shaikh, *Nanoimaging - Future of Precision Medicine*,
https://doi.org/10.1007/978-981-97-7292-6_3

3.2 History and Development of Nanotechnology

Richard Feynman is the man who is credited with the development of nanotechnology; this was when he delivered a lecture at the California Institute of Technology on December 29, 1959, entitled "There's Plenty of Room at the Bottom: An Invitation to Enter a New Field of Physics." In the 1970s, Eric Drexler was the pioneer for working on molecular manufacturing, and in 1986 he coined the term nanotechnology in his book Engines of Creation [6]. In 1992, Drexler did major work and published on Nanosystems [7], in which he showed the study of high-performance machines into molecular carbon lattices. Nanosystems was the first nanotechnology textbook in the world. The synthesis of basic materials like fullerene, or "buckyball," construction is mentioned in 1985, and this is seen as the important synthetic hollow geodesic sphere made up of 60 carbon atoms, thus resulting in the origin of the carbon nanotube in 1991 [8]. In 1996, the Nobel Prize in chemistry was awarded to Robert F. Curl, Jr.; Sir Harold W. Kroto; and Richard E. Smalley for their discovery of fullerenes [9]. The multifunctional nanoparticles are now being used and manufactured by various links, which can be different functional moieties like different radioisotopes, various chemotherapeutic agents, and also various photodynamic moieties for therapeutics.

3.3 Overview of Nanoparticles

The prefix "nano" means one-billionth; thus, a nanometer is one-billionth of a meter. The term nanotechnology refers to the creation of objects with nanoscale dimensions between 1.0 and 100.0 nm [1]. Nanoparticles are approximately 1/1000 the diameter of a hair and close to the size of a cellular organelle [10]. Nanoparticles due to their size are useful for molecular imaging and also for the delivery of the various therapeutics to molecular targets [2].

The use of the nanoparticles is also based on the adjustments and also the synergy between the various components of the nanoimaging. Imaging of subcellular components is also based on the cell compositions, nanoparticle biocompatibility, and cellular uptake mechanisms and side effects as shown that in most cases the toxicity/biocompatibility of nanoparticles depends on their concentration [11–19]. The differentiation of the conjugate and non-conjugates is seen in the various cellular uptake mechanisms [20–27]. The surface alterations are also seen in the various functionalities. The use of semiconductor material like TiO_2 is being used for both optically fluorescent agents and also for subcellular targeting. Newly emerging imaging techniques are photon microscopy [28], Raman scattering infrared technique [29, 30], X-ray fluorescence micro- and nano-probe imaging [31–35], and coherent X-ray diffraction imaging [36–38] will be more useful for the imaging of the nanoparticles in cells. Coherent X-ray diffraction imaging will also be useful for the evaluation of the frozen cells with the nanoparticles.

3.4 Nanoparticle Chemistry

Nanoparticles are mesostructures and have similar properties to atomic or molecular structures. When compared with them, these materials will be of the same size and have quantum properties that are similar to the semiconductor nanoparticles, having superparamagnetism, and many more. These nanoparticles are of three types: polymer/biomacromolecule nanoparticles, semiconductor nanoparticles, and metal nanoparticles.

The polymer/biomacromolecule nanoparticles are biocompatible nanomaterials, and despite this, they are added or combined with different other materials, which can increase

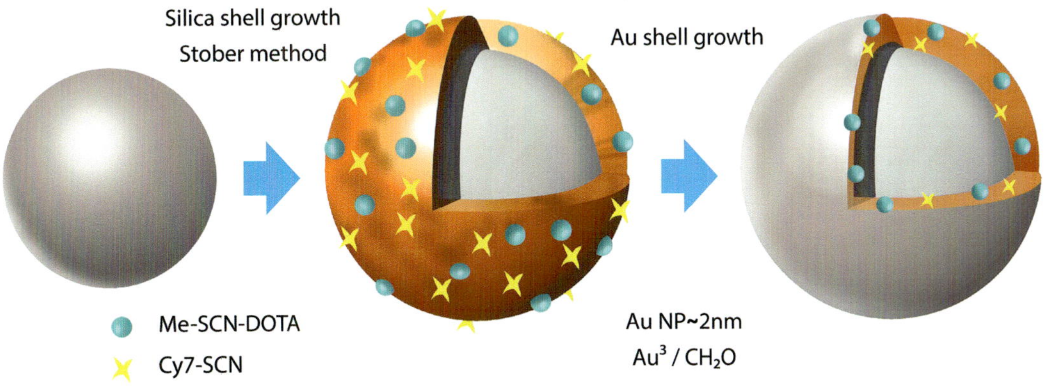

Diagram 3.1 Silica and gold shell growth

their overall biocompatibility or functionality. These nanoparticles are made up of various materials like polymer nanoparticles such as polyglycolic acid, polystyrene, polyethylene glycol, poly-coated calcium phosphate, lipids and lipoproteins, proteins condensed nanoparticles, non-nucleic acid polymers, DNA nanoparticles, pDNA complexes, and various fluorescent polymer nanoparticles [32–54]; Semiconductor nanoparticles are also part of the discussion on the quantum dots [28, 55–69], and other different semiconductor metal oxides like SiO_2, ZnO, Al_2O_3, CrO, SnO_2, and TiO_2 [20, 22, 24, 43, 47, 70–79]. These are located in groups II–VI, III–V, or IV–IV in the periodic table. These are inorganic salts or metal oxides and are less than 10 nm in size. These are based on the principle of photoelectron emission. The quantum dots will have the more "classical" organic fluorescent properties, which will have more photostability with a wider range of fluorescence emission wavelengths having different physical size, shape, and chemical components. TiO_2 nanoparticles can trap the electrons in the conjugated molecules by which the electrons from the molecules of water will be removed when in contact with the TiO_2 surface [80]. These will cause oxidation, and this will be useful for various therapeutic applications. These TiO_2 nanoparticles are less than 20 nm in size, which will form the "corner defects" about the bidentate ligands [81, 82]. These DNA composed oligonucleotides are capable of retaining the TiO_2-DNA oligonucleotide nanoconjugates [34, 76, 77]. Gold nanoparticles are the important nanomaterials that are used in nanoimaging [24, 29, 64, 79, 83–92] (Diagram 3.1). Silver, cobalt, and nickel nanoparticles are other nanomaterials after the gold to be used in nanoimaging [47]. These will be stable nanostructures, having less toxicity and capable of binding various targeting molecules. These conjugations of nucleic acids (DNA, RNA, LNAs, PNAs, polypeptides, or cellular ligands) will result in the conjugation and targeting of various cells having folate receptors [84]. Gold and silver have the optical properties of surface plasmon resonance based on their size and shape (Diagram 3.2).

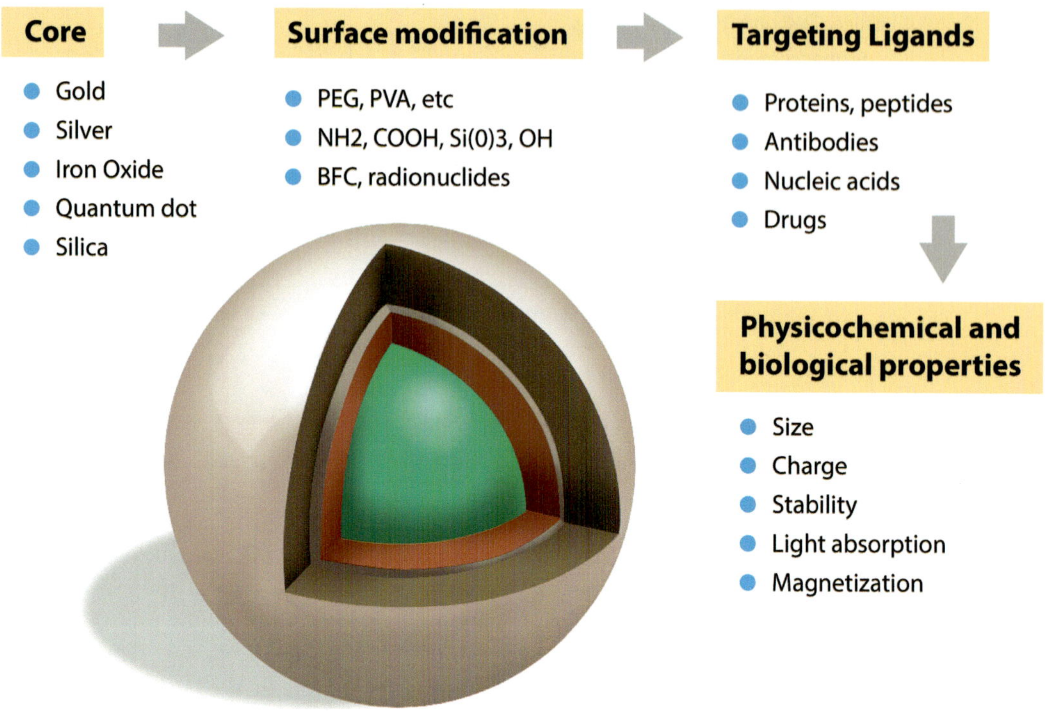

Diagram 3.2 Graphical representation of the surface modification and targeting ligands

3.5 Nanoparticles for Imaging in Fixed Cells

Various advancements are seen in nanotechnology, and this has led to significant development in nanoapplications. The modified quantum dots and gold nanoparticles are used more than the fluorescent dyes and proteins, which are multifunctional in nature. Li-Shishido et al. have shown false negative findings when there is increased time during which the quantum dots will get illuminated and seen before the fluorescence intensity [62]. In the illumination period, these will not show any fluorescence, but after 10 min there will be increased illumination in the range of 2- to 3-times. Quantum dots blinking was prevented by adding mercaptoethanol and glutathione in the samples [62]. The nanoparticles in the future will have more increased applications based on the more diverse and precise use of the cellular biomarkers [59] along with the water on the TiO2 surface [80]. There will be an oxidation process, which will result in more use for therapeutic purposes. The TiO_2 nanoparticles, which are smaller than 20 nm result in "corner defects" on the nanoparticle surface [81, 82]. Gold nanoparticles [24, 29, 64, 76, 77, 79, 83–92] along with silver, cobalt, and nickel nanoparticles are more widely used [47]. By increasing the time, these quantum dots are illuminated even before the fluorescence intensity [62]. Nanoparticles have thus significantly increased the number and diversity of properties in the analysis of various cellular biomarkers [59].

3.5.1 Magnetic Resonance Imaging

Magnetic resonance imaging is a non-invasive imaging modality and now has multiparametric imaging properties [93, 94]. Magnetic resonance imaging was first used clinically in 1984, and since then it has become the most preferred and best imaging modality [95]. The use of contrast will give better resolution, and the ideal contrast agent is the one with fewer adverse effects and can be excreted easily. Despite this, adverse reactions are seen, and they are various allergic reactions, nephrotoxicity, gadolinium deposition in the tissues, and various physiologic reactions [96]. The use of nanoparticles also has shown that the NPs in the future will have fewer side effects.

3.5.2 Gadolinium (Gd)

Gadolinium-based contrast agents have been commonly used in the diagnosis for the last 30 years, and now it is increasing the functional and improved applications. Gd shows increased relaxivity when it is in contact with Zn^{+2} ions [97]. Zn^{+2} ions are important for various enzyme catalyst reactions; they are used as the biomarker for insulin secretion in β-cells. One of the recent examples is that the prostate will have a very high volume of Zn^{+2}, leading to better MRI contrast. One of the examples is that collagen regulation is deranged in the diseased cell or cancer cells, and sometimes excessive production is seen in common liver conditions, which will cause alcohol and/or drug abuse. Thus, magnetic resonance imaging can be used for the excessive detection of collagen with Gd NP-based contrast agents. Due to the increased r1 relaxivity, Gd is being attached to the larger molecule, to the target molecule, with the increased permeation and retention effects, ultimately increasing the MRI contrast [98]. Gadolinium acts as a carrier as well as the contrast agent for the IL-13 liposome to bypass the blood–brain barrier and is used as the interleukin-13 receptor for the detection of glioma [99]. The use of dendrimers has many advantages in nanomedical imaging and MRI applications due to properties like rigidity, low polydispersity, and also the ease of surface modification. Dendrimers in MRI cause cell tracking, lymph node imaging, blood pool imaging, and tumor-targeted theragnostic. Gadolinium is a paramagnetic agent having the highest relaxivities due to the high rotation time of the large dendrimer molecules. The relaxivity per Gd (III) ion in the dendrimer is enhanced up to sixfold compared to that of a single Gd (III) chelate. So, the Dendrimer-based Gd contrast agents give excellent contrast on 3D time of flight MR angiograms. In one of the studies, the Arg-Gly-Asp-Phe-Lys (MPa)(RGD) peptide complex is being used as a targeting moiety along with the multimodal Gd dendrimer contrast agent and gold nanoparticles (Au NPs) as carriers. In neutron capture therapy, the dendrimers are irradiated with an external neutron beam, and the use of the dendrimer-bound Gd will generate the auger electrons, which will be very highly cytotoxic to tumor cells. This concept of the higher accumulation of Gd in the target cell is being experimented with in ovarian carcinomas. MRI and the manganese dendrimer are being used as the dual-mode agent for CT-MRI [100]. Gadolinium oxide with diethylene glycol polymer and magneto-liposome NPs in Hepa 1–6 cell lines can also be used as a positive MRI contrast agent and as the marker for cell tracking [101]. Gadolinium chelates are safe, but nowadays, there is the Gd-based contrast agent associated with the development of nephrogenic systemic fibrosis [102]. Some of the studies also have shown that Gd-based contrast agents can be deposited in human bone and brain tissue, despite the existing normal renal function [103, 104].

3.5.3 Superparamagnetic Iron Oxide Nanoparticles (SPIONs)

Magnetic resonance imaging contrast agents are either T1 (positive) or T2 (negative). T1 positive contrast is commonly used due to its better evaluation of the hemorrhage and air tissue. Gadolinium-based contrast agents are T1 contrast agents, having very good contrast enhancement, and despite this, they are not safe completely. Superparamagnetic iron oxide nanoparticles are now being used instead of Gd, and their size in diameter ranges from 1 to 100 nm. The larger SPIONs have features like the T2 contrast agents, and the smaller SPIONs will be functioning as T1 [105]. The superparamagnetic iron oxide nanoparticles are of similar size and made up of small crystals of iron oxide that are coated with the organic compounds.

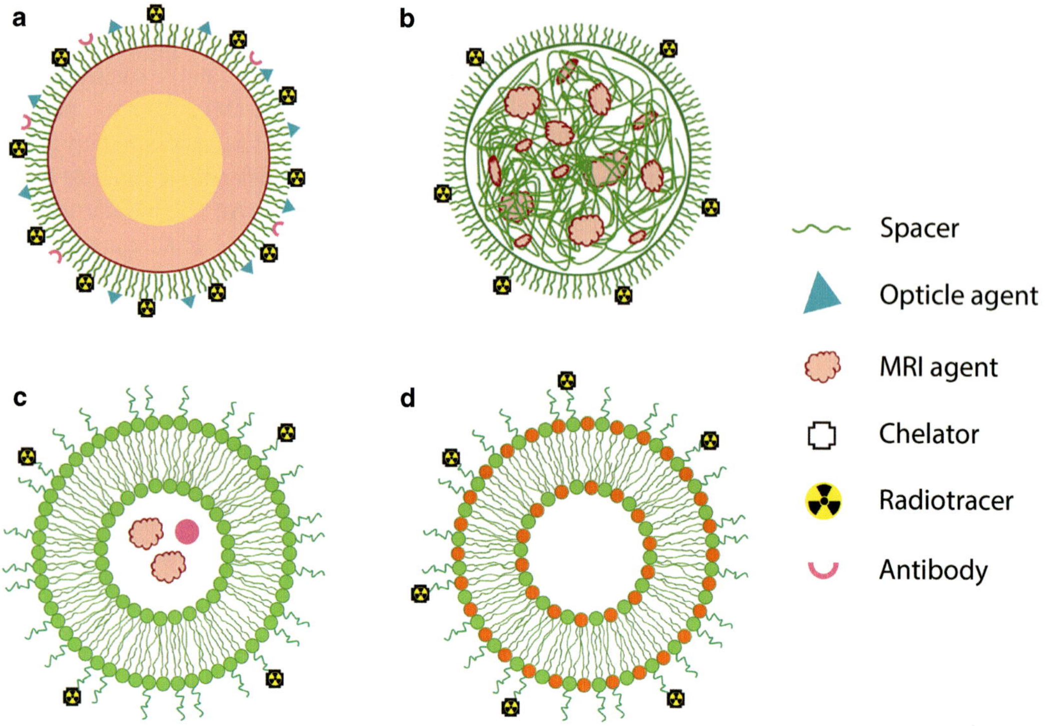

Diagram 3.3 Different forms of the nanospheres with different components that are attached and can be transferred to the cells

Superparamagnetic iron oxide nanoparticles can get conjugated into various particles, like antibodies, and can be seen as the drug carrier in cancer therapy [106]. SPION is also used in the stem cells, magnetic resonance imaging can monitor these stem cells, and cytotoxicity can be seen. Stem cell tracking can be seen in glioma-bearing mice in animal imaging [107]. Thus, superparamagnetic iron oxide nanoparticles can be used for both labeling and monitoring [108]. Shapewise, the spherical magnetic iron oxide NPs will have a better T2 relaxation rate with very strong relaxation properties of the NPs before and after NP administration. Thus, MRI can be used as a more specific agent in tumor imaging and also in whole-body imaging [109]. Superparamagnetic iron oxide nanoparticles have better T2 contrast agents but do not have as good contrast enhancement as Gd-based contrast agents despite the gadolinium-based contrast agents having toxicity concerns [110].

Magnetic resonance imaging is the most widely used for excellent spatial resolution. Inflammation is a process where the tissue healing mechanisms start. In various conditions like infection, ischemia, atherosclerosis, and metastasis, the associated imaging and diagnosis of the inflammation along with the damage done to the healthy tissue can be monitored (Diagram 3.3). Target inflammatory cells can be combined with the SPIONs by internalization of the NPs. Superparamagnetic iron oxide nanoparticle-tagged macrophages will have more affinity for the inflammatory processes. In tumor imaging and targeting, the macromolecular antibodies with cancer cell surface receptors are also seen as the various moieties of NPs due to their high specificity. The various nanoparticles with LHRH can be seen accumulating in primary tumor cells and metastatic cells. Folate receptors are overexpressed in cancerous tissues, and due to this, the SPIONs can be coated with different components: PEG, dextran, and 2-carboxyethyl phosphoric acid target-specific binding [111].

3.5.4 Carbon-13 (13C)

Carbon-13 (13C) is the backbone for all organic molecules. And relevant to various biological processes. MRI signal of carbon-13 is very low as it is widely seen (1.1%). The hyperpolarization of 13C will lead to a significant increase in the signal, which is seen more than 10,000-fold and due to this it will allow the dynamic evaluation of the various metabolism and physiologic processes. Pyruvate is also seen as the metabolic agent having various diseases like ischemia, inflammation, and cancer, monitoring response to therapies, energy metabolism, and changes related to hypoxia and oxidative stress [112].

Pyruvate is used to evaluate cancer progression, to characterize cancer biology, and is also useful as a biological marker [113]. The Warburg effect shows that the cancer cells will show increased glycolysis and lactic acid fermentation. Lactate dehydrogenase-mediated conversion of pyruvate to lactate is significantly elevated in malignant cells due to the Warburg effect. Increased glutathione results in the increased conversion of the 1–13C dehydroascorbate to 1–13C vitamin C [114]. NP-based pyruvate biosensors are already developed so that this can be easily detected with the total pyruvate level [115].

3.5.5 Nanodiamonds

Nanodiamond is a safer substrate being used for drug delivery and cellular tracking (fluorescent marker). The Overhauser effect is a proton–electron polarization transfer technique that can be used to increase the high contrast MRI of nanodiamonds in water in an ultra-low magnetic field. At ultra-low magnetic fields, the efficient Overhauser polarization transfer between electronic and nuclear spins is seen. Radiofrequency pulsing of the electron paramagnetic resonance transition between MRI signals will be transferring the spin polarization from the paramagnetic centers on the nanodiamond to 1H nuclei in the surrounding water. Thus, the nanodiamond in water will produce the enhancement in the 1H MRI signal, with a higher contrast sensitivity to the nanodiamond concentrations [116].

3.5.6 Carbon Nanotubes

Carbon nanotubes are single-walled or multi-walled having the diameters in the nm and length in the μm (Diagram 3.4). Carbon nanotubes are capable of being altered in living cells, and due to this, they are widely used in imaging and therapy. The electromagnetic properties of carbon nanotubes are very sensitive, even in photoacoustic molecular imaging and NIR imaging [117]. Photoacoustic imaging has better resolution and more depth than optical imaging. One of the preclinical studies will show eight times the photoacoustic signal in the tumor [118]. The single-walled carbon nanotubes combined with the sodium cholate are capable of producing high-resolution imaging having deep tissue penetration and lower autofluorescence in the NIR region beyond 1 μm [119] (Diagram 3.5).

3.5.7 Graphene

Graphene is a single layer of carbon atoms in the pattern of the 2D honeycomb lattice. They have excellent physicochemical and biological properties with variable size, lower toxicity, and excellent biocompatibility. This can also be seen based on the graphene-based nanomaterials, namely graphene–dye conjugates, graphene–antibody conjugates, graphene-NP–NP composites, and graphene quantum dots, which can be used as in vitro and in vivo imaging agents for molecular imaging [120]. Nanodots will have better visualization of tumors, and due to this, they can be used as an excellent optical imaging agent for oncological applications in deeper tissues [121].

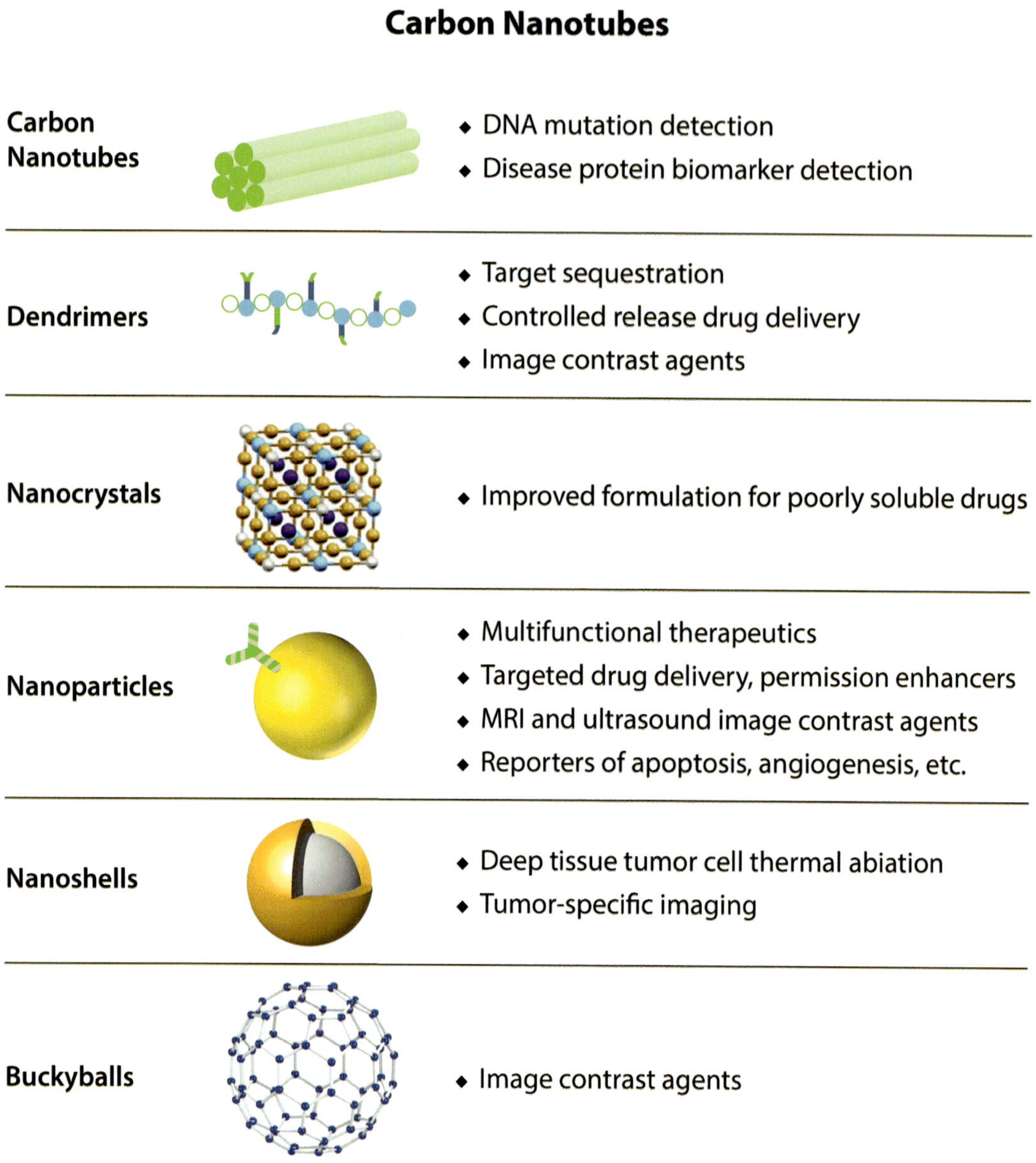

Diagram 3.4 Different types of nanoparticles shown

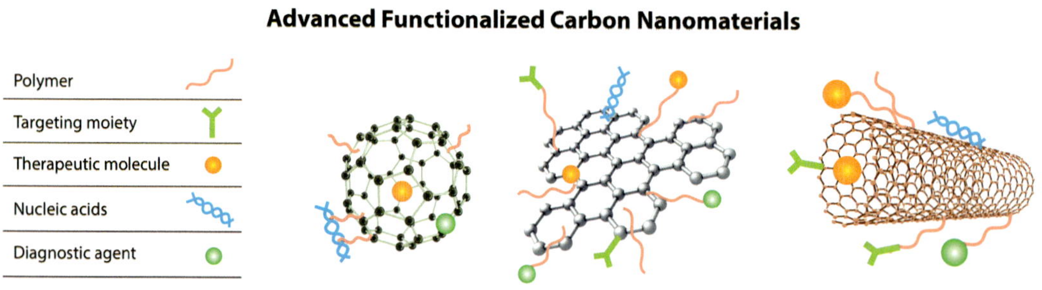

Diagram 3.5 Different types of nanoparticles with various functions

3.5.8 Manganese (Mn)

Manganese-based contrast agents have excellent biocompatibility with the short circulation time of Mn(II) ion chelate in the T1 weighted image. These NPs have very little toxicity and also good T1-weighted contrast effects. But when they are stored in the reticuloendothelial system, liver, and spleen, then they will have toxicity. The manganese oxide NPs will be excreted through the hepatobiliary pathway. The use of the polyethylene glycol coating will be able to reduce the toxicity with manganese oxide NPs [122].

3.5.9 Silicon (29Si)

Hyperpolarized silicon is also being used in various MRI applications. The large silicon particles will have a larger polarization than NPs. A silicon-based contrast agent shortens the nuclear spin-lattice relaxation time (T1) of the protons of nearby tissues and leads to long acquisition times. These can be resolved by the hyperpolarization techniques. The biggest advantage of silicon-based contrast agents is that when they are attached to the molecules on the surface of the particles, they will not reduce the nuclear magnetic resonance properties [123].

3.5.10 Peptide

Atherosclerosis is the leading cause of morbidity and mortality in the United States. Atherosclerosis is a chronic and inflammatory disease with the earliest identification of unstable plaques. Magnetic resonance imaging is an important imaging modality to image and characterize the blood vessel wall and plaque. Peptide-based NPs enhancing MRI images have important biodegradable properties and inherent biocompatibility. The peptide amphiphile micelles can be used in two types of amphiphilic molecules, which will contain a Gd chelator, and this enhances the image quality, and due to this, they are being used in dual optical imaging-MRI [124].

3.6 Computed Tomography

Computed tomography uses the X-ray source and a detector array to form images. It is capable of having a very high spatial and temporal resolution. It is capable of excellent anatomical details of the various tissues with the various cardiovascular, gastrointestinal, liver, and lung. The sensitivity towards the contrast agents is not precise, and due to this, newer contrast agents are available for CT [125].

3.6.1 Gold Nanoparticles

Gold nanoparticles use X-ray attenuation properties and can also be easily modified Au NPs can be functionalized with glucosamine as an excellent contrast agent [126]. Gold nanoparticles have a very high X-ray absorption coefficient, and this can be used for the imaging of various neoplasms as well as various pathologies where there is a more enhanced permeability and retention effect (EPR). In breast cancer, these can be combined with the PEG chains and various tumor biomarkers, and these can give more enhanced images in CT due to better resolution [127]. The use of mesenchymal stem cell, which is a type of adult stem cell that is being more commonly used in regenerative therapy and commonly used in various autoimmune, neurodegenerative, and cardiovascular diseases. The use of the mesenchymal cells, which will be using Au NPs as a marker, can provide better resolution and image quality [128]. Again, the various aspects are the porous Au NPs, which have a brighter contrast of 45 HU in comparison to the solid AuNPs, which show close to half contrast (26 HU) on the computed tomography scans [129]. CT-based nanoparticles are being developed, like the biodegradable poly-di (carboxylatophenoxy) phosphazene into gold nanospheres, which function as contrast agents, and thus, the contrast agents can give better contrast [130].

3.6.2 Iodine (131I)

Iodine-based NP contrast agents are more useful for the use of highly vascular lesions and tumor loading. These are also very small (~20 nm) in size, having better penetration into the tissues in comparison to the larger NPs. These contrast agents have a longer half-life (40 h), which will show significant tumor uptake and have better clearance in the liver in comparison to the Au NPs, and these will be more focused tumors, which will provide excellent contrast in the vascular tumor imaging [131]. In thyroid pathologies and radionuclide therapy, iodine is being used as it has a very high affinity towards the thyroid and will have a longer half-life (8.01 days). The property of the gamma emission makes it feasible for SPECT imaging, and with the concept of the beta minus decay, it will be useful for various therapeutic purposes. The functionalized polyethyleneimine (PEI)-entrapped Au NPs, combined with the targeting peptide, can be used in CT/SPECT imaging and also for the radionuclide therapy of glioma [132]. The sentinel lymph node evaluation with the tracer comprising of iodine and docetaxel for the evaluation of the locoregional chemotherapy for the lymphatic system [133]. The viability of the iodine is also seen in the dual-energy micro-CT with enhancing the various energy-integrating detectors with the photon-counting detector-based spectral micro-CT. And these detectors provided the background signals for better evaluation of the intratumoral NPs with the dual-energy micro-CT comprising of the energy-integrating detectors [134].

3.6.3 Bismuth

A hybrid cluster was prepared using the PEG 2000-DSPE, which contains hydrophobic bismuth (Bi2S3) NPs and quantum dots, and these are being used as the combined CT/fluorescence imaging components. The contrast enhancement in CT imaging of the liver and spleen was seen after 30 min and was persistent for more than 4 h. This phenomenon has shown that the probe has very good biocompatibility and will not disturb normal physiological function [135].

3.7 Positron Emission Tomography

Positron emission tomography (PET) is a hybrid imaging technique that uses radiotracers to produce images based on the radionuclide distribution. By the use of these radiotracers, information about the biological pathways can be evaluated non-invasively [136].

3.7.1 Gold Nanoparticles in PET

Gold nanoparticles are one of the commonly used NPs in PET. Due to this, the visualization of dendritic cell migration is seen more precisely for the biocompatible and stable imaging agents and their functioning. The evaluation of the dendritic cell migration is important for the evaluation of dendritic-based immunotherapy, good labeling efficiency, high radiosensitivity, and excellent chemical stability. It does not have any major reaction on the cell's biological function, which comprises the phenotype marker expression and proliferation showing the successful tracking of the tracking dendritic cell migration [137]. The changes in 18F–FDG uptake along with the apparent diffusion coefficient, will correlate with tumor survival, and based on this principle, it can be used for early evaluation of PTT [138]. A new 124I-labeled gold nanostar probe using PET is being used for the evaluation of sub-mm intracranial brain tumor detection, and this has become more sensitive for non-invasive imaging evaluation of tumors [139]. Positron emission tomography in combination with Cerenkov luminescent imaging has shown the tumor uptake after injection of NPs and signals will remain till 24 h [140].

3.7.2 Copper (64Cu)

Copper sulfide (CuS) NPs have a noncomplex chemistry, low toxicity, and small particle size, thus making them one of the most ideal contrast imaging agents. Radioactive [64Cu] CuS NPs are also used as radiotracers for PET and photothermal ablation therapy using near-infrared NIR laser irradiation and also for theranostic applications [141]. A study conjugated natriuretic peptide receptor-binding peptide (targeting entity) with a C-type atrial natriuretic factor to demonstrate the PET imaging agent to evaluate the atherosclerosis progression [142]. PET quantitative imaging of the macrophages in tumors is being pharmacokinetically developed for optimal visualization [143]. The use of these anti-prostate membrane antigens has shown a 1.6-fold increase in tumor targeting over the nontargeted lipid NPs and is being used as the source of the potential image enhancer [144].

3.7.3 Other Nanoparticles

The inflammatory macrophages will be destroying the tissue, which will lead to atherosclerosis, myocardial infarction, and also heart failure. The use of the monitoring macrophages as the criteria is also seen in the patients who are being seen for avoiding or early management of various diseases. 18F-Macroflor-modified polyglucose NPs are small in size, have high macrophage avidity, and are excreted really. The use of the Macroflor will enrich the cardiac and plaque macrophages with a significant increase in the PET signal [145]. The use of the active azide group in PET imaging shows its compatible properties for the use of the PET image-guided drug-delivery system [146]. The pH-sensitive pharmaceutical-grade carboxymethylcellulose-based NPs were introduced for white blood cells to be used in PET imaging. 68Ga3+ was used for labeling, which provides a more precise and greater spatial resolution, and it shows the advantage of PET over SPECT [147]. The use of the various polyphenol and poloxamer self-assembled supramolecular NPs will have better hydrogen bonding between tannic acid and Pluronic F-127 in combination with the hydrophobic reactions of poly (propylene oxide) chains that are being applied in NIRF and PET imaging. These agents have very good biocompatibility in different cell lines and in vitro, and the major component is that they do not cause hemolysis [148].

3.7.4 Cerium Oxide NPs Have a Unique Surface Chemistry

Cerium oxide NPs coated with 89Zr, are also being used clinically as the PET isotope, in PET imaging, and also for in vivo biodistribution, and this shows better resolution in all agents, which has great potential for use in PET imaging [149].

3.8 Single-Photon Emission Computerized Tomography

Single-photon emission computerized tomography is a nuclear imaging technique that uses gamma rays to evaluate the biochemical changes as the molecular target. SPECT is an already existing technique due to the 99mTc [150].

3.8.1 Gold Nanoparticles in SPECT

Gold nanoparticles have very high sensitivity for use in SPECT imaging and also as a contrast agent [151]. Technetium (99mTc)-radiolabeled Au NPs combined with the mannose are helpful for the tracking of the tracer in the lymph nodes, and this can be used as the SPECT contrast agent for lymphatic mapping [152].

3.8.2 Technetium (99mTc)

SPECT is an already established imaging modality for the evaluation of neoplastic lesions and also for metastases in the early stage. In one of the studies, it shows that cervical cancer patients evaluated the accuracy of 99mTc SPECT/MRI fusion that was being used for the selective sentinel lymph nodes (SLN) evaluation. In this study, the use of the size criteria and the absence of sharp demarcation for the presence of metastasis in cervical cancer patients without enlarged lymph nodes [153]. Thus, 99mTc is being used as the targeting agent in SPECT for mapping SLN in preoperative assessment in early-stage cervical cancer in women [154]. Single-photon emission computerized tomography also uses the 2D planer lymphoscintigraphy along with 99mTc

NP for dynamic sentinel lymph node biopsy in penile cancer patients and the detection rate of true tracer lymph nodes with more precise anatomic localization [155]. SPECT/CT is also used for the evaluation of the contralateral nodal failure for unilateral elective nodal irradiation in head-and-neck squamous cell carcinoma patients [156]. Sentinel lymph node biopsy can be done after the more precise evaluation of the intertumoral injection of 99mTc-labeled nanocolloid with SPECT/CT in renal tumors, which is feasible despite very high rejection rates [157].

3.9 Optical Imaging

Non-invasive optical imaging is an important imaging modality for the evaluation of the various macroscopic and microscopic dynamic levels. The use of optical imaging is seen in fluorescence, chemiluminescence, and Raman imaging, which can obtain non-invasive 2D or multidimensional image data. Fluorescence imaging is less time-consuming, and this is the reason researchers use it in biomedical imaging applications [158].

3.10 Fluorescence

Fluorescent NPs are more precisely used for the localization of molecules or various processes in living organisms or cell cultures. In this imaging, the excitation and absorption wavelengths should be in the NIR optical window, which is the optimal level of 700–750 nm excitation and 750–800 nm emission wavelengths [159]. Despite this, the fluorescence limitations are the longer time in biomarker analysis, immunoassays, and diagnostic imaging, as well as various fallacies like the wavelength range, photobleaching, and fluorescence self-quenching. This use will provide better stability, reduced photobleaching, and reduced toxicity [160]. To localize tumor pathology with a tumor to normal tissue ratio of 11.7 ± 1.3. one of the studies showed great potential in optical imaging application [161]. The use of the cross-linked functionalized layer will use 90% of the nanocomplex excretion within 2 weeks and with very little toxicity [162]. Indocyanine green (ICG) is an FDA-approved contrast and shows a significantly higher contrast-to-signal ratio for the evaluation of the vasculature for up to 4 h post-injection [163]. Fluorescent probes are very important for cancer diagnosis and surgery and are also being used as contrast agents. This will show the real-time evaluation of the cancer and adjacent normal tissue. Fluorescent Au NPs conjugated with diatriozoic acid and nucleolin targeted, the aptamer are being used as the molecular contrast agent for better tumor location in the, causing higher attenuation, good biocompatibility, high water solubility, strong X-ray attenuation, and visible fluorescence [164]. Fluorine MRI detects in vitro and vivo metabolism by tracking and monitoring immune cells and cancer cells. This shows the increased uptake in triple-negative breast cancer and ovarian cancer associated with minimum accumulation in off-target tissue [165]. Calcium-based NPs, which are fluorescent-based, will correlate the neuronal activity in the brain with improved spatial-temporal and chemical resolution, thus enabling the neurochemical circuit mapping with finer precision [166]. Glucocorticoids use in the inflammatory disorders show the NPs delivery, which will be controlled and monitored for the very small adverse reactions. In vivo glucocorticoid betamethasone phosphate, NPs, and fluorescent dye DY-647 can precisely improve the evaluation of the inflammation along with the monitoring of the drug delivery [167].

3.11 Quantum Dots

Fluorescence imaging is the evaluation of the real-time cellular function. Graphene quantum dots have good optical properties with unique structural features. These graphene quantum dots are on the surface of NPs like the core–shell NPs, and they have longer metabolism in the vascular system and are capable of monitoring the distribution and metabolism of NPs [168]. Mesoporous silica NPs with graphene quantum dots can simultaneously monitor real-time localization of the doxorubicin (DOX) as the carrier for drug targeting and also a fluorescent imaging agent. The other factors are pH and temperature-dependent release, which will be useful for the theranostic applications. The fluorescence properties will make them useful for the optical imaging with minimal cytotoxicity [169].

Quantum dots have better optical characteristics as compared to organic dyes. But they are toxic, and their potential toxicity is one of the disadvantages. In vivo, endoscopic imaging of human bladder cancer and the topical administration of quantum dots that are being conjugated to anti-CD47 are showing better results with no obvious acute toxicity till 7 days after installation [170]. The use of broad spectra and large stokes can be used for the multiple types of quantum dots for evaluation of the targeted imaging [171]. Lead sulfide quantum dots having 1100 nm emission peaks are being used in NIR fluorescence imaging of cerebral venous thrombosis. And important modality for the evaluation of the pathological state of cerebral blood vessels [172]. Indium arsenide quantum dots are high-resolution multicolor imaging structures that are easily modifiable, have better deep penetration, and faster acquisition speed, and also evaluate the metabolic turnover rates of lipoproteins in several organs at one time with 3D quantitative flow maps [173]. Carbon quantum dots (CDs) have excitation-dependent emissions with high fluorescence quantum yields, photostability, and long photoluminescence decay lifetime [174]. A versatile imaging probe with highly luminescent cadmium-free shell quantum dots conjugated to tumor-targeting peptide is being used due to the longer circulation time, excellent photostability, and minimal toxicity. They have very good potential for optical imaging [175] (Diagram 3.6a–d).

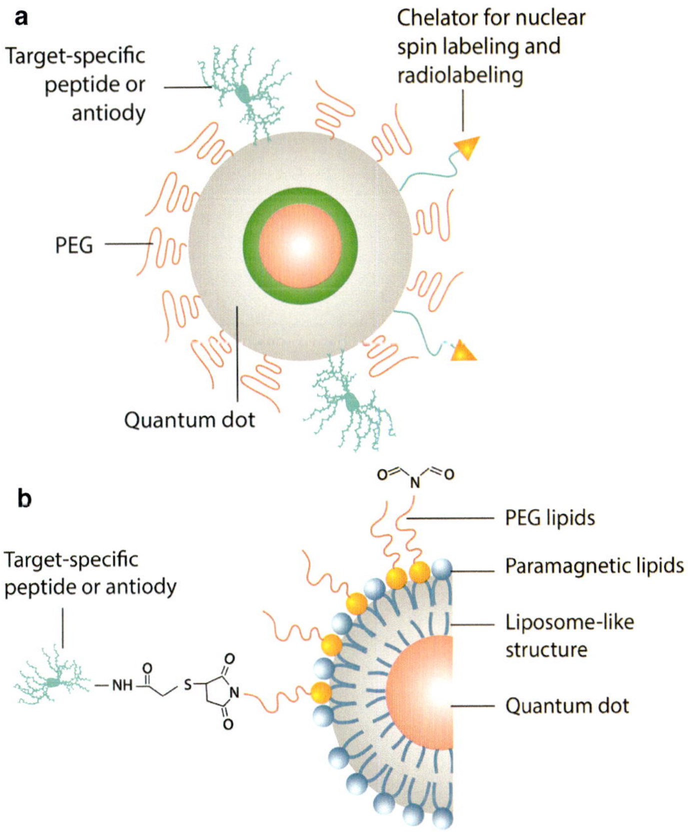

Diagram 3.6 Quantum dots different roles. (**a**) Different roles of the various targets as well as the different chelators by the use of the quantum dots. (**b**) Role of the various PEG ligands' use of the target's specific peptide and antibodies with the quantum dots. (**c**) Role of the quantum dots in the specific targeting peptides with the antibody with different hydrophobic components. (**d**) The role of the quantum dots in the antibody interaction

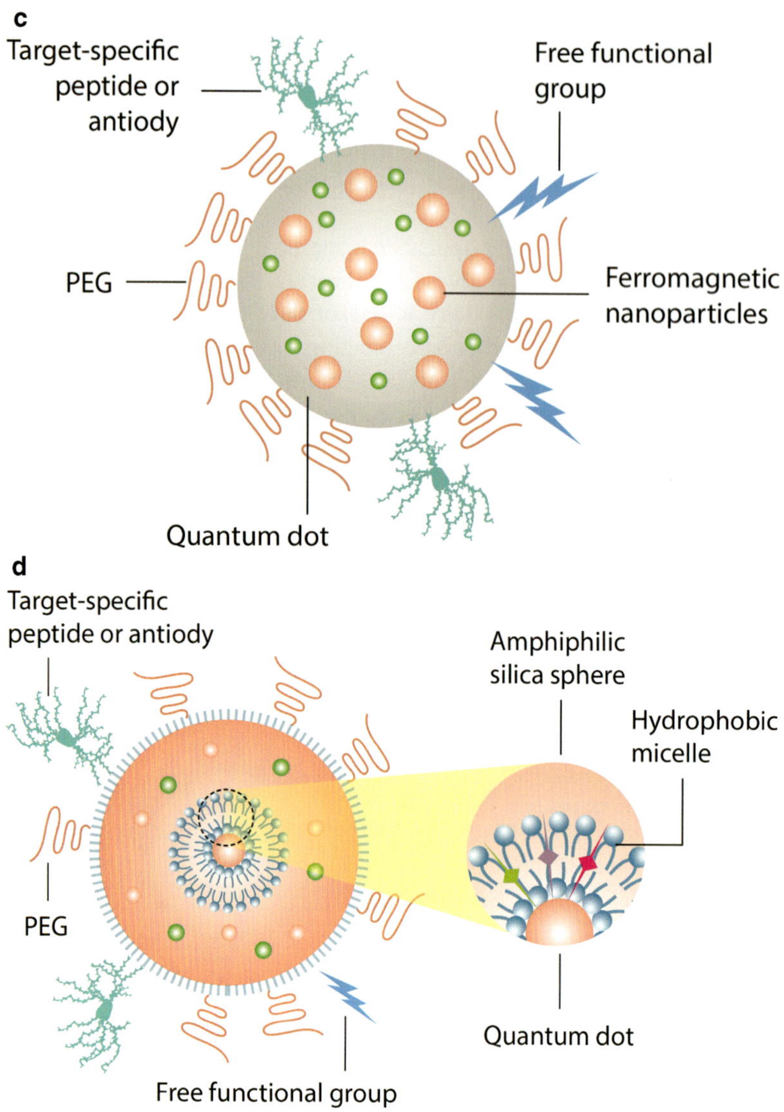

Fig. 3.6 (continued)

3.12 Gold Nanoparticles in mRNA

The identification and characterization of various disease-related mRNA in cells is of better relevance for the early diagnosis and treatment of numerous diseases. The oligonucleotide-functionalized Au NPs are being used as they have higher stability, high intracellular delivery efficiency by endocytosis, and high signal-to-background ratio for mRNA detection. Which are capable of detecting the intracellular mRNA level [176]. The immobilization of organic Cy5-NH2 dyes onto the Au NPs which are being used for the monitoring and cancer cells for determining their localization in cells by fluorescence microscopy by accumulation in the tumor after intravenous injection [177].

3.13 Persistent Luminescence NPs

Persistent luminescence NPs (PLNPs) are the newer imaging agents which are being developed as biomedical imaging agents based on the principle of the bioluminescence and fluorescence imaging as their optical properties are variable based on the chemical and physical properties like the composition, size, and surface nature. The PL mechanism is that when the PL materials are irradiated by light and the materials are charged until the excitation is stopped. Then, the PL materials emit light [178]. Preclinical studies on bioimaging were carried out to show that PLNPs have the advantages of high signal-to-noise ratio, high sensitivity, deep Nanomaterials 2020, 10, 1700 13 of 40 penetration, and no interference from tissue autofluorescence [179, 180]. For biocompatibility basis of the action is that the PLNPs, by using the zinc gallate (ZGO) in vivo when injected in mice show a direct proportionate of the weight gain in the different settings [181].

3.14 Ultrasound

Ultrasound is a non-invasive imaging technique used for the evaluation of the morphology, internal structure, orientation, and margins of the lesion multiplayer with high resolution [182]. Ultrasound-guided drug delivery using nanobubbles (NBs) has been used in recent years. These nanobubbles are made up of gas cores and stabilized shells and can cross the capillary wall and reach the targeted regions of interest. The use of the 5-fluorouracil with the NBs for hepatocellular carcinoma is seen. Nanobubbles-mediated DOX uptake and apoptosis have shown excellent drug-loading capability with better ultrasound enhancement [183] (Diagram 3.7). NBs are conjugated to poly (lactic-co-glycolic acid) and are capable of carrying the DOX as a cancer drug. The multifunctionality makes them the best theranostic agent with a drug-loading efficiency of 78.6% and an encapsulation efficiency of 7.4% [184]. Blood–brain barrier opening is also and tacked with the real-time evaluation of these NPs [185]. An ultrasound-responsive phosphatidylserine-based paclitaxel liposomes NBs conjugate that have a proapoptotic effect toward enhanced anticancer efficacy and image guidance was studied. The use of the significant reduction in tumor proliferation index [186]. One of the studies shows the intracellular delivery of the antibody in the cell was by using the Cy3-streptavidin with negligible toxicity [187].

Fluorescence up-conversion NPs are highly sensitive and can function as nanocarriers. They can be tumor-specific with NIR light. NIR has a lesser depth penetration of approximately 15 mm. The use of fluorescence with ultrasound can show the shortcoming of the high-resolution signal-to-noise ratio. The blend of the Nd3+-sensitized up-conversion NPs, graphic carbon nanodots, and NBs is being used for modality-based imaging

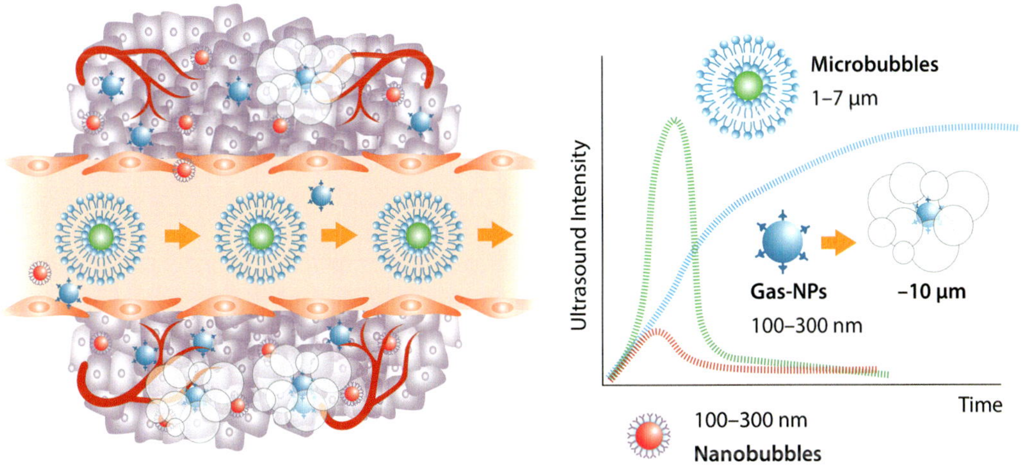

Diagram 3.7 Role of different ultrasound intensities with nanobubbles and microbubbles

[188]. Ultrasound-targeted and drug-loaded NBs can also see the desired destruction, selective growth inhibition, and apoptosis in HepG3 cells in vitro [189]. The tumor temperature is being used for the NB-liposome as compared to ultrasound causing tissue necrosis on ultrasound [190]. The use of the mitomycin-C showed significantly lesser tumor progression by the use of less than 50% lower concentration of chemotherapeutic drug [191]. An antitumor-targeted NBs conjugated are being used in prostatic cancer, which has binding capabilities to bind prostate-specific membrane antigens resulting in reduced efficiency, the use of cell cycle arrest, and cell apoptosis [192].

3.15 Photoacoustic Imaging

The photoacoustic imaging (PAI) concept is based on the photoacoustic effect. It generates ultrasound signals due to thermal expansion [193]. Photoacoustic imaging is also known as the optoacoustic imaging. This is based on the low cost, which will provide imaging of blood vessels with the higher spatial and temporal resolution and optimal depth [194].

3.15.1 Gold Nanoparticles in PAI

Metastases are a more important indicator of mortality than the primary lesion. Ultrasound-guided PAI is useful for the evaluation of the sentinel lymph node metastases by using the Au NPs (GC-Au NPs). This modality and technique can easily detect metastasis from non-metastatic lymph nodes using Au NPs [195]. This study has shown great interest in the murine sentinel lymph nodes [196]. Ultrasound-guided PAI and anti-epidermal growth factor (EGFR) antibodies ARE conjugated to gold nanorods which can effectively detect EGF-expressing primary tumors with lymph node metastases. The PAI signal in tumor and axillary lymph node metastases can be compared to the MCF-7 tumor and non-lymph node metastases [197]. Nanorod shows the surface plasmon resonance absorption which band covering the NIR range (~680–1100 nm). Gold nanorods showed good biocompatibility and stability with better contrast enhancement in PAI (3.1 times to the control group) [198]. As Au NPs contrast agents in PAI, the clearance is seen by Au NPs below the renal clearance (10 nm). Ultra-small 5 nm Au NPs are being used as molecular-activated plasmonic nano sensors for molecular-specific PAI [199]. The use of nanorods are 5–11 times smaller than regular gold nanorods, three times more thermally stable, and 3.5 times stronger PA signal under nanosecond pulsed laser illumination and verified in this study [200]. The use of gold nanoparticles which are being coated with the glycol chitosan (GC). The endocytosis of GC-Au NPs was also confirmed with dark microscopy, which is beneficial for minimizing toxicity [201]. Photoacoustic tomography (PAT) is the advanced emerging imaging technique for imaging cells by using contrast agents such as NPs and pigments. Astaxanthin-based Au NPs will have great applications in the future for PAI and therapy [202]. Blood–brain barrier (BBB) is the major drawback of the NP movement through the blood–brain barrier. This contrast agent had strong optical absorption, which allowed for visualization of the agent using ultrasound-guided PAI [203]. The in vivo imaging of Au MBs in a murine kidney model has also been shown as an alternative for the existing contrast agents for PAI [204].

3.15.2 Carbon Nanotube (CNT)

Carbon-based NPs have the unique physico-chemical properties [205]. In one study, a single-wall CNT complex with long circulation and capable of providing fluorescence and PAI of tumor evaluation in the therapeutic time window [206].

The NIR wavelengths will provide an excellent photoacoustic signal. In vivo, they accumulated in a mouse lymph node in sentinel lymph node detection [207]. The nanocomplex is made up of the various split fluorescent protein frag-

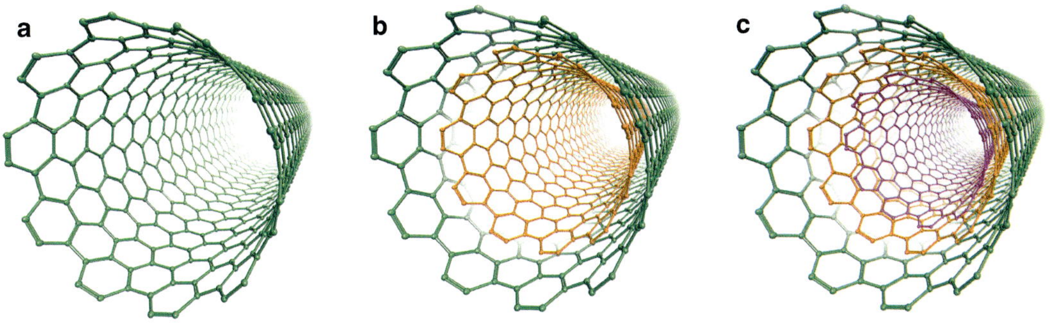

Diagram 3.8 Different types of nanoparticle mesh used, which can be altered

ments used as molecular glue and switchable Raman reporters used as the clusters. The fluorescent protein-driven assembly of metal colloids can be used as the yielding PA signal by the PAI agent [208]. A photoacoustic contrast agent will be allowing the PAI and analysis of prostate cancer [209]. Photoacoustic imaging has a penetration depth of a few cm and can generate better contrast from melanin and oxy−/deoxyhaemoglobin. ICG is a small molecule dye which is having very fast clearance, bleaching effects, and rapid protein binding, and it can be used for PAI. These NPs decrease their photobleaching and increase the retention of signals within the cells [210]. Core–shell silica PEG NPs are being developed with photothermal, photoacoustic, and nanomaterials having the NIR optical imaging properties and showing outstanding stability and enhanced photoacoustic signal [211]. Photoswitchable hybrid probes attached with the thermochromic dye and absorbing NPs sow the temperature-sensitive light–dark states and spectral shifts in absorption can be switched through controllable photothermal heating of doped NPs resulting in the higher contrast [212]. Hypoxia is proportional to tumor aggressiveness and poor treatment outcome. So the early evaluation of hypoxic tumor cells Hypoxia-activated NPs again can increase the photoacoustic intensity, fluorescence, and chemotherapy efficiencies. They activate in the hypoxic condition when they extravasate into the hypoxic tumor microenvironment. Azobenzene hypoxia-activated fluorescence nanoparticles have a high potential to be used in PAI [213].

The newer therapeutic nanoparticles are examples, Nanocage, Doxil (Janssen, Horsham, PA), Ambisome (Gilead, Foster City, CA), and Abraxane (Celgene, Summit, NJ), due to this the different applications can be done in preclinical and clinical trials [214–217] (Diagram 3.8). The nano-diagnostics must be excreted or metabolized in very local, oral, or parenteral administration and don't have any side effects. For the intravenously administered nanotherapeutics, the size of nanoparticles is proportional to the accumulation in the tissues without any side effects. The purpose of leaky neo vasculature of tumors and inflammation will lead to an increased permeability and retention (EPR) effect leading to increased therapeutic efficacy [214–216, 218–220]. The EPR is based on the long-circulating properties of nanoparticles which gradually increase over a period, ranging from 12 and 48 h. Another concept is the use of the contrast enhancement necessary for the diagnosis. The various aspects of the systemically administered nanoparticles for diagnosis are the poor tissue-penetrating properties due to large size, having limitations to the target cells which are usually away from the blood vessels, more prolonged time in the liver and spleen which will cause more side effects and the chemical composition like heavy metal-based quantum dots. The use of various nanoparticles can act as blood pool agents which act like contrast agents and these can be used for labeling and tracking the stem cells, movement of the nanoparticles and also felicitating the various interventional vascular procedures. This also can be an important entity for the various theranostic applications.

3.16 Conclusion

Thus, the role of nanoimaging is very important for more precise and correct diagnosis of the diseases and pathologies that can be detected at suboptimal levels. This has revolutionized the concept of molecular imaging for diagnostic and therapeutic options as well. Various nanomaterials, various applications, newer approaches of drug targeting, imaging, and theranostic approaches have revolutionized the era of imaging leading to the concept of the personalized imaging.

References

1. Ratner M, Ratner D. Size matters. In: Ratner M, Ratner D, editors. Nanotechnology. Upper Saddle River, NJ: Prentice Hall; 2003. p. 11–8.
2. Kim BY, Rutka JT, Chan WC. Nanomedicine. N Engl J Med. 2010;363(25):2434–43.
3. Power S, Slattery MM, Lee MJ. Nanotechnology and its relationship to interventional radiology. Part I: imaging. Cardiovasc Intervent Radiol. 2011;34(2):221–6.
4. Power S, Slattery MM, Lee MJ. Nanotechnology and its relationship to interventional radiology. Part II: drug delivery, thermotherapy, and vascular intervention. Cardiovasc Intervent Radiol. 2011;34(4):676–90.
5. Goldberg SN, Bonn J, Dodd G, et al. Society of Interventional Radiology Interventional Oncology Task Force: interventional oncology research vision statement and critical assessment of the state of research affairs. J Vasc Interv Radiol. 2005;16(10):1287–94.
6. Drexler KE. Engines of creation. New York, NY: Anchor; 1986.
7. Drexler KE. Nanosystems: molecular machinery, manufacturing, and computation. New York, NY: Wiley Interscience; 1992.
8. Qin LC, Zhao X, Hirahara K, Miyamoto Y, Ando Y, Iijima S. The smallest cardon nanotube. Nature. 2000;408(6808):50.
9. Bao L, Cui X, Chen C. Nanomedicine. In: Toxicology for nanotechnology. Singapore: Springer Nature; 2023. p. 157–77.
10. Thrall JH. Nanotechnology and medicine. Radiology. 2004;230(2):315–8.
11. Gojova A, Guo B, Kota RS, Rutledge JC, Kennedy IM, Barakat AI. Induction of inflammation in vascular endothelial cells by metal oxide nanoparticles: effect of particle composition. Environ Health Perspect. 2007;115(3):403–9.
12. Buniyamin I, Akhir RM, Asli NA, Khusaimi Z, Malek MF, Mahmood MR. Nanotechnology applications in biomedical systems. Curr Nanomater. 2022;7:167–80. https://doi.org/10.2174/2405461507666220301121135.
13. Kalita M, Payne MM, Bossmann SH. Glyco-nanotechnology: a biomedical perspective. Nanomedicine. 2022;42:102542. https://doi.org/10.1016/j.nano.2022.102542.
14. Medina C, Santos-Martinez MJ, Radomski A, Corrigan OI, Radomski MW. Nanoparticles: pharmacological and toxicological significance. Br J Pharmacol. 2007;150(5):552–8.
15. Nel A, Xia T, Madler L, Li N. Toxic potential of materials at the nanolevel. Science. 2006;311(5761):622–7.
16. Lal HM, Uthaman A, Thomas S. Polymer nanocomposites based on silver nanoparticles. In: Silver nanoparticle as an effective antiviral agent. Berlin/Heidelberg: Springer; 2021. p. 247–65.
17. Seleverstov O, Zabirnyk O, Zscharnack M, Bulavina L, Nowicki M, Heinrich JM, Yezhelyev M, Emmrich F, O'Regan R, Bader A. Quantum dots for human mesenchymal stem cells labeling. A size-dependent autophagy activation. Nano Lett. 2006;6(12):2826–32.
18. Tsuji JS, Maynard AD, Howard PC, James JT, Lam CW, Warheit DB, Santamaria AB. Research strategies for safety evaluation of nanomaterials, part IV: risk assessment of nanoparticles. Toxicol Sci. 2006;89(1):42–50.
19. Warheit DB, Hoke RA, Finlay C, Donner EM, Reed KL, Sayes CM. Development of a base set of toxicity tests using ultrafine TiO2 particles as a component of nanoparticle risk management. Toxicol Lett. 2007;171(3):99–110.
20. Chung TH, Wu SH, Yao M, Lu CW, Lin YS, Hung Y, Mou CY, Chen YC, Huang DM. The effect of surface charge on the uptake and biological function of mesoporous silica nanoparticles in 3T3-L1 cells and human mesenchymal stem cells. Biomaterials. 2007;28(19):2959–66.
21. Foged C, Brodin B, Frokjaer S, Sundblad A. Particle size and surface charge affect particle uptake by human dendritic cells in an in vitro model. Int J Pharm. 2005;298(2):315–22.
22. Kim JS, Yoon TJ, Yu KN, Noh MS, Woo M, Kim BG, Lee KH, Sohn BH, Park SB, Lee JK, Cho MH. Cellular uptake of magnetic nanoparticle is mediated through energy-dependent endocytosis in A549 cells. J Vet Sci. 2006;7(4):321–6.
23. Lai SK, Hida K, Man ST, Chen C, Machamer C, Schroer TA, Hanes J. Privileged delivery of polymer nanoparticles to the perinuclear region of live cells via a non-clathrin, non-degradative pathway. Biomaterials. 2007;28(18):2876–84.
24. Rothen-Rutishauser BM, Schurch S, Haenni B, Kapp N, Gehr P. Interaction of fine particles and nanoparticles with red blood cells visualized with advanced microscopic techniques. Environ Sci Technol. 2006;40(14):4353–9.
25. Walsh M, Tangney M, O'Neill MJ, Larkin JO, Soden DM, McKenna SL, Darcy R, O'Sullivan GC,

O'Driscoll CM. Evaluation of cellular uptake and gene transfer efficiency of pegylated poly-L-lysine compacted DNA: implications for cancer gene therapy. Mol Pharm. 2006;3(6):644–53.
26. Huang M, Ma Z, Khor E, Lim LY. Uptake of FITC-chitosan nanoparticles by A549 cells. Pharm Res. 2002;19(10):1488–94.
27. Mousavi SA, Malerod L, Berg T, Kjeken R. Clathrin-dependent endocytosis. Biochem J. 2004;377(Pt 1):1–16.
28. Medda R, Jakobs S, Hell SW, Bewersdorf J. 4Pi microscopy of quantum dot-labeled cellular structures. J Struct Biol. 2006;156(3):517–23.
29. Tang HW, Yang XB, Kirkham J, Smith DA. Probing intrinsic and extrinsic components in single osteosarcoma cells by near-infrared surface-enhanced Raman scattering. Anal Chem. 2007;79(10):3646.
30. Yang PH, Sun X, Chiu JF, Sun H, He QY. Transferrin-mediated gold nanoparticle cellular uptake. Bioconjug Chem. 2005;16(3):494–6.
31. Lai B, Maser J, Paunesku T, Woloschak GE. Report on the Workshop of Biological Applications of X-ray Microbeams. Int J Radiat Biol. 2002;78(8):749–52.
32. Lai B, Maser J, Vogt S, Paunesku T, Woloschak GE. Workshop on biological applications of X-ray microscopy and imaging. Advanced Photon Source, Argonne National Laboratory, Argonne, IL, USA, 28-29 April 2003. Int J Radiat Biol. 2004;80(6):459–61.
33. McRae R, Lai B, Vogt S, Fahrni CJ. Correlative microXRF and optical immunofluorescence microscopy of adherent cells labeled with ultrasmall gold particles. J Struct Biol. 2006;155(1):22–9.
34. Paunesku T, Vogt S, Maser J, Lai B, Woloschak G. X-ray fluorescence microprobe imaging in biology and medicine. J Cell Biochem. 2006;99(6):1489–502.
35. Yang L, McRae R, Henary MM, Patel R, Lai B, Vogt S, Fahrni CJ. Imaging of the intracellular topography of copper with a fluorescent sensor and by synchrotron x-ray fluorescence microscopy. Proc Natl Acad Sci USA. 2005;102(32):11179–84.
36. Pfeifer MA, Williams GJ, Vartanyants IA, Harder R, Robinson IK. Three-dimensional mapping of a deformation field inside a nanocrystal. Nature. 2006;442(7098):63–6.
37. Robinson IK, Vartanyants IA, Williams GJ, Pfeifer MA, Pitney JA. Reconstruction of the shapes of gold nanocrystals using coherent x-ray diffraction. Phys Rev Lett. 2001;87(19):195505.
38. Miao J, Charalambous P, Kirz J, Sayre D. Extending the methodology of X-ray crystallography to allow imaging of micrometre-sized non-crystalline specimens. Nature. 1999;400:342–4.
39. Davda J, Labhasetwar V. Characterization of nanoparticle uptake by endothelial cells. Int J Pharm. 2002;233(1–2):51–9.
40. Harush-Frenkel O, Debotton N, Benita S, Altschuler Y. Targeting of nanoparticles to the clathrin-mediated endocytic pathway. Biochem Biophys Res Commun. 2007;353:26–32.
41. Kim SH, Jeong JH, Chun KW, Park TG. Target-specific cellular uptake of PLGA nanoparticles coated with poly(L-lysine)-poly(ethylene glycol)-folate conjugate. Langmuir. 2005;21(19):8852–7.
42. Qaddoumi MG, Gukasyan HJ, Davda J, Labhasetwar V, Kim KJ, Lee VH. Clathrin and caveolin-1 expression in primary pigmented rabbit conjunctival epithelial cells: role in PLGA nanoparticle endocytosis. Mol Vis. 2003;9:559–68.
43. Geiser M, Rothen-Rutishauser B, Kapp N, Schurch S, Kreyling W, Schulz H, Semmler M, Im Hof V, Heyder J, Gehr P. Ultrafine particles cross cellular membranes by nonphagocytic mechanisms in lungs and in cultured cells. Environ Health Perspect. 2005;113(11):1555–60.
44. Reddy GR, Bhojani MS, McConville P, Moody J, Moffat BA, Hall DE, Kim G, Koo YE, Woolliscroft MJ, Sugai JV, Johnson TD, Philbert MA, Kopelman R, Rehemtulla A, Ross BD. Vascular targeted nanoparticles for imaging and treatment of brain tumors. Clin Cancer Res. 2006;12(22):6677–86.
45. Kakizawa Y, Furukawa S, Kataoka K. Block copolymer-coated calcium phosphate nanoparticles sensing intracellular environment for oligodeoxynucleotide and siRNA delivery. J Control Release. 2004;97(2):345–56.
46. Sokolova V, Kovtun A, Heumann R, Epple M. Tracking the pathway of calcium phosphate/DNA nanoparticles during cell transfection by incorporation of red-fluorescing tetramethylrhodamine isothiocyanate-bovine serum albumin into these nanoparticles. J Biol Inorg Chem. 2007;12(2):174–9.
47. Peters K, Unger RE, Kirkpatrick CJ, Gatti AM, Monari E. Effects of nano-scaled particles on endothelial cell function in vitro: studies on viability, proliferation and inflammation. J Mater Sci Mater Med. 2004;15(4):321–5.
48. Zheng G, Chen J, Li H, Glickson JD. Rerouting lipoprotein nanoparticles to selected alternate receptors for the targeted delivery of cancer diagnostic and therapeutic agents. Proc Natl Acad Sci USA. 2005;102:17757–62.
49. Arnedo A, Irache JM, Merodio M, Espuelas Millan MS. Albumin nanoparticles improved the stability, nuclear accumulation and anticytomegaloviral activity of a phosphodiester oligonucleotide. J Control Release. 2004;94(1):217–27.
50. Steinhauser I, Spankuch B, Strebhardt K, Langer K. Trastuzumab-modified nanoparticles: optimisation of preparation and uptake in cancer cells. Biomaterials. 2006;27(28):4975–83.
51. Brigger I, Dubernet C, Couvreur P. Nanoparticles in cancer therapy and diagnosis. Adv Drug Deliv Rev. 2002;54(5):631–51.

52. Feng M, Lee D, Li P. Intracellular uptake and release of poly(ethyleneimine)-co-poly(methyl methacrylate) nanoparticle/pDNA complexes for gene delivery. Int J Pharm. 2006;311(1–2):209.
53. Owens III DE, Peppas NA. Opsonization, biodistribution, and pharmacokinetics of polymeric nanoparticles. Int J Pharm. 2006;307:93–102. https://doi.org/10.1016/j.ijpharm.2005.10.010.
54. Vogt A, Combadiere B, Hadam S, Stieler KM, Lademann J, Schaefer H, Autran B, Sterry W, Blume-Peytavi U. 40 nm, but not 750 or 1,500 nm, nanoparticles enter epidermal CD1a+ cells after transcutaneous application on human skin. J Invest Dermatol. 2006;126(6):1316–22.
55. Bruchez M Jr, Moronne M, Gin P, Weiss S, Alivisatos AP. Semiconductor nanocrystals as fluorescent biological labels. Science. 1998;281(5385): 2013–6.
56. Chattopadhyay PK, Price DA, Harper TF, Betts MR, Yu J, Gostick E, Perfetto SP, Goepfert P, Koup RA, De Rosa SC, Bruchez MP, Roederer M. Quantum dot semiconductor nanocrystals for immunophenotyping by polychromatic flow cytometry. Nat Med. 2006;12(8):972–7.
57. Gao X, Chan WC, Nie S. Quantum-dot nanocrystals for ultrasensitive biological labeling and multicolor optical encoding. J Biomed Opt. 2002;7(4):532–7.
58. Gao X, Cui Y, Levenson RM, Chung LW, Nie S. In vivo cancer targeting and imaging with semiconductor quantum dots. Nat Biotechnol. 2004;22(8):969–76.
59. Giepmans BN, Adams SR, Ellisman MH, Tsien RY. The fluorescent toolbox for assessing protein location and function. Science. 2006;312(5771):217–24.
60. Vasudevan D, Ranganathan Gaddam R, Trinchi A, Cole I. Core-Shell quantum dots: Properties and applications. J Alloys Comp. 2015;636:395–404.
61. Kaul Z, Yaguchi T, Kaul SC, Wadhwa R. Quantum dot-based protein imaging and functional significance of two mitochondrial chaperones in cellular senescence and carcinogenesis. Ann N Y Acad Sci. 2006;1067:469–73.
62. Bilan R, Fleury F, Nabiev I, Sukhanova A. Quantum dot surface chemistry and functionalization for cell targeting and imaging. Bioconjug Chem. 2015;26:609–624.
63. Mittag A, Lenz D, Gerstner AOH, Tarnok A. Slide-based cytometry for cytomics—A minireview. Cytometry A. 2006;69A:555–62.
64. Prow T, Smith JN, Grebe R, Salazar JH, Wang N, Kotov N, Lutty G, Leary J. Construction, gene delivery, and expression of DNA tethered nanoparticles. Mol Vis. 2006;12:606–15.
65. Sapsford K, Pons T, Medintz I, Mattoussi H. Biosensing with luminescent semiconductor quantum dots. Sensors. 2006;6(8):925–53.
66. Nisman R, Dellaire G, Ren Y, Li R, Bazett-Jones DP. Application of quantum dots as probes for correlative fluorescence, conventional, and energy-filtered transmission electron microscopy. J Histochem Cytochem. 2004;52(1):13–8.
67. Tokumasu F, Dvorak J. Development and application of quantum dots for immunocytochemistry of human erythrocytes. J Microsc. 2003;211(Pt 3):256–61.
68. Wang HZ, Wang HY, Liang RQ, Ruan KC. Detection of tumor marker CA125 in ovarian carcinoma using quantum dots. Acta Biochim Biophys Sin Shanghai. 2004;36(10)
69. Efros AL, Brus LE. Nanocrystal quantum dots: from discovery to modern development. ACS Nano. 2021;15:6192–210.
70. Cha MH, Rhim T, Kim KH, Jang AS, Paik YK, Park CS. Proteomic identification of macrophage migration-inhibitory factor upon exposure to TiO_2 particles. Mol Cell Proteomics. 2007;6(1):56–63.
71. Herr JK, Smith JE, Medley CD, Shangguan D, Tan W. Aptamer-conjugated nanoparticles for selective collection and detection of cancer cells. Anal Chem. 2006;78(9):2918–24.
72. Jeng HA, Swanson J. Toxicity of metal oxide nanoparticles in mammalian cells. J. Environ. Sci. Health A Tox Hazard Subst Environ Eng. 2006;41(12):2699–711.
73. Liu J, Saponijc Z, Dimitrijevic NM, Luo S, Preuss D, Rajh T. Hybrid TiO_2 nanoparticles: an approach for developing site specific DNA cleavage. Proc SPIE. 2006;6096(48):2006.
74. Long TC, Saleh N, Tilton RD, Lowry GV, Veronesi B. Titanium dioxide (P25) produces reactive oxygen species in immortalized brain microglia (BV2): implications for nanoparticle neurotoxicity. Environ Sci Technol. 2006;40(14):4346–52.
75. Deng Y. Sensing mechanism and evaluation criteria of semiconducting metal oxides gas sensors. Semicond Metal Ox Gas Sens. Springer, Singapore, 2019;23–51.
76. Paunesku T, Rajh T, Wiederrecht G, Maser J, Vogt S, Stojicevic N, Protic M, Lai B, Oryhon J, Thurnauer M, Woloschak G. Biology of TiO_2-oligonucleotide nanocomposites. Nat Mater. 2003;2(5):343–6.
77. Paunesku T, Vogt S, Lai B, Maser J, Stojicevic N, Thurn KT, Osipo C, Liu H, Legnini D, Wang Z, Lee C, Woloschak GE. Intracellular distribution of TiO2-DNA oligonucleotide nanoconjugates directed to nucleolus and mitochondria indicates sequence specificity. Nano Lett. 2007;7(3):596–601.
78. Freund JB, Shapiro B. Transport of particles by magnetic forces and cellular blood flow in a model microvessel. Phys Fluids. 2012;24:51904.
79. Zhang HD, Williams PS, Zborowski M, Chalmers JJ. Binding affinities/avidities of antibody-antigen interactions: quantification and scale-up implications. Biotechnol Bioeng. 2006;95(5):812–29.
80. Fujishima A, Honda K. Electrochemical photolysis of water at a semiconductor electrode. Nature. 1972;238:37–8.
81. Michelmore A, Gong W, Jenkins P, Ralston J. The interaction of linear polyphosphates with tita-

nium dioxide surfaces. Phys Chem Chem Phys. 2000;2:2985–92.
82. Rajh T, Chen LX, Lukas K, Liu T, Thurnauer M, Tiede DM. Surface restructuring of nanoparticles: An efficient route for ligand-metal oxide crosstalk. J Phys Chem B. 2002;106(41):10543–52.
83. de la Fuente JM, Berry CC. Tat peptide as an efficient molecule to translocate gold nanoparticles into the cell nucleus. Bioconjug Chem. 2005;16:1176–80.
84. Dixit V, Van den Bossche J, Sherman DM, Thompson DH, Andres RP. Synthesis and grafting of thioctic acid-PEG-folate conjugates onto Au nanoparticles for selective targeting of folate receptor-positive tumor cells. Bioconjug Chem. 2006;17:603–9.
85. Herdt AR, Drawz SM, Kang Y, Taton TA. DNA dissociation and degradation at gold nanoparticle surfaces. Colloids Surf B Biointerfaces. 2006;51(2):130–9.
86. Kirpotin DB, Drummond DC, Shao Y, Shalaby MR, Hong K, Nielsen UB, Marks JD, Benz CC, Park JW. Antibody targeting of long-circulating lipidic nanoparticles does not increase tumor localization but does increase internalization in animal models. Cancer Res. 2006;66(13):6732–40.
87. Park SY, Lee JS, Georganopoulou D, Mirkin CA, Schatz GC. Structures of DNA-linked nanoparticle aggregates. J Phys Chem B Condens Matter Mater Surf Interfaces Biophys. 2006;110(25):12673–81.
88. Qin WJ, Yung LY. Efficient manipulation of nanoparticle-bound DNA via restriction endonuclease. Biomacromolecules. 2006;7(11):3047–51.
89. Rosi NL, Giljohann DA, Thaxton CS, Lytton-Jean AK, Han MS, Mirkin CA. Oligonucleotide-modified gold nanoparticles for intracellular gene regulation. Science. 2006;312(5776):1027–30.
90. Tkachenko AG, Xie H, Coleman D, Glomm W, Ryan J, Anderson MF, Franzen S, Feldheim DL. Multifunctional gold nanoparticle-peptide complexes for nuclear targeting. J Am Chem Soc. 2003;125(16):4700–1.
91. Tkachenko AG, Xie H, Liu Y, Coleman D, Ryan J, Glomm WR, Shipton MK, Franzen S, Feldheim DL. Cellular trajectories of peptide-modified gold particle complexes: comparison of nuclear localization signals and peptide transduction domains. Bioconjug Chem. 2004;15(3):482–90.
92. Salnikov V, Lukyanenko YO, Frederick CA, Lederer WJ, Lukyanenko V. Probing the outer mitochondrial membrane in cardiac mitochondria with nanoparticles. Biophys J. 2007;92(3):1058–71.
93. Loffroy R, Chevallier O, Moulin M, Favelier S, Genson PY, Pottecher P, et al. Current role of multiparametric magnetic resonance imaging for prostate cancer. Quant Imaging Med Surg. 2015;5:754–64.
94. Yousaf T, Dervenoulas G, Politis M. Advances in MRI methodology. Int Rev Neurobiol. 2018;141:31–76.
95. Hemond CC, Bakshi R. Magnetic resonance imaging in multiple sclerosis. Cold Spring Harb Perspect Med. 2018;8:1–21.
96. Behzadi AH, Farooq Z, Newhouse JH, Prince MR. MRI and CT contrast media extravasation. Medcine. 2018;97:e0055.
97. Esqueda AC, López JA, Andreu-de-Riquer G, Alvarado-Monzón JC, Ratnakar J, Lubag AJ, Sherry AD, De León-Rodríguez LM. A new gadolinium-based MRI zinc sensor. J Am Chem Soc. 2009;131:11387–91. [CrossRef] Nanomaterials 2020, 10, 1700 30 of 40
98. Lux J, Sherry AD. Advances in gadolinium-based MRI contrast agent designs for monitoring biological processes in vivo. Curr Opin Chem Biol. 2018;45:121–30.
99. Liu X, Madhankumar AB, Miller PA, Duck KA, Hafenstein S, Rizk E, Slagle-Webb B, Sheehan JM, Connor JR, Yang QX. MRI contrast agent for targeting glioma: Interleukin-13 labeled liposome encapsulating gadolinium-DTPA. Neuro-Oncology. 2016;18:691–9.
100. McMahon MT, Bulte JWM. Two decades of dendrimers as versatile MRI agents: a tale with and without metals. Wiley Interdiscip Rev Nanomed Nanobiotechnol. 2018;10:e1496.
101. Moghimi H, Zohdiaghdam R, Riahialam N, Behrouzkia Z. The assessment of toxicity characteristics of cellular uptake of paramagnetic nanoparticles as a new magnetic resonance imaging contrast agent. Iran J Pharm Res. 2019;18: 2083–92.
102. Rogosnitzky M, Branch S. Gadolinium-based contrast agent toxicity: a review of known and proposed mechanisms. Biometals. 2016;29:365–76.
103. Layne KA, Dargan PI, Archer JRH, Wood DM. Gadolinium deposition and the potential for toxicological sequelae—a literature review of issues surrounding gadolinium-based contrast agents. Br J Clin Pharmacol. 2018;84:2522–34.
104. Ramalho J, Semelka RC, Ramalho M, Nunes RH, AlObaidy M, Castillo M. Gadolinium-based contrast agent accumulation and toxicity: an update. Am J Neuroradiol. 2016;37:1192–8.
105. Wei H, Bruns OT, Kaul MG, Hansen EC, Barch M, Wiśniowska A, Chen O, Chen Y, Li N, Okada S, et al. Exceedingly small iron oxide nanoparticles as positive MRI contrast agents. Proc Natl Acad Sci USA. 2017;114:2325–30.
106. Dulińska-Litewka J, Łazarczyk A, Hałubiec P, Szafrański O, Karnas K, Karewicz A. Superparamagnetic iron oxide nanoparticles-current and prospective medical applications. Materials. 2019;12:617.
107. Kim SJ, Lewis B, Steiner MS, Bissa UV, Dose C, Frank JA. Superparamagnetic iron oxide nanoparticles for direct labelling of stem cells and in vivo MRI tracking. Contrast Media Mol Imaging. 2016;11:55–64.
108. Iyer SR, Xu S, Stains JP, Bennett CH, Lovering RM. Superparamagnetic iron oxide nanoparticles in musculoskeletal biology. Tissue Eng Part B Rev. 2017;23:373–85.

109. Kumar S, Kumar BSH, Khushu S. Increased transverse relaxivity in ultrasmall superparamagnetic iron oxide nanoparticles used as MRI contrast agent for biomedical imaging. Contrast Media Mol Imaging. 2016;11:350–61.
110. Hobson NJ, Weng X, Siow B, Veiga C, Ashford M, Thanh NTK, Schätzlein AG, Uchegbu IF. Clustering superparamagnetic iron oxide nanoparticles produces organ-targeted high-contrast magnetic resonance images. Nanomedicine. 2019;14:1135–52.
111. Shari S, Seyednejad H, Laurent S, Atyabi F. Superparamagnetic iron oxide nanoparticles for in vivo molecular and cellular imaging. Contrast Media Mol Imaging. 2015;10:329–55.
112. Wang ZJ, Ohliger MA, Larson PEZ, Gordon JW, Bok RA, Slater J, Villanueva-Meyer JE, Hess CP, Kurhanewicz J, Vigneron DB. Hyperpolarized 13C MRI: state of the art and future directions. Radiology. 2019;291:273–84.
113. Kurhanewicz J, Vigneron DB, Ardenkjaer-Larsen JH, Bankson JA, Brindle K, Cunningham CH, Gallagher FA, Keshari KR, Kjaer A, Laustsen C, et al. Hyperpolarized 13C MRI: path to clinical translation in oncology. Neoplasia. 2019;21:1–16.
114. Cho A, Lau JYC, Geraghty BJ, Cunningham CH, Keshari KR. Noninvasive interrogation of cancer metabolism with hyperpolarized 13C MRI. J Nucl Med. 2017;58:1201–6.
115. Malik M, Chaudhary R, Pundir CS. An improved enzyme nanoparticles based amperometric pyruvate biosensor for detection of pyruvate in serum. Enzym Microb Technol. 2019;123:30–8.
116. Waddington DEJ, Sarracanie M, Zhang H, Salameh N, Glenn DR, Rej E, Gaebel T, Boele T, Walsworth RL, Reilly DJ, et al. Nanodiamond-enhanced MRI via in situ hyperpolarization. Nat Commun. 2017;8:1–8.
117. Kostarelos K, Bianco A, Prato M. Promises, facts and challenges for carbon nanotubes in imaging and therapeutics. Nat Nanotechnol. 2009;4:627–33.
118. De La Zerda A, Zavaleta C, Keren S, Vaithilingam S, Bodapati S, Liu Z, Levi J, Smith BR, Ma TJ, Oralkan O, et al. Carbon nanotubes as photoacoustic molecular imaging agents in living mice. Nat Nanotechnol. 2008;3:557–62.
119. Welsher K, Liu Z, Sherlock SP, Robinson JT, Chen Z, Daranciang D, Dai H. A route to brightly fluorescent carbon nanotubes for near-infrared imaging in mice. Nat Nanotechnol. 2009;4:773–80.
120. Garg B, Sung CH, Ling YC. Graphene-based nanomaterials as molecular imaging agents. Wiley Interdiscip Rev Nanomed Nanobiotechnol. 2015;7:737–58.
121. Nurunnabi M, Khatun Z, Reeck GR, Lee DY, Lee YK. Photoluminescent graphene nanoparticles for cancer phototherapy and imaging. ACS Appl Mat Interf. 2014;6:12413–21.
122. Cai X, Zhu Q, Zeng Y, Zeng Q, Chen X, Zhan Y. Manganese oxide nanoparticles as mri contrast agents in tumor multimodal imaging and therapy. Int J Nanomedicine. 2019;14:8321–44.
123. Kwiatkowski G, Jähnig F, Steinhauser J, Wespi P, Ernst M, Kozerke S. Nanometer size silicon particles for hyperpolarized MRI. Sci Rep. 2017;7:1–6.
124. Yoo SP, Pineda F, Barrett JC, Poon C, Tirrell M, Chung EJ. Gadolinium-functionalized peptide amphiphile micelles for multimodal imaging of atherosclerotic lesions. ACS Omega. 2016;1:996–1003.
125. Dong YC, Hajfathalian M, Maidment PSN, Hsu JC, Naha PC, Si-Mohamed S, Breuilly M, Kim J, Chhour P, Douek P, et al. Effect of gold nanoparticle size on their properties as contrast agents for computed tomography. Sci Rep. 2019;9:1–13.
126. Silvestri A, Zambelli V, Ferretti AM, Salerno D, Bellani G, Polito L. Design of functionalized gold nanoparticle probes for computed tomography imaging. Contrast Media Mol Imaging. 2016;11:405–14.
127. Nakagawa T, Gonda K, Kamei T, Cong L, Hamada Y, Kitamura N, Tada H, Ishida T, Aimiya T, Furusawa N, et al. X-ray computed tomography imaging of a tumor with high sensitivity using gold nanoparticles conjugated to a cancer-specific antibody via polyethylene glycol chains on their surface. Sci Technol Adv Mater. 2016;17:387–97.
128. Mok PL, Leow SN, Koh AEH, Mohd Nizam HH, Ding SLS, Luu C, Ruhaslizan R, Wong HS, Halim WHWA, Ng MH, et al. Micro-computed tomography detection of gold nanoparticle-labelled mesenchymal stem cells in the rat subretinal layer. Int J Mol Sci. 2017;18:345.
129. Farooq Aziz AI, Nazir A, Ahmad I, Bajwa SZ, Rehman A, Diallo A, Khan WS. Novel route synthesis of porous and solid gold nanoparticles for investigating their comparative performance as contrast agent in computed tomography scan and effect on liver and kidney function. Int J Nanomedicine. 2017;12:1555.
130. Cheheltani R, Ezzibdeh RM, Chhour P, Pulaparthi K, Kim J, Jurcova M, Hsu JC, Blundell C, Litt HI, Ferrari VA, et al. Tunable, biodegradable gold nanoparticles as contrast agents for computed tomography and photoacoustic imaging. Biomaterials. 2016;102:87–97.
131. Hainfeld JF, Ridwan SM, Stanishevskiy Y, Smilowitz NR, Davis J, Smilowitz HM. Small, long blood half-life iodine nanoparticle for vascular and tumor imaging. Sci Rep. 2018;8:2–11.
132. Sun N, Zhao L, Zhu J, Li Y, Song N, Xing Y, Qiao W, Huang H, Zhao J. 131I-labeled polyethylenimine-entrapped gold nanoparticles for targeted tumor Spect/Ct imaging and radionuclide therapy. Int J Nanomedicine. 2019;14:4367–81.
133. Kim H, Jang EJ, Kim SK, Hyung WJ, Choi DK, Lim SJ, Lim JS. Simultaneous sentinel lymph node computed tomography and locoregional chemotherapy for lymph node metastasis in rabbit using an iodine-docetaxel emulsion. Oncotarget. 2017;8:27177–88.
134. Badea CT, Clark DP, Holbrook M, Srivastava M, Mowery Y, Ghaghada KB. Functional imaging of tumor vasculature using iodine and gadolinium-based nanoparticle contrast agents: a comparison of spectral micro-CT using energy integrating

and photon counting detectors. Phys Med Biol. 2019;64:65007.
135. Chen J, Yang XQ, Qin MY, Zhang XS, Xuan Y, Zhao YD. Hybrid nanoprobes of bismuth sulfide nanoparticles and CdSe/ZnS quantum dots for mouse computed tomography/fluorescence dual mode imaging. J Nanobiotechnol. 2015;13:1–10.
136. Santos BS, Ferreira MJ. Positron emission tomography in ischemic heart disease. Rev Port Cardiol. 2019;38:599–608.
137. Lee SB, Lee SW, Jeong SY, Yoon G, Cho SJ, Kim SK, Lee IK, Ahn BC, Lee J, Jeon YH. Engineering of radioiodine-labeled gold core-shell nanoparticles as efficient nuclear medicine imaging agents for trafficking of dendritic cells. ACS Appl Mater Interfaces. 2017;9:8480–9.
138. Simón M, Jørgensen JT, Norregaard K, Kjaer A. 18F-FDG positron emission tomography and diffusion-weighted magnetic resonance imaging for response evaluation of nanoparticle-mediated photothermal therapy. Sci Rep. 2020;10:1–9.
139. Liu Y, Carpenter AB, Pirozzi CJ, Yuan H, Waitkus MS, Zhou Z, Hansen L, Seywald M, Odion R, Greer PK, et al. Non-invasive sensitive brain tumor detection using dual-modality bioimaging nanoprobe. Nanotechnology. 2019;30:275101.
140. Lee SB, Kumar D, Li Y, Lee IK, Cho SJ, Kim SK, Lee SW, Jeong SY, Lee J, Jeon YH. PEGylated crushed gold shell-radiolabeled core nanoballs for in vivo tumor imaging with dual positron emission tomography and Cerenkov luminescent imaging. J Nanobiotechnol. 2018;16:1–12.
141. Cui L, Xiong C, Zhou M, Shi S, Chow DSL, Li C. Integrin αvβ3-targeted [64 Cu]CuS nanoparticles for PET/CT imaging and photothermal ablation therapy. Bioconjug Chem. 2018;29:4062–71.
142. Woodard PK, Liu Y, Pressly ED, Luehmann HP, Detering L, Sultan DE, Laforest R, McGrath AJ, Gropler RJ, Hawker CJ. Design and modular construction of a polymeric nanoparticle for targeted atherosclerosis positron emission tomography imaging: a story of 25% (64)Cu-CANF-Comb. Pharm Res. 2016;33:2400–10.
143. Kim HY, Li R, Ng T, Courties G, Rodell CB, Prytyskach M, Kohler RH, Pittet MJ, Nahrendorf M, Weissleder R, et al. Quantitative imaging of tumor associated macrophages and their response to therapy using (64)Cu-labeled macrin. ACS Nano. 2018;12:12015–29.
144. Wong P, Li L, Chea J, Delgado MK, Crow D, Poku E, Szpikowska B, Bowles N, Channappa D, Colcher D, et al. PET imaging of (64)Cu-DOTA-scFv-anti-PSMA lipid nanoparticles (LNPs): enhanced tumor targeting over anti-PSMA scFv or untargeted LNPs. Nucl Med Biol. 2017;47:62–8.
145. Keliher EJ, Ye YX, Wojtkiewicz GR, Aguirre AD, Tricot B, Senders ML, Groenen H, Fay F, Perez-Medina C, Calcagno C, et al. Polyglucose nanoparticles with renal elimination and macrophage avidity facilitate PET imaging in ischaemic heart disease. Nat Commun. 2017;8:1–12.
146. Sun J, Sun L, Li J, Xu J, Wan Z, Ouyang Z, Liang L, Li S, Zeng D. A multi-functional polymeric carrier for simultaneous positron emission tomography imaging and combination therapy. Acta Biomater. 2018;75:312–22.
147. Piras AM, Fabiano A, Sartini S, Zambito Y, Braccini S, Chiellini F, Cataldi AG, Bartoli F, de la Fuente A, Erba PA. pH-responsive carboxymethylcellulose nanoparticles for 68Ga-WBC labelling in PET imaging. Polymers. 2019;11:1615.
148. Wang X, Yan J, Pan D, Yang R, Wang L, Xu Y, Sheng J, Yue Y, Huang Q, Wang Y, et al. Polyphenol–poloxamer self-assembled supramolecular nanoparticles for tumor NIRF/PET imaging. Adv Healthc Mater. 2018;7:1–8.
149. McDonagh PR, Sundaresan G, Yang L, Sun M, Mikkelsen R, Zweit J. Biodistribution and PET imaging of 89-zirconium labeled cerium oxide nanoparticles synthesized with several surface coatings. Nanomedicine. 2018;14:1429–40.
150. Chakravarty R, Hong H, Cai W. Image-guided drug delivery with single-photon emission computed tomography: a review of literature. Curr Drug Targets. 2015;16:592–609.
151. Si-Mohamed S, Cormode DP, Bar-Ness D, Sigovan M, Naha PC, Langlois JB, Chalabreysse L, Coulon P, Blevis I, Roessl E, et al. Evaluation of spectral photon counting computed tomography K-edge imaging for determination of gold nanoparticle biodistribution in vivo. Nanoscale. 2017;9:18246–57.
152. Estudiante-Mariquez OJ, Rodríguez-Galván A, Ramírez-Hernández D, Contreras-Torres FF, Medina LA. Technetium-radiolabeled mannose-functionalized gold nanoparticles as nanoprobes for sentinel lymph node detection. Molecules. 2020;25:1982.
153. Hoogendam JP, Zweemer RP, Hobbelink MGG, Van Den Bosch MAAJ, Verheijen RHM, Veldhuis WB. 99mTc-nanocolloid SPECT/MRI fusion for the selective assessment of nonenlarged sentinel lymph nodes in patients with early-stage cervical cancer. J Nucl Med. 2016;57:551–6.
154. Collarino A, Vidal-Sicart S, Perotti G, Valdés Olmos RA. The sentinel node approach in gynaecological malignancies. Clin Translat Imaging. 2016;4:411–20.
155. Saad ZZ, Omorphos S, Michopoulou S, Gacinovic S, Malone P, Nigam R, Muneer A, Romanji J. Investigating the role of SPECT/CT in dynamic sentinel lymph node biopsy for penile cancers. Eur J Nucl Med Mol Imaging. 2017;44:1176–84.
156. De Veij Mestdagh PD, Schreuder WH, Vogel WV, Donswijk ML, Van Werkhoven E, Van Der Wal JE, Dirven R, Karakullukcu B, Sonke JJ, Van Den Brekel MWM, et al. Mapping of sentinel lymph node drainage using SPECT/CT to tailor elective nodal irradiation in head and neck cancer patients (SUSPECT-2): a single-center prospective trial. BMC Cancer. 2019;19:1–9.
157. De Veij Mestdagh PD, Schreuder WH, Vogel WV, Donswijk ML, Van Werkhoven E, Van Der Wal JE, Dirven R, Karakullukcu B, Sonke JJ, Van Den

Brekel MWM, et al. An analysis of SPECT/CT non-visualization of sentinel lymph nodes in renal tumors. EJNMMI Res. 2018;8:105.
158. Wang Y, Li Y, Wei F, Duan Y. Optical imaging paves the way for autophagy research. Trends Biotechnol. 2017;35:1181–93.
159. Boschi F, de Sanctis F. Overview of the optical properties of fluorescent nanoparticles for optical imaging. Eur J Histochem. 2017;61:245–8.
160. De-La-Cuesta J, González E, Pomposo JA. Advances in fluorescent single-chain nanoparticles. Molecules. 2017;22:1–14.
161. Wu W, Yang YQ, Yang Y, Yang YM, Wang H, Zhang KY, Guo L, Ge HF, Liu J, Feng H. An organic NIR-II nanofluorophore with aggregation-induced emission characteristics for in vivo fluorescence imaging. Int J Nanomedicine. 2019;14:3571–82.
162. Zhong Y, Ma Z, Wang F, Wang X, Yang Y, Liu Y, Zhao X, Li J, Du H, Zhang M, et al. In vivo molecular imaging for immunotherapy using ultra-bright near-infrared-IIb rare-earth nanoparticles. Nat Biotechnol. 2019;37:1322–31.
163. Bhavane R, Starosolski Z, Stupin I, Ghaghada KB, Annapragada A. NIR-II fluorescence imaging using indocyanine green nanoparticles. Sci Rep. 2018;8:1–10.
164. Li CH, Kuo TR, Su HJ, Lai WY, Yang PC, Chen JS, Wang DY, Wu YC, Chen CC. Fluorescence-guided probes of aptamer-targeted gold nanoparticles with computed tomography imaging accesses for in vivo tumor resection. Sci Rep. 2015;5:1–11.
165. Wallat JD, Czapar AE, Wang C, Wen AM, Wek KS, Yu X, Steinmetz NF, Pokorski JK. Optical and magnetic resonance imaging using fluorous colloidal nanoparticles. Biomacromolecules. 2017;18:103–12.
166. Kim EH, Chin G, Rong G, Poskanzer KE, Clark HA. Optical probes for neurobiological sensing and imaging. Acc Chem Res. 2018;51:1023–32.
167. Napp J, Andrea Markus M, Heck JG, Dullin C, Möbius W, Gorpas D, Feldmann C, Alves F. Therapeutic fluorescent hybrid nanoparticles for traceable delivery of glucocorticoids to inflammatory sites. Theranostics. 2018;8:6367–83.
168. Wang Y, Xu N, He Y, Wang J, Wang D, Gao Q, Xie S, Li Y, Zhang R, Cai Q. Loading graphene quantum dots into optical-magneto nanoparticles for real-time tracking in vivo. Materials. 2019;12:2191.
169. Flak D, Przysiecka Ł, Nowaczyk G, Scheibe B, Kościński M, Jesionowski T, Jurga S. GQDs-MSNs nanocomposite nanoparticles for simultaneous intracellular drug delivery and fluorescent imaging. J Nanopart Res. 2018;20:306.
170. Pan Y, Chang T, Marcq G, Liu C, Kiss B, Rouse R, Mach KE, Cheng Z, Liao JC. In vivo biodistribution and toxicity of intravesical administration of quantum dots for optical molecular imaging of bladder cancer. Sci Rep. 2017;7:1–9.
171. Wang YW, Reder NP, Kang S, Glaser AK, Liu JTC. Multiplexed optical imaging of tumor-directed nanoparticles: a review of imaging systems and approaches. Nano. 2017;1:369–88.
172. Imamura Y, Yamada S, Tsuboi S, Nakane Y, Tsukasaki Y, Komatsuzaki A, Jin T. Near-infrared emitting PbS quantum dots for in vivo fluorescence imaging of the thrombotic state in septic mouse brain. Molecules. 2016;21:80.
173. Bruns OT, Bischof TS, Harris DK, Franke D, Shi Y, Riedemann L, Bartelt A, Jaworski FB, Carr JA, Rowlands CJ, et al. Next-generation in vivo optical imaging with short-wave infrared quantum dots. Nat Biomed Eng. 2017;1:0056.
174. Mintz KJ, Zhou Y, Leblanc RM. Recent development of carbon quantum dots regarding their optical properties, photoluminescence mechanism, and core structure. Nanoscale. 2019;11: 4634–52.
175. Liu X, Braun GB, Zhong H, Hall DJ, Han W, Qin M, Zhao C, Wang M, She Z-G, Cao C, et al. Tumor-targeted multimodal optical imaging with versatile cadmium-free quantum dots. Adv Funct Mater. 2016;26:267–76.
176. Ou J, Zhou Z, Chen Z, Tan H. Optical diagnostic based on functionalized gold nanoparticles. Int J Mol Sci. 2019;20:4346.
177. Sánchez GJ, Maury P, Stefancikova L, Campion O, Laurent G, Chateau A, Hoch FB, Boschetti F, Denat F, Pinel S, et al. Fluorescent radiosensitizing gold nanoparticles. Int J Mol Sci. 2019;20:4618.
178. Brito HF, Hölsä J, Laamanen T, Lastusaari M, Malkamäki M, Rodrigues LC. Persistent luminescence mechanisms: human imagination at work. Opt Mater Exp. 2012;2:371–81.
179. Lecuyer T, Teston E, Maldiney T, Scherman D, Richard C. Physico-chemical characterizations of CR doped persistent luminescence nanoparticles. SPIE Bios. 2016;9722:97220X.
180. Sharma SK, Gourier D, Viana B, Maldiney T, Teston E, Scherman D, Richard C. Persistent luminescence in nanophosphors for long term in-vivo bio-imaging. SPIE Bios. 2015;9337:93370I.
181. Ramírez-García G, Gutiérrez-Granados S, Gallegos-Corona MA, Palma-Tirado L, d'Orlyé F, Varenne A, Mignet N, Richard C, Martínez-Alfaro M. Long-term toxicological effects of persistent luminescence nanoparticles after intravenous injection in mice. Int J Pharm. 2017;532:686–95.
182. Guo R, Lu G, Qin B, Fei B. Ultrasound imaging technologies for breast cancer detection and management: a review. Ultrasound Med Biol. 2018;44:37–70.
183. Zhou X, Guo L, Shi D, Duan S, Li J. Biocompatible chitosan nanobubbles for ultrasound-mediated targeted delivery of doxorubicin. Nanoscale Res Lett. 2019;14:24.
184. Meng M, Gao J, Wu C, Zhou X, Zang X, Lin X, Liu H, Wang C, Su H, Liu K, et al. Doxorubicin nanobubble for combining ultrasonography and targeted chemotherapy of rabbit with VX2 liver tumor. Tumour Biol. 2016;37:8673–80.

185. Cheng B, Bing C, Xi Y, Shah B, Exner AA, Chopra R. Influence of nanobubble concentration on blood–brain barrier opening using focused ultrasound under real-time acoustic feedback control. Ultrasound Med Biol. 2019;45:2174–87.
186. Chandan R, Banerjee R. Pro-apoptotic liposomes-nanobubble conjugate synergistic with paclitaxel: a platform for ultrasound responsive image-guided drug delivery. Sci Rep. 2018;8:1–15.
187. Thakur SS, Ward MS, Popat A, Flemming NB, Parat M, Barnett NL, Parekh HS. Stably engineered nanobubblesand ultrasound—an effective platform for enhanced macromolecular delivery to representative cells of the retina. PLoS One. 2017;12:1–17.
188. Chan MH, Pan YT, Chan YC, Hsiao M, Chen CH, Sun L, Liu RS. Nanobubble-embedded inorganic 808 nm excited upconversion nanocomposites for tumor multiple imaging and treatment. Chem Sci. 2018;9:3141–51.
189. Tian Y, Liu Z, Zhang L, Zhang J, Han X, Wang Q, Cheng W. Apatinib-loaded lipid nanobubbles combined with ultrasound-targeted nanobubble destruction for synergistic treatment of HepG2 cells in vitro. Onco Targets Ther. 2018;11:4785–95.
190. Suzuki R, Oda Y, Omata D, Nishiie N, Koshima R, Shiono Y, Sawaguchi Y, Unga J, Naoi T, Negishi Y, et al. Tumor growth suppression by the combination of nanobubbles and ultrasound. Cancer Sci. 2016;107:217–23.
191. Bhandari P, Novikova G, Goergen CJ, Irudayaraj J. Ultrasound beam steering of oxygen nanobubbles for enhanced bladder cancer therapy. Sci Rep. 2018;8:1–10.
192. Wu M, Zhao H, Guo L, Wang Y, Song J, Zhao X, Li C, Hao L, Wang D, Tang J. Ultrasound-mediated nanobubble destruction (UMND) facilitates the delivery of A10-3.2 aptamer targeted and siRNA-loaded cationic nanobubbles for therapy of prostate cancer. Drug Deliv. 2018;25:226–40.
193. Choi W, Park E-Y, Jeon S, Kim C. Clinical photoacoustic imaging platforms. Biomed Eng Lett. 2018;8:139–55.
194. Steinberg I, Huland DM, Vermesh O, Frostig HE, Tummers WS, Gambhir SS. Photoacoustic clinical imaging. Photoacoustics. 2019;14:77–98.
195. Dumani DS, Sun I-C, Emelianov SY. Ultrasound-guided immunofunctional photoacoustic imaging for diagnosis of lymph node metastases. Nanoscale. 2019;11:11649–59.
196. Nagaoka R, Tabata T, Yoshizawa S, Umemura S. Photoacoustics visualization of murine lymph vessels using photoacoustic imaging with contrast agents. Biochem Pharmacol. 2018;9:39–48.
197. Zhang M, Kim HS, Jin T, Yi A, Moon WK. Ultrasound-guided photoacoustic imaging for the selective detection of EGFR-expressing breast cancer and lymph node metastases. Biomed Opt Express. 2016;7:1920–31.
198. Li P, Wu Y, Li D, Su X, Luo C, Wang Y, Hu J, Li G, Jiang H, Zhang W. Seed-mediated synthesis of tunable-aspect-ratio gold nanorods for near-infrared photoacoustic imaging. Nanoscale Res Lett. 2018;13:313.
199. Han S, Bouchard R, Sokolov KV. Molecular photoacoustic imaging with ultra-small gold nanoparticles. Biomed Opt Express. 2019;10:3472–83.
200. Chen Y-S, Zhao Y, Yoon SJ, Gambhir SS, Emelianov S. Miniature gold nanorods for photoacoustic molecular imaging in the second near-infrared optical window. Nat Nanotechnol. 2019;14:465–72.
201. Sun I-C, Ahn C-H, Kim K, Emelianov S. Photoacoustic imaging of cancer cells with glycol-chitosan-coated gold nanoparticles as contrast agents. J Biomed Opt. 2019;24:121903.
202. Bharathiraja S, Manivasagan P, Bui NQ, Oh YO, Lim IG, Park S, Oh J. Cytotoxic induction and photoacoustic imaging of breast cancer cells using astaxanthin-reduced gold nanoparticles. Nanomaterials. 2016;6:78.
203. Hartman RK, Hallam KA, Donnelly EM, Emelianov SY. Photoacoustic imaging of gold nanorods in the brain delivered via microbubble-assisted focused ultrasound: a tool for in vivo molecular neuroimaging. Laser Phys Lett. 2019;16:025603.
204. Dixon AJ, Hu S, Klibanov AL, Hossack JA. Oscillatory dynamics and in vivo photoacoustic imaging performance of plasmonic nanoparticle-coated microbubbles. Small. 2015;11:3066–77.
205. Augustine S, Singh J, Srivastava M, Sharma M, Das A, Malhotra BD. Recent advances in carbon based nanosystems for cancer theranostics. Biomater Sci. 2017;5:901–52.
206. Xie L, Wang G, Zhou H, Zhang F, Guo Z, Liu C, Zhang X, Zhu L. Functional long circulating single walled carbon nanotubes for fluorescent/photoacoustic imaging-guided enhanced phototherapy. Biomaterials. 2016;103:219–28.
207. Fathi P, Knox HJ, Sar D, Tripathi I, Ostadhossein F, Misra SK, Esc MB, Chan J, Pan D. Biodegradable biliverdin nanoparticles for efficient photoacoustic imaging. ACS Nano. 2019;13:7690–704.
208. Köker T, Tang N, Tian C, Tian C, Zhang W, Wang X, Martel R, Pinaud F. Cellular imaging by targeted assembly of hot-spot SERS and photoacoustic nanoprobes using split-fluorescent protein scaffolds. Nat Commun. 2018;9:607.
209. Tan JWY, Murashov MD, Rosania GR, Wang X. Photoacoustic imaging of clofazimine hydrochloride nanoparticle accumulation in cancerous vs normal prostates. PLoS One. 2019;14(7):e0219655.
210. Swider E, Daoudi K, Staal AHJ, Koshkina O, Koen van Riessen N, van Dinther E, de Vries IJM, de Korte CL, Srinivas M. Clinically-applicable perfluorocarbon-loaded nanoparticles for in vivo photoacoustic,19f magnetic resonance and fluorescent imaging. Nano. 2018;2:258–68.
211. Biffi S, Petrizza L, Garrovo C, Rampazzo E, Andolfi L, Giustetto P, Nikolov I, Kurdi G, Danailov MB, Zauli G, et al. Multimodal near-infrared-emitting PluS silica nanoparticles with fluorescent, pho-

toacoustic, and photothermal capabilities. Int J Nanomedicine. 2016;11:4865–74.
212. Harrington WN, Haji MR, Galanzha EI, Nedosekin DA, Nima ZA, Watanabe F, Ghosh A, Biris AS, Zharov VP. Photoswitchable non-fluorescent thermochromic dye-nanoparticle hybrid probes. Sci Rep. 2016;6:1–11.
213. Wang Y, Shang W, Niu M, Tian J, Xu K. Hypoxia-active nanoparticles used in tumor theranostic. Int J Nanomedicine. 2019;14:3705–22.
214. Duncan R, Gaspar R. Nanomedicine(s) under the microscope. Mol Pharm. 2011;8(6):2101–41.
215. Lammers T, Aime S, Hennink WE, Storm G, Kiessling F. Theranostic nanomedicine. Acc Chem Res. 2011;44(10):1029–38.
216. Rizzo LY, Theek B, Storm G, Kiessling F, Lammers T. Recent progress in nanomedicine: therapeutic, diagnostic and theranostic applications. Curr Opin Biotechnol. 2013;24(6):1159–66.
217. Grimm J, Scheinberg DA. Will nanotechnology influence targeted cancer therapy? Semin Radiat Oncol. 2011;21(2):80–7.
218. Matsumura Y, Maeda H. A new concept for macromolecular therapeutics in cancer chemotherapy: mechanism of tumoritropic accumulation of proteins and the antitumor agent smancs. Cancer Res. 1986;46(12 Pt 1):6387–92.
219. Jain RK, Stylianopoulos T. Delivering nanomedicine to solid tumors. Nat Rev Clin Oncol. 2010;7(11):653–64.
220. Lammers T, Kiessling F, Hennink WE, Storm G. Drug targeting to tumors: principles, pitfalls and (pre-) clinical progress. J Control Release. 2012;161(2):175–87.

4 Nanocarriers in Imaging

4.1 Introduction

Nanotechnology has become an important modality and is being applied in various aspects of medicine. Nanocarriers are being used for carrying various therapeutic agents without having any impact on the surrounding healthy tissues. The delivery of the drugs by the nanocarriers is used in different target organs and also in the in vivo imaging. These different nanocarriers can be used for carrying the drugs thus having a more specific target to the various organs. The classical example is the blood–brain barrier (BBB) which can be crossed even by the larger molecules that will be entering into the brain. These BBBs have many facets and are comprised of the various growth factors, insulin, and transferrin and all these will result in increased nanocarrier targeting to the Brian. Few clinical conditions like stroke, tumors, and Alzheimer's disease have more specific organ-selective targeting, better bioaccumulation, and also the excretion of these agents. In vivo, imaging of nanomaterials is possible which will give real-time tracking of these nanocarriers. Thus, these nanocarriers are being used in various diagnostics, therapeutics, and molecular targeting of the various agents. The role is to have better functionalization, specific targeting, and also the live tracking of the drug delivery systems.

4.2 Types of Nanocarriers

The different types of nanocarriers are nanoparticles made up of inorganic components, and different polymeric and lipid nanocarriers. These nanocarriers are capable of BBB penetration in different conditions like schizophrenia, bipolar disorder, depression, anxiety disorder, and Alzheimer's disease. β-amyloid, or tau-protein targeted nanoparticles are used in Alzheimer's disease.

These nanoparticles are comprised of various mechanical and electrical properties which are like more fluorescent labels [1–3], drug and gene delivery agents [4, 5], pathogens that can detect diseases [6], detection of the protein [7], DNA structure evaluation [8], different tissue engineering [9, 10], and process of hyperthermia [11], biological molecules and cells can be differentiated and separated [12], contrast enhancement in magnetic resonance imaging (MRI) [13] and various studies of pharmacokinetics [14]. The basic principle is to interact and enter into the BBB [15], and also act as the monitoring passage preventing the higher-molecular weight molecules from passing from outside.

The nanomaterials can carry drugs, act as in vivo imaging markers, and act as optical imaging dyes, and based on this various fluorescence-based imaging techniques are seen in various

S. Shaikh, *Nanoimaging - Future of Precision Medicine*,
https://doi.org/10.1007/978-981-97-7292-6_4

biological processes, this fluorescence imaging can be used in both in vivo and in vitro despite of many limitations [16–18]. A few of the limitations are like trans-illumination as well as the various tomographic techniques that can be combined with the anatomical modalities leading to better sensitivity and higher resolution as seen in the optoacoustic molecular imaging and multi-spectral optoacoustic tomography (MSOT) [19, 20].

The use of the nanocarriers is based on various applications like the specific marker binding at the inflammation delivery site can be used as the conduit for the various drugs and probes. Another advantage is the excretion from the renal system, and/or reticuloendothelial system (RES), This will also be useful in preventing the biodegradation of different particles [16].

4.3 Nanoparticle Properties

The nanocarrier biodistribution and clearance will be based on the size and shape, the chemical composition, the surface charge of the particles, and the affinity for targeting the different moieties.

4.3.1 Size

Nanocarriers are nano-sized in the range of the 1–100 nanometer scale, and some of them will be of smaller size which will be less the 500 nm.

The nanoparticles will have a high surface area due to their smaller size and can target the drug/probe and also combine with the various drugs (Diagram 4.1). Due to this this will result in the changes in the circulation, pharmacokinetics, and permeability of contrast agents like MR contrast agents. The polyamidoamine (PAMAM) dendrimers will have less than 3 nm and will be crossing the barrier. The smaller-size dendrimers of size are from 3 to 6 nm and can be cleared rapidly however the dendrimers which will be larger than 12 nm will be showing increased uptake [17]. Frangioni's group has shown that there is a good relation between nanoparticle size and tissue distribution with 99mTc-labeled fluorescent quantum dots (QDs) by intraoperative fluorescence imaging and γ-scintigraphy [18].

4.3.2 Shape

The shape of the nanocarrier is also based on circulation, with different kinetics of the cells. The polymer particles which are elongated are more in circulation as compared to the spherical nanoparticles [19]. DeSimone's group has shown that the nano- and micron-sized particles are of similar size, shape, and surface chemistry [20]. The shape of the particle has more binding and phagocytosis by the macrophages. Ellipsoids shapes are larger in size as compared with spheres nearly four times [21]. The specific targeting can be seen more when there is no specific tissue uptake seen [22].

4.3.3 Surface Functionalization

The surface of the carrier's charge, its solubility, and the various affinity by the inclusion of various coating materials and also the reactive groups that will be targeting the surface. Nanoparticle

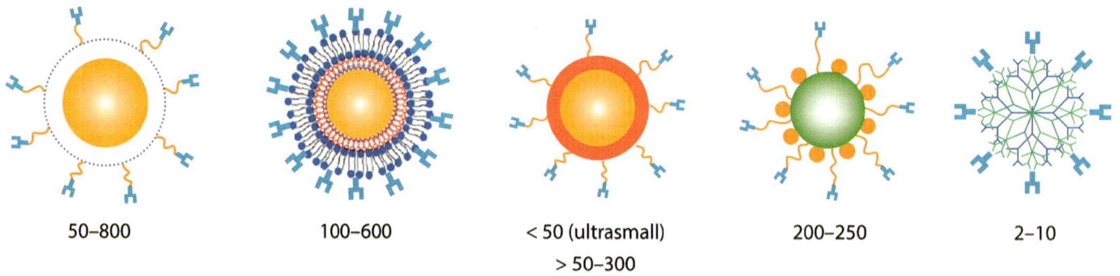

Diagram 4.1 Different types of nanoparticle sizes

charge on the surface can be used for the various protein binding and non-specific adhesion to cells. The neutral and negatively charged nanocarriers will have longer circulation times in the vascular system with more non-specific binding. The biocompatible hydrophilic polymer chains will be enhanced with the nanoparticle solubility, as well as the blood circulation time prolongation, with delayed clearance as compared to the ribosomes [23]. The "stealth" approach is used for better biocompatibility and pharmacokinetics control of the various carriers [24]. The linear-versus star-shaped polymer nanostructures of varied lengths like the radiolabeled 64Cu for PET imaging showed the star-shaped nanostructures retained for a longer time in the blood with lower hepatic and splenic uptake. The linear-shaped nanocarriers are rapidly excreted. The size (25–70 nm) and the core composition of the star-shaped structures have a lesser impact on the action of nanocarriers [25]. The various targeting moieties will increase the accessibility of the ligand to tissue antigens [23]. The various particle functionalization is also seen for the various chemical modifications of the surface. The accessing of the important sites leads to the various targets antibodies and peptides, various contrast agents like radionuclides [26], NIR-dyes [27], and MRI-active metal-chelates [28]; and various therapeutic agents. Various amines, carboxylic acids, and azides are also commonly used for nanocarrier coupling to various ligands [29, 30]. The use of the sulfhydryl and thiol groups also creates various proteins, peptides, and small molecules [31, 32].

Nanocarrier-based drug delivery systems are evolving more due to technological advancements. The nanocarriers need more designs that will be able to have more specific requirements and applications. Different nanocarriers are based on their composition and types like the carbon nanotubes, carbon dots, polymeric micelles, liposomes, iron oxide nanoparticles, nanogels, and dendrimers (Diagram 4.2).

4.3.4 Nanogels

Nanogels are nanoscale hydrogels that are made up of polymeric particles which are crosslinked [33]. They are hydrophilic, better biocompatible, and good content of water [33, 34]. The advantages of the nanogels are the encapsulation property, uniform, and moldable size, the ability to prepare easily, less toxicity, stability in the serum, and stimulus-response [35]. Different designs are available which will have a more focused delivery system [36]. These nanogels are capable of sustainable release of the activated components which can be degraded [36]. Other properties are

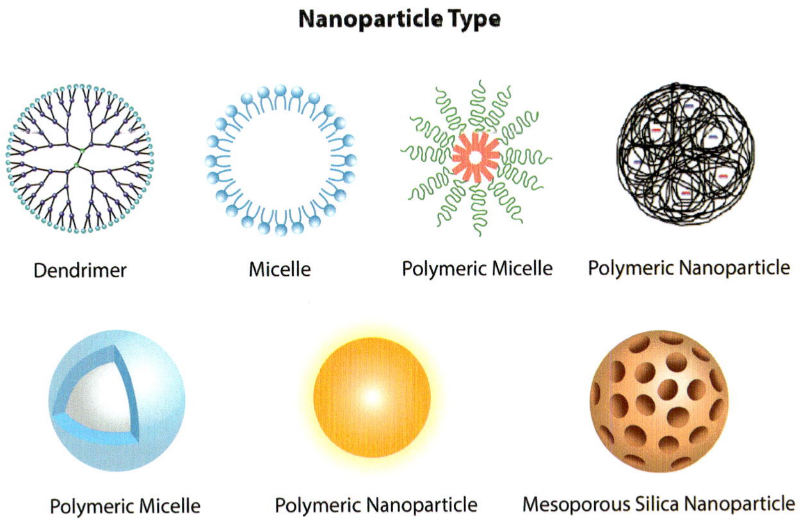

Diagram 4.2 Different types of nanoparticles

deformity, swell ability, capacity for loading drugs, better biodistribution, faster clearance, less toxicity, and scheduled release of the targeted drug and nano gels [34, 35]. One of the studies showed that nanogel which can be cross-linked can be changed to any shape thus targeting the focused delivery of the drugs [37].

4.3.5 Dendrimers

Dendrimers are three-dimensional structures that looks like branched polymers in nanoscale [38, 39]. Drug use can be seen which can have more dendrimer by different interactions or [40]. The modification of the surface can see that the release and entrapment of the drug can be achieved by modifying the surface of the dendrimers [40]. By attaching the various targeting ligands, the functionality can be achieved for better targeting [40]. The dendrimers have different properties due to their high molecular weight and this can be either hydrophilic or hydrophobic, having more surface-to-volume ratio thus enabling better and more precise gene therapy [39]. Nanocarriers have better solubility, stability, and oral bioavailability of many drugs [40]. The newly developed multidrug resistance (MDR)-related protein in the multidrug resistance can be seen [41]. These limitations of the cationic dendrimers can cause cell lysis [39]. The dendritic polymers can be helpful for lesser toxicity of dendrimers [42]. The important drawback of the dendrimers is that they are cleared faster having shorter circulation time, and all these are getting better and better due to advancements happening in this field [42].

4.3.6 Hybrid Nanocarriers

Hybrid nanocarriers will provide a better platform for drug delivery [43]. The fluorescent carbon dots are used as stabilizers and also provide stable signals for real-time imaging [44]. Liposomes or polymers are the hybrid nanocarriers that are being used for better biocompatibility of the nanocarriers [44]. Multiple combinations of nanomaterials are also used for the better drug delivery. The combination of the polymers and liposomes is an example like the addition of the hydrophobic HIV drugs to the PLGA polymeric cores which are added to the lipid bilayers [45]. This hybrid nanocarrier also added different antibodies to target CD4+ T cells [45]. This will increase the latency and decrease the toxicity of some HIV drugs [45].

4.4 Strategies to Enhance Drug Delivery Efficiency

Nanocarriers are using unique advantages, they will be facing the important challenges which will limit the passive accumulation and retention in the region of the interest [46]. This is still existing for better therapeutic outcomes, but it will be more precise for better control of drug release [47]. The principle used here is to have the passive diffusion of the nanocarriers accumulation in the diseased tissue [48]. In cancer therapy evaluation the various ligands on the surface are more specific to the receptors which will be overexpressed in certain cancer cells [32]. The smarter nanocarriers have better efficiency which is being used for better drug release and delivery, and these nanocarriers are under development [30]. The stimuli-responsive nanocarriers are triggered by various stimuli, like the temperature, magnetic fields, ultrasound intensity, light or electric pulses, or endogenous stimuli, such as changes in pH, enzyme concentration, or redox gradients [30]. Thus, the specific design of nanoscale stimuli-responsive systems will enable the controlled drug biodistribution in response to specific exogenous or endogenous stimuli [49]. These are responsiveness and on-demand drug release can be easily achieved [49].

4.5 Nanocarriers Based on Various Functions

4.5.1 Thermo-Responsive

Thermo-responsive nanocarriers are being used in various cancer therapies, especially for solid tumors [49]. The thermo-responsive nanocarriers will maintain the normal room temperature of 37 °C, the physiological body temperature, and this is seen as the tumor area (~40–42 °C) [49–52]. The use of the polymeric nanocarriers will exhibit the temperature limits for better control of drug release [53]. This technique is important for the various solid tumors in which the optimum temperature is 37 °C, and this can be done easily [53]. In this thermo-responsive technology, the thermo-responsive polymers to liquid crystals are being used and programmed for the desired drug release in the targeted tissues [54].

4.5.2 pH-Responsive

pH values are different in different settings and in various biological compartments, the pH- nanocarriers will have better control of site-specific drug release [55]. These are two important aspects of the pH-sensitive nanocarriers. The first one is for the polymers with ionizable groups which can have pH change and another one is to have acid-sensitive bonds which will be used [49, 56]. The concept here is the acidic microenvironment of tumor sites, where the pH-responsive nanocarriers will rise for targeted cancer therapy [57]. Here the intracellular pH values will differ from the extracellular matrix [58]. Nanocarriers are being widely used and equipped with pH-responsivity [59] Hepatitis B virus vaccines are developed using pH-responsive liposomes for better cytosolic drug release [60]. Polymeric micelles which are pH-responsive are under consideration for their role in chemotherapy [61, 62]. Drug-loading or drug-releasing rates are used to overcome the desired levels. In a few of the instances non-polymer pH-sensitive carbon dots are used for cancer therapy [63]. The pH-responsive has been developed for the delivery of anticancer drugs and genes [64].

4.5.3 Ultrasound-Triggered

The ultrasound-triggered drug-releasing is the most sought due to its non-invasiveness with regulation at the frequency, generated mechanical forces which can transform nanodroplets (such as liposomes) to nanobubbles [49]. The use of ultrasound-genic nanobubbles will increase the efficiency of drug delivery in the adjacent cells which are triggered to deliver the imaging agents and anticancer drugs, which will suggest the futuristic theranostic application of various ultrasound-triggered nanocarriers [65]. The trigger used for the drug release will differ from the various conventional settings for various clinical imaging purposes. Due to this, the chances of the mechanical-induced cell damage can be seen [66]. The ultrasound beam is being used for hard tissues and certain tissue movements [67].

4.5.4 Light-Responsive

Light-responsive drug release is by using these materials with the concept here is the use of the nanocarriers with variable wavelengths of light, and these are capable of the focused drug release in the deeper tissue penetration [49]. One of the studies by the Wang et al. showed the near-infrared (NIR) light-responsive polymeric nanocarrier by adding the selenium which will disintegrate on NIR light exposure due to reactive oxidation (ROS)-mediated selenium oxidation [68]. This irreversible disintegration of the nanocarriers will help in continuous drug release [68]. The light-responsive nanomaterials, the safety of light-responsive nanocarriers is not clear [69, 70]. And the irreversible change in these nanocarriers are raising the concerns of safety [69].

4.5.5 Redox-Responsive

The concept here is that the process of oxidative stress is seen in many diseases and the stimuli-responsive nanocarriers are of interest in these redox-responsive nanocarriers [71–73]. The tumor pathogenesis is based on the different microenvironment which is different from the normal tissues. Due to this process, there is a significant elevation in the concentrations of glutathione and ROS in tumors as compared to the normal tissues [74]. Due to this phenomenon, the redox-responsive polymers used as nanocarriers will felicitate the increased concentration of the drugs in the region of interest [74, 75].

Hyperglycemia induces cellular hypoxia is seen in mitochondrial ROS production [75, 76]. The hypoxia-sensitive vesicles provide fast glucose-responsive insulin delivery [77]. These nanoparticles are using H_2O_2-responsive polymeric vesicles attached to the glucose-mediated insulin delivery [78, 79]. This mechanism of the ROS and redox signaling is also important for the ischemia/reperfusion (I/R) injury, redox-responsive nanocarriers delivering antioxidants to the injured site [80–82]. Due to this ROS-responsive nanocarriers are being developed in various I/R injuries [83–87]. There are various degradable redox-responsive disulfide crosslinked nanogel drug carriers which will decrease cytotoxicity by increasing the increase drug uptake in areas with increased oxidative stress [88].

4.6 Magnetic Targeting

Nanocarriers are also impacted by the magnetic force. The therapeutic agents will be attached to the encapsulated magnetic nanocarriers, which are made up of polymers [89]. Among all the candidate nanomaterials, the biocompatible superparamagnetic iron oxide nanoparticles (SPIONs) with modifications are the most widely used as part of the magnetic nanocarriers [90]. Nanocarriers are being injected into the bloodstream near the target site [89]. After applying the magnetic fields to the target site, the magnetic force will lead to the accumulation and release of the payloads [89]. The phenomenon of magnetophoresis will result in the enhancement, accumulation, and penetration of nanocarriers into solid tumors [91]. However, this concept has lower efficacy with the uncontrollable magnetic nanoparticle distribution [92, 93]. The biggest drawback of this is that the magnetic force will be at the distance and the target sites are near to the surface of the body [89]. The process of implanting magnets in the body was also attempted for better tissue penetration tissue penetration [89]. However, the permanent magnets will not be possible for this [94, 95]. Real-time imaging with the magnetic force for more precise delivery is also a limitation for the use [94, 95].

4.7 Enzyme-Responsive Nanocarriers

Enzymes in the cellular system will be used as triggers for targeted drug release [96]. The various nanocarriers like polymeric micelles, liposomes, and dendrimers are often functionalized with the attachment of the cleavable peptides to the surface [97–100]. In the enzyme-responsive liposomes, the enzymes will directly impact by disturbing the lipid, lipopeptide, or lipopolymer which is used for the stabilization of the nanocarrier [98]. In addition, the various targeted enzymes will be causing the shielding of the polymers from the surface to increase cellular uptake, or by activating the various nanocarriers [98].

Different enzymes are being used to trigger local drug delivery. One important feature of the cancerous tissues is the overexpression of a type of extracellular proteolytic enzyme, called matrix metalloproteinases (MMPs) [101]. Due to this, the various MMP-responsive nanocarriers are developed for the targeted cancer treatment [102]. The targeted anti-inflammatory treatment, a protease secreted by neutrophils, called human neutrophil elastase (HNE) is used for the controlled drug release from nanocarriers which are well equipped with HNE-sensitive peptide linkers [103, 104]. This approach will increase the sensitivity of targeted anti-inflammatory treatment, and the neutrophils are the first cells to be present at the site of the inflammation [105]. Enzyme-responsive nanocarriers can also be applied to regulate coagulation locally [106, 107]. With the currently available imaging technologies, there will be controlled drug release in a precise manner [99].

4.8 Multimodal Nanocarriers

The various stimuli-responsive nanocarriers have improved drug delivery, with some limitations. Nanocarriers use ultrasound, light, heat, and magnetic forces, are for the exact target localization. Nanocarriers with the changes in the pH and oxidative stress will be released in the disease area as the nontargeted areas have elevated pH or oxidative stress, in various conditions of metabolic acidosis, vascular occlusion, or other inflammations. Thus, the multimodal nanocarriers increase the comprising of both thermo- and pH-responsiveness are used in certain cancers [108, 109]. The polymer library for stimuli-responsive nanocarriers is being developed for better delivery [110, 111]. Nanocarriers are being used for the response to the tumor microenvironment and can change the size and morphology in the acidic tumor microenvironment with near-infrared laser irradiation [112]. Jia et al. developed a smart nanodrug that can be able to change the size and morphology based on the tumor microenvironment which can be acidic, laser irradiation for tumor ablation, and also prevent tumor metastasis [112]. The acidic tumor microenvironment triggers the transformation of the nanodrug into net-like nanofibers [113]. The combination of gene and photothermal therapy and the different synergistic therapeutic effects are being seen as the advantages of multimodal nanomedicine [114]. On-demand drug release, Deng et al. have developed a new liposomal drug delivery platform that will have a payload that will be triggered by x-ray radiation [115]. This liposome will have various gold nanoparticles with a photosensitizer called verteporfin [116]. Under radiation, the photosensitizer will allow the payload to be released from the liposome, while the gold nanoparticles are used for radiation enhancement [115]. This will provide the synergistic effects of chemotherapy in combination with radiotherapy [115]. The addition of photosensitizer to the gold nanoparticle will cause cell damage [115]. Continued efforts are being used to make advancements in the theranostic nanocarriers. The surface of FDA-approved iron oxide nanoparticles with an imaging contrast agent and a peptide activatable by a tumor-specific enzyme new theranostic nanocarriers will be achieving the enzyme-specific drug delivery at the site of the tumor and simultaneous MRI imaging [116, 117]. Li et al. studied the carbon quantum dots which are functionalized by the large amino acids that can selectively accumulate at tumor sites for both imaging and therapeutic drug delivery purposes [118]. The development of different nanocarriers with various decorations (or functionalization) for the bioavailability, pharmacokinetics, and specificity of anticancer ruthenium-based drugs can be seen [119].

4.9 Bioinspired Nanocarriers

Besides using biodegradable and biocompatible synthetic materials for the development of nanocarriers, the use of natural materials is also being tried for drug delivery systems. Developing the biomimetic nanoparticles by use of the cell membrane as camouflage [120, 121]. Like the red blood cell (RBC) membrane vesicle integrated with the near-infrared persistent luminescence nano phosphors for the use of the nanocarriers showing the macrophage uptake and systemic clearance. Cancer membrane can be used as camouflage in the tumor nanocarrier which will have the various imaging and/or therapeutic moieties in the integrated newer drug delivery system [122, 123]. Cancer cell membranes are unstable and can be trapped as an autonomous drug delivery system without the support of other nanomaterials [124]. The cancer cell membrane material is combined with the porous silicon nanoparticles for the development of artificial organelles for supplementing cellular functions under oxidative stress [125, 126].

4.10 Strategies to Enhance Therapeutic Efficacy Using Different Payloads

Cell Replacement nanocarriers are capable of developing the payload to the cells to the diseased area thus applying to the cell replacement therapies. In type I diabetic patients, we all know that islet transplantation is the best form of treatment [127]. However, limitations are due to the shortage of donors and having significant side effects of immunosuppression [127]. The encapsulation of the islet cells is seen in immune-isolating membranes with properties of chemical modifications for transplantation [31, 127, 128].

The mesenchymal stem cells (MSCs) are cell-based drug delivery for cancer treatment having excellent results [129].

4.11 Gene Therapy

Nanocarrier-assisted gene therapy is the newer concept by the used of for making silence small interfering RNA (siRNA) to silence disease–causing genes [31]. With more than available 20 siRNA-based therapies in clinical trials [130]. But the pharmacokinetics is not well established [131]. Transporting siRNA across the cell membrane is difficult due to the anionic property [132]. siRNA has immunostimulatory effects and degraded in the bloodstream [132]. Various viral vectors are used to deliver siRNA but have drawbacks like immunogenic and cytotoxic [133]. Due to the non-viral in nature and having the target these nanocarriers have led to the emergence of RNAi therapeutics [130]. Various nanocarriers which are being applied in the delivery of siRNA, are nucleotides, lipids, and polymers [130, 132]. The biocompatibility and design flexibility of nanocarriers will allow better control of siRNA delivery for better knockdown efficiency [131–133]. Among all these the lipid nanocarriers are having good results clinical translation of siRNA therapy [113, 134–137]. Lipid nanoparticles prevent the damage of the siRNA from degradation along with the facilitation of endocytosis and endosomal escape [137]. The first nanoparticle-assisted targeted RNAi was done in humans in 2009 [138]. The nanocarriers can passively accumulate and enter the solid tumors [138, 139]. These nanocarriers have selective targeting molecules with tissue-specific gene editing [139]. Lipid nanocarriers can be helpful in the development and delivery of the various delivery of various components and liposome, called noisome, is being developed [140, 141]. The Niosomes are self-assembled vesicles comprising of single-chain non-ionic surfactants [141]. The majority of assays that are in use are for safety assessment of the therapeutics [142]. Few nanocarriers can enter the cell membrane due to their polarity and can enter the cell via endocytosis [143–145]. A safe entry of the cells is a crucial factor for the analysis of the nanomaterials [144]. The macrophages are presumed to be the first cells to take up the macrophages' first cells which will that take up nanoparticles, assessing the impact of nanocarriers on macrophages can be used to evaluate their immunocompatibility [146]. Mottas et al. have developed a rapid screening method for the evaluation of the impact of nanoparticles on the macrophages, for the new approach for the cytotoxicity for the quality control of nanoparticles [146]. The rapid screenings are also the process for the production of nanocarriers [146]. The interaction of nanoparticles with the blood components regarding the safety profile of the nanoparticles. Fornaguera et al. also show the various testing methods for the safety profile of a biocompatible nanomaterial, which will be interacting with blood components [147]. The fibrinogen aggregation is dependent on the surface charge of the various nanoparticles [147]. The complement is by the functionalization and concentration of nanoparticles [147]. These are based on results from their study, PLGA is about the embolism or cell lysis [147]. This is an important issue that was considered while developing new nanocarriers about the biocompatibility of a single component of the various nanocarriers [148]. Nanocarriers are composed of biocompatible materials which are made up of pro-inflammatory effects [148]. Various effects will be seen in major organs due to the size, shape, or charge of the nanocarriers [149]. Systemic effects of the nanocarriers, like the activation of complement, coagulation, or the platelets, and toxicity toward the clearing tissues (liver, kidney, lungs, etc.), can be seen further [148]. Various PEG-specific antibodies are seen for the administration of PEGylated liposomes, accelerating the clearing process of those liposomes [149]. Different antibodies will have different fragments or induce immune tolerance of the host using various approaches [149]. The safety of the nanocarriers is important is due to various reactions where the oxidation of the corona protein during the interaction between the nanoparticle and the physiological environment will induce oxidative stress to the cells, along with the chronic oxidative stress imbalance will exert harmful effects [146, 150–152].

4.12 Nanocarriers in Neurology

Different mechanisms by which the metabolites move in and out of the brain. The two important mechanisms are transport due to diffusion, by diffusion, or facilitated transport across aqueous channels. The primary concept is for the concentration gradient which is seen across the membranes between cells (i.e., paracellular) or across cells (i.e., transcellular). This passive diffusion movement is seen as accountable for the various means of transport of solutes by the cell membrane, which is dependent on size and lipophilicity [153, 154]. The active transport is being mediated by a carrier like the proteins. The transport is based on the affinity of the molecules and magnetic fields. Transports of solutes, drugs, and other particles. The cell migration, from the blood leukocytes is like monocytes/macrophages, and T cells which will be circulating through the capillary bed by the BBB driven by chemotaxis and modified by the various junctions [155] (Diagram 4.3).

Carrier-mediated transport (CMT) is a form of diffusion that can be passive or active, and can be unidirectional from the blood to the brain. Important carriers like glucose (GLUT1 glucose transporter), amino acids (the LAT1 large neutral amino acid transporter, the CAT1 cationic amino acid transporter), carboxylic acids (the MCT1 monocarboxylic acid transporter), and nucleosides (the CNT2 nucleoside transporter) will be there into the brain. The active efflux transport involves the movement of the drugs from the brain in the presence of the P-glycoprotein, multidrug resistance protein, breast cancer resistance protein, and other transporters [156]. The active

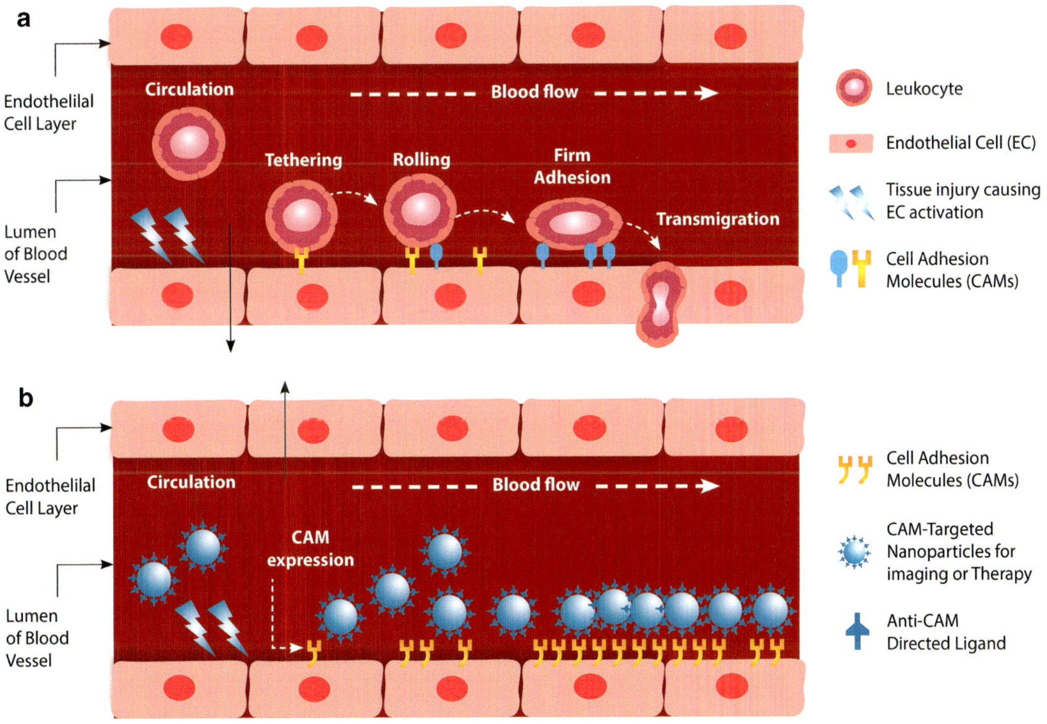

Diagram 4.3 Lumen of the blood vessel: (**a**) In the endothelial cell injury with tissue injury activation and cell adhesion molecules. (**b**) Showing the reaction of the cell expression with adhesion molecules with CAM-targeted nanoparticles for imaging and therapy and also activation of the anti-CAM ligand

Imaging of ischemic stroke hallmarks

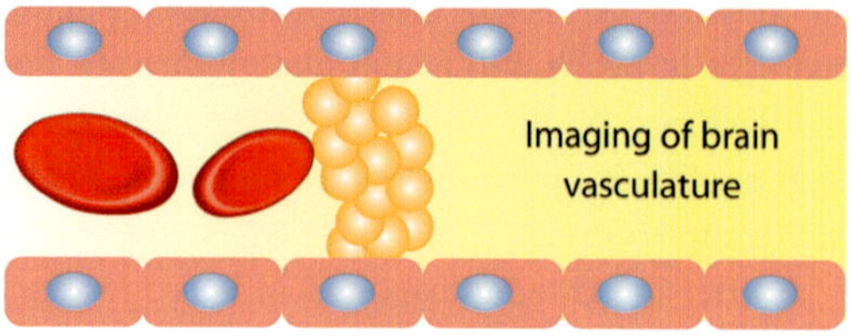

Diagram 4.4 Imaging of the brain vasculature in the stroke

efflux transport will cause the transport of the drug from the brain back to the blood. Receptor-mediated transport is also the transport of peptides and proteins across the BBB by conjugating with different ligands like lactoferrin, transferrin, and insulin [157–160]. This will be important for drug delivery. The conjugated NPs are entering the brain by the adsorptive mediated transport [161, 162]. The role of nanocarriers conjugated to ligands are capable of identifying the capillary endothelial cells and cerebral tumoral cells which will have better CNS drug delivery [163–166]. The BBB receptor-specific ligands are very effective for the transport of endogenous peptides like insulin and transferrin, albumin, and opioids [160, 167–180]. The receptor dissociates from the nanocarrier complex which degrades and releases the drug to the cell (Diagram 4.4).

The nanocarriers are conjugated for the different types of ligands of cell surface receptors based on the brain endothelial cells which are being modulated by the mechanism of receptor-mediated endocytosis. By direct and indirect conjugation of various peptides to nanocarriers or receptors of BBB, drug delivery is mediated [172–177, 180]. Three different mechanisms will be supporting the ligand-conjugated nanocarrier which are again based on the based transport of drugs (i) the adsorption of uptake which will promote apolipoproteins, (ii) the modulation of tight junctions, and (iii) the inhibition P-glycoprotein, playing a key role in drug resistance [181–187]. For the Alzheimer's disease the metal chelators which will be conjugated to NPs will play an important role in Alzheimer's disease. Thus, new channels and pathways are opened up for neurogenerative disease through the study of neural repair by using the conjugated nanocarrier system.

NP-mediated brain delivery systems due to their small size, customizable surface, improved solubility, targeted drug delivery, and multifunctionality, the NPs have emerged as potential drug delivery carriers in the body [188, 189]. The drug stability and its early degradation cannot be taken from the nanocarriers does not take place. Therefore, for drugs to be successfully delivered to their target, many factors such as their size, biocompatibility, target-specific affinity, avoidance of reticuloendothelial systems, stability in blood, or ability to facilitate controlled drug release need to be considered during the manufacture of the NPs. Ideal nanocarrier is particle diameter less than 100 nm, non-toxic, biodegradable and biocompatible, stable in blood (i.e., no opsonization by proteins), receptor-mediated endocytosis, no activation of neutrophils, non-inflammatory; no platelet aggregation; avoidance of the reticuloendothelial systems; prolonged circulation time; scalable and cost-effective concerning the manufacturing process; amenable to small molecules, peptides, proteins or nucleic acids; controlled drug release or should exhibit modulation of drug release profiles. The most prominent candidate biomolecules are peptides (CPP) such as SynB vectors, penetratin, and Tat for enhanced intracellular delivery [190–194], fluorescent dyes, tumoral markers for brain and gene therapeutic agents for genetic therapy such as siRNA [195–200]. The increase in the efficiency and kinetics of nanocarriers towards brains [201, 202], such as growth factors (e.g. epidermal growth factor [183], vascular endothelial growth factor [203], basic fibroblast growth factor [204], insulin-like growth factors (IGF-I and -II) [205]), biotin-binding proteins (avidin, streptavidin, or neutravidin) [206], insulin [184, 186], albumin [207–209], leptin [210, 211], lactoferrin [202, 212], iron-binding protein p97 (melanotransferrin) [213], transferrin [185, 214] and Angiopep-2 [187]. Some agents play a pivotal role in enhancing the permeability of nanoprobes through BBB [215].

Targeted nanocarriers are also used for the various targeted anti-inflammatory drugs for the reduction of vascular inflammation with increased

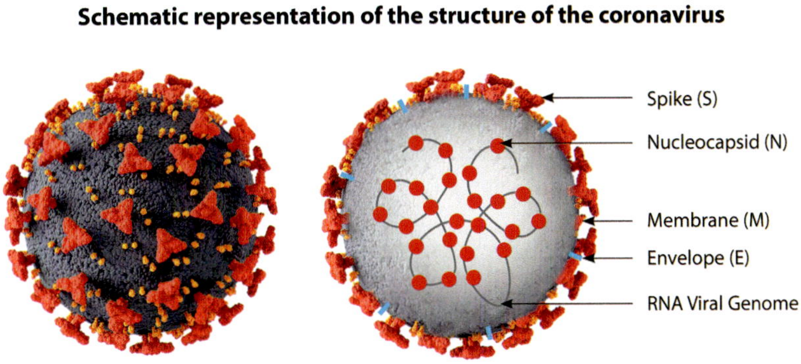

Diagram 4.5 Viral genome structure of the corona virus

therapeutic effect and also for reducing adverse effects (Diagram 4.5). Like the glucocorticoids are potent anti-inflammatory drugs, but adverse effects in nontargeted tissues include osteoporosis, obesity, and glaucoma [216–218]. Endothelial CAMs and the molecules that are being expressed in the inflamed vessels are being explored as the target therapeutics for the site of the anti-inflammatory agents, for the precise targeting of the antioxidants [219–221].

4.12.1 VCAM-1

VCAM-1 is an important therapeutic nanoparticle for the activated endothelium. The targeted liposomal treatment reversed atherosclerotic lesions, in mice has shown decreased accumulation in the inflammation, and the drug alone did not reverse atherosclerotic lesions [222].

4.12.2 ICAM-1

ICAM-1 is commonly seen in the vasculature, but more commonly seen in the sites of inflammation [216]. Thus, ICAM-1-targeted nanocarriers can be used for both diagnostic and therapeutic applications [217, 219]. Its use in the naïve mice, 200 nm PNCs targeted to ICAM-1 with anti-ICAM Ab showed nearly a 30-fold increase in endothelial uptake relative to control IgG control particles [217]. This delivery system is being used for to treat lysosomal storage diseases that are associated with vascular inflammation. ICAM-1-targeted PNCs will be used for the lysosomal replacement enzyme acid sphingomyelinase in vivo and indicated over a 20-fold increase in targeting to the endothelium versus untargeted free enzyme alone [223]. Interestingly, ICAM-1 molecules recycle back to the endothelial plasmalemma after internalization of ICAM-targeted nanocarriers, thereby providing sustained intracellular delivery [224]. Muro et al. have highlighted the targeting of nanocarriers to the endothelium via ICAM in vitro and in vivo by controlling the surface of Ab density [225]. Delivery of the stem cells to the site of injury can be used for the progression of inflammatory diseases like atherosclerosis. The ELIPs targeted to ICAM-1 are being used as the US contrast agent for imaging and also as the US-triggered controlled release delivery vehicle for vascular stem cells [226]. ELIPs loaded with vascular stem cells are seen in the vivo, in an atherosclerotic porcine model. The US-triggered sonoporation of the liposome caused double stem cell accumu lation. This also can cause enhanced stem cell release by binding the endothelium.

4.12.3 Selectins

Selectins are specific vascular targets for the onset of inflammation. Targeting of liposomes will cause a 3.6-fold greater localization in inflamed kidneys as compared to nontargeted IgG-coated liposomes. The targeted delivery of Dex will have fewer inflammatory markers by 60–70% for controlling the side effects [227]. Another selectin-targeted liposomal intervention by targeting of 100 nm liposomes with the selective accumulation in the inflamed site in comparison with the accumulation of untargeted liposomes [228]. Gene expression profiling shows that the genes are getting down-regulated with low-dose Dex-loaded sLex L as compared to high-dose Dex treatment alone.

4.13 Future Directions

The potential of the CAM-targeted nanocarriers for diagnostic agents and drug delivery vehicles is based on inflammation-driven diseases. Technological advancements, hybrid scanners, and the development of nanocarriers show favorable multifunctional nanoparticles. The discovery of new affinity ligands for various molecular signatures of inflammation will be the new imaging targets for both diagnosis and therapy. The dual nanoparticle-targeting of different CAM combinations for the increased adhesion to the target in comparison with the compared to single-epitope targeted particles [229, 230].

4.14 Conclusion

In this chapter, we discussed different types of nanocarriers for drug delivery. The use of nanocarriers can be used for the various nanomaterials and/or structures or in a hybrid fashion, for example, drug delivery. The functionalization of nanocarriers needs to have better stimuli (such as pH, heat, light, or oxidative stress) and increased chances for the delivery of different therapeutics.

In conclusion, nanocarriers show extensive biodistribution in the liver and kidney with a minimum concentration in other organs. The biodistribution study of nanocarriers experimented through radiolabeling showed the in vivo targeted ability of nanocarriers for efficient targeted drug delivery.

References

1. Bruchez M Jr, Moronne M, Gin P, Weiss S, Alivisatos AP. Semiconductor nanocrystals as fluorescent biological labels. Science. 1998;281(5385):2013–6.
2. Chan WC, Nie S. Quantum dot bioconjugates for ultrasensitive nonisotopic detection. Science. 1998;281(5385):2016–8.
3. Tian F, Prina-Mello A, Estrada G, Beyerle A, Moeller W, Schulz H, Kreyling W, Stoeger T. A novel assay for the quantification of internalized nanoparticles in macrophages. Nanotoxicology. 2008;2(4):232–42.
4. Cui D, Tian F, Coyer S, Wang J, Pan B, Gao F, He R, Zhang Y. Effects of antisense-myc-conjugated single-walled carbon nanotubes on HL-60 cells. J Nanosci Nanotechnol. 2007;7:1639–41.
5. Pantarotto D, Partidos CD, Hoebeke J, Brown F, Kramer E, Briand J-P, Muller S, Prato M, Bianco A. Immunization with peptide-functionalized carbon nanotubes enhances virus-specific neutralizing antibody responses. Chem Biol. 2003;10(10):961–6.
6. Edelstein RL, Tamanaha CR, Sheehan PE, Miller MM, Baselt DR, Whitman LJ, Colton RJ. The BARC biosensor applied to the detection of biological warfare agents. Biosens Bioelectron. 2000;14(10–11):805–13.
7. Klibanov AL, Hossack JA. Ultrasound in radiology: from anatomic, functional, molecular imaging to drug delivery and image-guided therapy. Investig Radiol. 2015;50:657–70.
8. Li B, Aid-Launais R, Labour MN, Zenych A, Juenet M, Choqueux C, Ollivier V, Couture O, Letourneur D, Chauvierre C. Functionalized polymer microbubbles as new molecular ultrasound contrast agent to target P-selectin in thrombus. Biomaterials. 2019;194:139–50.
9. de la Isla A, Brostow W, Bujard B, Estevez M, Rodriguez JR, Vargas S, Castaño VM. Nanohybrid scratch resistant coatings for teeth and bone viscoelasticity manifested in tribology. Mater Res Innov. 2003;7(2):110–4.
10. Ma J, Huifen W, Kong LB, Peng KW. Biomimetic processing of nanocrystallite bioactive apatite coating on titanium. Nanotechnology. 2003;14:619–23.
11. Shinkai M, Yanase M, Suzuki M, Hiroyuki H, Wakabayashi T, Yoshida J, Kobayashi T. Intracellular hyperthermia for cancer using magnetite cationic liposomes. J Magn Magn Mater. 1999;194(1–3):176–84.
12. Molday RS, Mackenzie D. Immunospecific ferromagnetic iron-dextran reagents for the labeling and magnetic separation of cells. J Immunol Methods. 1982;52(3):353–67.
13. Weissleder R, Elizondo G, Wittenberg J, Rabito CA, Bengele HH, Josephson L. Ultrasmall superparamagnetic iron oxide: characterization of a new class of contrast agents for MR imaging. Radiology. 1990;175(2):489–93.
14. Parak W, Boudreau R, Gros ML, Gerion D, Zanchet D, Micheel C, Williams S, Alivisatos AP, Larabell C. Cell motility and metastatic potential studies based on quantum dot imaging of phagokinetic tracks. Adv Mater. 2002;14(12):882–5.
15. Czerniawska A. Experimental investigations on the penetration of 198Au from nasal mucous mem-

brane into cerebrospinal fluid. Acta Otolaryngol. 1970;70(1):58–61.
16. Dziubla TD, Shuvaev VV, Hong NK, Hawkins BJ, Madesh M, Takano H, et al. Endothelial targeting of semi-permeable polymer nanocarriers for enzyme therapies. Biomaterials. 2008;29:215–27.
17. Kobayashi H, Brechbiel MW. Nano-sized MRI contrast agents with dendrimer cores. Adv Drug Deliv Rev. 2005;57:2271–86.
18. Choi HS, Liu W, Misra P, Tanaka E, Zimmer JP, Itty Ipe B, et al. Renal clearance of quantum dots. Nat Biotechnol. 2007;25:1165–70.
19. Geng Y, Dalhaimer P, Cai S, Tsai R, Tewari M, Minko T, et al. Shape effects of filaments versus spherical particles in flow and drug delivery. Nat Nanotechnol. 2007;2:249–55.
20. Gratton SE, Ropp PA, Pohlhaus PD, Luft JC, Madden VJ, Napier ME, et al. The effect of particle design on cellular internalization pathways. Proc Natl Acad Sci USA. 2008;105:11613–8.
21. Sharma G, Valenta DT, Altman Y, Harvey S, Xie H, Mitragotri S, et al. Polymer particle shape independently influences binding and internalization by macrophages. J Control Release. 2010;147:408–12.
22. Muro S, Garnacho C, Champion JA, Leferovich J, Gajewski C, Schuchman EH, et al. Control of endothelial targeting and intracellular delivery of therapeutic enzymes by modulating the size and shape of ICAM-1-targeted carriers. Mol Ther. 2008;16:1450–8.
23. Torchilin VP. Multifunctional nanocarriers. Adv Drug Deliv Rev. 2006;58:1532–55.
24. Otsuka H, Nagasaki Y, Kataoka K. PEGylated nanoparticles for biological and pharmaceutical applications. Adv Drug Deliv Rev. 2003;55:403–19.
25. Fukukawa K, Rossin R, Hagooly A, Pressly ED, Hunt JN, Messmore BW, et al. Synthesis and characterization of core-shell star copolymers for in vivo PET imaging applications. Biomacromolecules. 2008;9:1329–39.
26. Anderson CJ, Bulte JW, Chen K, Chen X, Khaw BA, Shokeen M, et al. Design of targeted cardiovascular molecular imaging probes. J Nucl Med. 2010;51(Suppl 1):3S–17S.
27. Chen W, Jarzyna PA, van Tilborg GAF, Nguyen VA, Cormode DP, Klink A, et al. RGD peptide functionalized and reconstituted high-density lipoprotein nanoparticles as a versatile and multimodal tumor targeting molecular imaging probe. FASEB J. 2010;24:1689–99.
28. Uppal R, Caravan P. Targeted probes for cardiovascular MR imaging. Future Med Chem. 2010;2:451–70.
29. Nobs L, Buchegger F, Gurny R, Allemann E. Current methods for attaching targeting ligands to liposomes and nanoparticles. J Pharm Sci. 2004;93:1980–92.
30. Wolinsky JB, Colson YL, Grinstaff MW. Local drug delivery strategies for cancer treatment: gels, nanoparticles, polymeric films, rods, and wafers. J Control Release. 2012;159:14–26.
31. Romero-Azogil L, Benito E, Iglesias N, Galbis E, De-Paz M-V, García-Martín MG. Chapter 11. Redox polymers for drug delivery. In: Casado N, Mecerreyes D, editors. Redox polymers for energy and nanomedicine, Polymer chemistry series. London: The Royal Society of Chemistry; 2020.
32. Summonte S, Racaniello GF, Lopedota A, Denora N, Bernkop-Schnürch A. Thiolated polymeric hydrogels for biomedical application: cross-linking mechanisms. J Control Release. 2021;330:470–82.
33. Oh JK, Drumright R, Siegwart DJ, Matyjaszewski K. The development of microgels/nanogels for drug delivery applications. Prog Polym Sci. 2008;33:448–77. https://doi.org/10.1016/j.progpolymsci.2008.01.002.
34. Soni KS, Desale SS, Bronich TK. Nanogels: an overview of properties, biomedical applications and obstacles to clinical translation. J Control Release. 2016;240:109–26. https://doi.org/10.1016/j.jconrel.2015.11.009.
35. Neamtu I, Rusu AG, Diaconu A, Nita LE, Chiriac AP. Basic concepts and recent advances in nanogels as carriers for medical applications. Drug Deliv. 2017;24:539–57. https://doi.org/10.1080/10717544.2016.1276232.
36. Eckmann DM, Composto RJ, Tsourkas A, Muzykantov VR. Nanogel carrier design for targeted drug delivery. J Mater Chem B. 2014;2:8085–97. https://doi.org/10.1039/C4TB01141D.
37. Myerson JW, McPherson O, Defrates KG, Towslee JH, Marcos-Contreras OA, Shuvaev VV, Braender B, Composto RJ, Muzykantov VR, Eckmann DM. Cross-linker-modulated nanogel flexibility correlates with tunable targeting to a sterically impeded endothelial marker. ACS Nano. 2019;13:11409–21. https://doi.org/10.1021/acsnano.9b04789.
38. Gupta U, Perumal O. Natural and synthetic biomedical polymers. In: Dendrimers and its biomedical applications. Amsterdam: Elsevier; 2014. p. 243–57.
39. Madaan K, Kumar S, Poonia N, Lather V, Pandita D. Dendrimers in drug delivery and targeting: drug-dendrimer interactions and toxicity issues. J Pharm Bioallied Sci. 2014;6:139–50. https://doi.org/10.4103/0975-7406.130965.
40. Chauhan AS. Dendrimers for drug delivery. Molecules. 2018;23:938. https://doi.org/10.3390/molecules23040938.
41. Pan J, Mendes LP, Yao M, Filipczak N, Garai S, Thakur GA, Sarisozen C, Torchilin VP. Polyamidoamine dendrimers-based nanomedicine for combination therapy with SiRNA and chemotherapeutics to overcome multidrug resistance. Eur J Pharm Biopharm. 2019;136:18–28. https://doi.org/10.1016/j.ejpb.2019.01.006.
42. Zhu Y, Liu C, Pang Z. Dendrimer-based drug delivery systems for brain targeting. Biomol Ther. 2019;9:790. https://doi.org/10.3390/biom9120790.
43. Wang H, Chen Q, Zhou S. Carbon-based hybrid nanogels: a synergistic nanoplatform for combined

biosensing, bioimaging, and responsive drug delivery. Chem Soc Rev. 2018;47:4198–232. https://doi.org/10.1039/C7CS00399D.
44. Li W, Liu Q, Zhang P, Liu L. Zwitterionic Nanogels crosslinked by fluorescent carbon dots for targeted drug delivery and simultaneous bioimaging. Acta Biomater. 2016;40:254–62. https://doi.org/10.1016/j.actbio.2016.04.006.
45. Cao S, Slack SD, Levy CN, Hughes SM, Jiang Y, Yogodzinski C, Roychoudhury P, Jerome KR, Schiffer JT, Hladik F, et al. Hybrid nanocarriers incorporating mechanistically distinct drugs for lymphatic CD4 + T cell activation and HIV-1 latency reversal. Sci Adv. 2019;5:eaav6322. https://doi.org/10.1126/sciadv.aav6322.
46. Blanco E, Shen H, Ferrari M. Principles of nanoparticle Design for Overcoming Biological Barriers to drug delivery. Nat Biotechnol. 2015;33:941–51. https://doi.org/10.1038/nbt.3330.
47. Needham D, Anyarambhatla G, Kong G, Dewhirst MW. A new temperature-sensitive liposome for use with mild hyperthermia: characterization and testing in a human tumor xenograft model. Cancer Res. 2000;60:1197–201.
48. Rizvi SAA, Saleh AM. Applications of nanoparticle systems in drug delivery technology. Saudi Pharm J. 2018;26:64–70. https://doi.org/10.1016/j.jsps.2017.10.012.
49. Mura S, Nicolas J, Couvreur P. Stimuli-responsive Nanocarriers for drug delivery. Nat Mater. 2013;12:991–1003. https://doi.org/10.1038/nmat3776.
50. Torchilin VP. Multifunctional, stimuli-sensitive nanoparticulate systems for drug delivery. Nat Rev Drug Discov. 2014;13:813–27. https://doi.org/10.1038/nrd4333.
51. Karimi M, Ghasemi A, Sahandi Zangabad P, Rahighi R, Moosavi Basri SM, Mirshekari H, Amiri M, Shafaei Pishabad Z, Aslani A, Bozorgomid M, et al. Smart micro/nanoparticles in stimulus-responsive drug/gene delivery systems. Chem Soc Rev. 2016;45:1457–501. https://doi.org/10.1039/C5CS00798D.
52. Sánchez-Moreno P, de Vicente J, Nardecchia S, Marchal JA, Boulaiz H. Thermo-sensitive nanomaterials: recent advance in synthesis and biomedical applications. Nano. 2018;8:935. https://doi.org/10.3390/nano8110935.
53. Bordat A, Boissenot T, Nicolas J, Tsapis N. Thermoresponsive polymer nanocarriers for biomedical applications. Adv Drug Deliv Rev. 2019;138:167–92. https://doi.org/10.1016/j.addr.2018.10.005.
54. Lin SY. Thermoresponsive gating membranes embedded with liquid crystal(s) for pulsatile transdermal drug delivery: an overview and perspectives. J Control Release. 2020;319:450–74. https://doi.org/10.1016/j.jconrel.2019.12.046.
55. Zhu YJ, Chen F. PH-responsive drug-delivery systems. Chem Asian J. 2015;10:284–305. https://doi.org/10.1002/asia.201402715.
56. Deirram N, Zhang C, Kermaniyan SS, Johnston APR, Such GK. PH-responsive polymer nanoparticles for drug delivery. Macromol Rapid Commun. 2019;40:1800917. https://doi.org/10.1002/marc.201800917.
57. Xiong H, Wu Y, Jiang Z, Zhou J, Yang M, Yao J. PH-activatable polymeric nanodrugs enhanced tumor chemo/antiangiogenic combination therapy through improving targeting drug release. J Colloid Interface Sci. 2019;536:135–48. https://doi.org/10.1016/j.jcis.2018.10.039.
58. Wu W, Luo L, Wang Y, Wu Q, Dai HB, Li JS, Durkan C, Wang N, Wang GX. Endogenous PH-responsive nanoparticles with programmable size changes for targeted tumor therapy and imaging applications. Theranostics. 2018;8:3038–58. https://doi.org/10.7150/thno.23459.
59. Goyal AK, Rath G, Faujdar C, Malik B. Applications of targeted nano drugs and delivery systems. In: Application and perspective of PH-responsive nano drug delivery systems. Amsterdam: Elsevier; 2019. p. 15–33.
60. Hu F, Yue H, Lu T, Ma G. Cytosolic delivery of HBsAg and enhanced cellular immunity by PH-responsive liposome. J Control Release. 2020;324:460–70. https://doi.org/10.1016/j.jconrel.2020.05.042.
61. Liao ZS, Huang SY, Huang JJ, Chen JK, Lee AW, Lai JY, Lee DJ, Cheng CC. Self-assembled PH-responsive polymeric micelles for highly efficient, noncytotoxic delivery of doxorubicin chemotherapy to inhibit macrophage activation: in vitro investigation. Biomacromolecules. 2018;19:2772–81. https://doi.org/10.1021/acs.biomac.8b00380.
62. Zhou XX, Jin L, Qi RQ, Ma T. Ph-responsive polymeric micelles self-assembled from amphiphilic copolymer modified with lipid used as doxorubicin delivery carriers. R Soc Open Sci. 2018;5:171654. https://doi.org/10.1098/rsos.171654.
63. Seo J, Lee J, Lee CB, Bae SK, Na K. Nonpolymeric PH-sensitive carbon dots for treatment of tumor. Bioconjug Chem. 2019;30:621–32. https://doi.org/10.1021/acs.bioconjchem.8b00813.
64. Cao Y, Huang HY, Chen LQ, Du HH, Cui JH, Zhang LW, Lee BJ, Cao QR. Enhanced lysosomal escape of PH-responsive polyethylenimine-betaine functionalized carbon nanotube for the codelivery of survivin small interfering RNA and doxorubicin. ACS Appl Mater Interfaces. 2019;11:9763–76. https://doi.org/10.1021/acsami.8b20810.
65. Prabhakar A, Banerjee R. Nanobubble liposome complexes for diagnostic imaging and ultrasound-triggered drug delivery in cancers: a theranostic approach. ACS Omega. 2019;4:15567–80. https://doi.org/10.1021/acsomega.9b01924.

66. Canavese G, Ancona A, Racca L, Canta M, Dumontel B, Barbaresco F, Limongi T, Cauda V. Nanoparticle-assisted ultrasound: a special focus on sonodynamic therapy against cancer. Chem Eng J. 2018;340:155–72. https://doi.org/10.1016/j.cej.2018.01.060.
67. Tharkar P, Varanasi R, Wong WSF, Jin CT, Chrzanowski W. Nano-enhanced drug delivery and therapeutic ultrasound for cancer treatment and beyond. Front Bioeng Biotechnol. 2019;7:324. https://doi.org/10.3389/fbioe.2019.00324.
68. Wang Y, Deng Y, Luo H, Zhu A, Ke H, Yang H, Chen H. Light-responsive nanoparticles for highly efficient cytoplasmic delivery of anticancer agents. ACS Nano. 2017;11:12134–44. https://doi.org/10.1021/acsnano.7b05214.
69. Linsley CS, Wu BM. Recent advances in light-responsive on-demand drug-delivery systems. Ther Deliv. 2017;8:89–107. https://doi.org/10.4155/tde-2016-0060.
70. Raza A, Hayat U, Rasheed T, Bilal M, Iqbal HMN. "Smart" materials-based near-infrared light-responsive drug delivery systems for cancer treatment: a review. J Mater Res Technol. 2019;8:1497–509. https://doi.org/10.1016/j.jmrt.2018.03.007.
71. Liguori I, Russo G, Curcio F, Bulli G, Aran L, Della-Morte D, Gargiulo G, Testa G, Cacciatore F, Bonaduce D, et al. Oxidative stress, aging, and diseases. Clin Interv Aging. 2018;13:757–72. https://doi.org/10.2147/CIA.S158513.
72. Huo M, Yuan J, Tao L, Wei Y. Redox-responsive polymers for drug delivery: from molecular design to applications. Polym Chem. 2014;5:1519–28. https://doi.org/10.1039/C3PY01192E.
73. Fang T, Zhang J, Zuo T, Wu G, Xu Y, Yang Y, Yang J, Shen Q. Chemo-photothermal combination cancer therapy with ROS scavenging, extracellular matrix depletion and tumor immune activation by telmisartan and diselenide-paclitaxel prodrug loaded nanoparticles. ACS Appl Mater Interfaces. 2020;12:31292–308. https://doi.org/10.1021/acsami.0c10416.
74. Guo X, Cheng Y, Zhao X, Luo Y, Chen J, Yuan WE. Advances in redox-responsive drug delivery systems of tumor microenvironment. J Nanobiotechnol. 2018;16:74. https://doi.org/10.1186/s12951-018-0398-2.
75. Kumari R, Sunil D, Ningthoujam RS. Hypoxia-responsive nanoparticle based drug delivery systems in cancer therapy: an up-to-date review. J Control Release. 2020;319:135–56. https://doi.org/10.1016/j.jconrel.2019.12.041.
76. Sada K, Nishikawa T, Kukidome D, Yoshinaga T, Kajihara N, Sonoda K, Senokuchi T, Motoshima H, Matsumura T, Araki E. Hyperglycemia induces cellular hypoxia through production of mitochondrial ROS followed by suppression of aquaporin-1. PLoS One. 2016;11:e0158619. https://doi.org/10.1371/journal.pone.0158619.
77. Yu J, Zhang Y, Ye Y, DiSanto R, Sun W, Ranson D, Ligler FS, Buse JB, Gu Z, Ho D. Microneedle-array patches loaded with hypoxia-sensitive vesicles provide fast glucose-responsive insulin delivery. Proc Natl Acad Sci USA. 2015;112:8260–5. https://doi.org/10.1073/pnas.1505405112.
78. Hu X, Yu J, Qian C, Lu Y, Kahkoska AR, Xie Z, Jing X, Buse JB, Gu Z. H2O2-responsive vesicles integrated with transcutaneous patches for glucose-mediated insulin delivery. ACS Nano. 2017;11:613–20. https://doi.org/10.1021/acsnano.6b06892.
79. Yu J, Qian C, Zhang Y, Cui Z, Zhu Y, Shen Q, Ligler FS, Buse JB, Gu Z. Hypoxia and H2O2 dual-sensitive vesicles for enhanced glucose-responsive insulin delivery. Nano Lett. 2017;17:733–9. https://doi.org/10.1021/acs.nanolett.6b03848.
80. Cadenas S. ROS and redox signaling in myocardial ischemia-reperfusion injury and cardioprotection. Free Radic Biol Med. 2018;117:76–89. https://doi.org/10.1016/j.freeradbiomed.2018.01.024.
81. Serracino-Inglott F, Habib NA, Mathie RT. Hepatic ischemia-reperfusion injury. Am J Surg. 2001;181:160–6. https://doi.org/10.1016/S0002-9610(00)00573-0.
82. Minutoli L, Puzzolo D, Rinaldi M, Irrera N, Marini H, Arcoraci V, Bitto A, Crea G, Pisani A, Squadrito F, et al. ROS-mediated NLRP3 inflammasome activation in brain, heart, kidney, and testis ischemia/reperfusion injury. Oxidative Med Cell Longev. 2016;2016:2183026. https://doi.org/10.1155/2016/2183026.
83. Lee D, Bae S, Hong D, Lim H, Yoon JH, Hwang O, Park S, Ke Q, Khang G, Kang PM. H2O2 -responsive molecularly engineered polymer nanoparticles as ischemia/reperfusion-targeted nanotherapeutic agents. Sci Rep. 2013;3:2233. https://doi.org/10.1038/srep02233.
84. Bae S, Park M, Kang C, Dilmen S, Kang TH, Kang DG, Ke Q, Lee SU, Lee D, Kang PM. Hydrogen peroxide-responsive nanoparticle reduces myocardial ischemia/reperfusion injury. J Am Heart Assoc. 2016;5:e003697. https://doi.org/10.1161/JAHA.116.003697.
85. Eshun D, Saraf R, Bae S, Jeganathan J, Mahmood F, Dilmen S, Ke Q, Lee D, Kang PM, Matyal R. Neuropeptide Y3-36 incorporated into PVAX nanoparticle improves functional blood flow in a murine model of hind limb ischemia. J Appl Physiol. 2017;122:1388–97. https://doi.org/10.1152/japplphysiol.00467.2016.
86. Mahmood E, Bae S, Chaudhary O, Feng R, Mahmood F, Robson S, Lee D, Kang PM, Matyal R. Neuropeptide Y3-36 incorporated into PVAX nanoparticle improves angiogenesis in a murine model of myocardial ischemia. Eur J Pharmacol. 2020;882:173261. https://doi.org/10.1016/j.ejphar.2020.173261.
87. Minami K, Bae S, Uehara H, Zhao C, Lee D, Iske J, Fanger MW, Reder J, Morrison I, Azuma H, et al. Targeting of intragraft reactive oxygen species by

APP-103, a novel polymer product, mitigates ischemia/reperfusion injury and promotes the survival of renal transplants. Am J Transplant. 2020;20:1527–37. https://doi.org/10.1111/ajt.15794.
88. Elkassih SA, Kos P, Xiong H, Siegwart DJ. Degradable redox-responsive disulfide-based nanogel drug carriers: via dithiol oxidation polymerization. Biomater Sci. 2019;7:607–17. https://doi.org/10.1039/C8BM01120F.
89. McBain SC, Yiu HHP, Dobson J. Magnetic nanoparticles for gene and Drug delivery. Int J Nanomedicine. 2008;3:169–80. https://doi.org/10.2147/ijn.s1608.
90. Mahmoudi M, Sant S, Wang B, Laurent S, Sen T. Superparamagnetic iron oxide nanoparticles (SPIONs): development, surface modification and applications in chemotherapy. Adv Drug Deliv Rev. 2011;63:24–46. https://doi.org/10.1016/j.addr.2010.05.006.
91. Liu JF, Lan Z, Ferrari C, Stein JM, Higbee-Dempsey E, Yan L, Amirshaghaghi A, Cheng Z, Issadore D, Tsourkas A. Use of oppositely polarized external magnets to improve the accumulation and penetration of magnetic nanocarriers into solid tumors. ACS Nano. 2020;14:142–52. https://doi.org/10.1021/acsnano.9b05660.
92. Zhou Z, Shen Z, Chen X. Tale of two magnets: an advanced magnetic targeting system. ACS Nano. 2020;14:7–11. https://doi.org/10.1021/acsnano.9b06842.
93. Price PM, Mahmoud WE, Al-Ghamdi AA, Bronstein LM. Magnetic drug delivery: where the field is going. Front Chem. 2018;6:619. https://doi.org/10.3389/fchem.2018.00619.
94. Shapiro B, Kulkarni S, Nacev A, Muro S, Stepanov PY, Weinberg IN. Open challenges in magnetic drug targeting. Wiley Interdiscip Rev Nanomed Nanobiotechnol. 2015;7:446–57. https://doi.org/10.1002/wnan.1311.
95. Liu YL, Chen D, Shang P, Yin DC. A review of magnet systems for targeted drug delivery. J Control Release. 2019;302:90–104. https://doi.org/10.1016/j.jconrel.2019.03.031.
96. De la Rica R, Aili D, Stevens MM. Enzyme-responsive nanoparticles for drug release and diagnostics. Adv Drug Deliv Rev. 2012;64:967–78. https://doi.org/10.1016/j.addr.2012.01.002.
97. Mu J, Lin J, Huang P, Chen X. Development of endogenous enzyme-responsive nanomaterials for theranostics. Chem Soc Rev. 2018;47:5554–73. https://doi.org/10.1039/C7CS00663B.
98. Fouladi F, Steffen KJ, Mallik S. Enzyme-responsive liposomes for the delivery of anticancer drugs. Bioconjug Chem. 2017;28:857–68. https://doi.org/10.1021/acs.bioconjchem.6b00736.
99. Shahriari M, Zahiri M, Abnous K, Taghdisi SM, Ramezani M, Alibolandi M. Enzyme responsive drug delivery systems in cancer treatment. J Control Release. 2019;308:172–89. https://doi.org/10.1016/j.jconrel.2019.07.004.
100. Zhang C, Pan D, Li J, Hu J, Bains A, Guys N, Zhu H, Li X, Luo K, Gong Q, et al. Enzyme-responsive peptide dendrimer-gemcitabine conjugate as a controlled-release drug delivery vehicle with enhanced antitumor efficacy. Acta Biomater. 2017;55:153–62. https://doi.org/10.1016/j.actbio.2017.02.047.
101. Shay G, Lynch CC, Fingleton B. Moving targets: emerging roles for MMPs in cancer progression and metastasis. Matrix Biol. 2015;44–46:200–6. https://doi.org/10.1016/j.matbio.2015.01.019.
102. Eskandari P, Bigdeli B, Porgham Daryasari M, Baharifar H, Bazri B, Shourian M, Amani A, Sadighi A, Goliaei B, Khoobi M, et al. Gold-capped mesoporous silica nanoparticles as an excellent enzyme-responsive nanocarrier for controlled doxorubicin delivery. J Drug Target. 2019;27:1084–93. https://doi.org/10.1080/1061186X.2019.1599379.
103. Korkmaz B, Horwitz MS, Jenne DE, Gauthier F. Neutrophil elastase, proteinase 3, and cathepsin G as therapeutic targets in human diseases. Pharmacol Rev. 2010;62:726–59. https://doi.org/10.1124/pr.110.002733.
104. Aimetti AA, Tibbitt MW, Anseth KS. Human neutrophil elastase responsive delivery from poly(ethylene glycol) hydrogels. Biomacromolecules. 2009;10:1484–9. https://doi.org/10.1021/bm9000926.
105. Rosales C. Neutrophil: a cell with many roles in inflammation or several cell types? Front Physiol. 2018;9:113. https://doi.org/10.3389/fphys.2018.00113.
106. Bhat R, Ribes À, Mas N, Aznar E, Sancen6n F, Marcos MD, Murguía JR, Venkataraman A, Martínez-Máñez R. Thrombin-responsive gated silica mesoporous nanoparticles as coagulation regulators. Langmuir. 2016;32:1195–200. https://doi.org/10.1021/acs.langmuir.5b04038.
107. Greineder CF, Howard MD, Carnemolla R, Cines DB, Muzykantov VR. Advanced drug delivery systems for antithrombotic agents. Blood. 2013;122:1565–75. https://doi.org/10.1182/blood-2013-03-453498.
108. Moreira AF, Dias DR, Costa EC, Correia IJ. Thermo- and PH-responsive nano-in-micro particles for combinatorial drug delivery to cancer cells. Eur J Pharm Sci. 2017;104:42–51. https://doi.org/10.1016/j.ejps.2017.03.033.
109. Hiruta Y, Kanda Y, Katsuyama N, Kanazawa H. Dual temperature-and PH-responsive polymeric micelle for selective and efficient two-step doxorubicin delivery. RSC Adv. 2017;7:29540–9. https://doi.org/10.1039/C7RA03579A.
110. Zhang P, Gao D, An K, Shen Q, Wang C, Zhang Y, Pan X, Chen X, Lyv Y, Cui C, et al. A programmable polymer library that enables the construction of stimuli-responsive nanocarriers containing logic gates. Nat Chem. 2020;12:381–90. https://doi.org/10.1038/s41557-020-0426-3.
111. Badeau BA, Comerford MP, Arakawa CK, Shadish JA, Deforest CA. Engineered modular biomaterial

logic gates for environmentally triggered therapeutic delivery. Nat Chem. 2018;10:251–8. https://doi.org/10.1038/nchem.2917.
112. Jia HR, Zhu YX, Liu X, Pan GY, Gao G, Sun W, Zhang X, Jiang YW, Wu FG. Construction of dually responsive nanotransformers with nanosphere-nanofiber-nanosphere transition for overcoming the size paradox of anticancer nanodrugs. ACS Nano. 2019;13:11781–92. https://doi.org/10.1021/acsnano.9b05749.
113. Coelho T, Adams D, Silva A, Lozeron P, Hawkins PN, Mant T, Perez J, Chiesa J, Warrington S, Tranter E, et al. Safety and efficacy of RNAi therapy for transthyretin amyloidosis. N Engl J Med. 2013;369:819–29. https://doi.org/10.1056/NEJMoa1208760.
114. Kim J, Kim J, Jeong C, Kim WJ. Synergistic nanomedicine by combined gene and photothermal therapy. Adv Drug Deliv Rev. 2016;98:99–112. https://doi.org/10.1016/j.addr.2015.12.018.
115. Deng W, Chen W, Clement S, Guller A, Zhao Z, Engel A, Goldys EM. Controlled gene and drug release from a liposomal delivery platform triggered by X-ray radiation. Nat Commun. 2018;9:2713. https://doi.org/10.1038/s41467-018-05118-3.
116. Ansari C, Tikhomirov GA, Hong SH, Falconer RA, Loadman PM, Gill JH, Castaneda R, Hazard FK, Tong L, Lenkov OD, et al. Development of novel tumor-targeted theranostic nanoparticles activated by membrane-type matrix metalloproteinases for combined cancer magnetic resonance imaging and therapy. Small. 2014;10:566–75. https://doi.org/10.1002/smll.201301456.
117. Sarkar S, Levi-Polyachenko N. Conjugated polymer nano-systems for hyperthermia, imaging and drug delivery. Adv Drug Deliv Rev. 2020;163-164:40–64. https://doi.org/10.1016/j.addr.2020.01.002.
118. Li S, Su W, Wu H, Yuan T, Yuan C, Liu J, Deng G, Gao X, Chen Z, Bao Y, et al. Targeted tumour theranostics in mice via carbon quantum dots structurally mimicking large amino acids. Nat Biomed Eng. 2020;4:704–16. https://doi.org/10.1038/s41551-020-0540-y.
119. Riccardi C, Musumeci D, Trifuoggi M, Irace C, Paduano L, Montesarchio D. Anticancer ruthenium(III) complexes and Ru(III)-containing nanoformulations: an update on the mechanism of action and biological activity. Pharmaceuticals. 2019;12:146. https://doi.org/10.3390/ph12040146.
120. Xia Q, Zhang Y, Li Z, Hou X, Feng N. Red blood cell membrane-camouflaged nanoparticles: a novel drug delivery system for antitumor application. Acta Pharm Sin B. 2019;9:675–89. https://doi.org/10.1016/j.apsb.2019.01.011.
121. Liu JM, Zhang DD, Fang GZ, Wang S. Erythrocyte membrane bioinspired near-infrared persistent luminescence nanocarriers for in vivo long-circulating bioimaging and drug delivery. Biomaterials. 2018;165:39–47. https://doi.org/10.1016/j.biomaterials.2018.02.042.
122. Jin J, Bhujwalla ZM. Biomimetic nanoparticles camouflaged in cancer cell membranes and their applications in cancer theranostics. Front Oncol. 2020;9:1560. https://doi.org/10.3389/fonc.2019.01560.
123. Xu C, Liu W, Hu Y, Li W, Di W. Bioinspired tumor-homing nanoplatform for co-delivery of paclitaxel and SiRNA-E7 to HPV-related cervical malignancies for synergistic therapy. Theranostics. 2020;10:3325–39. https://doi.org/10.7150/thno.41228.
124. Balasubramanian V, Poillucci A, Correia A, Zhang H, Celia C, Santos HA. Cell membrane-based nanoreactor to mimic the bio-compartmentalization strategy of a cell. ACS Biomater Sci Eng. 2018;4:1471–8. https://doi.org/10.1021/acsbiomaterials.7b00944.
125. Balasubramanian V, Correia A, Zhang H, Fontana F, Mäkilä E, Salonen J, Hirvonen J, Santos HA. Biomimetic engineering using cancer cell membranes for designing compartmentalized nanoreactors with organelle-like functions. Adv Mater. 2017;29:1605375. https://doi.org/10.1002/adma.201605375.
126. Zhang Z, Guan J, Jiang Z, Yang Y, Liu J, Hua W, Mao Y, Li C, Lu W, Qian J, et al. Brain-targeted drug delivery by manipulating protein corona functions. Nat Commun. 2019;10:3561. https://doi.org/10.1038/s41467-019-11593-z.
127. Ernst AU, Bowers DT, Wang LH, Shariati K, Plesser MD, Brown NK, Mehrabyan T, Ma M. Nanotechnology in cell replacement therapies for type 1 diabetes. Adv Drug Deliv Rev. 2019;139:116–38. https://doi.org/10.1016/j.addr.2019.01.013.
128. Warshauer JT, Bluestone JA, Anderson MS. New frontiers in the treatment of type 1 diabetes. Cell Metab. 2020;31:46–61. https://doi.org/10.1016/j.cmet.2019.11.017.
129. Krueger TEG, Thorek DLJ, Denmeade SR, Isaacs JT, Brennen WN. Concise review: mesenchymal stem cell-based drug delivery: the good, the bad, the ugly, and the promise. Stem Cells Transl Med. 2018;7:651–63. https://doi.org/10.1002/sctm.18-0024.
130. Dong Y, Siegwart DJ, Anderson DG. Strategies, design, and chemistry in SiRNA delivery systems. Adv Drug Deliv Rev. 2019;144:133–47. https://doi.org/10.1016/j.addr.2019.05.004.
131. Bumcrot D, Manoharan M, Koteliansky V, Sah DWY. RNAi therapeutics: a potential new class of pharmaceutical drugs. Nat Chem Biol. 2006;2:711–9. https://doi.org/10.1038/nchembio839.
132. Fernandes F, Kotharkar P, Chakravorty A, Kowshik M, Talukdar I. Nanocarrier mediated SiRNA delivery targeting stem cell differentiation. Curr Stem Cell Res Ther. 2019;15:155–72. https://doi.org/10.2174/1574888X14666191202095041.
133. Ramamoorth M, Narvekar A. Non viral vectors in gene therapy—An overview. J Clin Diagn

Res. 2015;9:GE01–6. https://doi.org/10.7860/JCDR/2015/10443.5394.
134. Wu YF, Wu HC, Kuan CH, Lin CJ, Wang LW, Chang CW, Wang TW. Multi-functionalized carbon dots as theranostic nanoagent for gene delivery in lung cancer therapy. Sci Rep. 2016;6:21170. https://doi.org/10.1038/srep21170.
135. Bates K, Kostarelos K. Carbon nanotubes as vectors for gene therapy: past achievements, present challenges and future goals. Adv Drug Deliv Rev. 2013;65:2023–33. https://doi.org/10.1016/j.addr.2013.10.003.
136. Nishiyama N, Bae Y, Miyata K, Fukushima S, Kataoka K. Smart polymeric micelles for gene and drug delivery. Drug Discov Today Technol. 2005;2:21–6. https://doi.org/10.1016/j.ddtec.2005.05.007.
137. Kulkarni JA, Witzigmann D, Chen S, Cullis PR, Van Der Meel R. Lipid nanoparticle technology for clinical translation of SiRNA therapeutics. Acc Chem Res. 2019;52:2435–44. https://doi.org/10.1021/acs.accounts.9b00368.
138. Davis ME. The first targeted delivery of SiRNA in humans via a self-assembling, cyclodextrin polymer-based nanoparticle: from concept to clinic. Mol Pharm. 2009;6:659–68. https://doi.org/10.1021/mp900015y.
139. Cheng Q, Wei T, Farbiak L, Johnson LT, Dilliard SA, Siegwart DJ. Selective organ targeting (SORT) nanoparticles for tissue-specific MRNA delivery and CRISPR–Cas gene editing. Nat Nanotechnol. 2020;15:313–20. https://doi.org/10.1038/s41565-020-0669-6.
140. Samaridou E, Heyes J, Lutwyche P. Lipid nanoparticles for nucleic acid delivery: current perspectives. Adv Drug Deliv Rev. 2020;154-155:37–63. https://doi.org/10.1016/j.addr.2020.06.002.
141. Riccardi C, Fàbrega C, Grijalvo S, Vitiello G, D'Errico G, Eritja R, Montesarchio D. AS1411-decorated niosomes AS effective nanocarriers for Ru(III)-based drugs in anticancer strategies. J Mater Chem B. 2018;6:5368–84. https://doi.org/10.1039/C8TB01563E.
142. Wolfram J, Zhu M, Yang Y, Shen J, Gentile E, Paolino D, Fresta M, Nie G, Chen C, Shen H, et al. Safety of nanoparticles in medicine. Curr Drug Targets. 2015;16:1671–81. https://doi.org/10.2174/1389450115666140804124808.
143. Ciappellano SG, Tedesco E, Venturini M, Benetti F. In vitro toxicity assessment of oral nanocarriers. Adv Drug Deliv Rev. 2016;106:381–401. https://doi.org/10.1016/j.addr.2016.08.007.
144. Foroozandeh P, Aziz AA. Insight into cellular uptake and intracellular trafficking of nanoparticles. Nanoscale Res Lett. 2018;13:339. https://doi.org/10.1186/s11671-018-2728-6.
145. Rothen-Rutishauser B, Kuhn DA, Ali Z, Gasser M, Amin F, Parak WJ, Vanhecke D, Fink A, Gehr P, Brandenberger C. Quantification of gold nanoparticle cell uptake under controlled biological conditions and adequate resolution. Nanomedicine (Lond). 2014;9:607–21. https://doi.org/10.2217/nnm.13.24.
146. Qi R, Wang Y, Bruno PM, Xiao H, Yingjie Y, Li T, Lauffer S, Wei W, Chen Q, Kang X, et al. Nanoparticle conjugates of a highly potent toxin enhance safety and circumvent platinum resistance in ovarian cancer. Nat Commun. 2017;8:2166. https://doi.org/10.1038/s41467-017-02390-7.
147. Fornaguera C, Calderó G, Mitjans M, Vinardell MP, Solans C, Vauthier C. Interactions of PLGA nanoparticles with blood components: protein adsorption, coagulation, activation of the complement system and hemolysis studies. Nanoscale. 2015;7:6045–58. https://doi.org/10.1039/C5NR00733J.
148. Howard M, Zern BJ, Anselmo AC, Shuvaev VV, Mitragotri S, Muzykantov V. Vascular targeting of nanocarriers: perplexing aspects of the seemingly straightforward paradigm. ACS Nano. 2014;8:4100–32. https://doi.org/10.1021/nn500136z.
149. Parhiz H, Khoshnejad M, Myerson JW, Hood E, Patel PN, Brenner JS, Muzykantov VR. Unintended effects of drug carriers: big Issues of small particles. Adv Drug Deliv Rev. 2018;130:90–112. https://doi.org/10.1016/j.addr.2018.06.023.
150. Mohammadpour R, Dobrovolskaia MA, Cheney DL, Greish KF, Ghandehari H. Subchronic and chronic toxicity evaluation of inorganic nanoparticles for delivery applications. Adv Drug Deliv Rev. 2019;144:112–32. https://doi.org/10.1016/j.addr.2019.07.006.
151. Jayaram DT, Runa S, Kemp ML, Payne CK. Nanoparticle-induced oxidation of corona proteins initiates an oxidative stress response in cells. Nanoscale. 2017;9:7595–601. https://doi.org/10.1039/C6NR09500C.
152. Lima T, Bernfur K, Vilanova M, Cedervall T. Understanding the lipid and protein corona formation on different sized polymeric nanoparticles. Sci Rep. 2020;10:1129. https://doi.org/10.1038/s41598-020-57943-6.
153. Zellnitz S, Müller M, Meindl C, Schröttner H, Fröhlich E. Impact of drug particle shape on permeability and cellular uptake in the lung. Eur J Pharm Sci. 2019;139:105065.
154. Zeb A, Rana I, Choi H-I, Lee C-H, Baek S-W, Lim C-W, Khan N, Arif ST, Sahar NU, Alvi AM, et al. Potential and applications of nanocarriers for efficient delivery of biopharmaceuticals. Pharmaceutics. 2020;12:1184.
155. Engelhardt B. Molecular mechanisms involved in T cell migration across the blood-brain barrier. J Neural Transm. 2006;113(4):477–85.
156. Ikumi Tamai AT. Transporter-mediated permeation of drugs across the blood-brain barrier. J Pharm Sci. 2000;89(11):1371–88.
157. Banks WA, Kastin AJ, Ehrensing CA. Endogenous peptide Tyr-Pro-Trp-GlyNH2 (Tyr-W-MIF-1) is transported from the brain to the blood by

peptide transport system-1. J Neurosci Res. 1993;35(6):690–5.
158. Maratos-Flier E, Kao CY, Verdin EM, King GL. Receptor-mediated vectorial transcytosis of epidermal growth factor by Madin-Darby canine kidney cells. J Cell Biol. 1987;105(4):1595–601.
159. King GL, Johnson SM. Receptor-mediated transport of insulin across endothelial cells. Science. 1985;227(4694):1583–6.
160. Roberts RL, Fine RE, Sandra A. Receptor-mediated endocytosis of transferrin at the blood-brain barrier. J Cell Sci. 1993;104(2):521–32.
161. Kumagai AK, Eisenberg JB, Pardridge WM. Absorptive-mediated endocytosis of cationized albumin and a beta-endorphin-cationized albumin chimeric peptide by isolated brain capillaries. Model system of blood-brain barrier transport. J Biol Chem. 1987;262(31):15214–9.
162. Sai Y, Kajita M, Tamai I, Wakama J, Wakamiya T, Tsuji A. Adsorptivemediated endocytosis of a basic peptide in enterocyte-like Caco-2 cells. Am J Physiol Gastrointest Liver Physiol. 1998;275(3):G514–20.
163. Mahajan SD, Aalinkeel R, Sykes DE, Reynolds JL, Bindukumar B, Adal A, Qi M, Toh J, Xu G, Prasad PN, et al. Methamphetamine alters blood brain barrier permeability via the modulation of tight junction expression: implication for HIV-1 neuropathogenesis in the context of drug abuse. Brain Res. 2008;1203:133–48.
164. Mahajan SD, Aalinkeel R, Sykes DE, Reynolds JL, Bindukumar B, Fernandez S, Chawda R, Shanahan T, Schwartz S. Tight junction regulation by morphine and HIV-1 tat modulates blood-brain barrier permeability. J Clin Immunol. 2008;28(5):528–41.
165. Choi JJ, Pernot M, Brown TR, et al. Spatio-temporal analysis of molecular delivery through the blood-brain barrier using focused ultrasound. Phys Med Biol. 2007;52(18):5509–30.
166. Béduneau A, Saulnier P, Benoit J-P. Active targeting of brain tumours using nanocarriers. Biomaterials. 2007;28(33):4947–67.
167. Bareford LM, Swaan PW. Endocytic mechanisms for targeted drug delivery. Adv Drug Deliv Rev. 2007;59(8):748–58.
168. Witt KA, Huber JD, Egleton RD, Davis TP. Insulin enhancement of opioid peptide transport across the blood-brain barrier and assessment of analgesic effect. J Pharmacol Exp Ther. 2000;295(3):972–8.
169. Tabernero A, Velasco A, Granda B, Lavado EM, Medina JM. Transcytosis of albumin in astrocytes activates the sterol regulatory element-binding protein-1, which promotes the synthesis of the neurotrophic factor oleic acid. J Biol Chem. 2002;277(6):4240–6.
170. Fiori A, Cardelli P, Negri L, Savi MR, Strom R, Erspamer V. Deltorphin transport across the blood brain barrier. Proc Natl Acad Sci USA. 1997;94(17):9469–74.
171. Thomas SA, Abbruscato TJ, Hau VS, Gillespie TJ, Zsigo J, Hruby VJ, Davis TP. Structure-activity relationships of a series of [D-Ala2] deltorphin I and II analogues; in vitro blood-brain barrier permeability and stability. J Pharmacol Exp Ther. 1997;281(2):817–25.
172. Liu H-M, Liu X-F, Yao J-L, Wang C-L, Yu Y, Wang R. Utilization of combined chemical modifications to enhance the blood-brain barrier permeability and pharmacological activity of endomorphin-1. J Pharmacol Exp Ther. 2006;319(1):308–16.
173. Dollinger S, Lober S, Klingenstein R, Korth C, Gmeiner P. A chimeric ligand approach leading to potent antiprion active acridine derivatives: design, synthesis, and biological investigations. J Med Chem. 2006;49(22):6591–5.
174. You JO, Auguste DT. Feedback-regulated paclitaxel delivery based on poly(N,N-dimethylaminoethyl methacrylate-co-2-hydroxyethyl methacrylate) nanoparticles. Biomaterials. 2008;29(12):1950–7.
175. Blasi P, Giovagnoli S, Schoubben A, Ricci M, Rossi C. Solid lipid nanoparticles for targeted brain drug delivery. Adv Drug Deliv Rev. 2007;59(6):454–77.
176. Kreuter J, Shamenkov D, Petrov V, Ramge P, Cychutek K, Koch-Brandt C, Alyautdin R. Apolipoprotein-mediated transport of nanoparticle-bound drugs across the blood-brain barrier. J Drug Target. 2002;10:317–25.
177. Kreuter J. Influence of the surface properties on nanoparticle-mediated transport of drugs to the brain. J Nanosci Nanotechnol. 2004;4:484–8.
178. Ahlawat J, Barroso GG, Asil SM, Alvarado M, Armendariz I, Bernal J, Carabaza X, Chavez S, Cruz P, Escalante V, et al. Nanocarriers as potential drug delivery candidates for overcoming the blood–brain barrier: challenges and possibilities. ACS Omega. 2020;5:12583–95.
179. Ohta S, Kikuchi E, Ishijima A, Azuma T, Sakuma I, Ito T. Investigating the optimum size of nanoparticles for their delivery into the brain assisted by focused ultrasound-induced blood–brain barrier opening. Sci Rep. 2020;10:18220.
180. Zhang L, Fan J, Li G, Yin Z, Fu BM. Transcellular model for neutral and charged nanoparticles across an in vitro blood–brain barrier. Cardiovasc Eng Technol. 2020;11:607–20.
181. Mishra B, Patel BB, Tiwari S. Colloidal nanocarriers: a review on formulation technology, types and applications toward targeted drug delivery. Nanomedicine. 2010;6(1):9–24.
182. Neubert RHH. Potentials of new nanocarriers for dermal and transdermal drug delivery. Eur J Pharm Biopharm. 2011;77(1):1–2.
183. How CW, Rasedee A, Manickam S, Rosli R. Tamoxifen-loaded nanostructured lipid carrier as a drug delivery system: characterization, stability assessment and cytotoxicity. Colloids Surf B Biointerfaces. 2013;112:393–9.
184. Honarbakhsh S, Guenther RH, Willoughby JA, Lommel SA, Pourdeyhimi B. Polymeric systems incorporating plant viral nanoparticles for tai-

lored release of therapeutics. Adv Healthc Mater. 2013;2(7):1001–7.
185. Li Y, Li N, Pan W, Yu Z, Yang L, Tang B. Hollow mesoporous silica nanoparticles with tunable structures for controlled drug delivery. ACS Appl Mater Interfaces. 2017;9(3):2123–9.
186. Desai D, Zhang J, Sandholm J, et al. Lipid bilayer-gated mesoporous silica nanocarriers for tumor-targeted delivery of zoledronic acid in vivo. Mol Pharm. 2017;14(9):3218–27.
187. Wu W, Zheng Y, Wang R, et al. Antitumor activity of folate-targeted, paclitaxel-loaded polymeric micelles on a human esophageal EC9706 cancer cell line. Int J Nanomedicine. 2012;7:3487–502.
188. Singh R, James W, Lillard J. Nanoparticle-based targeted drug delivery. Exp Mol Pathol. 2009;86(1):215–23.
189. Lockman PR, Mumper RJ, Khan MA, Allen DD. Nanoparticle technology for drug delivery across the blood-brain barrier. Drug Dev Ind Pharm. 2002;28(1):1–13.
190. Mortensen LJ, Oberdörster G, Pentland AP, DeLouise LA. In vivo skin penetration of quantum dot nanoparticles in the murine model: the effect of UVR. Nano Lett. 2008;8(9):2779–87.
191. Berry CC. Intracellular delivery of nanoparticles via the HIV-1 tat peptide. Nanomedicine. 2008;3:357–65.
192. de la Fuente JM, Berry CC. Tat peptide as an efficient molecule to translocate gold nanoparticles into the cell nucleus. Bioconjug Chem. 2005;16(5):1176–80.
193. Rao KS, Reddy MK, Horning JL, Labhasetwar V. TAT-conjugated nanoparticles for the CNS delivery of anti-HIV drugs. Biomaterials. 2008;29(33):4429–38.
194. Scherrmann J-M, Temsamani J. The use of pep: trans vectors for the delivery of drugs into the central nervous system. Int Congr Ser. 2005;1277:199–211.
195. Hu Q, Chen C, Yan J, Yang X, Shi X, Zhao J, Lei J, Yang L, Wang K, Chen L, et al. Therapeutic application of gene silencing MMP-9 in a middle cerebral artery occlusion-induced focal ischemia rat model. Exp Neurol. 2009;216(1):35–46.
196. Gonzalez-Alegre P. Therapeutic RNA interference for neurodegenerative diseases: from promise to progress. Pharmacol Ther. 2007;114(1):34–55.
197. Pardridge WM. shRNA and siRNA delivery to the brain. Adv Drug Deliv Rev. 2007;59(2–3):141–52.
198. Hoyer D, Dev KK. RNA interference as a therapeutic strategy for treating CNS disorders. Drug Discov Today. 2006;3(4):451–6.
199. Xie FY, Woodle MC, Lu PY. Harnessing in vivo siRNA delivery for drug discovery and therapeutic development. Drug Discov Today. 2006;11(1–2):67–73.
200. Leung RKM, Whittaker PA. RNA interference: from gene silencing to gene-specific therapeutics. Pharmacol Ther. 2005;107(2):222–39.
201. Sharpless NE, DePinho RA. The mighty mouse: genetically engineered mouse models in cancer drug development. Nat Rev Drug Discov. 2006;5(9):741–54.
202. Hu K, Li J, Shen Y, Lu W, Gao X, Zhang Q, Jiang X. Lactoferrin-conjugated PEG-PLA nanoparticles with improved brain delivery: in vitro and in vivo evaluations. J Control Release. 2009;134(1):55–61.
203. Ikeda E, Takubo K, Kodama T, Okada Y. Brain-specific expression of vascular endothelial growth factor 146 correlates with the blood-brain barrier induction in quail embryos. Dev Neurosci. 2008;30(5):331–9.
204. Deguchi Y, Okutsu H, Okura T, Yamada S, Kimura R, Yuge T, Furukawa A, Morimoto K, Tachikawa M, Ohtsuki S, et al. Internalization of basic fibroblast growth factor at the mouse blood-brain barrier involves perlecan, a heparan sulfate proteoglycan. J Neurochem. 2002;83(2):381–9.
205. Reinhardt RR, Bondy CA. Insulin-like growth factors cross the blood-brain barrier. Endocrinology. 1994;135(5):1753–61.
206. Townsend SA, Evrony GD, Gu FX, Schulz MP, Brown RH Jr, Langer R. Tetanus toxin C fragment-conjugated nanoparticles for targeted drug delivery to neurons. Biomaterials. 2007;28(34):5176–84.
207. Pardridge WM, Triguero D, Buciak J, Yang J. Evaluation of cationized rat albumin as a potential blood-brain barrier drug transport vector. J Pharmacol Exp Ther. 1990;255(2):893–9.
208. Lu W, Zhang Y, Tan Y-Z, Hu K-L, Jiang X-G, Fu S-K. Cationic albuminconjugated pegylated nanoparticles as novel drug carrier for brain delivery. J Control Release. 2005;107(3):428–48.
209. Lu W, Tan Y-Z, Hu K-L, Jiang X-G. Cationic albumin conjugated pegylated nanoparticle with its transcytosis ability and little toxicity against bloodbrain barrier. Int J Pharm. 2005;295(1–2):247–60.
210. Pan W, Barron M, Hsuchou H, Tu H, Kastin AJ. Increased leptin permeation across the blood-brain barrier after chronic alcohol ingestion. Neuropsychopharmacology. 2007;33(4):859–66.
211. Banks WA, DiPalma CR, Farrell CL. Impaired transport of leptin across the blood-brain barrier in obesity: relationship to plasma levels and adiposity in humans. Peptides. 1999;20:1341–5.
212. Fillebeen C, Descamps L, Dehouck M-P, Fenart L, Benaissa M, Spik G, Cecchelli R, Pierce A. Receptor-mediated transcytosis of lactoferrin through the blood-brain barrier. J Biol Chem. 1999;274(11):7011–7.
213. Karkan D, Pfeifer C, Vitalis TZ, Arthur G, Ujiie M, Chen Q, Tsai S, Koliatis G, Gabathuler R, Jefferies WA. A unique carrier for delivery of therapeutic compounds beyond the blood-brain barrier. PLoS One. 2008;3(6):2469.
214. Ulbrich K, Hekmatara T, Herbert E, Kreuter J. Transferrin- and transferrinreceptor-antibody-modified nanoparticles enable drug delivery across the blood-brain barrier (BBB). Eur J Pharm Biopharm. 2009;71(2):251–6.

215. Lumenta DB, Plesnila N, Kläsner B, Baethmann A, Pruneau D, SchmidElsaesser R, Zausinger S. Neuroprotective effects of a postischemic treatment with a bradykinin B2 receptor antagonist in a rat model of temporary focal cerebral ischemia. Brain Res. 2006;1069(1):227–34.
216. Barua S, Mitragotri S. Challenges associated with penetration of nanoparticles across cell and tissue barriers: a review of current status and future prospects. Nano Today. 2014;9:223–43.
217. Lin J, Miao L, Zhong G, Lin C-H, Dargazangy R, Alexander-Katz A. Understanding the synergistic effect of physicochemical properties of nanoparticles and their cellular entry pathways. Commun Biol. 2020;3:205.
218. Buttgereit F, Burmester GR, Lipworth BJ. Optimised glucocorticoid therapy: the sharpening of an old spear. Lancet. 2005;365:801–3.
219. Singhvi G, Rapalli VK, Nagpal S, Dubey SK, Saha RN. Nanocarriers as potential targeted drug delivery for cancer therapy. In: Daima HK, Pn N, Ranjan S, Dasgupta N, Lichtfouse E, editors. Nanoscience in medicine, vol. 1. Cham: Springer International Publishing; 2020. p. 51–88.
220. Muzykantov VR. Biomedical aspects of targeted delivery of drugs to pulmonary endothelium. Expert Opin Drug Deliv. 2005;2:909–26.
221. Carnemolla R, Shuvaev VV, Muzykantov VR. Targeting antioxidant and antithrombotic biotherapeutics to endothelium. Semin Thromb Hemost. 2010;36:332–42.
222. Homem de Bittencourt PI Jr, Lagranha DJ, Maslinkiewicz A, Senna SM, Tavares AM, Baldissera LP, et al. LipoCardium: endothelium-directed cyclopentenone prostaglandin-based liposome formulation that completely reverses atherosclerotic lesions. Atherosclerosis. 2007;193:245–58.
223. Garnacho C, Dhami R, Simone E, Dziubla T, Leferovich J, Schuchman EH, et al. Delivery of acid sphingomyelinase in normal and niemann-pick disease mice using intercellular adhesion molecule-1-targeted polymer nanocarriers. J Pharmacol Exp Ther. 2008;325:400–8.
224. Muro S, Gajewski C, Koval M, Muzykantov VR. ICAM-1 recycling in endothelial cells: a novel pathway for sustained intracellular delivery and prolonged effects of drugs. Blood. 2005;105:650–8.
225. Calderon AJ, Bhowmick T, Leferovich J, Burman B, Pichette B, Muzykantov V, et al. Optimizing endothelial targeting by modulating the antibody density and particle concentration of anti-ICAM coated carriers. J Control Release. 2010;150:37–44.
226. Herbst SM, Klegerman ME, Kim H, Qi J, Shelat H, Wassler M, et al. Delivery of stem cells to porcine arterial wall with echogenic liposomes conjugated to antibodies against CD34 and intercellular adhesion molecule-1. Mol Pharm. 2009;7:3–11.
227. Asgeirsdottir SA, Zwiers PJ, Morselt HW, Moorlag HE, Bakker HI, Heeringa P, et al. Inhibition of proinflammatory genes in anti-GBM glomerulonephritis by targeted dexamethasone-loaded AbEsel liposomes. Am J Physiol Ren Physiol. 2008;294:F554–61.
228. Hashida N, Ohguro N, Yamazaki N, Arakawa Y, Oiki E, Mashimo H, et al. Highefficacy site-directed drug delivery system using sialyl-Lewis X conjugated liposome. Exp Eye Res. 2008;86:138–49.
229. Ferrante EA, Pickard JE, Rychak J, Klibanov A, Ley K. Dual targeting improves microbubble contrast agent adhesion to VCAM-1 and P-selectin under flow. J Control Release. 2009;140:100–7.
230. Weller GE, Villanueva FS, Tom EM, Wagner WR. Targeted ultrasound contrast agents: in vitro assessment of endothelial dysfunction and multi-targeting to ICAM-1 and sialyl Lewisx. Biotechnol Bioeng. 2005;92:780–8.

5 Nanopores in Imaging

5.1 Introduction: Nanoparticle Design and Applications in Therapy and Diagnostics

The basic structure of the nanoparticles is made up of a core material, i.e., nanocrystalline materials or hydrophobic oils, or the shell interior is hollow and will be incorporated by the various payloads. The first type will have both diagnostic or therapeutic applications and in the second case, the interior is capable of carrying the diagnostic or therapeutic agents. The shell of a nanoparticle and core will have separate properties and there will be variations in some of the molecules which cannot be integrated in the nanoparticle lumen or core. As the surface of most of the nanoparticles is easily modified the diagnostic and therapeutic efficacy of the molecules can increase the biocompatibility and bio applicability. These coatings will include various hydrophilic polymers and polysaccharides for increasing the target-specific uptake, various ligands, like the various antibodies, antibody fragments, sugars, peptides, peptidomimetics, and small molecules which are capable of conjugation with the surface or also to the surface coating of nanoparticles.

5.2 Construction of PA-Based Multimodal Nanoprobes

The basic structure of the multimodal nanoprobes is to separate the various multifunctional agents into two types [1–3]. The first type is like "one for all" nanoprobes, for the preparation of the various intrinsic multimodal properties of the element into one single particle, like the gold-based nanoparticles [4–6], and various copper-based nanoparticles [7, 8], and tungsten-based nanoparticles and many [9]. The second type is labeled as "all in one" nanoprobes. The "all in one" multimodal nanoprobes can be prepared by the use of the techniques comprising the different components into core material for making it multifunctional second is the encapsulating or loading of different imaging components into one carrier, and the third is the conjugating different kinds of functional components with the various covalent or non-covalent linkage.

5.3 One for All

"One for all" multimodal nanoprobes are the two-dimensional nanomaterials (Diagram 5.1). Which can be used in the areas of energy storage and lithium-ion battery. The synthesis of the two-dimensional nanomaterials was widely used when it was introduced by the A. Geim and

S. Shaikh, *Nanoimaging - Future of Precision Medicine*,
https://doi.org/10.1007/978-981-97-7292-6_5

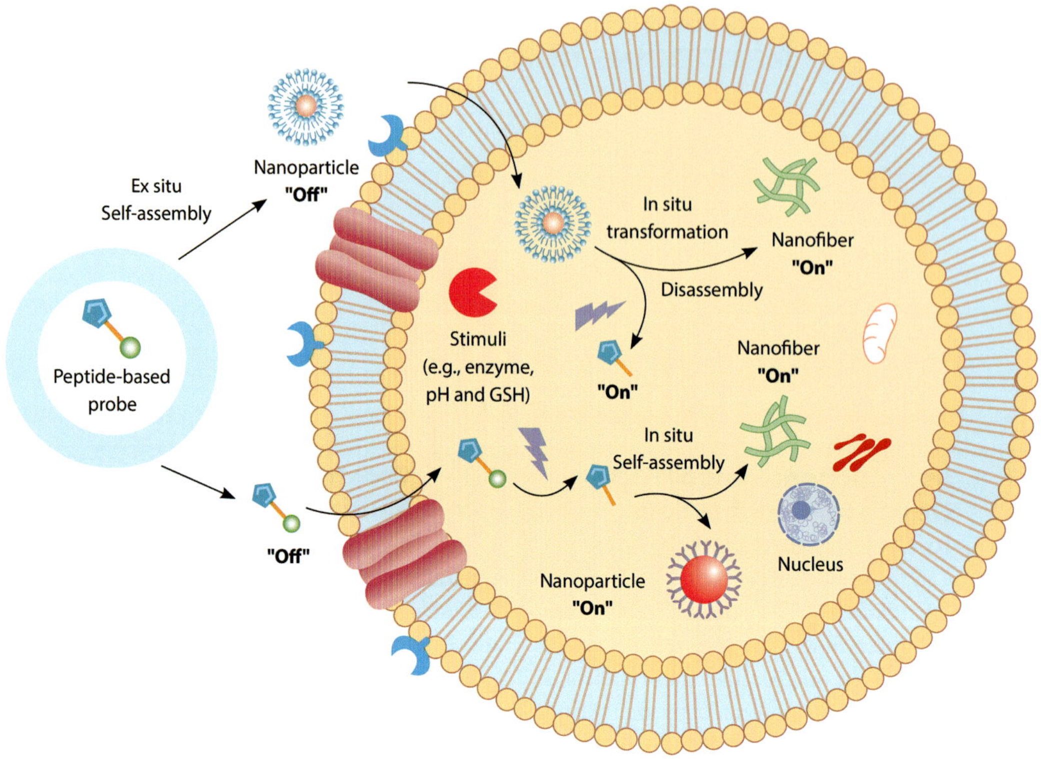

Diagram 5.1 shows in situ activation of the multimodal pathway of the peptide pathway of the nanoparticle

K. Novoselov's work in graphene [10–15]. Other than graphene, metal-based two-dimensional nanomaterials are being used in various aspects [16–18]. Liu Zhuang et al. have developed dual-modal tungsten sulfide nanosheets for use in computed tomography (CT) and PA imaging [19]. These nanosheets will be prepared by the "top-down" technique in which the N-butyllithium is used for the exfoliation of the bulk tungsten sulfide and to make it biocompatible and water-soluble, and thus the polyethylene glycol will be conjugated on the surface of the various nanosheets. The nanosheets are capable of very strong near-infrared absorption having a diameter of 50–100 nm with a thickness of about 1.6 nm and due to this, they are the ideal contrast agents for PA imaging. The tungsten-based nanomaterials can be added to the cobalt selenide nanoplates resulting in the multimodality probe [16]. The cobalt selenide nanoplates are capable of infrared absorption and also have significant magnetic properties. Thus, the cobalt selenide nanoplates are being used as the dual-modal PA imaging/magnetic resonance imaging (MRI) contrast agent as one for all multimodal nanoprobe copper-based nanoparticles. Copper-based nanoparticles are being widely used in PA imaging due to their very strong near-infrared absorption. Many copper-based nanoparticles are available like CuSe nanoparticles, CuS nanoparticles, and CuTe nanoparticles, and many more [19–22]. Recently, the copper sulfide nanoparticle has been used as the multimodal nanoprobe which has been discovered by Mou Juan in Chen Hangrong's group [7]. The CuS is different from the "top to down" method where the "bottom to up" method is being used to form the small molecular to nanoscale, which is also a more demanding chemical method to synthesize nanoparticles. The first step in this is the surface modification of the nanoprobes by making the water-dispersible so that it can be circulated faster in the normal physiological environment, this is a similar concept for increasing the circulation time in the blood circulation. The CuS nanoprobe is capable of many advantages after the range of the second near-infrared will be at about 1160 nm, in this process, the photons will be penetrating deeper into the biological tissues. The absorption of the CuS nanoprobe will make it an excellent contrast agent for PA imaging as well as infrared thermal imaging. The study by Chen Hangrong's group applied the as-synthesized CuS nanoprobe for MRI as the MRI ability of the CuS nanoprobe is like the multimodal nanoprobe which will be used for the nanoimaging. Another group from Toyo University has shown that the Cu2S is an effective X-ray probe [8]. The use of the Cu2S nanoprobe as the first imaging modality is suitable for the X-ray imaging signal and used as the X-ray contrast agent. Along with the two-dimensional and various copper-based multicomponent multimodal nanoprobes, single-component multimodal nanoprobes are being used. The gold-based multimodal nanoprobes like the gold Nanaoshell, gold cube, and gold nanorods will be used for PA imaging and X-ray imaging [4–6]. Carbon dots are also being used as multimodal imaging even though it is a single contrast agent. The carbon dots with red emitting are being constructed by Wang Pengfei's group [23], in which polythiophene phenyl propionic acid is the carbon source for the production of the various multimodal nanoprobes. The absorption and red emission of the various carbon dots will be used for the vivo tumor PA imaging and fluorescence imaging and can be the excellent imaging concept of the in vivo multimodal imaging. Paramagnetic metal ions like gadolinium ions, manganese ions, and their variants are being used in the MRI [24–26]. The various nanomaterials for PA imaging will have optical absorption where the energy can be excited with the laser and convert the energy into heat. Positron emission tomography will have the radiation in which the many small molecular dyes and quantum dots (QDs) can be used in fluorescence imaging.

The concept of bioimaging is the visualization of biological structures and processes. Various techniques are being used which have their advantages as well as disadvantages. Various imaging probes that can label the target molecules or organs can be used to provide the better acquisition of more anatomical and functional information [27–30]. Due to this, the use of imaging probes is an important component of biological research and diagnosis and due to this, there are many recent advances in the development of imaging probes capable of imaging at the subcellular or molecular level [31–33]. The organic probes are the majority used imaging probes used widely [34–36], and they are limited due to their intrinsic physical and physiological properties. The various fluorescent dyes will be used for optical imaging, and they will interact with the photobleaching [37]. The MRI contrast agents which are the Gd3 + -chelates will have a very weak contrast due to their low magnetic moment [38]. These small molecule-based probes have a very short circulation time in vivo, which will have an impact on targeting efficiency and enhancement [39]. Thus, nanotechnology has led to prepare various imaging probes which are having excellent performance [40–42]. Inorganics are the most commonly used nanoparticles due to their various unique physical and chemical properties [43]. These nanoparticles will be using magnetic, X-ray attenuation, and optical properties. The magnetic nanoparticles of the superparamagnetic iron oxide nanoparticles are being used due to the strong T2 MRI Gd3+- based MRI contrast agents, which will have very high sensitivity for the detection [44–46] Z elements like the gold [47, 48], bismuth [49–51], and tantalum [52, 53] will be studied for enhanced CT contrast agents due to the very high X-ray attenuation. The QDs also have better optical and chemical stability and are easily moldable for the emission wavelength in comparison with the fluorescent dyes which will be labeled as robust fluorescent tags [54–56]. The various inorganic nanoparticle-based imaging probes are having various disadvantages and can be used in various clinical settings, like the magnetic susceptibility artifacts of T2 MRI contrast agents [57]. The photoinduced tissue damage due to the ultraviolet (UV) excitation for the QDs [58], which have the potential toxicity of the heavy metal-containing nanoparticles [59, 60]. Many nanoparticle probes are being used and many efforts for the various inorganic nanoparticle imaging probes. Small iron oxide nanoparticles are being used in the T1 MRI contrast agents [61, 62]. The near-infrared (NIR) light will be used for the excitation of nanoparticle probes [63–66]. Various surface modifications are being done for the increased sensitivity of the biocompatibility and various functionalities like the various stimuli-based responsiveness, in the various targeted imaging, and also for the therapy [67]. There will be more increased inorganic nanoparticle probes which will be there for the evaluation of the MRI, CT, and anti-Stokes which will be based on shift-based optical imaging,

5.4 MRI Contrast Agents

MRI is the most commonly used imaging modality which is based on the principle of nuclear magnetic resonance (NMR) [68]. The hydrogen nuclei in the presence of the magnetic field will be absorbing the various resonant radiofrequency pulses, and this will be excited nuclei which will be returning to the original stage of the emission of the absorbed radio frequency energy. The contrast in the MRI is being produced by the various relaxation features of the hydrogen atoms in tissues which are being impacted by the presence of nearby magnetic materials. The use of the paramagnetic materials will enhance the longitudinal T1 relaxation processes which will produce more brighter MR signal. The superparamagnetic and ferromagnetic materials' transverse T2 relaxation processes will be in the hypointense MR signal. Due to these properties, the use of the paramagnetic gadolinium ions (Gd3+) and superparamagnetic iron oxide nanoparticles (SPIONs) are being used as the T1 and T2 contrast agents, respectively [69]. Gd(III) complexes are the most commonly used MRI contrast agents [70]. Along with this, the various safety, efficacy, and market shares of the nanoparticles need to be checked. First-generation commonly used magnetic nanoparticle-based T2 MRI contrast agents are being used to detect the various liver lesions, and second-generation agents will be used for the diagnosis of lymph metastases [57]. They were withdrawn from the market not for safety concerns, but rather due to their small market shares: T1 contrast agents are preferred in clinics due to bright MR images, and more importantly, Gd(III)-based T1 contrast agents can cover most organs including the liver [71]. Furthermore, the contrast effects of the early-generation magnetic nanoparticle-based contrast agents were not sufficiently strong owing to their small core size and low crystallinity [72]. Newly developed magnetic nanoparticles have a strong chance to compete with the Gd(III)-based contrast agents [40]. First, while a serious side effect of Gd(III) complexes, such as nephrogenic systemic fibrosis, is an issue of major concern [73], iron oxides are generally regarded as benign and biologically tolerable [74]. When intravenously injected, the iron oxide nanoparticles are typically degraded in the liver and spleen, and subsequently incorporated into iron metabolic pathways [75]. Indeed, although the early-generation SPION-based contrast agents for intravenous injection are no longer available in clinics, iron oxide nanoparticles are still used for the treatment of iron deficiency anemia [76] and MRI of the gastrointestinal tract via oral administration [77]. Second, nanoparticle syntheses based on the thermal decomposition of metal complexes yield high-quality nanoparticles with tunable size and superior crystallinity [46, 78]. As a result, the magnetic property of the nanoparticles can be controlled from nearly paramagnetic to ferromagnetic by tuning their size from a few to ~100 nm. Such modulation of nanoparticle size allows the magnetic nanoparticles to be used either as a nontoxic alternative to Gd(III)-based T1 contrast agents or as a highly sensitive T2 contrast agent. For example, extremely small-sized iron oxide nanoparticles (ESIONs) less than 3 nm in core size exhibit a large T1 contrast effect in high-resolution MR angiography.61 On the other hand, ferrimagnetic iron oxide nanoparticles (FIONs) with a diameter larger than 30 nm enable highly sensitive

T2-weighted MRI of individual cells due to their strong magnetic property and facile cellular uptake [79]. In addition, FIONs with an average core size of 22 nm exhibit ~7 times stronger T2 contrast effects than those of the first-generation SPION-based agents predicted by outer-sphere relaxation theory [44]. Such a strong contrast effect can be attributed to the balance between the magnetization and the diffusion rate of the 22 nm sized FIONs, which respectively are directly and inversely proportional to the nanoparticle size [80]. Moreover, it is also possible to control the MR contrast effect by changing the magnetic composition of nanoparticles. For instance, the addition of paramagnetic Gd3+ ions into iron oxide nanoparticles improves the T1 contrast effect due to the increased interactions between the Gd3+ ions and water molecules [81]. Likewise, manganese ferrite and zinc-doped ferrite nanoparticles show increased net magnetization, resulting in a much stronger T2 contrast effect [45, 82]. Third, while modification of Gd(III) complexes usually requires complicated multistep organic reactions, the surface of nanoparticles can be modified relatively easily using conventional bioconjugate chemistry with various functional molecules [83]. Since the interactions between biological tissues and nanoparticles are mainly determined by the surface characteristics of the nanoparticles, biodistribution and cellular uptake of the nanoparticles can be readily controlled by surface modification [67]. Furthermore, conjugation of targeting ligands allows more accurate diagnosis by providing information on the biological processes of interest [84] To date, various targeting ligands including antibodies [85], aptamers [86], folic acid [87], and Arg-Gly-Asp (RGD) peptide [88] have been studied for tumor.

As described above, there is still enormous potential in nanoparticle-based MRI contrast agents, and new trials for more sensitive MR imaging are in progress. One of the challenging issues in the development of MR contrast agents lies in overcoming the intrinsic limitations of MRI such as low sensitivity and artifact signals. For example, either hyperintense or hypointense signals can be generated from endogenous factors such as fat, air, bleeding, calcification, or metal deposition, and they are sometimes confused with MR signals generated by contrast agents [57]. To address this issue, T1–T2 dual-mode MRI contrast agents have been introduced by combining superparamagnetic nanoparticles with paramagnetic metal ions [89]. The dual-mode contrast agents generate bright and dark signals in T1- and T2-weighted MRI, respectively, enabling the intrinsic ambiguities to be overcome. In addition, the sensitivity and accuracy of MRI can be improved by obtaining complementary information using multimodal imaging [90]. Therefore, various methods of preparing multimodal imaging probes have been proposed, including the direct conjugation of fluorescent molecules or radioisotopes [91], the assembly of magnetic nanoparticles with QDs or upconversion nanoparticles (UCNPs) [92], and the doping of radioisotopes into magnetic nanoparticles [93] (Diagram 5.2). Another challenging issue is designing the way that MR contrast agents respond to the stimuli of surrounding environments such as pH, temperature, and spe-

Diagram 5.2 Role of the Nanoimaging as Tri modality imaging for detection and Phototherapy for therapy

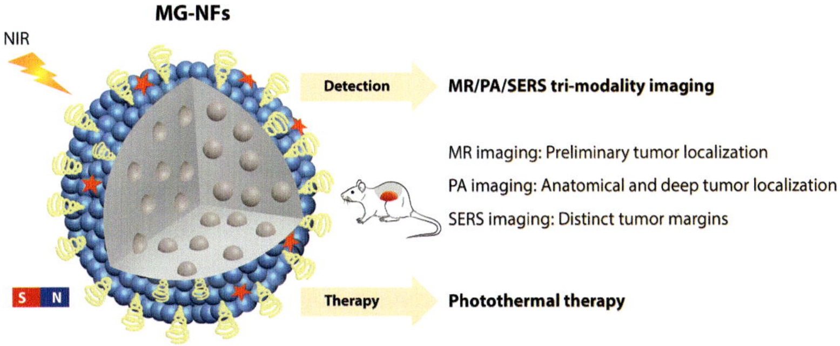

cific enzymes. For the case of Gd(III)-complex MR contrast agents, conformational changes of their chelate structures in response to various stimuli have been proposed [94]. In contrast, there have been scarce reports on the stimuli-responsive nanoparticle-based MR contrast agents because a magnetic field generated by superparamagnetic nanoparticles is not affected by a conformational change of ligands, making the contrast effect "always on." On the other hand, clustering of the magnetic nanoparticles can change the T2 relaxation rate, which is referred to as a magnetic relaxation switch (MRS) [95]. Because the aggregation of nanoparticles can be induced by specific interactions with specific target molecules, various small molecules including oligonucleotides, enzymes, and drugs are detected by MRS using MRI scanners and NMR spectrometers [84]. However, in vivo, MRS remains very challenging as signal attenuation depends on the nanoparticle concentration as well the degree of clustering. Recently, it has been shown that extremely small iron oxide nanoparticles assembled within pH-responsive polymers can activate the MR signals in acidic conditions [96]. When the nanoparticles are aggregated, a strong T2 contrast effect prevents the T1 contrast effect. However, the disassembly of the nanoparticles in acidic conditions leads to an increase in r1 and a decrease in r2, which results in signal enhancement in T1-weighted MRI. Although magnetic nanoparticles are not currently available as MR contrast agents for systemic delivery, much attention and effort have been made to develop superior nanoparticles. (a) X-ray attenuation factors of various elements. (b) Simulated attenuations of I, Yb, Ta, and Au against X-ray produced at 80 kVp (A), 100 kVp (B), 120 kVp (C), and 140 kVp (D). (c) Schematic illustration of RITC-doped tantalum oxide nanoparticles for multimodal imaging. (d) In vivo CT images (left) and optical images (right) of the sentinel lymph node of the rat 2 h after intradermal injection of tantalum oxide nanoparticles in both paws. Besides MR contrast agents based on iron oxide nanoparticles, lanthanide ion-doped nanoparticles are also strong candidates for novel MRI contrast agents like nanoparticles have been developed as multimodal imaging agents for T1-weighted MRI, CT, and upconversion imaging [97]. In addition, Dy3+ and Ho3+ ions exhibit unique magnetic characteristics such as short electronic relaxation time and large magnetic moment, which are suitable for high-field MRI [98, 99]. Although high-field MRI improves the resolution and sensitivity, the contrast effect of iron oxide nanoparticles is marginally increased because their magnetization is already saturated. On the other hand, the magnetization of Dy3+ and Ho3+ is not saturated at the high magnetic field, making NaHoF4 and NaDyF4 good candidates for a T2 contrast agent for high-field MRI. Furthermore, it is expected that an optimized contrast effect can be obtained by modulating particle size, surface coating, and magnetic field.

5.5 Multiphoton Fluorescence Imaging Probes

While whole-body imaging techniques such as MRI and CT play an important role in medical imaging owing to their high resolution and superior penetration depth, their long acquisition time prevents their practical use in real-time monitoring. Fluorescence imaging can be used to overcome such limitations. In general, fluorescence imaging is capable of obtaining high temporal and spatial resolution with good sensitivity [100, 101]. The utility of in vivo fluorescence imaging for live animals, however, has been hampered by the shallow penetration depth of light in tissues and decreased spatial resolution that comes from light scattering. For this reason, there have been demands for the development of innovative fluorescence imaging probes and techniques. One of the recent examples of progress in this field is utilizing the anti-Stokes emission process that generates emission light with a shorter wavelength than that of the excitation light [102, 103]. If combined with near-infrared (NIR) excitation sources, increased tissue penetration depth, as well as reduced background autofluorescence or light scattering, can be achieved [104]. Multiphoton absorption is a well-known anti-Stokes emission process that has the potential to reduce both the photoinduced damage of samples and the photobleaching of fluorophores [102]. Unfortunately, most small molecule-based multiphoton fluorescent dyes still suffer from their low photostability which prevents repeated excitation and prolonged imaging. Therefore, inorganic nanoparticle-based multiphoton fluorescence probes are studied as an alternative due to their improved resistance to photobleaching and relatively facile surface modification with functional molecules. In particular, semiconducting QDs are very attractive in that their emission spectra are tunable, and their multiphoton absorption cross sections are much larger than those of traditional fluorescent dyes [104, 105]. Several other issues need to be considered to fully make use of the potential of the QD-based multiphoton fluorescence probes in bioimaging, most notably safety and imaging efficiency. While cadmium-containing QDs such as CdSe/CdS/ZnS core–shell nanoparticles have been demonstrated as a two-photon imaging probe, potential toxicity from cadmium is a major concern. To address this issue, cadmium-free QDs have been studied [106–108]. For example, manganese-doped ZnS (ZnS: Mn) nanoparticles have been used in multiphoton imaging [107, 109–111]. Besides their low toxicity, the manganese dopants change the emission wavelength from ~430 nm to ~580 nm, allowing light to escape from the tissues. The large three-photon absorption cross-section of ZnS: Mn nanoparticles, which is 4 orders of magnitude larger than those of ultraviolet (UV) fluorescent dyes, enables the three-photon excitation by 920 nm NIR laser, allowing deeper tissue penetration compared with two-photon imaging. Spatial resolution is also much improved due to the reduced out-of-focus excitation and diminished background fluorescence. Other than the ZnS: Mn nanoparticles, InP/ZnS [112] or CuInS2/ZnS QDs [109] are also promising candidates for less-toxic probes (Diagram 5.3).

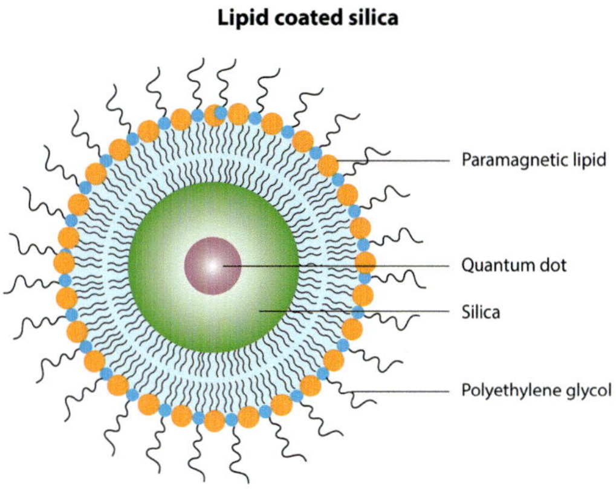

Diagram 5.3 Lipid-coated silica showing various components of the Nanoparticles

5.6 Luminescence Upconversion Imaging Probes

Upconversion is another mechanism of the anti-Stokes emission processes, and it has received much attention in recent years to develop novel luminescent probes. Similar to the multiphoton absorption, an NIR excitation source can be used for deeper light penetration and minimal background autofluorescence. Compared with multiphoton absorption, however, the upconversion mechanism involves photon absorption through real electronic intermediate states, resulting in a much higher emission efficiency and a longer luminescence lifetime of up to several hundred microseconds [111, 113]. Therefore, UCNPs can be excited at a low-power density using a continuous-wave laser diode. As such, image scanning by the focused pulsed laser is not necessary, and data acquisition can be performed much faster using wide-field microscopy [42, 114–116]. Unlike the QDs, the emission wavelength of lanthanide-doped inorganic UCNPs is not related to the quantum confinement effect but is dependent on the energy levels of individual lanthanide elements [117, 118]. Therefore, emission color tuning is achieved by controlling the elemental composition of the UCNPs [119, 120].

Luminescence lifetime is also tunable from several to thousands of microseconds by changing the type or the percentage of dopants [113], which allows multiplex imaging not only by different emission colors but also by different lifetimes. The long luminescence lifetime of the UCNPs is also beneficial to time-gated fluorescence imaging, where increased image contrast is obtained by separating the UCNP emission from light scattering [121]. Although the upconversion efficiency of the lanthanide-doped inorganic UCNPs is exceeded by those of the organic dye-based UCNPs that use triplet–triplet annihilation upconversion [122, 123], it has been shown that the photon collection efficiency can be enhanced by functionalizing the lanthanide-doped inorganic UCNPs with antenna materials such as NIR dyes [124–127], gold nanoshells [110], or QDs [128]. These antennas can also expand the range of absorption wavelength, allowing the flexible choice of excitation sources. In addition, more robust chemical stability and photostability of the inorganic UCNPs than those of the upconverting organic dyes in aqueous or air condition render them better suited for bioimaging applications. Further efforts to enhance the emission efficiency of the lanthanide-doped inorganic UCNPs include the controlled doping of core–shell structured UCNPs with different lanthanide dopants to facilitate the energy transfer, and the high-irradiance excitation of UCNPs, where the luminescence quenching of the dopant ions is alleviated by the strong excitation power [129]. While UCNPs are promising as a new luminescence imaging probe, it is important to point out the current limitations of the UCNPs. First of all, limited tunability of the emission wavelength can restrict the multiplex imaging. Because of the ladder-like energy levels of the lanthanide emitters, there are always multiple emission peaks. For example, erbium ions generate green and red emissions, and thulium ions are known to exhibit UV, blue, and NIR emissions [130]. Although there have been several reports on obtaining pure emission colors from UCNPs, especially for red, emission wavelength tuning is still a challenging issue that requires further exploration [131, 132]. Another issue is the heating effect by the 980 nm NIR laser, which is usually used for the excitation of many types of UCNPs. Since water molecules can absorb the photons of the incident 980 nm laser, their temperature may increase. This heating effect may not be noticeable in most cellular imaging situations [114]. However, the high-power laser used for in vivo imaging may induce a thermal change large enough to affect the upconversion luminescence properties of the UCNPs and possibly cause damage to tissues. To minimize the heating effect, alternative excitation wavelengths whose energies are less absorbed by water molecules have been sought [133]. In a recent study, Nd3+ ions were introduced to UCNPs as a new sensitizer dopant that can be excited at 800 nm with a marginal heating effect [134–136]. Further research on various combinations of sensitizer and host materials is needed to develop an optimized system for clinical use. UCNPs are not free from biosafety issues. Since an increasing number of UCNPs are now studied for in vivo applications, careful evaluation of their potential toxic effects is of great importance. To date, several systematic studies have been made to investigate the in vivo toxicity of UCNPs in mice [137–141], zebrafish embryos [142, 143], and *Caenorhabditis elegans* worms [144–146], and many results suggest little to no toxicity with small doses (e.g., UCNPs indeed can be regarded as safer than the cadmium-containing QDs, but they are not completely safe as an overdose of UCNPs still can induce severe toxicity. Consequently, the administration of UCNPs for bioimaging should be kept as minimal as possible,

5.7 Nanoscale Therapeutic Interventions

Nanoscale therapeutic interventions are increasingly important elements of cancer therapy [147, 148]. Nanoparticles [149, 150] can be effective delivery vehicles for toxic chemotherapeutic drugs, increasing delivery efficiency to the targeted tumor while reducing off-target delivery [151]. Liposomal anthracyclines were the first nanotherapeutic agents to be approved for clinical use as the first line for treatment of AIDS-related Kaposi sarcoma and relapsed ovarian cancer [152] and are under investigation in numerous clinical trials (128 active studies) for the treatment of many types of cancer, especially breast cancer, which accounts for 41 active clinical trials [153]. In addition to the cytotoxic effect of the drug at the molecular level, the success of systemically delivered nanotherapeutic agents for solid tumors is critically dependent on the access that these agents have to tumors through the so-called leaky vasculature of the tumor microvasculature network. This network consists of an immature blood microvessel system, with hypervascularization, abnormal vascular architecture, increased leakage through the vessel wall, and lack of lymphatic drainage [154, 155]. Nanoparticles preferentially accumulate in solid tumors using passive convective transport through leaky endothelium (extravasation) that is caused by pores varying from approximately 100–800 nm [156–158]. The phenomenon is termed the enhanced permeation and retention effect. To date to our knowledge, there exist no clinical tools to determine whether tumor blood vessels are permeable to nanoparticles in this fashion. For instance, the current clinical protocols for liposomal doxorubicin consist of a standard dose every 3–4 weeks [159]. No prior knowledge of tumor vessel status, especially leakiness, is considered for the dose schedule. However, it is well known that the degree of tumor vasculature leakiness differs not only among same-type tumors but even spatially in the same tumor [160–162]. Our purpose was to prospectively predict the effectiveness of a clinically used nano chemotherapeutic agent by detecting and measuring the intratumoral uptake of an X-ray contrast agent nanoprobe by using digital mammography. Materials and Methods The contrast agent formulation used in this study is similar to one that has been licensed to a start-up company, Marval Biosciences (Houston, Tex), founded by two authors (A.V.A. and R.V.B.). Fabrication of the Nanoscale X-ray Probe A highly concentrated iodine solution (650 mg of iodine per milliliter) was prepared by dissolving iodixanol powder (lyophilized from Visipaque 320; GE Healthcare, Milwaukee, Wis) in pure water, with stirring and heating at 70 °C. The rest of the procedures (E.K., with 7 years of experience in nanoparticle fabrication) were similar to those described previously [163]. The liposomal probe contained 72 mg/mL of lipids and 155 mg/mL of iodine, and 100% of the iodine was encapsulated within the liposomes. The mean diameter of the liposomes was 96 nm (standard deviation, 8 nm), a size known to prevent renal clearance. An in vitro leakage experiment against isotonic phosphate-buffered saline exhibited very low leakage of the encapsulated iodine (5% of the initial payload) for 3 days.

5.8 PA-Based Multimodal Biomedical Imaging

Various kinds of multimodal imaging nanoprobes are synthesized for biomedical imaging. They are tools that provide a chance to enhance the detection sensitivity, imaging contrast, and spatial resolution. In this review, in vivo lymph node imaging and tumor imaging were considered as the two representative cases involving PA-based multimodal imaging based on multifunctional nanoprobes. Abnormal lymph nodes are related to many diseases, especially at the early stage of cancer. Therefore, it is of great significance to image lymph nodes by the utilization of exogenously administered contrast-enhancing agents. MRI, fluorescence imaging, and positron emission tomography are involved in lymph node imaging. In recent years, PA imaging as a noninvasive imaging technique has been used for imaging lymph nodes. Multimodal nanoprobes-based PA imaging has also been established. Zeng Wenbin's group has detected the lymph node by fluorescence imaging and PA imaging with the usage of a multimodal nanoprobe which is synthesized by loading fluorescent dye into mesoporous silica nanoparticles [164]. As mentioned before, a multimodal nanoprobe integrated MRI and PA imaging has also been used for lymph node imaging. The multimodal nanoprobe has been synthesized through the doping method.

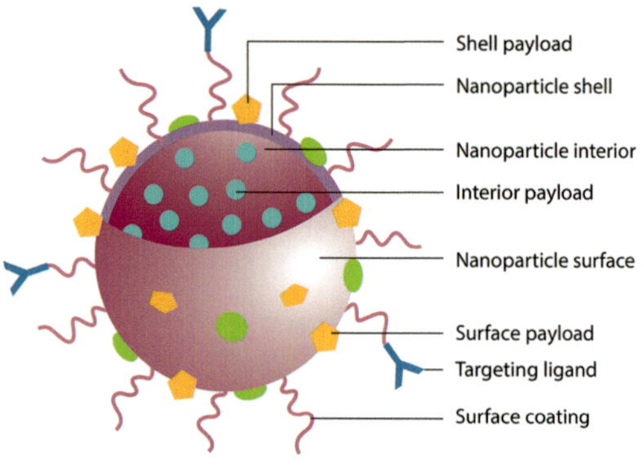

Diagram 5.4 Various interaction in the interior and exterior of the Nanoparticle actions

Imaging nanoprobes, optical nanoprobes capable of monitoring biological events or progress at the cellular and molecular levels have been developed for early detection, accurate diagnosis, and personalized image-guided treatment of diseases. The optical activities of nanoprobes can be tuned on demand for specific applications by engineering their size, surface nature, morphology, and composition. In addition, by the nanostructure, nanoprobes have displayed favorable pharmacokinetic features and target specificity reflecting clinical demands. In this review, we focus on typical approaches and recent trends in the development of nanoprobe-mediated optical imaging and their potential as a clinical diagnostic modality.

Multifunctional imaging nanoprobes have proven to be of great value in the research of pathological processes, as well as the assessment of the delivery, fate, and therapeutic potential of encapsulated drugs (Diagram 5.4). Moreover, such probes may potentially support therapy schemes by the exploitation of their physical properties, e.g., through thermal ablation. This review will present four classes of nanoparticulate imaging probes used in this area: multifunctional probes [165] that can be tracked with at least three different and complementary imaging techniques [166], that carry a drug and have bimodal imaging properties [167], that are employed for nucleic acid delivery and imaging, and [168] imaging probes with capabilities that can be used for thermal ablation. We will highlight several examples where the suitable combination of different (bio)materials like polymers, inorganic nanocrystals, fluorophores, proteins/peptides, and lipids can be tailored to manufacture multifunctional probes to accomplish nanomaterials of each of the aforementioned classes. Moreover, it will be demonstrated how multimodality imaging approaches improve our understanding of in vivo nanoparticle behavior and efficacy at different levels, ranging from the subcellular level to the whole body.

Moreover, multimodal imaging probes not only provide the means for complementary imaging of the same region of interest but also can enable the imaging of different regions by individual imaging techniques. As a result, a more comprehensive and reliable diagnosis is possible with smaller quantities of nanoparticle probes than those for separate imaging.

5.9 Conclusion

The use of multimodality imaging in nanoparticle imaging has come in a great way with many applications in various application in different fields. The use of the different modalities with recent technological advances has led to a greater extent with many few avenues in different areas of the body and the difficult applications.

References

1. Torchilin VP. Recent advances with liposomes as pharmaceutical carriers. Nat Rev Drug Discov. 2005;4:145–60.
2. Cormode DP, Skajaa T, Fayad ZA, Mulder WJ. Nanotechnology in medical imaging: probe design and applications. Arterioscler Thromb Vasc Biol. 2009;29:992–1000.
3. McCarthy JR, Kelly KA, Sun EY, Weissleder R. Targeted delivery of multifunctional magnetic nanoparticles. Nanomedicine. 2007;2:153–67.
4. Hembury M, Chiappini C, Bertazzo S, Kalber TL, Drisko GL, Ogunlade O, et al. Gold-silica quantum rattles for multimodal imaging and therapy. Proc Natl Acad Sci USA. 2015;112:1959–64.
5. Jin YD. Multifunctional compact hybrid au nanoshells: a new generation of nanoplasmonic probes for biosensing, imaging, and controlled release. Acc Chem Res. 2014;47:138–48.

6. Huang G, Yang S, Yuan Y, Xing D. Combining x-ray and photoacoustics for in vivo tumor imaging with gold nanorods. Appl Phys Lett. 2011;99:123701–4.
7. Mou J, Liu C, Li P, Chen Y, Xu H, Wei C, et al. A facile synthesis of versatile Cu2-xS nanoprobe for enhanced MRI and infrared thermal/photoacoustic multimodal imaging. Biomaterials. 2015;57:12–21.
8. Poulose AC, Veeranarayanan S, Mohamed MS, Nagaoka Y, Aburto RR, Mitcham T, et al. Multi-stimuli responsive Cu2S nanocrystals as trimodal imaging and synergistic chemo-photothermal therapy agents. Nanoscale. 2015;7:8378–88.
9. Cheng L, Liu J, Gu X, Gong H, Shi X, Liu T, et al. PEGylated WS2 nanosheets as a multifunctional theranostic agent for in vivo dual-modal CT/photoacoustic imaging guided photothermal therapy. Adv Mater. 2014;26:1886–93.
10. Tan C, Zhang H. Wet-chemical synthesis and applications of non-layer structured two-dimensional nanomaterials. Nat Commun. 2015;6:7873–86.
11. Tan C, Zhang H. Epitaxial growth of hetero-nanostructures based on ultrathin two-dimensional nanosheets. J Am Chem Soc. 2015;137:12162–74.
12. Matthaei JF, DiMaio F, Richards JJ, Pozzo LD, Baker D, Baneyx F. Designing two-dimensional protein arrays through fusion of multimers and Interface mutations. Nano Lett. 2015;15:5235–9.
13. Manzeli S, Allain A, Ghadimi A, Kis A. Piezoresistivity and strain-induced band gap tuning in atomically thin MoS2. Nano Lett. 2015;15:5330–5.
14. Dong R, Pfeffermann M, Liang H, Zheng Z, Zhu X, Zhang J, Feng X. Large-area, free-standing, two-dimensional supramolecular polymer single-layer sheets for highly efficient electrocatalytic hydrogen evolution. Angew Chem Int Ed. 2015;54:12058–63.
15. Dienel T, Kawai S, Söde H, Feng X, Müllen K, Ruffieux P, et al. Resolving atomic connectivity in graphene nanostructure junctions. Nano Lett. 2015;15:5185–90.
16. Song XR, Wang X, Yu SX, Cao J, Li SH, Li J, et al. Co9Se8 nanoplates as a new theranostic platform for photoacoustic/magnetic resonance dual-modal-imaging-guided chemo-photothermal combination therapy. Adv Mater. 2015;27:3285–91.
17. Liu J, Wang C, Wang X, Wang X, Cheng L, Li Y, Liu Z. Mesoporous silica coated single-walled carbon nanotubes as a multifunctional light-responsive platform for cancer combination therapy. Adv Funct Mater. 2015;25:384–92.
18. Liu Z, Liu J, Wang R, Du Y, Ren J, Qu X. An efficient nano-based theranostic system for multi-modal imaging-guided photothermal sterilization in gastrointestinal tract. Biomaterials. 2015;56:206–18.
19. Mou J, Li P, Liu C, Xu H, Song L, Wang J, et al. Ultrasmall Cu2-xS nanodots for highly efficient photoacoustic imaging-guided photothermal therapy. Small. 2015;11:2275–83. Int J Biol Sci 2017, Vol. 13: 412. http://www.ijbs.com
20. Zha Z, Zhang S, Deng Z, Li Y, Li C, Dai Z. Enzyme-responsive copper sulphide nanoparticles for combined photoacoustic imaging, tumor-selective chemotherapy and photothermal therapy. Chem Comm. 2013;49:3455–7.
21. Liu X, Law WC, Jeon M, Wang X, Liu M, Kim C, et al. Cu2-xSe nanocrystals with localized surface plasmon resonance as sensitive contrast agents for in vivo photoacoustic imaging: demonstration of sentinel lymph node mapping. Adv Healthc Mater. 2013;2:952–7.
22. Ku G, Zhou M, Song S, Huang Q, Hazle J, Li C. Copper sulfide nanoparticles as a new class of photoacoustic contrast agent for deep tissue imaging at 1064 nm. ACS Nano. 2012;6:7489–96.
23. Ge J, Jia Q, Liu W, Guo L, Liu Q, Lan M, et al. Red-emissive carbon dots for fluorescent, photoacoustic, and thermal theranostics in living mice. Adv Mater. 2015;27:4169–77.
24. Kunjachan S, Ehling J, Storm G, Kiessling F, Lammers T. Noninvasive imaging of nanomedicines and nanotheranostics: principles, progress, and prospects. Chem Rev. 2015;115:10907–37.
25. Kircher MF, Willmann JK. Molecular body imaging: MR imaging, CT, and US. Part I. principles. Radiology. 2012;263:633–43.
26. Kobayashi H, Longmire MR, Ogawa M, Choyke PL. Rational chemical design of the next generation of molecular imaging probes based on physics and biology: mixing modalities, colors and signals. Chem Soc Rev. 2011;40:4626.
27. Pan D, Roessl E, Schlomka J-P, Caruthers SD, Senpan A, Scott MJ, Allen JS, Zhang H, Hu G, Gaffney PJ, Choi ET, Rasche V, Wickline SA, Proksa R, Lanza GM. Computed tomography in color: NanoK-enhanced spectral CT molecular imaging. Angew Chem Int Ed. 2010;49:9635–9.
28. Barrow M, Taylor A, Murray P, Rosseinsky MJ, Adams DJ. Design considerations for the synthesis of polymer coated iron oxide nanoparticles for stem cell labelling and tracking using MRI. Chem Soc Rev. 2015;44:6733–48.
29. Ni D, Zhang J, Bu W, Xing H, Han F, Xiao Q, Yao Z, Chen F, He Q, Liu J, Zhang S, Fan W, Zhou L, Peng W, Shi J. Dual-targeting upconversion nanoprobes across the blood–brain barrier for magnetic resonance/fluorescence imaging of intracranial glioblastoma. ACS Nano. 2014;8:1231–42.
30. Rieffel J, Chen F, Kim J, Chen G, Shao W, Shao S, Chitgupi U, Hernandez R, Graves SA, Nickles RJ, Prasad PN, Kim C, Cai W, Lovell JF. Hexamodal imaging with porphyrin phospholipid-coated upconversion nanoparticles. Adv Mater. 2015;27:1785–90.
31. Cai E, Ge P, Lee SH, Jeyifous O, Wang Y, Liu Y, Wilson KM, Lim SJ, Baird MA, Stone JE, Lee KY, Davidson MW, Chung HJ, Schulten K, Smith AM, Green WN, Selvin PR. Stable small quantum dots for synaptic receptor tracking on live neurons. Angew Chem Int Ed. 2014;53:12484–8.

32. Motiei M, Dreifuss T, Betzer O, Panet H, Popovtzer A, Santana J, Abourbeh G, Mishani E, Popovtzer R. Differentiating between cancer and inflammation: a metabolic-based method for functional computed tomography imaging. ACS Nano. 2016;10:3469–77.
33. Ghosh D, Lee Y, Thomas S, Kohli AG, Yun DS, Belcher AM, Kelly KA. M13-templated magnetic nanoparticles for targeted in vivo imaging of prostate cancer. Nat Nanotechnol. 2012;7:677–82.
34. Aime S, Castelli DD, Crich SG, Gianolio E, Terreno E. Pushing the sensitivity envelope of lanthanide-based magnetic resonance imaging (MRI) contrast agents for molecular imaging applications. Acc Chem Res. 2009;42:822–31.
35. Singh J, Daftary A. Iodinated contrast media and their adverse reactions. J Nucl Med Technol. 2008;36:69–74. (quiz 76–7)
36. Choi HS, Nasr K, Alyabyev S, Feith D, Lee JH, Kim SH, Ashitate Y, Hyun H, Patonay G, Strekowski L, Henary M, Frangioni JV. Synthesis and in vivo fate of zwitterionic near-infrared fluorophores. Angew Chem Int Ed. 2011;50:6258–63.
37. Wu X, Liu H, Liu J, Haley KN, Treadway JA, Larson JP, Ge N, Peale F, Bruchez MP. Immunofluorescent labeling of cancer marker Her2 and other cellular targets with semiconductor quantum dots. Nat Biotechnol. 2003;21:41–6.
38. Hadjipanayis CG, Bonder MJ, Balakrishnan S, Wang X, Mao H, Hadjipanayis GC. Metallic iron nanoparticles for MRI contrast enhancement and local hyperthermia. Small. 2008;4:1925–9.
39. Mi P, Kokuryo D, Cabral H, Wu H, Terada Y, Saga T, Aoki I, Nishiyama N, Kataoka K. A pH-activatable nanoparticle with signal-amplification capabilities for non-invasive imaging of tumour malignancy. Nat Nanotechnol. 2016;11:724–30.
40. Lee N, Yoo D, Ling D, Cho MH, Hyeon T, Cheon J. Iron oxide based nanoparticles for multimodal imaging and magnetoresponsive therapy. Chem Rev. 2015;115:10637–89.
41. Lee N, Choi SH, Hyeon T. Nano-sized CT contrast agents. Adv Mater. 2013;25:2641–60.
42. Park YI, Lee KT, Suh YD, Hyeon T. Upconverting nanoparticles: a versatile platform for wide-field two-photon microscopy and multi-modal in vivo imaging. Chem Soc Rev. 2015;44:1302–17.
43. Giljohann DA, Seferos DS, Daniel WL, Massich MD, Patel PC, Mirkin CA. Gold nanoparticles for biology and medicine. Angew Chem Int Ed. 2010;49:3280–94.
44. Lee N, Choi Y, Lee Y, Park M, Moon WK, Choi SH, Hyeon T. Water-dispersible ferrimagnetic iron oxide nanocubes with extremely high R2 relaxivity for highly sensitive in vivo MRI of tumors. Nano Lett. 2012;12:3127–31.
45. Jang J-T, Nah H, Lee J-H, Moon SH, Kim MG, Cheon J. Critical enhancements of MRI contrast and hyperthermic effects by dopant-controlled magnetic nanoparticles. Angew Chem Int Ed. 2009;48:1234–8.
46. Chen R, Christiansen MG, Sourakov A, Mohr A, Matsumoto Y, Okada S, Jasanoff A, Anikeeva P. High-performance ferrite nanoparticles through nonaqueous redox phase tuning. Nano Lett. 2016;16:1345–51.
47. Hainfeld JF, Slatkin DN, Focella TM, Smilowitz HM. Gold nanoparticles: a new X-ray contrast agent. Br J Radiol. 2006;79:248–53.
48. Kim D, Park S, Lee JH, Jeong YY, Jon S. Antibiofouling polymer-coated gold nanoparticles as a contrast agent for in vivo X-ray computed tomography imaging. J Am Chem Soc. 2007;129:7661–5.
49. Rabin O, Perez JM, Grimm J, Wojtkiewicz G, Weissleder R. An X-ray computed tomography imaging agent based on long circulating bismuth sulphide nanoparticles. Nat Mater. 2006;5:118–22.
50. Kinsella JM, Jimenez RE, Karmali PP, Rush AM, Kotamraju VR, Gianneschi NC, Ruoslahti E, Stupack D, Sailor MJ. X-ray computed tomography imaging of breast cancer by using targeted peptide-labeled bismuth sulfide nanoparticles. Angew Chem Int Ed. 2011;50:12308–11.
51. Ai K, Liu Y, Liu J, Yuan Q, He Y, Lu L. Large-scale synthesis of Bi2S3 nanodots as a contrast agent for in vivo X-ray computed tomography imaging. Adv Mater. 2011;23:4886–91.
52. Oh MH, Lee N, Kim H, Park SP, Piao Y, Lee J, Jun SW, Moon WK, Choi SH, Hyeon T. Large-scale synthesis of bioinert tantalum oxide nanoparticles for X-ray computed tomography imaging and bimodal image-guided sentinel lymph node mapping. J Am Chem Soc. 2011;133:5508–15.
53. Bonitatibus PJ Jr, Torres AS, Goddard GD, FitzGerald PF, Kulkarni AM. Synthesis, characterization, and computed tomography imaging of a tantalum oxide nanoparticle imaging agent. Chem Commun. 2010;46:8956–8.
54. Gao X, Cui Y, Levenson RM, Chung LWK, Nie S. In vivo cancer targeting and imaging with semiconductor quantum dots. Nat Biotechnol. 2004;22:969–76.
55. Bruchez M Jr, Moronne M, Gin P, Weiss S, Alivisatos AP. Semiconductor nanocrystals as fluorescent biological labels. Science. 1998;281:2013–6.
56. Kim S, Lim YT, Soltesz EG, De Grand AM, Lee J, Nakayama A, Parker JA, Mihaljevic T, Laurence RG, Dor DM, Cohn LH, Bawendi MG, Frangioni JV. Near-infrared fluorescent type II quantum dots for sentinel lymph node mapping. Nat Biotechnol. 2004;22:93–7.
57. Bulte JWM, Kraitchman DL. Iron oxide MR contrast agents for molecular and cellular imaging. NMR Biomed. 2004;17:484–99.
58. Mortensen LJ, Oberdoerster G, Pentland AP, Delouise LA. In vivo skin penetration of quantum dot nanoparticles in the murine model: the effect of UVR. Nano Lett. 2008;8:2779–87.
59. Tsoi KM, Dai Q, Alman BA, Chan WCW. Are quantum dots toxic? exploring the discrepancy between

cell culture and animal studies. Acc Chem Res. 2013;46:662–71.
60. Ye L, Yong K-T, Liu L, Roy I, Hu R, Zhu J, Cai H, Law W-C, Liu J, Wang K, Liu J, Liu Y, Hu Y, Zhang X, Swihart MT, Prasad PN. A pilot study in non-human primates shows no adverse response to intravenous injection of quantum dots. Nat Nanotechnol. 2012;7:453–8.
61. Kim BH, Lee N, Kim H, An K, Park YI, Choi Y, Shin K, Lee Y, Kwon SG, Na HB, Park J-G, Ahn T-Y, Kim Y-W, Moon WK, Choi SH, Hyeon T. Large-scale synthesis of uniform and extremely small-sized iron oxide nanoparticles for high resolution T1 magnetic resonance imaging contrast agents. J Am Chem Soc. 2011;133:12624–31.
62. Shen Z, Wu A, Chen X. Iron oxide nanoparticle based contrast agents for magnetic resonance imaging. Mol Pharm. 2017;14:1352–64.
63. Weissleder R, Ntziachristos V. Shedding light onto live molecular targets. Nat Med. 2003;9:123–8.
64. Hong G, Lee JC, Robinson JT, Raaz U, Xie L, Huang NF, Cooke JP, Dai H. Multifunctional in vivo vascular imaging using near-infrared II fluorescence. Nat Med. 2012;18:1841–6.
65. Naczynski DJ, Tan MC, Zevon M, Wall B, Kohl J, Kulesa A, Chen S, Roth CM, Riman RE, Moghe PV. RareEarth-doped biological composites as in vivo shortwave infrared reporters. Nat Commun. 2013;4:2199.
66. Li C, Zhang Y, Wang M, Zhang Y, Chen G, Li L, Wu D, Wang Q. In vivo real-time visualization of tissue blood flow and angiogenesis using Ag2S quantum dots in the NIR-II window. Biomaterials. 2014;35:393–400.
67. Biju V. Chemical modifications and bioconjugate reactions of nanomaterials for sensing, imaging, drug delivery and therapy. Chem Soc Rev. 2014;43:744–64.
68. Brown MA, Semelka RC. MRI: basic principles and applications. Wiley; 2011.
69. Modo MMJ, Bulte JWM. Molecular and cellular MR imaging. CRC Press; 2007.
70. Wang Y-XJ. Superparamagnetic iron oxide based MRI contrast agents: current status of clinical application. Quant Imaging Med Surg. 2011;1:35–40.
71. Kiessling F, Mertens ME, Grimm J, Lammers T. Nanoparticles for imaging: top or flop? Radiology. 2014;273:10–28.
72. Jun YW, Huh Y-M, Choi JS, Lee JH, Song HT, Kim S, Yoon S, Kim KS, Shin JS, Suh JS, Cheon J. Nanoscale size effect of magnetic nanocrystals and their utilization for cancer diagnosis via magnetic resonance imaging. J Am Chem Soc. 2005;127:5732–3.
73. Penfield JG, Reilly RF. What nephrologists need to know about gadolinium. Nat Clin Pract Nephrol. 2007;3:654–68.
74. Arami H, Khandhar A, Liggitt D, Krishnan KM. In vivo delivery, pharmacokinetics, biodistribution and toxicity of iron oxide nanoparticles. Chem Soc Rev. 2015;44:8576–607.
75. Kolosnjaj-Tabi J, Javed Y, Lartigue L, Volatron J, Elgrabli D, Marangon I, Pugliese G, Caron B, Figuerola A, Luciani N, Pellegrino T, Alloyeau D, Gazeau F. The one year fate of iron oxide coated gold nanoparticles in mice. ACS Nano. 2015;9:7925–39.
76. Schwenk MH. Ferumoxytol: a new intravenous iron preparation for the treatment of iron deficiency anemia in patients with chronic kidney disease. Pharmacotherapy. 2010;30:70–9.
77. Giovagnoni A, Fabbri A, Maccioni F. Oral contrast agents in MRI of the gastrointestinal tract. Adbom Imaging. 2002;27:367–75.
78. Park J, An K, Hwang YS, Park JG, Noh HJ, Kim JY, Park JH, Hwang NM, Hyeon T. Ultra-large-scale syntheses of monodisperse nanocrystals. Nat Mater. 2004;3:891–5.
79. Lee N, Kim H, Choi SH, Park M, Kim D, Kim H-C, Choi Y, Lin S, Kim BH, Jung HS, Kim H, Park KS, Moon WK, Hyeon T. Magnetosome-like ferrimagnetic iron oxide nanocubes for highly sensitive MRI of single cells and transplanted pancreatic islets. Proc Natl Acad Sci USA. 2011;108:2662–7.
80. Gillis P, Moiny F, Brooks RA. On T2-shortening by strongly magnetized spheres: a partial refocusing model. Magn Reson Med. 2002;47:257–63.
81. Zhou Z, Wang L, Chi X, Bao J, Yang L, Zhao W, Chen Z, Wang X, Chen X, Gao J. Engineered iron-oxide-based nanoparticles as enhanced T1 contrast agents for efficient tumor imaging. ACS Nano. 2013;7:3287–96.
82. Lee J-H, Huh Y-M, Jun Y-W, Seo J-W, Jang J-T, Song H-T, Kim S, Cho E-J, Yoon H-G, Suh J-S, Cheon J. Artificially engineered magnetic nanoparticles for ultra-sensitive molecular imaging. Nat Med. 2007;13:95–9.
83. Laurent S, Forge D, Port M, Roch A, Robic C, Vander Elst L, Muller RN. Magnetic iron oxide nanoparticles: synthesis, stabilization, vectorization, physicochemical characterizations, and biological applications. Chem Rev. 2008;108:2064–110.
84. Tassa C, Shaw SY, Weissleder R. Dextran-coated iron oxide nanoparticles: a versatile platform for targeted molecular imaging, molecular diagnostics, and therapy. Acc Chem Res. 2011;44:842–52.
85. Kievit FM, Stephen ZR, Veiseh O, Arami H, Wang T, Lai VP, Park JO, Ellenbogen RG, Disis ML, Zhang M. Targeting of primary breast cancers and metastases in a transgenic mouse model using rationally designed multifunctional Spions. ACS Nano. 2012;6:2591–601.
86. Park J, Park S, Kim S, Lee I-H, Saw PE, Lee K, Kim Y-C, Kim Y-J, Farokhzad OC, Jeong YY, Jon S. Her2-specific aptide conjugated magneto-nanoclusters for potential breast cancer imaging and therapy. J Mater Chem B. 2013;1:4576.
87. Landmark KJ, DiMaggio S, Ward J, Kelly C, Vogt S, Hong S, Kotlyar A, Myc A, Thomas TP, Penner-

Hahn JE, Baker JR, Holl MMB, Orr BG. Synthesis, characterization, and in vitro testing of superparamagnetic iron oxide nanoparticles targeted using folic acid-conjugated dendrimers. ACS Nano. 2008;2:773–83.
88. Zhang F, Huang X, Zhu L, Guo N, Niu G, Swierczewska M, Lee S, Xu H, Wang AY, Mohamedali KA, Rosenblum MG, Lu G, Chen X. Noninvasive monitoring of orthotopic glioblastoma therapy response using RGD-conjugated iron oxide nanoparticles. Biomaterials. 2012;33:5414–22.
89. Shin T-H, Choi J-S, Yun S, Kim I-S, Song H-T, Kim Y, Park KI, Cheon J. T1 and T2 dual-mode MRI contrast agent for enhancing accuracy by engineered nanomaterials. ACS Nano. 2014;8:3393–401.
90. Shin T-H, Choi Y, Kim S, Cheon J. Recent advances in magnetic nanoparticle-based multi-modal imaging. Chem Soc Rev. 2015;44:4501–16.
91. Yang X, Hong H, Grailer JJ, Rowland IJ, Javadi A, Hurley SA, Xiao Y, Yang Y, Zhang Y, Nickles RJ, Cai W, Steeber DA, Gong S. cRGD-functionalized, dox-conjugated, and 64Cu-labeled superparamagnetic iron oxide nanoparticles for targeted anticancer drug delivery and PET/MR imaging. Biomaterials. 2011;32:4151–60.
92. Cheng L, Yang K, Li Y, Chen J, Wang C, Shao M, Lee S-T, Liu Z. Facile preparation of multifunctional upconversion nanoprobes for multimodal imaging and dual-targeted photothermal therapy. Angew Chem Int Ed. 2011;50:7385–90.
93. Chakravarty R, Valdovinos HF, Chen F, Lewis CM, Ellison PA, Luo H, Meyerand ME, Nickles RJ, Cai W. Intrinsically germanium-69-labeled iron oxide nanoparticles: synthesis and in-vivo dual-modality PET/MR imaging. Adv Mater. 2014;26:5119–23.
94. Yu J, Martins AF, Preihs C, Clavijo Jordan V, Chirayil S, Zhao P, Wu Y, Nasr K, Kiefer GE, Sherry AD. Amplifying the sensitivity of zinc(II) responsive MRI contrast agents by altering water exchange rates. J Am Chem Soc. 2015;137:14173–9.
95. Perez JM, Josephson L, O'Loughlin T, Högemann D, Weissleder R. Magnetic relaxation switches capable of sensing molecular interactions. Nat Biotechnol. 2002;20:816–20.
96. Ling D, Park W, Park S-J, Lu Y, Kim KS, Hackett MJ, Kim BH, Yim H, Jeon YS, Na K, Hyeon T. Multifunctional tumor pH-sensitive self-assembled nanoparticles for bimodal imaging and treatment of resistant heterogeneous tumors. J Am Chem Soc. 2014;136:5647–55.
97. Ma D, Meng L, Chen Y, Hu M, Chen Y, Huang C, Shang J, Wang R, Guo Y, Yang J. NaGdF4:Yb3+/Er3+@ NaGdF4:Nd3+@sodium-gluconate: multifunctional and biocompatible ultrasmall core−shell nanohybrids for UCL/MR/CT multimodal imaging. ACS Appl Mater Interfaces. 2015;7:16257–65.
98. Cormode DP, Sanchez-Gaytan BL, Mieszawska AJ, Fayad ZA, Mulder WJ. Inorganic nanocrystals as contrast agents in MRI: synthesis, coating and introduction of multifunctionality. NMR Biomed. 2013;26:766–80.
99. Bruns OT, Ittrich H, Peldschus K, Kaul MG, Tromsdorf UI, Lauterwasser J, Nikolic MS, Mollwitz B, Merkel M, Bigall NC, et al. Real-time magnetic resonance imaging and quantification of lipoprotein metabolism in vivo using nanocrystals. Nat Nanotechnol. 2009;4:193–201.
100. Giepmans BNG, Adams SR, Ellisman MH, Tsien RY. The fluorescent toolbox for assessing protein location and function. Science. 2006;312:217–24.
101. Kobayashi H, Ogawa M, Alford R, Choyke PL, Urano Y. New strategies for fluorescent probe design in medical diagnostic imaging. Chem Rev. 2010;110:2620–40.
102. Gamelin DR, Gü del H, U. Design of luminescent inorganic materials: new photophysical processes studied by optical spectroscopy. Acc Chem Res. 2000;33:235–42.
103. Auzel F. Upconversion and anti-stokes processes with F and D ions in solids. Chem Rev. 2004;104:139–73.
104. Weissleder R. A clearer vision for in vivo imaging. Nat Biotechnol. 2001;19:316–7.
105. Larson DR, Zipfel WR, Williams RM, Clark SW, Bruchez MP, Wise FW, Webb WW. Water-soluble quantum dots for multiphoton fluorescence imaging in vivo. Science. 2003;300:1434–6.
106. Deng ZT, Tong L, Flores M, Lin S, Cheng JX, Yan H, Liu Y. High-quality manganese-doped zinc sulfide quantum rods with tunable dual-color and multiphoton emissions. J Am Chem Soc. 2011;133:5389–96.
107. Yu JH, Kwon S-H, Petrašek Z, Park OK, Jun SW, Shin K, Choi M, Park YI, Park K, Na HB, Lee N, Lee DW, Kim JH, Schwille P, Hyeon T. High-resolution three-photon biomedical imaging using doped ZnS nanocrystals. Nat Mater. 2013;12:359–66.
108. Zhao B, Yao Y, Yang K, Rong P, Huang P, Sun K, An X, Li Z, Chen X, Li W. Mercaptopropionic acid-capped Mn2+:ZnSe/ZnO quantum dots with both downconversion and upconversion emissions for bio-imaging applications. Nanoscale. 2014;6:12345–9.
109. Cichy B, Wawrzynczyk D, Bednarkiewicz A, Samoc M, Strek W. Optical nonlinearities and two-photon excited time resolved luminescence in colloidal quantum-confined CuInS2/ZnS heterostructures. RSC Adv. 2014;4:34065–72.
110. Pan D, Schmieder AH, Wickline SA, Lanza GM. Manganese-based MRI contrast agents: past, present, and future. Tetrahedron. 2011;67:8431.
111. Wang F, Deng R, Wang J, Wang Q, Han Y, Zhu H, Chen X, Liu X. Tuning upconversion through energy migration in core-shell nanoparticles. Nat Mater. 2011;10:968–73.
112. Bharali DJ, Lucey DW, Jayakumar H, Pudavar HE, Prasad PN. Folate-receptor-mediated delivery of InP quantum dots for bioimaging using confocal and two-photon microscopy. J Am Chem Soc. 2005;127:11364–71.

113. Lu Y, Zhao J, Zhang R, Liu Y, Liu D, Goldys EM, Yang X, Xi P, Sunna A, Lu J, Shi Y, Leif RC, Huo Y, Shen J, Piper JA, Robinson JP, Jin D. Tunable lifetime multiplexing using luminescent nanocrystals. Nat Photonics. 2014;8:32–6.
114. Nam SH, Bae YM, Park YI, Kim JH, Kim HM, Choi JS, Lee KT, Hyeon T, Suh YD. Long-term real-time tracking of lanthanide ion doped upconverting nanoparticles in living cells. Angew Chem Int Ed. 2011;50:6093–7.
115. Bae YM, Park YI, Nam SH, Kim JH, Lee K, Kim HM, Yoo B, Choi JS, Lee KT, Hyeon T, Suh YD. Endocytosis, intracellular transport, and exocytosis of lanthanide-doped upconverting nanoparticles in single living cells. Biomaterials. 2012;33:9080–6.
116. Jo HL, Song YH, Park J, Jo E-J, Goh Y, Shin K, Kim M-G, Lee KT. Fast and background-free three-dimensional (3D) live-cell imaging with lanthanide-doped upconverting nanoparticles. Nanoscale. 2015;7:19397–402.
117. Carnall WT, Fields PR, Rajnak K. Spectral intensities of the trivalent lanthanides and actinides in solution. II. Pm3+, Sm3+, Eu3+, Gd3+, Tb3+, Dy3+, and Ho3+. J Chem Phys. 1968;49:4412–23.
118. Bunzli J-CG, Piguet C. Taking advantage of luminescent lanthanide ions. Chem Soc Rev. 2005;34:1048–77.
119. Zhang F, Shi Q, Zhang Y, Shi Y, Ding K, Zhao D, Stucky GD. Fluorescence upconversion microbarcodes for multiplexed biological detection: nucleic acid encoding. Adv Mater. 2011;23:3775–9.
120. Lee J, Bisso PW, Srinivas RL, Kim JJ, Swiston AJ, Doyle PS. Universal process-inert encoding architecture for polymer microparticles. Nat Mater. 2014;13:524–9.
121. Zheng X, Zhu X, Lu Y, Zhao J, Feng W, Jia G, Wang F, Li F, Jin D. High-contrast visualization of upconversion luminescence in mice using time-gating approach. Anal Chem. 2016;88:3449–54.
122. Liu Q, Yang T, Feng W, Li F. Blue-emissive upconversion nanoparticles for low-power-excited bioimaging in vivo. J Am Chem Soc. 2012;134:5390–7.
123. Liu Q, Yin B, Yang T, Yang Y, Shen Z, Yao P, Li F. A general strategy for biocompatible, high-effective upconversion nanocapsules based on triplet-triplet annihilation. J Am Chem Soc. 2013;135:5029–37.
124. Zou W, Visser C, Maduro JA, Pshenichnikov MS, Hummelen JC. Broadband dye-sensitized upconversion of near-infrared light. Nat Photonics. 2012;6:560–4.
125. Chen G, Damasco J, Qiu H, Shao W, Ohulchanskyy TY, Valiev RR, Wu X, Han G, Wang Y, Yang C, Ågren H, Prasad PN. Energy-cascaded upconversion in an organic dye-sensitized core/shell fluoride nanocrystal. Nano Lett. 2015;15:7400–7.
126. Lee J, Yoo B, Lee H, Cha GD, Lee H-S, Cho Y, Kim SY, Seo H, Lee W, Son D, Kang M, Kim HM, Park YI, Hyeon T, Kim D-H. Ultra-wideband multi-dye-sensitized upconverting nanoparticles for information security application. Adv Mater. 2017;29:1603169.
127. Wu X, Zhang Y, Takle K, Bilsel O, Li Z, Lee H, Zhang Z, Li D, Fan W, Duan C, Chan EM, Lois C, Xiang Y, Han G. Dye-sensitized Core/active shell upconversion nanoparticles for optogenetics and bioimaging applications. ACS Nano. 2016;10:1060–6.
128. Wu M, Congreve DN, Wilson MWB, Jean J, Geva N, Welborn M, Van Voorhis T, Bulovic V, Bawendi MG, Baldo MA. Solid-state infrared-to-visible upconversion sensitized by colloidal nanocrystals. Nat Photonics. 2016;10:31–4.
129. Fischer S, Frohlich B, Kramer KW, Goldschmidt JC. Relation between excitation power density and Er3+ doping yielding the highest absolute upconversion quantum yield. J Phys Chem C. 2014;118:30106–14.
130. Wang F, Liu X. Upconversion multicolor fine-tuning: visible to near-infrared emission from lanthanide-doped NaYF4 nanoparticles. J Am Chem Soc. 2008;130:5642–3.
131. Tian G, Gu Z, Zhou L, Yin W, Liu X, Yan L, Jin S, Ren W, Xing G, Li S, Zhao Y. Mn2+ dopant-controlled synthesis of NaYF4:Yb/Er upconversion nanoparticles for in vivo imaging and drug delivery. Adv Mater. 2012;24:1226–31.
132. Wu Z, Lin M, Liang S, Liu Y, Zhang H, Yang B. HotInjection synthesis of manganese-ion-doped NaYF4:Yb,Er nanocrystals with red up-converting emission and tunable diameter. Part Part Syst Charact. 2013;30:311–5.
133. Zhan Q, Qian J, Liang H, Somesfalean G, Wang D, He S, Zhang Z, Andersson-Engels S. Using 915 nm laser excited Tm3+/Er3+/Ho3+-doped NaYbF4 upconversion nanoparticles for in vitro and deeper in vivo bioimaging without overheating irradiation. ACS Nano. 2011;5:3744–57.
134. Xie X, Gao N, Deng R, Sun Q, Xu Q-H, Liu X. Mechanistic investigation of photon upconversion in Nd3+-sensitized core-shell nanoparticles. J Am Chem Soc. 2013;135:12608–11.
135. Wang Y-F, Liu G-Y, Sun L-D, Xiao J-W, Zhou J-C, Yan C-H. Nd3+-sensitized upconversion nanophosphors: efficient in vivo bioimaging probes with minimized heating effect. ACS Nano. 2013;7:7200–6.
136. Shen J, Chen G, Vu A-M, Fan W, Bilsel OS, Chang C-C, Han G. Engineering the upconversion nanoparticle excitation wavelength: cascade sensitization of tri-doped upconversion colloidal nanoparticles at 800 nm. Adv Opt Mater. 2013;1:644–50.
137. Xiong L, Yang T, Yang Y, Xu C, Li F. Long-term in vivo biodistribution imaging and toxicity of polyacrylic acid-coated upconversion nanophosphors. Biomaterials. 2010;31:7078–85.
138. Cheng L, Yang K, Shao M, Lu X, Liu Z. In vivo pharmacokinetics, long-term biodistribution and toxicology study of functionalized upconversion nanoparticles in mice. Nanomedicine. 2011;6:1327–40.

139. Park HS, Nam SH, Kim J, Shin HS, Suh YD, Hong KS. Clear-cut observation of clearance of sustainable upconverting nanoparticles from lymphatic system of small living mice. Sci Rep. 2016;6:27407. ACS Central Science Outlook https://doi.org/10.1021/acscentsci.7b00574 ACS Cent Sci. 2018, 4, 324–336 335.
140. Gao Y, Zhu X, Zhang Y, Chen X, Wang L, Feng W, Huang C, Li F. In vivo biodistribution and passive accumulation of upconversion nanoparticles in colorectal cancer models via intraperitoneal injection. RSC Adv. 2017;7:31588–96.
141. Yu J, Yin W, Peng T, Chang YN, Zu Y, Li J, He X, Ma X, Gu Z, Zhao Y. Biodistribution, excretion, and toxicity of polyethyleneimine modified NaYF4: Yb, Er upconversion nanoparticles in mice via different administration routes. Nanoscale. 2017;9:4497–507.
142. Wang K, Ma J, He M, Gao G, Xu H, Sang J, Wang Y, Zhao B, Cui D. Toxicity assessments of near-infrared upconversion luminescent LaF3:Yb,Er in early development of zebrafish embryos. Theranostics. 2013;3:258–66.
143. Jayakumar MKG, Bansal A, Li BN, Zhang Y. Mesoporous silica-coated upconversion nanocrystals for near infrared light-triggered control of gene expression in zebrafish. Nanomedicine. 2015;10:1051–61.
144. Lim SF, Riehn R, Ryu WS, Khanarian N, Tung C-K, Tank D, Austin RH. In vivo and scanning electron microscopy imaging of upconverting nanophosphors in caenorhabditis elegans. Nano Lett. 2006;6:169–74.
145. Chen J, Guo C, Wang M, Huang L, Wang L, Mi C, Li J, Fang X, Mao C, Xu S. Controllable synthesis of NaYF4: Yb,Er upconversion nanophosphors and their application to in vivo imaging of caenorhabditis elegans. J Mater Chem. 2011;21:2632–8.
146. Bansal A, Liu H, Jayakumar MKG, Andersson-Engels S, Zhang Y. Quasi-continuous wave near-infrared excitation of upconversion nanoparticles for optogenetic manipulation of celegans. Small. 2016;12:1732–43.
147. Ferrari M. Cancer nanotechnology: opportunities and challenges. Nat Rev Cancer. 2005;5(3):161–71.
148. Service RF. Materials and biology: nanotechnology takes aim at cancer. Science. 2005;310(5751):1132–4.
149. Gradishar WJ, Tjulandin S, Davidson N, et al. Phase III trial of nanoparticle albuminbound paclitaxel compared with polyethylated castor oil-based paclitaxel in women with breast cancer. J Clin Oncol. 2005;23(31):7794–803.
150. Lasic DD. Doxorubicin in sterically stabilized liposomes. Nature. 1996;380(6574):561–2.
151. Yezhelyev MV, Gao X, Xing Y, Al-Hajj A, Nie S, O'Regan RM. Emerging use of nanoparticles in diagnosis and treatment of breast cancer. Lancet Oncol. 2006;7(8):657–67.
152. Lasic DD, Papahadjopoulos D. Liposomes revisited. Science. 1995;267(5202):1275–6.
153. U.S. National Institutes of Health. ClinicalTrials.gov. http://www.clinicaltrials.gov/. Accessed 12 Dec 2007.
154. Folkman J. Angiogenesis in cancer, vascular, rheumatoid and other disease. Nat Med. 1995;1(1):27–31.
155. Carmeliet P, Jain RK. Angiogenesis in cancer and other diseases. Nature. 2000;407(6801):249–57.
156. Maeda H. SMANCS and polymer-conjugated macromolecular drugs: advantages in cancer chemotherapy. Adv Drug Deliv Rev. 2001;46(1–3):169–85.
157. Maeda H, Wu J, Sawa T, Matsumura Y, Hori K. Tumor vascular permeability and the EPR effect in macromolecular therapeutics: a review. J Control Release. 2000;65(1–2):271–84.
158. Yuan F, Leunig M, Huang SK, Berk DA, Papahadjopoulos D, Jain RK. Microvascular permeability and interstitial penetration of sterically stabilized (stealth) liposomes in a human tumor xenograft. Cancer Res. 1994;54(13):3352–6.
159. Wolff AC. Liposomal anthracyclines and new treatment approaches for breast cancer. Oncologist. 2003;8(suppl 2):25–30.
160. Fukumura D, Jain RK. Tumor microenvironment abnormalities: causes, consequences, and strategies to normalize. J Cell Biochem. 2007;101(4):937–49.
161. Hobbs SK, Monsky WL, Yuan F, et al. Regulation of transport pathways in tumor vessels: role of tumor type and microenvironment. Proc Natl Acad Sci USA. 1998;95(8):4607–12.
162. Yuan F, Chen Y, Dellian M, Safabakhsh N, Ferrara N, Jain RK. Time-dependent vascular regression and permeability changes in established human tumor xenografts induced by an anti-vascular endothelial growth factor/vascular permeability factor antibody. Proc Natl Acad Sci USA. 1996;93(25):14765–70.
163. Mukundan S Jr, Ghaghada KB, Badea CT, et al. A liposomal nanoscale contrast agent for preclinical CT in mice. AJR Am J Roentgenol. 2006;186(2):300–7.
164. Liu Z, Rong P, Yu L, Zhang X, Yang C, Guo F, et al. Dual-modality noninvasive mapping of sentinel lymph node by photoacoustic and near-infrared fluorescent imaging using dye-loaded mesoporous silica nanoparticles. Mol Pharm. 2015;12:3119–28.
165. Prabhakar U, Maeda H, Jain RK, et al. Challenges and key considerations of the enhanced permeability and retention effect for nanomedicine drug delivery in oncology. Cancer Res. 2013;73(8):2412–7.
166. Das A, Mukherjee P, Singla SK, Guturu P, Frost MC, Mukhopadhyay D, Shah VH, Patra CR. Fabrication and characterization of an inorganic gold and silica nanoparticle mediated drug delivery system for nitric oxide. Nanotechnology. 2010;21:305102.
167. Ali Z, Abbasi AZ, Zhang F, Arosio P, Lascialfari A, Casula MF, et al. Multifunctional nanoparticles for dual imaging. Anal Chem. 2011;83:2877–82.
168. Sun Y, Zhu X, Peng J, Li F. Core-shell lanthanide upconversion nanophosphors as four-modal probes for tumor angiogenesis imaging. ACS Nano. 2013;7:11290–300.

6 Ultrasound Nanoimaging I

6.1 Introduction

Ultrasound has established that it is the most common medical imaging modality [1]. Ultrasound contrast agents work by enhancing acoustic signals to differentiate between normal and abnormal tissues. Microbubbles (MBs) are the commonest ultrasound contrast agents, having sizes ranging from 1 to 8 μm. These agents generate blood pool signals, providing contrast against smaller, less stable circulations [2]. Enough acoustic reflections based on the size of the ultrasound contrast nanoparticles are generated, which are larger than CT or MRI contrast nanoparticle sizes ranging from hundreds to thousands of nanometers [3, 4].

6.2 Nanoparticles in Ultrasound Imaging Applications: Advantages and Disadvantages

Commonly used US contrast nanoparticles in clinical use are in the range of 200 nm [1, 5]. These nanoparticles will be labeled along with the surface ligands, which will have more focused target molecules. These nanoparticles passively accumulate in tumors due to cellular uptake mechanisms. One of the examples is the folate in the silica nanoparticles associated with perfluoro pentane for prostate cancer [6]. Three types of nanoparticle contrast agents are commonly used in US imaging. The first one is the gas (microbubble), which will create very strong acoustic reflections. A few of these examples of MBs are nitrogen, carbon dioxide, perfluorocarbon, and sulfur hexafluoride [7]. These nanoparticles will be using various types, like liposomes, silica nanoparticles, and polymer nanoparticles [4, 8, 9]. The solid-based nanoparticle is the second type of nanoparticle, which has higher scattering acoustic signals and is capable of changing the nanoparticle shape [10]. The third type of nanoparticles, gas-based nanoparticles, are small and can accumulate in target lesions via cellular uptake, the EPR effect, or receptor binding. Mesoporous silica nanostructure is also having a very powerful ultrasound signal in tumors [11]. The liquid-based nanoparticle perfluorooctyl bromide (PFOB) nanoparticles will generate the acoustic signal due to the lower speed of sound transmission in comparison to the water, and the shell of this nanoparticle can have the intrinsic acoustic reflection [12]. The gas and liquid-based nanoparticles will have the core–shell structure, similar to that of the gas and liquid in the core.

S. Shaikh, *Nanoimaging - Future of Precision Medicine*,
https://doi.org/10.1007/978-981-97-7292-6_6

6.3 Application of the Nanoparticle in Biomedical Imaging

The nanoparticle ultrasound contrast agents are very small, and due to this physical property, the targeting of the lesion will be easier [1]. These nanoparticles generate strong blood pool signals, facilitating early and enhanced detection of tumors, stem cells, inflammation, and enabling drug delivery [13].

Various gas nanoparticles are nitrogen, air, perfluorocarbon, and sulfur hexafluoride [4, 8]. And the most commonly used is the liquid gas. When liquid is converted into gas, it produces strong acoustic reflections in ultrasound imaging. This is seen with gold nanoparticle-coated mesoporous silica nanocapsules. Carbonate is also used to produce gas by the chemical reaction [2]. The carbonate is encapsulated in nanoparticles as it has neutral PH. The acidic tumor environment will be reacting to the carbonate core to generate carbon dioxide gas by creating better acoustic contrast. Tumor imaging [14] using the ammonium bicarbonate acoustic imaging [15].

The use of micro-nanomaterials can also lead to various biological effects that will have an impact on better drug delivery to the tumor sites, which is due to the EPR effects and MNBs and the enhancement of these effects. This smart bubble drug delivery system can be used for future cancer theronostics. The various EPR acoustic transducers can generate different transverse or longitudinal waves of ultrasound frequencies greater than 20 kHz. The distribution of various transducers is seen in various forms of the continuous and pulsed ultrasound waveform, comprising of the pulse length and pulse repetition frequency. The acoustic densitometry of ultrasound is defined as the time-averaged rate of sonic energy flow in the given unit area [16]. The sonic intensity of more than 5 W/cm^2 is high intensity, and the intensity between is 0.5 and 1 W/cm^2 low-intensity ultrasound. The output of the ultrasound equipment can be calculated by the mechanical index (MI), defined by the FDA as the ratio of in situ peak negative pressure (PnP) to the square root of the center frequency (Fc). The best output levels can be up to the MI of 1.9 in the diagnostic instruments [17]. Various drug carriers like micelles and liposomes cannot be compressed, having a maximum MI of 1.9, and MBs that are compressible will have an MI of 0.8 [18]. This will be used for diagnosis as well as therapeutic applications, where the acoustic waves enhance the EPR by drug release and bioeffects. In the process of the drug release, the carrier of the ultrasound increases the distribution concentration in the tumor. This process can be controlled by acoustic parameters like ultrasound frequency, power density, and pulse duration [19, 20]. Alexander et al. [21] studied the impact of continuous wave (CW) and pulsed ultrasound on the drug doxorubicin (DOX) uptake by HL-60 cells is increased as compared with CW ultrasound (10 s pulse). The high-frequency ultrasound will have better results as compared to the low-frequency ultrasound. As we all know, low-frequency ultrasound is more penetrating as compared to high-frequency ultrasound. The low-frequency ultrasound can penetrate deeper structures, while the high-frequency ultrasound can penetrate superficial tumors [22]. Again in [19], the drug release increases with the increasing power density. With the various bioeffects, the acoustic energy with EPR will concentrate better on the enhancement of cell membrane permeability [18]. Liu et al. [23] study showed that ultrasonic exposure at 1 MHz can cause better platelet penetration of gold nanoparticles (GNPs). Different acoustic waves will interact with the drug carriers, body tissues, and cell membranes due to the various thermal and mechanical effects [20]. The acoustic waves will interact with the EPR by the use of the two different forms of the energies, mechanical and thermal energy. The low-ultrasound waves with low intensity will not have any thermal properties, and due to this, there will be clear demarcation between the mechanical and thermal effects.

6.4 Mechanical Effects

The mechanical effects seen in the use of ultrasound and MNB-assisted ultrasound are due to the various acoustic forces due to the radiation as well as the different bioeffects mechanically. The acoustic radiation force is due to the various exchanges happening at the different momentum, resulting between the object and sound field having an acoustic wave with the different forces known as the acoustic radiation force (ARF) [24]. The air bubble will be acted by the different forces in a sound field, and these are known as the Bjerknes forces [25]. These forces are due to two mechanisms, namely primary Bjerknes forces and secondary Bjerknes forces. The forces that will be having an impact primarily are known as the primary Bjerknes force [2]. The EPR effect will be significantly increased in the case where the radiotracer is located in a particular region; as a result, the velocity of the passing agent will cause reduced forces. Dayton et al. [26] concluded that the radiation force of ultrasound can cause the microbubble contrast agent to collide with the wall of a 50-mm blood vessel, which can result in reversible aggregation. This microbubble aggregation seen in the large vessel will be influencing the acoustic radiation force by Wang et al. [27]. Thus, the use of the intravascular ultrasound transducer in the common carotid artery can potentially result in the increased efficiency of intravascular MBs and drug delivery [28].

6.5 Acoustic Streaming

The acoustic streaming is due to the radiation of the force by different reflectors and scatterers, which will cause the displacements of the various particles due to the impact of the fluid currents in the ultrasound field. This same property is being used in the blood by the displacements of the various particles [29] due to the motion of blood flow [30]. These will cause an impact on the release of drugs from various carriers, thus impacting the drug movement in the particularly designated tissues, which are targeted [31] by the property of vitro low-frequency sonophoresis [32].

6.6 Barriers for Nanoparticle Drug Delivery to Tumor

There are a lot of technical developments, due to which there is a lot of research going on for the drug delivery systems, and the best example is the chemotherapeutic agent administration and delivery in the management of various cancers. This has resulted in more dedicated planning and better delivery of the drugs to the target organs. This is possible due to nanoparticle attachment to the drugs in the targeted cancer cells. A few of the receptors are endothelial growth factor receptor (VEGFR), epidermal growth factor receptor (EGFR integrin receptor vascular), folate receptor (FR), and human epidermal growth factor receptor 2 (HER2) [33]. This has resulted in better bioavailability, biodistribution, and better targeting of the cells. Still, the drug targeting in the tumors is very less than $\sim 1\%$ of nanoparticles [34]. Various physiological barriers are causing hurdles to this less accumulation in the tumors [35]. Nanoparticles have early excretion from the blood circulation before reaching the targeted tissues, and proteins in the blood, which are opsonized in the various cells of the mononuclear phagocyte system (MPS), are excreted from the blood circulation later. Tumor tissue vasculature will be more abnormal having a high density of extracellular matrix (ECM). The incomplete or delayed delivery of therapeutic nanoparticle agents to tumors due to the abnormal and more complex tumor networks, tumor complexity. This has resulted in nanoparticle delivery delay, nanoparticle accumulation, and diffusion of nanoparticles in tumors, leading to lesser antitumor activity [36, 37].

Ultrasound is also an important modality for the drug delivery of various nanoparticles, and this modality will also result in less stimulation of drug release, which needs to be addressed.

Alongside there are different processes of sonoporosis, cavitation, and hyperthermia having different interactions of ultrasound with tumor cells, and this also applies to the nanoparticles. Ultrasound-assisted drug delivery is an important entity for the reduced side effects comprising of drug toxicity and non-targeted delivery [38].

As a modality, we all know that it has many advantages, like the higher sensitivity of the evaluation in the various structural morphologies of the lesion in different scan planes [39].

Ultrasound-guided drug delivery is the most challenging and demanding aspect of nanoimaging in the use of various nanobubbles (NBs) (Diagram 6.1). The composition of NBs is very complex, and this will cause the movement of these bubbles to cross the capillary wall. This concept will be useful for the use of the 5-fluorouracil-loaded NBs in the management of hepatocellular carcinoma. These NBs will be mediated by the different patterns of uptake and apoptosis, which can be quantified by the concept of the drug loading capability and pattern of enhancement in the ultrasound [40]. This is seen with the efficiency of 78.6% in the enhancement, and an encapsulation efficiency is measured at 7.4% [41]. The blood–brain barrier is also seen in different settings with feedback [42]. The permeation of the micromolar enhancement is being noted as the better ultrasound-responsive NBs, which have very negligible toxicity with better internalization [43, 44]. Another important agent is the fluorescence NPs, which are seen as the nanocarriers having the better high-resolution signal-to-noise ratio. These NBs can be used as the dual imaging seen in the mouse model [45]. One of the studies has shown that the novel GPC3-targeted and drug-loaded NBs are being used in hepatocellular carcinoma cells [46]. Another important particle is NB-liposome as one of the antitumor agents infused intratumorally, resulting in the necrosis of the tissue on ultrasound exposure [47]. Another important agent is the oxygen-encapsulated agent, which has anti-chemotherapeutic properties [48] and is capable of increased transfection efficiency, causing cell cycle arrest with cell apoptosis, by the principle of the FoxM1 expression reduction [49].

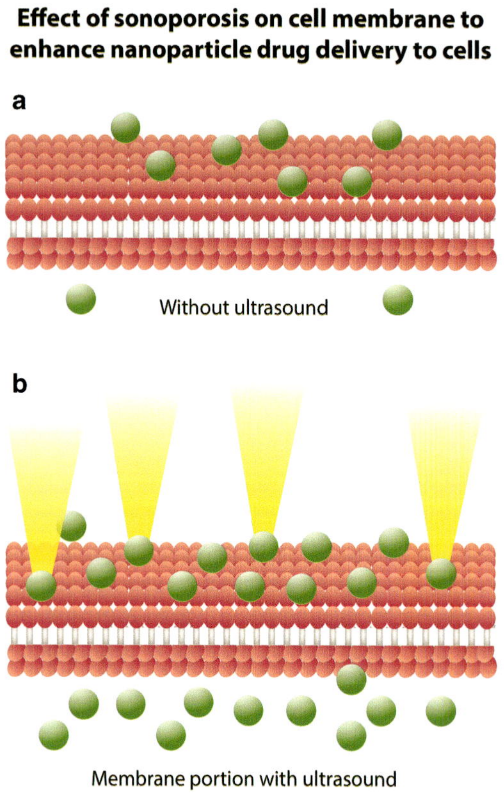

Diagram 6.1 Effect of the sonoporosis for the increase drug delivery in the cell membrane. A without ultrasound and B with ultrasound

NBs are very small bubbles having a diameter range of 150–500 nm [50–52], with better antitumor properties, which can cause more cancer aggregation and retention of these agents as compared with MBs [53, 54]. The properties of the synthesis and characterization of NBs are done by the process of sonication along with mechanical agitation [55–58]. These NBs will have more and better stability [55, 59, 60], as confirmed in various applications [61, 62]. These NBs will have more US contrast augmentation, which will result in better diagnostic [63] and therapeutic effects [64]. These will also have better transmission of medications, chromosomes, and gas [65, 66]. NBs also have more extravasation in the surrounding tissues, causing better transmission capacity and localization at the tumor target cells [67] and also resulting in the more retention and permeability effect on the tissues [68] with the

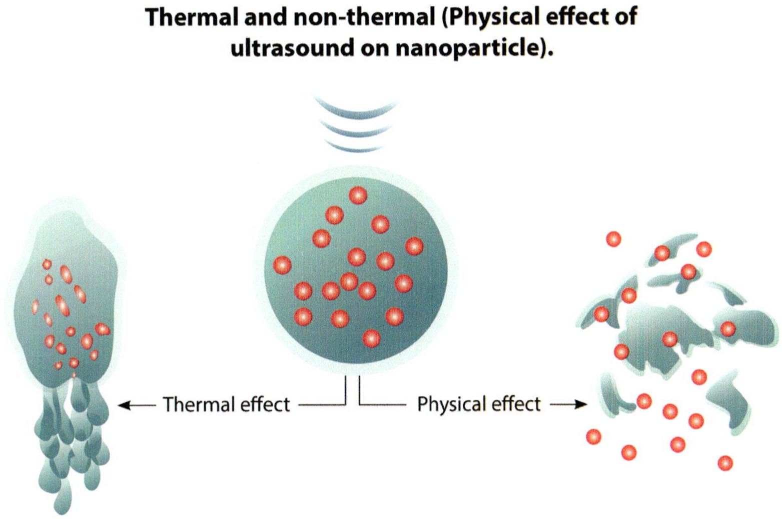

Diagram 6.2 Thermal and non-thermal effects of the ultrasound on the Nanoparticles

concept of targeting, or attaching these antibodies to their surfaces [69], resulting in the better and increased intracellular uptake [70] and better delivery and distribution of these medications by the increased ultrasonic imaging [71] (Diagram 6.2). Thus, these NBs are more biocompatible with biodegradable substances [72, 73]. The NB shell will have better stability due to better surface tension of the wall as walls are made up of various polymers or lipids [74]. The thickness and elasticity of the nanoshell will impact the durability of NBs [75]. NB shell will reduce the surface tension, and this is an important cause of the stability. Due to the property of the long acyl chains, it will have a better and higher shelf life [76–78]. The effect of external polymer coating on nanobubbles was studied by Rapoport et al. [77]. Crosslinking of the nanoshell, by various means, like chemical or physical properties, will provide better stability and elasticity [79]. Glycol chitosan polymer is an important NB that will have better stability with time [80, 81].

6.7 Biological Effects

With the various sorts of ultrasonic exposure in combination with the various NBs, the ultrasonic delivery will be dependent on the various biophysical processes of cell poration and this combination of the ultrasound with NBs is known as sonoporation (Diagram 6.3). The NBs have an important property of "collapsing" under insonation when the ultrasound energy is being used, causing changes in the permeability of the cell membrane [82]. The use of ultrasonic energy can cause the effect of "ultrasonic cavitation." The concept of ultrasound cavitation will result in liquids vibrating and growing, resulting in the energy from the acoustic field, which will collapse till some threshold point of the energy levels is reached. The basis of the irradiation of the ultrasound will be causing the cavitation, further resulting in the cell membrane leading to the increased permeability of blood vessels [83]. In addition, this will also cause the thermal and mechanical effects which will be used for therapy. Dyson's study revealed that increased temperature mostly from 40 C to 45 C will cause more therapeutic temperature, resulting in an increased temperature of 0.86 C min 1 (1 MHz, 1 W cm^2). The various biological effects of radiation by ultrasound will lead to increased permeability of the cell membrane, causing gene transfer, increased drug concentration of the embolized tumors, nourishing vessels, and better therapeutic results [84, 85]. Ultrasound is a valuable modality for delivering various medications, particularly anticancer drugs. Suzuki et al. [86] showed the use of NBs with the US led to an increase in the antitumor drugs. The oxygen-trapping NBs will cause increased targeting and

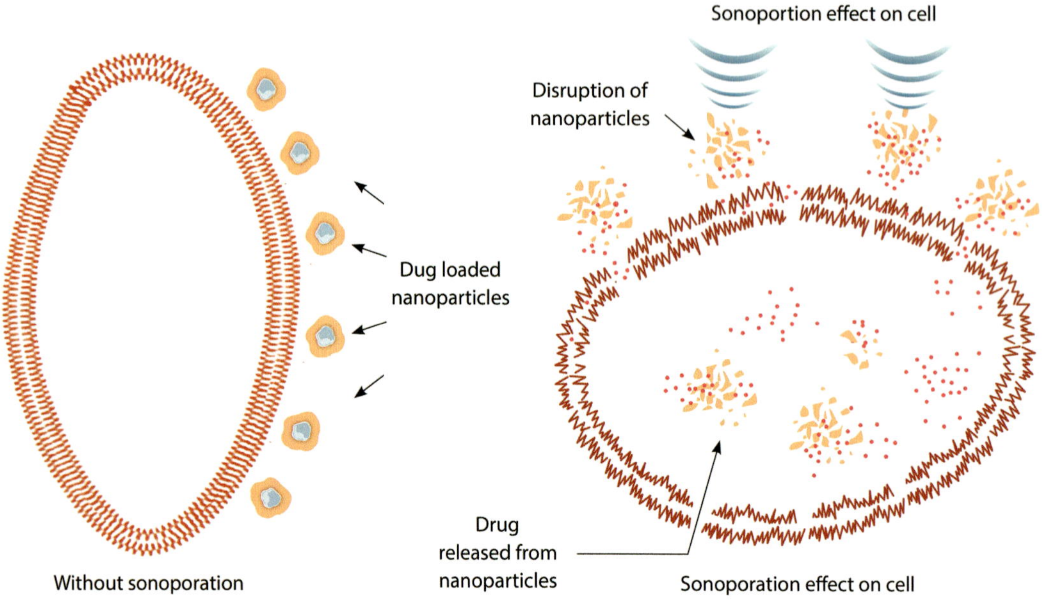

Diagram 6.3 Without and with sonoporation effect loaded with the nanoparticles on the cells

effectiveness of the hypoxia in non-muscle invasive bladder cancer (NMIBC) tumors. This noninvasive and stable ultrasonic will be useful for a wide variety of therapeutic applications [87]. With the use of ultrasound, there will be more precise apoptosis along with the NBs labelled therapeutic drugs and this will result in newer chemotherapeutic avenues for the management of cancers [88]. NBs carrying chemotherapeutic agents will have greater potential for killing cancer cells. Thus, these NBs will be the best material for the non-invasive and selective drug delivery of various chemotherapeutic agents [89].

Perfluorocarbon NBs are a more advanced form of nanocarriers that have the property to transform by using micron-scale gas bubbles. These bubbles will interact with the tissues, resulting in a better evaluation of the tissues, especially tumors by Gao et al. [90] created the NB polymeric device, showing the better and higher sensitivity of chemotherapeutic drugs when used along with the. Rapoport et al. showed that when there is an increased temperature to 37 C from the nanodrops ultimately, it leads to bigger bubbles. This will cause the tumor irradiation to be larger as compared to the nanodrops along with the thermodynamic effects leading to restraining tumor proliferation [81]. In another study, the presence of the amphiphilic chain along with the more transporters will have a more precise effect, causing significant resolution of the tumor.

There is a definite advantage of NBs over MBs as the NBs will cause more ultrasonic energy for the drug/gene release in different target tissues or cells [91–93]. This is due to better cell permeability for the therapeutic agents, and also the disadvantage is damaging the various biological tissues (Diagram 6.3). The use of ultrasound irradiation will result in more ultrasonic heating and damaging the tissues. Again, the important aspect is that the use of the different ultrasound intensities will be useful for various purposes [94]. Zhang et al. [95] showed that NBs are made up of phospholipids, which are the low-toxicity materials and thus safer to use [96, 97]. Despite this, there is a lot of scope for more future research [96]. With this, the use of NBs is more worthwhile for various clinical applications [91]. The use of the NBs combined along with the US for tumor therapy is in the preliminary stages. But their unique structure and multifunctional characteristics, along with the higher degree of curative effect, the important aspect here is that a significant amount of the sufficient number of drugs/genes can be delivered and released in tumor tissues. The ultrasonic-response materials like gaseous, PFC, or gas-generation MBs, NBs, and droplets can reduce the drug/gene content in many tissues, resulting in more therapeutic effects [96, 98].

6.8 Targeting of Nanobubbles

NBs have a more important role in ultrasonic molecular imaging and potential therapy functions, such as passive targeting mechanisms known as enhanced permeability and retention effects (EPR) [99]. NBs as they are smaller in size can penetrate the vascular wall into tumor tissues. Thus, this will be more important for the tumor-targeted imaging and therapy [100] (Diagram 6.4).

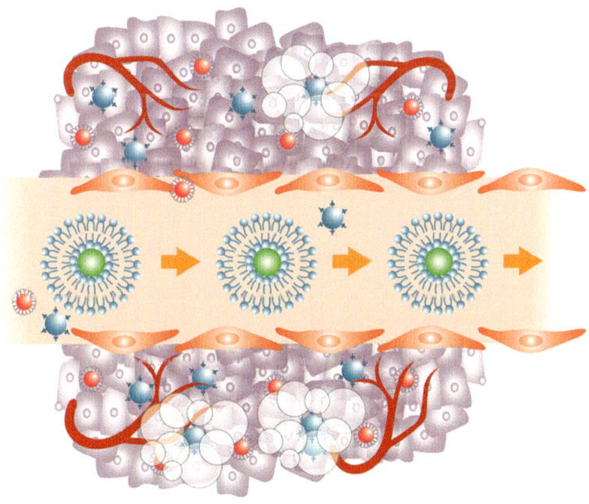

Diagram 6.4 Basic structure of the tumor tissue with increased vascularization

6.9 Conclusions

NBs combined with US will have more applications, especially for innovative therapeutic devices in specific clinical environments. This will create more sophisticated and targeted distribution strategies for anticancer agents, resulting in better therapeutic sensitivity of anticancer drugs while minimizing their off-target side effects. The reason is that ultrasound is a well-established, non-invasive, and cheap diagnostic imaging procedure, and with this background can increase. US combined with NBs may improve therapeutic efficacy, enhancing clinical applications. Current research suggests that nano-drug delivery systems hold significant potential in enhancing treatment efficacy.

References

1. Klibanov AL, Hossack JA. Ultrasound in radiology: from anatomic, functional, molecular imaging to drug delivery and image-guided therapy. Investig Radiol. 2015;50:657–70.
2. Wang X, Chen H, Zheng Y, Ma M, Chen Y, Zhang K, Zeng D, Shi J. Au-nanoparticle coated mesoporous silica nanocapsule-based multifunctional platform for ultrasound mediated imaging, cytoclasis and tumor ablation. Biomaterials. 2013;34:2057–68.
3. Sakamoto JH, Smith BR, Xie B, Rokhlin SI, Lee SC, Ferrari M. The molecular analysis of breast cancer utilizing targeted nanoparticle based ultrasound contrast agents. Technol Cancer Res Treat. 2005;4:627–36.
4. Köse G, Darguzyte M, Kiessling F. Molecular ultrasound imaging. Nanomaterials. 2020;10:1935.
5. Min KH, Min HS, Lee HJ, Park DJ, Yhee JY, Kim K, Kwon IC, Jeong SY, Silvestre OF, Chen X, Hwang YS, Kim EC, Lee SC. pH-controlled gas-generating mineralized nanoparticles: a theranostic agent for ultrasound imaging and therapy of cancers. ACS Nano. 2015;9:134–45.
6. Wang J, Barback CV, Ta CN, Weeks J, Gude N, Mattrey RF, Blair SL, Trogler WC, Lee H, Kummel AC. Extended lifetime in vivo pulse stimulated ultrasound imaging. IEEE Trans Med Imaging. 2018;37:222–9.

7. Mackman N, Bergmeier W, Stouffer GA, Weitz JI. Therapeutic strategies for thrombosis: new targets and approaches. Nat Rev Drug Discov. 2020;19:333–52.
8. Wang CW, Yang SP, Hu H, Du J, Li FH. Synthesis, characterization and in vitro and in vivo investigation of C3F8-filled poly (lactic-co-glycolic acid) nanoparticles as an ultrasound contrast agent. Mol Med Rep. 2015;11:1885–90.
9. Park SH, Yoon YI, Moon H, Lee GH, Lee BH, Yoon TJ, Lee HJ. Development of a novel microbubble-liposome complex conjugated with peptide ligands targeting IL4R on brain tumor cells. Oncol Rep. 2016;36:131–6.
10. Chen F, Ma M, Wang J, Wang F, Chern SX, Zhao ER, Jhunjhunwala A, Darmadi S, Chen H, Jokerst JV. Exosome-like silica nanoparticles: a novel ultrasound contrast agent for stem cell imaging. Nanoscale. 2017;9:402–11.
11. Zhang K, Chen H, Guo X, Zhang D, Zheng Y, Zheng H, Shi J. Double-scattering/reflection in a single nanoparticle for intensified ultrasound imaging. Sci Rep. 2015;5:8766.
12. Şen T, Tüfekçioğlu O, Koza Y. Mechanical index. Anatol J Cardiol. 2015;15:334–6.
13. Kim GW, Kang C, Oh YB, Ko MH, Seo JH, Lee D. Ultrasonographic imaging and anti-inflammatory therapy of muscle and tendon injuries using polymer nanoparticles. Theranostics. 2017;7:2463–76.
14. Min HS, Son S, You DG, Lee TW, Lee J, Lee S, Yhee JY, Lee J, Han MH, Park JH, Kim SH, Choi K, Park K, Kim K, Kwon IC. Chemical gas-generating nanoparticles for tumor-targeted ultrasound imaging and ultrasound-triggered drug delivery. Biomaterials. 2016;108:57–70.
15. Xia J, Feng G, Xia X, Hao L, Wang Z. NH4HCO3 gas-generating liposomal nanoparticle for photoacoustic imaging in breast cancer. International journal of nanomedicine. Int J Nanomedicine. 2017;12:1803–13.
16. Kim YS, Rhim H, Choi MJ, Lim HK, Choi D. High-intensity focused ultrasound therapy: an overview for radiologists. Korean J Radiol. 2008;9:291–302.
17. Dalecki D. Mechanical bioeffects of ultrasound. Annu Rev Biomed Eng. 2004;6:229–48.
18. Sirsi SR, Borden MA. State-of-the-art materials for ultrasound-triggered drug delivery. Adv Drug Deliv Rev. 2014;72:3–14.
19. Husseini GA, Myrup GD, Pitt WG, Christensen DA, Rapoport NAY. Factors affecting acoustically triggered release of drugs from polymeric micelles. J Control Release. 2000;69:43–52.
20. Ahmed SE, Moussa HG, Martins AM, Abbas Y, Al-Sayah MH, Husseini GA. Factors affecting the acoustic in vitro release of calcein from pegylated liposomes. J Nanosci Nanotechnol. 2019;19:6899–906.
21. Li B, Aid-Launais R, Labour MN, Zenych A, Juenet M, Choqueux C, Ollivier V, Couture O, Letourneur D, Chauvierre C. Functionalized polymer microbubbles as new molecular ultrasound contrast agent to target P-selectin in thrombus. Biomaterials. 2019;194:139–50.
22. Marin A, Sun H, Husseini GA, Pitt WG, Christensen DA, Rapoport NY. Drug delivery in pluronic micelles: effect of high-frequency ultrasound on drug release from micelles and intracellular uptake. J Control Release. 2002;84:39–47.
23. Liu T, Li M, Tang J, Li J, Zhou Y, Liu Y, et al. An acoustic strategy for gold nanoparticle loading in platelets as biomimetic multifunctional carriers. J Mater Chem B. 2019;7:2138–44.
24. Li Y, Qiu C, Xu S, Ke M, Liu Z. Theoretical study of large angle bending transport of microparticles by 2D acoustic half-bessel beams. Sci Rep. 2015;5:13063.
25. Crum LA. Bjerknes forces on bubbles in a stationary sound field. J Acoust Soc Am. 1975;57:1363–70.
26. Dayton P, Klibanov A, Brandenburger G, Ferrara K. Acoustic radiation force in vivo: a mechanism to assist targeting of microbubbles. Ultrasound Med Biol. 1999;25:1195–201.
27. Wang S, Wang CY, Unnikrishnan S, Klibanov AL, Hossack JA, Mauldin FW. Optical verification of microbubble response to acoustic radiation force in large vessels with in vivo results. Investig Radiol. 2015;50:772–84.
28. Kilroy JP, Klibanov AL, Wamhoff BR, Hossack JA. Intravascular ultrasound catheter to enhance microbubble-based drug delivery via acoustic radiation force. IEEE Trans Ultrason Ferroelectr Freq Control. 2012;59:2156–66.
29. Zhang X, Tian X, Li P, Zhu H, Zhou N, Fang Z, Yang Y, Jing Y, Yuan J. Ultrasound-targeted microbubble destruction promotes myocardial angiogenesis and functional improvements in rat model of diabetic cardiomyopathy. BMC Cardiovasc Disord. 2021;21:21.
30. Solovchuk MA, Thiriet M, Sheu TWH. Computational study of acoustic streaming and heating during acoustic hemostasis. Appl Therm Eng. 2017;124:1112–22.
31. Rapoport N. Ultrasound-mediated micellar drug delivery. Int J Hyperth. 2012;28:374–85.
32. Peng H, Zhu P, Lu P. Acoustic streaming simulation and analyses in in vitro low frequency sonophoresis. Sens Actuators A Phys. 2017;263:744–53.
33. Ko YJ, Kim WJ, Kim K, Kwon IC. Advances in the strategies for designing receptor-targeted molecular imaging probes for cancer research. J Control Release. 2019;305:1–17. https://doi.org/10.1016/j.jconrel.2019.04.030.
34. Wilhelm S, Tavares AJ, Dai Q, Ohta S, Audet J, Chan WCW. Analysis of nanoparticle delivery to tumours. Nat Rev Mater. 2016;1:16014. https://doi.org/10.1038/natrevmats.2016.14.
35. Rosenblum D, Joshi N, Tao W, Karp JM, Peer D. Progress and challenges towards targeted delivery of cancer therapeutics. Nat Commun. 2018;9:1410. https://doi.org/10.1038/s41467-018-03705-y.

36. Sriraman SK, Aryasomayajula B, Torchilin VP. Barriers to drug delivery in solid tumors. Tissue Barriers. 2014;2:e29528. https://doi.org/10.4161/tisb.29528.
37. Zhang YR, Lin R, Li HJ, He WL, Du JZ, Wang J. Strategies to improve tumor penetration of nanomedicines through nanoparticle design. Wiley Interdiscip Rev Nanomed Nanobiotechnol. 2019;11:e1519. https://doi.org/10.1002/wnan.1519.
38. Mullick Chowdhury S, Lee T, Willmann JK. Ultrasound-guided drug delivery in cancer. Ultrasonography. 2017;36:171–84. https://doi.org/10.14366/usg.17021.
39. Guo R, Lu G, Qin B, Fei B. Ultrasound imaging technologies for breast cancer detection and management: a review. Ultrasound Med Biol. 2018;44:37–70.
40. Zhou X, Guo L, Shi D, Duan S, Li J. Biocompatible chitosan nanobubbles for ultrasound-mediated targeted delivery of doxorubicin. Nanoscale Res Lett. 2019;14:1–9.
41. Meng M, Gao J, Wu C, Zhou X, Zang X, Lin X, Liu H, Wang C, Su H, Liu K, et al. Doxorubicin nanobubble for combining ultrasonography and targeted chemotherapy of rabbit with VX2 liver tumor. Tumour Biol. 2016;37:8673–80.
42. Cheng B, Bing C, Xi Y, Shah B, Exner AA, Chopra R. Influence of nanobubble concentration on blood–brain barrier opening using focused ultrasound under real-time acoustic feedback control. Ultrasound Med Biol. 2019;45:2174–87.
43. Chandan R, Banerjee R. Pro-apoptotic liposomes-nanobubble conjugate synergistic with paclitaxel: a platform for ultrasound responsive image-guided drug delivery. Sci Rep. 2018;8:1–15.
44. Thakur SS, Ward MS, Popat A, Flemming NB, Parat M, Barnett NL, Parekh HS. Stably engineered nanobubblesand ultrasound—An effective platform for enhanced macromolecular delivery to representative cells of the retina. PLoS One. 2017;12:1 17.
45. Chan MH, Pan YT, Chan YC, Hsiao M, Chen CH, Sun L, Liu RS. Nanobubble-embedded inorganic 808 nm excited upconversion nanocomposites for tumor multiple imaging and treatment. Chem Sci. 2018;9:3141–51.
46. Tian Y, Liu Z, Zhang L, Zhang J, Han X, Wang Q, Cheng W. Apatinib-loaded lipid nanobubbles combined with ultrasound-targeted nanobubble destruction for synergistic treatment of HepG2 cells in vitro. Onco Targets Ther. 2018;11:4785–95.
47. Suzuki R, Oda Y, Omata D, Nishiie N, Koshima R, Shiono Y, Sawaguchi Y, Unga J, Naoi T, Negishi Y, et al. Tumor growth suppression by the combination of nanobubbles and ultrasound. Cancer Sci. 2016;107:217–23.
48. Bhandari P, Novikova G, Goergen CJ, Irudayaraj J. Ultrasound beam steering of oxygen nanobubbles for enhanced bladder cancer therapy. Sci Rep. 2018;8:1–10.
49. Wu M, Zhao H, Guo L, Wang Y, Song J, Zhao X, Li C, Hao L, Wang D, Tang J. Ultrasound-mediated nanobubble destruction (UMND) facilitates the delivery of A10-3.2 aptamer targeted and siRNA-loaded cationic nanobubbles for therapy of prostate cancer. Drug Deliv. 2018;25:226–40.
50. Klibanov AL, et al. Ultrasound-triggered release of materials entrapped in microbubble-liposome constructs: a tool for targeted drug delivery. J Controlled Release. 2010;148(1):13–7. 12924 I RSC Adv., 2021, 11, 12915–12928
51. Michailidi ED, et al. Chapter 4—Fundamentals and applications of nanobubbles. In: Kyzas GZ, Mitropoulos AC, editors. Interface science and technology. Elsevier; 2019. p. 69–99.
52. Bosca F, et al. Porphyrin-loaded Pluronic Nanobubbles: a new US-activated agent for future Theranostic applications. Bioconjug Chem. 2018;29(2):234–40.
53. Wu H, et al. Time-intensity-curve analysis and tumor extravasation of nanobubble ultrasound contrast agents. Ultrasound Med Biol. 2019;45(9):2502–14.
54. Nittayacharn P, et al. Enhancing tumour drug distribution with ultrasound-triggered nanobubbles. J Pharm Sci. 2019;108(9):3091–8.
55. Peyman SA, et al. On-chip preparation of nanoscale contrast agents towards high-resolution ultrasound imaging. Lab Chip. 2016;16(4):679–87.
56. Wu M, et al. Ultrasound-mediated nanobubble destruction (UMND) facilitates the delivery of A10-3.2 aptamer targeted and siRNA-loaded cationic nanobubbles for therapy of prostate cancer. Drug Deliv. 2018;25(1):226–40.
57. Xing Z, et al. The fabrication of novel nanobubble ultrasound contrast agent for potential tumor imaging. Nanotechnology. 2010;21(14):145607.
58. Azevedo RSS, et al. In situ immune response and mechanisms of cell damage in central nervous system of fatal cases microcephaly by Zika virus. Sci Rep. 2018;8(1):1.
59. Perera RH, et al. Improving performance of nanoscale ultrasound contrast agents using N,N-diethylacrylamide stabilization. Nanomedicine. 2017;13(1):59–67.
60. Hernandez C, et al. Sink or float? Characterization of shellstabilized bulk nanobubbles using a resonant mass measurement technique. Nanoscale. 2019;11(3):851–5.
61. Tan BH, An H, Ohl CD. How bulk nanobubbles might survive. Phys Rev Lett. 2020;124(13):134503.
62. Alheshibri M, et al. A history of nanobubbles. Langmuir. 2016;32(43):11086–100.
63. Pellow C, et al. Threshold-dependent nonlinear scattering from porphyrin nanobubbles for vascular and extravascular applications. Phys Med Biol. 2018;63(21):215001.
64. Cavalli R, et al. New chitosan nanobubbles for ultrasoundmediated gene delivery: preparation and in vitro characterization. Int J Nanomedicine. 2012;7:3309–18.

65. Deng Y, Zhang X, Shen H, He Q, Wu Z, Liao W, Yuan M. Application of the nano-drug delivery system in treatment of cardiovascular diseases. Front Bioeng Biotechnol. 2020;7:489.
66. O'Neill BE, Rapoport N. Phase-shift, stimulirespon-sive drug carriers for targeted delivery. Ther Deliv. 2011;2(9):1165–87.
67. Cavalli R, Bisazza A, Lembo D. Micro- and nanobubbles: a versatile non-viral platform for gene delivery. Int J Pharm. 2013;456(2):437–45.
68. Fokong S, et al. Image-guided, targeted and triggered drug delivery to tumors using polymer-based microbubbles. J Control Release. 2012;163(1):75–81.
69. Zhang C-B, et al. Enhancement effect of ultrasoundinduced microbubble cavitation on branched polyethylenimine-mediated VEGF165 transfection with varied N/P ratio. Ultrasound Med Biol. 2013;39(1):161–71.
70. Cai X, et al. Ultrasound-responsive materials for drug/gene delivery. Front Pharmacol. 2020;10:1650.
71. Kim C, et al. Multifunctional microbubbles and nanobubbles for photoacoustic and ultrasound imaging. J Biomed Opt. 2010;15(1):010510.
72. Li H, et al. Characteristics of micro-nano bubbles and potential application in groundwater bioremediation. Water Environ Res. 2014;86(9):844–51.
73. Khan MS, et al. Oxygen-carrying micro/nanobubbles: composition, synthesis techniques and potential prospects in photo-triggered theranostics. Molecules. 2018;23(9):2210.
74. du Toit LC, et al. Investigating the effect of polymeric approaches on circulation time and physical properties of nanobubbles. Pharm Res. 2011;28(3):494–504.
75. Delalande A, et al. Ultrasound and microbubble-assisted gene delivery: recent advances and ongoing challenges. Ther Deliv. 2012;3(10):1199–215.
76. Kothapalli SVVN, et al. Assessment of the viscoelastic and oscillation properties of a nano-engineered multimodality contrast agent. Ultrasound Med Biol. 2014;40(10):2476–87.
77. Rapoport N, et al. Ultrasound-mediated tumor imaging and nanotherapy using drug loaded, block copolymer stabilized perfluorocarbon nanoemulsions. J Control Release. 2011;153(1):4–15.
78. Abou-Saleh RH, Peyman SA, Johnson BRG. The infuence of intercalating perfluorohexane into lipid shells on nano and microbubble stability. Soft Matter. 2016;12(34):7223–30.
79. Capece S, et al. A general strategy for obtaining biodegradable polymer shelled microbubbles as theranostic devices. Chem Commun. 2013;49(51):5763–5.
80. Cavalli R, Soster M, Argenziano M. Nanobubbles: a promising efficient tool for therapeutic delivery. Ther Deliv. 2016;7(2):117–38.
81. Zhang N, Li J, Hou R, Zhang J, Wang P, Liu X, Zhang Z. Bubble-generating nano-lipid carriers for ultrasound/CT imaging-guided efficient tumor therapy. Int J Pharm. 2017;534(1-2):251–62.
82. Elhelf IAS, Albahar H, Shah U, Otoe A, Cressman E, Almekkawy M. High intensity focused ultrasound: the fundamentals, clinical applications, and research trends. Diagn Interv Imaging. 2018;99:349–59.
83. Duan S, Guo L, Shi D, Shang M, Meng D, Li J. Development of a novel folate-modified nanobubbles with improved targeting ability to tumour cells. Ultrason Sonochem. 2017;37:235–43.
84. Bhandari P, et al. Ultrasound beam steering of oxygen nanobubbles for enhanced bladder cancer therapy. Sci Rep. 2018;8(1):3112.
85. Tian Y, et al. Apatinib-loaded lipid nanobubbles combined with ultrasound-targeted nanobubble destruction for synergistic treatment of HepG2 cells in vitro. Onco Targets Ther. 2018;11:4785–95.
86. Zhou X, et al. Biocompatible chitosan nanobubbles for ultrasound-mediated targeted delivery of doxorubicin. Nanoscale Res Lett. 2019;14(1):24.
87. Alphandéry E. Biodistribution and targeting properties of iron oxide nanoparticles for treatments of cancer and iron anemia disease. Nanotoxicology. 2019;13:573–96.
88. Yamamoto N, Renfrew AK, Kim BJ, Bryce NS, Hambley TW. Dual targeting of hypoxic and acidic tumor environments with a cobalt(III) chaperone complex. J Med Chem. 2012;55(24):11013–21. https://doi.org/10.1021/jm3014713.
89. Lv W, Yang T, Yu Q, et al. A phosphorescent iridium(III) complex-modified nanoprobe for hypoxia bioimaging via time-resolved luminescence microscopy. Adv Sci (Weinh). 2015;2(10):1500107.
90. Adhikari U, Goliaei A, Berkowitz ML. Nanobubbles, cavitation, shock waves and traumatic brain injury. Phys Chem Chem Phys. 2016;18(48):32638–52.
91. Zhang J, et al. The optimized fabrication of a novel nanobubble for tumor imaging. Front Pharmacol. 2019;31(10):610.
92. Cong H, et al. Ultrasound-mediated nanobubble destruction (UMND) facilitates the delivery of VEGFR2- targeted CD-TK-loaded cationic nanobubbles in the treatment of bladder cancer. J Cancer Res Clin Oncol. 2020;146(6):1415–26.
93. Gao S, et al. Lipid nanobubbles as an ultrasound-triggered artesunate delivery system for imaging-guided, tumortargeted chemotherapy. Onco Targets Ther. 2019;12:1841–50.
94. Cheng B, et al. Influence of nanobubble concentration on blood-brain barrier opening using focused ultrasound under real-time acoustic feedback control. Ultrasound Med Biol. 2019;45(8):2174–87.
95. Fabiilli ML, et al. Delivery of chlorambucil using an acoustically-triggered perfluoropentane emulsion. Ultrasound Med Biol. 2010;36(8):1364–75.
96. Feng Z, Tao P, Zou L, et al. Hyperbranched phosphorescent conjugated polymer dots with iridium(III) complex as the core for hypoxia imaging and photodynamic therapy. ACS Appl Mater Interfaces. 2017;9(34):28319–30.
97. Liu J, Liu Y, Bu W, et al. Ultrasensitive nanosensors based on upconversion nanoparticles for selective

hypoxia imaging in vivo upon near-infrared excitation. J Am Chem Soc. 2014;136(27):9701–9.
98. Zhou X, Liang H, Jiang P, et al. Multifunctional phosphorescent conjugated polymer dots for hypoxia imaging and photodynamic therapy of cancer cells. Adv Sci (Weinh). 2016;3(2):1500155.
99. Maeda H, Sawa T, Konno T. Mechanism of tumortargeted delivery of macromolecular drugs, including the EPR effect in solid tumor and clinical overview of the prototype polymeric drug SMANCS. J Control Release. 2001;74(1–3):47–61.
100. Cai WB, et al. The optimized fabrication of nanobubbles as ultrasound contrast agents for tumor imaging. Sci Rep. 2015;5:13725.

Ultrasound Nanoimaging II

7

7.1 Introduction

The normal phenomenon of these barriers is to protect the cells from external agents. With this the use of ultrasounds will show the favor for the diffusion like in the stratum corneum in the skin, this will increase the efficiency by using the low-frequency ultrasounds which are capable of transmission through the skin [1–4]. These various transmissions are based on the various combinations comprising cavitation [5, 6], thermal force [7, 8], radiation force, convection [9], and lipid extraction [10] which will have various effects due to the various applications of the ultrasound. These nanosystems are made up of siRNAs and liposomes and are capable of being delivered to the skin tumors in the mouse including both dermal and epidermal layers, and this needs the high-frequency ultrasound probes of around 20 kHz ultrasound frequency. With this management, there will be a significant reduction in the melanoma tumors that were growing under the skin of the treated mice [11]. The use of low-frequency ultrasounds will cause disruption of the blood-brain barrier (BBB) causing better and more precise diffusion in the brain parenchyma. This will have more precision having a resolution that will be lower than 1 mm, causing increased temperature increases, causing BBB permeability in a more synergic manner [12]. Thus, liposomes will cross the blood-brain barrier leading to better delivery to the brain parenchyma [13]. Another concept of tumor treatment is tumor ablation, causing increased blood vessel/tissue permeability and facilitating more drug release [14]. In this scenario, the nanosystems (NS) can cause better ultrasound antitumor treatment by using heat based on the increased heat production by the ultrasounds [15]. This concept will lead to a better mechanism due to the presence of nanosystems in the tumor, which will cause increased ultrasound attenuation with higher ultrasound intensity and frequency [16]. After injecting these nanoparticles (NPs) there was a significant increase in the temperatures from 15% to 40% in the NP tissues as compared with the NP-free tissue, for example, there was an increased heating rate of 1.6 °C/min for Au NP when it is dispersed in water as compared to the free water, i.e. 0.4 °C/min [16]. The other parameters that need to be discussed and explained are the size, composition, concentration of the NPs, ultrasound intensity, and frequency, this will result in better thermal efficacy, but these parameters are not yet fully studied. These NPs are made up of phospholipid or polymer membranes which are capable of temperature maintenance and helpful in tumor targeting, imaging, and treatment. Again, the optimum temperature needs to be there for the stability of the configuration like temperature, to maintain drug capsulation, and stability at a higher temperature by the ultrasound application,

S. Shaikh, *Nanoimaging - Future of Precision Medicine*,
https://doi.org/10.1007/978-981-97-7292-6_7

so that drug release can be done at an optimal level. The maintenance of the temperature will be in the range of (40–43 °C) which is slightly more and is known as the transition temperature. Various agents are liposomes that are thermosensitive and made up of phosphatidylcholines [17], PEGylated phospholipids associated with DPPC/DSPC [18], and porous lisolipid membranes [19]. All these are capable of the antitumor activity by the anti-cancer drug release/activation and mild hyperthermia [17]. The major drawback is that these liposomes are in the instability zone and they will be destroyed before reaching the tumor [20], to counter this many methods are developed to prevent the lack of stability and they don't leak at normal body temperature [21]. Cavitation is the mechanical effect that is generated for the antitumor activity under ultrasound. Cavitation and thermal mechanisms are closely related to each other, i.e. cavitation will cause increased temperature and temperature variation will impact the cavitation. NS will increase the level of cavitation resulting from ultrasound application. There are two types of cavitation stable and inertial ones. Stable cavitation, is the process where gas pockets are formed, and they will oscillate by the mechanism of acoustic compression and decompression cycles, this process will cause the generation of fluid streaming and mechanical stresses, leading to the destruction of the cancer cells, inertial cavitation is also an unstable phenomenon where the nanobubbles will expand and collapse in presence of the ultrasound thus having more higher temperatures (>5000 K), pressures (>800 atm), and ROS production. The cavitation phenomenon will have nanobubbles or microbubbles, and these can be easily identified. These two types of micro/nanobubbles are endogenous and exogenous and can be easily distinguished based on the results of the ultrasound application. The basis of the existence and nature of cavitation in the nanosystem depends on the various ultrasound parameters like ultrasound frequency, pressure, intensity, pulse sequence, or duration of application. The negative acoustic pressure will cause the nano/microbubbles to lead to growth or expansion. The favorable cavitation will be with low negative acoustic pressure and higher negative acoustic pressure will lead to the destruction of the nano/microbubble disruption and unstable cavitation [22]. MI expression is the ratio between the negative acoustic pressure and the square root of the ultrasound frequency. The parameters that will not impact are (1) the medium in which ultrasound travels, like water and blood [23], or (2) the pulsation length of the acoustic media [24]. The high ultrasound frequency will reduce MI [24]. One of the classical examples is the use of the polymeric NP having porphyrins when they are exposed to ultrasounds of frequencies which are well below 20 MHz and with high-pressure of amplitude up to 120 MPa seen in vitro neuroblastoma model [25]. Ultrasound MI is capable of various effects causing destruction of cells, creating porosity between the cells, and endocytosis enhancement [26]. The visualization or measurement of the bubbles can be a little difficult task [27]. However, some studies are already cavitation monitoring [28]. This cavitation is being used in many clinical trials [29].

7.2 Sonoluminescence

Sonoluminescence (SL) is the process where the flashes of light are emitted by cavitation bubbles [30]. This will result in the different electron/hole pairs leading to the generation of ROS species. With the use of ultrasound, the nonthermal cavitation will result in the sonoluminescent light exciting C-doped TiO_2 NP. The decaying of these NPs will lead to energy transfer which can lead to cell death in tumors [31]. Sonoluminescence can also activate various light-sensitive drugs (LSD) leading to the triggering of these substances. Sonoporation will cause the cavitation to be separated for clarity, and this sonoporation will be labeled as sonopermeation, in which the pores are formed in cell membranes when they are exposed to the ultrasound causing cavitation [32]. The diffusion of NS will cause the movement through the cell membrane, which is known as sonoporation [33–35] (Figs. 7.1 and 7.2). The various mechanisms will lead to the various (1) intracellular interactions happening between

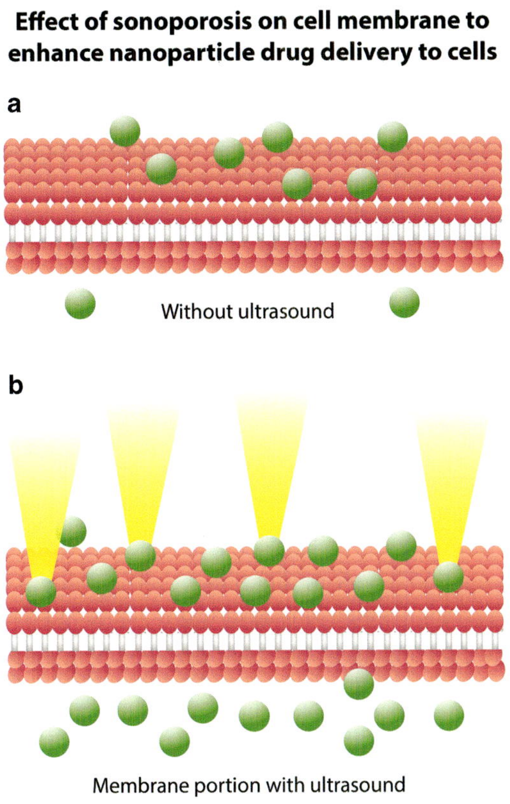

Fig. 7.1 Sonoporosis nanoparticle delivery to the cells (**a**) without ultrasound and (**b**) with ultrasound on the cells

microbubbles and cell membranes, (2) micro-jetting, (3) micro-streaming, (4) shockwaves, and (5) diffusion of microbubbles through cells. Various intracellular interactions will be seen by different directions of movement of the microbubbles, causing the various mechanical forces that can cause better permeability (Fig. 7.3). Various aspects happen the interaction of the jets, streams, and shockwaves occur under ultrasound applications when microbubbles are not there [36]. Another process known as micro-streaming is seen along with the various fluid movements surrounding the bubbles leading to various effects like [37], and better permeability in blood vessels [38–40]. Another concept of the description is cytotoxicity and the mechanical destruction of the cell membrane which can be seen in the presence of NP [41, 42], due to NP interactions with the cell membrane [43, 44]. The different ultrasound frequencies will have different intensities in the range of 0.02 and 6 MHz [45, 46], and low intensities below 1 W/cm^2 [46], resulting in better use of this technique clinically [47], with better ultrasound parameters. Another aspect is the use of the various drugs/bio-active molecules that can be released from polymer-based NS [48],

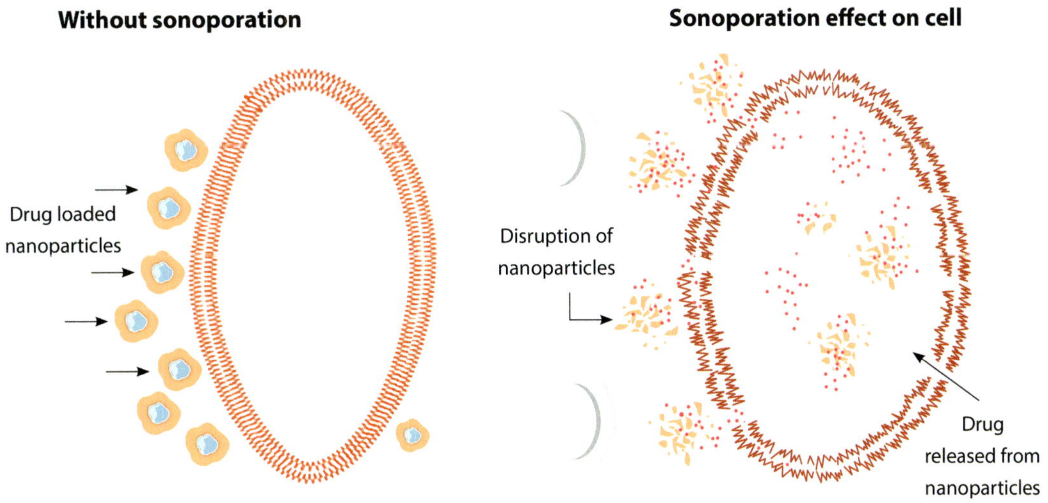

Fig. 7.2 With and without sonoporation on the cells

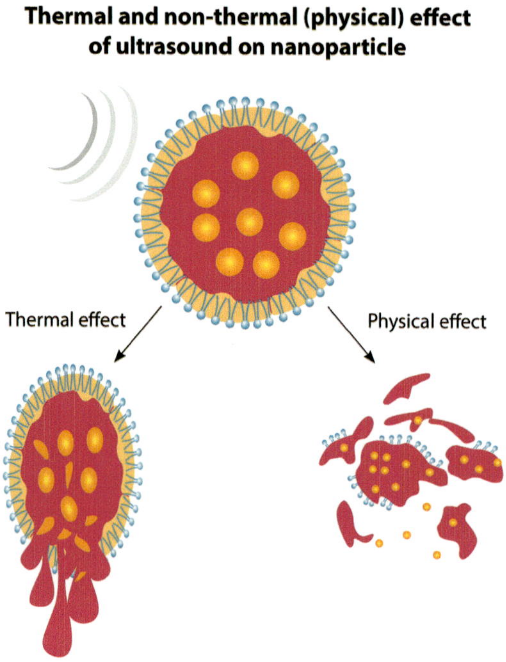

Fig. 7.3 Thermal and nonthermal effect of the ultrasound on the nanoparticle

hydrogels [49], ultrasound excitation leading to the microbubble excitation in various tumor tissues leading to the inhibition of the tumor growth [50]. These have many advantages including continuous release of the drug by using ultrasound [49]. Thus, there is prevention of permanent damages due to the use of mild ultrasound parameters and also prevention of drug release when no ultrasound application is there. Thus, the toxicity of the chemotherapeutic agents is reduced. Also reducing the leakage of the drug in the blood circulation. Thus, there controlled release of the drug in the tumor, having better exposure as required [51–54]. There are some of the gases which can be used as antitumor activity [55]. They can be inserted within the nano/microbubbles, thus controlling the release in the tumor by using the ultrasound application. Secondly, some of the inactive gases may become inactive like the conversion of H_2O_2 to O_2. The tumor microenvironment can have higher levels of H_2O_2, which can be converted to O_2 by the catalase enzymes or inorganic systems looking like the mimicking enzyme activity, another one is the catalase-associated iron oxide NPs which will produce the O_2 bubbles in tumors under HIFU application [56]. The various gases will act in different ways in the tumor or against the tumor. Oxygen (O_2) helps overcome the tumor hypoxia, nitric oxide (NO) produces highly reactive ROS, i.e. peroxynitrites ($ONOO^-$) [57–59], hydrogen sulfide (H_2S) is capable of causing the selective ROS activation [60], sulfur dioxide (SO_2) regulates redox balance in a tumor [61], CO_2 will be favoring the drug release [62]. Technical problems will lead to low solubility and irritability due to the encapsulation in nano/microbubbles. An important aspect of the nanosystems is the capability of the gases for their selective release in the tumor microenvironment (TME) [63, 64]. These gases will help to visualize and release drugs [65], starving the tumor cells by the CO generation which will be blocking nutrients like, e.g., glucose or O_2 [66], and also for creating the embolization by using nanobubbles filled with SF6 and thrombin exposed to ultrasounds [67]. The important concept of the use of ultrasound and nanospheres is the size of the particles like increasing the size from the nanometer to micrometer or can be in the reverse order like micrometer to nanometer [68]. This is possible due to the mechanical nature of the acoustic wave, where the application of the various positive/compressive and/or negative/expansive pressures on these nanospheres. There can be the conversion of the liquid to the gaseous state of the nanodroplets can be due to vaporization leading to the expansion [69, 70]. The different mechanisms can be there the first one is based on the acoustic principles and another one is due to the generation of radiations like lasers [71, 72]. This dual principle of the NS sizes which will cause the bubbles/droplets to combine a good echogenicity due to the micrometric sizes leading to the diffuse tumor via the EPR effect from their dimensions [73]. The nanodroplets will have different properties as (1) Vaporization will lead to the microbubble transformation as an ultrasound contrast agent, (2) Can extravasate in the tumor when injected and transported towards the tumor, (3) releasing into the tumor resulting in the better tumor chemotherapeutic efficacy [74]. The process of the transition of MB from micrometric to nanometric sizes. Conversion will lead to more amount of porphyrin accumulation at the tumor site [75]. Another aspect is the MB loaded with siRNA is being converted into NP for ultrasound exposure, for better efficacy of antitumor efficacy [76]. For the best results ultrasound should be applied when the bubbles/droplets have reached the tumor, and this also the strongest interaction between them will lead to the destruction [77]. There will be a lot of impact of the ultrasound treatment with biodistribution for ultrasound tumor imaging. The commonest used ultrasound contrast is microbubble agents [78], which will have the best resolution ultrasound microscopy [79]. The blood vessels irrigating tumors also could serve as ultrasound (US) contrast agents for better resolution, having better diffraction parts [80]. The laser light can cause thermal expansion and an acoustic wave, which can be detected by the process called photoacoustic (PA) imaging and this can be used for the deeper tumors [81]. This can be a combination of laser-based tumor treatment management like PDT and PTT. One of the preclinical animal imaging will show the mice with tumor when injected intravenously DPP-TPA NP and this along with the passive tumor targeting of the NS which is being highlighted as PA imaging will show the tumor destruction via PDT and PTT treatments [82]. Third, the combination of echography with other imaging methods such as CT or magnetic resonance imaging (MRI) can be done by NS [83]. MB is the main component of in vivo tumor imaging, which has resulted the use of the better use of the newer NS which will be useful as US contrast agents, for example, silica-based micro/nanoplatforms [84], NPs that will stabilize the gas [85], some of the re-chargeable nanobubbles like ZnO nanocrystals, polytetrafluoroethylene NPs [86], here cavitation process will lead to the better ultrasound contrast. In vitro, anti-tumor efficacy Combined NS/US treatments were tested in vitro on different cell lines to determine the various aspects like NS concentration, NS incubation time, ultrasound intensity and frequency for better tumor cell destruction, newer pore formation, and endocytosis [87]. The amount of tumor cell death will increase with the increasing amount of NPs in the range of 0.004 and 0.4 μg/mL, 0.05 and 8 μg/mL, 0.06 and 12 μM [88], or 12 and 200 μg/mL, this will have better and longer incubation time, which will be based on the time between the 24 and 72 h [70], also increasing ultrasound intensities from 1 and 6 W. Thus the ultrasound frequency will affect the penetration depth, and also the production of the ROS.

Ultrasounds are capable of the mechanical displacements of NS or the bubbles/gas which can be used for the evaluation of the various mechanisms of tumor cell death that are being done by the NS/US systems, like the NS internalization, more specific tumor cell targeting in the chemotherapeutic drug release [55], apoptosis of the cells [89, 90]. The level of cavitation by the use of the US application can lead to damage at the cellular level [91]. The various gels will cause the various tissues will be exposed to the NS and also to the HIFU, which can be seen in the various types of measurements.

One of the studies showed that the presence of the NS will cause the ablation which will link the cavitation and tumor tissue [91]. Another concept of the nanoscalpel effect will cause acoustic shock waves, and this will be associated with various mechanical damages when there is an application in the adenocarcinoma or leukemia cancer cells in the presence of various metal oxide NPs [92–94].

7.3 In Vivo Antitumor Efficacy

Antitumor activity of various nanosystems has also been identified on various tumors, i.e. glioblastoma U87, fibrosarcoma HT-1080 [95], different sizes of the tumors in the range of 5–300 mm^3 and injected with 0.1–2 mg of NS. The ultrasound frequency is calculated by the frequency/intensity of the sonoporation [96], or drug release [97], frequency/intensity which is directly related to the temperature as required [98]. As mentioned, the ultrasound frequency and intensity are important for ultrasound penetration, based on size, number, type of bubbles, or amplitude of temperature increase, and all these will again be based on the depth of the ultrasound wave traveled, tumor, and transducer surface distance, transducer ultrasound trajectory pattern, and also the NS injection and different ultrasound application patterns. Antitumor activity is dependent on the injection dose having an impact on the tumor growth retardation during the treatment. Antitumor activity will be planned and more scheduled drug release in the tumor [99]. The ROS production, release, or generation of gases in the tumor, increased tumor temperature, antitumor immune reactions, or any combination of all these mechanisms [100]. In vivo detection of cavitation when the smaller gas is used which can be seen in vivo. In a couple of the studies, ultrasounds are used for imaging and another concept is the use of other irradiations like a laser for PDT, or PTT, One of the important examples is that subcutaneous tumors were grown on two mouse fanks will have an immune modulator with NS and this tumor which will be exposed to ultrasound and radiofrequency thermal ablation, i.e. tumor temperature above ~60 °C. This mechanism will induce the important aspect by the activation of dendritic and long-term immune memory T cells and simultaneously de-activation of Treg cells, this has caused a significant reduction in the size of the primary tumor, another aspect is that the tumor cell apoptosis [101]. Another bigger advantage of ultrasounds is like cost-effectiveness and compactness of the modality [102]. Due to this, many ultrasound clinical applications are being done for better tumor detection by using methods like mammography [103, 104], to either replace or complement each other; (1) the imaging of tumors in pretherapy settings [105]; (2) recurrent nodal involvement in post-treatment scenarios in head and neck cancers [106], this will be useful for better monitoring the tumor recurrence; (3) for the guidance of surgery or cryoablation which will be seen during the pre-operative analysis for the identification of the right location [101, 107]; (4) blood-brain barrier disruption for better chemotherapeutic drug delivery to brain tumor [108]; (5) heat production in the tumors by using HIFU, in ablation of the prostate [109], breast [110], or brain tumors [111]; (6) endoscopic ultrasound use for the guidance of the various pancreatic cancer treatment [112]; and (7) for enhancing the release and delivery of the drugs in the tumor [14]. Newer ultrasound contrast agents (UCA) are made up of the proteins (Optison) or phospholipids (Definity or Sonovue), gas acting as a contrast agent, like C3F8 for Option and Definity or SF6 for Sonovue [113]. The use of various contrast agents will improve the efficacy of anti-cancer drugs when used in the US for enhancing the antitumor activity [114]. The disruption of the BBB in the presence of ultrasounds will cause better delivery of DOX to brain tumors. Thermosensitive liposomes encapsulating DOX (Termo DOX) which are being followed by ultrasound application will cause increased DOX accumulation in the tumor [115]. Due to this ultrasounds in combination with the nanomaterials in cancer are in a more advanced phase where ultrasound as a therapeutic modality is being used as HIFU in the management of liver cancers [115], thyroid [116], breast cancer [110], or low-intensity ultrasound which has better permeability in BBB will result in the better management of the glioblastoma [116]. Different nanomaterials are being used in clinical imaging in humans like iron-based NP and liposomes [117]. The role of the various ultrasound contrast agents along with the therapeutic options is seen by the use of the MB to HIFU for pancreatic tumors [114], and facilitating the release of the drug from the liposomes in tumors [118]. Nanomaterials combined with therapeutic ultrasounds are the newer concept with promising results due to the tumor heating, and specific tumor targeting with localized treatment at TME location. Such an approach should yield a more efficient and less toxic cancer treatment compared with the use of ultrasounds or NM alone (Fig. 7.4).

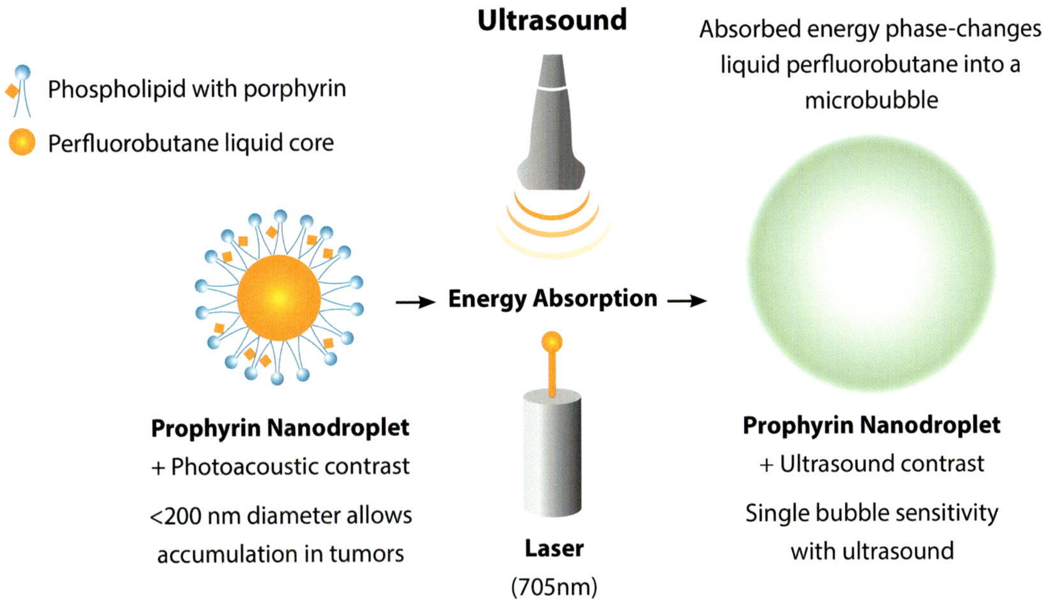

Fig. 7.4 Role of the ultrasound on the energy absorption by the photoacoustic and ultrasound contrast

7.4 Monitoring of Kidney Function and Structure

NPs are being used by CT imaging for the analysis of renal fibrosis. One of the studies using gold NPs with anti-collagen-I antibodies (Co-I-AuNPs) in mice showed that the suspicion of renal artery stenosis was demonstrated by the increased retention of Co-I-AuNPs as compared with no renal artery stenosis. Fibrosis will cause more ischaemic damage to the kidney due to renal artery stenosis. And for this invasive biopsy is the option. After injecting the gold NPs there will not be much change in the hematology, electrolyte, hepatic, and renal functions seen in the mice after 24 h of injection of gold NPs, showing safety which is termed [119–121]. Again, it is the most important area of interest showing the amount of renal-clearable luminescent metal NPs in renal diseases, cancer, and antimicrobial activity. This is being done by the kidney clearance kinetics (KCK) assessment.

Non-invasive fluorescence imaging is not very sensitive due to the conventional fluorophores in the skin, and this can be done with better renal-clearable of the metal NPs with NIR excitation and emission. The use of GS-AuNPs the high-contrast and non-invasive fluorescence imaging is the most promising functional marker of renal function [122–124]. Ordikhani et al. [125] designed NPs that interact with proximal tubule epithelial cells in the selective manner used in renal cancers predominantly arising from the proximal tubular epithelial cells [126–133].

Nanocarriers are of different categories and a few of them are PEGylated polylactic-coglycolic acid which has lambda light chains on the surface and also perfluorocarbons will react with the membrane protein which is the megalin (Fig. 7.5). Megalin has a very high sensitivity for the evaluation of subclinical disease. Ultra-small gold nanoclusters (AuNCs) are the components which are having renal clearance in better ways so that there will be better tumor accumulation as well as renal clearance properties. Loynachan et al. [134, 135] studied and demonstrated the multiple protease nanosensors that will have the better peroxidase-mimicking activity of AuNCs, and this diagnosis is being done by the calorimetric catalytic activity in mice urine. One of the study [136] studied and identified the use of NPs in anti-fibrotic therapies, particularly in damaged

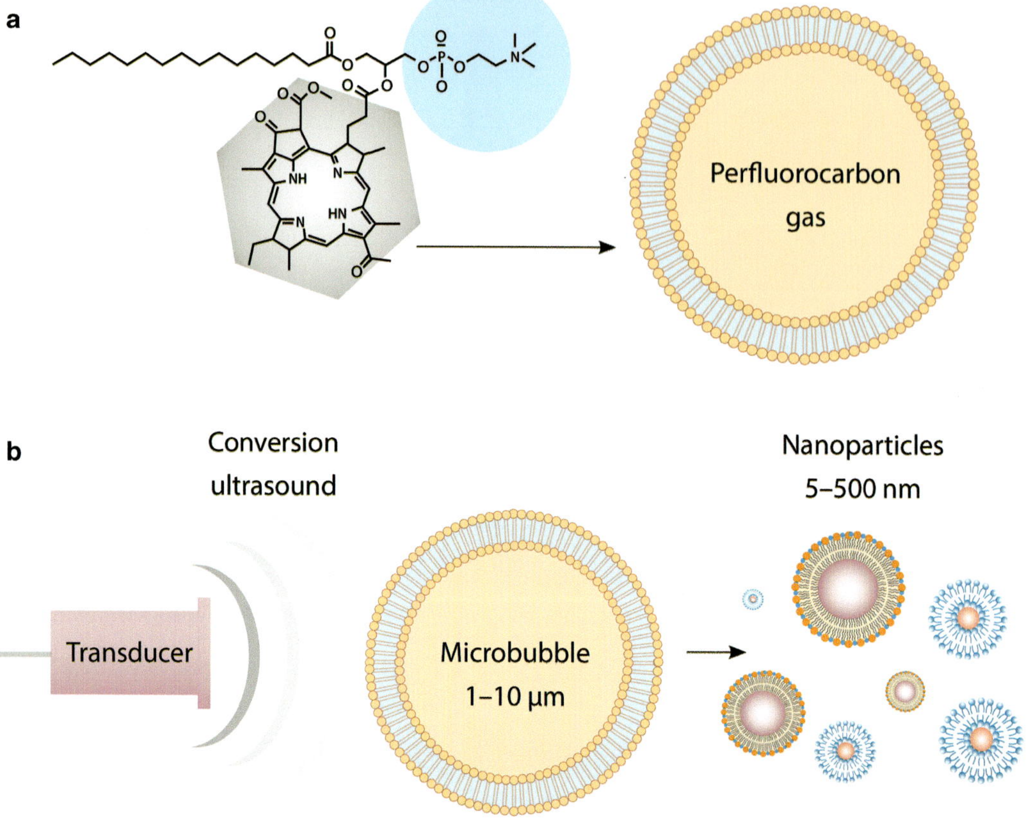

Fig. 7.5 (**a**) Structure of the perfluorocarbon gas. (**b**) Ultrasound conversion of the microbubble to the nanoparticle contrast

cells and tissues. Another important entity is that the NPs can have the property of improving renal function, especially in the clinical scenarios of chronic kidney disease, which is progressing, and this is being done by the removal of collagen fibers, reversing fibrosis, and regenerating renal tubules. Here the use of the gene delivery NP therapy is better than the cytokine therapy [137]. Another emerging role of the NPs with kidney disease antibodies will have a significant impact on the medical management of various diseases. One of the important uses is by using the iron oxide NPs which will coordinate the anti-vascular cell adhesion molecule-1 (anti-VCAM-1) antibody by binding the endothelium which will deliver the region of interest. Vascular cell adhesion molecule-1 (VCAM-1) is seen in the renal ischemia-reperfusion injury [122]. Kidney ischemia-reperfusion injury has significant importance which is seen in transplantation. The different molecules that will be seen on the vascular endothelium will remove the cessation of ischemia, highlighting the injury [138]. Focal segmental glomerulosclerosis (FSGS) is an important condition in which there will be a significant loss of renal podocytes leading to the scarring of the glomeruli and this plays an important role in chronic kidney disease. NPas that are based on various parameters will have drug delivery systems that will be based on the size, composition, charge, and ligands such as peptides, antibodies, and aptamers of the NP. In glomerular pathologies where there will be alteration in the kinetics leading to various stages of injury and inflammation. There will be more compression of the receptors on the surface of the blood vessels which are being engineered for better release of the drugs at the diseased site [139]. Williams et al. [123] showed better conduction in the mice that were administered mesoscale NPs (MNPs).

Nanoencapsulation is the advanced technique in which there will be the encapsulation of various drugs like anti-inflammatory drugs, and chemotherapeutics and this will not be harmful. These NPs releases are scheduled in different phasic patterns which will be targeted to the renal cells. The fluorophores are based on the mechanism of the fluorescence, and this will be helpful in renal cancer cells. This can be done by combining the cancer tumor cells with the antibody on which the antigens will be binding on the surface of the cancer cells [140–143].

Cerium oxide NPs (CNPs) have already mentioned the toxicity to the kidneys and there will be a reduction in the systemic side effects [144–147]. The various antibody-conjugated NP drug carriers which will be targeted to the renal vasculature during NMP. The NPs targeted on CD31 are capable of the increased level of the NPs in the tissues having very good blood perfusion. This concept will show the possibility of delivering various drugs to a very specific organ [148]. DiRito et al. [149] showed that the NPs are capable of accumulating the microvascular plugs due to decreased perfusion mainly in the renal cortex. In one of the studies, the authors showed results with 39 human kidneys, where the low temperatures will cause more accumulation of fibrinogen in microvascularity and will cause the formation of the plugs. These plugs need not be used by using plasminogen and tissue plasminogen activators during NMP. NPs are also used for the assessment of machine perfusion. Wood et al. [120], showed that the analysis of NPs, in the donor-derived extracellular vehicles (EVs) seen in perfusion fluid, will show the kidney status before transplantation. These EVs will show the condition of the tissues from which they are derived. They will be seen in the blood or urine so that the status of the function of the organ after transplantation can be assessed. These NPs containing EVs will be showing the transplant quality before, and the analysis of perfusion fluid can be useful before transplantation. The existing most perfect form of assessing renal function is histopathological analysis, which is invasive, tedious, cost-effective, and takes time. Feng et al. [150] showed that the assessment of kidney function and also to access the damage can be done by surface-enhanced Raman spectroscopy (SERS). SERS spectra is a newer technique based on the silver NPs which are measured in urine. This also shows the various dominant vibrational bands which are due to various strong interactions between urine substances and silver colloids. The urine components which are absorbed from the surface of the silver NPs, and Raman scattering in seen by various optical methods and causing the significant strong intensity enhancement. This can easily identify minute structural changes. The qualification of patients for transplantation is challenging when the donors with AKI have seen serious effects and died. Raman analyses use the principal component analysis (PCA) where the sample volume is not labeled and has very low sensitivity. The analysis of urine using surface-enhanced Raman spectroscopy is also based on various applying iodide-modified silver NPs (Ag-I NPs). The use of the NP in the analysis will show significant improvement as per the SERS spectra quality which will protect the protein from denaturation. SERS analysis is a non-invasive having very high sensitivity (almost 90%) and also for the early assessment of the transplant from dead donors [151]. These are highly specific and sensitive biomarkers that will be monitoring the function of the kidney transplant. There is also a significant relationship between the exosomal miRNA, and endogenous NPs, and also renal function. Chen et al. [152] studied and showed the creation of various circulating exosomal miRNA panels for the post-renal transplant assessment. The exosomal miRNA panel also has various advantages which will show different advantages, like non-invasive procedures, and in this, there will be miRNA stability and lower susceptibility to external factors, and also identify the difference between the identification of chronic graft dysfunction and normal function.

7.5 Monitoring Patients After KTx the Kidney Disease

Improving Global Outcomes (KDIGO) Transplant Work Group has recommended the continuation of the immunosuppressive therapy for 8–16 weeks after transplantation to formulate no rejection as per the recommendation the drug tacrolimus is the drug of choice for the prevention of the rejection.. Another drug, calcineurin inhibitor (CI), can be started before or during KTx. In acute rejection settings, corticosteroids will be recommended as the drug of choice [153–155].

7.6 NPs for US Therapy

NPs can also be used for the US therapy, like HIFU, US-triggered drug release, US-triggered gene transfection, and US-irradiated SDT.

7.6.1 HIFU Ablation

HIFU ablation is based on the principle of inducing hyperthermia and tissue penetrating US by HIFU, this will lead to the issue of coagulative necrosis by various thermal, mechanical, and cavitation effects [156–158]. These NPs are seen in the HIFU ablation therapy, which has significantly increased therapeutic potential.

The different processes of the bimodal imaging techniques where two modalities will be combined like US/MRI for the guided tumor PTT process which can be done by using PGS-SP nanocapsules. In vivo US contrast imaging before and after the administration of PGS-SP nanocapsules or nanobubbles in for the evaluation of the rabbit kidney (Fig. 7.6). Also, for the evaluation of the various changes in the tumor growth rate by using the different therapeutic methods [157, 159].

Contrast agents for HIFU are being used for the differentiation of the various tumors from normal tissues. The use of the Silica NPs is seen to have better HIFU ablation enhancement. The use of the mesoporous Prussian blue NPs which are encapsulated with DOX/PFH are capable of the exhibition by simultaneous increase in the efficacy of the HIFU treatment by decreasing the different cavitation thresholds [156]. The use of the various MSNs which will show the functionalized component associated with paramagnetic MnOx NPs will be able to demonstrate more effective MR imaging and also the enhancement of the HIFU ablation capability [46, 160]. The use of the MSNC along with the Au-PFH-PEG is capable of increasing the US-guided HIFU enhancement and thus increasing the efficiency

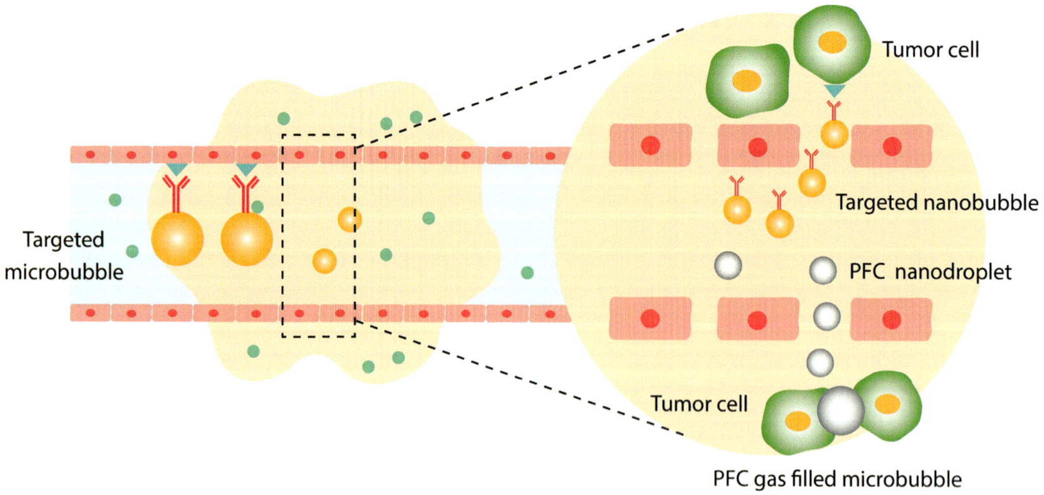

Fig. 7.6 The targeted microbubble to perfluorocarbon gas-filled microbubble in the tumor cells

[161–163]. Phospholipid-coated MSNs are one of the important US agents which are capable of increasing cancer cells and also for better HIFU efficacy leading to better biocompatibility and excellent stability [164]. This also along with the tri-phase transition with L-menthol is seen as enhancing the HIFU efficacy [165]. MR thermometry with Au NPs are also inducing increased efficiency by increasing the temperature of the HIFU [166, 167].

7.7 US-Stimulated Drug Release and US-Enhanced Gene Transfection

This is the newer concept of managing the toxicity levels of any chemotherapeutic agents without any effect on efficiency. These NPs in the drug delivery systems will protect the drug by preventing its degradation and covering it till it is released in the tumor site which is under internal and external stimuli [168]. Ultrasound-targeted nanobubble destruction can also be a parameter for the better-controlled release of the drug which is targeted at the particular location leading to better drug levels at the region of the interest. Different US stimuli-responsive NPs are like polymerases, liposomes, MSNs, and Au-based NPs. The role of the polymeric NPs is by the amphiphilic pathway leading to better drug delivery [169]. Other agents are liposomal nanocarriers, they have better efficiency as compared to the polymer-coated materials in the release of the anti-cancer drugs at the desired site [170]. Paclitaxel-loaded PEGylated immunoliposomes are the agents that are being used in the local drug delivery for Her2-overexpressing breast cancers [171]. This carrier is showing the higher cellular uptake efficiency with the longer circulating times compared to the non-PEGylated system. The polymeric micelles are also performing as US-mediated drug delivery [172]. Pluronic-based micelles are the most commonly used agents which are used for drug delivery purposes [173, 174]. These agents are having low stability and capable of decomposing below the level of the critical micellar concentration [175–177]. US-responsive PLGA NP agents which are the aggregates loaded are being used for the better use for the drug release and also for enhanced efficacy of chemotherapy by US irradiation [178]. Polymeric micelles and US which is being used for drug delivery is being reviewed by Tanbour et al. [179] and it was shown that the NPs based on polymers can also be used for non-viral gene delivery leading to immunogenic responses [180, 181]. FA (folic acid)–PEG–COL (chitosan oligosaccharide lactate) NPs when they are encapsulated with siRNA which will be targeting hypoxia-inducible factor-1a has shown that there is effective gene therapy of targeted ovarian cancer [182]. The US-sensitive nanobubbles with the polymeric micelles and liposomes which will be carrying siRNA or both siRNA and PTX are also showing the anti-cancer effects for hepatocellular carcinoma [183, 184]. Yang et al. also showed that the PLGA NPs modified with PEI loaded with PTX and shRNA can be used as the theranostic agents for breast cancer thus exhibiting excellent US imaging and anti-cancer effects [185]. Liposomes are excellent drug and siRNA carriers due to their physical properties of the lower immunogenicity, good biocompatibility, and degradability. These liposomes have better-enhancing imaging signal intensities as compared to the polymeric NPs [186]. The use of ultrasound will be helpful for the release of materials that are encapsulated in liposomes and containing the vaporizable emulsion droplets [187]. The PEG-PLA NPs along with the cationic lipids will be delivering the CDK1 which is targeted for silencing of triple-negative breast cancer by the s c-Myc overexpression [188]. These results will be showing that these systems will be inhibiting the expression of CDK1 in the ex vivo and also causing the inhibition of the tumor growth which will be at the dose of 2 mg/kg and causing no systemic toxicity in vivo. This system will be carrying siRNA which is the targeting glucose transporter 3 (GLUT3) which will be achieving the silencing of GLUT3 in vitro and causing the inhibition of the tumor growth in vivo [189]. MSNs in the form of inorganic form are seen as carriers based on their stability and better high drug-loading capacity [190]. The organic-inorganic hybrid HMONs can be also loaded with the anti-cancer drug DOX and seen as the HIFU-responsive drug release and also felicitating the drug behavior [191]. Hybrid MSNs are capable of preventing the premature release of the drugs that are grafted on the surface thus stimulating better drug delivery [192]. Hydrophobic-hydrophilic phase transformation when they are irradiated will prevent premature drug release. PLGA Nanocapsules are coated with ultrathin silicon and then loaded along with the anti-cancer drug camptothecin (CPT) and show excellent hydrophobic drug-loading capacity which will also be causing the accelerated release which will be triggered by HIFU [193]. To coordinate and better control of the drug for transportation and also for the release of the anti-cancer agents in the subcellular site, and these are based on the US-degradable MSNs which will cause the increase the drug accumulation and cause the better drug retention time at the level of the target action [77]. This along with the Au NPs encapsulated along with the phase-change materials and will be leading the better enhancement of the PA contrast imaging and also for the better release of drugs [194, 195]. The traditional sonosensitizers which are seen as synthesized from the photosensitizers, and causing less stability and phototoxicity [196, 197]. Si-based NPs are being used along with the aSDTs there is excellent biocompatibility along with the very good SDT effects [198]. Various MSNs which are being porous silicon NPs and another one which is ion-bound protoporphyrin-PEG will be having more efficient for the MRI-guided SDT and these will be excited by US [199–201]. Graphene oxide nanosheet can be seen along the magnetic-targeted efficacy for the treatment and also seen as better SDT seen by the US irradiation [200]. Porous silicon NPs will have lesser cytotoxicity as compared to that of the various NPs which are not coated with dextran and these also will exhibit better antitumor effects and these SDT of cancer [201]. Porous silicon nanowires are also better sonosensitizers for the SDT of HepG2 cancer cells [202–204]. TiO_2 NPs are also seen as the better activators for the US as photosensitizers [205]. Despite this, there are fewer quantum yield of ROS for the faster recombination of electrons and holes, and this will be an important limitation for pure TiO_2 [206]. To prevent this limitation the TiO_2 NPs will be combined with the noble metals like Au forming the Au-TiO_2 nanocomposites [207]. These nanocomposites are better for the increased SDT efficiency and also for prolonging systemic circulation. And these composites with Au will be doing better analysis [208]. Another classical example of the Au NPs combination is the antitumor efficacy and one of these is documented for colorectal tumors [209]. Carbon NPs can also be used as better sonosensitizers for SDT [210]. Organic NPs, like organic porphyrin, are being used importantly as the sonosensitizers for SDT as it is having very good biocompatibility and biodegradability, like hematoporphyrin [25] and protoporphyrin IX [211]. Another example is that poly(methyl methacrylate) core-shell NPs will be seen enhancing with SDT effect under US radiation with shockwaves [212] (Fig. 7.6).

7.8 Biodegradation and Biosafety

As the role of NPs is seen increasing significantly for both diagnostic and therapeutic purposes, but the important issue is the various side effects like the biological effects and also biosafety of the various agents. The organic NPs which are made up of polymers or lipids are being excreted easily and safely from the body [213, 214]. However the role of the various inorganic NPs is not clear as the biodegradation is not clearly defined. Various NPs which are made up of metal oxides, carbon, silica, and different combustion products will be difficult to biodegrade having a longer degradation time and various cytotoxic effects [215, 216]. These different components like the homogeneous organic-inorganic hybrids which contain disulfide bonds (–S-S–) will cause better biocompatibility and biodegradation as compared to the routine MSNs [217]. The use of the metal ion-doping by adding the –Mn-O– bonds into the silica framework, is capable of accelerating the biodegradation rates of MSNs [218]. Ca^{2+} or iron (Fe) can also be used for the improvement of biodegradability [219]. Silica-based NPs and carbon-based nanosystems can also facilitate biodegradation by enzymatic oxidation [220, 221]. A few of the NPs are having toxicity like carbon-based nanosystems [221, 222]. Different Au NP sizes also show features of toxicity with sizes ranging from 1.4 nm are toxic and 15 nm are non-toxic [223]. Iron oxide particles also show cytotoxic effects [224]. This can be done by surface modifications [225]. Gadolinium fullerene particles with hydroxyl groups are also capable of reducing the side effects by preventing the generation of ROS.

7.9 Conclusion

Nanosystems are being widely used for the evaluation of US-based diagnosis and therapy and showing significant high biocompatibility and low toxicity. Selective targeting of tumors and other diseased tissues can be better. US-based agents can be used for both the use for both US contrast imaging and therapy as multifunctional theranostic agents by using different US parameters. Nanobubbles are excellent agents for evaluation of the low acoustic intensities, and these bubbles will undergo cavitation resulting in the US sonication with high acoustic intensities. Thus, these are having different applications from MRI, CT, PET, and PA. Here we discussed the use of the NPs for US-based diagnosis, therapy, and theronostics based on the characteristic features of these materials. Thus, the use of the NPs will have to have a new methodology for the US-based theranostics, along with various applications.

References

1. AIUM curriculum for fundamentals of ultrasound physics and instrumentation. J Ultrasound Med. 2019;38(8):1933–5.
2. Fischetti AJ, Scott RC. Basic ultrasound beam formation and instrumentation. Clin Tech Small Anim Pract. 2007;22(3):90–2.
3. Prabhu SJ, Kanal K, Bhargava P, Vaidya S, Dighe MK. Ultrasound artifacts: classification, applied physics with illustrations, and imaging appearances. Ultrasound Q. 2014;30(2):145–57.
4. Rosenblum D, Joshi N, Tao W, Karp JM, Peer D. Progress and challenges towards targeted delivery of cancer therapeutics. Nat Commun. 2018;9:1410.
5. Bhatnagar S, Schifter H, Coussios C-C. Exploitation of acoustic cavitation-induced microstreaming to enhance molecular transport. J Pharm Sci. 2014;103:1903–12.
6. Bhatnagar S, Kwan JJ, Shah AR, Coussios CC, Carlisle RC. Exploitation of sub-micron cavitation nuclei to enhance ultrasound-mediated transdermal transport and penetration of vaccines. J Control Release. 2016;238:22–30.
7. Mitragotri S. Transdermal drug delivery using low-frequency sonophoresis. In: Ferrari M, Desai T, Bhatia S, editors. BioMEMS and biomedical nanotechnology, Therapeutic micro/nanotechnology, vol. III. Boston: Springer; 2007. p. 223–36.
8. Boucaud A, Montharu J, Machet L, Arbeille B, Machet MC, Patat F, Vaillant L. Clinical histologic, and electron microscopy study of skin exposed to low-frequency ultrasound. Anat Rec. 2001;264:114–9.
9. Simonin JP. On the mechanisms of in vitro and in vivo phonophoresis. J Control Release. 1995;33:125–41.
10. Alvarez-Román R, Merino G, Kalia YN, Naik A, Guy RH. Skin permeability enhancement by low frequency sonophoresis: lipid extraction and transport pathways. J Pharm Sci. 2003;92:1138–46.
11. Tran MA, Gowda R, Sharma A, Park EJ, Adair J, Kester M, Smith NB, Robertson GP. Targeting V600EB-Raf and Akt3 using nanoliposomal-small interfering RNA inhibits cutaneous melanocytic lesion development. Cancer Res. 2008;68:7638–49.
12. Ljubimova JY, Sun T, Mashouf L, Ljubimov AV, Israel LL, Ljubimov VA, Falahatian V, Holler E. Covalent nanodelivery systems for selective imaging and treatment of brain tumors. Adv Drug Deliv Rev. 2017;113:177–200.
13. Treat LH, McDannold N, Vykhodtseva N, Zhang Y, Tam K, Hynynen K. Targeted delivery of doxorubicin to the rat brain at therapeutic levels using MRI-guided focused ultrasound. Int J Cancer. 2007;121:901–7.
14. Tharkar P, Varanasi R, Wong WSF, Jin CT, Chrzanowski W. Nanoenhanced drug delivery and therapeutic ultrasound for cancer treatment and beyond. Front Bioeng Biotechnol. 2019; https://doi.org/10.3389/fbioe.2019.00324.
15. Beik J, Abed Z, Ghoreishi FS, Hosseini-Nami S, Mehrzadi S, ShakeriZadeh A, Kamran Kamrava S. Nanotechnology in hyperthermia cancer therapy: from fundamental principles to advanced applications. J Control Release. 2016;235:205–21.
16. Sviridov AP, Andreev VG, Ivanova EM, Osminkina LA, Tamarov K, Timoshenko VY. Porous silicon nanoparticles as sensitizers for ultrasonic hyperthermia. Appl Phys Lett. 2013;103:193110.

17. Landon CD, Park JY, Needham D, Dewhirst MW. Nanoscale drug delivery and hyperthermia: the materials design and preclinical and clinical testing of low temperature-sensitive liposomes used in combination with mild hyperthermia in the treatment of local cancer. Open Nanomed J. 2011;1:38–64.
18. Unezaki S, Maruyama K, Takahashi N, et al. Enhanced delivery and antitumor activity of doxorubicin using long-circulating thermosensitive liposomes containing amphipathic polyethylene glycol in combination with local hyperthermia. Pharm Res. 1994;11:1180–5. https://doi.org/10.1023/A:1018949218380.
19. Needham D, Anyarambhatla G, Kong G, Dewhirst MW. A new temperature-sensitive liposome for use with mild hyperthermia: characterization and testing in a human tumor xenograft model. Can Res. 2000;60:1197–201.
20. Banno B, Ickenstein LM, Chiu GNC, Bally MB, Thewalt J, Brief E, Wasan EK. The functional roles of poly(ethylene glycol)-lipid and lysolipid in the drug retention and release from lysolipid-containing thermosensitive liposomes in vitro and in vivo. J Pharm Sci. 2010;99(5):2295–308.
21. Marzban E, Alavizadeh SH, Ghiadi M, Khoshangosht M, Khashayarmanesh Z, Abbasi A, Jaafari MR. Optimizing the therapeutic efficacy of cisplatin PEGylated liposomes via incorporation of different DPPG ratios: in vitro and in vivo studies. Colloids Surf B Biointerfaces. 2015;136:885–91. https://doi.org/10.1016/j.colsurfb.2015.10.046.
22. Şen T, Tüfekçioğlu O, Koza Y. Mechanical index. Anatol J Cardiol. 2015;15:334–6.
23. Chung YE, Kim KW. Contrast-enhanced ultrasonography: advance and current status in abdominal imaging. Ultrasonography. 2015;34:3–18.
24. Yang X, Jo J. Enhanced cavitation by using two consecutive ultrasound waves at different frequencies. Appl Phys Lett. 2014;105:193701.
25. Canaparo R, Varchi G, Ballestri M, Foglietta F, Sotgiu G, Guerrini A, Francovich A, Civera P, Frairia R, Serpe L. Polymeric nanoparticles enhance the sonodynamic activity of meso-tetrakis (4-sulfonatophenyl) porphyrin in an in vitro neuroblastoma model. Int J Nanomedicine. 2013;8:4247–63.
26. Kooiman L, Roovers S, Langeveld S, Kleven RT, Dewitte H, O'Reilly MA, Escofre JM, Bouakaz A, Verweij M, Hynynen K, Lentacker I, Stride E, Holland CK. Ultrasound-responsive cavitation nuclei for therapy and drug delivery. Ultrasound Med Biol. 2020;46:1296–325.
27. Coussios CC, Farny CH, Ter Haar G, Roy RA. Role of acoustic cavitation in the delivery and monitoring of cancer treatment by high-intensity focused ultrasound (HIFU). Int J Hyperth. 2007;23:105–20.
28. Wu SY, Sanchez CS, Samiotaki G, Buch A, Ferrera VP, Konofagou EE. Characterizing focused ultrasound mediated drug delivery to the heterogeneous primate brain in vivo with acoustic monitoring. Sci Rep. 2016;6:37094. https://doi.org/10.1038/srep37094.
29. Wischhusen J, Padilla F. Ultrasound-targeted microbubble destruction (UTMD) for localized drug delivery into tumor tissue. IRBM. 2019;40:10–5.
30. Beguin E, Shrivastava S, Dezhkunov NV, McHale AP, Callan JF, Stride E. Direct evidence of multibubble sonoluminescence using therapeutic ultrasound and microbubbles. ACS Appl Mater Interfaces. 2019;11:19913–9.
31. Yang CC, Wang CX, Kuan CY, Chi CY, Chen CY, Lin YY, Chen GS, Hou CG, Lin FH. Using C-doped TiO_2 nanoparticles as a novel sonosensitizer for cancer treatment. Antioxidants. 2020;9:880.
32. Snipstad S, Sulheim E, Davies CDL, Moonen C, Storm G, Kiessling F, Schmid R, Lammers T. Sonopermeation to improve drug delivery to tumors: from fundamental understanding to clinical translation. Expert Opin Drug Deliv. 2018;15:1249–61.
33. Delalande A, Kotopoulis S, Postema M, Midoux P, Pichon C. Sonoporation: mechanistic insights and ongoing challenges for gene transfer. Gene. 2013;525:191–9.
34. Chowdhury SM, Lee T, Willmann JK. Ultrasound-guided drug delivery in cancer. Ultrasonography. 2017;36:171–84.
35. Postema M, Gilja OH. Ultrasound-directed drug delivery. Curr Pharm Biotechnol. 2007;8(6):355–61. https://doi.org/10.2174/138920107783018453.
36. Fan Z, Kumon RE, Deng CX. Mechanisms of microbubble-facilitated sonoporation for drug and gene delivery. Ther Deliv. 2014;5:467–86. https://doi.org/10.4155/tde.14.10.
37. Khokhlova TD, Haider Y, Hwang JH. Therapeutic potential of ultrasound microbubbles in gastrointestinal oncology: recent advances and future prospects. Ther Adv Gastroenterol. 2015;8:384–94.
38. Tay LM, Xu C. Coating microbubbles with nanoparticles for medical imaging and drug delivery. Nanomedicine. 2017;12:91–4.
39. Lee H, Kim H, Han H, Lee M, Lee S, Yoo H, Chang JH, Kim H. Microbubbles used for contrast enhanced ultrasound and theragnosis: a review of principles to applications. Biomed Eng Lett. 2017;7:59–69.
40. Gu M, Wang X, Toh TB, Chow EKH. Applications of stimuli-responsive nanoscale drug delivery systems in translational research. Drug Discov Today. 2018;23:1043–52.
41. Blomley MJK, Cooke JC, Unger EC, Monaghan MJ, Cosgrove DO. Microbubble contrast agents: a new era in ultrasound. BMJ. 2001;322:1222–5.
42. Guimaraes D, Cavaco-Paulo A, Nogueira E. Design of liposomes as drug delivery system for therapeutic applications. Int J Pharm. 2021;601:120571.
43. Ke H, Wang J, Dai Z, Jin Y, Qu E, Xing Z, Guo C, Yue X, Liu J. Gold-nanoshelled microcapsules: atheranostic agent for ultrasound contrast imaging and photothermal therapy. Angew Chem Int Ed. 2011;50:3017–21.

44. Ke H, Yue X, Wang J, Xing S, Zhang Q, Dai Z, Tian J, Wang S, Jin Y. Gold nanoshelled liquid perfuorocarbon nanocapsules for combined dual modal ultrasound/CT imaging and photothermal therapy of cancer. Small. 2014;6:1220–7.
45. Forbes MM, Steinberg RL, O'Brien WD. Frequency-dependent evaluation of the role of definity in producing sonoporation of Chinese hamster ovary cells. J Ultrasound Med. 2011;30:61–9.
46. Wang X, Chen H, Zheng Y, Ma M, Chen Y, Zhang K, Zeng D, Shi J. Au-nanoparticle coated mesoporous silica nanocapsule-based multifunctional platform for ultrasound mediated imaging, cytoclasis and tumor ablation. Biomaterials. 2013;34:2057–68.
47. Castle J, Kotopoulis S, Forsberg F. Sonoporation for augmenting chemotherapy of pancreatic ductal adenocarcinoma. Methods Mol Biol. 2020;2059:191–205. https://doi.org/10.1007/978-1-4939-9798-5_9.
48. Cao Y, Langer R. Optimizing the delivery of cancer drugs that block angiogenesis. Sci Transl Med. 2010;2:1513.
49. Huebsch N, Kearney CJ, Zhao X, Kim J, Cezar CA, Suo Z, Mooney DJ. Ultrasound-triggered disruption and self-healing of reversibly crosslinked hydrogels for drug delivery and enhanced chemotherapy. Proc Natl Acad Sci USA. 2014;111:9762–7.
50. Li P, Zheng Y, Ran H, Tan J, Lin Y, Zhang Q, Ren J, Wang Z. Ultrasound triggered drug release from 10-hydroxycamptothecin-loaded phospholipid microbubbles for targeted tumor therapy in mice. J Control Release. 2012;162:349–54.
51. Yoon YI, Yoon T-J, Lee HJ. Optimization of ultrasound parameters for microbubble-nanoliposome complex-mediated delivery. Ultrasonography. 2015;34:297–303.
52. Hesham GM, Ana MM, Ghaleb AH. Review on triggered liposomal drug delivery with a focus on ultrasound. Curr Cancer Drug Targets. 2015;15:282–313.
53. Ahmed SE, Martins AM, Husseini GA. The use of ultrasound to release chemotherapeutic drugs from micelles and liposomes. J Drug Target. 2015;23:16–42.
54. Kosheleva OK, Lai TC, Chen NG, Hsiao M, Chen CH. Selective killing of cancer cells by nanoparticle-assisted ultrasound. J Nanobiotechnol. 2016;14:46. https://doi.org/10.1186/s12951-016-0194-9.
55. Cheng CA, Deng T, Lin FC, Cai Y, Zink JI. Supramolecular nanomachines as stimuli-responsive gatekeepers on mesoporous silica nanoparticles for antibiotic and cancer drug delivery. Theranostics. 2019;9:3341–64.
56. Liu T, Zhang N, Wang Z, Wu M, Chen Y, Ma M, Chen H, Shi J. Endogenous catalytic generation of O_2 bubbles for in situ ultrasound-guided high intensity focused ultrasound ablation. ACS Nano. 2017;11:9093–102.
57. Deng L, O'Reilly MA, Jones RM, An R, Hynynen K. A multi-frequency sparse hemispherical ultrasound phased array for microbubble mediated transcranial therapy and simultaneous cavitation mapping. Phys Med Biol. 2016;61:8476–501.
58. Xiang HJ, Deng Q, An L, Guo M, Yang SP, Liu JG. Tumor cell specific and lysosome-targeted delivery of nitric oxide for enhanced photodynamic therapy triggered by 808 nm near-infrared light. Chem Commun. 2016;52:148–51.
59. Wan SS, Zeng JY, Cheng H, Zhang XZ. ROS-induced NO generation for gas therapy and sensitizing photodynamic therapy of tumor. Biomaterials. 2018;185:51–62.
60. Ohsawa I, Ishikawa M, Takahashi K, Watanabe M, Nishimaki K, Yamagata K, Katsura K, Katayama Y, Asoh S, Ohta S. Hydrogen acts as a therapeutic antioxidant by selectively reducing cytotoxic oxygen radicals. Nat Med. 2007;13:688–94.
61. Zhang Y, Wan Y, Chen Y, Blum NT, Lin J, Huang P. Ultrasound-enhanced chemo-photodynamic combination therapy by using albumin "nanoglue"-based nanotheranostics. ACS Nano. 2020;14:5560–6556.
62. Liu J, Ma H, Wei T, Liang XJ. CO_2 gas induced drug release from pH-sensitive liposome to circumvent doxorubicin resistant cells. Chem Commun. 2012;48:4869–71.
63. Min HS, Son S, You DG, Lee TW, Lee J, Lee S, Yhee JY, Lee J, Han MH, Park JH, Kim SH, Choi K, Park K, Kim K, Kwon IC. Chemical gas-generating nanoparticles for tumor-targeted ultrasound imaging and ultrasound-triggered drug delivery. Biomaterials. 2016;108:57–70.
64. Mo R, Gu Z. Tumor microenvironment and intracellular signal-activated nanomaterials for anticancer drug delivery. Mater Today. 2016;19:274–83.
65. Hernandez C, Gulati S, Fioravanti G, Stewart PL, Exner AA. Cryo-EM visualization of lipid and polymer-stabilized perfluorocarbon gas nanobubbles—a step towards nanobubble mediated drug delivery. Sci Rep. 2017;7:13517.
66. Wang X, Wang J, Pan J, Zhao F, Kan D, Cheng R, Zhang X, Sun SK. Rhenium sulfide nanoparticles as a biosafe spectral CT contrast agent for gastrointestinal tract imaging and tumor theranostics in vivo. ACS Appl Mater Interfaces. 2019;11:33650–8.
67. Shao Y, Guo L, Li A, Zhang K, Liu W, Shi J, Liu J, Zhang Z. US-triggered ultra-sensitive "thrombus constructor" for precise tumor therapy. J Control Release. 2020;318:136–44.
68. Huynh E, Rajora MA, Zheng G. Multimodal micro, nano, and size conversion ultrasound agents for imaging and therapy. WIREs Nanomed Nanobiotechnol. 2016;8:796–813.
69. Wilson K, Homan K, Emelianov S. Biomedical photoacoustics beyond thermal expansion using triggered nanodroplet vaporization for contrast-enhanced imaging. Nat Commun. 2012;3:618.
70. Cao Y, Chen Y, Yu T, Guo Y, Liu F, Yao Y, Li P, Dong W, Wang Z, Chen Y, Ran H. Drug release from phase-changeable nanodroplets triggered by low-intensity focused ultrasound. Theranostics. 2018;8:1327–39.

71. Loskutova K, Grishenkov D, Ghorbani M. Review on acoustic droplet vaporization in ultrasound diagnostics and therapeutics. Biomed Res Int. 2019;2019:1.
72. Choo S, He B, Duan F. Evaporation of droplet with and without laser excitation. Appl Therm Eng. 2015;88:341–6.
73. Rapoport N. Phase-shift, stimuli-responsive perfluorocarbon nanodroplets for drug delivery to cancer. WIREs Nanomed Nanobiotechnol. 2012;4:492–510.
74. Gao Z, Kennedy AM, Christensen DA, Rapoport NY. Drug-loaded nano/microbubbles for combining ultrasonography and targeted chemotherapy. Ultrasonics. 2008;48:260–70.
75. Huynh E, Leung BYC, Helfeld BL, Shakiba M, Gandier JA, Jin CS, Master ER, Wilson BC, Goertz DE, Zheng G. In situ conversion of porphyrin microbubbles to nanoparticles for multimodality imaging. Nat Nanotechnol. 2015;10:325–32.
76. Wang P, Yin T, Li J, Zheng B, Wang X, Wang Y, Zheng J, Zheng R, Shuai X. Ultrasound-responsive microbubbles for sonography-guided siRNA delivery. Nanomed Nanotechnol Biol Med. 2016;12:1139–49.
77. Ferrara K, Pollard R, Borden M. Ultrasound microbubble contrast agents: fundamentals and application to gene and drug delivery. Annu Rev Biomed Eng. 2007;9:415–47.
78. Chong WK, Papadopoulou V, Dayton PA. Imaging with ultrasound contrast agents: current status and future. Abdominal Radiol. 2018;43:762–72.
79. Christensen-Jefries K, Couture O, Dayton PA, Eldar YC, Hynynen K, Kiessling F, O'Reilly M, Pinton GF, Schmitz G, Tang MX, Tanter M, Van Sloun RJG. Super-resolution ultrasound imaging. Ultrasound Med Biol. 2020;46:865–91.
80. Tang S, Song P, Trzasko JD, Lowerison M, Huang C, Gong P, Lok UW, Manduca A, Chen S. Kalman filter-based microbubble tracking for robust super-resolution ultrasound microvessel imaging. IEEE Trans Ultrason Ferroelectr Freq Control. 2020;67:1738–51.
81. Wang S, Lin J, Wang T, Chen X, Huang P. Recent advances in photoacoustic imaging for deep-tissue biomedical applications. Therasonics. 2016;6:2394–413.
82. Cai Y, Liang P, Tang Q, Yang X, Si W, Huang W, Zhang Q, Dong X. Diketopyrrolopyrrole–triphenylamine organic nanoparticles as multifunctional reagents for photoacoustic imaging-guided photodynamic/photothermal synergistic tumor therapy. ACS Nano. 2017;11:1054–63.
83. Li X, Zhang XN, Li XD, Chang J. Multimodality imaging in nanomedicine and nanotheranostics. Cancer Biol Med. 2016; https://doi.org/10.20892/j.issn.2095-3941.2016.0055.
84. Zhou Y, Han X, Jing X, Chen Y. Construction of silica-based micro/nanoplatforms for ultrasound theranostic biomedicine. Adv Health Mater. 2017;6:1700646.
85. Kwan JJ, Graham S, Myers R, Carlisle R, Stride E, Coussios CC. Ultrasound-induced inertial cavitation from gas-stabilizing nanoparticles. Phys Rev E. 2015;92:023019.
86. Jin Q, Kang ST, Chang YC, Zheng H, Yeh CK. Inertial cavitation initiated by polytetrafluoroethylene nanoparticles under pulsed ultrasound stimulation. Ultrason Sonochem. 2016;32:1–7.
87. Yang Y, Li Q, Guo X, Tu J, Zhang D. Mechanisms underlying sonoporation: interaction between microbubbles and cells. Ultrasonics Sonochem. 2020;67:105096.
88. Chen M, Liang X, Gao C, Zhao R, Zhang N, Wang S, Chen W, Zhao B, Wang J, Dai Z. Ultrasound triggered conversion of porphyrin/camptothecin-fluoroxyuridine triad microbubbles into nanoparticles overcomes multidrug resistance in colorectal cancer. ACS Nano. 2018;12:7312–26.
89. Chen XJ, Zhang XQ, Liu Q, Zhang J, Zhou G. Nanotechnology: a promising method for oral cancer detection and diagnosis. J Nanobiotechnol. 2018;16:52.
90. Chen S, Liu Y, Zhu S, Chen C, Xie W, Xiao L, Zhu L, Hao L, Wang Z, Sun J, Chang S. Dual-mode imaging and therapeutic effects of drug-loaded phase-transition nanoparticles combined with near-infrared laser and low-intensity ultrasound on ovarian cancer. Drug Deliv. 2018;25:1683–93. https://doi.org/10.1080/10717544.2018.1507062.
91. Yildirim A, Shi D, Roy S, Blum NT, Chattaraj R, Cha JN, Goodwin AP. Nanoparticle-mediated acoustic cavitation enables high intensity focused ultrasound ablation without tissue heating. ACS Appl Mater Interfaces. 2018;10:36786–95.
92. Osminkina LA, Nikolaev AL, Sviridov AP, Andronova NV, Tamarov KP, Gongalsky MB, Kudryavtsevd AA, Treshalinac HM, Timoshenkoa YV. Porous silicon nanoparticles as efficient sensitizers for sonodynamic therapy of cancer. Microporous Mesoporous Mater. 2015;210:169–75.
93. Racca L, Limongi T, Vighetto V, Dumontel B, Ancona A, Canta M, Canavese G, Garino N, Cauda V. Zinc oxide nanocrystals and high-energy shock waves: a new synergy for the treatment. Front Bioeng Biotechnol Nanobiotechnol. 2020;8:577.
94. Vighetto V, Racca L, Canta M, Matos JC, Dumontel B, Gonçalves MC, Cauda V. Smart shockwave responsive titania-based nanoparticles for cancer treatment. Pharmaceutics. 2021;13:1423.
95. Xie X, Lin W, Li M, Yang Y, Deng J, Liu H, Chen Y, Fu X, Liu H, Yang Y. Efficient siRNA delivery using novel cell-penetrating peptide-siRNA conjugate-loaded nanobubbles and ultrasound. Ultrasound Med Biol. 2016;42:1362–74.
96. Liu Q, Jiang J, Tang L, Chen M. The effect of low frequency and low intensity ultrasound combined with microbubbles on the sonoporation efficiency of MDA-MB-231. Ann Transl Med. 2020;8:298.

97. Schroeder A, Avnir Y, Weisman S, Najajreh Y, Gabizon A, Talmon Y, Kost J, Barenholz Y. Controlling liposomal drug release with low frequency ultrasound: mechanism and feasibility. Langmuir. 2007;23:4019–25.
98. Bove T, Zawada T, Serup J, Jessen A, Poli M. High-frequency (20-MHz) high-intensity focused ultrasound (HIFU) system for dermal intervention: preclinical evaluation in skin equivalents. Skin Res Technol. 2019;25:217–28.
99. Min HS, You DG, Son S, Jeon S, Park JH, Lee S, Kwon IC, Kim K. Echogenic glycol chitosan nanoparticles for ultrasound-triggered cancer theranostics. Theranostics. 2015;5:1402–18.
100. Tung YS, Vlachos F, Choi JJ, Defeux T, Selert K, Konofagou EE. In vivo transcranial cavitation threshold detection during ultrasound-induced blood–brain barrier opening in mice. Phys Med Biol. 2010;55:6141.
101. Arko D, Sikošeka NC, Kozar N, Sobocanb M, Taka I. The value of ultrasound-guided surgery for breast cancer. Eur J Obstet Gynecol Reprod Biol. 2017;216:198–203.
102. Gummadi S, Eisenbrey J, Li J, Li Z, Forsberg F, Lyshchik A, Liu JB. Advances in modern clinical ultrasound. Adv Ultrasound Diagn Ther. 2018;02:051–63.
103. Rebolj M, Assi V, Brentnall A, Parmar D, Dufy SW. Addition of ultrasound to mammography in the case of dense breast tissue: systematic review and meta-analysis. Br J Cancer. 2018;118:1559–70.
104. Sood R, Rositch AF, Shakoor D, Ambinder E, Pool KL, Pollack E, Mollura DJ, Mullen LA, Harvey SC. Ultrasound for breast cancer detection globally: a systematic review and meta-analysis. J Global Oncol. 2019;5:1–17.
105. Giannella L, Carpini GD, Montik N, Verdecchia V, Puccio F, Di Giuseppe J, Tsiroglou D, Goteri G, Ciavattini A. Ultrasound features of a uterine perivascular epithelioid cell tumor (PEComa): case report and literature review. Diagnostics. 2020;10:553.
106. Lin CM, Wang CP, Chen CN, Lin CY, Li TY, Chou CH, Hsu YC, Kuo PY, Yang TL, Lou PJ, Ko JY, Chen TC. The application of ultrasound in detecting lymph nodal recurrence in the treated neck of head and neck cancer patients. Sci Rep. 2017;7:3958.
107. Ward RC, Lourenco AP, Mainiero MB. Ultrasound-guided breast cancer cryoablation. AJR. 2019;213:716–22.
108. Beccaria K, Canney M, Bouchoux G, Puget S, Grill J, Carpentier A. Blood-brain barrier disruption with low-intensity pulsed ultrasound for the treatment of pediatric brain tumors: a review and perspectives. Neurosurg Focus. 2020;48:E10.
109. Chaussy CG, Thurof S. High-intensity focused ultrasound for the treatment of prostate cancer: a review. J Endourol. 2017;31:S30–9.
110. Peek MCL, Wu F. High-intensity focused ultrasound in the treatment of breast tumours. Ecancer. 2018;12:794.
111. Prada F, Kalani MYS, Yagmurlu K, Norat P, Del Bene M, DiMeco F, Kassell NF. Applications of focused ultrasound in cerebrovascular diseases and brain tumors. Neurotherapeutics. 2019;16:67–87.
112. Testoni SGG, Healey AJ, Dietrich CF, Arcidiacono PG. Systematic review of endoscopy ultrasound-guided thermal ablation treatment for pancreatic cancer. Endosc Ultrasound. 2020;9:83–100.
113. Borden MA, Song KH. Reverse engineering the ultrasound contrast agent. Adv Colloid Interf Sci. 2018;262:39–49.
114. Dimcevski G, Kotopoulis S, Bjånes T, Hoem D, Schjøtt J, Gjertsen BT, Biermann M, Molven A, Sorbye H, McCormack E, Postema M, Gilja OH. A human clinical trial using ultrasound and microbubbles to enhance gemcitabine treatment of inoperable pancreatic cancer. J Control Release. 2016;243:172–81.
115. Lyon PC, Grifths LF, Lee J, Chung D, Carlisle R, Wu F, Middleton MR, Gleeson FV, Coussios CC. Clinical trial protocol for TARDOX: a phase I study to investigate the feasibility of targeted release of lysothermosensitive liposomal doxorubicin (ThermoDox®) using focused ultrasound in patients with liver tumours. J Therap Ultrasound. 2017;5:28.
116. Lang BH, Woo YC, Chiu KWH. High intensity focused ultrasound (HIFU) ablation of benign thyroid nodule is safe and efficacious in patients who continue taking an anti-coagulation or antiplatelet agent in the treatment period. Int J Hyperth. 2019;36:185–9.
117. Wu SK, Tsai CL, Huang Y, Hynynen K. Focused ultrasound and microbubbles-mediated drug delivery to brain tumor. Pharmaceutics. 2021;13:15.
118. Gray MD, Lyon PC, Mannaris C, Folkes LK, Stratford M, Dip LC, Chung DYF, Scott S, Anderson M, Goldin R, Carlisle R, Wu F, Middleton MR, Gleeson FV, Coussios CC. Focused ultrasound hyperthermia for targeted drug release from thermosensitive liposomes: results from a phase I trial. Radiology. 2019;291:232–8.
119. Freestone I, Meeks N, Sax M, Higgitt C. The Lycurgus Cup—a Roman nanotechnology. Gold Bull. 2007;40:270–7.
120. Bhattacharya R, Mukherjee P. Biological properties of "naked" metal nanoparticles. Adv Drug Deliv Rev. 2008;60:1289–306.
121. Arvizo RR, Bhattacharyya S, Kudgus RA, Giri K, Bhattacharya R, Mukherjee P. Intrinsic therapeutic applications of noble metal nanoparticles: past, present and future. Chem Soc Rev. 2012;41:2943–70.
122. Fakruddin M, Hossain Z, Afroz H. Prospects and applications of nanobiotechnology: a medical perspective. J Nanobiotechnol. 2012;10:31.
123. Surendiran A, Sandhiya S, Pradhan SC, Adithan C. Novel applications of nanotechnology in medicine. Indian J Med Res. 2009;130:689–701.
124. Xue Y, Yu G, Shan Z, Li Z. Phyto-mediated synthesized multifunctional Zn/CuO NPs hybrid nanopar-

ticles for enhanced activity for kidney cancer therapy: a complete physical and biological analysis. J Photochem Photobiol B Biol. 2018;186:131–6.
125. Minardi S, Shah S, Luo X. Biomimetic nanoparticles for transplantation tolerance. Curr Opin Organ Transplant. 2018;23:15–21.
126. Vemuri C, Upadhya GA, Arif B, Jia J, Lin Y, Gaut JP, Fazal J, Pan H, Wickline SA, Chapman WC. Antithrombin perfluorocarbon nanoparticles improve renal allograft function in a murine deceased criteria donor model. Transplant Direct. 2018;4:e384.
127. Uehara M, Bahmani B, Jiang L, Jung S, Banouni N, Kasinath V, Solhjou Z, Zhao J, Ordikhani F, Bae M, et al. Nanodelivery of mycophenolate mofetil to the organ improves transplant vasculopathy. ACS Nano. 2019;13:12393–407.
128. Zhang Y, Pan J, Li H, Yu D, Wu T, Wang L, Wang Y, Zhou L, Zheng S. Albumin based nanomedicine for enhancing tacrolimus safety and lymphatic targeting efficiency. J Biomed Nanotechnol. 2019;15:1313–24.
129. Stead SO, Kireta S, McInnes SJP, Kette FD, Sivanathan KN, Kim J, Cueto-Diaz EJ, Cunin F, Durand J-O, Drogemuller CJ, et al. Murine and non-human primate dendritic cell targeting nanoparticles for in vivo generation of regulatory T-cells. ACS Nano. 2018;12:6637–47.
130. Brasile L, Henry N, Stubenitsky B. Underlying mechanisms of protection involved in immunocloak. Transplantation. 2017;101:e49–56.
131. Xie D, Xu Y, Jing W, Juxiang Z, Hailun L, Yu H, Zheng D-H, Lin Y-T. Berberine nanoparticles protects tubular epithelial cells from renal ischemia-reperfusion injury. Oncotarget. 2017;8: 24154–62.
132. Stephen Inbaraj B, Chen B-H. An overview on recent in vivo biological application of cerium oxide nanoparticles. Asian J Pharm Sci. 2020;15:558–75.
133. Loynachan CN, Soleimany AP, Dudani JS, Lin Y, Najer A, Bekdemir A, Chen Q, Bhatia SN, Stevens MM. Renal clearable catalytic gold nanoclusters for in vivo disease monitoring. Nat Nanotechnol. 2019;14:883–90.
134. Midgley AC, Wei Y, Zhu D, Gao F, Yan H, Khalique A, Luo W, Jiang H, Liu X, Guo J, et al. Multifunctional natural polymer nanoparticles as antifibrotic gene carriers for CKD therapy. J Am Soc Nephrol. 2020;31:2292–311.
135. Kamaly N, He JC, Ausiello DA, Farokhzad OC. Nanomedicines for renal disease: current status and future applications. Nat Rev Nephrol. 2016;12:738–53.
136. Akhtar AM, Schneider JE, Chapman SJ, Jefferson A, Digby JE, Mankia K, Chen Y, McAteer MA, Wood KJ, Choudhury RP. In vivo quantification of Vcam-1 expression in renal ischemia reperfusion injury using non-invasive magnetic resonance molecular imaging. PLoS One. 2010;5:e12800.
137. Liu GW, Pippin JW, Eng DG, Lv S, Shankland SJ, Pun SH. Nanoparticles exhibit greater accumulation in kidney glomeruli during experimental glomerular kidney disease. Physiol Rep. 2020;8:e14545.
138. Soriano ML, Rodríguez-Benot A, Valcárcel M. Bases nanotecnológicas de una «nueva» Nefrología. Nefrología. 2018;38:368–78.
139. Martuszewski A, Paluszkiewicz P, Król M, Banasik M, Kepinska M. Donor-derived cell-free DNA in kidney transplantation as a potential rejection biomarker: a systematic literature review. J Clin Med. 2021;10:193.
140. Stanimirova I, Banasik M, Ząbek A, Dawiskiba T, Kościelska-Kasprzak K, Wojtowicz W, Krajewska M, Janczak D, Młynarz P. Serum metabolomics approach to monitor the changes in metabolite profiles following renal transplantation. Sci Rep. 2020;10:17223.
141. Terasaki PI. Humoral theory of transplantation: humoral theory of transplantation. Am J Transplant. 2003;3:665–73.
142. Klinger M, Banasik M. Immunological characteristics of the elderly allograft recipient. Transplant Rev. 2015;29:219–23.
143. Banasik M, Boratyńska M, Kościelska-Kasprzak K, Kamińska D, Bartoszek D, Zabińska M, Myszka M, Zmonarski S, Protasiewicz M, Nowakowska B, et al. The influence of non-HLA antibodies directed against angiotensin II type 1 receptor (AT1R) on early renal transplant outcomes. Transpl Int. 2014;27:1029–38.
144. Tietjen GT, Hosgood SA, DiRito J, Cui J, Deep D, Song E, Kraehling JR, Piotrowski-Daspit AS, Kirkiles-Smith NC, Al-Lamki R, et al. Nanoparticle targeting to the endothelium during normothermic machine perfusion of human kidneys. Sci Transl Med. 2017;9:eaam6764.
145. Hosgood SA, Hoff M, Nicholson ML. Treatment of transplant kidneys during machine perfusion. Transpl Int. 2021;34:224–32.
146. DiRito JR, Hosgood SA, Tietjen GT, Nicholson ML. The future of marginal kidney repair in the context of normothermic machine perfusion. Am J Transpl. 2018;18:2400–8.
147. DiRito JR, Hosgood SA, Reschke M, Albert C, Bracaglia LG, Ferdinand JR, Stewart BJ, Edwards CM, Vaish AG, Thiru S, et al. Lysis of cold-storage-induced microvascular obstructions for ex vivo revitalization of marginal human kidneys. Am J Transpl. 2021;21:161–73.
148. Feng S, Zhou L, Lin D, Zhao J, Guan Q, Zheng B, Wang K, Li H, Chen R, Zeng H, et al. Assessment of treatment efficacy using surface-enhanced Raman spectroscopy analysis of urine in rats with kidney transplantation or kidney disease. Clin Exp Nephrol. 2019;23:880–9.
149. Chi J, Ma Y, Weng FL, Thiessen-Philbrook H, Parikh CR, Du H. Surface-enhanced Raman scattering analysis of urine from deceased donors as a prognostic

tool for kidney transplant outcome. J Biophotonics. 2017;10:1743–55.
150. Chen Y, Han X, Sun Y, He X, Xue D. A circulating exosomal microRNA panel as a novel biomarker for monitoring post-transplant renal graft function. J Cell Mol Med. 2020;24:12154–63.
151. Kasiske BL, Zeier MG, Chapman JR, Craig JC, Ekberg H, Garvey CA, Green MD, Jha V, Josephson MA, Kiberd BA, et al. KDIGO clinical practice guideline for the care of kidney transplant recipients: a summary. Kidney Int. 2010;77:299–311.
152. Eckardt K-U, Kasiske BL, Zeier MG. Special issue: KDIGO clinical practice guideline for the care of kidney transplant recipients. Am J Transplant. 2009;9:S1–S155.
153. Josephson MA. Monitoring and managing graft health in the kidney transplant recipient. Clin J Am Soc Nephrol. 2011;6:1774–80.
154. Jenssen T, Hartmann A. Post-transplant diabetes mellitus in patients with solid organ transplants. Nat Rev Endocrinol. 2019;15(3):172–88. https://doi.org/10.1038/s41574-018-0137-7.
155. Edgar L, Pu T, Porter B, Aziz JM, La Pointe C, Asthana A, et al. Regenerative medicine, organ bioengineering and transplantation. Br J Surg. 2020;107(7):793–800. https://doi.org/10.1002/bjs.11686.
156. Kennedy JE. High-intensity focused ultrasound in the treatment of solid tumours. Nat Rev Cancer. 2005;5:321.
157. You Y, Wang Z, Ran H, Zheng Y, Wang D, Xu J, Wang Z, Chen Y, Li P. Nanoscale. 2016;8:4324.
158. Chen Y, Chen H, Sun Y, Zheng Y, Zeng D, Li F, Zhang S, Wang X, Zhang K, Ma M, He Q, Zhang L, Shi J. Angew Chem Int Ed Engl. 2011;50:12505.
159. Wang X, Chen H, Chen Y, Ma M, Zhang K, Li F, Zheng Y, Zeng D, Wang Q, Shi J. Adv Mater. 2012;24:785.
160. Wang X, Chen H, Zhang K, Ma M, Li F, Zeng D, Zheng S, Chen Y, Jiang L, Xu H, Shi J. Small. 2014;10:1403.
161. Niu D, Wang X, Li Y, Zheng Y, Li F, Chen H, Gu J, Zhao W, Shi J. Adv Mater. 2013;25:2686.
162. Yildirim A, Chattaraj R, Blum NT, Shi D, Kumar K, Goodwin AP. Adv Healthc Mater. 2017;6:1700514.
163. Zhang K, Chen H, Li F, Wang Q, Zheng S, Xu H, Ma M, Jia X, Chen Y, Mou J, Wang X, Shi J. Biomaterials. 2014;35:5875.
164. Devarakonda SB, Myers MR, Lanier M, Dumoulin C, Banerjee RK. Nano Lett. 2017;17:2532.
165. Feng G, Hao L, Xu C, Ran H, Zheng Y, Li P, Cao Y, Wang Q, Xia J, Wang Z. Int J Nanomedicine. 2017;12:4647.
166. Lim EK, Jang E, Lee K, Haam S, Huh YM. Pharmaceutics. 2013;5:294.
167. Discher DE, Eisenberg A. Science. 2002;297:967.
168. de Assis DN, Mosqueira VC, Vilela JM, Andrade MS, Cardoso VN. Int J Pharm. 2008;349:152.
169. Yang T, Choi MK, Cui FD, Kim JS, Chung SJ, Shim CK, Kim DD. J Control Release. 2007;120:169.
170. Chen W, Meng F, Cheng R, Zhong Z. J Control Release. 2010;142:40.
171. Adams ML, Lavasanifar A, Kwon GS. J Pharm Sci. 2003;92:1343.
172. Husseini GA, Pitt WG. J Nanosci Nanotechnol. 2008;8:2205.
173. Husseini GA, Pitt WG. J Pharm Sci. 2009;98:795.
174. Husseini GA, Pitt WG, Martins AM. Colloids Surf B Biointerfaces. 2014;123:364.
175. Eggen S, Fagerland SM, Morch Y, Hansen R, Sovik K, Berg S, Furu H, Bohn AD, Lilledahl MB, Angelsen A, Angelsen B, de Lange Davies C. J Control Release. 2014;187:39.
176. Papa AL, Korin N, Kanapathipillai M, Mammoto A, Mammoto T, Jiang A, Mannix R, Uzun O, Johnson C, Bhatta D, Cuneo G, Ingber DE. Biomaterials. 2017;139:187.
177. Tanbour R, Martins AM, Pitt WG, Husseini GA. Curr Pharm Des. 2016;22:2796.
178. Wang TY, Choe JW, Pu K, Devulapally R, Bachawal S, Machtaler S, Chowdhury SM, Luong R, Tian L, KhuriYakub B, Rao J, Paulmurugan R, Willmann JK. J Control Release. 2015;203:99.
179. Yang XZ, Dou S, Sun TM, Mao CQ, Wang HX, Wang J. J Control Release. 2011;156:203.
180. Li TS, Yawata T, Honke K. Eur J Pharm Sci. 2014;52:48.
181. Yin T, Wang P, Li J, Zheng R, Zheng B, Cheng D, Li R, Lai J, Shuai X. Biomaterials. 2013;34:4532.
182. Yin T, Wang P, Li J, Wang Y, Zheng B, Zheng R, Cheng D, Shuai X. Biomaterials. 2014;35:5932.
183. Yang H, Deng L, Li T, Shen X, Yan J, Zuo L, Wu C, Liu Y. J Biomed Nanotechnol. 2015;11:2124.
184. Marxer EE, Brussler J, Becker A, Schummelfeder J, Schubert R, Nimsky C, Bakowsky U. Eur J Pharm Biopharm. 2011;77:430.
185. Lattin JR, Pitt WG, Belnap DM, Husseini GA. Ultrasound Med Biol. 2012;38:2163.
186. Liu Y, Zhu YH, Mao CQ, Dou S, Shen S, Tan ZB, Wang J. J Control Release. 2014;192:114.
187. Xu CF, Liu Y, Shen S, Zhu YH, Wang J. Biomaterials. 2015;51:1.
188. Chen Y, Chen H, Shi J. Adv Mater. 2013;25:3144.
189. Chen Y, Meng Q, Wu M, Wang S, Xu P, Chen H, Li Y, Zhang L, Wang L, Shi J. J Am Chem Soc. 2014;136:16326.
190. Paris JL, Cabanas MV, Manzano M, Vallet Regi M. ACS Nano. 2015;9:11023.
191. Ma M, Xu H, Chen H, Jia X, Zhang K, Wang Q, Zheng S, Wu R, Yao M, Cai X, Li F, Shi J. Adv Mater. 2014;26:7378.
192. Kang B, Zheng MB, Song P, Chen AP, Wei JW, Xu JJ, Shi Y, Chen HY. Small. 2017;13:1604228.
193. Moon GD, Choi SW, Cai X, Li W, Cho EC, Jeong U, Wang LV, Xia Y. J Am Chem Soc. 2011;133:4762.
194. Konan YN, Gurny R, Allemann E. J Photochem Photobiol B. 2002;66:89.
195. Vargas A, Pegaz B, Debefve E, Konan-Kouakou Y, Lange N, Ballini JP, van den Bergh H, Gurny R, Delie F. Int J Pharm. 2004;286:131.

196. Osminkina LA, Kudryavtsev AA, Zinovyev SV, Sviridov AP, Kargina YV, Tamarov KP, Nikiforov VN, Ivanov AV, Vasilyev AN, Timoshenko VY. Bull Exp Biol Med. 2016;161:296.
197. Huang P, Qian X, Chen Y, Yu L, Lin H, Wang L, Zhu Y, Shi J. J Am Chem Soc. 2017;139:1275.
198. Chen YW, Liu TY, Chang PH, Hsu PH, Liu HL, Lin HC, Chen SY. Nanoscale. 2016;8:12648.
199. Sviridov AP, Osminkina LA, Kharin AY, Gongalsky MB, Kargina JV, Kudryavtsev AA, Bezsudnova YI, Perova TS, Geloen A, Lysenko V, Timoshenko VY. Nanotechnology. 2017;28:105102.
200. Osminkina LA, Nikolaev AL, Sviridov AP, Andronova NV, Tamarov KP, Gongalsky MB, Kudryavtsev AA, Treshalina HM, Timoshenko VY. Nanoscale Res Lett. 2014;9:463.
201. Allain SYP, Aoued S, Quintin-Poulon A, Goune M, Danoix F, Hell JC, Bouzat M, Soler M, Geandier G. Mater Sci Eng A. 2018;11:245.
202. Mano SS, Kanehira K, Sonezaki S, Taniguchi A. Int J Mol Sci. 2012;13:3703.
203. Yamaguchi S, Kobayashi H, Narita T, Kanehira K, Sonezaki S, Kudo N, Kubota Y, Terasaka S, Houkin K. Ultrason Sonochem. 2011;18:1197.
204. Ozawa K, Emori M, Yamamoto S, Yukawa R, Yamamoto S, Hobara R, Fujikawa K, Sakama H, Matsuda I. J Phys Chem Lett. 1953;2014:5.
205. Deepagan VG, You DG, Um W, Ko H, Kwon S, Choi KY, Yi GR, Lee JY, Lee DS, Kim K, Kwon IC, Park JH. Nano Lett. 2016;16:8b00495.
206. Shanei A, Sazgarnia A, Tayyebi Meibodi N, Eshghi H, Hassanzadeh-Khayyat M, Esmaily H, Attaran Kakhki N. Iran J Basic Med Sci. 2012;15:759.
207. Beik J, Shiran MB, Abed Z, Shiri I, Ghadimi-Daresajini A, Farkhondeh F, Ghaznavi H, Shakeri-Zadeh A. Med Phys. 2018;45:4306.
208. Pan X, Bai L, Wang H, Wu Q, Wang H, Liu S, Xu B, Shi X, Liu H. Adv Mater. 2018;30:e1800180.
209. Yu T, Zhang Y, He H, Zhou S, Liu Y, Huang P. Mol Pharm. 2011;8:1408.
210. Chen H, Zhou X, Gao Y, Zheng B, Tang F, Huang J. Drug Discov Today. 2014;19:502.
211. Zhang L, Gu FX, Chan JM, Wang AZ, Langer RS, Farokhzad OC. Clin Pharmacol Ther. 2008;83:761.
212. Cattaneo AG, Gornati R, Sabbioni E, Chiriva-Internati M, Cobos E, Jenkins MR, Bernardini G. J Appl Toxicol. 2010;30:730.
213. Maggini L, Cabrera I, Ruiz-Carretero A, Prasetyanto EA, Robinet E, De Cola L. Nanoscale. 2016;8:7240.
214. Prasetyanto EA, Bertucci A, Septiadi D, Corradini R, Castro-Hartmann P, De Cola L. Angew Chem Int Ed Engl. 2016;55:3323.
215. Huang P, Chen Y, Lin H, Yu L, Zhang L, Wang L, Zhu Y, Shi J. Biomaterials. 2017;125:23.
216. Yu L, Chen Y, Wu M, Cai X, Yao H, Zhang L, Chen H, Shi J. J Am Chem Soc. 2016;138:9881.
217. Pohaku Mitchell KK, Liberman A, Kummel AC, Trogler WC. J Am Chem Soc. 2012;134:13997.
218. Kurapati R, Backes C, Menard-Moyon C, Coleman JN, Bianco A. Angew Chem Int Ed Engl. 2016;55:5506.
219. Kotchey GP, Allen BL, Vedala H, Yanamala N, Kapralov AA, Tyurina YY, Klein-Seetharaman J, Kagan VE, Star A. ACS Nano. 2011;5:2098.
220. Zhao Y, Xing G, Chai Z. Nat Nanotechnol. 2008;3:191.
221. Pan Y, Neuss S, Leifert A, Fischler M, Wen F, Simon U, Schmid G, Brandau W, Jahnen-Dechent W. Small. 1941;2007:3.
222. Zhu MT, Wang B, Wang Y, Yuan L, Wang HJ, Wang M, Ouyang H, Chai ZF, Feng WY, Zhao YL. Toxicol Lett. 2011;203:162.
223. Yan L, Zhao F, Li S, Hu Z, Zhao Y. Nanoscale. 2011;3:362.
224. Meng H, Xing G, Sun B, Zhao F, Lei H, Li W, Song Y, Chen Z, Yuan H, Wang X, Long J, Chen C, Liang X, Zhang N, Chai Z, Zhao Y. ACS Nano. 2010;4:2773.
225. Zhao F, Zhao Y, Liu Y, Chang X, Chen C, Zhao Y. Small. 2011;7:1322.

8 CT Nanoimaging

8.1 Introduction

Nano-computed tomography (nano-CT) is the most recent and emerging high-resolution imaging modality, and now it has been getting micro-CT technology. The focal spot size in the X-ray tube is reduced to less than 400 nm. These imaging protocols, with specialized detectors and examination methods, achieve superior spatial resolution and are enabled by micro-CT systems. Nano-CT imaging can be used to identify and characterize atherosclerotic plaques, which may contain intraplaque hemorrhage and calcifications, as observed in research using a murine model of atherosclerosis (ApoE(−/−)/LDLR(−/−)). Nano-CT applications can be done for the imaging of the cerebral microcirculation, lung, and trabecular microstructure in comparison to micro-CT imaging.

Bio- and medical imaging are the main components for the very early detection and diagnosis of diseases. Early detection, diagnosis, and treatment are important for diseases like cancer and other diseases.

Nanoparticles (NPs) have an advantage over CT contrast agents due to their longer blood-pool times for cell tracking and targeted imaging. This nanoparticle CT imaging has an advantage due to the contrast reactions for these iodinated contrast agents. Micelles and lipoproteins, a micelle class of NPs, are used as CT contrast agents.

CT is based on the principle of differential tissue X-ray attenuation and thickness by which images are acquired and reconstructed in three-dimensional images. CT advantage is faster speed, less cost, more efficiency, and higher spatial resolution for clinical imaging; thus, CT is more commonly used as compared to conventional X-rays despite radiation.

8.2 Nanoparticle Contrast

NPs are now important contrast agents that have a better diagnosis, and this is compatible with various tissue characterizations. NPs have more biological interactions, various clearance routes, and the use of a coating of NPs for various functions.

NPs are small particles, typically ranging in size from 1 to 100 nm [1–4]. Their common shape is a sphere but different shapes, like rods, cages, or stars are also there [5–7] (Fig. 8.1). NPs are widely used in various clinical applications, a field known as nanomedicine [8]. Different therapeutic applications like the Doxil use of the doxorubicin and FDA-approved agents for the treatment of various cancers [9]. Other therapeu-

S. Shaikh, *Nanoimaging - Future of Precision Medicine*,
https://doi.org/10.1007/978-981-97-7292-6_8

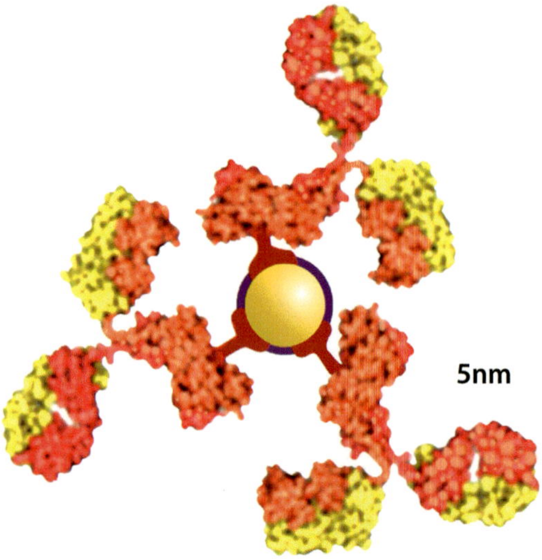

Fig. 8.1 Nanosphere for use in various applications

tic applications for NPs the RNA delivery or thermal ablation [10, 11]. Different NPs are used as tracers in various contrast agents, with modalities like PET, magnetic resonance imaging (MRI), photoacoustics, and fluorescence imaging. The use of these NPs as contrast agents in computed tomography (CT) is well-established [12, 13]. Common contrast agents are iodine [14], gold [15], or bismuth [16], bromine [17], tantalum [18], platinum [19], ytterbium [20], yttrium [21], gadolinium [22, 23], tungsten [1], and others have been reported. The core is coated with soluble and biocompatible polymer, lipid, protein, silica, or other compounds of similar characteristics. The coating is used for targeting different moieties like various antibodies, proteins, peptides, aptamers, etc. [24–27]. Different imaging techniques, like fluorophores, will be used for fluorescent imaging or gadolinium chelates for MR imaging when the nanoparticle is loaded [15, 28]. “Theranostic” nanoparticle applications can be added by coating the various NPs with the diagnosis and therapeutics [29, 30]. Nanoparticle contrast agents for CT are emulsions, liposomes, micelles, lipoproteins, polymeric NPs, solid metal NPs, and many more. Micelles and lipoproteins form good CT contrast agents [15, 31–35]. NPs will have circulation half-lives longer than 15 h, in comparison with the iodinated contrast agents [36]. CT contrast agents will have more circulation time [37] and more injections are required. These NPs can track the cells in vivo [38]. These targeted nanoparticle CT contrast agents can detect different cell types in tissues, and one of the important uses is the atherosclerotic plaque evaluation [31]. NPs can be used in spectral CT for the detection of exogenous contrast agents [39]. NP’s role is multifunctional, and different imaging modalities are used [15]; also, there is a therapeutic impact on the CT contrast. Conventional iodinated contrast media will result in renal insufficiency or allergic responses [40]. Micelle and lipoprotein-based nanoparticle CT contrast agents, liposomes, emulsions, micelles, or lipoproteins [2, 3, 33, 37, 41–43], solid metal, metal alloys, or metal salts [16, 19, 36] or a combination of any of these [15, 31, 35]. Lipid-based structures can be used for various biomedical applications due to their compatibility [12]. Liposomes can be loaded in the core or encapsulated in a lipid coating, making the core itself functional. Micelles and lipoproteins are used as CT contrast agents as they contain lipids and also other elements like iodine [33], coated metal/metal salt, gold, or bismuth sulfide [15, 35]. A mixture of the lipid with iodinated contrast can achieve better payloads and concentrations, leading to better CT contrast [37]. This will be more compatible with any toxic effects. Liposomal CT contrast agents have been developed for many years [44, 45], and this is formed by the hydration of lipids or other organic solutions of lipids, which will have an iodinated contrast agent [37, 46]. Nanoparticle CT contrast agents may contain platinum [19], gadolinium [5, 21], ytterbium [6], yttrium [21], and tantalum [7–9, 18]. This despite the least toxicity by gadolinium, which has significant toxicity [11]. Gadolinium has multifunctional NPs that are capable of producing contrast for CT, MRI, and fluorescence [5, 21]. These NPs are similar to CT contrast properties, like the in vivo imaging capabilities, efficacy, pharmacokinetics, biodistribution, and physiochemical properties of all these agents [7–9]. Tantalum is a cheap and common element and is being used in various trials [47].

8.3 Targeted Nanoparticles for CT Imaging

Targeted imaging along the CT will represent the various challenges due to its low sensitivity for contrast media. Krause has defined the minimum detectable signal change for the CT to be 30 HU [4]. The attenuation rate of gold is 5.1 HU 25–120/mM, and this is the minimum detectable concentration difference between target and background of 5.9 mM. Despite the variations in the properties of the CT contrast media, the concentration of any other agent is closer to this value. Achieving the accumulation of these concentrations at the target sites in the contrast media is challenging. Despite the high expression of CT contrast media on various targets, targeted imaging is considered feasible with CT [15]. In certain clinical scenarios like atherosclerosis in the coronary arteries, CT is a better modality as compared to the alternative imaging methods (MRI, PET). With this, the molecular imaging of the evaluation of the atherosclerosis by CT will be the best modality for imaging. CT contrast agents can also be targeted for the imaging of various thrombi and are helpful for the evaluation of heart attacks, strokes, and various cardiovascular events. Thrombi are made up of densely packed fibrin, which will be helpful for the CT evaluation by the fibrin targeting. The use of the nanoemulsions when combined with the perfluorooctyl bromide or an iodinated oil will be modified and synthesized when attached to the surface. These studies were done by using the clots that are formed from human plasma in vitro and then incubated in the targeted NPs. The targeting where incubation along with the excess of free anti-fibrin antibodies will cause a significant increase in the contrast-to-noise ratio in the clot [48]. A similar pattern of targeting can be pursued for performing in vitro imaging of clots, and the example is that the micelle-based targeted nanoparticle was reported by Kinsella et al. [35] Bismuth sulfide NPs are made by a combination of the sulfur to bismuth acetate and oleic acid in boiling octadecene, and this has resulted in the hydrophobically coated (oleic acid) 10 nm core NPs. These NPs are being made biocompatible by mixing the aPEG-phospholipids and also forming a lipid film. These aPEG-phospholipids are being used with maleimide and amines to preferentially target breast cancer cells [21]. Various in vitro assays and biodistribution analyses have shown that the targeted NPs are being accumulated more in the cancer cells as compared to the non-targeted NPs. Comparison of CT images acquired before and after 24 h post-injection of LyP-1 targeted NPs showed a significant amount of contrast in the tumors. Lipoproteins are being considered as the form of micelles. Cormode et al. studies showed that the gold core high-density lipoprotein nanoparticle when targeted to the various macrophages in atherosclerotic plaque, which will be accumulated in the aorta of a mouse model, can also be detected with micro-CT imaging [32]. Hyafil et al. also targeted different macrophages having iodinated polymeric NPs [31]. NPs used in spectral CT, as reported by Schlomka et al., demonstrated the development of a newer CT scanner called spectral or multicolor CT, developed by Philips [39]. This CT scanner can differentiate materials in the field of view, such as gold, iodine, bismuth, ytterbium, gadolinium, calcified tissue, and soft tissue. The differences are depicted using color coding, hence the name multicolor CT [20, 27, 31]. The important difference between this scanner and a conventional CT scanner is based on the detectors [39]. The spectral CT scanner has detectors that are sensor-based and made up of cadmium telluride, and this allows the better detection of the energy of incident X-rays. When these X-rays hit the detector, a current pulse is generated. Thus, the magnitude of the current is directly proportional to the X-ray energy, and these pulses are recorded depending on their magnitude. These channels are freely adjustable and are capable of altering the K-edges of the contrast agents used in each experiment. One of the classical examples is that there are different energy levels from 25 to 34, 34 to 51, 51 to 80, 80 to 91, 91 to 110, and 110 to 130 keV when imaging the iodine and gold-based contrast agents in the same experiment. The first and third energy boundaries match with the K-edges of iodine and gold, which will facilitate the detection of these elements.

Despite recent advancements in multidetector CT for improved temporal resolution, the maximum spatial resolution remains in the range of 240–600 μm. The in vivo scanners are also called the high-resolution peripheral quantitative CT scanners, that have a spatial resolution which is between the range between 105 and 55 μm at 10% MTF, allowing visualization of the finer structures of peripheral bone [49]. Still, histopathology is the gold standard for the visualization and quantification of structures with a size of less than 200 μm. Micro-CT is well-established and complementary to the histological examination, which is a non-destructive three-dimensional imaging technique on the microscopic level. There is no standard definition of the term "micro-CT" [50]. There is the suggestion of a spatial resolution of at least 100 μm. The CT scanners that have a higher spatial resolution are known as micro-CT scanners. The isotropic voxel geometry has an average spatial resolution in the range of 5–5 μm, which can be quantified by histometry. This technique was used in the evaluation of the morphometry of compact and spongy bones [51]. The development of the newer contrast enhancement techniques will be used for the expansion of the various newer possible applications. These intravascular polymerizing contrast agents will be used for better accuracy and quantification for the identification of any vascular changes [52]. In vivo micro-CT is being used as an important functional diagnostic method for the evaluation of pulmonary and cardiac performances in different parameters in smaller animal models [53, 54]. One of the important applications is micro-CT, seen in both in vivo and ex vivo, which has limited spatial resolution. This is important for the better visualization of smaller vessels, atherosclerotic plaques, and cellular lacunae in calcified tissues including the bones. Nano-CT systems are different from micro-CT systems based on the spatial resolution, tube voltage, and tube power. By altering the focal spot size and suitable detectors and different examination protocols like geometric enlargement, angular scanning, and noise reduction by repetitive image acquisition, these systems will be in the micrometer range, and this is how the difference between the CT systems can be identified and differentiated.

8.4 Basics of Nano-CT Technology

X-ray generation: In the high-resolution CT of the submicrometric range of synchrotron-based micro-computed tomography (synMCT) [55, 56]. This is achievable due to the high photon density and the use of monoenergetic X-ray radiation; the resulting image quality is the gold standard due to the very high signal-to-noise ratio and without beam-hardening artifacts. The nano-CT is a high-resolution device, which results in good compact individual machines having standalone technology. In this technology, the X-ray tubes do not have a uniform vacuum. A vacuum of 6–9 × 10^{-4} Pa is created using a suction pump to maintain stability. This source is known as the "open pumped type X-ray source." Here the coil of the cathode filament is made of lanthanum hexaboride (LaB_6). Thus, two horizontally and vertically condenser/objective electron lenses are connected and are focusing on the emitted electrons. The anode is made up of the tungsten-vaporized beryllium glass plate in which the X-ray radiation is generated at various acceleration voltages of 20–80 kVp with the tube current in the range of 100–200 μA. Without collimation, the generated beam is ineffective. The aperture angle of the radiation is approximately 170°. The divergence of the radiation in horizontal and vertical patterns will cause the coverage of the detector, which will have the flat detector in the cone beam CT. The focal spot sizes will be in the range between <400 and 600 nm, which can be targeted with the different spatial resolution and magnification having the voxel side, which will be in the length in the submicrometer range with the smallest possible focus (<400 nm). In comparison with the X-ray tubes, the classic reflection of the direct beam tubes, which will be in the micro-CT, and the transmission tubes, which will have the X-ray window and focal spot, and this focal spot will be smaller as compared to the various conventional micro-CT, which will have the better spatial resolution in the submicrometer range. The thermal energy resulting from the various X-ray generations means that the focal spot cannot be deflected, which results in the tube current being lower than that in micro-CT. This will also be different between the focal spots of the different sizes. Due to this reason, the photon density is lower in nano-CT as compared to micro-CT. Image noise will be bigger where the focus is the smallest. The greater enhancement of the detector signal may be due to the lower space-charge effect in the X-ray tube when the voltage is lower, which causes increased noise.

As we have seen, contrast is a crucial aspect of bio- and medical imaging, enabling early detection and diagnosis of diseases. With the development of the nanocontrast agents, this problem of better soft tissue resolution between the various tissues will be more precise [57, 58]. Available contrast agents have many limitations for their applications [59, 60]. Different reasons for the lesser sensitivities are shorter blood half-life, nonspecific biodistribution of the contrast, faster clearance, renal toxicity, and poor contrast in fatty patients [61].

NPs in the nano-CT are the novel contrast agents that are being used for non-invasive diagnosis, and the compatibility of the various contrast agents is that (a) the surface can be used for one or more target molecules having different densities; (b) plasma circulation time can be altered for their physicochemical properties; and (c) various contrast agents and drugs can be aligned for their surfaces [62–64].

8.5 Biological Interactions and Clearance Routes

The different biological interactions where there can be various accumulations or interference of nanomaterials that are being used for various diagnostic applications. Issue of the slow renal clearance and their nonspecific accumulation in different systems like the mononuclear phagocyte system (MPS), which can be seen in the liver, spleen, and lymph nodes [65, 66]. This is based on various factors like the size, shape, and surface properties [67]. It is usually presumed that the larger particles will have rapid clearance as compared to the smaller particles. NPs larger than 200 nm are sequestered by MPS in the spleen, while NPs smaller than 10 nm are found in the reticuloendothelial system and liver [68, 69]. Most of these are cleared by renal route, preferably the NPs smaller than 5.5 nm [70, 71]. Accumulation in the RES is influenced by the size and surface properties of the particles, as well as improved absorption and opsonization [72, 73]. The nonspecific binding and uptake, selective binding on the cell surface receptors, along with better elimination from the body with minimal or without toxicity of NPs, is an important concern. The biodistribution pattern and the partial clearance will also lead to various risks, acute or chronic toxicity, immunotoxicity, and carcinogenicity due to various biological interactions that will be happening in the body. Some of these will be broken down into highly toxic components, and they are not considered safe [74–76]. Rapid biodistribution and complete clearance from the body enhance biocompatibility and make biodegradable, targeted contrast agents more useful for clinical applications [77].

8.6 Coating and Targeting

The surface of the various NPs will also determine their fate and their interaction in the clinical application. Coating can be used effectively for the differentiation of the NPs from MPS for better protein adsorption [78]. These coatings can also cause the neutral and hydrophilic surface, which will be resistant to the aggregation and agglomeration by the contact of the proteins. Here the challenge is that the surface coating material needs to be made that will have better colloidal stability and biocompatibility under various physiological conditions. This stability of NPs in CT should also have the stability with better biocompatibility of NPs and the interactions in the different biological environments [64, 67,

76]. In the in vivo applications, different classes of polymeric coating materials are being used like polyethylene glycol (PEG), poly L-lysine (PLL), polyvinyl alcohol (PVA), polyacrylic acid (PAA), alginate, dextran, chitosan, and starch [67, 79]. PEG, one of the most common coating agents, is inexpensive, versatile, and FDA-approved and is being used in many preclinical studies for medical imaging. This addition of PEG to the NP surface will reduce the proteins and small-molecule interactions and simultaneously increase the permeability and retention (EPR) effects and solubility. The controllable bioconjugation of other molecules, such as antibodies, small peptides or molecules, lectins, aptamers, and engineered proteins; proper selection of surface coating agents is also an important concept [64, 80]. The CT NP-based contrast agents are valuable and have potentially transformative tools and many imaging modalities such as PET, MRI, ultrasound (US), CT, and single-photon emission computed tomography (SPECT). Ryvolova et al. [81] and Michael Farle [82] focused on molecular imaging applying NPs. CT is one of the important diagnostic modalities in medical imaging and has various advantages like cost, efficiency, and availability. Due to the different atomic numbers, electron density, and X-ray attenuation, the coefficient of different tissues in the body is different, and here the CT can distinguish between different tissues for the production of the images for different body structures and tissues. The concept of the mass attenuation coefficient is to determine the loss of X-ray photon intensity due to different interactions with matter. The concept here is that the denser tissues and electron-rich elements will absorb the higher amounts of X-ray beam, and vice versa [83]. CT images are seen in the grayscale, seen as the coefficient attenuation in the different tissues of the body [84, 85]. Thus, to differentiate the surrounding tissues between the normal and abnormal tissues and also for the fat or muscle, the mass attenuation coefficients are being used [57]. Available intravenously injected iodine-based or oral barium sulfate suspensions, which are being used as contrast agents, are seen as having different blood half-life, nonspecific biodistribution, fast clearance, slight renal toxicity, and poor contrast in fat patients [57, 86, 87]. Now, with the availability of nanoparticulate contrast agents for CT, many limitations are being overcome. This is possible due to the higher payload of contrast, which will generate the increased absorption of X-ray as compared to iodine-based contrast agents, which will be exposed to the lesser doses of X-ray. Based on the different pharmacokinetic properties, these agents will provide a better and wider range for perfusion imaging [88, 89]. Different types of NPs can be used as the better novel contrast-enhancing agents, which will include different emulsions, liposomes, micelles, lipoproteins, polymeric NPs, and gold NPs for several types of soft tissue. Various modifications and coatings of NPs as contrast agents are seen as important components of the biological media and biocompatibility [84, 90, 91]. Nanomaterial due to its unique properties of long-circulating half-life, simple surface modification, and capability of passive targeting. Many nanoparticulate CT contrast agents will be assessed in vivo applications of these nanomaterials. A novel Yb-based nanoparticulate agent shows better contrast efficacy compared to routine CT contrast agents. In the study conducted by Yanlan et al., they integrated both Ba and Yb [86]. They demonstrated that the Ba and Yb have higher contrast as compared to the iodinated agents, having prolonged in vivo circulation time with very low toxicity. Myoung et al. used bioinert tantalum oxide NPs as contrast agents [87]. Studies have shown that these NPs perform better in in vivo X-ray CT angiography and bimodal image-guided lymph node mapping.

Now, with recent advances, the use of gold nanoparticles (GNPs) as contrast agents in the CT is becoming important [92–94]. Thus, GNPs, with their better biocompatibility and unique physical and chemical properties [95–100], are more attractive as contrast agents for various imaging modalities, including CT [101–107]. The best example is the gold particle having a higher atomic number as compared to the stronger X-ray attenuation, making it the best GNP ideal contrast agent for CT as an important modality of biomedical imaging, especially in

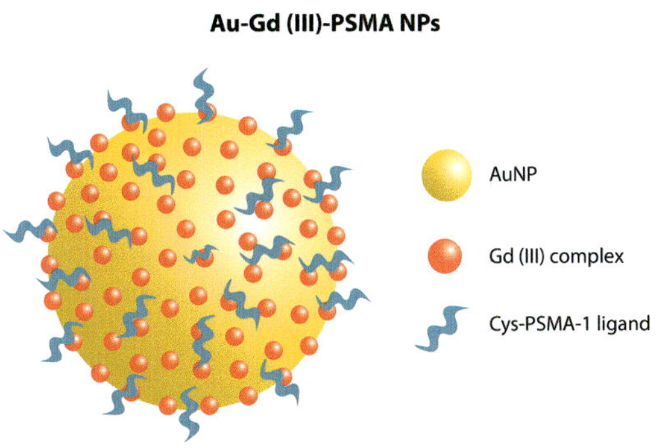

Fig. 8.2 Gold nanoparticle with Gadolinium and PSMA complex

cancers like prostatic cancers, and can be used with PSMA ligand NP (Fig. 8.2). CT has higher temporal and spatial resolutions and a better role in terms of availability, efficiency, and costs. With these properties, the combination of GNPs and CT imaging in non-invasive cell tracking and also for cell therapy can be used. CT3 technology for non-invasive monitoring of mesenchymal stem cells in the transplanted rat model for depression [93], for cancer immunotherapy treatment with cancer-specific T cells [92]. In this process, the optimum balance between the therapeutic capabilities and abilities of the injected cells along with the maximum cell loading is achieved for the sufficient contrast of the cells by CT [94].

8.7 Conclusions

Nano-CT has become an important modality for addressing the limitations of previous CT technologies. The excellent resolution for the evaluation of the visualization of soft cellular lacunae in calcified tissues and the information regarding the interaction between cells and an extracellular matrix in the three-dimensional space. This, along with the development and research for the new contrast agents, will be useful for the differentiation of cells and the extracellular matrix in various soft tissues, especially those that have a lower contrast range. The differentiation of different types of plaques, like atherosclerotic plaques, can be evaluated due to the significantly improved resolution of the nano-CT.

Improved blood-pool agents are being developed, along with the use of new elements like gold, bismuth, and tantalum. Targeting and cell tracking agents are now possible. Another range of the agents are long-circulating iodine-loaded liposomes, and small, really excretable gold or tantalum oxide NPs are under research. These newer elements will have better formulations, which can be useful in compromised renal function patients and those who are hypersensitive to the iodine contrast agents.

The different pharmacokinetic properties of these smaller molecular agents will provide a wider time window for better perfusion imaging. Targeted agents need to be further validated in animal models and optimized for their physicochemical properties. Evaluation of CT contrast properties should be standardized.

References

1. Yu S-B, Watson AD. Metal-based X-ray contrast media. Chem Rev. 1999;99:2353–77.
2. Violante MR, Dean PB, Fischer HW, Mahoney JA. Particulate contrast media for computed tomographic scanning of the liver. Investig Radiol. 1980;15(6 Suppl):S171–5.
3. Longino MA, Weichert JP, Schwendner SW, Szabo SM, Counsell RE, Glazer GM. Biodistribution of a new lipid-soluble CT contrast agent. Evaluation of cholesteryl iopanoate in the rabbit. Investig Radiol. 1983;18(3):275–8.
4. Krause W. Delivery of diagnostic agents in computed tomography. Adv Drug Deliv Rev. 1999;37:159–73.
5. Jana NR, Gearheart L, Murphy CJ. Wet chemical synthesis of high aspect ratio cylindrical gold nanorods. J Phys Chem B. 2001;105(19):4065–7.
6. Nehl CL, Liao H, Hafner JH. Optical properties of star-shaped gold nanoparticles. Nano Lett. 2006;6(4):683–8.
7. Song KH, Kim C, Cobley CM, Xia Y, Wang LV. Near-infrared gold nanocages as a new class of tracers for photoacoustic sentinel lymph node mapping on a rat model. Nano Lett. 2009;9:183–8.
8. Kim BYS, Rutka JT, Chan WCW. Nanomedicine. N Engl J Med. 2010;363(25):2434–43.
9. Peer D, Karp JM, Hong S, Farokhzad OC, Margalit R, Langer R. Nanocarriers as an emerging platform for cancer therapy. Nat Nanotechnol. 2007;2:751–60.

10. Frank-Kamenetsky M, Grefhorst A, Anderson NN, Racie TS, Bramlage B, Akinc A, Butler D, Charisse K, Dorkin R, Fan Y, Gamba-Vitalo C, Hadwiger P, Jayaraman M, John M, Jayaprakash KN, Maier M, Nechev L, Rajeev KG, Read T, Rohl I, Soutschek J, Tan P, Wong J, Wang G, Zimmermann T, de Fougerolles A, Vornlocher HP, Langer R, Anderson DG, Manoharan M, Koteliansky V, Horton JD, Fitzgerald K. Therapeutic RNAi targeting PCSK9 acutely lowers plasma cholesterol in rodents and LDL cholesterol in nonhuman primates. Proc Natl Acad Sci USA. 2008;105(33):11915–20.
11. von Maltzahn G, Park JH, Agrawal A, Bandaru NK, Das SK, Sailor MJ, Bhatia SN. Computationally guided photothermal tumor therapy using long-circulating gold nanorod antennas. Cancer Res. 2009;69(9):3892–900.
12. Mulder WJM, Strijkers GJ, Van Tilborg GAF, Cormode DP, Fayad ZA, Nicolay K. Nanoparticulate assemblies of amphiphiles and diagnostically active materials for multimodality imaging. Acc Chem Res. 2009;42(7):904–14.
13. Kim D, Park S, Lee JH, Jeong YY, Jon S. Antibiofouling polymer-coated gold nanoparticles as a contrast agent for in vivo X-ray computed tomography imaging. J Am Chem Soc. 2007;129(24):7661–5.
14. Yordanov AT, Lodder AL, Woller EK, Cloninger MJ, Patronas N, Milenic D, Brechbiel MW. Novel iodinated dendritic nanoparticles for computed tomography (CT) imaging. Nano Lett. 2002;2(6):595–9.
15. van Schooneveld MM, Cormode DP, Koole R, van Wijngaarden JT, Calcagno C, Skajaa T, Hilhorst J, Hart DC, Fayad ZA, Mulder WJM, Meijerink A. A fluorescent, paramagnetic and PEGylated gold/silica nanoparticle for MRI, CT and fluorescence imaging. Contrast Media Mol Imaging. 2010;5(4):231–6.
16. Rabin O, Perez JM, Grimm J, Wojtkiewicz G, Weissleder R. An X-ray computed tomography imaging agent based on long-circulating bismuth sulphide nanoparticles. Nat Mater. 2006;5:118–22.
17. Caride VJ, Sostman HD, Twickler J, Zacharis H, Orphanoudakis SC, Jaffe CC. Brominated radiopaque liposomes: contrast agent for computed tomography of liver and spleen a preliminary report. Investig Radiol. 1982;17:381–5.
18. Oh MH, Lee N, Kim H, Park SP, Piao Y, Lee J, Jun SW, Moon WK, Choi SH, Hyeon T. Largescale synthesis of bioinert tantalum oxide nanoparticles for X-ray computed tomography imaging and bimodal image-guided sentinel lymph node mapping. J Am Chem Soc. 2011;133:5508–15.
19. Chou S-W, Shau Y-H, Wu P-C, Yang Y-S, Shieh D-B, Chen C-C. In vitro and in vivo studies of FePt nanoparticles for dual modal CT/MRI molecular imaging. J Am Chem Soc. 2010;132(38):13270–8.
20. Pan D, Schirra CO, Senpan A, Schmieder AH, Stacy AJ, Roessl E, Thran A, Wickline SA, Proska R, Lanza GM. An early investigation of ytterbium nanocolloids for selective and quantitative "multicolor" spectral CT imaging. ACS Nano. 2012;6(4):3364–70.
21. Xing H, Bu W, Zhang S, Zheng X, Li M, Chen F, He Q, Zhou L, Peng W, Hua Y, Shi J. Multifunctional nanoprobes for upconversion fluorescence, MR and CT trimodal imaging. Biomaterials. 2012;33:1079–89.
22. Reginoa CAS, Walbridge S, Bernardo M, Wong KJ, Johnson D, Lonser R, Oldfield EH, Choyke PL, Brechbiel MW. A dual CT-MR dendrimer contrast agent as a surrogate marker for convection-enhanced delivery of intracerebral macromolecular therapeutic agents. Contrast Media Mol Imaging. 2008;3:2–8.
23. Carrascosa P, Capunay C, Deviggiano A, Bettinotti M, Goldsmit A, Tajer C, Carrascosa J, Garcia MJ. Feasibility of 64-slice gadolinium-enhanced cardiac CT for the evaluation of obstructive coronary artery disease. Heart. 2010;96(19):1543–9.
24. Javier DJ, Nitin N, Levy M, Ellington A, Richards-Kortum R. Aptamer-targeted gold nanoparticles as molecular-specific contrast agents for reflectance imaging. Bioconjug Chem. 2008;19(6):1309–12.
25. van Tilborg GAF, Vucic E, Strijkers GJ, Cormode DP, Mani V, Skajaa T, Reutelingsperger CPM, Fayad ZA, Mulder WJM, Nicolay K. Annexin A5-functionalized bimodal nanoparticles for MRI and fluorescence imaging of atherosclerotic plaques. Bioconjug Chem. 2010;21:1794.
26. Nahrendorf M, Jaffer FA, Kelly KA, Sosnovik DE, Aikawa E, Libby P, Weissleder R. Noninvasive vascular cell adhesion molecule-1 imaging identifies inflammatory activation of cells in atherosclerosis. Circulation. 2006;114:1504–11.
27. Pan D, Roessl E, Schlomka J-P, Caruthers SD, Senpan A, Scott MJ, Allen JS, Zhang H, Hu G, Gaffney PJ, Choi ET, Rasche V, Wickline SA, Proksa R, Lanza GM. Computed tomography in color: nanoK-enhanced spectral CT molecular imaging. Angew Chem Int Ed. 2010;49:9635–9.
28. Alric C, Taleb J, Le Duc G, Mandon C, Billotey C, Le Meur-Herland A, Brochard T, Vocanson F, Janier M, Perriat P, Roux S, Tillement O. Gadolinium chelate coated gold nanoparticles as contrast agents for both X-ray computed tomography and magnetic resonance imaging. J Am Chem Soc. 2008;130(18):5908–15.
29. Lee J-S, Green JJ, Love KT, Sunshine J, Langer R, Anderson DG. Gold, poly(beta-amino ester) nanoparticles for small interfering RNA delivery. Nano Lett. 2009;9(6):2402–6.
30. Libutti SK, Paciotti GF, Byrnes AA, Alexander HR, Gannon WE, Walker M, Seidel GD, Yuldasheva N, Tamarkin L. Phase I and pharmacokinetic studies of CYT-6091, a novel PEGylated colloidal gold-rhTNF nanomedicine. Clin Cancer Res. 2010;16(24):6139–49.
31. Cormode DP, Roessl E, Thran A, Skajaa T, Gordon RE, Schlomka JP, Fuster V, Fisher EA, Mulder WJM, Proksa R, Fayad ZA. Atherosclerotic plaque compo-

sition: analysis with multicolor CT and targeted gold nanoparticles. Radiology. 2010;256(3):774–82.
32. Cormode DP, Skajaa T, van Schooneveld MM, Koole R, Jarzyna P, Lobatto ME, Calcagno C, Barazza A, Gordon RE, Zanzonico P, Fisher EA, Fayad ZA, Mulder WJM. Nanocrystal core high-density lipoproteins: a multimodal molecular imaging contrast agent platform. Nano Lett. 2008;8(11):3715–23.
33. Torchilin VP, Frank-Kamenetsky MD, Wolf GL. CT visualization of blood pool in rats by using long-circulating, iodine-containing micelles. Acad Radiol. 1999;6(1):61–5.
34. Trubetskoy VS, Gazelle GS, Wolf GL, Torchilin VP. Block-copolymer of polyethylene glycol and polylysine as a carrier of organic iodine: design of long-circulating particulate contrast medium for X-ray computed tomography. J Drug Targ. 1997;4(6):381–8.
35. Kinsella JM, Jimenez RE, Karmali PP, Rush AM, Kotamraju VR, Gianneschi NC, Ruoslahti E, Stupack D, Sailor MJ. X-ray computed tomography imaging of breast cancer by using targeted peptide-labeled bismuth sulfide nanoparticles. Angew Chem Int Ed. 2011;50(51):12308–11.
36. Cai QY, Kim SH, Choi KS, Kim SY, Byun SJ, Kim KW, Park SH, Juhng SK, Yoon KH. Colloidal gold nanoparticles as a blood-pool contrast agent for X-ray computed tomography in mice. Investig Radiol. 2007;42(12):797–806.
37. Mukundan S, Ghaghada KB, Badea CT, Kao C-Y, Hedlund LW, Provanzale JM, Johnson GA, Chen E, Bellamkonda RV, Annapragada A. A liposomal nanoscale contrast agent for preclinical CT in mice. Am J Roentgenol. 2006;186:300–7.
38. Arifin DR, Long CM, Gilad AA, Alric C, Roux S, Tillement O, Link TW, Arepally A, Bulte JWM. Trimodal gadolinium-gold microcapsules containing pancreatic islet cells restore normoglycemia in diabetic mice and can be tracked by using US, CT, and positive-contrast MR imaging. Radiology. 2011;260(3):790 8.
39. Schlomka JP, Roessl E, Dorscheid R, Dill S, Martens G, Istel T, Baumer C, Herrmann C, Steadman R, Zeitler G, Livne A, Proksa R. Experimental feasibility of multi-energy photoncounting K-edge imaging in pre-clinical computed tomography. Phys Med Biol. 2008;53(15):4031–47.
40. Stacul F, van der Molen AJ, Reimer P, Webb JAW, Thomsen HS, Morcos SK, Almen T, Aspelin P, Bellin MF, Clement O, Heinz-Peer G, Contrast Media Safety Committee of European Society of Urogenital Radiology (ESUR). Contrast induced nephropathy: updated ESUR Contrast Media Safety Committee guidelines. Eur Radiol. 2011;21(12):2527–41.
41. Galper MW, Saung MT, Fuster V, Roessl E, Thran A, Proksa R, Fayad ZA, Cormode DP. Effect of computed tomography scanning parameters on gold nanoparticle and iodine contrast. Investig Radiol. 2012;47(8):475–81.
42. Hill ML, Corbin IR, Levitin RB, Cao WG, Mainprize JG, Yaffe MJ, Zheng G. In vitro assessment of poly-iodinated triglyceride reconstituted low-density lipoprotein: initial steps toward CT molecular imaging. Acad Radiol. 2010;17(11):1359–65.
43. Mattrey RF. Perfluorooctylbromide—a new contrast agent for CT, sonography, and MR imaging. Am J Roentgenol. 1989;152(2):247–52.
44. http://www.marvalbiosciences.com.
45. Ghaghada K. Telephone interview, 16 Aug 2012.
46. Havron A, Seltzer SE, Davis MA, Shulkin P. Radiopaque liposomes—a promising new contrast material for computed-tomography of the spleen. Radiology. 1981;140(2):507–11.
47. Cormode DP, Skajaa T, Fayad ZA, Mulder WJM. Nanotechnology in medical imaging: probe design and applications. Arterioscler Thromb Vasc Biol. 2009;29:992–1000.
48. Oh MH, Lee N, Kim H, Park SP, Piao Y, Lee J, Jun SW, Moon WK, Choi SH, Hyeon T. Large-scale synthesis of bioinert tantalum oxide nanoparticles for X-ray computed tomography imaging and bimodal image-guided sentinel lymph node mapping. J Am Chem Soc. 2011;133:5508–15.
49. Kocijan R, Finzel S, Englbrecht M, et al. Decreased quantity and quality of the periarticular and nonperiarticular bone in patients with rheumatoid arthritis: a cross-sectional HR-pQCT study. J Bone Miner Res. 2014;29:1005–14.
50. Kalender WA. Special applications. In: Kalender WA, editor. Computed tomography fundamentals, system technology, image quality, applications. 3rd ed. Erlangen: Publicis Corp Publ; 2011. p. 272.
51. Engelke K, Karolczak M, Lutz A, et al. Micro-CT. Technology and application for assessing bone structure. Radiologe. 1999;39:203–12.
52. Ortiz MC, Garcia-Sanz A, Bentley MD, et al. Microcomputed tomography of kidneys following chronic bile duct ligation. Kidney Int. 2000;58:1632–40.
53. Badea C, Hedlund LW, Johnson GA. Micro-CT with respiratory and cardiac gating. Med Phys. 2004;31:3324–9.
54. Lee CL, Min H, Befera N, et al. Assessing cardiac injury in mice with dual energy-microCT, 4D-microCT, and microSPECT imaging after partial heart irradiation. Int J Radiat Oncol Biol Phys. 2014;88:686–93.
55. Schneider P, Stauber M, Voide R, et al. Ultrastructural properties in cortical bone vary greatly in two inbred strains of mice as assessed by synchrotron light based micro- and nano-CT. J Bone Miner Res. 2007;22:1557–70.
56. Pacureanu A, Langer M, Boller E, et al. Nanoscale imaging of the bone cell network with synchrotron X-ray tomography: optimization of acquisition setup. Med Phys. 2012;39:2229–38.
57. He W, Ai K, Lu L. Nanoparticulate X-ray CT contrast agents. Sci China Chem. 2015;58:753–60.

58. Cormode DP, Jarzyna PA, Mulder WJ, et al. Modified natural nanoparticles as contrast agents for medical imaging. Adv Drug Deliv Rev. 2010;62:329–38.
59. Du Y, Lai PT, Leung CH, et al. Design of superparamagnetic nanoparticles for magnetic particle imaging (MPI). Int J Mol Sci. 2013;14:18682–710.
60. Lee SH, Kim BH, Na HB, et al. Paramagnetic inorganic nanoparticles as T1 MRI contrast agents. Wires Nanomed Nanobiotechnol. 2014;6:196–209.
61. Razi M, Dehghani A, Beigi F, et al. The peep of nanotechnology in reproductive medicine: a mini-review. Int J Med Lab. 2015;2:1–15.
62. Janib SM, Moses AS, MacKay JA. Imaging and drug delivery using theranostic nanoparticles. Adv Drug Deliv Rev. 2010;62:1052–63.
63. Massoud TF, Gambhir SS. Molecular imaging in living subjects: seeing fundamental biological processes in a new light. Genes Dev. 2003;17:545–80.
64. Wang X, Yang L, Chen Z, et al. Application of nanotechnology in cancer therapy and imaging. CA Cancer J Clin. 2008;58:97–110.
65. Mohammadian F, Pilehvar-Soltanahmadi Y, Mofarrah M, et al. Down regulation of miR-18a, miR-21 and miR-221 genes in gastric cancer cell line by chrysin-loaded PLGA-PEG nanoparticles. Artif Cells Nanomed Biotechnol. 2016;44:1972–8.
66. Sadeghzadeh H, Pilehvar-Soltanahmadi Y, Akbarzadeh A, et al. The effects of nanoencapsulated curcumin-Fe_3O_4 on proliferation and hTERT gene expression in lung cancer cells. Anti Cancer Agents Med Chem. 2017;17:1363–73. https://doi.org/10.2174/1871520617666170213115756.
67. Rosen JE, Yoffe S, Meerasa A, et al. Nanotechnology and diagnostic imaging: new advances in contrast agent technology. J Nanomed Nanotechnol. 2011;2:115.
68. Mohammadian F, Pilehvar-Soltanahmadi Y, Zarghami F, et al. Upregulation of miR-9 and Let-7a by nanoencapsulated chrysin in gastric cancer cells. Artif Cells Nanomed Biotechnol. 2017;45:1201–6.
69. Amirsaadat S, Pilehvar-Soltanahmadi Y, Zarghami F, et al. Silibinin-loaded magnetic nanoparticles inhibit hTERT gene expression and proliferation of lung cancer cells. Artif Cells Nanomed Biotechnol. 2017; https://doi.org/10.1080/21691401.2016.1276922.
70. Mohammadian F, Abhari A, Dariushnejad H, et al. Effects of chrysin-PLGA-PEG nanoparticles on proliferation and gene expression of miRNAs in gastric cancer cell line. Iran J Cancer Prev. 2016;9:e4190.
71. Mohammadian F, Abhari A, Dariushnejad H, et al. Upregulation of Mir-34a in AGS gastric cancer cells by a PLGA-PEG-PLGA chrysin nano formulation. Asian Pac J Cancer Prev. 2015;16:8259–63.
72. Sun C, Lee JS, Zhang M. Magnetic nanoparticles in MR imaging and drug delivery. Adv Drug Deliv Rev. 2008;60:1252–65.
73. Farajzadeh R, Pilehvar-Soltanahmadi Y, Dadashpour M, et al. Nano-encapsulated metformin-curcumin in PLGA/PEG inhibits synergistically growth and hTERT gene expression in human breast cancer cells. Artif Cells Nanomed Biotechnol. 2018; https://doi.org/10.1080/21691401.2017.1347879.
74. Cormode DP, Skajaa T, Fayad ZA, et al. Nanotechnology in medical imaging probe design and applications. Arterioscler Thromb Vasc Biol. 2009;29:992–1000.
75. Lee JH, Park G, Hong GH, et al. Design considerations for targeted optical contrast agents. Quant Imaging Med Surg. 2012;2:266–73.
76. Surendiran A, Sandhiya S, Pradhan S, et al. Novel applications of nanotechnology in medicine. Indian J Med Res. 2009;130:689–701.
77. Ahn S, Jung SY, Lee SJ. Gold nanoparticle contrast agents in advanced X-ray imaging technologies. Molecules. 2013;18:5858–90.
78. Michalet X, Pinaud F, Bentolila L, et al. Quantum dots for live cells, in vivo imaging, and diagnostics. Science. 2005;307:538–44.
79. Rhee I, Hong S, Chang Y. Chitosan-coated ferrite (Fe_3O_4) nanoparticles as a T2 contrast agent for magnetic resonance imaging. J Korean Phys Soc. 2010;56:868–73.
80. Jokerst JV, Lobovkina T, Zare RN, et al. Nanoparticle PEGylation for imaging and therapy. Nanomedicine. 2011;6:715–28.
81. Ryvolova M, Chomoucka J, Drbohlavova J, et al. Modern micro and nanoparticle-based imaging techniques. Sensors. 2012;12:14792–820.
82. Farle M. Imaging techniques: nanoparticle atoms pinpointed. Nature. 2017;542:35–6.
83. Popovtzer R, Agrawal A, Kotov NA, et al. Targeted gold nanoparticles enable molecular CT imaging of cancer. Nano Lett. 2008;8:4593–6.
84. Cormode DP, Naha PC, Fayad ZA. Nanoparticle contrast agents for computed tomography: a focus on micelles. Contrast Media Mol Imaging. 2014;9:37–52.
85. Cho EC, Glaus C, Chen J, et al. Inorganic nanoparticle-based contrast agents for molecular imaging. Trends Mol Med. 2010;16:561–73.
86. Lee N, Choi SH, Hyeon T. Nano-sized CT contrast agents. Adv Mater. 2013;25:2641–60.
87. Rand D, Ortiz V, Liu Y, et al. Nanomaterials for X-ray imaging: gold nanoparticle enhancement of X-ray scatter imaging of hepatocellular carcinoma. Nano Lett. 2011;11:2678–83.
88. Naha PC, Al Zaki A, Hecht E, et al. Dextran coated bismuth–iron oxide nanohybrid contrast agents for computed tomography and magnetic resonance imaging. J Mater Chem B. 2014;2:8239–48.
89. Mukundan S Jr, Ghaghada KB, Badea CT, et al. A liposomal nanoscale contrast agent for preclinical CT in mice. AJR Am J Roentgenol. 2006;186:300–7.
90. Chien CC, Chen HH, Lai SF, et al. Gold nanoparticles as high-resolution X-ray imaging contrast agents for the analysis of tumor-related microvasculature. J Nanobiotechnol. 2012;10:10.
91. Wathen CA, Foje N, Avermaete TV, et al. In vivo X-ray computed tomographic imaging of soft tissue

with native, intravenous, or oral contrast. Sensors. 2013;13:6957–80.
92. Meir R, et al. Nanomedicine for cancer immunotherapy: tracking cancer-specific T-cells in vivo with gold nanoparticles and CT imaging. ACS Nano. 2015;9:6363–72.
93. Betzer O, et al. Nanoparticle-based CT imaging technique for longitudinal and quantitative stem cell tracking within the brain: application in neuropsychiatric disorders. ACS Nano. 2014;8:9274–85.
94. Betzer O, et al. In-vitro optimization of nanoparticle-cell labeling protocols for in-vivo cell tracking applications. Sci Rep. 2015;5:15400.
95. Cho EC, Zhang Q, Xia Y. The effect of sedimentation and diffusion on cellular uptake of gold nanoparticles. Nat Nanotechnol. 2011;6:385–91.
96. Chithrani BD, Ghazani AA, Chan WCW. Determining the size and shape dependence of gold nanoparticle uptake into mammalian cells. Nano Lett. 2006;6:662–8.
97. Chithrani BD, Stewart J, Allen C, Jaffray DA. Intracellular uptake, transport, and processing of nanostructures in cancer cells. Nanomed Nanotechnol Biol Med. 2009;5:118–27.
98. Khan MS, Vishakante GD, Siddaramaiah H. Gold nanoparticles: a paradigm shift in biomedical applications. Adv Colloid Interf Sci. 2013;199–200:44–58.
99. Bao C, et al. A promising road with challenges: where are gold nanoparticles in translational research? Nanomedicine (Lond). 2014;9:2353–70.
100. Liu Y, et al. Human induced pluripotent stem cells for tumor targeted delivery of gold nanorods and enhanced Photothermal therapy. ACS Nano. 2016;10:2375–85.
101. Nam SY, Ricles LM, Suggs LJ, Emelianov SY. In vivo ultrasound and photoacoustic monitoring of mesenchymal stem cells labeled with gold nanotracers. PLoS One. 2012;7:e37267.
102. Bhatnagara P, et al. Imaging of genetically engineered T cells by PET using gold nanoparticles complexed to Cooper-64. Integr Biol (Camb). 2014;5:231–8.
103. Ricles LM, Nam SY, Sokolov K, Emelianov SY, Suggs LJ. Function of mesenchymal stem cells following loading of gold nanotracers. Int J Nanomedicine. 2011;6:407–16.
104. Astolfo A, et al. In vivo visualization of gold-loaded cells in mice using x-ray computed tomography. Nanomedicine. 2013;9:284–92.
105. Long CM, Alric C, Tillement O, Link TW. Trimodal gadolinium-gold pancreatic islet cells restore normoglycemia in diabetic mice and can be tracked by using US, CT, and positive-contrast MR imaging. Radiology. 2011;260:790.
106. Jokerst JV, Thangaraj M, Kempen PJ, Sinclair R, Sanjiv S. Photoacoustic imaging of mesenchymal stem cells in living mice via silica-coated gold nanorods. ACS Nano. 2013;6:5920–30.
107. Ankri R, et al. Gold nanorods based air scanning electron microscopy and diffusion reflection imaging for mapping tumor margins in squamous cell carcinoma. ACS Nano. 2016;10:2349–56.

9 MRI Nanoimaging: Part I

9.1 Introduction

The most recent major and most important advances that happened in nanoscale magnetic resonance imaging (nano-MRI) are due to the concept of personalized medicine (PM). With the advent of the technology, nano-MRI is being developed as an important modality as compared to the existing traditional magnetic resonance [1, 2]. Nano-MRI is seen as potentially important for the evaluation of the various techniques, and these are not seen in various molecular imaging techniques, which has increased the sensitivity of the nano-MRI as an important imaging modality. In this chapter, the work of the nano-MRI will be discussed as MRI has higher sensitivity, which will lead to better diagnosis and result in more precise personalized management [3, 4]. The sensitivity of the magnetic nanoprobes is higher as compared to the routine conventional MRI, and this is based on the nitrogen-vacancy (NV) centers.

T2 contrast agents, also known as negative contrast agents, will significantly decrease the MR signal intensity, resulting in the hypointense signals in T2 weighted sequences, and this will look darker. This will result in more heterogeneity in the magnetic field that will be surrounding the NP, and in this process, the water molecules will diffuse, by the process of diffusion resulting in the induction of the dephasing of the proton magnetic moments, and this entire process will cause T2 shortening [5, 6]. Because of this, the T2 contrast agents are called susceptibility agents due to their effects on the magnetic field. The T2 shortening will be less effective as compared to the T1 shortening process where the close interaction of the different water molecules and T1 agents will happen. The T1 contrast agents, when in the higher magnetic fields, the R2, the relaxation rate, will have the positive constant. The commonly used T2 contrast agents are iron particles, and the iron oxides will behave like the ferromagnetic or superparamagnetic properties, which are based on the size of the core of the NP. The common iron oxides that are being used for various applications are magnetite (Fe_3O_4), resulting in the stable form of maghemite (γ-Fe_2O_3). The upper size limit threshold of the superparamagnetism is approximately 25 nm for magnetite and 30 nm for maghemite. There are two components: (1) chemical stability will have normal physiological settings, (2) low toxicity, and (3) sufficiently high magnetic moments. As these two iron oxides will have superparamagnetic behavior, the net magnetization due to the loss will result in the external magnetic field limits, which will be for self-aggregation and will have good biological response [1]. Due to this, Van der Waals forces, there will be the natural aggregation of the particles, which will be helpful for the stabilization of the particles and can be further stabilized by using the polymers for the stabilizing materials, which will be of two types

S. Shaikh, *Nanoimaging - Future of Precision Medicine*,
https://doi.org/10.1007/978-981-97-7292-6_9

of hydrophilic or amphiphilic, neutral, or charged, homopolymers or copolymers. The different polymers like poly(ethylene glycol) (PEG), dextran and its derivatives, alginate, chitosan, starch, polyvinyl alcohol, albumin, poly(ethylene imine), organic siloxane, sulfonated styrene-divinylbenzene, and bioactive molecules and structures such as liposomes are being used [7–9]. The overall action of the nanoparticles (NPs) and their biodistribution will be determined by the size of the NPs (Fig. 9.1).

The various sizes of the NPs are (1) ultrasmall superparamagnetic iron oxide nanoparticles (USPIONs), which will have a diameter less than 50 nm, (2) superparamagnetic iron oxide nanoparticles (SPIONs) with sizes of hundreds of nanometers, and (3) micron-sized particles of iron oxide (MPIO) having diameters greater than 1 μm.

The first two sizes will be administered intravenously due to their size, and the larger particles will be taken orally as they will be traversing through GIT. The other components are monocrystalline iron oxide particles (MION), which are like cross-linked iron oxides (CLIO) [3]. SPION formulations will be used by intravenous or oral administration and approved by EMEA and FDA [10]. SPION will be used for oral administration and is FDA-approved; SPIONs are dextran and carboxydextran. The NPs will be coated by the polymers, and this will be based on the type of coating and thickness. The thickness of the coating is inversely proportional as the coating increases, the ratio r2/r1 decreases [11]. These will result in the outer hydrophilic PEG layer by allowing water into the diffuse coating zone [12]. The pharmacokinetics and biodistribution of the NPs are based on the size of the particle. The different NPs having an average size of 5.5 nm will have renal clearance better [13, 14]. The SPIONs will be taken by the phagocytic cells having a diameter larger than 200 nm and accumulating in the monocyte phagocyte system (MPS), and this will be seen in the liver and spleen macrophages. These are administered intravenously and will be accumulated in the liver 80%, 5–10% in the spleen, and the plasma half-life will be less than 10 min [2]. Because of this, SPIONs will result in the reduction of the signal of the liver and spleen post-injection in a

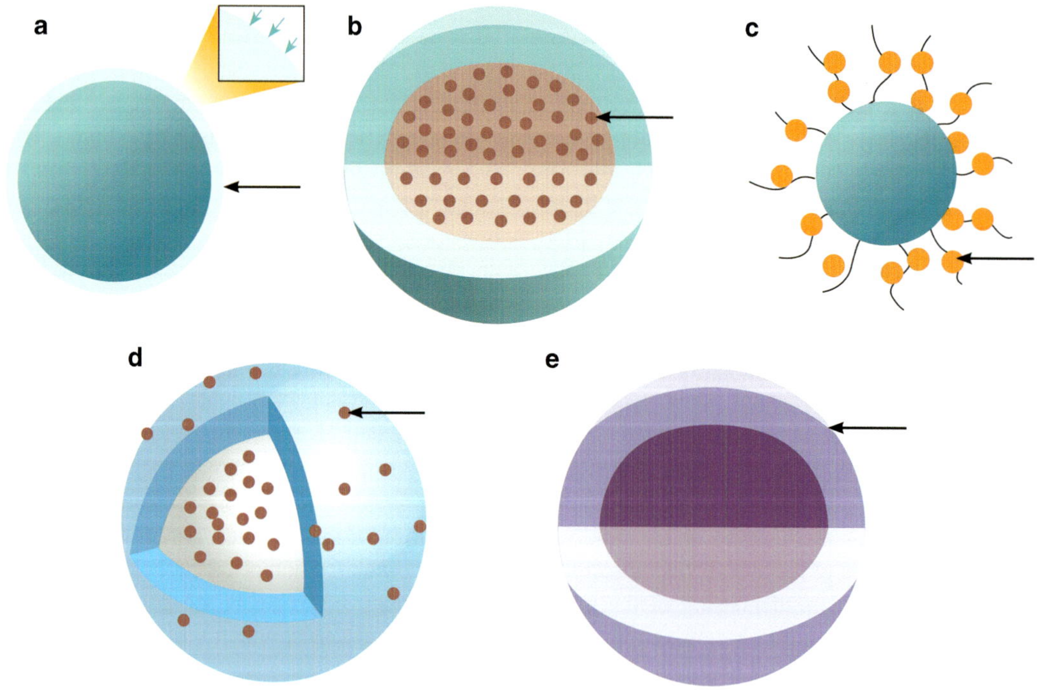

Fig. 9.1 Structure of the MR nanoparticle showing (**a**) basic structure, (**b**) core–shell structure, (**c**) a vector structure, (**d**) mixed structure, and (**e**) a core–shell structure

couple of minutes. Kupffer cells will not be seen in the malignant tumors or metastases, and due to this, the Kupffer cells and a smaller amount of the NPs will result in a higher contrast difference, especially in normal and abnormal tissues on T2-weighted sequences. The USPIONs will not take MPS uptake, and this will increase their blood half-life of 2 h. This phenomenon of increased blood circulation will result in the SPIONs reaching the target tissues. The metabolism of the SPIONs and USPIONs is by the use of the superparamagnetic iron, which is soluble in form, and to this iron pool added, e.g., ferritin, hemosiderin, and hemoglobin in a couple of days [3]. Due to the smaller size of the NP, SPIONs cannot see small solid tumors and also metastatic cells. This is due to the passive targeting of the enhanced permeation and retention (EPR) effects. This EPR effect will help NPs to cross the leaky vasculature into the interstitium [15]. Due to this, there will be poor lymphatic drainage by the particles that will be entrapped in these solid tumors. The SPIONs will use negative contrast for liver imaging, and USPIONs will use negative contrast in lymph node imaging. The USPIONs are also blood-pool agents that will be easily distributed in the intravascular extracellular space. These USPIONs will be used as contrast agents for lymphography [16, 17] and angiography [18, 19], bone marrow contrast, and perfusion agents in the brain and kidney. The iron oxide NPs, which will be targeting macrophages and also other target cells, will need different approaches [20]. By this phenomenon, the contrast agents will be using the different targets, which will increase the sensitivity by increasing the local SPION concentration [21–23]. For better targeting of the MRI cancer cells in vitro, the hyaluronic acid will be targeted with the iron oxide NPs seen as the efficient probes for the xenografted tumor model in vivo. Various ligands will be different markers like antibodies, small peptides, lectins, aptamers, engineered proteins, and protein fragments for use. Based on the size criteria, the USPIONs which will be less than 10 nm in diameter are being used for the tumor-specific MRI targeting [24]. These USPIONs are stabilized when added to the Mannich reaction [25]. These peptides will bind to the RGD, which will have capacity to bind by the cell adhesion molecule in tumor vasculature and also by the various invasive tumor cells that are expressed too far, and due to this, there will be the over-expression of this molecule [26]. MRI signals in the tumor will decrease when the RGD is injected into the body. Another important drawback of the SPIONs is seen when there is a reduction of the signal in the negative contrast agents. These will be reduced or dark in signal, mimicking like the pathologies distorting the normal tissue signals due to the higher susceptibility of T2 contrast. And this distortion will be seen as a susceptibility artifact or "blooming effect" [27]. On the other side, there will be higher signal intensity seen in the magnetic susceptibility in the lung and hemorrhagic areas [28]. NPs at this point will have different iron oxides, with different magnetic cores showing better signal sensitivity [29]. Transverse relaxivity will be based on saturation magnetization (Ms), creating a better way of the magnetic NPs with higher MRI sensitivity. The higher the magnetization, the iron NPs will have the higher T2 relaxivity [30]. T2 contrast agents are better with higher and better relaxivities with nanomaterials with alloys [31]. The better suitability of the Mn-ferrites as MRI contrast agents will increase the nonmagnetic ions replacing the Fe ions [32]. The combination of the Fe^{2+} with nonmagnetic Zn^{2+} will cause the increased net magnetization of the NPs [33]. Gold-iron oxide (Au-Fe_3O_4) NPs, which are metallic ion NPs, are more helpful in the use of the MRI with hollow Fe_3O_4 NPs and Fe-based alloy NPs. In this scenario, the biological solutions will have the combined iron–cobalt (FeCo) and iron–platinum (FePt) NPs as they will be more reactive. There will also be the increased stability of these metallic NPs for better contrast in MRI. Technically, the FePt NPs will be better contrast agents in comparison with the FeCo NPs [34–36]. Dy^{3+} as a chelate can be used (e.g., Dy^{3+}-DTPA) or as NPs (e.g., Dy_2O_3). Due to the outer-sphere relaxation theory [37], the smaller the particles the better the T1-shortening agents as compared to the larger ones. Because of their larger sizes and magnetic moments, the SPIONs are more used as T2-agents. And the newer USPION agents, which are less than 10 nm in diameter, have excellent

T1-enhancing properties [4, 38]. Again, in comparison with the paramagnetic ions, the SPIONs have better and higher molar relaxivities, showing as the varied blood-pool and tissue-specific agents [39]. The SPIONs cause the dark or negative contrast at the target sites by the blooming effect from magnetite susceptibility artifacts. The persistent T2 effects will result in delaying the SPION circulation after 24–72 h post-injection [40, 41]. Inversion recovery ON-resonant water suppression (IRON)-MRI is again another positive contrast agent [42]. These will remove the signal loss or magnetic susceptibility artifacts causing the brighter images even in delayed phases. Due to this delay, there will be a delay in the treatment and imaging caused by the MPIO, which are in the range of 0.76–1.63 μm in size [43]. Due to the larger size, the iron oxide particles will get aggregated for the pulmonary entrapment, and leukocyte mimicking homing ligand can be seen [44].The colloidal iron oxide nanoparticle platform (CION), when mixed with the oleate-coated magnetite particles in the hydrophobic matrix, which is made up of vegetable oil and partially cross-linked phospholipids, results in decreased T2 effects, causing positive T1-weighted contrast detection.

9.2 Dual (T1 and T2) Contrast Agents

MRI contrast agents in our conventional methods are better as single mode, either T1 or T2. The combination of both the T1 and T2 contrast will result in a single contrast medium for more MRI differentiation of the normal versus the diseased areas [45]. These contrast agents' availability and also preparing them are difficult and challenging [46]. USPIONs that are smaller than 10 nm in diameter will produce a positive contrast in T1–T1-weighted images when they are administered in average concentrations [47, 48]. Positive T1 contrast, which is seen in the USPIONs, can also have T2 effects [49, 50]. For better results, the combination of all these agents iron oxides, SPIONs and USPIONs, will be the better option. Still, there will be issues as both the T1 and T2 contrast agents will be in close vicinity [24]. A core–shell-type T1–T2 dual-mode NP as the contrast will be seen as the T1 contrast effects; and the superparamagnetic T2 contrast material, $MnFe_2O_4$ of 15 nm, will be as the core, which will be inducing the magnetic field in the relaxation of water molecules. As these two materials separated by SiO_2 will be adjusted by the thickness of the SiO_2, the magnetic coupling between these two will be controlled [51, 52]. The iron core with the nickel ions will form the nickel ferrite shell NP, and this will be reacted with the dopamine-PEG, which has dual-mode T1 and T2 contrast agents [53]. Gd chelate can be mixed with the polymer coating of SPIONs also [54, 55] and will result in a dual contrast agent. The Gd chelates like Gd–DTPA cannot cross the blood–brain barrier (BBB) [56], resulting in limited brain tumor imaging (Fig. 9.2). SPIONs have a larger surface area and can cross the BBB, and due to this, they can be used for non-invasive brain tumor imaging.

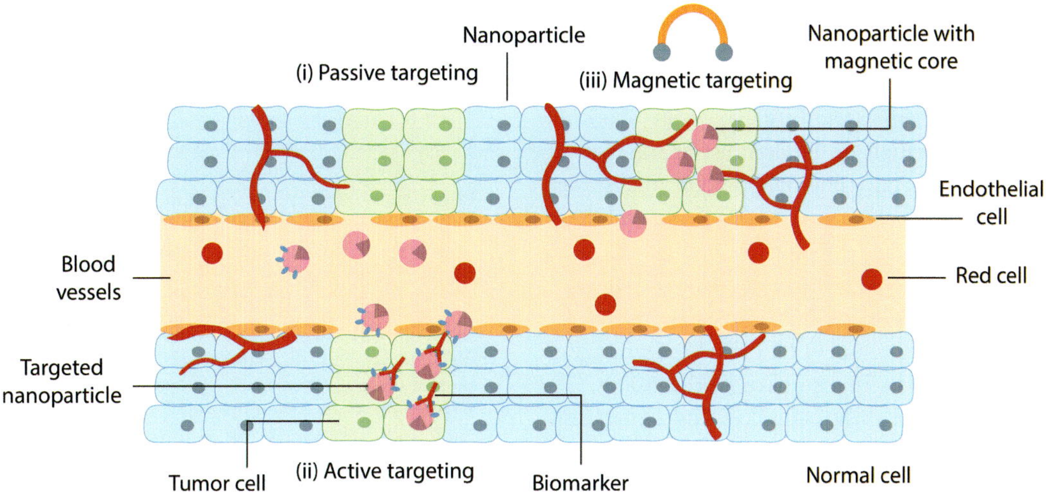

Fig. 9.2 The role of MR targeting in the tumor tissues along with the active and passive targeting that happens in the tumors and as compared to the normal cells

9.3 Paramagnetic/ Superparamagnetic Liposomes

These are the MRI probes, which are liposomes and have several advantages [57]. One of the important advantages is that they are capable of encapsulating the different hydrophilic substances that will be there in the aqueous inner core. Important advantages will be the compatibility and no side effects on the biological systems. These will have advantages, resulting in the liposomes as carriers and capable of being used as both therapeutics or diagnostics in vivo. These will be capable of having the Gd-based contrast agents, which are the paramagnetic liposomes or Fe-based contrast agents, superparamagnetic liposomes [57]. The toxicity of the Gd chelates is better for renal clearance [3]. Nanoprobes use will be more helpful in identifying the specific molecular marker with the pathological approach for better diagnosis as well as therapy [58]. For this, the amphiphilic lipid will be added as a bilayer, having increased sensitivity [58]. The relaxivity rates will be based on the liposomes as their content of cholesterol will affect the contrast agent concentration in the local tissues [59]. Encapsulation of a Gd chelate will have the freer Gd chelate [60]. Nano-MRI research in the newer concepts like magnetic resonance force microscopy (MRFM), optically detected biomedicines, and also related to the various applications of the nanotechnologies. Nano-MRI is made up of single-biomolecule imaging in vivo at room-temperature conditions [61]. The various nuclear spins with the spin-lattice relaxation time and hyperpolarized media will be effective. The adjacent spins will be used as (RF) pulses [62, 63]. The difference between the sync and anti-geometries is also seen in the J-couplings. Magnetic nanoprobes [64] will be used as better contrast agents for MRI, especially in the imaging of specific tumors.

9.4 MR Force Microscopy

MR force microscopy [65] is also a newer concept in which the shift of solids with 1-μm accuracy in a bi-dimensional spectrum will be there.

Hadamard encoding. Medical diagnosis can be arrived at by different modalities, the one that can visualize the in vivo internal tissues, as seen in the functional MRI (fMRI) [28].

The different NPs and complexes will be acting as the MRI contrast agents, which will be of great use in medicine. The transition or lanthanide metals, ion gadolinium (Gd^{3+}), can use the T1 contrast agents, which will have the better longitudinal relaxation times. With this, the newer generation of T1 contrast agents will be like the materials (like nanoporous silicas, dendrimers, perfluorocarbon NPs, and nanotubes).

The dual-mode agents will have strong T1–T2 contrast effects, and this is very challenging, despite the fact that this will have better and improve the accuracy of biomedical imaging. Nanomaterials will also have MRI artifacts. The NP can be seen as the cationic lipids, fluorescent-lipid conjugates, targeting ligands, drugs, and PEG, comprising the single nanosystems. The most notable limitations associated with the use of such contrast agents are the current detection limits and the lack of tissue specificity. Current detection limits need to be improved for the successful translation of NPs.

9.5 Conclusion

The use of various nanomaterials in imaging has shown a significant change in the use of the MRI by different algorithms, which has led to earlier and more precise diagnosis of the diseases at the molecular levels.

References

1. Lam T, Pouliot P, Avti PK, et al. Superparamagnetic iron oxide based nanoprobes for imaging and theranostics. Adv Colloid Interf Sci. 2013;199–200:95–113.
2. Mornet S, Vasseur S, Grasset F, et al. Magnetic nanoparticle design for medical diagnosis and therapy. J Mater Chem. 2004;14:2161–75.
3. Geraldes CF, Laurent S. Classification and basic properties contrast agents for magnetic resonance imaging. Contrast Media Mol Imaging. 2009;4:1–23.
4. Senpan A, Caruthers SD, Rhee I, et al. Conquering the dark side: colloidal iron oxide nanoparticles. ACS Nano. 2009;3:3917–26.
5. Park JY, Choi ES, Baek MJ, et al. Water-soluble ultra small paramagnetic or superparamagnetic metal oxide nanoparticles for molecular MR Imaging. Eur J Inorg Chem. 2009;9:2477–81.
6. Filippousi M, Angelakeris M, Sikini M, et al. Surfactant effects on the structural and magnetic properties of iron oxide nanoparticles. J Phys Chem C. 2014;118:16209–17.
7. Shokrollahi H. Structure, synthetic methods, magnetic properties and biomedical applications of ferrofluids. Mater Sci Eng C Mater Biol Appl. 2013;33:2476–87.
8. Pflipsen C, Forge D, Benali S, et al. Improved stability and relaxivity of a commercial magnetic ferrofluid. J Phys Chem C. 2013;117:20919–26.
9. Bulte JW, Kraitchman DL. Iron oxide MR contrast agents for molecular and cellular imaging. NMR Biomed. 2004;17:484–99.
10. Bulte JW. In vivo MRI cell tracking: clinical studies. Am J Roentgenol. 2009;193:314–25.
11. LaConte LE, Nitin N, Zurkiya O, et al. Coating thickness of magnetic iron oxide nanoparticles affects R2 relaxivity. J Magn Reson Imaging. 2007;26:1634–41.
12. Hu F, Joshi HM, Dravid VP, et al. High-performance nanostructured MR contrast probes. Nanoscale. 2010;2:1884–91.
13. Choi HS, Liu W, Misra P, et al. Renal clearance of nanoparticles. Nat Biotechnol. 2007;25:1165–70.
14. Longmire M, Choyke PL, Kobayashi H. Clearance properties of nanosized particles and molecules as imaging agents: considerations and caveats. Nanomedicine. 2008;3:703–17.
15. Maeda H, Wu J, Sawa T, et al. Tumor vascular permeability and the EPR effect in macromolecular therapeutics: a review. J Control Release. 2000;65:271–84.
16. Hamm B. Iron-oxide-enhanced MR lymphography: just a new toy or a breakthrough? Eur Radiol. 2002;12:957–8.
17. Lind K, Kresse M, Debus NP, Muller RH. A novel formulation for superparamagnetic iron oxide (SPIO) particles enhancing MR lymphography: comparison of physicochemical properties and the in vivo behavior. J Drug Target. 2002;10:221–30.
18. Maes RM, Lewin JS, Duerk JL, et al. A new type of susceptibility artefact-based magnetic resonance angiography: intra-arterial injection of superparamagnetic iron oxide particles (SPIO) A Resovist in combination with TrueFisp imaging: a feasibility study. Contrast Media Mol Imaging. 2006;1:189–95.
19. Schmitz SA, Albrecht T, Wolf KJ. MR angiography with superparamagnetic iron oxide: feasibility study. Radiology. 1999;213:603–7.
20. Morawski A, Winter P, Crowder K, et al. Targeted nanoparticles for quantitative imaging of sparse molecular epitopes with MRI. Magn Reson Med. 2004;51:480–6.
21. Cai H, An X, Cui J, et al. Facile hydrothermal synthesis and surface functionalization of polyethyleneimine-coated iron oxide nanoparticles for biomedical applications. ACS Appl Mater Interfaces. 2013;5:1722–31.
22. Li J, He Y, Sun W, et al. Hyaluronic acid-modified hydrothermally synthesized iron oxide nanoparticles for targeted tumor MR imaging. Biomaterials. 2014;35:3666–77.
23. Zhang H, Li J, Sun W, et al. Hyaluronic acid-modified magnetic iron oxide nanoparticles for MR imaging of

surgically induced endometriosis model in rats. PLoS One. 2014;9:e94718.
24. Gupta AK, Gupta M. Synthesis and surface engineering of iron oxide nanoparticles for biomedical applications. Biomaterials. 2005;26:3995–4021.
25. Xie J, Chen K, Lee HY, et al. Ultrasmall c(RGDyK)-coated Fe_3O_4 nanoparticles and their specific targeting to integrin alpha(v)-beta(3)-rich tumor cells. J Am Chem Soc. 2008;126:11458–9.
26. Desgrosellier JS, Cheresh DA. Integrins in cancer: biological implications and therapeutic opportunities. Nat Rev Cancer. 2010;10:9–22.
27. Bin Na H, Chan Song I, Hyeon T. Inorganic nanoparticles for MRI contrast agents. Adv Mater. 2009;21:2133–48.
28. Terreno E, Castelli DD, Viale A, et al. Challenges for molecular magnetic resonance imaging. Chem Rev. 2010;110:3019–42.
29. Yoo D, Lee JH, Shin TH, et al. Theranostic magnetic nanoparticles. Acc Chem Res. 2011;44:863–74.
30. Hadjipanayis CJ, Bonder MJ, Balakrishnan S, et al. Metallic iron nanoparticles for MRI contrast enhancement and local hyperthermia. Small. 2008;4:1925–9.
31. Yang H, Zhang C, Shi X, et al. Water-soluble superparamagnetic manganese ferrite nanoparticles for magnetic resonance imaging. Biomaterials. 2010;31:3667–73.
32. Ho D, Sun X, Sun S. Monodisperse magnetic nanoparticles for theranostic applications. Acc Chem Res. 2011;44:875–82.
33. Jang JT, Nah H, Lee JH, et al. Critical enhancements of MRI contrast and hyperthermic effects by dopant-controlled magnetic nanoparticles. Angew Chem Int Ed. 2009;48:1234–8.
34. Das GK, Johnson NJ, Cramen J, et al. NaDyF4 nanoparticles as T2 contrast agents for ultrahigh field magnetic resonance imaging. J Phys Chem Lett. 2012;3:524–9.
35. Bulte JW, Wu C, Brechbiel MW, et al. Dysprosium-DOTA-PANAM dendrimers as macromolecular T2 contrast agents: preparation and relaxometry. Investig Radiol. 1998;33:841–5.
36. Das GK, Zhang Y, Silva LD, et al. Single-phase Dy_2O_3:Tb^{3+} nanocrystals as dual-modal contrast agent for field high magnetic resonance and optical imaging. Chem Mater. 2011;23:2439–46.
37. Gueron M. Nuclear relaxation in macromolecules by paramagnetic ions: a novel mechanism. J Magn Reson. 1975;19:58–66.
38. Taboada E, Rodríguez E, Roig A, et al. Relaxometric and magnetic characterization of ultrasmall iron oxide nanoparticles with high magnetization. Evaluation as potential T1 magnetic resonance imaging contrast agents for molecular imaging. Langmuir. 2007;23:4583–7.
39. Shubayev VI, Pisanic TR II, Jin S. Magnetic nanoparticles for theragnostics. Adv Drug Deliv Rev. 2009;61:467–77.
40. Cunningham CH, Arai T, Yang PC, et al. Positive contrast magnetic resonance imaging of cells labeled with magnetic nanoparticles. Magn Reson Med. 2005;53:999–1005.
41. Huang J, Zhong X, Wang L, et al. Improving the magnetic resonance imaging contrast and detection methods with engineered magnetic nanoparticles. Theranostics. 2012;2:86–102.
42. Stuber M, Gilson WD, Schär M, et al. Positive contrast visualization of iron oxide-labeled stem cells using inversion-recovery with ON-resonant water suppression (IRON). Magn Reson Med. 2007;58:1072–7.
43. McAteer MA, Sibson NR, von zur Muhlen C, et al. In vivo magnetic resonance imaging of acute brain inflammation using microparticles of iron oxide. Nat Med. 2007;10:1253–8.
44. McAteer MA, Schneider JE, Ali ZA, et al. Magnetic resonance imaging of endothelial adhesion molecules in mouse atherosclerosis using dual targeted microparticles of iron oxide. Arterioscler Thromb Vasc Biol. 2008;28:77–83.
45. von zur Muhlen C, von Elverfeldt D, Moeller JA, et al. Magnetic resonance imaging contrast agent targeted toward activated platelets allows in vivo detection of thrombosis and monitoring of thrombolysis. Circulation. 2008;118:258–67.
46. Corot C, Robert P, Idèe JM, et al. Recent advances in iron oxide nanocrystal technology for medical imaging. Adv Drug Deliv Rev. 2006;58:1471–504.
47. Park JY, Daksha P, Lee GH, et al. Highly dispersible PEG surface modified ultra-small superparamagnetic iron oxide nanoparticles useful for target-specific biomedical applications. Nanotechnology. 2008;19:365603.
48. Di Marco M, Sadun C, Port M, et al. Physicochemical characterization of ultrasmall superparamagnetic iron oxide particles (USPIO) for biomedical application as MRI contrast agents. Int J Nanomedicine. 2007;2:609–22.
49. Gossuin Y, Gillis A, Hocq A, et al. Magnetic resonance relaxation properties of superparamagnetic particles. Wiley Interdiscip Rev Nanomed Nanobiotechnol. 2009;1:299–310.
50. Jun YW, Huh YM, Choi JS, et al. Nanoscale size effect of magnetic nanocrystals and their utilization for cancer diagnosis via magnetic resonance imaging. J Am Chem Soc. 2005;127:5732–3.
51. Choi JS, Lee JH, Shin TH, et al. Self-confirming, "AND" logic nanoparticles for fault-free MRI. J Am Chem Soc. 2010;132:11015–7.
52. Bonmatí MO, Gossuin Y, Morales M, et al. Comparative analysis of the 1H NMR relaxation enhancement produced by iron oxide and core-shell iron–iron oxide nanoparticles. Magn Reson Imaging. 2007;25:1437–41.
53. Shultz MD, Calvin S, Fatouros PP, et al. Enhanced ferrite nanoparticles as MRI contrast agents. J Magn Mater. 2007;311:464–8.
54. Bae KY, Kim YB, Lee Y, et al. Bioinspired synthesis and characterization of gadolinium-labeled magnetite nanoparticles for dual contrast T1- and T2-weighted

magnetic resonance imaging. Bioconjug Chem. 2010;21:505–12.
55. Yang H, Zhuang Y, Sun Y, et al. Targeted dual contrast T1- and T2-weighted magnetic resonance imaging of tumors using multifunctional gadolinium-labeled superparamagnetic iron oxide nanoparticles. Biomaterials. 2011;32:4584–93.
56. Ding D, Kanaly CW, Cummings TJ, et al. Long-term safety of combined intracerebral delivery of free gadolinium and targeted chemotherapeutic agent PRX321. Neurol Res. 2010;32:810–5.
57. Torchilin VP. Liposomes as delivery agents for medical imaging. Mol Med Today. 1996;2:242–9.
58. Kamaly N, Miller AD. Paramagnetic liposome nanoparticles for cellular and tumour imaging. Int J Mol Sci. 2010;11:1759–76.
59. Victor KG, Korb JP, Bryant RG. Translational dynamics of water at the phospholipid interface. J Phys Chem B. 2013;117:12475–8.
60. Jung K-H, Lee K-H. Molecular imaging in the era of personalized medicine. J Pathol Transl Med. 2015;49:5–12.
61. Boretti A, Rosa L, Castelletto S. Towards single biomolecule imaging via optical nanoscale magnetic resonance imaging. Small. 2015;11:4229–36.
62. DeVience SJ, Walsworth RL, Rosen MS. Preparation of nuclear spin singlet states using spin-lock induced crossing. Phys Rev Lett. 2013;111:173002.
63. DeVience SJ, Walsworth RL, Rosen MS. Probing scalar coupling differences via long-lived singlet states. J Magn Reson. 2016;262:42–9.
64. Gao Z, Ma T, Zhao E, Docter D, Yang W, Stauber RH, Gao M. Small is smarter: nano MRI contrast agents—advantages and recent achievements. Small. 2016;12:556–76.
65. Joss R, Tomka IT, Eberhardt KW, van Beek JD, Meier BH. Chemical-shift imaging in micro- and nano-MRI. Phys Rev B. 2011;84:104435.

10 MRI Nanoimaging: Part II

10.1 Introduction

The radiofrequency pulse in the magnetic resonance imaging (MRI) will be resonating at the frequency of 5–100 MHz, and this can transfer the energy to the protons. This phenomenon leads to the magnetic moments shifting the axis in various phases, and this can be seen in the flip angle. When this is reduced, the magnetic moments of the protons will cause the relaxation of the equilibrium [1]. This happens in MRI, in a quick succession of various RF pulses. The time in which the magnetic moments will be returned to the original alignment of the magnetic field is known as the relaxation time, which is tissue-dependent. This relaxation is seen in two different ways: (1) longitudinal relaxation (which is identified by the parameter T1) and (2) transverse relaxation (which is identified by the parameter T2). T1 is also known as the spin-lattice relaxation time, which will relate the magnetization parallel to the static magnetic field recovering the process. These protons will be relaxing in the form of short T1, leading to the full magnetization along the longitudinal axis. Full magnetization along the longitudinal axis will not be seen before the RF pulses, and this will also be seen as the lower signal intensity [1]. T2 signal intensity will show the fast magnetization that is seen in the perpendicular plane to the static magnetic field losing coherence. During this RF pulse, the proton nuclei spin seen in each spin will be interacting with the nuclei and the energy will be exchanged. Due to this, the nuclei will lose phase coherence, and this will randomly lead to the spin [2]. T2 decay is due to the exchange of energy between spinning protons, which is also known as spin–spin relaxation. These two longitudinal and transverse relaxation processes will be seen independently and together as T2 will be much shorter than T1, and due to this, the tissues can be differentiated easily [3]. The combination of these intrinsic molecular interactions and the extrinsic magnetic field inhomogeneities causes the transverse relaxation time, T2.

10.2 Superparamagnetic Iron Oxide Nanoparticles (SPION)

Superparamagnetic iron oxide nanoparticles (SPION) has now become the most important MRI contrast agent for the evaluation of tumors, which has better efficacy and safety. Research is going on to improve the SPION, which is tailor-made for adding the therapeutic aspects. The research is going on for the dual-modal contrast applications of SPION. The development of multifunctional nanoparticles (NPs) will have both diagnostic and therapeutic properties, and this will be more helpful in the evaluation of the various cancers. Here the physiochemical properties

S. Shaikh, *Nanoimaging - Future of Precision Medicine*,
https://doi.org/10.1007/978-981-97-7292-6_10

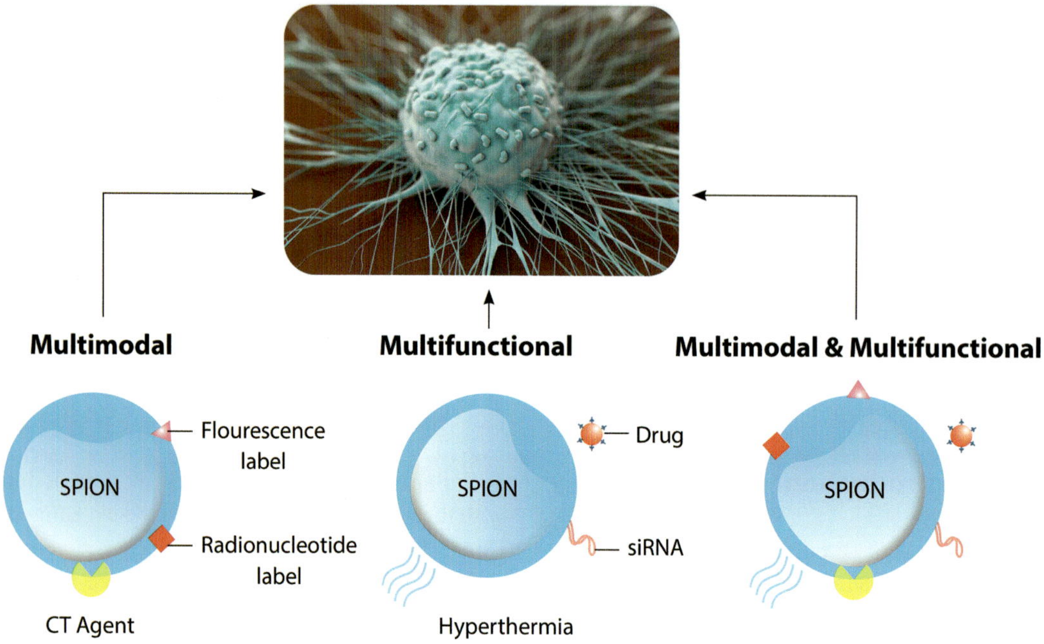

Fig. 10.1 The multimodal, multifunctional, and both multifunctional and multimodal roles of SPION

and various methods for the synthesis of the SPION for the multimodal contrast agents are seen (Fig. 10.1).

The early detection and treatment of cancer will be the important factors for better cancer diagnosis seen in the nanotechnology, which is the emerging field. Tremendous efforts in biomedical research have been devoted to improving the sensitivity and accuracy of diagnosing cancer using early detection methods and to improving the efficacy of treatment methods. The non-invasive diagnosis of patients is a powerful tool for diagnosing the abnormal state of the body and also for the monitoring of the biological target sites. Non-invasive imaging techniques are computed tomography (CT), MRI, nuclear imaging of positron emission tomography (PET), single-photon emission computed tomography (SPECT), and optical (or fluorescence) imaging (Fig. 10.2). These imaging modalities are now capable of evaluating both in vivo for a better depiction of the anatomy and physiology in both animals and humans [4–9]. MRI has a very high sensitivity for the high soft tissue resolution and for the unlimited penetration into tissues [7, 10]. The use of nanotechnology has led to the newer concept of theranostics, where therapy and imaging can be done simultaneously [11, 12]. This nanotechnology is also seen in molecular imaging in the form of probes, which are capable of enhancing the sensitivity of the image and also the specificity toward the target tissue. The imaging probes will consist of NPs, which will be conjugated for targeting ligands actively [13, 14]. Another means of tumor targeting is enhanced permeability and retention (EPR) [15, 16]. SPION have a superparamagnetic iron core, thus making them more useful as T2 contrast agents for MRI. SPION is being detected with higher sensitivity for iron and polymer components of SPION, which are biocompatible and degradable [17]. The excretion of the iron oxide NPs is based on the target cellular uptake and also on the elimination from the body. NPs having more than 200 nm in diameter and sizes lower 10 nm will be accumulated in the liver and spleen, and these will be excreted by the excretory renal systems [18]. The properties of the NPs are being modified to better accommodate the targeting agent [19]. With the increased use of multimodal imag-

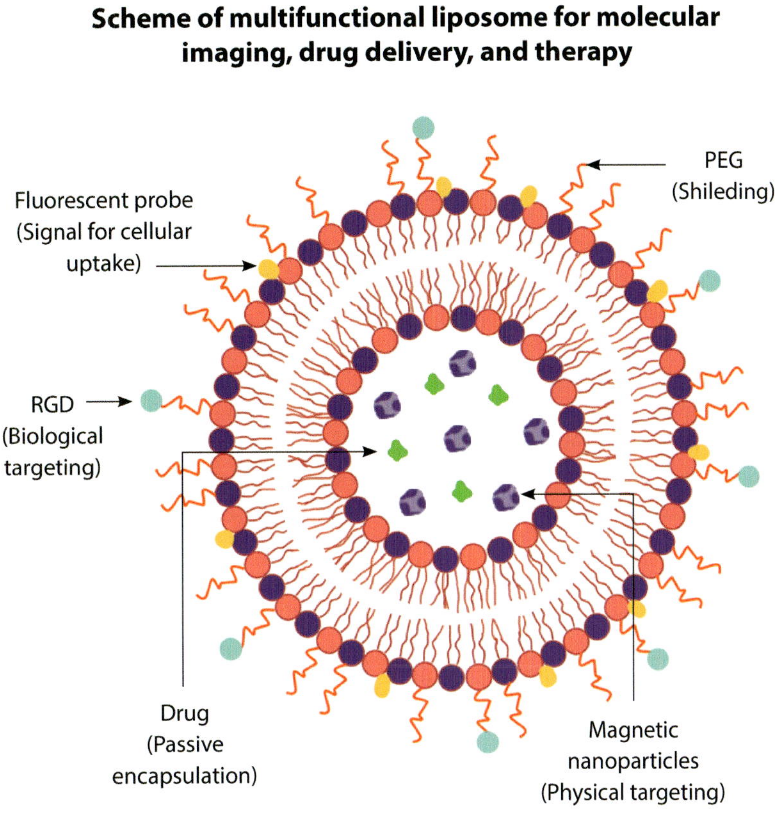

Fig. 10.2 Different multifunctional roles of molecular imaging in the drug delivery and therapy

ing, there is more development of the various novel probes, which can be seen for the all-in-one probes as well as the multimodal capability as well as the therapeutic functionality [20]. Multifunctional NPs will be more useful in combination with more imaging modalities for the use of theronostics. These techniques will be used more as compared to routine imaging diagnosis as well as therapy, leading to the concept of optimized therapy, which is known as "personalized medicine" [9]. Various NPs are drug conjugates and complexes, micelles, dendrimers, vesicles, core–shell particles, microbubbles, and carbon nanotubes [21]. Among these NPs, the magnetic NPs will execute the best integrated design for cell targeting, imaging, and therapy. Due to this concept, the NPs will be useful for theronostics. Magnetic NPs will also be useful for multimodality preclinical imaging, leading to the combinations of various MRI, nuclear, and optical imaging [18, 22].

10.3 Synthesis of Iron Oxide Nanoparticles

SPION is one of the newest MRI contrast agents due to the T2 relaxivity of a SPION-based agent, which is higher than the gadolinium agents. The physiochemical properties of SPION are based on the charge, size, and surface chemistry, impacting the biodistribution, stability, and metabolism [23]. The surface coating has to be compatible with very good biocompatibility and biodegradability, and SPION is also an immune response with serum protein adsorption. The important deciding factor for the determination of the colloidal nature of NPs is the surface charge and biocompatible as compared to the charged surfaces. The surface charged is by plasma protein binding, which will impact directly by affecting the biodistribution and nanoparticle clearance from the body [24]. Classical SPION is comprised of the magnetite (Fe_3O_4) or maghemite (Fe_2O_3), with the coatings for maintaining the stability. Different mechanisms are there for the synthesis of the SPION by various processes, including co-precipitation, thermal decomposition, and microemulsion. Microwaving-assisted synthesis, hydrolysis method, and also microfluidic synthesis are other there [25–27]. Thermal decomposition is a common method for better use of NPs due to better physiological properties like high crystallinity, high magnetization, and less biodistribution. These NPs are hydrophobic and need more coating for biomedical applications [28]. Co-precipitation is used for the synthesis of the SPION is the co-precipitation method. In this, there will be additional use of the divalent or trivalent ferrous salt solutions which, will be polydisperse in the range of 2–15 nm [18, 29]. These will lead to the better aggregate and causing faster growth [30]. The microemulsion technique has an important advantage in maintaining the size of NPs. SPION formation by the use of the water-in-oil or oil-in-water emulsion having the size of 2 and 12 nm [18]. The superparamagnetism is having single-domain nanoparticle, with a net magnetic dipole. In the presence of the magnetic field, the NPs will be realigning themselves, which will be similar to the paramagnetic substances, and this will be much higher as compared to the paramagnetic substances [17, 18]. SPION functions in the MRI are contrasted by different nuclear spin relaxation of water protons, which has dark signal intensity. The biosafety of iron oxide NPs (Fe_3O_4 and $MnFe_3O_4$) is seen at the 200 μg/mL concentration [19]. Higher dosages form more ROS and impact the normal cells' functioning, which may cause apoptosis or cell death [24, 31].

Routine MRI contrast agents are capable of having relaxation times, which will enhance the positive signal on T1-weighted images, and the T2 SPION-based contrast agents will cause decrease in the signal intensity on T2-weighted images. The use of multimodal imaging has many benefits along with many drawbacks. The analysis of the contrast MRI in the targeting of the cells and tissues is also an important role in the diagnosis of the early phase of the disease (Fig. 10.3). This will lead to specific limitations for the specific modality, and the combinations will increase the strengths [32, 33]. PET imaging has a high sensitivity to the many biological and functional aspects of cancer. On the contrary, the CT and MRI will give better anatomical information. Thus, these two modalities will have higher sensitivity and resolution along with detailed anatomical or biological information about cancer. Multimodal imaging is also being used for cancer theronostics when used with the advantages of NPs [34, 35]. Thus iron oxide NPs are important for multimodal imaging, which will have comfortable pharmacokinetics and surface modification.

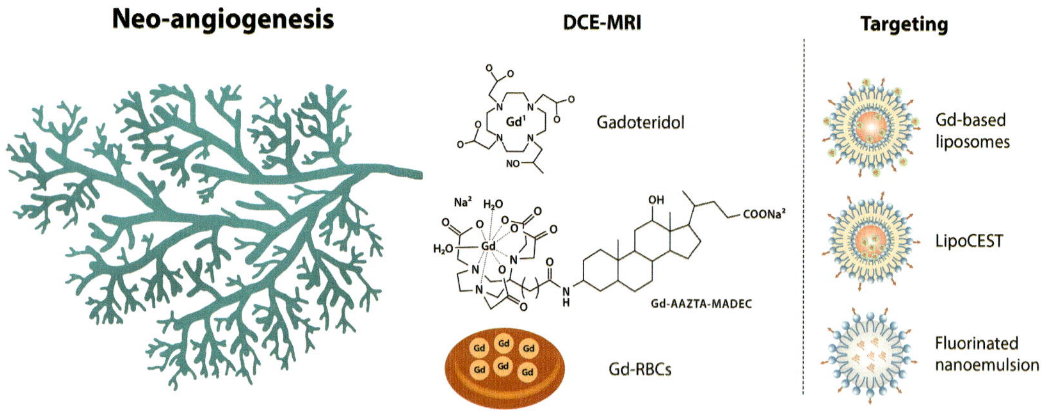

Fig. 10.3 The relevance of the dynamic contrast enhancement in the angiogenesis of the tumors. The different targeting of the various contrast agents like the Gadolinium, LipoCEST and fluorinated nanoemulsions. These can be tagged to the Gadolinium with the red blood cells

10.4 MRI/Optical Imaging

With the research going on the MRI and NIRF dye conjugated contrast agents, a lot of the MRI/optical dual imaging protocols are being developed [36, 37]. MRI is not conducive to measuring molecular events like protease activity and gene expression. However, NIRF imaging is capable of producing the finer functional details in the molecular event as these combined techniques will have improved imaging quality [38]. In the group of NIRF dyes, the cyanine dye (Cy5.5) is commonly used SPION. The combination of MRI and optical imaging as multimodal imaging is associated with NIRF-based molecular imaging probes and has emission wavelengths that will range from 650 to 900 nm [39].

MRI/CT imaging again is the concept of multimodality anatomical imaging, and neither MRI nor CT are strong in functional imaging, as they are excellent modalities for anatomical imaging. An MRI/CT dual-modal agent is seen as a combination of two anatomic modalities that will be capable of functional imaging features [40]. Only the difference involved is the MRI and CT physics of the respective imaging acquisition. CT uses X-ray radiation, and the image is formed by the absorption and attenuation features of the contrast agents. MRI is based on the strong magnetic field, where the image analysis is based on the T1 or T2 contrast enhancement, which will be induced by NPs [41, 42]. Gold nanoparticles (AuNPs) are the most commonly used contrast agents in CT. This SPION will combine with the AuNPs to form the gold iron oxide nanoparticles (GION), and this will act as the dual-contrast agent for MRI and CT [43–46]. Gold is acting as the CT contrast moiety, which will have lesser spatial resolution and sensitivity in comparison with the MRI. Another dual-contrast agent in combination with the iron oxide core as well as the gold-layered shell with PEG coating on the surface will be able to provide the combined synergistic effects in both CT and MRI modalities. This PEG-coated GION when it is dispersed in the water, will have an average hydrodynamic size of 47 nm. This intensity of the CT images for PEG-coated GION is always greater than iodine-based counterparts when used in similar concentrations. In this process, the T2 signal efficiency will be decreased as compared to oleic acid/oleylamine-coated SPION. One of the studies reported [47] that the synthesis of hybrid NPs with the thermal decomposition of Au-oleylamine and Fe-oleate mixtures will be followed by subsequent coating in the amphiphilic poly(DMA-r-mPEGMA-r-MA) by nanoemulsion. These NPs will show water dispersion and biocompatibility effectively. This in vivo CT contrast will have better efficiency with the improved contrast enhancement in the 1 h post-injection in a murine hepatoma model.

10.5 MRI/PET Imaging

MRI/PET imaging is the newest form of imaging and is now one of the most preferred imaging modalities wherever it is available. MRI/PET has the biggest advantage of not having the radiation. MRI has significantly enhanced spatial resolution, which will have the partial-volume effects caused by PET imaging. Here in multimodal PET/MRI imaging, the radionucleotide conjugated with an MRI agent, SPION, can be used as a nanocarrier [43, 44]. The tumor will have the integrin αvβ3 expression targeted using PEGylated SPION by cyclic Arg-Gly-Asp-D-Phe-Cys (cRGD) peptides as dual-modality imaging by using PET ^{64}Cu chelators for PET imaging. PET imaging in the nanoimaging will be better evaluated for the capability of cRGD-conjugated SPION, which will be accumulated in tumors. One of the studies also shows that the use of the αvβ3-integrin in the tumor was being targeted by the use of the iron oxide NPs coated with polyaspartic acid (PASP) [45]. This RGD peptide is also seen as the integrin αvβ3 targeting agent and 1,4,7,10-tetraazacyclododecane-*N,N′,N″*,-tetraacetic acid (DOTA), where the chelators can be labeled by ^{64}Cu for PET. The use of the DOTA IO-RGD conjugates, which will be bound by the specific integrin αvβ3 in vitro. Thus, the MRI/PET imaging has shown the use of the integrin-specific delivery for the conjugated RGD-PASP-IO (Arg-Gly-Asp-D-Phe-Cys-

Polyaspartic acid-iron oxide) with significant uptake of the NPs in the reticuloendothelial system (RES).

10.6 MRI/SPECT Imaging

The use of the SPION by the SPECT probes in the MRI/SPECT dual-modality imaging is seen as the molecular processes by the use of the radiolabels, which needs to be specific. SPECT will also be used for the identification of the biodistribution noninvasively of the radiotracers, which will be tagged to the different particles in vivo as well as non-invasively by the picomolar. SPECT has limitations in the anatomical details as well as the spatial resolution. Thus, MRI along with the SPECT can be used for the evaluation of the anatomical images, having both the structural and functional evaluation of the dynamic imaging. MRI/SPECT is associated with multiple sessions due to the scan times that are needed for separate imaging procedures [48]. One of the studies showed that the ^{99m}Tc-labeled PEG-coated iron oxide can be used as the multimodal contrast agent in the evaluation of the sentinel lymph node (SLN) [46]. Thus, the SLN is the receiver of the lymphatic drainage from a malignant tumor, and this evaluation of the SLN will provide the pre-surgical information irrespective of the different locations and various characteristics of the lymph nodes. The MR images will also show us the uptake of the ^{99m}Tc-SPION, and this will be useful for breast cancer and malignant melanoma imaging.

10.7 Multifunctional Magnetic Nanoparticle for Cancer Theranostics

The multifunctional capabilities of the NPs where the combination of diagnosis as well as therapeutic applications is known as theranostics. Theranostics will have different agents that will have the more specific targeted molecular biomarkers of a disease and also will be contributing to the fate of personalized medicine (Fig. 10.4). These SPION also will have significant potential in the future for the development of theranostic nanomedicines like contrast agents for diagnosis as well as therapeutic agents for delivery as anticancer agents siRNA, covering all the focused disease sites. In addition, the SPION is capable of

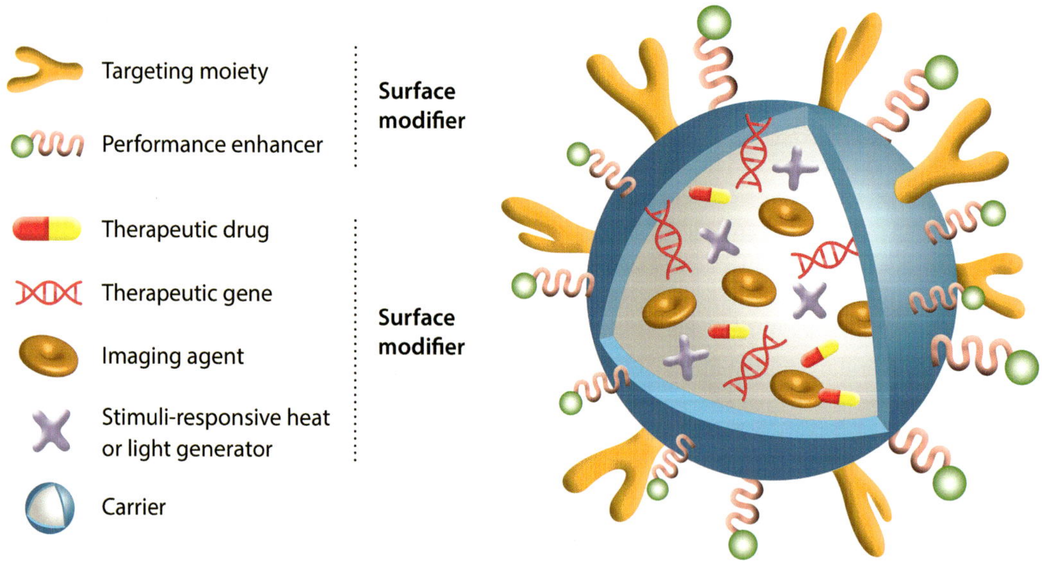

Fig. 10.4 Nanoparticle surface modifier with different roles

emitting heat when it is exposed to the external magnetic field (AMF), which will be alternating due to the conversion of the electromagnetic energy into heat energy [49–54]. The multifunctionality of a SPION-based nanosystem is being used by various drugs and other imaging modalities. The use of the biocompatible polymer-coated SPION is seen loaded [55]. Due to this, the tumor size in the lung cancer seen in the MRI and NIRF is seen as the diagnostic information [55, 56]. The SPION (TCL-SPION) is being linked and synthesized after coating the negatively charged anti-biofouling polymer and Cy5.5 dye. By different electrostatic actions, these NPs will then be loaded with the positively charged doxorubicin (DOX). After injecting these particles into the lung cancer, there was a significant reduction in the size of the cancer. This encapsulation is carried out by the NIR dye, an important anticancer drug having the hydrophobic region of polyacrylic acid (PAA), which can be done by another solvent technique, and SPION is also used for diagnosing the carcinoma lung [57–61]. These SPION particles are synthesized and then coated by the gold and PEI as the intermediate layer [61–64].

MRI's newer applications in nanoimaging include the blood flow and physiochemical traits of different tissues and organs in the body, leading to increased sensitivity [65–69]. Metal NPs can also be a better contrast agent, which can be used in MRI scanning [70]. CT contrast agents have very short circulation half-lives [71] and also will cause more potential tissue damage [72]. Due to this reason, the metal NPs can be used in CT imaging [73, 74]. AuNPs are better as studied by Kojima et al., where the AuNPs are conjugated with a PEGylated dendrimer (PEG-AuNPs), which is made up of the better contrast agent in vitro and also in the X-ray computed tomography, in comparison with the iodine contrast agent iopamidal [75]. The PEG-AuNPs had better contrast efficiency as compared to the commercially available iopamidol, which has very rapid excretion from the body [76]. Due to these properties, these contrast agents can enable photothermal therapy and due to this the imaging tool can be used for atherosclerosis. When these AuNPs are applied, this medical imaging is called SPECT [77]. This imaging is very similar to a gamma camera and is capable of providing true 3D images that can be sliced, rotated, and manipulated to achieve more precise analytical techniques [77]. These modified NPs will be able to target the specifically targeted atherosclerosis plaques, which will have apoptotic macrophages, which shows the more invasive and accurate detection of atherosclerosis plaques [77]. These AuNPs will be demonstrating photoacoustic imaging (PA), which will have higher spatial resolution and sensitivity [78, 79]. The PA will rely on the detection of ultrasonic waves emitted from various tissues when exposed to non-ionizing pulsed laser irradiation [80, 81]. This intensity/magnitude of the ultrasonic emission can produce the image contrast, where they can absorb the laser pulses in the heat, which will increase the magnitude of the ultrasonic emission, and AuNPs possess the ability to do both of these [82, 83]. Lung cancer cells can be treated by the AuNPs having the target epidermal growth factor receptor, which are having the bright areas in the NPs showing the different locations in the cells [84, 85]. The nanoparticle delivery will make it ideal for drug delivery as it will have a higher surface area-to-volume ratio as compared to the bulk material [86]. Along with this, it can interact with the immune system in different ways [87, 88]. The increased hydrophobicity of NPs/subgroups conjugated to the NPs can cause an increased immune response by measuring cytokine mRNA levels in mice [87]. On the other side, the various NPs will be conjugated along with different ligands, which will activate the immune system in the body and cause the destruction of the tumor cells [89] or also the accumulation in the liver or spleen where the immunity is generated [88]. AuNP applications like the AuNPs for the transmucosal delivery of insulin. These AuNPs will be synthesized in the presence of polymer chitosan [90]. These NPs are administered both nasally and orally when added to the insulin. More recently, the "smart" AuNPs are also seen in the PA [79]. The size of the NPs is in the range of 10 nm in diameter, and due to this property, they can interact with the citraconic amide moieties and get converted into the positively charged primary amino acids. This will result in better adaptation of the molecules having negative charges at normal physiological pH [79]. These two properties will combine with the "smart" NPs, which will have positive and negative charges leading to the electrostatic attraction. These NPs are labeled as "smart" particles due to their cancer-specific properties, which will be accumulating in the cancer tissues as compared to the normal tissues [91]. AuNPs act also as a delivery system for nucleic acids [92], for example, oligonucleotides [93, 94] and small interfering RNA (siRNA) [95]. Various methods will be developed for the functionalizing of the AuNPs with nucleic acids; one of the examples is the AuNPs, which will be modified by thiocholine, as these will be bound to DNA and will be fused like the wire-like structures throughout the DNA [96]. Sandström et al. demonstrated the binding of the nucleic acids onto AuNPs [93], and this can be done by Rosi et al., where tetrathiol-modified antisense oligonucleotides were bound to 13-nm AuNPs [94].

10.8 Type of Nanoparticle Size (nm)

Various scanning types of Ref PEG-AuNPs 3–8 CT [76] will be there with the modified AuNPs 17–23 SPECT/CT [77] AuNPs 130–147 PA [78] AuNPs with citraconic amide moieties 10 PA [79]. These AuNPs will be in combination with radiotherapy in dual-energy CT [80]. Nanoscale will show the deficiency in that protein, which will not produce the protein itself [92, 95–97]. AuNPs, which are modified with DNA, can identify the various cancer cells [98]. Paciotti et al. had shown that the PEGylated AuNPs as a carrier for tumor necrosis factor (TNF) can be used as the cell-signaling protein, which will have the ability for the apoptosis induction in the healthy cells [99]. The doxorubicin is an important cancer therapeutic agent having limited ability and having various degrees of cardiac toxicity. The combination of the AuNPs and doxorubicin concluded the cardiotoxicity to mice while being able to treat cancer [100]. Dixit et al. showed that the selective delivery of folic acid coated AuNPs into folate receptor (FR) positive cancer cells can be used to compare the anti-tumor activity [101]. Here the important drawback is the nanoparticle drug delivery to the target area [102–104]. Different methods are being used like the metal nanoparticle which will be targeting the various antibodies [105–107] and also the peptides [108, 109]. An important limitation is that the various target cells need to pass through different barriers, like blood vessels and the blood-brain barrier [110]. This can be overcome by the use of magnetic NPs [111]. The different magnetic NPs will be targeting the NPs to a target site by the use of the external magnetic field, which can be seen by the magnetic anisotropy of the nanoparticle [112]. The concept here is that the change in anisotropy will change the efficacy of hypothermia [113]. Superparamagnetic metal NPs will have this tendency (they only present magnetic properties that can be seen due to the presence of a magnetic field) [114–116]. The iron oxides will be showing the few adverse effects when they are labeled in the cell labeling [117–119]. By coating the iron, this problem of the potential toxicity can be overcome [120]. Different coating materials are being used like silica [121–123], polymers [124, 125], gold [126], or silver [127]. Gold has lower pharmaceutical activity [128], and silver is also used in different biomedical applications for many years [129, 130]. The combination of a superparamagnetic core when added to the inert and safe metal coating NPs will have better characteristics than non-magnetic metal particles [131]. This will reduce the toxicity, and also the coating can be conjugated of functionalized molecules on the surface, like the drugs and biomolecules for application in the medical field. The core–shell nanoparticle will have the superparamagnetic core–shell which will be directing the nanoparticle in the body [132]. The use of the AuNPs conjugates can deliver the lower nanoparticle sizes (nm) [90]. The AuNPs are conjugated to an oligonucleotide, which will be modified with thiol, and this will be acting as the better mode for the gene therapy [93]. AuNPs conjugates will be showing the antisense oligonucleotide, which will be modified with tetra-thiol groups.

10.9 Current Medicinal Uses of Gold-Coated Iron Oxide Nanoparticles

The core–shell superparamagnetic NPs can be assessed for their biomedical uses [133] with the majority of research is on the use of gold as a shell for the NPs due to their biocompatibility and ease of binding to a variety of materials, and the magnetic carrier is one of the examples for the [133–137]. Kayal and Ramanujan have tested an in vitro apparatus simulating the human circulatory system as a test for the magnetic delivery of gold-coated iron oxide NPs (Au-Fe_3O_4) loaded with doxorubicin [135]. In this process, many NPs are being captured in the magnetic fields, signifying the magnetic NPs used in drug delivery. Bhana et al. have shown that the core–shell system, when used with the combination therapy for the different cancer cell lines, has different cancer cells like head and neck (KB-3-1) and breast (SK-BR-3) when they are combined with the photothermal and photodynamic therapy [138]. In the photothermal therapy, the AuNPs will be coated with ligands like PEG [83]; these NPs, when they are irradiated with the laser, the wavelength will match with the UV-vis λ-max of the AuNPs [135]. These NPs will vibrate with the laser's frequency, which will cause the death of the surrounding tissues [139]. This introduction is with the superparamagnetic targeting in the targeting of the therapy. One of the studies shows that gold hybrid NPs in colorectal cancer in photothermal therapy led to cellular apoptosis after the therapy, which will show increased uptake, thus reducing the threshold therapeutic levels [140]. This use of core–shell NPs in the photothermal therapy of cancer is also documented [141, 142]. Metal NPs like the core–shell NPs in T1- and T2-weighted imaging in MRI [143], are also used as biosensors [143]. The concurrent use of the magnetic carrier, which will be using the photothermal therapy, will show the various multifunctional NPs when combined with photothermal therapy [144]. The core–shell metal NPs will also show the impact on the enzyme prodrug therapy (DEPT) [111, 132, 145, 146]. The department can be the prodrug-activating enzyme leading tumor site accumulation, and here the enzyme can activate the prodrug [145, 147] or antibodies [148, 149], rather than an active targeting system for enzyme delivery. A novel approach to the therapy will be the conjugation of a genetically modified prodrug-activating enzyme on the surface of nanoparticle conjugates, which will have the anticancer therapy [150].

Different conjugates that will be targeting the cancer cells by the property of cell apoptosis [99], like folic acid, can be used as the PEG linker in nucleic acid production [101]. The use of the AuNPs with doxorubicin can be used in tumor targeting/therapy [100]. Lymphoma cell drug delivery will also have AuNPs where the conjugated cellular uptake peptides to the lymphoma cells [102].

10.10 Conclusion

Different subsets of the use of the various NPs with the MRI can be more specific in diagnosis as well as the theronostics aspects. This will be the futuristic era in the concept of personalized medicine.

References

1. Stephen ZR, Kievit FM, Zhang M. Magnetic nanoparticles for medical MR imaging. Mater Today. 2011;14:330–8.
2. Mornet S, Vasseur S, Grasset F, et al. Magnetic nanoparticle design for medical diagnosis and therapy. J Mater Chem. 2004;14:2161–75.
3. Rümenapp C, Gleich B, Haase A. Magnetic nanoparticles in magnetic resonance imaging and diagnostics. Pharm Res. 2012;29:1165–79.
4. Kabasawa H. MR imaging in the 21st century: technical innovation over the first two decades. Magn Reson Med Sci. 2022;21:71–82.
5. Soher BJ, Dale BM, Merkle EM. A review of MR physics: 3 T versus 1.5 T. Magn Reson Imaging Clin N Am. 2007;15(3):277–90.
6. Hawkes RC, Holland GN, Moore WS, Worthington BS. Nuclear magnetic resonance (NMR) tomography of the brain: a preliminary clinical assessment with demonstration of pathology. J Comput Assist Tomogr. 1980;4(5):577–86.
7. Soher BJ, Dale BM, Merkle EM. A review of MR physics: 3T versus 1.5T. Magn Reson Imaging Clin N Am. 2007;15(3):277–90, v.
8. Chavhan GB, Babyn PS, Thomas B, Shroff MM, Haacke EM. Principles, techniques, and applications

of T2*-based MR imaging and its special applications. Radiographics. 2009;29(5):1433–49.
9. Jung BA, Weigel M. Spin echo magnetic resonance imaging. J Magn Reson Imaging. 2013;37(4): 805–17.
10. Gautam A, Kapoor P, Chaudhary K, Kumar R, Raghava G, Consortium SDD. Tumor homing peptides as molecular probes for cancer therapeutics, diagnostic s and theranostics. Curr Med Chem. 2014;21:2367–91.
11. Zhi D, Yang T, Yang J, Fu S, Zhang S. Targeting strategies for superparamagnetic iron oxide nanoparticles in cancer therapy. Acta Biomater. 2020;102:13–34.
12. Price PM, Mahmoud WE, Al-Ghamdi AA, Bronstein LM. Magnetic drug delivery: where the field is going. Front Chem. 2018;6:619.
13. Sjogren CE, Briley-Saebo K, Hanson M, Johansson C. Magnetic characterization of iron oxides for magnetic resonance imaging. Magn Reson Med. 1994;31(3):268–72.
14. Filippousi M, Angelakeris M, Sikini M, et al. Surfactant effects on the structural and magnetic properties of iron oxide nanoparticles. J Phys Chem C. 2014;118:16209–17.
15. Michel MM, Chen Y, Geisler WS, Seidemann E. An illusion predicted by V1 population activity implicates cortical topography in shape perception. Nat Neurosci. 2013;16(10):1477–83.
16. Kaim AH, Jundt G, Wischer T, et al. Functional-morphologic MR imaging with ultrasmall superparamagnetic particles of iron oxide in acute and chronic soft-tissue infection: study in rats. Radiology. 2003;227:169–74.
17. Alam SR, Shah AS, Richards J, et al. Ultrasmall superparamagnetic particles of iron oxide in patients with acute myocardial infarction: early clinical experience. Circ Cardiovasc Imaging. 2012;5:559–65.
18. Kumar P, Agnihotri S, Roy I. Synthesis of dox drug conjugation and citric acid stabilized superparamagnetic iron-oxide nanoparticles for drug delivery. Biochem Physiol. 2016;5:2.
19. Lassenberger A, Scheberl A, Stadlbauer A, Stiglbauer A, Helbich T, Reimhult E. Individually stabilized, superparamagnetic nanoparticles with controlled shell and size leading to exceptional stealth properties and high relaxivities. ACS Appl Mater Interfaces. 2017;9:3343–53. https://doi.org/10.1021/acsami.6b12932.
20. Mu K, Zhang S, Ai T, Jiang J, Yao Y, Jiang L, Zhou Q, Xiang H, Zhu Y, Yang X, et al. Monoclonal antibody-conjugated superparamagnetic iron oxide nanoparticles for imaging of epidermal growth factor receptor-targeted cells and gliomas. Mol Imaging. 2015;14:2–12. https://doi.org/10.2310/7290.2015.00002.
21. Tadic M, Panjan M, Damnjanovic V, Milosevic I. Magnetic properties of hematite (α-Fe_2O_3) nanoparticles prepared by hydrothermal synthesis method. Appl Surf Sci. 2014;320:183–7.
22. Petri-Fink A, Hofmann H. Superparamagnetic iron oxide nanoparticles (SPIONs): from synthesis to in vivo studies—a summary of the synthesis, characterization, in vitro, and in vivo investigations of SPIONs with particular focus on surface and colloidal properties. IEEE Trans Nano Biosci. 2007;6:289–97.
23. Kania G, Sternak M, Jasztal A, Chlopicki S, Błażejczyk A, Nasulewicz-Goldeman A, Wietrzyk J, Jasiński K, Skórka T, Zapotoczny S, et al. Uptake and bioreactivity of charged chitosan-coated superparamagnetic nanoparticles as promising contrast agents for magnetic resonance imaging. Nanomedicine. 2018;14:131–40.
24. Unterweger H, Dézsi L, Matuszak J, Janko C, Poettler M, Jordan J, Bäuerle T, Szebeni J, Fey T, Boccaccini AR, et al. Dextran-coated superparamagnetic iron oxide nanoparticles for magnetic resonance imaging: evaluation of size-dependent imaging properties, storage stability and safety. Int J Nanomedicine. 2018;13:1899–915.
25. Yu S, Gonzales-Moragas L, Milla M, Kolovou A, Santarella-Mellwig R, Schwab Y, Laromaine A, Roig A. Bio-identity and fate of albumin-coated SPIONs evaluated in cells and by the *C. elegans* model. Acta Biomater. 2016;43:348–57.
26. Sakulkhu U, Mahmoudi M, Maurizi L, Salaklang J, Hofmann H. Superparamagnetic iron oxide nanoparticles with various physico-chemical properties and coatings. Sci Rep. 2014;4:5020.
27. Yu S, Perálvarez-Marín A, Minelli C, Faraudo J, Roig A, Laromaine A. Albumin-coated SPIONs: an experimental and theoretical evaluation of protein conformation, binding affinity and competition with serum proteins. Nanoscale. 2016;8:14393–405.
28. Shevtsov MA, Nikolaev BP, Ryzhov VA, Yakovleva LY, Marchenko YY, Parr MA, Rolich VI, Mikhrina AL, Dobrodumov AV, Pitkin E, et al. Ionizing radiation improves glioma-specific targeting of superparamagnetic iron oxide nanoparticles conjugated with cmHsp70.1 monoclonal antibodies (SPION–cmHsp70.1). Nanoscale. 2015;7:20652–64.
29. Jeon M, Halbert MV, Stephen ZR, Zhang M. Iron oxide nanoparticles as t1 contrast agents for magnetic resonance imaging: fundamentals, challenges, applications, and prospectives. Adv Mater. 2020;33:1906539. https://doi.org/10.1002/adma.201906539.
30. Vermeij EA, Koenders MI, Bennink MB, Crowe LA, Maurizi L, Vallée J-P, Hofmann H, Van den Berg VB, Van Lent PLEM, Van de Loo FAJ. The in-vivo use of superparamagnetic iron oxide nanoparticles to detect inflammation elicits a cytokine response but does not aggravate experimental arthritis. PLoS One. 2015;10:e0126687.
31. Hendee WR, Morgan CJ. Magnetic resonance imaging. Part 1—Physical principles. West J Med. 1984;141:491–500.

32. Nahrendorf M, Zhang H, Hembrador S, Panizzi P, Sosnovik DE, Aikawa E, Libby P, Swirski FK, Weissleder R. Nanoparticle PET-CT imaging of macrophages in inflammatory atherosclerosis. Circulation. 2008;117:379–87.
33. Sosnovik DE, Nahrendorf M, Weissleder R. Molecular magnetic resonance imaging in cardiovascular medicine. Circulation. 2007;115:2076–86.
34. Cai H, Dai X, Wang X, Tan P, Gu L, Luo Q, et al. A nanostrategy for efficient imaging-guided antitumor therapy through a stimuli-responsive branched polymeric prodrug. Adv Science. 2020;7:1903243.
35. de Moura CL, Gallo J, García-Hevia L, Pessoa ODL, Ricardo NMPS, Bañobre-López M. Magnetic hybrid wax nanocomposites as externally controlled theranostic vehicles: high MRI enhancement and synergistic magnetically assisted thermo/chemo therapy. Chem Eur J. 2020;26:4531–8.
36. Park JH, von Maltzahn G, Ruoslahti E, Bhatia SN, Sailor MJ. Micellar hybrid nanoparticles for simultaneous magnetofluorescent imaging and drug delivery. Angew Chem Int Ed Engl. 2008;47:7284–8.
37. Josephson L, Kircher MF, Mahmood U, Tang Y, Weissleder R. Near-infrared fluorescent nanoparticles as combined MR/optical imaging probes. Bioconjug Chem. 2002;13:554–60.
38. Cha EJ, Jang ES, Sun IC, Lee IJ, Ko JH, Kim YI, Kwon IC, Kim K, Ahn CH. Development of MRI/NIRF 'activatable' multimodal imaging probe based on iron oxide nanoparticles. J Control Release. 2011;155:152–8.
39. He X, Gao J, Gambhir SS, Cheng Z. Near-infrared fluorescent nanoprobes for cancer molecular imaging: status and challenges. Trends Mol Med. 2010;16:574–83.
40. Hasebroock KM, Serkova NJ. Toxicity of MRI and CT contrast agents. Expert Opin Drug Metab Toxicol. 2009;5:403–16.
41. Wang YX. Superparamagnetic iron oxide based MRI contrast agents: current status of clinical application. Quant Imag Med Surg. 2011;1:35–40.
42. Kim D-K, Kim J-W, Jeong Y-Y, Jon S-Y. Antibiofouling polymer coated gold@iron oxide nanoparticle (gion) as a dual contrast agent for CT and MRI. Bull Kor Chem Soc. 2009;30:1855–7.
43. Cherry SR. Multimodality imaging: beyond PET/CT and SPECT/CT. Semin Nucl Med. 2009;39:348–53.
44. Yang X, Hong H, Grailer JJ, Rowland IJ, Javadi A, Hurley SA, Xiao Y, Yang Y, Zhang Y, Nickles RJ, et al. CRGD-functionalized, dox-conjugated, and (6)(4) Cu-labeled superparamagnetic iron oxide nanoparticles for targeted anticancer drug delivery and PET/MR imaging. Biomaterials. 2011;32:4151–60.
45. Lee HY, Li Z, Chen K, Hsu AR, Xu C, Xie J, Sun S, Chen X. PET/MRI dual-modality tumor imaging using arginine-glycine-aspartic (RGD)-conjugated radiolabeled iron oxide nanoparticles. J Nucl Med. 2008;49:1371–9.
46. Madru R, Kjellman P, Olsson F, Wingardh K, Ingvar C, Stahlberg F, Olsrud J, Latt J, Fredriksson S, Knutsson L, et al. ^{99m}Tc-labeled superparamagnetic iron oxide nanoparticles for multimodality SPECT/MRI of sentinel lymph nodes. J Nucl Med. 2012;53:459–63.
47. Kim D, Yu MK, Lee TS, Park JJ, Jeong YY, Jon S. Amphiphilic polymer-coated hybrid nanoparticles as CT/MRI dual contrast agents. Nanotechnology. 2011;22:155101.
48. Szpak A, Fiejdasz S, Prendota W, Strączek T, Kapusta C, Szmyd J, Nowakowska M, Zapotoczny S. T1–T2 dual-modal MRI contrast agents based on superparamagnetic iron oxide nanoparticles with surface attached gadolinium complexes. J Nanopart Res. 2014;16:2678.
49. Santra S, Kaittanis C, Grimm J, Perez JM. Drug/dye-loaded, multifunctional iron oxide nanoparticles for combined targeted cancer therapy and dual optical/magnetic resonance imaging. Small. 2009;5:1862–8.
50. Drake P, Cho H-J, Shih P-S, Kao C-H, Lee K-F, Kuo C-H, Lin X-Z, Lin Y-J. Gd-doped iron-oxide nanoparticles for tumour therapy via magnetic field hyperthermia. J Mater Chem. 2007;17:4914.
51. Silva AC, Oliveira TR, Mamani JB, Malheiros SM, Malavolta L, Pavon LF, Sibov TT, Amaro E Jr, Tannus A, Vidoto EL, et al. Application of hyperthermia induced by superparamagnetic iron oxide nanoparticles in glioma treatment. Int J Nanomedicine. 2011;6:591–603.
52. Zhao Q, Wang L, Cheng R, Mao L, Arnold RD, Howerth EW, Chen ZG, Platt S. Magnetic nanoparticle-based hyperthermia for head & neck cancer in mouse models. Theranostics. 2012;2:113–21.
53. Laurent S, Dutz S, Hafeli UO, Mahmoudi M. Magnetic fluid hyperthermia: focus on superparamagnetic iron oxide nanoparticles. Adv Colloid Interf Sci. 2011;166:8–23.
54. Lee JH, Lee K, Moon SH, Lee Y, Park TG, Cheon J. All-in-one target-cell-specific magnetic nanoparticles for simultaneous molecular imaging and siRNA delivery. Angew Chem Int Ed Engl. 2009;48:4174–9.
55. Quan Q, Xie J, Gao H, Yang M, Zhang F, Liu G, Lin X, Wang A, Eden HS, Lee S, et al. HSA coated iron oxide nanoparticles as drug delivery vehicles for cancer therapy. Mol Pharm. 2011;8:1669–76.
56. Yu MK, Jeong YY, Park J, Park S, Kim JW, Min JJ, Kim K, Jon S. Drug-loaded superparamagnetic iron oxide nanoparticles for combined cancer imaging and therapy in vivo. Angew Chem Int Ed Engl. 2008;47:5362–5.
57. Hoskins C, Min Y, Gueorguieva M, McDougall C, Volovick A, Prentice P, Wang Z, Melzer A, Cuschieri A, Wang L. Hybrid gold-iron oxide nanoparticles as a multifunctional platform for biomedical application. J Nanobiotechnol. 2012;10:27.
58. Ray S, Li Z, Hsu CH, et al. Dendrimer-and copolymer-based nanoparticles for magnetic resonance cancer theranostics. Theranostics. 2018;8:6322–49.
59. Mahajan UM, Teller S, Sendler M, et al. Tumour-specific delivery of siRNA-coupled superparamag-

netic iron oxide nanoparticles, targeted against PLK1, stops progression of pancreatic cancer. Gut. 2016;65:1838–49.
60. Liang Y, Xie J, Yu J, Zheng Z, Liu F, Yang A. Recent advances of high performance magnetic iron oxide nanoparticles: Controlled synthesis, properties tuning and cancer theranostics. Nano Sel. 2021;2:216–50. https://doi.org/10.1002/nano.202000169.
61. Zhao S, Yu X, Qian Y, Chen W, Shen J. Multifunctional magnetic iron oxide nanoparticles: an advanced platform for cancer theranostics. Theranostics. 2020;10:6278–309.
62. Revia RA, Stephen ZR, Zhang M. Theranostic nanoparticles for RNA-based cancer treatment. Acc Chem Res. 2019;52:1496–506.
63. Kwon KC, Jo E, Kwon YW, et al. Superparamagnetic gold nanoparticles synthesized on protein particle scaffolds for cancer theragnosis. Adv Mater. 2017;29:1701146.
64. Beik J, Asadi M, Khoei S, et al. Simulation-guided photothermal therapy using MRI-traceable iron oxide-gold nanoparticle. J Photochem Photobiol B Biol. 2019;199:111599.
65. Royal College of Radiologists, Padhani A, Allen C, Carey B. Recommendations for cross-sectional imaging in cancer management: computed tomography—CT magnetic resonance imaging—MRI positron emission tomography—PET-CT. 2014. www.rcr.ac.uk.
66. McRobbie DW, Moore EA, Graves MJ, Prince MR. MRI from picture to proton. Cambridge: Cambridge University Press; 2006.
67. Murphy KJ, Brunberg JA, Cohan RH. Adverse reactions to gadolinium contrast media: a review of 36 cases. Am J Roentgenol. 1996;167:847–9.
68. Rhee CM, Bahn I, Alexander EK, Brunelli SM. Association between iodinated contrast media exposure and incident hyperthyroidism and hypothyroidism. Arch Intern Med. 2012;172:153.
69. Lee M, Lee G, Park S-S, et al. Synthesis of TiO_2/SiO_2 nanoparticles in a water-in-carbon-dioxide microemulsion and their photocatalytic activity. Res Chem Intermediates. 2005;31:379–89.
70. Blasiak B, Van Veggel FCJM, Tomanek B. Applications of nanoparticles for MRI cancer diagnosis and therapy. J Nanomater. 2013;2013:13–5.
71. Stacul F, van der Molen AJ, Reimer P, et al. Contrast induced nephropathy: updated ESUR contrast media safety committee guidelines. Eur Radiol. 2011;21:2527–41.
72. Li W, Chen X. Gold nanoparticles for photoacoustic imaging. Nanomedicine. 2015;10:299–320.
73. van Schooneveld MM, Cormode DP, Koole R, et al. A fluorescent, paramagnetic and PEGylated gold/silica nanoparticle for MRI, CT and fluorescence imaging. Contrast Media Mol Imaging. 2010;5:231–6.
74. Cheheltani R, Ezzibdeh RM, Chhour P, et al. Tunable, biodegradable gold nanoparticles as contrast agents for computed tomography and photoacoustic imaging. Biomaterials. 2016;102:87–97.
75. Kojima C, Umeda Y, Ogawa M, et al. X-ray computed tomography contrast agents prepared by seeded growth of gold nanoparticles in PEGylated dendrimer. Nanotechnology. 2010;21:245104.
76. Umeda Y, Kojima C, Harada A, et al. PEG-attached PAMAM dendrimers encapsulating gold nanoparticles: growing gold nanoparticles in the dendrimers for improvement of their photothermal properties. Bioconjug Chem. 2010;21:1559–64.
77. Li X, Wang C, Tan H, et al. Gold nanoparticles-based SPECT/CT imaging probe targeting for vulnerable atherosclerosis plaques. Biomaterials. 2016;108:71–80.
78. Wang Y, Xie X, Wang X, et al. Photoacoustic tomography of a nanoshell contrast agent in the in vivo rat brain. Nano Lett. 2004;4:1689–92.
79. Song J, Kim J, Hwang S, et al. "Smart" gold nanoparticles for photoacoustic imaging: an imaging contrast agent responsive to the cancer microenvironment and signal amplification via pH-induced aggregation. Chem Commun. 2016;52:8287–90.
80. Ashton JR, Castle KD, Qi Y, et al. Dual-energy CT imaging of tumor liposome delivery after gold nanoparticle-augmented radiation therapy. Theranostics. 2018;8:1782–97.
81. Xu M, Wang LV. Photoacoustic imaging in biomedicine. Rev Sci Instrum. 2006;77:2115–20.
82. Huang X, El-Sayed IH, Qian W, El-Sayed MA. Cancer cell imaging and photothermal therapy in the near-infrared region by using gold nanorods. J Am Chem Soc. 2006;128:2115–20.
83. Hirsch LR, Stafford RJ, Bankson JA, et al. Nanoshell-mediated near-infrared thermal therapy of tumors under magnetic resonance guidance. Proc Natl Acad Sci. 2003;100:13549–54.
84. Woodford O, Harriman A, McFarlane W, Wills C. Dramatic effect of solvent on the rate of photobleaching of organic pyrrole-BF 2 (BOPHY) dyes. ChemPhotoChem. 2017;1:317–25.
85. Ashton JR, Gottlin EB, Patz EF, et al. A comparative analysis of EGFR-targeting antibodies for gold nanoparticle CT imaging of lung cancer. PLoS One. 2018;13:1–20.
86. Buzea C, Pacheco II, Robbie K. Nanomaterials and nanoparticles: sources and toxicity. Biointerphases. 2007;2:MR17–71.
87. Moyano DF, Goldsmith M, Solfiell DJ, et al. Nanoparticle hydrophobicity dictates immune response. J Am Chem Soc. 2012;134:3965–7.
88. Boraschi D, Duschl A. Nanoparticles and the immune system: safety and effects. Oxford, UK: Academic; 2013.
89. Lammers T, Kiessling F, Hennink WE, Storm G. Drug targeting to tumors: principles, pitfalls and (pre-) clinical progress. J Control Release. 2012;161:175–87.
90. Bhumkar DR, Joshi HM, Sastry M, Pokharkar VB. Chitosan reduced gold nanoparticles as novel carriers for transmucosal delivery of insulin. Pharm Res. 2007;24:1415–26.

91. Nam J, La WG, Hwang S, et al. pH-responsive assembly of gold nanoparticles and "spatiotemporally concerted" drug release for synergistic cancer therapy. ACS Nano. 2013;7:3388–402.
92. Ding Y, Jiang Z, Saha K, et al. Gold nanoparticles for nucleic acid delivery. Mol Ther. 2014;22: 1075–83.
93. Sandström P, Boncheva M, Åkerman B. Nonspecific and thiol-specific binding of DNA to gold nanoparticles. Langmuir. 2003;19:7537–43.
94. Rosi NL, Giljohann DA, Thaxton CS, et al. Oligonucleotide-modified gold nanoparticles for intracellular gene regulation. Science. 2006;312:1027–30.
95. Giljohann DA, Seferos DS, Prigodich AE, et al. Gene regulation with polyvalent siRNA–nanoparticle conjugates. J Am Chem Soc. 2009;131:2072–3.
96. Yonezawa T, Onoue S, Kimizuka N. Metal coating of DNA molecules by cationic, metastable gold nanoparticles. Chem Lett. 2002;31:1172–3.
97. Barron N, Piskareva O, Muniyappa M. Targeted genetic modification of cell lines for recombinant protein production. Cytotechnology. 2007;53:65–73.
98. Niidome T, Nakashima K, Takahashi H, Niidome Y. Preparation of primary amine-modified gold nanoparticles and their transfection ability into cultivated cells. Chem Commun. 2004; https://doi.org/10.1039/B406189F.
99. Paciotti GF, Myer L, Weinreich D, et al. Colloidal gold: a novel nanoparticle vector for tumor directed drug delivery. Drug Deliv. 2004;11:169–83.
100. Du Y, Xia L, Jo A, et al. Synthesis and evaluation of doxorubicin-loaded gold nanoparticles for tumor-targeted drug delivery. Bioconjug Chem. 2018;29:420–30.
101. Dixit V, Van Den Bossche J, Sherman DM, et al. Synthesis and grafting of thioctic acid-PEG-folate conjugates onto Au nanoparticles for selective targeting of folate receptor-positive tumor cells. Bioconjug Chem. 2006;17:603–9.
102. Kalimuthu K, Lubin B C, Bazylevich A, et al. Gold nanoparticles stabilize peptide-drug-conjugates for sustained targeted drug delivery to cancer cells. J Nanobiotechnol. 2018;16:34.
103. Gao H. Progress and perspectives on targeting nanoparticles for brain drug delivery. Acta Pharm Sin B. 2016;6:268–86.
104. Rosenblum D, Joshi N, Tao W, et al. Progress and challenges towards targeted delivery of cancer therapeutics. Nat Commun. 2018;9:1–12.
105. Friedman AD, Claypool SE, Liu R. The smart targeting of nanoparticles. Curr Pharm Des. 2013;19:6315–29.
106. Fay F, Scott CJ. Antibody-targeted nanoparticles for cancer therapy. Immunotherapy. 2011;3:381–94.
107. Carter T, Mulholland P, Chester K. Antibody-targeted nanoparticles for cancer treatment. Immunotherapy. 2016;8:941–58.
108. Okur AC, Erkoc P, Kizilel S. Targeting cancer cells via tumor-homing peptide CREKA functional PEG nanoparticles. Colloids Surf B Biointerfaces. 2016;147:191–200.
109. King A, Ndifon C, Lui S, et al. Tumor-homing peptides as tools for targeted delivery of payloads to the placenta. Sci Adv. 2016;2:1–16.
110. Pardridge WM. Blood–brain barrier delivery. Drug Discov Today. 2007;12:54–61.
111. Gwenin VV, Gwenin CD, Kalaji M. Colloidal gold modified with a genetically engineered nitroreductase: toward a novel enzyme delivery system for cancer prodrug therapy. Langmuir. 2011;27:14300–7.
112. Obaidat I, Issa B, Haik Y. Magnetic properties of magnetic nanoparticles for efficient hyperthermia. Nanomaterials. 2015;5:63–89.
113. Bødker F, Mørup S, Linderoth S. Surface effects in metallic iron nanoparticles. Phys Rev Lett. 1994;72:282–5.
114. Mody VV, Cox A, Shah S, et al. Magnetic nanoparticle drug delivery systems for targeting tumor. Appl Nanosci. 2014;4:385–92.
115. Donaldson JD, Beyersmann D. Cobalt and cobalt compounds. In: Ullmann's encyclopedia of industrial chemistry. Weinheim: Wiley-VCH Verlag GmbH & Co. KGaA; 2005. p. 467–97.
116. Crossgrove J, Zheng W. Manganese toxicity upon overexposure. NMR Biomed. 2004;17:544–53.
117. Karlsson HL, Gustafsson J, Cronholm P, Möller L. Size-dependent toxicity of metal oxide particles—a comparison between nano- and micrometer size. Toxicol Lett. 2009;188:112–8.
118. Soto K, Garza KM, Murr LE. Cytotoxic effects of aggregated nanomaterials. Acta Biomater. 2007;3:351–8.
119. Brunner TJ, Wick P, Manser P, et al. In vitro cytotoxicity of oxide nanoparticles: comparison to asbestos, silica, and the effect of particle solubility. Environ Sci Technol. 2006;40:4374–81.
120. Chia SL, Leong DT. Reducing ZnO nanoparticles toxicity through silica coating. Heliyon. 2016;2:1–18.
121. Mine E, Yamada A, Kobayashi Y, et al. Direct coating of gold nanoparticles with silica by a seeded polymerization technique. J Colloid Interface Sci. 2003;264:385–90.
122. Cerruti MG, Sauthier M, Leonard D, et al. Gold and silica-coated gold nanoparticles as thermographic labels for DNA detection. Anal Chem. 2006;78:3282–8.
123. Kobayashi Y, Nagasu R, Shibuya K, et al. Synthesis of a colloid solution of silica-coated gold nanoparticles for X-ray imaging applications. J Nanopart Res. 2014;16:1–13.
124. Tural B, Özkan N, Volkan M. Preparation and characterization of polymer coated superparamagnetic magnetite nanoparticle agglomerates. J Phys Chem Solids. 2009;70:860–6.
125. Wang Y, Dave RN, Pfeffer R. Polymer coating/encapsulation of nanoparticles using a supercritical anti-solvent process. J Supercrit Fluids. 2004;28: 85–99.

126. Chen D, Li J, Shi C, et al. Properties of core-shell Ni-Au nanoparticles synthesized through a redox-transmetalation method in reverse microemulsion. Chem Mater. 2007;19:3399–405.
127. Jana NR. Silver coated gold nanoparticles as new surface enhanced Raman substrate at low analyte concentration. Analyst. 2003;128:954.
128. Dierks S. Gold “MSDS”. Electronic Space Products International; 2005. p. 1–6.
129. Alexander JW. History of the medical use of silver. Surg Infect. 2009;10:289–92.
130. Marx DE, Barillo DJ. Silver in medicine: the basic science. Burns. 2014;40:S9–S18.
131. Shubayev VI, Pisanic TR, Jin S. Magnetic nanoparticles for theragnostics. Adv Drug Deliv Rev. 2009;61:467–77.
132. Gwenin VV, Gwenin CD, Kalaji M. Gold coated magnetic particles enabling nitroreductase delivery to cancer cells. European Union Patent PCT/EP2010/062871. 2009.
133. McBain SC, Yiu HHP, Dobson J. Magnetic nanoparticles for gene and drug delivery. Int J Nanomedicine. 2008;3:169–80.
134. Veiseh O, Gunn JW, Zhang M. Design and fabrication of magnetic nanoparticles for targeted drug delivery and imaging. Adv Drug Deliv Rev. 2010;62:284–304.
135. Kayal S, Ramanujan RV. Anti-cancer drug loaded iron–gold core–shell nanoparticles (Fe@Au) for magnetic drug targeting. J Nanosci Nanotechnol. 2010;10:5527–39.
136. Dobson J. Magnetic nanoparticles for drug delivery. Drug Dev Res. 2006;67:55–60.
137. Arruebo M, Fernández-Pacheco R, Ibarra MR, Santamaría J. Magnetic nanoparticles for drug delivery. Nanotoday. 2007;2:22–32.
138. Bhana S, Lin G, Wang L, et al. Near-infrared-absorbing gold nanopopcorns with iron oxide cluster core for magnetically amplified photothermal and photodynamic cancer therapy. ACS Appl Mater Interfaces. 2015;7:11637–47.
139. Abadeer NS, Murphy CJ. Recent progress in cancer thermal therapy using gold nanoparticles. J Phys Chem C. 2016;120:4691–716.
140. Kirui DK, Rey DA, Batt CA. Gold hybrid nanoparticles for targeted phototherapy and cancer imaging. Nanotechnology. 2010;21:1–10.
141. Lal S, Clare SE, Halas NJ. Photothermal therapy: impending clinical impact. Acc Chem Res. 2008;41:1842–51.
142. Sotiriou GA, Starsich F, Dasargyri A, et al. Photothermal killing of cancer cells by the controlled plasmonic coupling of silica-coated Au/Fe_2O_3 nanoaggregates. Adv Funct Mater. 2014;24:2818–27.
143. Cho S-J, Jarrett BR, Louie AY, Kauzlarich SM. Gold-coated iron nanoparticles: a novel magnetic resonance agent for T1 and T2 weighted imaging. Nanotechnology. 2006;17:640–4.
144. Cheng L, Yang K, Li Y, et al. Multifunctional nanoparticles for upconversion luminescence/MR multimodal imaging and magnetically targeted photothermal therapy. Biomaterials. 2012;33:2215–22.
145. Searle PF, Chen MJ, Hu L, et al. Nitroreductase: a prodrug-activating enzyme for cancer gene therapy. Clin Exp Pharmacol Physiol. 2004;31:811–6.
146. Patel P, Young JG, Mautner V, et al. A phase I/II clinical trial in localized prostate cancer of an adenovirus expressing nitroreductase with CB1984. Mol Ther. 2009;17:1292–9.
147. Palmer DH, Mautner V, Mirza D, et al. Virus-directed enzyme prodrug therapy: intratumoral administration of a replication-deficient adenovirus encoding nitroreductase to patients with resectable liver cancer. J Clin Oncol. 2004;22:1546–52.
148. Bagshawe KD. Antibody-directed enzyme prodrug therapy professor. In: Valentino S, Borchardt R, Hageman M, et al., editors. Prodrugs: challenges and rewards. New York: Springer; 2005. p. 526–36.
149. Sharma SK, Chester KA, Bagshawe KD. Antibody-directed enzyme prodrug therapy (ADEPT). In: Handbook of therapeutic antibodies, vol. 1. 2nd ed; 2014. p. 475–86.
150. Gwenin CD, Kalaji M, Williams PA, Jones RM. The orientationally controlled assembly of genetically modified enzymes in an amperometric biosensor. Biosens Bioelectron. 2007;22:2869–75.

11 PET-CT Nanoimaging

11.1 Introduction

Different radiolabeled inorganic nanoparticles are being developed and evaluated for their potential, which includes using probes in PET with more applications in cancer. Inorganic nanoparticles are also important in the paradigm of molecular imaging probe design, allowing them to be used in different imaging modalities, with various targeting ligands and therapeutic payloads. Developing disease-specific nanoparticles is important so they will be more useful and help in the future planning of radiolabeling strategies. These radiolabeled nanoparticles can be used in both in vitro and in vivo applications, providing greater stability and enhanced sensitivity in the early detection of disease. They offer optimized pharmacokinetics with reduced non-specific organ uptake, leading to focused targeting of organs for better efficacy. Despite the many challenges and issues, the only radiolabeled nanoparticle known as "C-dots" (Cornell dots) has progressed to multiple clinical trials.

Nanoparticles are size-dependent with physical and chemical properties, and due to this, they are very useful in clinical oncology [11, 12]. They have different surface functions, which result in the use of personalized cancer management [11, 13]. Nanoparticle-based distribution by various carriers can be seen in the body after administration. Radiolabeling is the process of labeling specific radioisotopes to the nanoparticles [12, 15, 16]. These radiotracers will circulate in the body and can be imaged non-invasively by various molecular imaging techniques like single photon emission computed tomography (SPECT) [17], PET [18], Cerenkov luminescence (CL) [19], Cerenkov resonance energy transfer (CRET) [20], etc. These are important imaging modalities for cancer imaging in both preclinical and clinical stages. Radiolabeled nanoparticles used in molecular imaging have many applications, which are increasing day by day [16]. These radiolabeled nanoparticles act as signal amplifiers, resulting in better contrast indices as well as better sensitivities compared to conventional radiotracers. The larger surface area of nanoparticles can result in the attachment of more ligands to the surface, which can be used for the evaluation of various cancers. With the availability of multimodality and hybrid imaging techniques, these nanoparticles will yield better results. The third advantage is that the nanoparticles are capable of both diagnostic and therapeutic capabilities in a single nanoplatform, resulting in the emerging concept of "image-guided drug delivery" [17, 18]. Along with this, different therapeutic agents like chemotherapeutic drugs and therapeutic radiolabeled isotopes can also be incorporated into a similar nanoplatform for better multimodality targeted therapy for cancer management. These nanoparticles, which will be used in cancer research, are

S. Shaikh, *Nanoimaging - Future of Precision Medicine*,
https://doi.org/10.1007/978-981-97-7292-6_11

categorized into four groups: (1) inorganic nanoparticles like quantum dots, metal nanoparticles, iron oxide nanoparticles, silica nanoparticles, metal sulfide nanoparticles, and upconversion nanophosphors; (2) polymer nanoparticles such as core-shell dendrimers and amphiphilic nanoparticles; (3) lipid nanoparticles such as liposomes and solid lipid nanoparticles; and (4) carbon-based nanoparticles, such as carbon nanotubes, graphene oxide, and nanodiamonds. The different methods for radiolabeling nanoparticles have been described along with their pros and cons. The successful translation of the radiolabeled nanoparticles faces many barriers, but there will be better clinical translation and analysis in the future.

11.2 PET Imaging Using Radiolabeled Nanoparticles

With the advent of molecular imaging of cancer, various radiolabeled nanoparticles are being used for early detection, characterization, staging, and also for monitoring the response to drugs [12, 16]. PET imaging is highly sensitive and provides better quantitative information, which plays an important role in earlier disease detection and improved therapeutic decision-making [21, 22]. In this process, the radioisotope will be added to the nanoparticle, and its decay will result in the emission of positrons interacting with nearby electrons within a short distance (~1 mm) while traveling within the body [22]. The "coincidence events" of the positron emission will help reconstruct the spatial distribution of these radiolabeled nanoparticles, showing their amount and distribution in the body. The four important parameters that are needed are: (1) Radioisotope choice; (2) Type of nanoplatform; (3) Radiolabeling method; and (4) Stability of the radiolabeled agent.

11.2.1 Different Choices of the Radioisotope

The radioisotope should be suitable for radiolabeling with the nanoparticles based on these criteria: (1) the decay time of the radioisotope; (2) the production ease and availability of the radioisotopes; and (3) the radiolabeling strategy [12, 22]. It is known that radioisotopes with low positron energy and a high $\beta+$ branching ratio are favorable for PET imaging. When radioisotopes have higher positron energy, the $\beta+$ particles travel slightly longer before positron annihilation occurs, resulting in a loss of spatial resolution. Radioisotopes with a low $\beta+$ branching ratio, combined with longer scan times, can make PET images noisy. In molecular imaging of cancer by PET, the half-life of the radioisotopes should be in concurrence with the imaging procedure. Also, the decay time should be as short as possible to minimize radiation exposure concerns. In these nanoplatforms, radioisotopes must be compatible with different chemical compositions, functional and surface characteristics, including aspects such as hydrodynamic diameter, diagnostic and therapeutic agent addition, ease of conjugation with ligands for active targeting, and colloidal stability in physiological medium. Nanoparticles can be used for multimodality imaging to achieve precise cancer treatment [15]. A nanoparticle can serve as an imaging agent, in addition to the attached radioisotope [15]. Fluorescent emissions from quantum dots can serve as inorganic nanoparticles for imaging [23]; another example is superparamagnetic iron oxide nanoparticles (SPIONs) used as MRI contrast agents [24, 25]. Dyes can also be used with functional nanoparticles (such as silica nanoparticles) for PET/optical dual-modality imaging [26]. Nanoparticles can also be used in PET-guided drug delivery, where the nanoplatform needs to be chosen for precise targeting, accurate dose delivery, and minimal toxicity to enhance therapeutic efficacy [18]. The use of radiolabeled inorganic nanoparticles can offer an advantage by choosing biocompatible nanoparticles (such as silica, iron oxide, or gold nanoparticles) [14], all of which are already established for their safety, efficacy, and reliability to a great extent. Another important aspect is the biodegradability and renal clearance behavior of the nanoparticles. The radiolabeling method for this should be robust, rapid, and safe, without altering the original characteristics of the nanoparticles, which should remain efficient. This process should adhere to the principle of "as low as reasonably achievable (ALARA)" for radioactivity handling processes. Other aspects include higher temperatures and prolonged reaction times associated with various nanoparticle properties.

The important properties are the radiolabeling strategies, which are four in number: (1) The complexation reaction of radiometal ions with chelators using different chemistries; (2) The direct bombardment of nanoparticles by hadronic projectiles; (3) The synthesis of intrinsically radiolabeled nanoparticles from a mixture of radioactive and non-radioactive precursors; (4) Chelator-free post-synthetic radiolabeling. Each radiolabeling method has its advantages and limitations, and the choice can depend on the specific radioisotope, research purpose, and feasibility. Various options include the use of 1,4,7-triazacyclononane-1,4,7-triacetic acid (NOTA), 1,4,7,10-tetraazacyclododecane-1,4,7,10-tetraacetic acid (DOTA), desferrioxamine (Df), diethylene triamine pentaacetic acid (DTPA), etc., which can be selected as the chelator of choice based on the chemistry of the radiometal ion used for radiolabeling [27–30]. This strategy is easier to operate, as it requires significant structural/surface modifications that will affect the pharmacokinetics of nanoparticles for improved targeting and in vivo distribution. Other limitations are related to factors such as the radiometal chelator in high protein concentrations in the bloodstream or transchelation of the radiometal from the chelator complex, which can lead to erroneous interpretations of PET imaging results. Due to these limitations, conventional chelator-based radiolabeling remains a foundation for significant advancements in nanomedicine. The direct irradiation of various nanoparticles is performed in nuclear reactors or cyclotrons to generate radiolabeled nanoparticles. One example is direct proton irradiation applied to ^{18}O-enriched Al_2O_3 nanoparticles for the production of ^{18}F-labeled Al_2O_3 nanoparticles via the ^{18}O (p, n) ^{18}F nuclear reaction [31]. When the radioisotope is added to the nanoparticles, it reflects the distribution of the nanoparticles without leaking. The second method for incorporating radioisotopes into nanoparticles is the use of radioactive precursors for the synthesis of intrinsically radiolabeled nanoparticles [32]. In this method, radioactive precursors (which depend on the nuclear reaction by which the radioisotope is produced) are mixed with non-radioactive analogues. This alignment of radioisotopes with the crystal lattice of the nanocrystals leads to improved radiochemical stability [32]. Commonly used intrinsically radiolabeled inorganic nanoparticles for PET imaging include ^{64}CuS nanoparticles, synthesized by the reaction of $^{64}CuCl_2$, $CuCl_2$, and Na_2S at 95 °C for 1 h [33, 59]. Due to this, there is always the risk of radioactive contamination and radiation exposure to the radiation workers. For this, the entire process should be well shielded from the facility to prevent exposure and minimize radiation hazards. Recently, the post-synthesis radiolabeling of nanoparticles based on a chelator-free protocol is gaining priority [32]. This process involves various physical or chemical interactions between the radionuclide ions and nanoparticles, allowing the radioisotopes to be incorporated without changing the nanoparticle's characteristics. The chelator-free radioisotopes will show uptake based on (1) bonding and specific trapping of radionuclide ions into the nanoparticles and (2) different ion-exchange mechanisms. Different methods of trapping are seen with various radioisotopes (such as ^{64}Cu [34], ^{89}Zr [35], ^{72}As [36], and ^{69}Ge [37], etc.) and also with nanoparticles. Ion-exchange-based approaches are used in the radiolabeling of lanthanide-doped upconversion nanoparticles, which include ^{18}F [38] and ^{64}Cu [39] for PET imaging. The use of different chelator-free post-synthetic radiolabeling methods is faster and more specific, with a higher radiolabeling yield under mild reaction conditions. The stability of radiolabeled nanoparticles is of two types: (1) colloidal stability and (2) radiochemical stability [22]. Colloidal stability is an important factor in the use of different nanoplatforms for molecular imaging applications. The colloidal stability of nanoparticles can be measured by the hydrodynamic diameter using the dynamic light scattering (DLS) technique [40] or by monitoring the magnetic T2 relaxation time [41] over a specific period. Along with colloidal stability, radiochemical stability is also crucial for radiolabeled nanoparticles. The dissociation of radioisotopes can cause the uptake of radioactivity in non-target organs, resulting in PET imaging that does not show the exact distribution of radiolabeled nanoparticles in vivo. This is also influenced by the different components of the nanoparticles, such as size, shape, surface charge, and multiple attachments [41]. After in vivo administration, the radiochemical stability of these radioisotopes can be studied by injecting them into mice, where they are excreted via the renal route, and this will show a stronger radioactivity signal in the urinary bladder.

11.3 Targeting of Radiolabeled Inorganic Nanoparticles to the Tumor Tissue

The delivery of radiolabeled nanoparticles to target tissues can be achieved in two ways: passive and active targeting [42]. Passive targeting occurs due to the permeability of tumor tissue [42]. As cancer cells exhibit rapid vascularization, the capillary endothelium in cancerous tissue becomes more permeable to nanoplatforms in abnormal cells compared to the endothelium in normal tissues. If radiolabeled nanoparticles can remain in blood circulation for a longer time, they will lead to greater enrichment of nanoplatforms in tumor tissues. Since the lymphatic system is not fully developed in tumor tissue, these nanoparticles will remain in the interstitial space within the tumor. This entire process, in which the accumulation of nanoplatforms in tumor tissues occurs due to their property of enhanced permeability and retention (EPR) effect, is well established [42, 43]. The EPR phenomenon is influenced by various factors, including particle size, surface charge, hydrophobicity, immunogenicity, and tumor characteristics, all of which contribute to better optimization of passive targeting [42]. However, the specificity of passive targeting toward the tumor and the suboptimal chemotherapeutic drug concentrations at the tumor site often result in poor therapeutic efficacy [42]. This issue can be overcome by active targeting through the conjugation of functionalized nanoplatforms with ligands such as aptamers, peptides, or proteins, which promote preferential accumulation of the drug in tumor tissue [42]. Combining the effects of both passive and active targeting can achieve maximal therapeutic efficacy [42]. Biocompatible hydrophilic polymer chains, such as polyethylene glycol (PEG), enhance circulation times compared to uncoated nanoplatforms, leading to higher tumor uptake [44]. Radiolabeled quantum dots are being explored for in vivo imaging due to their properties, including high quantum yield, resistance to photobleaching, narrow emission peaks, and tunable emission wavelengths [23, 45–54]. The radiolabeling of quantum dots with positron-emitting radioisotopes adds PET functionality, enabling dual-modality PET/optical imaging. Schipper et al. have shown the comparative biodistribution between PEGylated and non-PEGylated CdSe quantum dots of two different sizes (12 nm and 21 nm diameter) in mice when injected [48]. These quantum dots, when combined with DOTA chelators, were radiolabeled with ^{64}Cu and then injected into normal mice. In vivo PET imaging studies revealed that the size of the quantum dots did not impact the biodistribution, and significant rapid uptake was seen in the liver (27.4–38.9%ID/g) and spleen (8.0–12.4%ID/g). As there was no clearance of the radiolabeled quantum dots from these organs, the quantum dots require further modifications to be suitable as molecular imaging probes [49]. The smaller the size of the quantum dots (2-nm diameter), the easier they can be cleared by both renal and hepatobiliary routes [50]. These radiolabeled quantum dots exhibit very high radiochemical stability. In vivo PET imaging showed that the clearance of the radiolabeled agent occurred via the renal and hepatobiliary routes. These modified quantum dots are suitable for use as molecular probes for imaging [51]. An amine-functionalized quantum dot (QD705; emission maximum, 705 nm; Invitrogen) was conjugated with the RGD peptide for integrin $\alpha v\beta 3$ targeting, as well as with the DOTA chelator for ^{64}Cu labeling. There was a better correlation between in vivo PET imaging and ex vivo infrared fluorescence imaging. Various components can be used for dual-modality PET/near-infrared fluorescence imaging in prostate (PC-3) tumor-bearing mice [52]. Tissue penetration limitation is one of the important components for evaluating fluorescence imaging [53]. Cerenkov luminescence imaging is another modality in which high-energy charged particles (β+, β−, or α particles) are produced, and they travel through a medium (e.g., water) at the speed of light in that medium. Cerenkov luminescence spectra are more continuous and have higher frequencies (ultraviolet/blue), and because of this, the ultraviolet/blue wavelengths experience more attenuation by tissues [54]. In the quantum dot-CRET system, Cerenkov luminescence imaging shows the energy transferred by the quantum dots in proximity. Due to this property, the blue-weighted Cerenkov luminescence can be converted to the longer quantum dot emission wavelengths, enabling the use of optical imaging in animals [55]. Sun et al. showed that the synthesis of self-illuminating quantum dots can demonstrate different efficient CRET processes [56]. This shows the significant advantage of CRET-quantum dot systems for multimodality cancer imaging.

11.4 Radiolabeled Metal Nanoparticles

Radiolabeled gold nanoparticles are commonly used in molecular imaging due to their unique size-dependent physicochemical and optical properties, better biocompatibility, easier characterization, and excellent surface chemistry [57]. Xie et al. demonstrated the use of gold nanoshells conjugated with a DOTA-based chelator for radiolabeling with ^{64}Cu [58]. The radiolabeled nanoparticles showed lower tumor uptake in PET imaging.

11.4.1 Radiolabeled Metal Sulfide Nanoparticles

The use of CuS nanoparticles in photothermal ablation therapy has yielded very good results in various cancers [61]. CuS nanoparticles have advantages over gold nanoparticles: (1) CuS has stable near-infrared absorption, and (2) it is low-cost and advantageous for theranostics [65]. The synthesis of CuS nanoparticles involves a metathesis reaction of radioactive $^{64}CuCl_2$, $CuCl_2$, and Na_2S materials. When these ^{64}CuS nanoparticles are formed, PEGylated particles are created. These PEGylated and non-PEGylated ^{64}CuS nanoparticles can be injected for in vivo PET imaging, with the uptake of non-PEGylated nanoparticles being three times higher compared to ^{64}CuS-PEG nanoparticles. These radiolabeled ^{64}CuS nanoparticles can also be used as theranostic agents [66–68] and for sentinel lymph node mapping [69]. Folic acid conjugates with ^{64}CuS-PEG nanoparticles can be used in PET imaging and photothermal ablation therapy as receptors [70]. Multicomponent nanoparticles, with a single structure comprising different particle components, can be used synergistically in various cancer theranostics applications [64]. When the nanoparticle is added to the DOTA chelator for ^{64}Cu labeling, it functions in dual-modality (PET/CT) imaging in squamous cell carcinoma head and neck cancer (HNSCC) xenografts [60] (Fig. 11.1). Much research is ongoing, especially on silica nanoparticles (Cornell dots) [74], which are radiolabeled with inorganic nanoparticles and can be used because of their distinct pharmacokinetic prop erties, such as radiochemical stability, toxicity, biodegradation, and elimination.

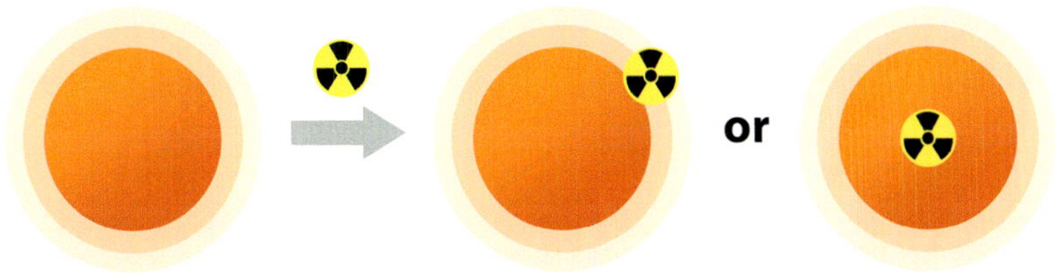

Fig. 11.1 PET + CT = PET − CT

In these key challenges, the radiolabeled nanoparticle formulations will be discussed based on the different clinical or translational points of view.

11.5 Biological Challenges

The biodistribution of radiolabeled nanoparticles will be within the compartment in which they are administered. These will be cleared by lymphatic drainage when they are injected locally. If they are given orally, these nanoparticles will be excreted in the feces. Tumor targeting is lower compared to the radiolabeled nanoparticles that are administered orally. The common mode of administration of the radiolabeled nanoparticles is by intravenous injection, and after injection, they will be based on the hydrodynamic diameter, charge, and surface characteristics [63]. The uptake of the radiolabeled nanoparticles will occur in various tumors due to the fenestrated endothelium and increased permeability of vasculature at the sites of inflammation [72]. The important limitation is to have a high tumor-to-background ratio, as there will be non-specific uptake of radiolabeled nanoparticles in healthy organs, as seen in the liver and spleen. These organs will be made up of various mononuclear phagocyte systems (MPS), such as phagocytic cells, and these will accumulate in different cells of the body, like macrophages in the liver, spleen, and lymph nodes, after injection [71, 72]. This will result in significant early and rapid uptake in these organs. This process of sequestration is due to the opsonization of nanoparticles by the adsorption of plasma proteins on the surface of radiolabeled nanoparticles [72]. The protein coating of the nanoparticles is based on various factors, such as nanoparticle size, surface charge, hydrophobicity, and surface chemistry. After protein adsorption, these radiolabeled nanoparticles will attach to the receptors on the surface of phagocytes. After this, they are internalized by the radiolabeled nanoparticles, and these will be transported to phagosomes and fused with lysosomes [72, 73]. Due to this, PEG is widely used for surface modification of various nanoparticles, which is reported for its "stealth" properties, and the different biocompatibilities of PEG will increase the circulation times of the nanoparticles and cause a reduction in opsonization, resulting in the increased uptake of radiolabeled nanoparticles in tumors by both passive targeting (EPR effect) and active targeting [74]. PEGylation will cause reduced uptake of the nanoparticles by target cells, causing better-enhanced serum protein binding, which will result in a better immune response for improved clearance of the nanoparticles [74]. The size, shape, and surface charge of the radiolabeled nanoparticles will alter the biodistribution in different organs, including the lungs, liver, spleen, and kidneys. The size of the nanoparticles will also cause a cut-off in circulation half-lives, causing extravasation in the vasculature and resulting in macrophage uptake [72]. The EPR effect in tumors also enhances nanoparticle-based tumor targeting [62], which will vary depending on the extent of tumor vascularity [72, 75]. Another important difference between human and animal experiments is the better use of targeting ligands on a nanoparticle. Another important problem is the amount of biological chelators, which will have more precise biological fate and long-term biocompatibility. As already mentioned, the size, shape, charge, and porosity will guide the biocompatibility and biodegradability of nanoparticles in vivo [72, 75] (Fig. 11.2).

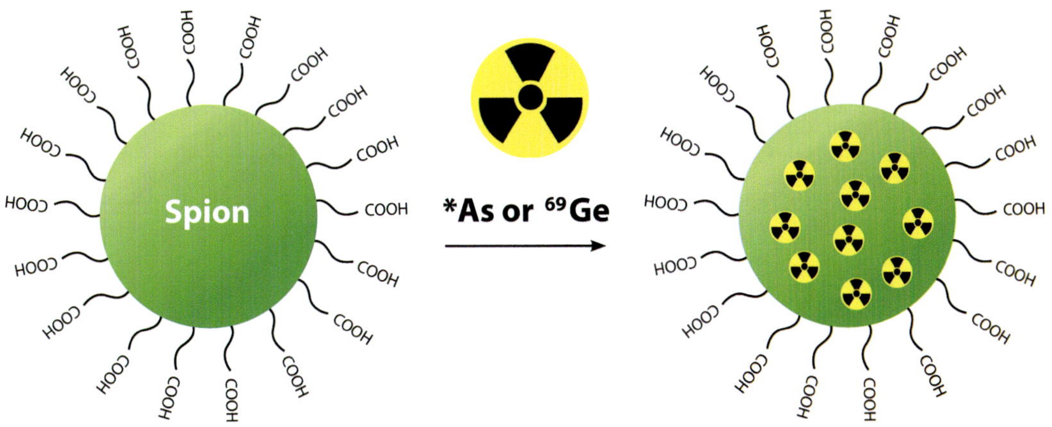

Fig. 11.2 Role of SPION in the use of radionuclides nanoimaging

11.6 Conclusion

Inorganic nanoparticles that are radiolabeled with positron-emitting radioisotopes for PET imaging are used in the detection of various cancers in different phases of preclinical and clinical settings. PET is commonly used with various modalities in which nanomaterials are employed for early-stage disease detection, as well as monitoring disease progression, regression, and recurrence. Due to advancements in science and technology, multifunctional nanoparticles now have different functions. Thus, these nanomaterial radiolabeling methods require radioisotopes to be incorporated onto nanoparticles with appreciably high radiochemical stability and minimal impact on their original biological behavior. Although great strides have been made in the use of radiolabeled inorganic nanoparticles for PET imaging of cancer in preclinical settings, most of these studies are in the 'proof-of-principle' stage, and more research efforts are needed to translate radiolabeled nanoparticles to the clinical phase in the near future. As such, when a new strategy moves forward toward clinical applications, several types of biological, technological, and regulatory challenges emerge that need to be surmounted. Besides these scientific issues, several socioeconomic and political factors might also impede the clinical translation of radiolabeled inorganic nanoplatforms for PET imaging. Additionally, complicated bureaucratic protocols in some countries, the initially limited potential market for novel molecular imaging agents, lobbying by the manufacturers of other approved imaging probes, and lack of reimbursement strategies by insurance companies for new agents, among other factors, might affect the clinical translation process. Despite these initial hurdles, the clinical translation of radiolabeled nanoparticles for the benefit of millions of cancer patients all over the world is feasible with increased interdisciplinary collaboration and knowledge exchange among various stakeholders in this field, including funding agencies, regulatory authorities, clinicians, and scientists.

References

1. Gambhir SS, Czernin J, Schwimmer J, Silverman DH, Coleman RE, Phelps ME. A tabulated summary of the FDG PET literature. J Nucl Med. 2001;42:1S–93S.
2. Jadvar H, Parker JA. PET radiotracers. In: Clinical PET and PET/CT. 1st ed. Philadelphia: Springer; 2005. p. 45–67.
3. Karakatsanis NA, Lodge MA, Tahari AK, Zhou Y, Wahl RL, Rahmim A. Dynamic whole-body PET parametric imaging. Part 1. Concept, acquisition protocol optimization and clinical application. Phys Med Biol. 2013;58:7391–418.
4. Hillner BE, Siegel BA, Hanna L, et al. Impact of ^{18}F-FDG PET used after initial treatment of cancer: comparison of the National Oncologic PET Registry 2006 and 2009 cohorts. J Nucl Med. 2012;53:831–7.

5. Herrmann K, Buck AK. Proliferation imaging with ^{18}F-fluorothymidine PET/computed tomography: physiologic uptake, variants, and pitfalls. PET Clin. 2014;9:331–8.
6. Acuff SN, Jackson AS, Subramaniam RM, Osborne D. Practical considerations for integrating PET/CT into radiation therapy planning. J Nucl Med Technol. 2018;46:343–8.
7. Unterrainer M, Ruzicka M, Fabritius MP, et al. PET/CT imaging for tumour response assessment to immunotherapy: current status and future directions. Eur Radiol Exp. 2020;4:63.
8. Hirmas N, Leyh C, Sraieb M, Barbato F, Schaarschmidt B, Umutlu L, Nader M, Wedemeyer H, Ferdinandus J, Rischpler C, et al. Ga-68-PSMA-11 PET/CT improves tumor detection and impacts management in patients with hepatocellular carcinoma (HCC). Nuklearmedizin. 2021;60:V55.
9. Liang G, Wang H, Shi H, Wang H, Zhu M, Jing A, Li J, Li G. Recent progress in the development of upconversion nanomaterials in bioimaging and disease treatment. J Nanobiotechnol. 2020;18:154. https://doi.org/10.1186/s12951-020-00713-3.
10. Brandt M, Cardinale J, Aulsebrook ML, Gasser G, Mindt TL. An overview of PET radiochemistry, Part 2: Radiometals. J Nucl Med. 2018;59:1500–6. https://doi.org/10.2967/jnumed.117.190801.
11. Toy R, Bauer L, Hoimes C, Ghaghada KB, Karathanasis E. Targeted nanotechnology for cancer imaging. Adv Drug Deliv Rev. 2014;76:79–97.
12. Goel S, England CG, Chen F, Cai W. Positron emission tomography and nanotechnology: a dynamic duo for cancer theranostics. Adv Drug Deliv Rev. 2017;113:157–76.
13. Min Y, Caster JM, Eblan MJ, Wang AZ. Clinical translation of nanomedicine. Chem Rev. 2015;115:11147–90.
14. Anselmo AC, Mitragotri S. A review of clinical translation of inorganic nanoparticles. AAPS J. 2015;17:1041–54.
15. Xing Y, Zhao J, Conti PS, Chen K. Radiolabeled nanoparticles for multimodality tumor imaging. Theranostics. 2014;4:290–306.
16. De Barros AB, Tsourkas A, Saboury B, Cardoso VN, Alavi A. Emerging role of radiolabeled nanoparticles as an effective diagnostic technique. EJNMMI Res. 2012;2:39.
17. Chakravarty R, Hong H, Cai W. Image-guided drug delivery with single-photon emission computed tomography: a review of literature. Curr Drug Targets. 2015;16:592–609.
18. Chakravarty R, Hong H, Cai W. Positron emission tomography image-guided drug delivery: current status and future perspectives. Mol Pharm. 2014;11:3777–97.
19. Thorek DL, Das S, Grimm J. Molecular imaging using nanoparticle quenchers of Cerenkov luminescence. Small. 2014;10:3729–34.
20. Du J, Wang Y, Zhang W. Label-free, non-derivatization cret detection platform for 6-mercaptopurine based on the distance-dependent optical properties of gold nanoparticles. Chemistry. 2012;18:8540–6.
21. Srivatsan A, Chen X. Recent advances in nanoparticle-based nuclear imaging of cancers. Adv Cancer Res. 2014;124:83–129.
22. Sun X, Cai W, Chen X. Positron emission tomography imaging using radiolabeled inorganic nanomaterials. Acc Chem Res. 2015;48:286–94.
23. Cai W, Hong H. In a "nutshell": intrinsically radio-labeled quantum dots. Am J Nucl Med Mol Imaging. 2012;2:136–40.
24. Ai F, Ferreira CA, Chen F, Cai W. Engineering of radiolabeled iron oxide nanoparticles for dualmodality imaging. Wiley Interdiscip Rev Nanomed Nanobiotechnol. 2016;8:619–30.
25. Bouziotis P, Psimadas D, Tsotakos T, Stamopoulos D, Tsoukalas C. Radiolabeled iron oxide nanoparticles as dual-modality spect/MRI and pet/MRI agents. Curr Top Med Chem. 2012;12:2694–702.
26. Nahrendorf M, Keliher E, Marinelli B, Waterman P, Feruglio PF, Fexon L, et al. Hybrid pet-optical imaging using targeted probes. Proc Natl Acad Sci USA. 2010;107:7910–5.
27. Ehlerding EB, Chen F, Cai W. Biodegradable and renal clearable inorganic nanoparticles. Adv Sci (Weinh). 2016;3:1500223.
28. Ametamey SM, Honer M, Schubiger PA. Molecular imaging with pet. Chem Rev. 2008;108:1501–16.
29. Deng H, Wang H, Li Z. Matching chelators to radiometals for positron emission tomography imaging-guided targeted drug delivery. Curr Drug Targets. 2015;16:610–24.
30. Price EW, Orvig C. Matching chelators to radiometals for radiopharmaceuticals. Chem Soc Rev. 2014;43:260–90.
31. Perez-Campana C, Gomez-Vallejo V, Puigivila M, Martin A, Calvo-Fernandez T, Moya SE, et al. Biodistribution of different sized nanoparticles assessed by positron emission tomography: a general strategy for direct activation of metal oxide particles. ACS Nano. 2013;7:3498–505.
32. Goel S, Chen F, Ehlerding EB, Cai W. Intrinsically radiolabeled nanoparticles: an emerging paradigm. Small. 2014;10:3825–30.
33. Goel S, Chen F, Cai W. Synthesis and biomedical applications of copper sulfide nanoparticles: from sensors to theranostics. Small. 2014;10:631–45.
34. Sun X, Huang X, Yan X, Wang Y, Guo J, Jacobson O, et al. Chelatorfree (64)Cu-integrated gold nanomaterials for positron emission tomography imaging guided photothermal cancer therapy. ACS Nano. 2014;8:8438–46.
35. Shaffer TM, Wall MA, Harmsen S, Longo VA, Drain CM, Kircher MF, et al. Silica nanoparticles as substrates for chelator-free labeling of oxophilic radioisotopes. Nano Lett. 2015;15:864–8.
36. Chen F, Ellison PA, Lewis CM, Hong H, Zhang Y, Shi S, et al. Chelator-free synthesis of a dualmodality PET/MRI agent. Angew Chem Int Ed Engl. 2013;52:13319–23.

37. Chakravarty R, Valdovinos HF, Chen F, Lewis CM, Ellison PA, Luo H, et al. Intrinsically germanium-69-labeled iron oxide nanoparticles: synthesis and in-vivo dual-modality PET/MR imaging. Adv Mater. 2014;26:5119–23.
38. Liu Q, Sun Y, Li C, Zhou J, Li C, Yang T, et al. ^{18}F-labeled magnetic-upconversion nanophosphors via rare-earth cation-assisted ligand assembly. ACS Nano. 2011;5:3146–57.
39. Sun X, Huang X, Guo J, Zhu W, Ding Y, Niu G, et al. Self-illuminating ^{64}Cu-doped CdSe/ZnS nanocrystals for in vivo tumor imaging. J Am Chem Soc. 2014;136:1706–9.
40. Hassan PA, Rana S, Verma G. Making sense of Brownian motion: colloid characterization by dynamic light scattering. Langmuir. 2015;31:3–12.
41. Chakravarty R, Chakraborty S, Ningthoujam RS, Vimalnath Nair KV, Sharma KS, Ballal A, et al. Industrial-scale synthesis of intrinsically radiolabeled ^{64}CuS nanoparticles for use in positron emission tomography (PET) imaging of cancer. Ind Eng Chem Res. 2016;55:12407–19.
42. Bertrand N, Wu J, Xu X, Kamaly N, Farokhzad OC. Cancer nanotechnology: the impact of passive and active targeting in the era of modern cancer biology. Adv Drug Deliv Rev. 2014;66:2–25.
43. Kunjachan S, Pola R, Gremse F, Theek B, Ehling J, Moeckel D, et al. Passive versus active tumor targeting using RGD- and NGR-modified polymeric nanomedicines. Nano Lett. 2014;14:972–81.
44. Jokerst JV, Lobovkina T, Zare RN, Gambhir SS. Nanoparticle pegylation for imaging and therapy. Nanomedicine (Lond). 2011;6:715–28.
45. Biju V, Itoh T, Ishikawa M. Delivering quantum dots to cells: bioconjugated quantum dots for targeted and nonspecific extracellular and intracellular imaging. Chem Soc Rev. 2010;39:3031–56.
46. Li ZH, Peng J, Chen HL. Bioconjugated quantum dots as fluorescent probes for biomedical imaging. J Nanosci Nanotechnol. 2011;11:7521–36.
47. Volkov Y. Quantum dots in nanomedicine: recent trends, advances and unresolved issues. Biochem Biophys Res Commun. 2015;468:419–27.
48. Schipper ML, Cheng Z, Lee SW, Bentolila LA, Iyer G, Rao J, et al. Micropet-based biodistribution of quantum dots in living mice. J Nucl Med. 2007;48:1511–8.
49. Schipper ML, Iyer G, Koh AL, Cheng Z, Ebenstein Y, Aharoni A, et al. Particle size, surface coating, and PEGylation influence the biodistribution of quantum dots in living mice. Small. 2009;5:126–34.
50. Tu C, Ma X, House A, Kauzlarich SM, Louie AY. PET imaging and biodistribution of silicon quantum dots in mice. ACS Med Chem Lett. 2011;2:285–8.
51. Cai W, Chen K, Li ZB, Gambhir SS, Chen X. Dual-function probe for PET and near-infrared fluorescence imaging of tumor vasculature. J Nucl Med. 2007;48:1862–70.
52. Hu K, Wang H, Tang G, Huang T, Tang X, Liang X, et al. In vivo cancer dual-targeting and dualmodality imaging with functionalized quantum dots. J Nucl Med. 2015;56:1278–84.
53. Liu H, Ren G, Miao Z, Zhang X, Tang X, Han P, et al. Molecular optical imaging with radioactive probes. PLoS One. 2010;5:e9470.
54. Xu Y, Liu H, Cheng Z. Harnessing the power of radionuclides for optical imaging: Cerenkov luminescence imaging. J Nucl Med. 2011;52:2009–18.
55. Dothager RS, Goiffon RJ, Jackson E, Harpstrite S, Piwnica-Worms D. Cerenkov radiation energy transfer (CRET) imaging: a novel method for optical imaging of PET isotopes in biological systems. PLoS One. 2010;5:e13300.
56. Guo W, Sun X, Jacobson O, Yan X, Min K, Srivatsan A, et al. Intrinsically radioactive [^{64}Cu]CuInS/ZnS quantum dots for PET and optical imaging: improved radiochemical stability and controllable Cerenkov luminescence. ACS Nano. 2015;9:488–95.
57. Alex S, Tiwari A. Functionalized gold nanoparticles: synthesis, properties and applications—a review. J Nanosci Nanotechnol. 2015;15:1869–94.
58. Xie H, Wang ZJ, Bao A, Goins B, Phillips WT. In vivo PET imaging and biodistribution of radiolabeled gold nanoshells in rats with tumor xenografts. Int J Pharm. 2010;395:324–30.
59. Guerrero S, Herance JR, Rojas S, Mena JF, Gispert JD, Acosta GA, et al. Synthesis and in vivo evaluation of the biodistribution of a ^{18}F-labeled conjugate gold-nanoparticle-peptide with potential biomedical application. Bioconjug Chem. 2012;23:399–408.
60. Schwaminger SP, Fraga-García P, Blank-Shim SA, Straub T, Haslbeck M, Muraca F, et al. Magnetic one-step purification of His-tagged protein by bare iron oxide nanoparticles. ACS Omega. 2019;4(2):3790–9.
61. de Almeida M, Susnik E, Drasler B, Taladriz-Blanco P, Petri-Fink A, Rothen-Rutishauser B. Understanding nanoparticle endocytosis to improve targeting strategies in nanomedicine. Chem Soc Rev. 2021;50:5397.
62. Janßen HC, Angrisani N, Kalies S, Hansmann F, Kietzmann M, Warwas DP, et al. Biodistribution, biocompatibility and targeted accumulation of magnetic nanoporous silica nanoparticles as drug carrier in orthopedics. J Nanobiotechnol. 2020;18(1):14.
63. Reifenrath J, Janßen HC, Warwas DP, Kietzmann M, Behrens P, Willbold E, et al. Implant based direction of magnetic nanoporous silica nanoparticles: influence of macrophage depletion and infection. Nanomedicine. 2020;30:102289.
64. Janßen HC, Warwas DP, Dahlhaus D, Meißner J, Taptimthong P, Kietzmann M, et al. In vitro and in vivo accumulation of magnetic nanoporous silica nanoparticles on implant materials with different magnetic properties. J Nanobiotechnol. 2018;16(1):96.
65. Zhou M, Zhang R, Huang M, Lu W, Song S, Melancon MP, et al. Achelator-free multifunctional [^{64}Cu]CuS nanoparticle platform for simultaneous micro-PET/CT imaging and photothermal ablation therapy. J Am Chem Soc. 2010;132:15351–8.
66. Zhou M, Chen Y, Adachi M, Wen X, Erwin B, Mawlawi O, et al. Single agent nanoparticle for radio-

therapy and radio-photothermal therapy in anaplastic thyroid cancer. Biomaterials. 2015;57:41–9.
67. Zhou M, Zhao J, Tian M, Song S, Zhang R, Gupta S, et al. Radiophotothermal therapy mediated by a single compartment nanoplatform depletes tumor initiating cells and reduces lung metastasis in the orthotopic 4T1 breast tumor model. Nanoscale. 2015;7:19438–47.
68. Zhou M, Melancon M, Stafford RJ, Li J, Nick AM, Tian M, et al. Precision nanomedicine using dual PET and MR temperature imaging guided photothermal therapy. J Nucl Med. 2016;57:1778–83.
69. Liu Q, Zhou M, Li P, Ku G, Huang G, Li C, et al. ^{64}CuS-labeled nanoparticles: a new sentinel-lymph-node-mapping agent for PET-CT and photoacoustic tomography. Contrast Media Mol Imaging. 2016;11:475–81.
70. Zhou M, Song S, Zhao J, Tian M, Li C. Theranostic CuS nanoparticles targeting folate receptors for PET image-guided photothermal therapy. J Mater Chem B Mater Biol Med. 2015;3:8939–48.
71. Phillips E, Penate-Medina O, Zanzonico PB, Carvajal RD, Mohan P, Ye Y, et al. Clinical translation of an ultrasmall inorganic optical-PET imaging nanoparticle probe. Sci Transl Med. 2014;6:260ra149.
72. Blanco E, Shen H, Ferrari M. Principles of nanoparticle design for overcoming biological barriers to drug delivery. Nat Biotechnol. 2015;33:941–51.
73. Li J, Chang X, Chen X, Gu Z, Zhao F, Chai Z, et al. Toxicity of inorganic nanomaterials in biomedical imaging. Biotechnol Adv. 2014;32:727–43.
74. Verhoef JJ, Anchordoquy TJ. Questioning the use of PEGylation for drug delivery. Drug Deliv Transl Res. 2013;3:499–503.
75. Choi H, Lee Y-S, Hwang Do W, Lee Dong S. Translational radionanomedicine: a clinical perspective. Eur J Nanomed. 2016;8:71.

12 SPECT CT Nanoimaging

12.1 Introduction

Hybrid imaging, as we all know, is the combination of functional and anatomical imaging modalities for better-enhanced localization and of molecular aspects with structural abnormalities. In this, two or more multimodality probes and multimodality probes will be used in multimodality imaging, thus reducing the time required. This is an important synergistic component that results in better identification of biological targets in terms of structure and function, which cannot be achieved by a single modality [1]. PET/computed tomography (CT) and single-photon emission CT (SPECT)/CT are commonly used in human patients as routine imaging modalities (Fig. 12.1). On the contrary, PET/magnetic resonance (MR) imaging is a newer, important hybrid imaging modality, and SPECT/MR is also an important newer modality. The use of PET/CT, SPECT/CT, and PET/MR has increased in recent years [2–6]. In this chapter, the newer developments in SPECT/MR hybrid imaging and the development of new imaging multimodal nanoprobes are also considered. An important advantage of nuclear medicine modalities like PET and SPECT is their ability to detect signals at picomolar concentrations of radiotracers without interrupting normal physiology. Due to this, various physiological and

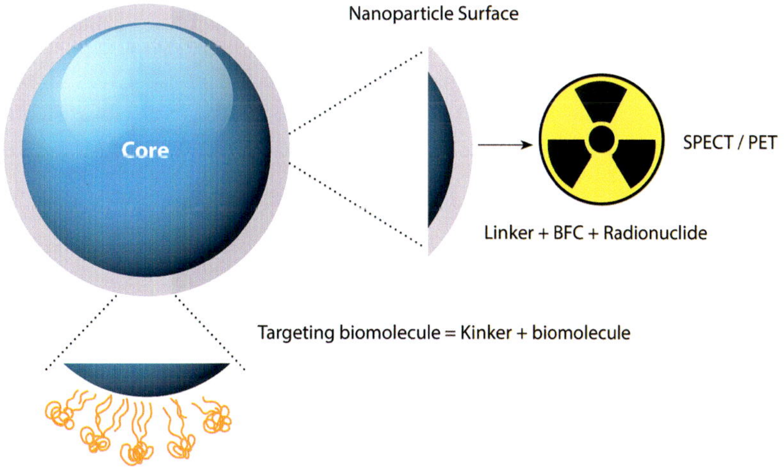

Fig. 12.1 Role of nanoparticle surface used in radionuclide SPECT/PET and targeting biomolecule

S. Shaikh, *Nanoimaging - Future of Precision Medicine*,
https://doi.org/10.1007/978-981-97-7292-6_12

pathological aspects can be easily identified at the molecular level, identifying micro-quantities of radiolabeled probes. Positron-emitting radionuclides are composed of short-lived isotopes, such as fluorine (^{18}F), carbon (^{11}C), nitrogen (^{13}N), and oxygen (^{15}O), as well as longer-lived isotopes, such as copper (^{64}Cu), gallium (^{68}Ga), and zirconium (^{89}Zr), which enable the radiolabeling of biological components [7]. SPECT is based on various radiolabeled components, such as indium (^{111}In), technetium (^{99m}Tc), and radioactive iodine isotopes ^{123}I and ^{131}I. The half-lives of these radiotracers used in SPECT in comparison with PET tracers are better in terms of kinetics, as they are slower. The range of SPECT radiopharmaceuticals allows for the production of kits onsite, thus preventing the need for an expensive on-site cyclotron/radiochemistry production as required for PET tracers. The SPECT tracers are advantageous over PET tracers as the PET tracers are lower in cost and easily available. Both these SPECT and PET imaging techniques are used for functional imaging of various molecular processes. One disadvantage is the reduced sensitivity for anatomic information, which results in very low sensitivity. For this reason, CT and MR imaging are two important anatomical imaging techniques used with SPECT and PET. Due to these new hybrid imaging techniques, more and better sensitivities lead to a newer terminology 'new-clear medicine' [8]. Between these two modalities, MR imaging has advantages over CT, as it does not involve ionizing radiation, has very high soft tissue contrast, and better sensitivity to tissue alterations. MR can use contrast agents like gadolinium and iron oxide-based agents, leading to excellent soft tissue resolutions. However, MR imaging is less sensitive to millimolar vs. picomolar concentrations as compared to PET or SPECT, as there will be lower quantum energy involved. MR imaging also provides higher spatial resolution than various nuclear imaging methods, such as micrometer vs. millimeter [9, 10]. The MR image contrast is based on basic parameters like spin-lattice relaxation time (T1) and spin–spin relaxation time (T2), which are depended on the chemical structure of the molecules being imaged, thus offering advantages for molecular imaging [11]. MR imaging, which is fast-growing, is used in different imaging tissues, cells, and molecules, along with dynamic contrast-enhanced MR imaging, diffusion-weighted imaging, and (MR) spectroscopy. The transendothelial transport of the MR contrast agent, which is conjugated with albumin by dynamic contrast-enhanced MR imaging, serves as a marker for angiogenesis in tumors [12]. Different cells labeled with superparamagnetic iron oxide nanoparticles (SPIONs) contrast agents are being developed [13], allowing greater efficiency in tracking of in vivo stem cells, progenitor cells, or the cells expressing transgenes, even at the single-cell level [14]. The concept of combining SPECT and MR as a hybrid imaging modality is that the high spatial resolution of MR and the high sensitivity of SPECT will yield better results [15]. SPECT/MR imaging also enables dynamic imaging with radiotracers and different MR imaging contrast agents, facilitating better monitoring of treatment, pharmacokinetics, and metabolism of drugs non-invasively. The use of hybrid molecular imaging can also fast-track the evaluation of various molecular abnormalities [16], growth kinetics [17], angiogenesis [18], and tumor cell markers [19] and provides the potential for earlier detection and characterization of disease and evaluation of treatment. The use and assessment of tumor response to therapy can be either unidimensional or bidimensional based on the size of the tumor, as per the Response Evaluation Criteria in Solid Tumors (RECIST) classification [20]. Imaging with hybrid techniques like SPECT/MR improves therapeutic efficiency at the molecular level, even before various phenotypic changes occur. Despite various technical advances, combining MR and nuclear medicine instrumentation presents challenges, as the radiofrequency radiating components of PET and SPECT can interfere with the MRI. PET and SPECT detectors are based on scintillators combined with photomultiplier tubes, which are very sensitive to magnetic fields [21]. Newer developments include X-ray detectors based on avalanche photodiodes, such as lutetium oxyorthosilicate, a fast sensitive material that is fully magnetic-field-insensitive, and high-performance PET detectors used in the PET/MR imaging scanners [22, 23].

12.2 SPECT Radionuclides

SPECT CT radiotracers should have a very comfortable physical half-life, with an energy range (100–300 keV), and stability in the daughter radionuclide; these are the criteria for radionuclide used in SPECT imaging [24]. ^{99m}Tc is the most commonly used radiotracer having a physical half-life of 6 h and having better sensitivity and also preventing excessive radiation dosage to the patient [25]. This enables optimal imaging with a major g line at 140 keV [24]. Along with ^{99m}Tc, ^{111}In is another isotope that is very commonly used in nuclear medicine. ^{111}In has a half-life of 68 h and the energy lines at 171 and 245 keV, with no β-emission. The advantage of ^{111}In is its use with probes that have slower kinetics, as its longer half-life reduces background uptake. ^{111}In is more sensitive due to the less scatter and internal shielding provided by the longer half-life of radionuclides (^{123}I and ^{125}I). ^{123}I and ^{125}I have half-lives of 13.2 h and 60.1 days with energies of 159 and 35.5 keV, respectively, and are used in SPECT imaging; however, their clinical applications is limited due to high cost, low availability, and less stability [24–28]. Thallium (^{201}TI) [29, 30] and holmium (^{166}Ho) are radiotracers that are less used now [31]. Another range incudes lanthanide isotopes like ytterbium (^{169}Yb), samarium (^{153}Sm), dysprosium (^{165}Dy), and holmium (^{166}Ho), all of which emit β-electrons and are useful brachytherapy applications where the g component is more useful [32, 33]. Technetium and rhenium in ^{188}Re and ^{186}Re are effective g radioisotopes used with therapeutic β component [29, 34].

12.3 MR Imaging Contrast Agents

The two important agents used in MR imaging are T1 and T2 contrast agents. All these contrast agents produce both T1 (longitudinal) and T2 (transverse) relaxation, and the type of relaxation produced to a greater extent can classify them. MR contrast agents produce nuclear magnetization, causing relaxation from a non-equilibrium state to equilibrium, analyzed in terms of relaxivity. This relaxivity value is a parameter that define the enhancement of the water proton relaxation rate (1/T1 or 1/T2) in various solutions containing a 1 mM concentration of the magnetic solute. Due to the low concentration of biological targets, these contrast agents have higher sensitivity, which is required for molecular MR imaging. Due to this, the majority of the MR contrast agents are not sensitive. The use of hybrid SPECT/MR imaging T2 agents shows the transverse relaxation of contrast agents as the relaxation rate of protons surrounding water molecules, causing T2 shortening and hypointense (dark) contrast [30]. SPIONs can shorten T2, and particularly T2* due to their contribution to both transverse relaxation and local magnetic field nonuniformities of water protons and are used as MR imaging probes for molecular imaging experiments [35]. Various antibodies are conjugated with SPIONs, which can image HER-2-expressing cells or tissues in both in vitro and in vivo MR imaging [36]. ^{64}Cu-labeled SPION probes produce strong MR and PET signals at 24 h at 37 °C in mouse serum, which is used in dual-modality PET/MR imaging bio-probes [37]. When ^{64}Cu (half-life of 12.7 h) is replaced with ^{67}Cu (half-life of 2.58 days), it provides a longer half-life for the nanoprobe in circulation and can be imaged with SPECT and MR imaging. The uptake of SPIONs by phagocytic cells results in target focusing, facilitating delivery to tumor cells or other cells like cardiac myocytes and endothelial cells [38–40]. Iron oxide-based nanoparticles utilizing bimetallic cores (e.g., iron-cobalt [FeCo] or manganese ferrite [$MnFe_2O_4$]) aid in developing T2 contrast agents with very higher relaxivity [41–43], though all are in the preclinical stage and need detailed further assessment of their toxicity. SPIONs are structurally made up of the core of magnetite (Fe_3O_4) and/or maghemite (Fe_2O_3) [44, 45], having important pharmacological and magnetic properties of SPIONs. These particles of monodisperse crystal cores with a very high degree of crystallinity provide better relaxivity characteristics in imaging probes with higher sensitivity. SPIONs are in the range of 20–50 nm having higher r^2 (relaxivity due to transverse relaxation) values. In the intracellular space within a cell, the size of 5–10 nm is better as compared to circulating nanoparticles in the bloodstream. Nanoparticles targeting vascular and extravascular (tumors) structures in sizes of 30–100 nm seem to be useful [46]. In vivo applications require these nanoparticles to be coated with biocompatible materials, preventing aggregation and sedimentation for better stability. Surface coating is important in defining the pharmacokinetic behavior of the particle, as nanoparticles are like foreign bodies and will be taken by the phagocytic cells, depositing in the spleen and liver. When coated with polyethylene glycol (PEG), it will reduce the ability of serum proteins to recognize the particle prolonging circulation times in the blood pool [37]. Other than this PEG, materials such as dextran, starch, and silica can also be used as coating materials for SPIONs [47–49]. SPIONs are also used clinically for the detection of liver and lymph nodes [50, 51]. In hybrid imaging, Bligh et al. first tested radioactive imaging and MR imaging procedures as combined hybrid modalities [52]. They prepared hybrid imaging probes by conjugating Fe_3O_4 particles of 0.1–0.5 μm in diameter with either ^{99m}Tc or ^{111}In. Biodistribution and g-camera studies showed that the main targets will be the liver and lung. Due to this, Lewin et al. showed the hybrid imaging probe, which is capable of imaging three modalities, MR imaging, fluorescence imaging, and SPECT by conjugating cross-linked iron oxide (CLIO) nanoparticles with HIV-Tat peptide and labeling with fluorescence (fluorescein isothiocyanate) and the radioisotope ^{111}In [53]. CLIO nanoparticles having a diameter of 45 nm made up of the monocrystalline, superparamagnetic iron oxide core, are stabilized by crosslinking aminated dextran. Radiolabeling with ^{111}In can be done by conjugating a diethylenetriamine penta acetic acid (DTPA) chelator to CLIO nanoparticles by the dextran coating. These authors further demonstrated that triple-labeled CLIO–Tat probes (hybrid imaging probes) tagged to hematopoietic cells (CD34+) and neural progenitor cells can track the distribution and differentiation of progenitor and stem cells with high-resolution in vivo imaging techniques. One of the studies showed a new dual-modality imaging agent will be conjugating radiolabeled bisphosphonates (BPs) to SPIONs with a high binding affinity of BPs to iron, labeling with the iron oxide liver MR imaging contrast agent, with ^{99m}Tc-dipicolylamine-alendronate, which is the SPECT imaging agent for bone imaging [23]. Bimodal imaging of these nanoparticles was confirmed by MR imaging and nano-SPECT CT imaging, which concluded that the ^{99m}Tc, and Endorem will be seen in the liver and spleen. In another study, Kaufner et al. developed SPIONs of Fe_2O_3 coated with poly(ethylene oxide)-block-poly(glutamic acid) (PEO–PGA) with a hydrodynamic diameter of 60 nm [54]. Nanoparticles combined with tumor-specific targeting ligands and antibodies are used for diagnosing tumors and detecting peripheral metastases [55, 56]. The hybrid imaging nanoprobe composed of Fe_3O_4 conjugated to radioisotope ^{125}I and anti-gastric cancer monoclonal antibody 3H11 (mAb 3H11) was designed by Liu et al. [57]. Various in vivo experiments have shown the importance of probes for detecting xenograft tumors in nude mice by MR imaging and g-imaging techniques. Otsuji et al. have shown the importance of radiolabeled monoclonal antibody Mab A7 against human colonic carcinoma by using the ^{125}I chloramine-T method [58–62]. The use of the ^{125}I in the nano-construct will make the hybrid imaging more relevant with SPECT/MR. The dual design of dual-modality SPECT/MR imaging nanoprobes will help in imaging cancers like mesotheliomas, pancreatic and ovarian cancers, which are challenging to diagnose and have low survival rates. Available diagnostic methods have limitations for the diagnosis of these cancers [63–66]. The imaging technique for the detection of the molecular markers in these cancers can potentially help identify early tumors. This will result in the identification and staging of the tumor-node metastasis for the status of their resectability before surgery [67]. Tumor site imaging will help in identifying more accurate, better-targeted drug or radiation delivery in managing these cancers. Tumor-specific molecular imaging bioprobe are developed and can be conjugated with radiolabeled antibodies to carboxymethyl dextran-coated SPIONs via carbodiimide coupling [68]. The mAb MB specifically targets the cell surface antigen known as 'mesothelin,' which is overexpressed in mesotheliomas and pancreatic and ovarian cancers, and is radiolabeled with ^{111}In [63, 64, 69]. In tumor-bearing mice, ^{111}In-labeled mAbMB antibody (^{111}In-mAbMB) shows specific uptake in mesothelin-expressing A431K5 tumors, as seen in SPECT imaging [65]. ^{111}In-mAbMB is then conjugated with SPIONs with an average diameter of 70 nm, resulting in the ^{111}In-mAbMB-SPIONs' uptake in mesothelin-expressing A431K5 tumors, causing change in the MR signal. These radioactive hybrid nanoprobes are useful in determining biodistribution and in vivo characteristics. The higher relaxivity and specific targeting of bioprobes enhance MR contrast in mesothelin-expressing tumors. The intra-tumor distribution of the hybrid nanoprobe is helpful in the identification of in vivo detection and characterization, providing both functional and anatomical information from MR imaging and SPECT to aid in early diagnosis and monitoring of mesothelin-expressing cancers.

12.4 Hybrid SPECT/MR Imaging T1 Agents

The paramagnetic complexes of metal ions and their ground states will have the ability to catalyze the relaxation process of adjacent water protons, which are identified as longitudinal relaxivity (r1) in T1 contrast imaging. Paramagnetic substances like gadolinium Gd(III) and manganese Mn(II) have hyperintensity (bright contrast) in T1-weighted MR images due to the longitudinal relaxation process. Lijowski et al. designed and characterized dual-modality αvβ3-targeted nanoparticles that are sensitive to nuclear detection and provide high-resolution MR for tumor angiogenesis characterization [66]. Gadolinium (T1 contrast agent) was added to the outer phospholipid layer of these nanoparticles, resulting in a single nanoparticle structure suitable for both SPECT and MR imaging. With the use of the SPECT/CT imaging techniques, αvβ3-targeted ^{99m}Tc gadolinium nanoparticles can effectively demonstrate tumor neovascular imaging. After administration of the αvβ3-targeted gadolinium nanoparticles, SPECT scan was done for MR imaging, showing the angiogenic regions in the tumor mass. High-resolution MR molecular nanoprobes eliminated 95% of the nanoprobes by kidneys within 18 days, a pharmacokinetic behavior observable through both SPECT and MR imaging. Zielhuis et al. developed liposomal nanodevices measuring 130 nm, which can be used in hybrid imaging and also for radiotherapy [70]. The liposomes were created by the thin film hydration technique and radiolabeled with ^{166}Ho (both a β- and γ-emitter and also highly paramagnetic) or ^{99m}Tc, having very high radiolabeling efficiency and radiochemical stability by incorporating the amphiphilic molecule DTPA-lipid into the liposomal bilayer [71]. Paramagnetic gadolinium co-loaded into the bilayer reached concentrations up to 20 mol%. This study demonstrated the feasibility of liposomal nanodevices for the development of the hybrid theranostic nanoprobes suitable for SPECT/MR imaging and radiotherapy. Low-molecular-weight contrast agents in dynamic contrast-enhanced imaging show the features of tumor perfusion and extravasation [72, 73]. A higher-molecular-weight contrast agent remains intravascular, even within tumor vasculature, and this can be used for the identification of the physiological parameters of permeability and perfusion along with the normal versus abnormal tissues (Fig. 12.2). A macromolecular platform for multimodal imaging as unimolecular, nano-sized, long-circulating, biocompatible, and non-toxic macromolecule (high molecular-weight hyperbranched polyglycerols [HPG]) is equipped by MR, using the radioactive and fluorescence markers [74]. This HPG is further modified with probes for better use of the SPECT and MR imaging. The ligand (chelator) used (DOTA) was suitable to bind In^{3+} and Gd^{3+} metal ions. When DOTA is added to HPG, the molecule chemistry, water solubility, and charge are minimally altered,

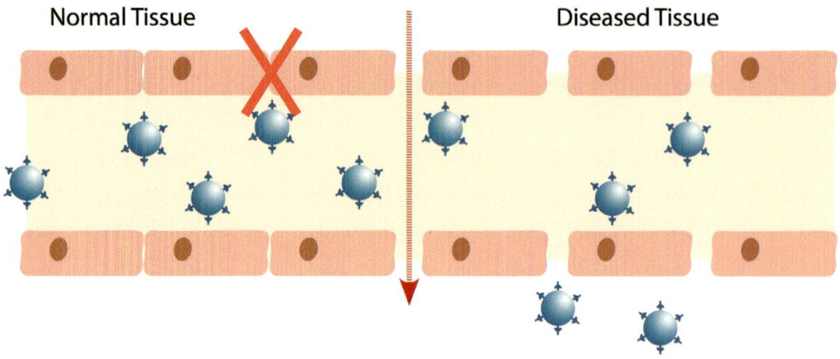

Fig. 12.2 Cell surface in the normal and second one with cell membrane breach, leading to movement of various components in and out of the cell

leading to no change in the behavior and biodistribution as compared to the original HPG molecule. Alexa 647 can also be used for fluorescent imaging, resulting in a trimodal imaging agent. The fluorescent tag confirms tumor distribution and also extravasation by ex vivo imaging. The relaxivity of the HPG-Gd system is 1075 $mM^{-1}\ s^{-1}$, much higher than that of commercially available gadolinium-labeled agents.

12.5 Conclusion

The use of hybrid imaging modalities has been more useful for the precise diagnosis of various diseases at a very early stage. The combinations of hybrid imaging modalities with SPECT CT is playing a significant role by using biomarkers that enhance better physiological aspects of disease evaluation.

References

1. Jennings LE, Long NJ. 'Two is better than one'—probes for dual-modality molecular imaging. Chem Commun (Camb). 2009;24:3511–24.
2. Delbeke D, Schoder H, Martin WH, Wahl RL. Hybrid imaging (SPECT/CT and PET/CT): improving therapeutic decisions. Semin Nucl Med. 2009;39(5):308–40.
3. Bockisch A, Freudenberg LS, Schmidt D, Kuwert T. Hybrid imaging by SPECT/CT and PET/CT: proven outcomes in cancer imaging. Semin Nucl Med. 2009;39(4):276–89.
4. Antoch G, Bockisch A. Combined PET/MRI: a new dimension in whole-body oncology imaging? Eur J Nucl Med Mol Imaging. 2009;36(Suppl. 1):S113–20.
5. Pichler BJ, Judenhofer MS, Wehrl HF. PET/MRI hybrid imaging: devices and initial results. Eur Radiol. 2008;18(6):1077–86.
6. Fatemi-Ardekani A, Samavati N, Tang J, Kamath MV. Advances in multimodality imaging through a hybrid PET/MRI system. Crit Rev Biomed Eng. 2009;37(6):495–515.
7. Cook GJ. Oncological molecular imaging: nuclear medicine techniques. Br J Radiol. 2003;76(Spec No. 2):S152–8.
8. Patel CN, Chowdhury FU, Scarsbrook AF. Hybrid SPECT/CT: the end of 'unclear' medicine. Postgrad Med J. 2009;85(1009):606–13.
9. Shapiro EM, Skrtic S, Sharer K, Hill JM, Dunbar CE, Koretsky AP. MRI detection of single particles for cellular imaging. Proc Natl Acad Sci USA. 2004;101(30):10901–6.
10. Zhou SA, Brahme A. Development of high-resolution molecular phase-contrast stereoscopic X-ray imaging for accurate cancer diagnostics. Radiat Prot Dosim. 2010;139(1–3):334–8.
11. Gallagher FA. An introduction to functional and molecular imaging with MRI. Clin Radiol. 2010;65(7):557–66.
12. Dafni H, Landsman L, Schechter B, Kohen F, Neeman M. MRI and fluorescence microscopy of the acute vascular response to VEGF165: vasodilation, hyper-permeability and lymphatic uptake, followed by rapid inactivation of the growth factor. NMR Biomed. 2002;15(2):120–31.
13. Weissleder R, Moore A, Mahmood U, et al. In vivo magnetic resonance imaging of transgene expression. Nat Med. 2000;6(3):351–5.
14. Dodd SJ, Williams M, Suhan JP, Williams DS, Koretsky AP, Ho C. Detection of single mammalian cells by high-resolution magnetic resonance imaging. Biophys J. 1999;76(1 Pt 1):103–9.
15. Hamamura MJ, Ha S, Roeck WW, et al. Development of an MR-compatible SPECT system (MRSPECT) for simultaneous data acquisition. Phys Med Biol. 2010;55(6):1563–75.
16. Thakur ML. Genomic biomarkers for molecular imaging: predicting the future. Semin Nucl Med. 2009;39(4):236–46.
17. Mishani E, Hagooly A. Strategies for molecular imaging of epidermal growth factor receptor tyrosine kinase in cancer. J Nucl Med. 2009;50(8):1199–202.
18. Dobrucki LW, De Muinck ED, Lindner JR, Sinusas AJ. Approaches to multimodality imaging of angiogenesis. J Nucl Med. 2010;51(Suppl. 1):S66–79.

19. Dunn BK, Wagner PD, Anderson D, Greenwald P. Molecular markers for early detection. Semin Oncol. 2010;37(3):224–42.
20. Eisenhauer EA, Therasse P, Bogaerts J, et al. New response evaluation criteria in solid tumours: revised RECIST guideline (version 1.1). Eur J Cancer. 2009;45(2):228–47.
21. Catana C, Procissi D, Wu Y, et al. Simultaneous in vivo positron emission tomography and magnetic resonance imaging. Proc Natl Acad Sci USA. 2008;105(10):3705–10.
22. Wehrl HF, Judenhofer MS, Wiehr S, Pichler BJ. Preclinical PET/MR: technological advances and new perspectives in biomedical research. Eur J Nucl Med Mol Imaging. 2009;36(Suppl. 1):S56–68.
23. Torres Martin De Rosales R, Tavare R, Glaria A, Varma G, Protti A, Blower PJ. ^{99m}Tcbisphosphonate-iron oxide nanoparticle conjugates for dual-modality biomedical imaging. Bioconjug Chem. 2011;22(3):455–65.
24. Liu Y, Wu C. Radiolabeling of monoclonal antibodies with metal chelates. Pure Appl Chem. 1991;63(2):427–63.
25. Mahmood A, Jones AG. Technetium radiopharmaceuticals. In: Welch MJ, Redvanly CS, editors. Handbook of radiopharmaceuticals: radiochemistry and applications. West Sussex, UK: Wiley; 2003. p. 323–62.
26. Halpern SE. The advantages and limits of indium-111 labeling of antibodies. Experimental studies and clinical applications. Int J Rad Appl Instrum B. 1986;13(2):195–201.
27. Shiga H, Taki J, Yamada M, et al. Evaluation of the olfactory nerve transport function by SPECT-MRI fusion image with nasal thallium-201 administration. Mol Imaging Biol. 2011;13(6):1262–6.
28. Goetz C, Breton E, Choquet P, Israel-Jost V, Constantinesco A. SPECT low-field MRI system for small-animal imaging. J Nucl Med. 2008;49(1):88–93.
29. Van Gog FB, Visser GW, Stroomer JW, Roos JC, Snow GB, Van Dongen GA. High dose rhenium 186 labeling of monoclonal antibodies for clinical application: pitfalls and solutions. Cancer. 1997;80(Suppl. 12):2360–70.
30. Aime S, Nano R, Grandi M. A new class of contrast agents for magnetic resonance imaging based on selective reduction of water-T2 by chemical exchange. Investig Radiol. 1988;23(Suppl. 1):S267–70.
31. Gianolio E, Maciocco L, Imperio D, et al. Dual MRI-SPECT agent for pH-mapping. Chem Commun (Camb). 2011;47(5):1539–41.
32. Cutler CS, Smith CJ, Ehrhardt GJ, Tyler TT, Jurisson SS, Deutsch E. Current and potential therapeutic uses of lanthanide radioisotopes. Cancer Biother Radiopharm. 2000;15:531–45.
33. Liu S, Edwards DS. Bifunctional chelators for therapeutic lanthanide radiopharmaceuticals. Bioconjug Chem. 2001;12:7–34.
34. Knapp FF Jr. Rhenium-188—a generator-derived radioisotope for cancer therapy. Cancer Biother Radiopharm. 1998;13(5):337–49.
35. Laurent S, Forge D, Port M, et al. Magnetic iron oxide nanoparticles: synthesis, stabilization, vectorization, physicochemical characterizations, and biological applications. Chem Rev. 2008;108(6):2064–110.
36. Kinoshita M, Yoshioka Y, Okita Y, Hashimoto N, Yoshimine T. MR molecular imaging of HER-2 in a murine tumor xenograft by SPIO labeling of anti-HER-2 affibody. Contrast Media Mol Imaging. 2010;5(1):18–22.
37. Glaus C, Rossin R, Welch MJ, Bao G. In vivo evaluation of (64)Cu-labeled magnetic nanoparticles as a dual-modality PET/MR imaging agent. Bioconjug Chem. 2010;21(4):715–22.
38. Weissleder R, Kelly K, Sun EY, Shtatland T, Josephson L. Cell-specific targeting of nanoparticles by multivalent attachment of small molecules. Nat Biotechnol. 2005;23(11):1418–23.
39. Weissleder R, Lee AS, Fischman AJ, et al. Polyclonal human immunoglobulin G labeled with polymeric iron oxide: antibody MR imaging. Radiology. 1991;181(1):245–9.
40. Weissleder R, Lee AS, Khaw BA, Shen T, Brady TJ. Antimyosin-labeled monocrystalline iron oxide allows detection of myocardial infarct: MR antibody imaging. Radiology. 1992;182(2):381–5.
41. Lee JH, Sherlock SP, Terashima M, et al. High-contrast in vivo visualization of microvessels using novel FeCo/GC magnetic nanocrystals. Magn Reson Med. 2009;62(6):1497–509.
42. Yang H, Zhang C, Shi X, et al. Water-soluble superparamagnetic manganese ferrite nanoparticles for magnetic resonance imaging. Biomaterials. 2010;31(13):3667–73.
43. Tromsdorf UI, Bigall NC, Kaul MG, et al. Size and surface effects on the MRI relaxivity of manganese ferrite nanoparticle contrast agents. Nano Lett. 2007;7(8):2422–7.
44. Tanimoto A, Kuribayashi S. Application of superparamagnetic iron oxide to imaging of hepatocellular carcinoma. Eur J Radiol. 2006;58(2):200–16.
45. Jain TK, Foy SP, Erokwu B, Dimitrijevic S, Flask CA, Labhasetwar V. Magnetic resonance imaging of multifunctional pluronic stabilized iron-oxide nanoparticles in tumor-bearing mice. Biomaterials. 2009;30(35):6748–56.
46. Berry CC, Curtis ASG. Functionalisation of magnetic nanoparticles for applications in biomedicine. J Phys D Appl Phys. 2003;36:R198–206.
47. Kobukai S, Baheza R, Cobb JG, et al. Magnetic nanoparticles for imaging dendritic cells. Magn Reson Med. 2010;63(5):1383–90.
48. Lind K, Kresse M, Debus NP, Muller RH. A novel formulation for superparamagnetic iron oxide (SPIO) particles enhancing MR lymphography: comparison of physicochemical properties and the in vivo behavior. J Drug Target. 2002;10(3):221–30.
49. Guardia P, Perez N, Labarta A, Batlle X. Controlled synthesis of iron oxide nanoparticles over a wide size range. Langmuir. 2010;26(8):5843–7.
50. Wen G, Zhang XL, Chang RM, Xia Q, Cang P, Zhang Y. Superparamagnetic iron oxide (Feridex)-enhanced

MRI in diagnosis of focal hepatic lesions. Di Yi Jun Yi Da Xue Bao. 2002;22(5):451–2.
51. Wang YX, Hussain SM, Krestin GP. Superparamagnetic iron oxide contrast agents: physicochemical characteristics and applications in MR imaging. Eur Radiol. 2001;11(11):2319–31.
52. Bligh SW, Sadler PJ, Marriott JA, Latham IA, Kelly JD. Characterization and in vivo distribution of ^{99m}Tc- and ^{111}In-labeled magnetite. Int J Rad Appl Instrum A. 1989;40(9):751–7.
53. Lewin M, Carlesso N, Tung CH, et al. Tat peptide-derivatized magnetic nanoparticles allow in vivo tracking and recovery of progenitor cells. Nat Biotechnol. 2000;18(4):410–4.
54. Kaufner L, Cartier R, Wüstneck R, et al. Poly(ethylene oxide)-block-poly(glutamic acid) coated maghemite nanoparticles: in vitro characterization and in vivo behavior. Nanotechnology. 2007;18(11):115710.
55. Baio G, Fabbi M, Salvi S, et al. Two-step in vivo tumor targeting by biotin-conjugated antibodies and superparamagnetic nanoparticles assessed by magnetic resonance imaging at 1.5 T. Mol Imaging Biol. 2010;12(3):305–15.
56. Gao X, Cui Y, Levenson RM, Chung LW, Nie S. In vivo cancer targeting and imaging with semiconductor quantum dots. Nat Biotechnol. 2004;22(8):969–76.
57. Liu S, Jia B, Qiao R, et al. A novel type of dual-modality molecular probe for MR and nuclear imaging of tumor: preparation, characterization and in vivo application. Mol Pharm. 2009;6(4):1074–82.
58. Otsuji E, Kuriu Y, Okamoto K, et al. Monoclonal antibody A7 coupled to magnetic particles as a contrast enhancing agent for magnetic resonance imaging of human colorectal carcinoma. Cancer Immunol Immunother. 2006;55(6):728–33.
59. West SD, Lee YC. Management of malignant pleural mesothelioma. Clin Chest Med. 2006;27(2):335–54.
60. Renshaw AA, Dean BR, Antman KH, Sugarbaker DJ, Cibas ES. The role of cytologic evaluation of pleural fluid in the diagnosis of malignant mesothelioma. Chest. 1997;111(1):106–9.
61. Pakzad F, Groves AM, Ell PJ. The role of positron emission tomography in the management of pancreatic cancer. Semin Nucl Med. 2006;36(3):248–56.
62. Buy JN, Ghossain MA, Sciot C, et al. Epithelial tumors of the ovary: CT findings and correlation with US. Radiology. 1991;178(3):811–8.
63. Muminova ZE, Strong TV, Shaw DR. Characterization of human mesothelin transcripts in ovarian and pancreatic cancer. BMC Cancer. 2004;4:19.
64. Yokokawa J, Palena C, Arlen P, et al. Identification of novel human CTL epitopes and their agonist epitopes of mesothelin. Clin Cancer Res. 2005;11(17):6342–51.
65. Misri R, Saatchi K, Ng SSW, Kumar U, Häfeli UO. Evaluation of ^{111}In labeled antibodies for SPECT imaging of mesothelin expressing tumors. Nucl Med Biol. 2011;38(6):885–96.
66. Lijowski M, Caruthers S, Hu G, et al. High sensitivity: high-resolution SPECT-CT/MR molecular imaging of angiogenesis in the Vx2 model. Investig Radiol. 2009;44(1):15–22.
67. Patz EF Jr, Rusch VW, Heelan R. The proposed new international TNM staging system for malignant pleural mesothelioma: application to imaging. AJR Am J Roentgenol. 1996;166(2):323–7.
68. Misri R, Meier D, Yung A, Kozlowski P, Hafeli UO. Development and evaluation of a dual-modality (MRI/SPECT) molecular imaging bioprobe. Nanomedicine. 2011; https://doi.org/10.1016/j.nano.2011.10.013.
69. Argani P, Iacobuzio-Donahue C, Ryu B, et al. Mesothelin is overexpressed in the vast majority of ductal adenocarcinomas of the pancreas: identification of a new pancreatic cancer marker by serial analysis of gene expression (SAGE). Clin Cancer Res. 2001;7(12):3862–8.
70. Zielhuis SW, Seppenwoolde JH, Mateus VA, et al. Lanthanide-loaded liposomes for multimodality imaging and therapy. Cancer Biother Radiopharm. 2006;21(5):520–7.
71. Koning GA, Fretz MM, Woroniecka U, Storm G, Krijger GC. Targeting liposomes to tumor endothelial cells for neutron capture therapy. Appl Radiat Isot. 2004;61(5):963–7.
72. O'Connor JP, Jackson A, Parker GJ, Jayson GC. DCE-MRI biomarkers in the clinical evaluation of antiangiogenic and vascular disrupting agents. Br J Cancer. 2007;96(2):189–95.
73. Tofts PS, Brix G, Buckley DL, et al. Estimating kinetic parameters from dynamic contrast-enhanced T(1)-weighted MRI of a diffusable tracer: standardized quantities and symbols. J Magn Reson Imaging. 1999;10(3):223–32.
74. Saatchi K, Soema P, Gelder N, et al. Hyperbranched polyglycerols as trimodal imaging agents: design, biocompatibility and tumor uptake. Bioconjug Chem. 2011;23(3):372–81.

13 PET-MR Nanoimaging

13.1 Introduction

The combined positron emission tomography (PET)/magnetic resonance imaging (MRI) modality has revolutionized various cancer diagnoses, treatments, and follow-ups. Again, this is considered one of the best multimodality imaging techniques, combining both functional and morphological information at biological and cellular levels, along with tracking various biological pathways. Nanoparticle-based contrast agents (CAs) are the best examples, as they offer better probe biocompatibility and biodistribution, along with more specific target accumulation and non-toxicity. The CAs are capable of delivering drugs, both general and specific therapeutic agents, for better diagnostic and therapeutic effects, enhancing cancer diagnosis and treatment. This is expected to pave the way for personalized medicine in the near future. In this chapter, we discuss the principles, characteristics, applications, and disadvantages of nanoparticle (NP)-based PET/MRI CAs.

The whole-body PET/MRI system incorporates MRI-compatible photodiodes (avalanche photodiodes) and various MR-based attenuation corrections. MRI itself is a functional imaging modality using different sequences like diffusion-weighted imaging (DWI), blood oxygen level-dependent (BOLD) imaging in functional MRI (fMRI), T1/T2 mapping, perfusion imaging, spectroscopy, and dynamic contrast-enhanced (DCE) imaging. The technical advantage of MRI is that anatomical data, needed for correcting the partial volume effect caused by PET [1], can also be useful in motion correction. Additionally, the arterial input function can aid in the characterization of PET kinetic modeling and PET iterative reconstruction. PET provides valuable molecular information and highly sensitive quantification [2]. fMRI data and metabolic PET information, when used simultaneously, can identify the coupling between metabolic demand and functional brain activity, as oxygen and glucose metabolism are related to cerebral blood flow, which delivers O_2 and glucose to tissues [3]. MRI has the advantage of high resolution and high tissue contrast, and along with multiparametric, functional, and quantitative imaging, supplies complementary information for imaging of breast, head, and neck [4], liver, musculoskeletal, and brain tumors [5], and heart [6]. Thus, hybrid imaging is complementary for molecular imaging development, in which molecular imaging represents the “in vivo” visualization, characterization, and measurement of various biological processes at molecular and cellular levels [7, 8]. So far, different molecular imaging modalities have been studied for disease diagnosis, pathology stratification, and treatment assessment [9], along with image-guided therapy. Molecular imaging involves the administration of imaging probes for detecting signals in various

S. Shaikh, *Nanoimaging - Future of Precision Medicine*,
https://doi.org/10.1007/978-981-97-7292-6_13

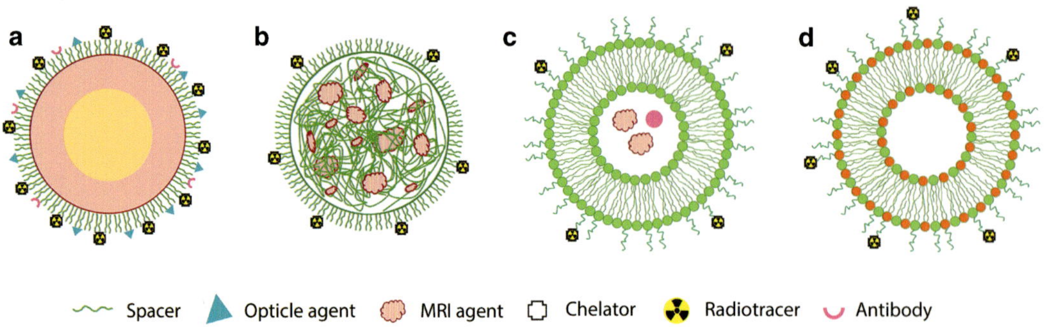

Fig. 13.1 **a** is an antibody with spacer and radiotracer. **b** is Spacer, MRI agent and radiotracer. **c** is also Spacer, MRI agent and radiotracer. **d** is comprising of spacer and radiotracer

probes [10], playing an important role in understanding the pathophysiological principles of diseases. The concept of personalized medicine is aimed at identifying better management strategies and achieving improved therapeutic efficacy. Different imaging probes are being developed, which poses challenges for imaging scientists [11, 12]. The use of nanoparticle (NP)-based PET/MRI multimodal tracers in clinical oncological imaging is an important step, with many preclinical studies showing promising results for tumor detection, staging, and grading (Fig. 13.1).

13.2 Nanoparticle-Based PET/MRI Probes Overview

MRI contrast agents and PET tracers are widely used for hybrid PET/MRI. One of the early agents includes superparamagnetic iron oxide (IO) and ^{18}F-Fluorodeoxyglucose, which provide enhanced information from each modality [13–15]. Since both MRI and PET are used to evaluate different physiological processes, sequential imaging is needed. The combination of PET and MRI agents in a single probe allows better temporal and spatial correlation of these imaging modalities. The basic concept is that biological systems are different in various aspects, one of which is their dynamic nature, which responds differently to various contrast agents. The integration of these agents in a single platform prevents differential biodistribution from various imaging agents in the body, allowing simultaneous monitoring [16]. Nanoparticles have great potential as imaging probes due to their excellent physical and chemical properties [17]. Compared to molecule-based probes such as gadolinium(III) complexes, which are smaller in size and have fewer moieties or linkages with drugs, nanoparticles have a larger surface area to volume ratio, allowing multiple surfaces for imaging, targeting, and therapeutic purposes [13]. Their size and surface modifications also lead to increased circulation time in the blood and reduced uptake by the reticuloendothelial system (RES) [18]. Important factors include probe design, physical properties, pharmacokinetics, solubility, lipophilicity, efficacy, and material stability [19, 20]. The kinetics of the nanoparticle surface depend on size, surface charge, and coating functionality, all of which are used in the development of nanoparticle-based PET/MRI probes. The size of nanoparticles influences cell uptake and clearance from the body. An ideal size range is 20–100 nm in diameter, which allows longer circulation times and reduces renal clearance by the liver and spleen [21]. These particles accumulate in tumor tissues due to increased permeability and retention in affected tissues, such as in macrophages found in atherosclerotic plaques and ischemic myocardium. The exact size can be measured by transmission electron microscopy (TEM), while hydrodynamic diameter is measured by dynamic light scattering. Stabilization of the colloidal system is achieved by adsorbed or grafted polymers or electrostatic repulsion [22]. Steric hindrance methods use hydrophilic polymers like PEG, where direct contact between the particles is prevented by the polymer chains extending into the medium, forming "polymer brushes" [22]. By coating the particles with polymer chains, blood protein adsorption is reduced and rapid clearance by the reticuloendothelial system is also minimized [23]. Higher zeta values (positive values greater than 40 mV and negative values lower than −40 mV) suggest strong electrostatic repulsion and a lower likelihood of aggregation due to Brownian motion [24, 25]. The stabilizing function along with the surface charge also leads to the interactions between nanoparticles and their biological targets, such as polyionic molecules [26, 27]. This signifies the role of the cellular interactions with nanoparticles and their cellular responses in diagnosis and therapeutics [27]. There will be a decrease in relaxivity when the surface is coated with the IO nanoparticles [28]. Additionally, Cu^{2+} ions can be chelated on the surface of on Fe_3O_4 nanoparticles [29]. IO-based dual-modality probes exhibit higher transverse relaxivity compared to other contrast agents [30]. Gd^{3+}-based nanoparticles, such as $^{124}I\text{-}Gd_3N@C_{80}$, can load more Gd^{3+} ions [31, 32]. Enhanced contrast is achieved with high r1 per particle from Gd^{3+} centers, and particles with high payloads may have superior sensitivity compared to those with only high r1 per Gd^{3+}. Since PET uses small molecules conjugated to nanoparticles, there is minimal impact on the physical properties of the nanoprobe. Multimodal nanoparticle-based PET/MRI probes are typically composed of separate signal-generating components. Most PET/MRI probes feature nanoparticle carriers with MRI contrast agents, targeting various scaffolds, and PET tracers. These PET tracers, added to PET/MRI, bind radiometals to create stable complexes with radioisotopes, forming coordinate bonds on the nanoparticle surface. Such surface modifications enhance targeting specific biomarkers for lymph nodes and tumors. Coating materials are selected for colloidal stability, circulation time in the blood, clearance (serum protein adsorption during circulation that promotes rapid RES recognition) rates, and water solubility. The MRI component of nanoparticle-based PET/MRI probes enhances contrast in the chemical surroundings of nearby water protons [29, 33] and through transverse relaxation in IO nanoparticles. Nanoparticle-based PET/MRI probes include IO, sentinel ferrite, MFe_2O_4 (M = Co, Mn), and gadolinium(III) ions used in metallofullerenes and as codopants in nanophosphors. Superparamagnetic IO nanoparticles are now commonly used as MRI contrast components in PET/MRI probes due to their non-toxicity, biodegradability, and cost-effective precursors. Synthetic properties allow better size distribution, control, easy purification, and affordability. These nanoparticles can be combined with metals like cobalt [34, 35] and manganese [36, 37], forming ferrites with higher magnetic anisotropy than IO. This increased magnetic anisotropy enhances contrast efficiency [38] and reduces nanoparticle size, extending blood circulation time [39]. Lanthanide-based nanomaterials, combined with luminescence and magnetic properties, offer superior multimodal probe capabilities compared to others. Paramagnetic Gd^{3+} is used as an MRI contrast agent in metallofullerene Gd_3N with C80 nanoprobes, which serve as effective T1 contrast agents [40]. Hydroxylated Gd_3N $C_{80}(OH)$ has low cytotoxicity [40]. These probes combine the paramagnetic properties of Gd^{3+} with the luminescence of other lanthanides, facilitating high sensitivity and spatial resolution [40] for fluorescence molecular tomography in animal imaging [12, 41]. Gd^{3+} MRI contrast agents combined with ^{18}F-labeled nanophosphors allow PET/MR whole-body imaging and in vivo luminescence imaging of living cells [41]. These nanophosphors exhibit low cytotoxicity at 3 T [42].

13.3 PET Component

PET tracers and chelators are essential components of nanoparticle-based PET/MRI probes. PET tracers, which include radiohalides and radiometals, emit positrons. These positrons collide with electrons, resulting in annihilation and the emission of γ-rays, which can be detected by a γ detector. Smith et al. studied optimal PET radionuclides, identifying key factors for effective imaging, such as maximum positron energy, half-life, nuclide availability, and ease of use [43]. PET tracers generally possess low maximum positron energy, which can lead to increased absorption and reduced spatial resolution [44]. Consequently, ^{18}F and ^{64}Cu are commonly used due to their lower maximum positron energy and favorable half-lives. ^{64}Cu requires ^{64}Ni and protons in its purification process, which is conducted through anion exchange chromatography [45]. Gallium-68 is obtained by a $^{68}Ge/^{68}Ga$ generator [45], and an important aspect is to note its smaller half-life. In triple modal PET/MRI/optical imaging, ^{124}Iodine [31, 46] and ^{68}Ga are used as radioisotopes because they emit high-energy positrons emission, which result in better Cerenkov luminescence imaging [36, 47, 48]. The radiometals used in nanoparticle-based PET/MRI probes don't require chelators [49], a direct labeling method [48] with bonding with the material [50]. The majority of PET/MRI nanoprobes require the chelation of radiometals [51]. Common chelators used in PET/MRI nanoprobes include polyazacarboxylate macrocycle like DOTA; pentadecatriene-*S*-4triacetic acid PCTA [52]; 1,4,7-triazacyclododecane-*N*,*N′*,*N″*-triacetic acid (NOTA) [19, 35, 53, 54]; and a derivative of NOTA with one of the carboxylic arms modified to ethylenediamine [45]. These chelators can coordinately saturate Cu^{2+} and Ga^{3+} to their coordination number of six and have been demonstrated to form stable complexes with Cu^{2+} and Ga^{3+} (log KML for Cu^{2+}DOTA, Cu^{2+}NOTA, and Ga^{3+}NOTA = 20–30) [54–56]. NOTA is used for easier conjugation of NOTA to amine-terminated nanoparticles [29, 33, 52, 53, 57–60]. Ga^{3+} and DOTA can cause transmetallation when using DOTA as a ligand for $^{68}Ga^{3+}$ radionuclides [61, 62]. Chelating agents used for stabilizing radiometals iclude deferoxamine for ^{89}Zr [63] and the acyclic bis(dithiocarbamatebisphosphonate) [37, 64] for chelating ^{64}Cu. Different strategies are considered as the important component of the synthesis and functionalization of different nanoparticles [16, 20, 65, 66]. Metal cores of PET/MRI probes are typically created by coprecipitation [52, 57] or thermal decomposition [19, 29, 33, 37, 41, 53, 54]. Nanoparticles produced by thermal decomposition are hydrophobic, requiring additional steps for water solubility; however, this method provides better control over size, dispersity, and crystallinity [67–71]. Synthesized nanoparticle-based PET/MRI probes are often coated with biocompatible materials like polyaspartic acid [57], serum albumin [36, 58], PEG [19, 34, 35, 45, 48, 50, 53, 54, 59, 60, 64, 72], dextran [15, 64], citrate [41], poly(acrylic acid) [49], and silica [35]. Neutral polymers like dextran and PEG resist opsonization, while poly(aspartic acid) promotes electrostatic interactions with plasma proteins. Hydrophilic coatings for PET/MRI probes can be applied through ligand exchange on the nanoparticle surface, enhancing water solubility and improving biocompatibility.

Nanoparticle-based PET/MRI probes have demonstrated utility in preclinical research for imaging various target tumors or macrophages. They are particularly useful for imaging lymph nodes, tumors, and cardiovascular disease.

13.4 Lymph Node Imaging

Lymph node imaging is an important part of imaging for the characterization of specific lymph nodes (LNs) as well as the tumor pathways for metastasis to various organs via the lymphatic route. Thus, evaluation and staging of tumor invasion are important parts of cancer staging. These need to be differentiated between reactive nodes and metastatic nodes [47]. The use of nanoparticle-based PET/MRI probes allows for better evaluation with greater stability, with a hydrodynamic size in the range of 5–50 nm [47]. This size criterion is also very important due to the mobility of nanoparticles in the lymphatic tract [36]. The nanoparticle-based PET/MRI probes can also be coated for better stability. Different studies describe how nanoparticles delineate the brachial, axillary, popliteal, and iliac lymph nodes in normal mice [36, 38, 63, 64], as well as in tumor-bearing mice [47, 63]. Mn-doped IO (nonspecific) nanoparticles with a size of 32 nm are also useful for evaluating the brachial and axillary lymph nodes in healthy mice [63]. The MR probe is useful for reconstructing surgical guidance images, while PET complements MRI with higher resolution.

Tumor imaging using nanoparticle-based PET/MRI probes can show the passive or active accumulation of these nanoparticles in tumors.

13.5 Cardiovascular Diseases

Cardiovascular imaging is useful for the evaluation of coronary artery disease, atherosclerosis, and heart failure. Challenges include the small sample volume in the coronaries and carotid arteries, as well as significant movement of blood vessels. Multimodal imaging can overcome these limitations in sensitivity and spatial resolution for different structures and various physiological parameters, including coronary anatomy, cardiac volumes, infarct size, and plaque inflammation [18]. Some new PET/MRI probes for imaging macrophages can be seen in atherosclerotic plaques. Mal-BSA-DOTA-Gd/Cu (PET component: ^{64}Cu, MRI component: Gd^{3+}, chelator: DOTA, targeting ligand: maleylated bovine serum albumin) and MDIO-^{64}Cu-DOTA (PET component: ^{64}Cu; MRI component: IO; chelator: DOTA; targeting ligand: maleylated dextran) are being used for the specific targeting of the scavenger receptor A seen on activated macrophages, one of the inflammation biomarkers in vulnerable atherosclerotic plaques [73, 74].

13.6 Cell Tracking

Cell-based therapy systems are also seen as important advancements in pharmaceutics [17]. Monitoring cell activity is critical for assessing the therapeutic potential of these cells. Molecular imaging is also regarded as a targeted nanoparticle delivery system to deliver to a greater number of cells in tissues. These cell tracking methods will be primarily labeled for easy detection in the body. These multimodal PET/MRI probes will be aligned for cell migration tracking and evaluation of drug response at the molecular level.

The mixture of MRI and PET probes, combined as imaging agents, poses a higher risk for the patient [75, 76]. Despite this, the spatial and temporal resolutions of these two imaging modalities differ based on the biodistribution, pharmacodynamic, and pharmacokinetic properties of the imaging agent. For this reason, nanoparticles (NPs) are selected differently for each bimodal probe to monitor both modalities. NPs are defined as particles sized between 1 and 100 nm [77, 78]. Different types of NPs are available, including proteins, polymers, dendrimers, micelles, liposomes, viral capsids, metal oxides (iron-oxide NPs), zeolites, and mesoporous silicas, which are used for their various shapes, such as spheres, cylinders (nanorods), and tubes [79]. In these combined probes, the MRI component can carry iron-oxide or gadolinium-oxide NPs, or they can be combined into nanosized systems such as apoferritin cages [80] and low-density lipoprotein (LDL) particles [81]. The particle size, charge, core and surface properties, shape, and multivalency influence distribution, reducing NP toxicity [82]. Size will determine parameters of renal excretion, where larger particle sizes (up to 100 nm) will be taken up by macrophages [83, 84]. NP shapes are important for cell tracking and labeling, with internalization rates observed in spherical particles [85]. The internalization properties similar to rod-like bacterium are seen when they are coated by plasma proteins in a process known as opsonization. These NPs are recognized by plasma membrane receptors on monocytes and macrophages and are taken up by the reticuloendothelial system (RES), also known as the mononuclear phagocyte system (MPS). The liver, spleen, and bone marrow are rich in macrophages, allowing NP access to these organs [86]. Preventing nonspecific uptake by the RES is known as the "stealth effect" [86–88]. Blood components like serum proteins are optimal for minimizing opsonization and clearance [89, 90]. Polymers without different functional groups (amine, carboxyl, or hydroxyl) for ligand linkage will not be activated. Various hydrophilic bifunctional materials, such as bisphosphonate [91] or aluminum hydroxide [92], are used as coatings. Coating layers should be biocompatible, such as polyethylene glycol (PEG), dendrimers, polysaccharides (dextran and chitosan), and polypeptides (serum albumin), for enhanced properties, including better stability against agglomeration, biocompatibility, and water solubility, along with low toxicity. Commonly used coating polymers are dextran, chitosan, and PEG [80]. PEG is a hydrophilic, water-soluble, biocompatible polymer that reduces opsonization by increasing circulation time from seconds or minutes to hours [93]. This affects the superparamagnetic properties of iron-oxide NPs [94]. In particular, the nature and thickness of the coating influence relaxivity. A more hydrophilic coating material retains water molecules for interacting with magnetic centers, while a thicker coating causes more protons to be shielded from the magnetic field [95]. NP delivery to malignant cells is achieved by passive and active targeting. Passive targeting relies on the enhanced permeability and retention (EPR) effect, where the tumor vessels have larger fenestrations, resulting in higher permeability and allowing easier NP extravasation in tumor tissue. Poor lymphatic drainage results in more NP retention in the tumor's interstitial space [96, 97]. The high NP surface-to-volume ratio presents some limitations due to the NP surface's target-specific moieties, enabling more active permeation and retention in tumor cells.

13.7 Radiolabeled Nanoparticles

Commonly used tracers utilize labeled positron emitters with shorter half-lives ranging from 2.037 to 109.8 min. These radionuclides are produced via a cyclotron and used for labeling. The cyclotron produces a beam of accelerated protons and deuterons, which is used to irradiate the target (e.g., $^{14}N_2$ gas, ^{20}Ne gas, ^{18}O water or gas), forming the desired radioisotope through a nuclear reaction.

Nanoparticles (NPs) are increasingly used at the preclinical level as imaging probes for dual MRI/PET tumor imaging. There are two types of NPs: inorganic and organic [98]. Inorganic NPs are becoming more widely used due to their chemical inertness, good stability, and ease of surface functionalization, making them suitable for imaging malignant tumors via endocytosis, although this can lead to higher toxicity due to accumulation in endo-lysosomal compartments [99]. Inorganic NPs have superior physical and chemical properties like inertness, stability, and better surface functionalization, which aid in differentiating malignant cells. However, higher toxicity remains a concern.

Magnetic iron-oxide NPs, including magnetite Fe_3O_4 and maghemite (γ Fe_2O_3), are used in MRI imaging, particularly for the liver, spleen, and bone marrow, due to their shortened T2 and T2* relaxation times. These NPs are categorized by size: micrometer-sized paramagnetic iron oxide (MPIO) (several micrometers), superparamagnetic iron oxide (SPIO) (hundreds of nanometers), ultrasmall superparamagnetic iron oxide (USPIO) (below 50 nm) [100], and monocrystalline iron oxide (MION), a subset of USPIO ranging from 10 to 30 nm [98, 101, 102]. The most common method for SPIO and USPIO production is synthesizing ferrous and ferric salts in a basic aqueous medium [103–105]. These NPs tend to form polydisperse, poorly crystalline structures, though other methods, such as thermal decomposition and microwave synthesis, are also used [60, 106]. To improve colloidal and chemical stability, different polymeric coatings are applied, including dextran, carboxymethylated dextran, chitosan, starch, PEG, heparin, albumin, arabinogalactan, glycosaminoglycan, sulfonated styrene–divinylbenzene, organic siloxane, polyvinyl alcohol, poloxamers, and polyoxamines [107, 108]. These coatings protect the iron-oxide NPs and reduce cytotoxicity [109]. The coprecipitation process can be performed during NP synthesis, with coating either occurring simultaneously or post-synthesis [110, 111]. Surface coating plays an important role in NP bioconjugation with biological ligands, such as peptides or antibodies, increasing potential for cancer imaging [112, 113]. These nanoparticles provide negative contrast, with a loss of signal compared to T1 contrast agents, resulting in high susceptibility with distortion artifacts and a reduced contrast-to-noise ratio [100]. Gd-based T1 agents are commonly used. Alloy materials can also be used to obtain more efficient T2 contrast agents with higher magnetic anisotropy [3], crystallinity, and relaxivity. Various bimetallic ferrite NPs, known as magnetic engineered iron-oxide NPs, such as $CoFe_2O_4$, $MnFe_2O_4$, and $NiFe_2O_4$, have been tested [114].

Iron-oxide NPs like magnetite Fe_3O_4 and maghemite (γ Fe_2O_3) are also used in MRI imaging, especially for the liver, spleen, and bone marrow, due to shorter T2 and T2* relaxation times. Based on their size, these are categorized into MPIO (several micrometers), SPIO (hundreds of nanometers), USPIO (below 50 nm) [100], and MION, a subset of USPIO ranging from 10 to 30 nm [98, 102]. SPIO and USPIO are typically synthesized by the reduction and coprecipitation reactions of ferrous and ferric salts in a basic aqueous medium [103–105, 107, 108]. The polymeric corona protects iron-oxide NPs, preventing erosion at acidic pH levels and lowering cytotoxicity [109]. The coating process may be conducted during coprecipitation or post-synthesis [110–113]. Surface coating is an important factor for NP bioconjugation with biological ligands, such as peptides or antibodies, for cancer imaging. Alloys are used to improve crystallinity and relaxivity, with bimetallic ferrite NPs, such as $CoFe_2O_4$, $MnFe_2O_4$, and $NiFe_2O_4$, and the dextran. Dextran-like Feridex and Endorem, which are alkali-treated low-molecular-weight carboxydextran, are used for Resovist [115, 116]. Various PET/MRI probes are used in cancer imaging, employing both passive (e.g., lymph node mapping) and active targeting strategies [117]. Radiolabeled ^{89}Zr iron-oxide NPs (ferumoxytol) are used for axillary and brachial lymph node evaluation in healthy wild-type mice. The iron-oxide core is covered by a semisynthetic polysaccharide coating of polyglucose sorbitol carboxymethylether and deferoxamine, which acts as a chelator. In a study by Madru et al. [118], chelator-free conjugation of ^{64}Cu provided efficient identification and characterization of sentinel lymph nodes (SLNs) in C57BL/6J mice. Lymph node metastases are crucial markers for cancer staging and treatment. Xie and colleagues [119] encapsulated iron-oxide NPs, modified with dopamine, into human serum albumin (HSA) matrices and labeled them with Cy5.5 dye and ^{64}Cu-DOTA. In this model, PET and NIRF imaging provided a higher signal-to-noise ratio compared to MRI due to higher sensitivity. MRI showed an inhomogeneous distribution pattern due to better spatial resolution, confirmed histopathologically. Lee and coworkers [120] developed an active targeting probe by combining RGD-conjugated ^{64}Cu radiolabeled iron-oxide NPs. Imaging on a U87MG mouse model confirmed NP accumulation mediated by $\alpha v\beta 3$ integrin binding. Kim and colleagues [121] injected ^{68}Ga-labeled iron-oxide NPs into BALB/c nude mice bearing HT-29 colon cancer cells, with oleanolic acid as a tumor-targeting molecule.

13.8 Silica-Based Nanoparticles

Silica-based nanoparticles (NPs) are widely used in drug delivery, bio-imaging, and cell targeting due to their biocompatibility and ease of integration into imaging probes. There are two main classes of silica-based NPs: solid silica nanoparticles (SiNPs) and mesoporous silica nanoparticles (MSNs). SiNPs are primarily used as optical imaging agents, while MSNs are applied in CT, MRI, PET, molecular, and multimodal imaging. MSNs are typically prepared using the surfactant-templated sol–gel method [122] and offer a large surface area, tunable structure in terms of size, morphology, and porosity, as well as easy functionalization through synthetic approaches [123]. In PET/MRI imaging, MSNs serve as coating materials for metallic NPs and as carriers for MRI contrast agents (CAs) and PET radioisotopes. A porous silica shell can enhance MRI contrast due to improved interaction between water molecules and iron-oxide NPs [124], making it useful for cancer imaging applications. Burke and colleagues [125] discussed the advantages of silica nanorods over nanospheres, noting their superior T2 MRI contrast and liver uptake through phagocytosis. Additionally, Huang and coworkers [126] explored the use of mesoporous silica-based, triple-modal imaging nanoprobes for tracking tumor metastatic sentinel lymph nodes (T-SLNs). The PET and MRI imaging probes are located on both the surface and within the mesoporous channels of these NPs. Compared to normal SLNs, T-SLNs demonstrate faster and higher uptake rates of multifunctional MSN probes.

13.9 Organic Nanoparticles

Organic NPs, such as dendrimers, polymeric micelles, liposomes, and proteins, are used in various applications for cancer diagnosis. These organic NPs are capable of carrying radionuclides, showing great potential for tumor diagnosis [98]. Liposomes are spherical phospholipid bilayers similar to cell membranes. Phospholipids are amphiphilic molecules, with a hydrophilic head group and two hydrophobic tails. Liposomes are classified by size and the number of bilayers or multilayers. The multilamellar liposomes have an onion-like structure where each bilayer of phospholipids is separated by an adjacent water layer [104]. Liposomes are prepared based on their size and number of layers, which is determined by sonication and extrusion [104]. They are biocompatible, non-toxic, biodegradable, and commonly used for drug delivery. Through PEGylation, the blood circulation time of liposomes is prolonged, allowing sustained release or targeted delivery of imaging and therapeutic agents [106]. Liposomes utilize the EPR effect for active targeting with antibodies, peptides, and vitamins to reach cancer cells [125–127]. Liposomal formulations are approved for cancer therapy, including those combined with doxorubicin, and are also used to treat infections such as fungal infections [128]. In MRI imaging, liposomes serve as a coating material to prevent iron-oxide NPs from aggregating and targeting tumor cells, contributing to various multimodal imaging and theranostic applications. In MRI and PET imaging, the magnetic properties of liposomes and glucose enable a dual tumor-targeting mechanism. Li and colleagues [129] demonstrated a multifunctional theranostic liposomal drug delivery system with liposomes encapsulating doxorubicin and conjugated with Gd-DOTA for MRI and IR dye for near-infrared fluorescence. Additionally, the liposomes were radiolabeled with ^{99m}Tc and ^{64}Cu for SPECT and PET imaging. After intratumoral injection, high-resolution MR images displayed the tumor, highlighting its potential as an important chemotherapeutic agent. Mitchell et al. [130] developed liposomal formulations for multifunctional imaging using the chelator (DOTA) to chelate Gd^{3+} for MRI, ^{111}In for SPECT, and ^{64}Cu for PET. Abou and colleagues [131] showed radiolabeled preformed paramagnetic (Gd) liposomes with ^{89}Zr (positron emitter). MR and PET images also show the accumulation of octreotide liposomes in SSTr2-expressing cells compared to control liposomes.

Micelles are composed of non-ionic surfactants (surfactant micelles) or amphiphilic block copolymers (polymeric micelles). In polymeric micelles, the length of the hydrophilic block is greater than the hydrophobic one [112]. Polymeric micelles are more capable of delivering hydrophobic drugs, even though they are smaller in size than liposomes [132, 133]. Polymeric micelles can be synthesized using water-soluble copolymers with lipids as the hydrophobic blocks (such as polyethylene glycol–phosphatidyl ethanolamine, PEG–PE). Size is an important component of PEG–lipid micelles used for diagnostic imaging [112], particularly due to the size limitations of liposomes, which need to have a diameter larger than a certain minimum (usually 70–100 nm) [134]. Trubetskoy and colleagues [135] developed a vector with Gd-DTPA-PE and ^{111}In-DTPA-SA for lymph node gamma scintigraphy and MRI imaging in rabbits. A recent study by Starmans et al. [136] demonstrated PET/MRI dual imaging for micelles functionalized with ^{89}Zr deferoxamine and Fe^{3+} deferoxamine. Both liposomes and micelles, however, tend to be unstable in the presence of serum [118, 119].

Dendrimers are highly branched, spherical polymers with tree-like internal structures, characterized by an inner core surrounded by branches known as generations. Depending on the number of generations, their size and molecular weight vary. Drugs can be encapsulated in the inner spaces or anchored on the external terminations

[18]. Dendrimers can serve as dual-modal agents in MRI and fluorescence imaging [137–139], optical imaging and nuclear medicine [113], CT and MRI [114], and positron-emitting radionuclide-labeled dendrimers have also been reported [140–142]. However, achieving PET/MRI or SPECT/MRI agents with radioisotope loading through specific chelating moieties remains challenging [143].

Various biological polymers, such as alginate, hyaluronic acid, and chitosan, as well as proteins, antibodies, enzymes, lipoproteins, and viral capsids (protein cages), are increasingly used for diagnostic and therapeutic applications [144]. Maham et al. synthesized different drug delivery systems based on various types and shapes (NPs, microspheres, films, mini rods, hydrogels) using gelatin, albumin, collagen, elastin, ferritin/apoferritin, gliadin, casein, zein (corn protein), whey protein, and soy protein [3], highlighting their potential and challenges. Vecchione et al. [145] developed a fully biocompatible platform for dual MRI/PET imaging with distinct relaxometric properties. Core–shell nanocarriers made of chitosan and hyaluronic acid entrapped Gd-DTPA, offering enhanced relaxometric properties up to five times better, and served as carriers for ^{18}F-FDG without altering the FDA-approved CAs for various cellular pathways [146].

13.10 Conclusion

The use of molecular imaging and multimodal imaging is under research in many aspects by researchers and scientists. Since there is not a single modality that can achieve the best results, this leads to a comprehensive overview of various morphological, functional, and metabolic processes underlying tumors. For this reason, the use of the hybrid imaging modality PET and MRI is advantageous. The nanoparticles (NPs) used in PET/MRI can provide better anatomical and pathological information about their size, shape, charge, coating, and multivalency. This approach is also useful in the theranostic field related to cancer imaging, diagnosis, characterization, as well as information related to drug release, efficient therapy, and monitoring of treatment response. The use of multicomponent imaging probes can achieve higher sensitivity and high resolution, providing complementary information.

References

1. Garcia J, Tang T, Louie AY. Nanoparticle-based multimodal PET/MRI probes. Nanomedicine. 2015;10:1343–59.
2. Nensa F, Beiderwellen K, Heusch P, Wetter A. Clinical applications of PET/MRI: current status and future perspectives. Diagn Interv Radiol. 2014;20:438–47.
3. Aiello M, Salvatore E, Cachia A, Pappata S, Cavaliere C, Prinster A, Nicolai E, Salvatore M, Baron J-C, Quarantelli M. Relationship between simultaneously acquired resting-state regional cerebral glucose metabolism and functional MRI: a PET/MR hybrid scanner study. NeuroImage. 2015;113:111–21.
4. Covello M, Cavaliere C, Aiello M, Cianelli MS, Mesolella M, Iorio B, Rossi A, Nicolai E. Simultaneous PET/MR head-neck cancer imaging: preliminary clinical experience and multiparametric evaluation. Eur J Radiol. 2015;84:1269–76.
5. Pace L, Nicolai E, Aiello M, Catalano OA, Salvatore M. Whole-body PET/MRI in oncology: current status and clinical applications. Clin Translat Imaging. 2013;1:31–44.
6. Nappi C, Altiero M, Imbriaco M, Nicolai E, Giudice CA, Aiello M, Diomiaiuti C, Pisani A, Spinelli L, Cuocolo A. First experience of simultaneous PET/MRI for the early detection of cardiac involvement in patients with Anderson-Fabry disease. Eur J Nucl Med Mol Imaging. 2015;42:1025–31.
7. Weissleder R, Mahmood U. Molecular imaging. Radiology. 2001;219:316–33.
8. Chen K, Chen X. Design and development of molecular imaging probes. Curr Top Med Chem. 2010;10:1227–36.
9. Cai W, Chen X. Multimodality molecular imaging of tumor angiogenesis. J Nucl Med. 2008;49:113S–28S.
10. Chen K, Conti PS. Target-specific delivery of peptide-based probes for PET imaging. Adv Drug Deliv Rev. 2010;62:1005–22.
11. Bouziotis P, Psimadas D, Tsotakos T, Stamopoulos D, Tsoukalas C. Radiolabeled iron oxide nanoparticles as dual-modality SPECT/MRI and PET/MRI agents. Curr Top Med Chem. 2012;12:2694–702.
12. Luker GD, Luker KE. Optical imaging: current applications and future directions. J Nucl Med. 2008;49(1):1–4.

13. Huang W-Y, Davis JJ. Multimodality and nanoparticles in medical imaging. Dalton Trans. 2011;40(23):6087–103.
14. Higuchi T, Nekolla SG, Jankaukas A, et al. Characterization of normal and infarcted rat myocardium using a combination of small-animal PET and clinical MRI. J Nucl Med. 2007;48(2):288–94.
15. Chapon C, Jackson JS, Aboagye EO, Herlihy AH, Jones WA, Bhakoo KK. An in vivo multimodal imaging study using MRI and PET of stem cell transplantation after myocardial infarction in rats. Mol Imaging Biol. 2009;11(1):31–8.
16. Xie J, Liu G, Eden HS, Ai H, Chen X. Surface-engineered magnetic nanoparticle platforms for cancer imaging and therapy. Acc Chem Res. 2011;44(10):883–92.
17. Hwang DW, Youn H, Lee DS. Molecular imaging using PET/MRI particle. Open Nucl Med J. 2010;2:186–91.
18. Lee DE, Koo H, Sun IC, Ryu JH, Kim K, Kwon IC. Multifunctional nanoparticles for multimodal imaging and theragnosis. Chem Soc Rev. 2012;41(7):2656–72.
19. Kim S-M, Chae MK, Yim MS, et al. Hybrid PET/MR imaging of tumors using an oleanolic acid-conjugated nanoparticle. Biomaterials. 2013;34(33):8114–21.
20. Yan K, Li P, Zhu H, et al. Recent advances in multifunctional magnetic nanoparticles and applications to biomedical diagnosis and treatment. R Soc Chem Adv. 2013;3(27):10598–618.
21. Thomas R, Park I-K, Jeong YY. Magnetic iron oxide nanoparticles for multimodal imaging and therapy of cancer. Int J Mol Sci. 2013;14(8):15910–30.
22. Araki J, Wada M, Kuga S. Steric stabilization of a cellulose microcrystal suspension by poly (ethylene glycol) grafting. Langmuir. 2001;17(1):21–7.
23. Gref R, Domb A, Quellec P, et al. The controlled intravenous delivery of drugs using PEG-coated sterically stabilized nanospheres. Adv Drug Deliv Rev. 1995;16(2):215–33.
24. Xu R. Progress in nanoparticles characterization: sizing and zeta potential measurement. Particuology. 2008;6(2):112–5.
25. Obeidat WM, Schwabe K, Müller RH, Keck CM. Preservation of nanostructured lipid carriers (NLC). Eur J Pharm Biopharm. 2010;76(1):56–67.
26. Altankov G, Richau K, Groth T. The role of surface zeta potential and substratum chemistry for regulation of dermal fibroblasts interaction. Materwiss Werksttech. 2003;34(12):1120–8.
27. Zhang Y, Yang M, Portney NG, et al. Zeta potential: a surface electrical characteristic to probe the interaction of nanoparticles with normal and cancer human breast epithelial cells. Biomed Microdev. 2008;10(2):321–8.
28. Laconte LE, Nitin N, Zurkiya O, et al. Coating thickness of magnetic iron oxide nanoparticles affects R2 relaxivity. J Magn Reson Imaging. 2007;26(6):1634–41.
29. Patel D, Kell A, Simard B, Xiang B, Lin HY, Tian G. The cell labeling efficacy, cytotoxicity and relaxivity of copper-activated MRI/PET imaging contrast agents. Biomaterials. 2011;32(4):1167–76.
30. Yang X, Grailer JJ, Rowland IJ, et al. Multifunctional SPIO/DOX-loaded wormlike polymer vesicles for cancer therapy and MR imaging. Biomaterials. 2010;31(34):9065–73.
31. Luo J, Wilson JD, Zhang J, et al. A dual PET/MR imaging nanoprobe: ^{124}I labeled $Gd_3N@C_{80}$. Appl Sci. 2012;2(2):465–78.
32. Goyen M, Lauenstein TC, Herborn CU, Debatin JF, Bosk S, Ruehm SG. 0.5 M Gd chelate (Magnevist®) versus 1.0 M Gd chelate (Gadovist®): dose-independent effect on image quality of pelvic three-dimensional MR-angiography. J Magn Reson Imaging. 2001;14(5):602–7.
33. Huang J, Xie J, Chen K, et al. HSA coated MnO nanoparticles with prominent MRI contrast for tumor imaging. Chem Commun. 2010;46(36):6684–6.
34. Ko HY, Lee JH, Kang H, et al. A nucleolin-targeted multimodal nanoparticle imaging probe for tracking cancer cells using an aptamer. J Nucl Med. 2010;51(1):98–105.
35. Hwang DW, Ko HY, Kim SK, Kim D, Lee DS, Kim S. Development of a quadruple imaging modality by using nanoparticles. Chemistry. 2009;15(37):9387–93.
36. Choi JS, Park JC, Nah H, et al. A hybrid nanoparticle probe for dual-modality positron emission tomography and magnetic resonance imaging. Angew Chem Int Ed. 2008;47(33):6259–62.
37. Cui X, Belo S, Krüger D, et al. Aluminium hydroxide stabilised $MnFe_2O_4$ and Fe_3O_4 nanoparticles as dual-modality contrasts agent for MRI and PET imaging. Biomaterials. 2014;35(22):5840–6.
38. Kim JS, Kim Y-H, Kim JH, et al. Development and in vivo imaging of a PET/MRI nanoprobe with enhanced NIR fluorescence by dye encapsulation. Nanomedicine. 2012;7(2):219–29.
39. Baldi G, Bonacchi D, Innocenti C, Lorenzi G, Sangregorio C. Cobalt ferrite nanoparticles: the control of the particle size and surface state and their effects on magnetic properties. J Magn Magn Mater. 2007;311(1):10–6.
40. Fernández-Suárez M, Ting AY. Fluorescent probes for super-resolution imaging in living cells. Nat Rev Mol Cell Biol. 2008;9(12):929–43.
41. Zhou J, Yu M, Sun Y, et al. Fluorine-18-labeled Gd^{3+}/Yb^{3+}/Er^{3+} co-doped $NaYF_4$ nanophosphors for multimodality PET/MR/UCL imaging. Biomaterials. 2011;32(4):1148–56.
42. Garcia J, Neelavalli J, Haacke EM, Allen MJ. Eu(II)-containing cryptates as contrast agents for ultra-high field strength magnetic resonance imaging. Chem Commun. 2011;47(48):12858–60.
43. Smith SV, Jones M, Holmes V. Production and selection of metal PET radioisotopes for molecular imaging. In: Singh N, editor. Radioisotopes. Rijeka, Croatia: InTech Open; 2011. p. 199–224.

44. Ametamey SM, Honer M, Schubiger PA. Molecular imaging with PET. Chem Rev. 2008;108(5):1501–16.
45. Locatelli E, Gil L, Israel LL, et al. Biocompatible nanocomposite for PET/MRI hybrid imaging. Int J Nanomed. 2011;7:6021–33.
46. Higuchi T, Anton M, Dumler K, et al. Combined reporter gene PET and iron oxide MRI for monitoring survival and localization of transplanted cells in the rat heart. J Nucl Med. 2009;50(7):1088–94.
47. Park JC, Yu MK, An GI, et al. Facile preparation of a hybrid nanoprobe for triple-modality optical/PET/MR imaging. Small. 2010;6(24):2863–8.
48. Madru R, Tran TA, Axelsson J, et al. ^{68}Ga-labeled superparamagnetic iron oxide nanoparticles (SPIONs) for multi-modality PET/MR/Cherenkov luminescence imaging of sentinel lymph nodes. Am J Nucl Med Mol Imaging. 2014;4(1):60.
49. Chen F, Ellison PA, Lewis CM, et al. Chelator-free synthesis of a dual-modality PET/MRI agent. Angew Chem Int Ed. 2013;52(50):13319–23.
50. Stelter L, Pinkernelle JG, Michel R, et al. Modification of aminosilanized superparamagnetic nanoparticles: feasibility of multimodal detection using 3T MRI, small animal PET, and fluorescence imaging. Mol Imaging Biol. 2010;12(1):25–34.
51. Wadas TJ, Wong EH, Weisman GR, Anderson CJ. Coordinating radiometals of copper, gallium, indium, yttrium, and zirconium for PET and SPECT imaging of disease. Chem Rev. 2010;110(5):2858–902.
52. Zolata H, Afarideh H, Abbasi-Davani F. Radioimmunoconjugated, Dox-loaded, surface-modified superparamagnetic iron oxide nanoparticles (SPIONs) as a bioprobe for breast cancer tumor theranostics. J Radioanal Nucl Chem. 2014;301(2):451–60.
53. Yang X, Hong H, Grailer JJ, et al. cRGD-functionalized, DOX-conjugated, and 64 Cu-labeled superparamagnetic iron oxide nanoparticles for targeted anticancer drug delivery and PET/MR imaging. Biomaterials. 2011;32(17):4151–60.
54. Yang M, Cheng K, Qi S, et al. Affibody modified and radiolabeled gold–iron oxide heteronanostructures for tumor PET, optical and MR imaging. Biomaterials. 2013;34(11):2796–806.
55. Delgado R, Sun Y, Motekaitis RJ, Martell AE. Stabilities of divalent and trivalent metal ion complexes of macrocyclic triazatriacetic acids. Inorg Chem. 1993;32(15):3320–6.
56. Clarke ET, Martell AE. Stabilities of the alkaline earth and divalent transition metal complexes of the tetraazamacrocyclic tetraacetic acid ligands. Inorg Chim Acta. 1991;190(1):27–36.
57. Lee H-Y, Li Z, Chen K, et al. PET/MRI dual-modality tumor imaging using arginine-glycine-aspartic (RGD)–conjugated radiolabeled iron oxide nanoparticles. J Nucl Med. 2008;49(8):1371–9.
58. Xie J, Chen K, Huang J, et al. PET/NIRF/MRI triple functional iron oxide nanoparticles. Biomaterials. 2010;31(11):3016–22.
59. Glaus C, Rossin R, Welch MJ, Bao G. In vivo evaluation of ^{64}Cu-labeled magnetic nanoparticles as a dual-modality PET/MR imaging agent. Bioconjug Chem. 2010;21(4):715–22.
60. Aryal S, Key J, Stigliano C, Landis MD, Lee DY, Decuzzi P. Positron emitting magnetic nanoconstructs for PET/MR imaging. Small. 2014;10(13):2688–96.
61. Fani M, André JP, Maecke HR. ^{68}Ga-PET: a powerful generator-based alternative to cyclotron-based PET radiopharmaceuticals. Contrast Media Mol Imaging. 2008;3(2):53–63.
62. Clarke ET, Martell AE. Stabilities of trivalent metal ion complexes of the tetraacetate derivatives of 12-, 13- and 14-membered tetraazamacrocycles. Inorg Chim Acta. 1991;190(1):37–46.
63. Thorek DL, Ulmert D, Diop N-FM, et al. Non-invasive mapping of deep-tissue lymph nodes in live animals using a multimodal PET/MRI nanoparticle. Nat Commun. 2014;5:3097.
64. Torres Martin De Rosales R, Tavaré R, Paul RL, et al. Synthesis of $^{64}Cu^{II}$–Bis (dithiocarbamatebisphosphonate) and its conjugation with superparamagnetic iron oxide nanoparticles: in vivo evaluation as dual-modality PET–MRI agent. Angew Chem Int Ed. 2011;50(24):5509–13.
65. Xie J, Huang J, Li X, Sun S, Chen X. Iron oxide nanoparticle platform for biomedical applications. Curr Med Chem. 2009;16(10):1278–94.
66. Hao R, Xing R, Xu Z, Hou Y, Gao S, Sun S. Synthesis, functionalization, and biomedical applications of multifunctional magnetic nanoparticles. Adv Mater. 2010;22(25):2729–42.
67. Wang WW, Zhu YJ, Ruan ML. Microwave-assisted synthesis and magnetic property of magnetite and hematite nanoparticles. J Nanopart Res. 2007;9(3):419–26.
68. Sreeja V, Joy P. Microwave–hydrothermal synthesis of γ-Fe_2O_3 nanoparticles and their magnetic properties. Mater Res Bull. 2007;42(8):1570–6.
69. Liao X, Zhu J, Zhong W, Chen H-Y. Synthesis of amorphous Fe_2O_3 nanoparticles by microwave irradiation. Mater Lett. 2001;50(5):341–6.
70. Osborne EA, Atkins TM, Gilbert DA, Kauzlarich SM, Liu K, Louie AY. Rapid microwave-assisted synthesis of dextran-coated iron oxide nanoparticles for magnetic resonance imaging. Nanotechnology. 2012;23(21):215602.
71. Wong RM, Gilbert DA, Liu K, Louie AY. Rapid size-controlled synthesis of dextran-coated, ^{64}Cu-doped iron oxide nanoparticles. ACS Nano. 2012;6(4):3461–7.
72. Chen K, Xie J, Xu H, et al. Triblock copolymer coated iron oxide nanoparticle conjugate for tumor integrin targeting. Biomaterials. 2009;30(36):6912–9.
73. Tu C, Ng TS, Jacobs RE, Louie AY. Multimodality PET/MRI agents targeted to activated macrophages. J Biol Inorg Chem. 2014;19(2):247–58.
74. Jarrett BR, Correa C, Ma KL, Louie AY. In vivo mapping of vascular inflammation using multimodal imaging. PLoS One. 2010;5(10):e13254.

75. Uppal R, Catana C, Ay I, Benner T, Sorensen AG, Caravan P. Bimodal thrombus imaging: simultaneous PET/MR imaging with a fibrin-targeted dual PET/MR probe—feasibility study in rat model. Radiology. 2011;258(3):812–20.
76. Rosales R. Potential clinical applications of bimodal PET-MRI or SPECT-MRI agents. J Label Compd Radiopharm. 2014;57:298–303.
77. Kim BY, Rutka JT, Chan WC. Nanomedicine. N Engl J Med. 2010;363:2434–43.
78. Chen GY, Roy I, Yang CH, Prasad PN. Nanochemistry and nanomedicine for nanoparticle-based diagnostics and therapy. Chem Rev. 2016;116:2826–85.
79. Shi X, Shen L. Integrin αvβ3 receptor targeting PET/MRI dual-modal imaging probe based on the 64 Cu labeled manganese ferrite nanoparticles. J Inorg Biochem. 2018;186:257–63.
80. Aime S, Frullano L, Crich SG. Compartmentalization of a gadolinium complex in the apoferritin cavity: a route to obtain high relaxivity contrast agents for magnetic resonance imaging. Angew Chem Int Ed. 2002;41:1017.
81. Crich SG, Lanzardo S, Alberti D, Belfiore S, Ciampa A, Giovenzanaz GB, Lovazzano C, Pagliarin R, Aime S. Magnetic resonance imaging detection of tumor cells by targeting low-density lipoprotein receptors with Gd-loaded low-density lipoprotein particles. Neoplasia. 2007;9:1046–56.
82. Baetke SC, Lammers T, Kiessling F. Applications of nanoparticles for diagnosis and therapy of cancer. Br J Radiol. 2015;88:20150207.
83. Peer D, Karp JM, Hong S, FaroKhzad OC, Margalit R, Langer R. Nanocarriers as an emerging platform for cancer therapy. Nat Nanotechnol. 2007;2:751–60.
84. Sun C, Lee JSH, Zhang M. Magnetic nanoparticles in MR imaging and drug delivery. Adv Drug Deliv Rev. 2008;60:1252–65.
85. Grimm J, Scheinberg DA. Will nanotechnology influence targeted cancer therapy? Semin Radiat Oncol. 2011;21:80–7.
86. Qiao R, Yang C, Gao M. Superparamagnetic iron oxide nanoparticles: from preparations to in vivo MRI applications. J Mater Chem. 2009;19:6274–93.
87. Berry CC, Curtis ASG. Functionalisation of magnetic nanoparticles for applications in biomedicine. J Phys D Appl Phys. 2003;36:R198–206.
88. Blanco E, Shen H, Ferrari M. Principles of nanoparticle design for overcoming biological barriers to drug delivery. Nat Biotechnol. 2015;33:941–51.
89. Neuberger T, Schopf B, Hofmann H, Hofmann M, von Rechenberg B. Superparamagnetic nanoparticles for biomedical applications: possibilities and limitations of a new drug delivery system. J Magn Magn Mater. 2005;293:483–96.
90. Choi HS, Liu W, Misra P, Tanaka E, Zimmer JP, Ipe BI, Bawendi MG, Frangioni JV. Renal clearance of quantum dots. Nat Biotechnol. 2007;25:1165–70.
91. de Rosales RTM, Tavaré R, Paul RL, Jauregui-Osoro M, Protti A, Glaria A, Varma G, Szanda I, Blower PJ. Synthesis of $^{64}Cu^{II}$–bis (dithiocarbamatebisphosphonate) and its conjugation with superparamagnetic iron oxide nanoparticles: in vivo evaluation as dual-modality PET–MRI agent. Angew Chem Int Ed. 2011;50:5509–13.
92. Cui X, Belo S, Krueger D, Yan Y, de Rosales RTM, Jauregui-Osoro M, Ye H, Su S, Mathe D, Kovács N, et al. Aluminium hydroxide stabilised $MnFe_2O_4$ and Fe_3O_4 nanoparticles as dual-modality contrasts agent for MRI and PET imaging. Biomaterials. 2014;35:5840–6.
93. Owens DE, Peppas NA. Opsonization, biodistribution, and pharmacokinetics of polymeric nanoparticles. Int J Pharm. 2006;307:93–102.
94. Mikhaylova M, Kim DK, Bobrysheva N, Osmolowsky M, Semenov V, Tsakalakos T, Muhammed M. Superparamagnetism of magnetite nanoparticles: dependence on surface modification. Langmuir. 2004;20:2472–7.
95. LaConte LEW, Nitin N, Zurkiya O, Caruntu D, O'Connor CJ, Hu X, Bao G. Coating thickness of magnetic iron oxide nanoparticles affects R-2 relaxivity. J Magn Reson Imaging. 2007;26:1634–41.
96. Jain RK, Stylianopoulos T. Delivering nanomedicine to solid tumors. Nat Rev Clin Oncol. 2010;7:653–64.
97. Stylianopoulos T, Jain RK. Design considerations for nanotherapeutics in oncology. Nanomed Nanotechnol Biol Med. 2015;11:1893–907.
98. Xing Y, Zhao J, Conti PS, Chen K. Radiolabeled nanoparticles for multimodality tumor imaging. Theranostics. 2014;4:290–306.
99. Saftig P, Klumperman J. Lysosome biogenesis and lysosomal membrane proteins: trafficking meets function. Nat Rev Mol Cell Biol. 2009;10:623–35.
100. Bulte JWM, Kraitchman DL. Iron oxide MR contrast agents for molecular and cellular imaging. NMR Biomed. 2004;17:484–99.
101. Estelrich J, Sánchez-Martín MJ, Busquets MA. Nanoparticles in magnetic resonance imaging: from simple to dual contrast agents. Int J Nanomedicine. 2015;10:1727–41.
102. Thorek D, Czupryna J, Chen A, Tsourkas A. Molecular imaging of cancer with superparamagnetic iron oxide particles. Cancer Imaging Instrum Appl. 2008;2:85–95.
103. Jung CW, Jacobs P. Physical and chemical-properties of superparamagnetic iron-oxide MR contrast agents—ferumoxides, ferumoxtran, ferumoxsil. Magn Reson Imaging. 1995;13:661–74.
104. Wang YXJ, Hussain SM, Krestin GP. Superparamagnetic iron oxide contrast agents: physicochemical characteristics and applications in MR imaging. Eur Radiol. 2001;11:2319–31.
105. Wunderbaldinger P, Josephson L, Weissleder R. Crosslinked iron oxides (CLIO): a new platform for the development of targeted MR contrast agents. Acad Radiol. 2002;9:S304–6.
106. Tu C, Ng TSC, Jacobs RE, Louie AY. Multimodality PET/MRI agents targeted to activated macrophages topical issue on metal-based MRI contrast agents.

Guest editor: Valerie C. Pierre. J Biol Inorg Chem. 2014;19(2):247–58.
107. Mornet S, Vasseur S, Grasset F, Duguet E. Magnetic nanoparticle design for medical diagnosis and therapy. J Mater Chem. 2004;14:2161–75.
108. Laurent S, Forge D, Port M, Roch A, Robic C, Elst LV, Muller RN. Magnetic iron oxide nanoparticles: synthesis, stabilization, vectorization, physicochemical characterizations, and biological applications. Chem Rev. 2008;108:2064–110.
109. Chen Y, Gibson SB. Three dimensions of autophagy in regulating tumor growth: cell survival/death, cell proliferation, and tumor dormancy. Biochim Biophys Acta Mol basis Dis. 2021;1867:166265.
110. Griffiths SM, Singh N, Jenkins GJ, Williams PM, Orbaek AW, Barron AR, Wright CJ, Doak SH. Dextran coated ultrafine superparamagnetic iron oxide nanoparticles: compatibility with common fluorometric and colorimetric dyes. Anal Chem. 2011;83:3778–85.
111. Babic M, Horák D, Trchová M, Jendelová P, Glogarová K, Lesný P, Herynek V, Hájek M, Syková E. Poly(L-lysine)-modified iron oxide nanoparticles for stem cell labeling. Bioconjug Chem. 2008;19:740–50.
112. Maeng JH, Lee D-H, Jung KH, Bae Y-H, Park I-S, Jeong S, Jeon Y-S, Shim C-K, Kim W, Kim J, et al. Multifunctional doxorubicin loaded superparamagnetic iron oxide nanoparticles for chemotherapy and magnetic resonance imaging in liver cancer. Biomaterials. 2010;31:4995–5006.
113. Liao Z, Wang H, Lv R, Zhao P, Sun X, Wang S, Su W, Niu R, Chang J. Polymeric liposomes-coated superparamagnetic iron oxide nanoparticles as contrast agent for targeted magnetic resonance imaging of cancer cells. Langmuir. 2011;27:3100–5.
114. Lee J-H, Huh Y-M, Jun Y-W, Seo J-W, Jang J-T, Song H-T, Kim S, Cho E-J, Yoon H-G, Suh J-S, et al. Artificially engineered magnetic nanoparticles for ultra-sensitive molecular imaging. Nat Med. 2007;13:95–9.
115. Lawaczeck R, Menzel M, Pietsch H. Superparamagnetic iron oxide particles: contrast media for magnetic resonance imaging. Appl Organomet Chem. 2004;18:506–13.
116. Blasiak B, van Veggel FC, Tomanek B. Applications of nanoparticles for MRI cancer diagnosis and therapy. J Nanomater. 2013; https://doi.org/10.1155/2013/148578.
117. Thorek DLJ, Ulmert D, Diop N-FM, Lupu ME, Doran MG, Huang R, Abou DS, Larson SM, Grimm J. Non-invasive mapping of deep-tissue lymph nodes in live animals using a multimodal PET/MRI nanoparticle. Nat Commun. 2014;5:3097.
118. Madru R, Budassi M, Benveniste H, Lee H, Smith SD, Schlyer DJ, Vaska P, Knutsson L, Strand S-E. Simultaneous preclinical positron emission tomography-magnetic resonance imaging study of lymphatic drainage of chelator-free Cu-64-labeled nanoparticles. Cancer Biother Radiopharm. 2018;33:213–20.
119. Xie J, Chen K, Huang J, Lee S, Wang J, Gao J, Li X, Chen X. PET/NIRF/MRI triple functional iron oxide nanoparticles. Biomaterials. 2010;31:3016–22.
120. Lee H-Y, Li Z, Chen K, Hsu AR, Xu C, Xie J, Sun S, Chen X. PET/MRI dual-modality tumor imaging using arginine-glycine-aspartic (RGD)—conjugated radiolabeled iron oxide nanoparticles. J Nucl Med. 2008;49:1371–9.
121. Kim S-M, Chae MK, Yim MS, Jeong IH, Cho J, Lee C, Ryu EK. Hybrid PET/MR imaging of tumors using an oleanolic acid-conjugated nanoparticle. Biomaterials. 2013;34:8114–21.
122. Vivero-Escoto JL, Huxford-Phillips RC, Lin W. Silica-based nanoprobes for biomedical imaging and theranostic applications. Chem Soc Rev. 2012;41:2673–85.
123. Nakamura T, Sugihara F, Matsushita H, Yoshioka Y, Mizukami S, Kikuchi K. Mesoporous silica nanoparticles for ^{19}F magnetic resonance imaging, fluorescence imaging, and drug delivery. Chem Sci. 2015;6:1986–90.
124. Patel D, Kell A, Simard B, Deng J, Xiang B, Lin H-Y, Gruwel M, Tian G. Cu^{2+}-labeled, SPION loaded porous silica nanoparticles for cell labeling and multifunctional imaging probes. Biomaterials. 2010;31:2866–73.
125. Burke BP, Baghdadi N, Kownacka AE, Nigam S, Clemente GS, Al-Yassiry MM, Domarkas J, Lorch M, Pickles M, Gibbs P, et al. Chelator free gallium-68 radiolabelling of silica coated iron oxide nanorods via surface interactions. Nanoscale. 2015;7:14889–96.
126. Huang X, Zhang F, Lee S, Swierczewska M, Kiesewetter DO, Lang L, Zhang G, Zhu L, Gao H, Choi HS, et al. Long-term multimodal imaging of tumor draining sentinel lymph nodes using mesoporous silica-based nanoprobes. Biomaterials. 2012;33:4370–8.
127. Elbayoumi TA, Torchilin VP. Current trends in liposome research. Methods Mol Biol. 2010;605:1–27.
128. Puri A, Loomis K, Smith B, Lee J-H, Yavlovich A, Heldman E, Blumenthal R. Lipid-based nanoparticles as pharmaceutical drug carriers: from concepts to clinic. Crit Rev Ther Drug Carrier Syst. 2009;26:523–80.
129. Li S, Goins B, Zhang L, Bao A. Novel multifunctional theranostic liposome drug delivery system: construction, characterization, and multimodality MR, near-infrared fluorescent, and nuclear imaging. Bioconjug Chem. 2012;23:1322–32.
130. Mitchell N, Kalber TL, Cooper MS, Sunassee K, Chalker SL, Shaw KP, Ordidge KL, Badar A, Janes SM, Blower PJ, et al. Incorporation of paramagnetic, fluorescent and PET/SPECT contrast agents into liposomes for multimodal imaging. Biomaterials. 2013;34:1179–92.

131. Abou DS, Thorek DLJ, Ramos NN, Pinkse MWH, Wolterbeek HT, Carlin SD, Beattie BJ, Lewis JS. Zr-89-labeled paramagnetic octreotide-liposomes for PET-MR imaging of cancer. Pharm Res. 2013;30:878–88.
132. Savic R, Luo LB, Eisenberg A, Maysinger D. Micellar nanocontainers distribute to defined cytoplasmic organelles. Science. 2003;300:615–8.
133. Trubetskoy VS. Polymeric micelles as carriers of diagnostic agents. Adv Drug Deliv Rev. 1999;37:81–8.
134. Enoch HG, Strittmatter P. Formation and properties of 1000-A-diameter, single-bilayer phospholipid vesicles. Proc Natl Acad Sci USA. 1979;76:145–9.
135. Trubetskoy VS, Frank-Kamenetsky MD, Whiteman KR, Wolf GL, Torchilin VP. Stable polymeric micelles: lymphangiographic contrast media for gamma scintigraphy and magnetic resonance imaging. Acad Radiol. 1996;3:232–8.
136. Starmans LW, Hummelink MA, Rossin R, Kneepkens E, Lamerichs R, Donato K, Nicolay K, Grüll H. ^{89}Zr- and Fe-labeled polymeric micelles for dual modality PET and T1-weighted MR imaging. Adv Healthc Mater. 2015;4:2137–45.
137. Xu H, Regino CA, Koyama Y, Hama Y, Gunn AJ, Bernardo M, Kobayashi H, Choyke PL, Brechbiel MW. Preparation and preliminary evaluation of a biotin-targeted, lectin-targeted dendrimer-based probe for dual-modality magnetic resonance and fluorescence imaging. Bioconjug Chem. 2007;18:1474–82.
138. Koyama Y, Talanov VS, Bernardo M, Hama Y, Regino CA, Brechbiel MW, Choyke PL, Kobayashi H. A dendrimer-based nanosized contrast agent dual-labeled for magnetic resonance and optical fluorescence imaging to localize the sentinel lymph node in mice. J Magn Reson Imaging. 2007;25:866–71.
139. Kobayashi H, Koyama Y, Barrett T, Hama Y, Regino CA, Shin IS, Jang B-S, Le N, Paik CH, Choyke PL, et al. Multimodal nanoprobes for radionuclide and five-color near-infrared optical lymphatic imaging. ACS Nano. 2007;1:258–64.
140. Wen S, Li K, Cai H, Chen Q, Shen M, Huang Y, Peng C, Hou W, Zhu M, Zhang G, et al. Multifunctional dendrimer-entrapped gold nanoparticles for dual mode CT/MR imaging applications. Biomaterials. 2013;34:1570–80.
141. Almutairi A, Rossin R, Shokeen M, Hagooly A, Ananth A, Capoccia B, Guillaudeu S, Abendschein D, Anderson CJ, Welch MJ, et al. Biodegradable dendritic positron-emitting nanoprobes for the non-invasive imaging of angiogenesis. Proc Natl Acad Sci USA. 2009;106:685–90.
142. Kobayashi H, Wu C, Kim M-K, Paik CH, Carrasquillo JA, Brechbiel MW. Evaluation of the in vivo biodistribution of indium-111 and yttrium-88 labeled dendrimer-1B4M-DTPA and its conjugation with anti-Tac monoclonal antibody. Bioconjug Chem. 1999;10:103–11.
143. Kumar A, Zhang SR, Hao GY, Hassan G, Ramezani S, Sagiyama K, Lo S-T, Takahashi M, Sherry AD, Öz OK, et al. Molecular platform for design and synthesis of targeted dual modality imaging probes. Bioconjug Chem. 2015;26:549–58.
144. MaHam A, Tang Z, Wu H, Wang J, Lin Y. Protein-based nanomedicine platforms for drug delivery. Small. 2009;5:1706–21.
145. Vecchione D, Aiello M, Cavaliere C, Nicolai E, Netti PA, Torino E. Hybrid core shell nanoparticles entrapping Gd-DTPA and F-18-FDG for simultaneous PET/MRI acquisitions. Nanomedicine. 2017;12:2223–31.
146. Gratton SEA, Ropp PA, Pohlhaus PD, Luft JC, Madden VJ, Napier ME, DeSimone JM. The effect of particle design on cellular internalization pathways. Proc Natl Acad Sci USA. 2008;105:11613–8.

14 Radionuclide Nanoimaging

14.1 Introduction

The amalgamation of radiochemistry for use in imaging and therapy has a promising future in various biomedical applications. The possibility of using various nanoparticles can be seen in the form of theranostic agents based on their biocompatibility. The radionuclide-imaging modalities are positron emission tomography (PET), single-photon emission computed tomography (SPECT), and Cerenkov luminescence (CL), in which the radiolabeled NP will be used for imaging. PET and SPECT can enable NP design for multimodal imaging, with radiotracers showing promise in both imaging and therapy. CL uses high-energy electrons (β^-) for imaging purposes.

The basis of radio-nanomedicine is the multifunctional use of radiolabeled nanoparticles for various theranostic applications. Radio-nanomedicine is also a similar theranostic approach for the diagnosis and therapy of various diseases [1, 2]. Radiolabeled nano colloids are used in liver and bone marrow scans, and lymphoscintigraphy imaging in nuclear medicine is another investigation [3, 4]. The principle behind these techniques is the size-dependent in vivo distribution or lymphatic clearance of radiolabeled nano colloids. Additionally, radiolabeled nanoparticles are capable of delivering drugs for both diagnosis and therapy. This includes the use of various nanoparticles in a multifunctional approach for both diagnostic and therapeutic effects in evaluating targeted tissues using radionuclides, which is known as internal radiotherapy. This can be achieved by modifying different components of the radiolabeling with various nanoparticles, resulting in more effective and novel applications.

Radio-nanomedicine is now an important component of clinical nanomedicine, raising significant concerns regarding the safety of nanomaterials. Despite this, various safety issues persist related to different mechanisms of toxicity and nanoparticles [5]. Different cellular and subcellular interactions with nanoparticles can result in various toxic side effects. The innate immune response of cells, with increased reactive oxygen species (ROS), can also cause cellular damage [6]. Phagocytosis of different metal oxide nanoparticles causes oxidative stress due to autocatalysis of ROS formation and cytokine production [7]. Oxidative stress induced by nanoparticles can affect DNA integrity, cell growth, subcellular signaling, and even carcinogenesis [8]. The use of nanoparticles may also lead to lysosomal dysfunction. Nanoparticles with a positive zeta potential sequester protons, which leads to increased activation of the proton pump, potentially causing lysosomal rupture [9]. The generation of ROS, followed by cellular signaling, involving inorganic nanomaterials like quantum dots, depends on the body's response to various

S. Shaikh, *Nanoimaging - Future of Precision Medicine*,
https://doi.org/10.1007/978-981-97-7292-6_14

toxic substances [10]. Gold nanoparticles, which have better biocompatibility, are still not free from cellular and systemic toxicity due to various surface ligands and/or associated phagocytic immune responses [11, 12]. The plausible in vivo toxicity of various biomaterial nanomaterials, such as cationic micelles and liposomes, has already been documented [7]. Toxicity may be influenced by factors such as chemical composition, size, shape, and surface molecules of nanoparticles [5], but the most critical factor is nanoparticle dosage [8]. Radio-nanomedicine generally uses very small amounts of nanomaterials, which are less than pharmacologic doses (Fig. 14.1). This suggests that radiolabeled nanomaterials can be used in various clinical settings, despite some risk of toxicity. Nuclear imaging is minimally invasive, such as using silica nanoparticles targeting αvβ3 integrin with ^{124}I and optical probes in small animal tumor models, for tumor targeting and efficient renal clearance [13], showing accumulation of nanoparticles in various tumor lesions [14]. PET or SPECT provide pharmacokinetic insights into nanomaterials in living organisms. This information aids in the engineering and surface modification of nanomaterials for in vivo applications [15, 16]. Nanomedicine applies these modifications for biomedical uses. For instance, polyethylene glycol (PEG) is introduced to reduce opsonin protein binding in the blood, significantly reducing RES sequestration and increasing tumor accumulation [15, 17]. Smaller quantum dots are excreted by the kidneys, showing a higher lesion-to-background accumulation ratio [18]. PET and SPECT can be engineered with nanoparticles to achieve high target lesion uptake, minimal critical organ accumulation, and rapid clearance. The use of the excretion patterns with the physiologic accumulation of nanomaterials also shows the toxicity with longer exposure. Nanoparticles excreted through the hepatobiliary system can reach high serum concentrations if hepatic function is impaired [12]. In vivo kinetics can be tracked with radiolabeled nanoparticles to assess toxicity in preclinical imaging. Phase 0 trials serve as platforms for target modulation and selection, facilitating early drug development with biomarkers to evaluate drug effects and pharmacokinetics in humans [19]. These trials reduce clinical trial costs, encouraging early-stage development [20, 21]. The theranostic approach with radiolabeled nanoparticles for specific diagnosis and therapy forms an essential concept of molecular theranostics. An example is ^{131}I, used in thyroid cancer for both internal radiotherapy and post-therapy evaluation for metastasis [22]. ^{90}Y-ibritumomab tiuxetan (Zevalin) is used in radioimmunotherapy for non-Hodgkin's lymphoma [23]. Therapeutic radionuclides are often

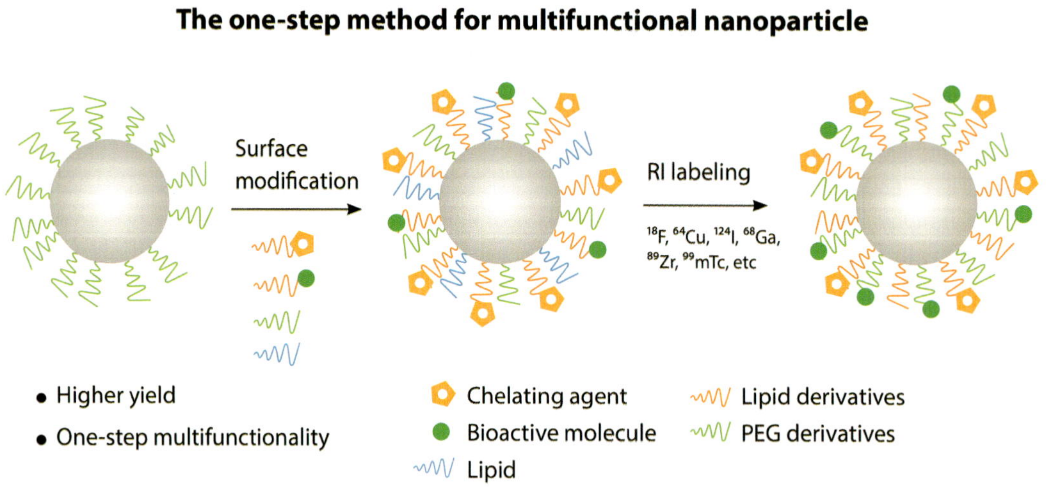

Fig. 14.1 Multifunctional nanoparticle showing surface modification with radiolabelling with different functions as described

combined with monoclonal antibodies for targeted molecular therapy. Radio microsphere therapy with ^{90}Y-labeled microspheres targets tumors by delivering them into the hepatic artery, exploiting the high tumor-to-normal liver ratio. The theranostic approach, with ^{99m}Tc-labeled macroaggregated albumin, enables SPECT imaging based on microsphere distribution [24]. Resin microspheres are used in hepatic metastases [25, 26] and hepatocellular carcinoma [27, 28]. PEGylated liposomes, combined with the chelator DTPA, can be labeled with ^{99m}Tc for SPECT imaging, ^{68}Ga for PET imaging, and ^{177}Lu for therapeutic purposes [29]. This image-guided theranostic approach provides a high radiation dose to tumor tissues [30]. Nanocarriers can also be used for drug delivery and imaging, which are biocompatible mesoporous silica labeled with ^{64}Cu, and can be used in image-guided drug delivery systems [31]. Gold nanorods which have the capability for multifunctional properties can be labeled with PET agents for chemotherapy, photothermal therapy, and PET imaging [32]. Tumor heterogeneity is an important aspect in which multiple molecular targets and ligands can be used for successful cancer therapy [33, 34]. These nanomaterials are capable of multifunctional platforms for radiolabeling, supporting simultaneous imaging and therapy for personalized radiotherapy. Surface modifications allow nanomaterials to carry multiple targeting ligands with multiple radioisotopes for both diagnostic and therapeutic purposes [1, 35]. This multifunctionality and flexibility promote broader use of radionuclide theranostics. Extrinsically labeled nanoparticles can carry chelators on their surfaces when they are labeled [36, 37]. However, intrinsic radiolabeled nanoparticles are limited as compared to extrinsically labeled ones. For instance, Cd-based quantum dots use various radioisotopes, such as ^{109}Cd [38], while ^{64}Cu incorporated into CdSe/ZnS quantum dots enables self-illuminating effects for tumor targeting in animal models [39]. Extrinsic radiolabeling involves chemical chelation or physical encapsulation of radiotracers, such as ^{99m}Tc or ^{111}In in liposomes [40–43], or by using positron-emitters of ^{124}I and ^{18}F attached to lipid molecules in the membrane [44, 45]. The extrinsic radiolabeling which is seen on the surfaces of nanomaterials will be in the range of inorganic nanomaterials like gold nanoparticles or quantum dots to organic nanoparticles including liposomes [36, 46]. The use of the nanoparticles ranges from inorganic to organic types, with different chelators linked to the nanoparticles, enabling the use of different isotopes, which can be changing [36, 47]. The toxicity of various nanoparticles is seen with different radionuclides, each having different capabilities of imaging and different doses. These nanoparticles can be matched with proper radionuclides for extrinsic labeling, which will exhibit both properties. ^{99m}Tc is one of the important radiotracers, with a 6 h half-life; however, it can take more time for the active targeting, which can be mediated by functional biomolecules. For better accumulation and longer activity, radioisotopes with longer half-lives are preferred. ^{141}Ce-labeled cerium oxide nanoparticles, with a half-life of 32.5 days, are seen in the liver and spleen, and are also excreted slowly via the intestines within 6 days [48]. ^{111}In-labeled HER-2 targeted gold nanoparticles (half-life 2.8 days) are other nanoparticles that accumulate in tumor lesions for more than 2 days [49]. Another important aspect to remember is that a longer half-life increases the radiation dose.

14.2 Types of Targeting

The modification strategy for the clinical application of these particles is based on permeability and can be categorized as "passive targeting" and "active targeting." Passive targeting relies on enhanced permeability and retention (EPR), where inflammation or cancer tissue leakage leads to better accumulation of nano-sized materials in these lesions [50, 51]. Theranostic nanoparticles can be modified by adding various targeting ligands on their surfaces, such as monoclonal antibodies [52]. The relatively large size of antibodies increases nanoparticle diameter, potentially altering the immunogenicity of these nanoparticles [28]. Antigen-binding proteins, which are part of monoclonal antibodies, and fragments are single surface modifications used for active targeting with lower immunogenicity [53]. Affibodies, aptamers, engineered peptides, and small proteins are also used for active targeting. Size is an important consideration, as smaller molecules are preferable for active targeting due to easier conjugation [54]. Most nanoparticles undergo opsonization and phagocytosis, requiring modification to increase bioavailability for successful active targeting [55, 56].

14.3 Dosimetry of Radio-Nanomedicine

Radiation dosimetry is an important aspect of radiolabeled nanoparticles for both therapy and diagnosis. The radiation doses absorbed by specific tissues can be identified based on the kinetics of radiopharmaceuticals and the physical characteristics of radionuclides, such as half-life, emission types, and retention time [57]. The calculation of the radiation doses along with the cumulative total can be depicted on the time-radioactivity curve. The absorbed radiation doses in target organs are influenced by radiation from various target organs and neighboring organs. This amount of radiation can be calculated using software like OLINDA/EXM [58]. The pharmacokinetics of radiolabeled nanoparticles, which have different anatomical distributions, are important for dosimetry. In one of the preclinical studies, the dosimetry for radiolabeled liposomes was performed to identify different liposomes for better dose delivery of the optimal dose without excessive radiation to the bone marrow [59–61]. Radioiodine-labeled nanoparticles are capable of causing iodine-related thyroid damage; therefore, radiation doses to the thyroid gland must be carefully planned. The radiation dosimetry of nanoparticles is similar to that of conventional radiopharmaceuticals. However, assessing radiation dosimetry is challenging for nanoparticles labeled with multiple radionuclides intended for internal radiotherapy. Different beta-emitters can be selected for internal radiotherapy based on tumor size and the uptake of nanoparticles [62, 63]. Different therapeutic radionuclides with beta-ray energy, such as ^{90}Y and ^{177}Lu, can be combined in personalized optimal ratios according to tumor properties [62] (Fig. 14.2). Future advancements in radiation dosimetry will support the use of radiolabeled nanoparticles in analyzing internal radiation therapy for tumors. The ^{186}Re-labeled liposomal doxorubicin allows simultaneous application for localized radiation and chemotherapy of tumor tissues [64]. The dosimetry, calculated based on the accumulation of radiolabeled liposomes, determines both radiation and chemotherapy doses for management. The study of radio-nanomedicine using endogenous nanoparticles is crucial for understanding their biomedical applications and associated safety concerns. Endogenous nanoparticles range from unicellular organisms for the evaluation of the complex multicellular organisms like the extracellular vesicles involved in intercellular communication [65]. Extracellular vesicles are composed of endogenous nanomaterials, while liposomes are made of exogenous organic nanomaterials, and both types can be modified and used for diagnostic and therapeutic purposes. Extracellular vesicles, found in serum, urine, and saliva, are emerging as important therapeutic agents in regenerative medicine [66–68]. Extracellular vesicles consist of lipid bilayer proteins, cytosolic components, and nucleic acids, which carry biological information for intercellular communication [69]. These nanocarriers are capable of transporting genetic materials, proteins, and drugs [66, 70]. The ease of loading lipophilic drugs (doxorubicin or curcumin) and hydrophilic molecules (therapeutic nucleic acids) makes exosomes a valuable delivery channel. Exosome-derived anti-tumor immune responses are being explored as potential cancer therapies. Exosome-based cancer immunotherapy has been studied in Phase I clinical trials with dendritic cells [71–73]. Radiolabeled extracellular vesicles, used to quantify systemic distribution, offer improved nanoparticle safety [74, 75]. The radiolabeled extracellular vesicles will have better biodistribution of systemically administered extracellular vesicles, leading to exosomes or exosome-mimetic nanovesicles. These vesicles,

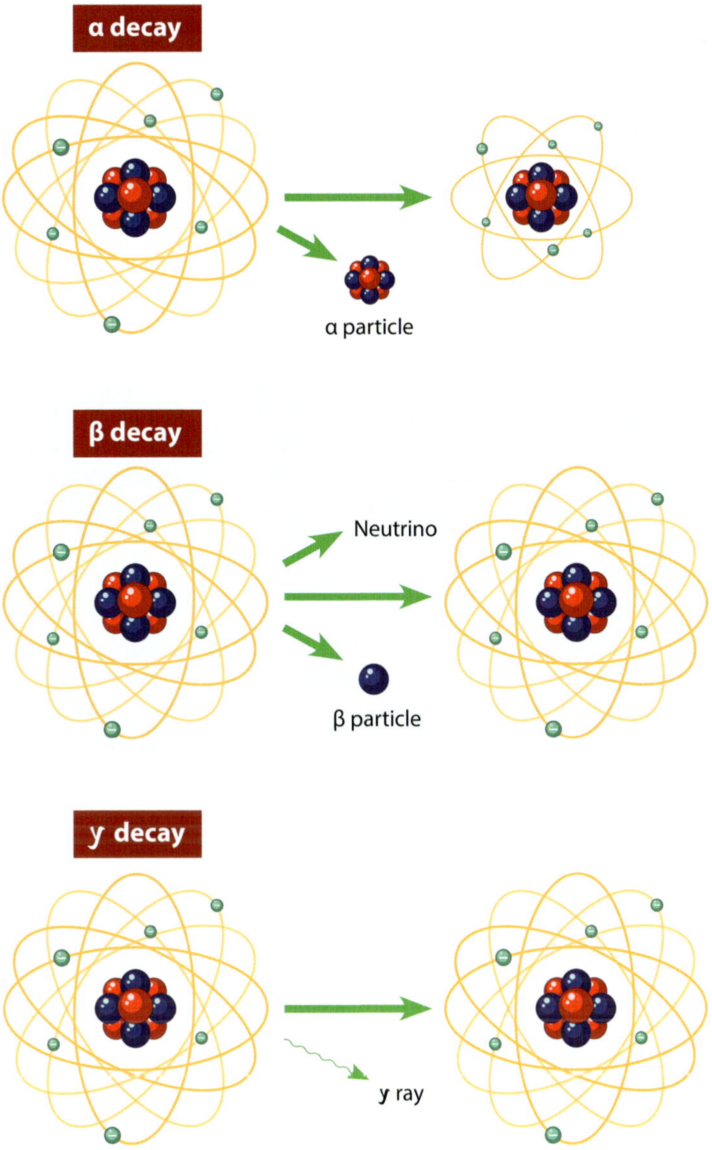

Fig. 14.2 Showing alpha, beta and gamma decays due to radiation

equipped with membrane-bound streptavidin and ^{125}I, can be labeled through norbiotinamide conjugation, allowing elucidation of their biodistribution [76]. Different radiotracers, such as ^{111}In-oxine [77] or ^{99m}Tc-HMPAO [56], can convert hydrophobic to hydrophilic compounds within the cell cytosol, offering biocompatible options for radiolabeling under physiological conditions for extracellular vesicles and radiotracers. SPECT imaging in animals uses ^{99m}Tc-HMPAO-labeled extracellular vesicles [56]. Exosomes also possess target-specific homing capabilities [78, 79]. Various radiolabeled exosomes can be targeted to specific tissues based on their physiological behaviors within the body. The exosomes from the cultured cells will limit the evaluation of the various therapeutic usability of extracellular vesicles for labeling and radiolabeling varies among studies. The vesicles are secreted by cells and isolated through processes such as ultracentrifugation, filtration, and immunoaffinity isolation from culture media or biological fluids [80]. However, large-scale production for in vivo applications remains challenging. A new method derived from the whole cells can be used for the exosome-mimetic nanovesicles for nanoimaging [81]. As the exogenous nanoparticles are being metabolized and degraded in the body, these extracellular vesicles similarly undergo rapid degradation [76, 82]. Thin-layer chromatography can differentiate radiolabeled extracellular vesicles from degraded ones, aiding in understanding the in vivo stability of the distribution of the radiolabeled extracellular vesicles [83]. In vivo studies demonstrate extracellular vesicles' distributions based on cell sources or administration routes [84]. Additionally, the cross-species distribution of extracellular vesicles from human and mouse origins reveals cell origin-dependent tissue tropism in mouse xenograft models [84].

14.4 Cancer Therapy

In cancer therapy, a few radiolabeled agents (e.g., ^{90}Y or ^{177}Lu) exhibit additional therapeutic properties in the form of drugs or radionuclides. These nanoparticles are not only radiotracers but also therapeutic agents. One of the processes of the micropinocytosis mechanism provides a window for imaging and therapy optimization in tumors [85]. Functionalizing unlabeled NPs allows them to align with imaging radioisotopes, making them easily trackable in vivo via various imaging methods like PET, SPECT, or CL. The properties of NPs are important for the biodistribution of particular imaging agents, and improper target delivery can lead to toxicity. Small molecules, which show uptake and biodistribution based on various physiochemical parameters described as "Lipinski's rule of Five" [86], include fewer than 5 hydrogen bond donors, fewer than 10 hydrogen bond acceptors, a molecular mass less than 500 Da, and a partition coefficient greater than 5. Lipinski's rules aim to achieve drugs with sufficient absorption, distribution, metabolism, and excretion (ADME) to limit off-target effects and extreme delivery methods. On the contrary, NPs are of larger magnitude, and their biodistribution depends on various properties like size, aspect ratio, charge, stiffness, and surface chemistry (which determines the properties of the NP coating) [87]. Particle opsonization is the elimination mechanism for NPs, seen in plasma-bound small molecules by renal excretion and metabolism.

14.5 Nanoparticles Used in Nuclear Imaging

Various lipidic, polymeric, carbonaceous, and inorganic NP platforms can be used for radiolabeling through chemical conjugation by direct methods or chelation. The choice of radionuclide is based on the radionuclide's intrinsic decay properties. These radionuclides have different synthetic routes, which include ^{11}C, ^{18}F, ^{76}Br, and ^{124}I, as well as other radionuclides like ^{64}Cu, ^{68}Ga, ^{89}Zr, ^{90}Y, ^{99m}Tc, and ^{177}Lu, which are attached by methods such as chelation or chelator-free surface adsorption. Morgan and coworkers pioneered the use of multilamellar vesicles for ^{99m}Tc by reducing pertechnetate in the presence of liposomes in 1981 [88]. "Cornell Dots" (C dots) are synthesized and can be fluorescent [89] or radiolabeled [90]. Inorganic NPs, like gold and iron oxide, are commonly used for surface-enhanced Raman scattering (SERS) [91] and as MR-weighted contrast agents [92].

The radiolabeling biodistribution and pharmacokinetics of NPs depend on certain criteria, including clearance times, as with SNPs [93] and radioisotopes with short half-lives, such as ^{68}Ga ($t_{1/2} = 68$ min). Imaging of NPs requires intrinsic particle properties that differ from those of the target tissue, such as magnetic resonance contrast or a fluorophore. Radiolabeling uses the high-energy decay from the parent radionuclide as the imaging source, and the contrast agent concentrations in MR, CT, PET, and SPECT provide lower spatial resolution due to collimator rejection for the identification of the angle of incidence [94]. Radionuclides used for imaging have higher specific activities, as in picomolar radiotracers, to detect potential harmful radiation effects on the organism. Multimodal imaging is possible when NPs have different intrinsic imaging properties and functions based on the various radionuclides—for instance, IONPs labeled with a radiotracer enable MR/PET imaging [95].

14.6 Nanoparticle Radiolabeling

The type of radionuclide and NP ultimately determines the outcome of the disease process. Conventional methods for radiolabeling an NP often use chelators such as 1,4,7,10-tetraazacyclo dodecane-1,4,7,10-tetraacetic acid (DOTA) or deferoxamine (DFO) for radiometal chelation [96], and tyrosine residues of proteins are modified by chloramine-T oxidation or the IODOGEN method [90] on the NP surface in different radiolabeling processes. Three important methods used for chelate-free radiolabeling are hot-plus-cold synthesis, ionic exchange, and cyclotron bombardment [97]. Radiolabeling of IONPs [98, 99], SNPs, silica-coated NPs [100], and quantum dots involves the pH-dependent addition of the radiometal to the NP at elevated temperatures [101]. ^{198}Au-NP can be synthesized from cold AuNP precursors [102], and another method using ^{64}Cu-alloyed AuNPs shows greater radiochemical stability compared with chelator and chelator-free radiolabeling methods [103]. An onsite cyclotron is required for this labeling [104]. One radiolabeling approach involves determining if the CNTs contain ions that can be identified for chelator-free labeling [105]. CNTs serve as hallmark imaging moieties; for instance, IR optical imaging [106] is possible through direct CNT radiolabeling methods.

14.7 Conclusion

The use of various nanoparticles is employed in disease diagnosis and therapeutic applications. This will be more useful for the future approach to precision medicine.

References

1. Thakor AS, Jokerst JV, Ghanouni P, Campbell JL, Mittra E, Gambhir SS. Clinically approved nanoparticle imaging agents. J Nucl Med. 2016;57(12):1833–7.
2. Lee DE, Koo H, Sun IC, Ryu JH, Kim K, Kwon IC. Multifunctional nanoparticles for multimodal imaging and theragnosis. Chem Soc Rev. 2012;41:2656–72.
3. Alazraki NP, Eshima D, Eshima LA, Herda SC, Murray DR, Vansant JP, et al. Lymphoscintigraphy, the sentinel node concept, and the intraoperative gamma probe in melanoma, breast cancer, and other potential cancers. Semin Nucl Med. 1997;27:55–67.
4. Strand SE, Bergqvist L. Radiolabeled colloids and macromolecules in the lymphatic system. Crit Rev Ther Drug Carrier Syst. 1989;6:211–38.
5. Nel A, Xia T, Madler L, Li N. Toxic potential of materials at the nanolevel. Science. 2006;311:622–7.
6. Xia T, Kovochich M, Brant J, Hotze M, Sempf J, Oberley T, et al. Comparison of the abilities of ambient and manufactured nanoparticles to induce cellular toxicity according to an oxidative stress paradigm. Nano Lett. 2006;6:1794–807.
7. Knudsen KB, Northeved H, Kumar PE, Permin A, Gjetting T, Andresen TL, et al. In vivo toxicity of cationic micelles and liposomes. Nanomedicine. 2015;11:467–77.
8. Buzea C, Pacheco II, Robbie K. Nanomaterials and nanoparticles: sources and toxicity. Biointerphases. 2007;2:MR17–71.
9. Nel AE, Mädler L, Velegol D, Xia T, Hoek EMV, Somasundaran P, et al. Understanding biophysicochemical interactions at the nano-bio interface. Nat Mater. 2009;8:543–57.
10. Kirchner C, Liedl T, Kudera S, Pellegrino T, Muñoz Javier A, Gaub HE, et al. Cytotoxicity of colloidal CdSe and CdSe/ZnS nanoparticles. Nano Lett. 2005;5:331–8.
11. Boisselier E, Astruc D. Gold nanoparticles in nanomedicine: preparations, imaging, diagnostics, therapies and toxicity. Chem Soc Rev. 2009;38:1759–82.
12. Khlebtsov N, Dykman L. Biodistribution and toxicity of engineered gold nanoparticles: a review of in vitro and in vivo studies. Chem Soc Rev. 2011;40:1647–71.
13. Benezra M, Penate-Medina O, Zanzonico PB, Schaer D, Ow H, Burns A, et al. Multimodal silica nanoparticles are effective cancer-targeted probes in a model of human melanoma. J Clin Invest. 2011;121:2768–80.
14. Phillips E, Penate-Medina O, Zanzonico PB, Carvajal RD, Mohan P, Ye Y, et al. Clinical translation of an ultrasmall inorganic optical-PET imaging nanoparticle probe. Sci Transl Med. 2014;6:260ra149.
15. Storm G, Belliot SO, Daemen T, Lasic DD. Surface modification of nanoparticles to oppose uptake by the mononuclear phagocyte system. Adv Drug Deliv Rev. 1995;17:31–48.
16. Roser M, Fischer D, Kissel T. Surface-modified biodegradable albumin nano- and microspheres. II: effect of surface charges on in vitro phagocytosis and biodistribution in rats. Eur J Pharm Biopharm. 1998;46:255–63.
17. Otsuka H, Nagasaki Y, Kataoka K. PEGylated nanoparticles for biological and pharmaceutical applications. Adv Drug Deliv Rev. 2003;55:403–19.
18. Schipper ML, Iyer G, Koh AL, Cheng Z, Ebenstein Y, Aharoni A, et al. Particle size, surface coating, and

PEGylation influence the biodistribution of quantum dots in living mice. Small. 2009;5:126–34.
19. Kummar S, Kinders R, Rubinstein L, Parchment RE, Murgo AJ, Collins J, et al. Compressing drug development timelines in oncology using phase '0' trials. Nat Rev Cancer. 2007;7:131–9.
20. Marchetti S, Schellens JH. The impact of FDA and EMEA guidelines on drug development in relation to phase 0 trials. Br J Cancer. 2007;97:577–81.
21. LoRusso PM. Phase 0 clinical trials: an answer to drug development stagnation? J Clin Oncol. 2009;27:2586–8.
22. Aktolun C, Goldsmith SJ. Nuclear medicine therapy: principles and clinical applications. New York: Springer Science and Business Media; 2012.
23. Goldsmith SJ. Radioimmunotherapy of lymphoma: Bexxar and Zevalin. Semin Nucl Med. 2010;40:122–35.
24. Garin E, Lenoir L, Rolland Y, Edeline J, Mesbah H, Laffont S, et al. Dosimetry based on 99mTc-macroaggregated albumin SPECT/CT accurately predicts tumor response and survival in hepatocellular carcinoma patients treated with 90Y-loaded glass microspheres: preliminary results. J Nucl Med. 2012;53:255–63.
25. Saxena A, Bester L, Shan L, Perera M, Gibbs P, Meteling B, et al. A systematic review on the safety and efficacy of yttrium-90 radioembolization for unresectable, chemorefractory colorectal cancer liver metastases. J Cancer Res Clin Oncol. 2014;140:537–47.
26. Sharma RA, Van Hazel GA, Morgan B, Berry DP, Blanshard K, Price D, et al. Radioembolization of liver metastases from colorectal cancer using yttrium-90 microspheres with concomitant systemic oxaliplatin, fluorouracil, and leucovorin chemotherapy. J Clin Oncol. 2007;25:1099–106.
27. Geschwind JFH, Salem R, Carr BI, Soulen MC, Thurston KG, Goin KA, et al. Yttrium-90 microspheres for the treatment of hepatocellular carcinoma. Gastroenterology. 2004;127:S194–205.
28. Salem R, Lewandowski RJ, Mulcahy MF, Riaz A, Ryu RK, Ibrahim S, et al. Radioembolization for hepatocellular carcinoma using Yttrium-90 microspheres: a comprehensive report of long-term outcomes. Gastroenterology. 2010;138:52–64.
29. Helbok A, Decristoforo C, Dobrozemsky G, Rangger C, Diederen E, Stark B, et al. Radiolabeling of lipid-based nanoparticles for diagnostics and therapeutic applications: a comparison using different radiometals. J Liposome Res. 2010;20:219–27.
30. Phillips WT, Bao A, Brenner AJ, Goins BA. Image-guided interventional therapy for cancer with radiotherapeutic nanoparticles. Adv Drug Deliv Rev. 2014;76:39–59.
31. Chen F, Hong H, Zhang Y, Valdovinos HF, Shi S, Kwon GS, et al. In vivo tumor targeting and image-guided drug delivery with antibody-conjugated, radiolabeled mesoporous silica nanoparticles. ACS Nano. 2013;7:9027–39.
32. Xiao Y, Hong H, Matson VZ, Javadi A, Xu W, Yang Y, et al. Gold nanorods conjugated with doxorubicin and cRGD for combined anticancer drug delivery and PET imaging. Theranostics. 2012;2:757–68.
33. Gerlinger M, Rowan AJ, Horswell S, Larkin J, Endesfelder D, Gronroos E, et al. Intratumor heterogeneity and branched evolution revealed by multiregion sequencing. N Engl J Med. 2012;366:883–92.
34. Fahey F, Zukotynski K, Capala J, Knight N, Organizing Committee, Contributors, and Participants of NCI/SNMMI Joint Workshop on Targeted Radionuclide Therapy. Targeted radionuclide therapy: proceedings of a joint workshop hosted by the National Cancer Institute and the Society of Nuclear Medicine and Molecular Imaging. J Nucl Med. 2014;55:337–48.
35. Lee YK, Jeong JM, Hoigebazar L, Yang BY, Lee YS, Lee BC, et al. Nanoparticles modified by encapsulation of ligands with a long alkyl chain to affect multispecific and multimodal imaging. J Nucl Med. 2012;53:1462–70.
36. Xing Y, Zhao J, Conti PS, Chen K. Radiolabeled nanoparticles for multimodality tumor imaging. Theranostics. 2014;4:290–306.
37. Goel S, Chen F, Ehlerding EB, Cai W. Intrinsically radiolabeled nanoparticles: an emerging paradigm. Small. 2014;10:3825–30.
38. Sun M, Hoffman D, Sundaresan G, Yang L, Lamichhane N, Zweit J. Synthesis and characterization of intrinsically radiolabeled quantum dots for bimodal detection. Am J Nucl Med Mol Imaging. 2012;2:122–35.
39. Sun X, Huang X, Guo J, Zhu W, Ding Y, Niu G, et al. Self-illuminating 64Cu-doped CdSe/ZnS nanocrystals for in vivo tumor imaging. J Am Chem Soc. 2014;136:1706–9.
40. Suresh M, Cao Y. A simple and efficient method for radiolabeling of preformed liposomes. J Pharm Sci. 1998;1:31–7.
41. Phillips WT, Rudolph AS, Goins B, Timmons JH, Klipper R, Blumhardt R. A simple method for producing a technetium-99m labeled liposome which is stable in vivo. Int J Rad Appl Instrum B. 1992;19:539–47.
42. Awasthi VD, Goins B, Klipper R, Phillips WT. Dual radiolabeled liposomes: biodistribution studies and localization of focal sites of infection in rats. Nucl Med Biol. 1998;25:155–60.
43. Bao A, Goins B, Klipper R, Negrete G, Phillips WT. Direct 99mTc labeling of pegylated liposomal doxorubicin (Doxil) for pharmacokinetic and non-invasive imaging studies. J Pharmacol Exp Ther. 2004;308:419–25.
44. Mougin-Degraef M, Jestin E, Bruel D, Remaud-Le Saëc P, Morandeau L, Faivre-Chauvet A, et al. High-activity radio-iodine labeling of conventional and stealth liposomes. J Liposome Res. 2006;16:91–102.
45. Marik J, Tartis MS, Zhang H, Fung JY, Kheirolomoom A, Sutcliffe JL, et al. Long-circulating liposomes

radiolabeled with [18F]fluorodipalmitin ([18F] FDP). Nucl Med Biol. 2007;34:165–71.
46. Morales-Avila E, Ferro-Flores G, Ocampo-García BE, de María Ramírez F. Radiolabeled nanoparticles for molecular imaging. In: Schaller B, editor. Molecular imaging. InTech; 2012. https://doi.org/10.5772/31109. http://www.intechopen.com/books/molecular-imaging/radiolabeled-nanoparticles-formolecular-imaging.
47. Locatelli E, Gil L, Israel LL, Passoni L, Naddaka M, Pucci A, et al. Biocompatible nanocomposite for PET/MRI hybrid imaging. Int J Nanomedicine. 2012;7:6021–33.
48. Yang L, Sundaresan G, Sun M, Jose P, Hoffman D, McDonagh PR, et al. Intrinsically radiolabeled multifunctional cerium oxide nanoparticles for in vivo studies. J Mater Chem B. 2013;1:1421–31.
49. Chattopadhyay N, Fonge H, Cai Z, Scollard D, Lechtman E, Done SJ, et al. Role of antibody-mediated tumor targeting and route of administration in nanoparticle tumor accumulation in vivo. Mol Pharm. 2012;9:2168–79.
50. Jain RK, Stylianopoulos T. Delivering nanomedicine to solid tumors. Nat Rev Clin Oncol. 2010;7:653–64.
51. Smith BR, Kempen P, Bouley D, Xu A, Liu Z, Melosh N, et al. Shape matters: intravital microscopy reveals surprising geometrical dependence for nanoparticles in tumor models of extravasation. Nano Lett. 2012;12:3369–77.
52. Brannon-Peppas L, Blanchette JO. Nanoparticle and targeted systems for cancer therapy. Adv Drug Deliv Rev. 2004;56:1649–59.
53. Alexis F, Pridgen E, Molnar LK, Farokhzad OC. Factors affecting the clearance and biodistribution of polymeric nanoparticles. Mol Pharm. 2008;5:505–15.
54. Kamaly N, Xiao Z, Valencia PM, Radovic-Moreno AF, Farokhzad OC. Targeted polymeric therapeutic nanoparticles: design, development and clinical translation. Chem Soc Rev. 2012;41:2971–3010.
55. Bertrand N, Wu J, Xu X, Kamaly N, Farokhzad OC. Cancer nanotechnology: the impact of passive and active targeting in the era of modern cancer biology. Adv Drug Deliv Rev. 2014;66:2–25.
56. Kumari P, Ghosh B, Biswas S. Nanocarriers for cancer-targeted drug delivery. J Drug Target. 2016;24:179 91.
57. Fisher DR. Internal dosimetry for systemic radiation therapy. Semin Radiat Oncol. 2000;10:123–32.
58. Stabin MG, Sparks RB, Crowe E. OLINDA/EXM: the second-generation personal computer software for internal dose assessment in nuclear medicine. J Nucl Med. 2005;46:1023–7.
59. Emfietzoglou D, Kostarelos K, Sgouros G. An analytic dosimetry study for the use of radionuclide-liposome conjugates in internal radiotherapy. J Nucl Med. 2001;42:499–504.
60. Kostarelos K, Emfietzoglou D. Tissue dosimetry of liposomeradionuclide complexes for internal radiotherapy: toward liposome-targeted therapeutic radiopharmaceuticals. Anticancer Res. 2000;20:3339–45.
61. Chen LC, Wu YH, Liu IH, Ho CL, Lee WC, Chang CH, et al. Pharmacokinetics, dosimetry and comparative efficacy of 188Reliposome and 5-FU in a CT26-luc lung-metastatic mice model. Nucl Med Biol. 2012;39:35–43.
62. Walrand S, Hanin F-X, Pauwels S, Jamar F. Tumour control probability derived from dose distribution in homogeneous and heterogeneous models: assuming similar pharmacokinetics, 125Sn–177Lu is superior to 90Y–177Lu in peptide receptor radiotherapy. Phys Med Biol. 2012;57:4263.
63. Lucas S, Feron O, Gallez B, Masereel B, Michiels C, Vander BT. Monte Carlo calculation of radioimmunotherapy with 90 Y-, 177 Lu-, 131 I-, 124 I-, and 188 re-nanoobjects: choice of the best radionuclide for solid tumour treatment by using TCP and NTCP concepts. Comput Math Methods Med. 2015;2015:284–360.
64. Soundararajan A, Dodd GD III, Bao A, Phillips WT, McManus LM, Prihoda TJ, et al. Chemoradionuclide therapy with 186Re-labeled liposomal doxorubicin in combination with radiofrequency ablation for effective treatment of head and neck cancer in a nude rat tumor xenograft model. Radiology. 2011;261:813–23.
65. Kim DK, Kang B, Kim OY, Choi DS, Lee J, Kim SR, et al. EVpedia: an integrated database of high-throughput data for systemic analyses of extracellular vesicles. J Extracell Vesicles. 2013;2:20384.
66. van Dommelen SM, Vader P, Lakhal S, Kooijmans SA, van Solinge WW, Wood MJ, et al. Microvesicles and exosomes: opportunities for cell-derived membrane vesicles in drug delivery. J Control Release. 2012;161:635–44.
67. Ohno S, Takanashi M, Sudo K, Ueda S, Ishikawa A, Matsuyama N, et al. Systemically injected exosomes targeted to EGFR deliver antitumor microRNA to breast cancer cells. Mol Ther. 2013;21:185–91.
68. Biancone L, Bruno S, Deregibus MC, Tetta C, Camussi G. Therapeutic potential of mesenchymal stem cell-derived microvesicles. Nephrol Dial Transplant. 2012;27:3037–42.
69. Kowal J, Tkach M, Thery C. Biogenesis and secretion of exosomes. Curr Opin Cell Biol. 2014;29:116–25.
70. Sun D, Zhuang X, Xiang X, Liu Y, Zhang S, Liu C, et al. A novel nanoparticle drug delivery system: the anti-inflammatory activity of curcumin is enhanced when encapsulated in exosomes. Mol Ther. 2010;18:1606–14.
71. Escudier B, Dorval T, Chaput N, André F, Caby MP, Novault S, et al. Vaccination of metastatic melanoma patients with autologous dendritic cell (DC) derived-exosomes: results of thefirst phase I clinical trial. J Transl Med. 2005;3:10.
72. Morse MA, Garst J, Osada T, Khan S, Hobeika A, Clay TM, et al. A phase I study of dexosome immunotherapy in patients with advanced non-small cell lung cancer. J Transl Med. 2005;3:9.

73. Dai S, Wei D, Wu Z, Zhou X, Wei X, Huang H, et al. Phase I clinical trial of autologous ascites-derived exosomes combined with GM-CSF for colorectal cancer. Mol Ther. 2008;16:782–90.
74. Mignot G, Roux S, Thery C, Segura E, Zitvogel L. Prospects for exosomes in immunotherapy of cancer. J Cell Mol Med. 2006;10:376–88.
75. Gyorgy B, Hung ME, Breakefield XO, Leonard JN. Therapeutic applications of extracellular vesicles: clinical promise and open questions. Annu Rev Pharmacol Toxicol. 2015;55:439–64.
76. Morishita M, Takahashi Y, Nishikawa M, Sano K, Kato K, Yamashita T, et al. Quantitative analysis of tissue distribution of the B16BL6-derived exosomes using a streptavidinlactadherin fusion protein and iodine-125-labeled biotin derivative after intravenous injection in mice. J Pharm Sci. 2015;104:705–13.
77. Smyth T, Kullberg M, Malik N, Smith-Jones P, Graner MW, Anchordoquy TJ. Biodistribution and delivery efficiency of unmodified tumor-derived exosomes. J Control Release. 2015;199:145–55.
78. Hood JL, San RS, Wickline SA. Exosomes released by melanoma cells prepare sentinel lymph nodes for tumor metastasis. Cancer Res. 2011;71:3792–801.
79. Zhuang X, Xiang X, Grizzle W, Sun D, Zhang S, Axtell RC, et al. Treatment of brain inflammatory diseases by delivering exosome encapsulated anti-inflammatory drugs from the nasal region to the brain. Mol Ther. 2011;19:1769–79.
80. Witwer KW, Buzas EI, Bemis LT, Bora A, Lässer C, Lötvall J, et al. Standardization of sample collection, isolation and analysis methods in extracellular vesicle research. J Extracell Vesicles. 2013;2:20360.
81. Jang SC, Kim OY, Yoon CM, Choi DS, Roh TY, Park J, et al. Bioinspired exosome-mimetic nanovesicles for targeted delivery of chemotherapeutics to malignant tumors. ACS Nano. 2013;7:7698–710.
82. Takahashi Y, Nishikawa M, Shinotsuka H, Matsui Y, Ohara S, Imai T, et al. Visualization and in vivo tracking of the exosomes of murine melanoma B16-BL6 cells in mice after intravenous injection. J Biotechnol. 2013;165:77–84.
83. Colombo M, Moita C, van Niel G, Kowal J, Vigneron J, Benaroch P, et al. Analysis of ESCRT functions in exosome biogenesis, composition and secretion highlights the heterogeneity of extracellular vesicles. J Cell Sci. 2013;126:5553–65.
84. Wiklander OP, Nordin JZ, O'Loughlin A, Gustafsson Y, Corso G, Mäger I, et al. Extracellular vesicle in vivo biodistribution is determined by cell source, route of administration and targeting. J Extracell Vesicles. 2015;4:26316.
85. Hertz B. Dr. Saul Hertz (1905–1950) discovers the medical uses of radioactive iodine: the first targeted cancer therapy. In: Ahmadzadehfar H, editor. Thyroid cancer-advances in diagnosis and therapy. IntechOpen; 2016. p. 1.
86. Swierczewska M, Liu G, Lee S, Chen X. High-sensitivity nanosensors for biomarker detection. Chem Soc Rev. 2012;41(7):2641–55.
87. Zhang F, Huang X, Zhu L, et al. Noninvasive monitoring of orthotopic glioblastoma therapy response using RGD-conjugated iron oxide nanoparticles. Biomaterials. 2012;33(21):5414–22.
88. Morgan JR, Williams KE, Davies RL, Leach K, Thomson M, Williams LAP. Localisation of experimental staphylococcal abscesses by 99MTC-technetium-labelled liposomes. J Med Microbiol. 1981;14:213–7.
89. Ow H, Larson DR, Srivastava M, Baird BA, Webb WW, Wiesner U. Bright and stable core-shell fluorescent silica nanoparticles. Nano Lett. 2005;5:113–7. https://doi.org/10.1021/nl0482478.
90. Benezra M, Penate-Medina O, Zanzonico PB, Schaer D, Ow H, Burns A, DeStanchina E, Longo V, Herz E, Iyer S, et al. Multimodal silica nanoparticles are effective cancer-targeted probes in a model of human melanoma. J Clin Invest. 2011;121:2768–80.
91. Kircher MF, de la Zerda A, Jokerst JV, Zavaleta CL, Kempen PJ, Mittra E, Pitter K, Huang R, Campos C, Habte F, et al. A brain tumor molecular imaging strategy using a new triple-modality MRI-photoacoustic-Raman nanoparticle. Nat Med. 2012;18:829–34.
92. Kaittanis C, Shaffer TM, Ogirala A, Santra S, Perez JM, Chiosis G, Li Y, Josephson L, Grimm J. Environment-responsive nanophores for therapy and treatment monitoring via molecular MRI quenching. Nat Commun. 2014;5:3384.
93. Van Schooneveld MM, Vucic E, Koole R, Zhou Y, Stocks J, Cormode DP, Tang CY, Gordon RE, Nicolay K, Meijerink A, et al. Improved biocompatibility and pharmacokinetics of silica nanoparticles by means of a lipid coating: a multimodality investigation. Nano Lett. 2008;8:2517–25.
94. Rahmim A, Zaidi H. PET versus SPECT: strengths, limitations and challenges. Nucl Med Commun. 2008;29:193–207.
95. Thorek DLJ, Ulmert D, Diop N-FM, Lupu ME, Doran MG, Huang R, Abou DS, Larson SM, Grimm J. Non-invasive mapping of deep-tissue lymph nodes in live animals using a multimodal PET/MRI nanoparticle. Nat Commun. 2014;5:3097.
96. Zeglis BM, Houghton JL, Evans MJ, ViolaVillegas N, Lewis JS. Underscoring the influence of inorganic chemistry on nuclear imaging with radiometals. Inorg Chem. 2014;53:1880–99.
97. Lee S, Choi KY, Chung H, et al. Real time, high resolution video imaging of apoptosis in single cells with a polymeric nanoprobe. Bioconjug Chem. 2011;22(2):125–31.
98. Boros E, Bowen AM, Josephson L, Holland JP. Chelate-free metal ion binding and heat-induced radiolabeling of iron oxide nanoparticles. Chem Sci. 2015;6:225–36.
99. Burke BP, Baghdadi N, Kownacka A, Nigam S, Clemente G, Al-Yassiry M, Domarkas J, Lorch M, Pickles M, Gibbs P, et al. Chelator free gallium-68 radiolabelling of silica coated iron oxide nanorods via surface interactions: evaluation as multimodal PET/MR liver imaging

agents. Nanoscale. 2015;7:14889–96. https://doi.org/10.1039/C5NR02753E.

100. Shaffer TM, Wall MA, Harmsen S, Longo VA, Drain CM, Kircher MF, Grimm J. Silica nanoparticles as substrates for chelator-free labeling of oxophilic radioisotopes. Nano Lett. 2015;15:864–8.

101. Zeng J, Jia B, Qiao R, Wang C, Jing L, Wang F, Gao M. In situ 111In-doping for achieving biocompatible and non-leachable 111In-labeled Fe3O4 nanoparticles. Chem Commun. 2014;50:2170–2.

102. Shukla R, Chanda N, Zambre A, Upendran A, Katti K, Kulkarni RR, Nune SK, Casteel SW, Smith CJ, Vimal J, et al. Laminin receptor specific therapeutic gold nanoparticles (198AuNP-EGCg) show efficacy in treating prostate cancer. Proc Natl Acad Sci USA. 2012;109:12426–31.

103. Zhao Y, Sultan D, Detering L, Cho S, Sun G, Pierce R, Wooley KL, Liu Y. Copper-64-alloyed gold nanoparticles for cancer imaging: improved radiolabel stability and diagnostic accuracy. Angew Chem Int Ed. 2014;53:156–9.

104. Pérez-Campaña C, Gómez-Vallejo V, Puigivila M, Martín A, Calvo-Fernández T, Moya SE, Ziolo RF, Reese T, Llop J. Biodistribution of different sized nanoparticles assessed by positron emission tomography: a general strategy for direct activation of metal oxide particles. ACS Nano. 2013;7:3498–505.

105. Matson ML, Villa CH, Ananta JS, Law JJ, Scheinberg DA, Wilson LJ. Encapsulation of alphaparticle-emitting 225Ac3+ ions within carbon nanotubes. J Nucl Med. 2015;56:897–900.

106. Bachilo SM, Strano MS, Kittrell C, Hauge RH, Smalley RE, Weisman RB. Structure-assigned optical spectra of single-walled carbon nanotubes. Science. 2002;298:2361–6.

Nanoimaging in Cancer Part I 15

15.1 Introduction

Molecular imaging (MI) is an important and emerging biomedical research modality based on the visual representation, characterization, and quantification of biological processes at the cellular and/or molecular level in living organisms [1–3]. Different MI modalities are being used, including fluorescence and bioluminescence imaging, targeted ultrasound (US), molecular magnetic resonance imaging (MRI), magnetic resonance spectroscopy (MRS), single-photon-emission computed tomography (SPECT), and positron emission tomography (PET) (Fig. 15.1). The basis of these imaging modalities is the use of probes or beacons, which are observed in the region of interest and can be evaluated by imaging. One of the important drawbacks is the many limitations due to very poor spatial resolution, low sensitivity, or poor signal penetration through tissues. Optical imaging is not sensitive due to its limited depth-penetration capability and poorer resolution; on the contrary, MRI is capable of imaging deep tissues but is still not sensitive. With the use of nanoparticles (NPs) as newer MI agents, these limitations are being overcome by various research advancements. NPs are more precise due to passive accumulation and various active-targeting approaches [4]. NPs are capable of carrying a higher capacity, have higher signal intensity, and offer better stability. NPs can also achieve higher concentrations of imaging agents in the target region. NPs are much smaller in comparison to cells [5], and similar in size to molecules and proteins [6].

S. Shaikh, *Nanoimaging - Future of Precision Medicine*,
https://doi.org/10.1007/978-981-97-7292-6_15

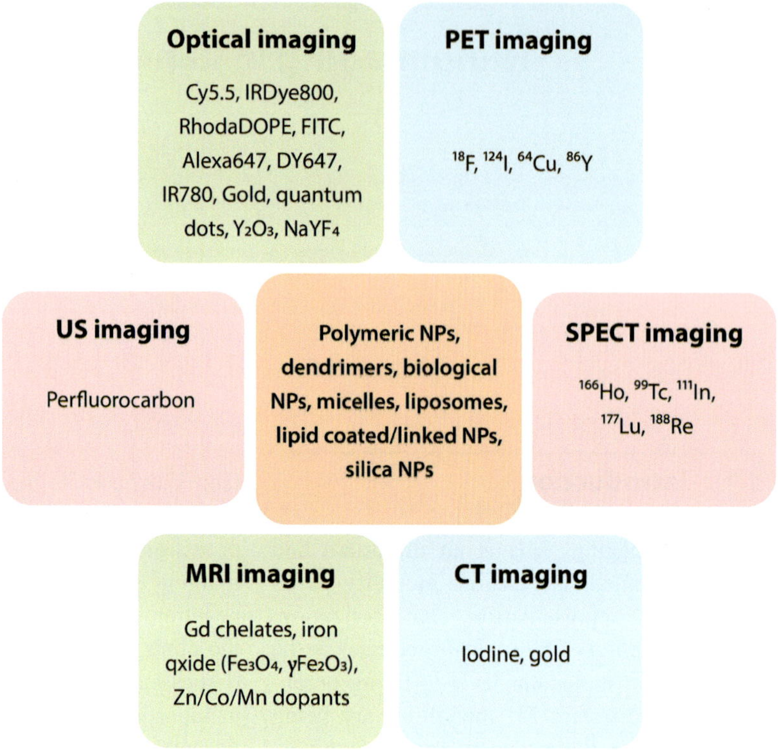

Fig. 15.1 Different types of molecular imaging as shown

15.2 Nanoplatforms

Different nanoplatforms can be used for different imaging modalities. As we all know, the different NPs are based on various compositions, sizes, shapes, and structures made up of different materials [7–9]. These are based on various materials and different shapes, such as spheres, rods, cubes, and structures resembling snowflakes, flowers, thorns, hemispheres, worms, discoids, and chains. All these are used in various imaging modalities, including optical imaging, MRI, US, and/or nuclear imaging, often combined with diagnostic and therapeutic abilities known as "theranostics." We have already discussed that these NPs are of two types, which target tumors based on two different pathways: passive- and/or active-targeting pathways. When there is leakage through capillary barriers via the lymphatic drainage system in different tumor tissues, these NP platforms accumulate in the tumor tissues, a process known as the enhanced permeability and retention (EPR) effect (Fig. 15.2). These NPs can recognize different tumor cells via receptor-mediated endocytosis, such as tumor-targeting moieties from antibodies, nucleic acids, proteins, or other ligands.

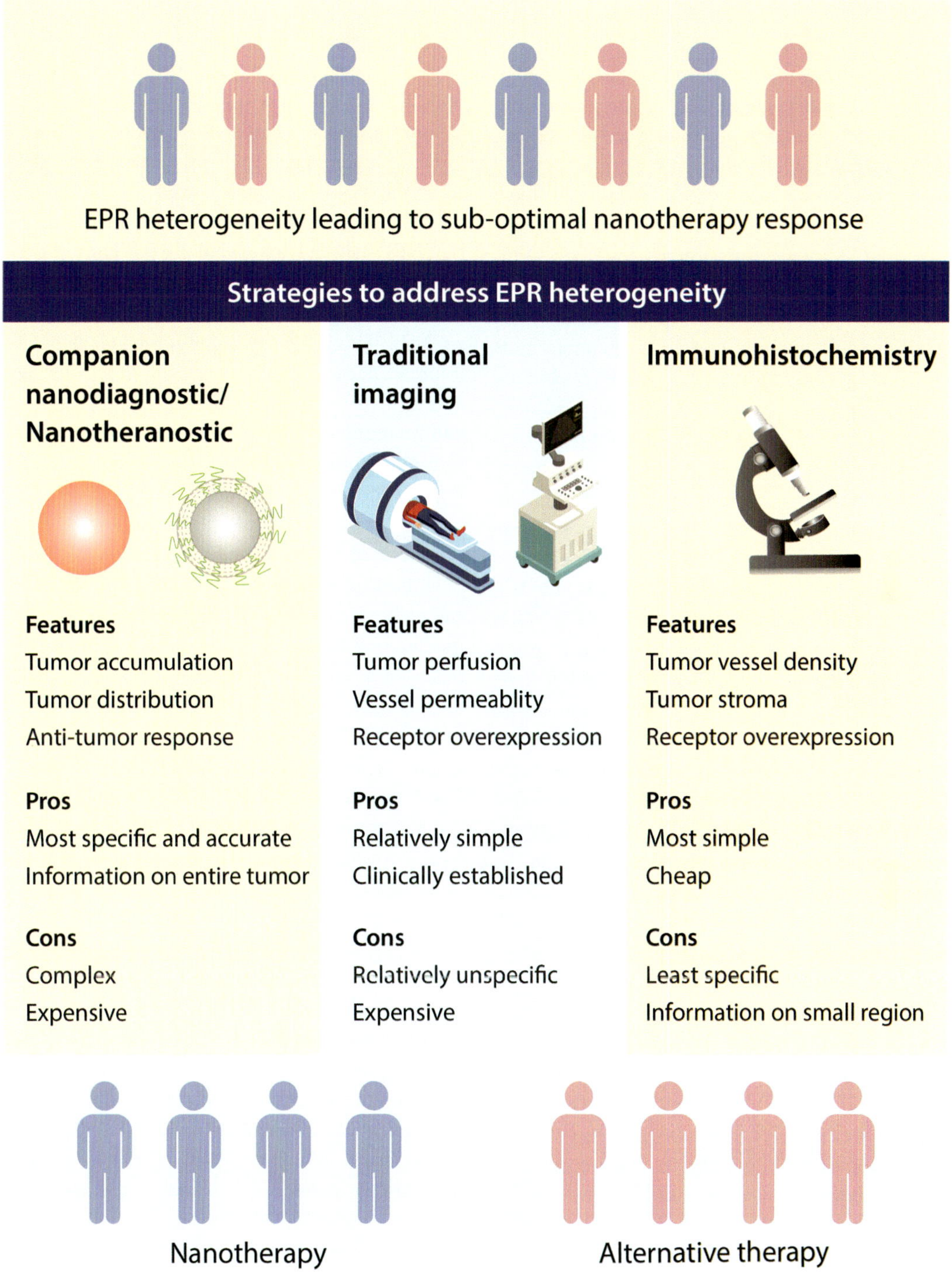

Fig. 15.2 Strategies of the EPR heterogeneity showing different aspects of nanodiagnosis and therapeutics versus traditional imaging and immunohistochemistry

15.3 Optical Nanoparticles

Optical imaging is very important due to its high sensitivity, better cost-effectiveness, and nonionizing radiation, and it can be used for various small-animal studies. However, optical imaging is not completely quantitative for living-subject imaging studies. For this reason, many probes, including synthetic fluorophores, various semiconductor fluorescent crystals, and probes made from lanthanide rare-earth ions, are used for small-animal imaging [10–12]. Quantum dots (QDs) are commonly used NPs for the evaluation of preclinical optical imaging. Compared with fluorophores, QDs have many superior properties for biological imaging, such as strong resistance to photobleaching and chemical degradation, higher quantum yields, continuous absorption spectra from ultraviolet (UV) to near-infrared (NIR), and larger effective Stokes shifts [13–16]. The Arginine-glycine-aspartic acid (RGD)-QDs bind the luminal endothelium in mouse tumor neovasculature [17]. Due to their smaller size, these QDs show less retention in the tumor, with easier washout in the bloodstream [16]. The possibility of toxicity and lack of deep-tissue imaging, along with the inability to quantify, are a few of the drawbacks of QDs [18]. The small-animal PET, which uses radiolabeled QDs, can be observed in the reticuloendothelial system (RES), liver, and spleen [19, 20]. The uptake of QDs can be reduced in the RES, and human serum albumin (HSA) serves as the exploring factor for the QD800-mercaptopropionic acid (MPA) NPs. This QD800-MPA-HSA can be identified in macrophage cells through fluorescence imaging in vivo [21]. The Affibody-modified QDs exhibit higher specificity for evaluating the targeting of human epidermal growth factor receptor 2 (HER2)-expressing cells and tumors [22]. Indocyanine dyes can be encapsulated and added to polymeric NPs made from indocyanine green (ICG) [23], the NIR region fluorescent cyanine 7 (Cy7) [24], and dialkylcarbocyanine fluorophores [25], which are FDA-approved. This nanoplatform strategy is an important method of optical imaging for overcoming the drawbacks of free and small dyes [26, 27].

QDs are being made capable of overcoming the light penetration issue and are also designed for deep-tissue imaging, for in vivo multiplexed optical imaging, in a similar way to the natural bioluminescence resonance energy transfer (BRET) technique, in which chemical energy is converted into photons to excite the QDs [28, 29]. The photocaged up-conversion nanoparticles (UCNPs) take advantage of the photon-up conversion of NIR light to UV light to trigger the release of D-luciferin from D-luciferin-conjugated UCNPs. This released D-luciferin enhances the fluorescence and bioluminescence signals in vitro and in vivo by allowing greater light penetration and reducing cellular damage in the region [30]. Cerenkov luminescence imaging is not an imaging modality like bioluminescence imaging, as it is based on the principle of visible photons that are captured when emitted by Cerenkov radiation [31]. Radioluminescent nanophosphors (RLNPs) can be combined with MI probes for the development of combined X-ray/optical imaging modalities, such as X-ray luminescence computed tomography (XLCT) [32–34]. These PEGylated Eu^{3+}-doped nanophosphors emit luminescence when excited by ^{18}F radioisotope or X-rays. These capabilities provide more potential for the use of various novel imaging modalities, such as XLCT or deep-tissue Cerenkov luminescence imaging [35].

15.4 Raman Spectroscopy

This technique is based on the phenomenon of light scattering, which differs from absorption/emission in fluorescence, as these Raman-active molecules are more photostable compared to fluorophores, which are photobleached. The use of surface-enhanced Raman spectroscopy (SERS), where active NPs, along with single-wall nanotubes (SWNTs), are used for noninvasive imaging in smaller living subjects by Raman spectroscopy, is gaining popularity [36, 37]. Gold silica NPs can be used in Raman MI for the evaluation of colon cancer in mouse models, aided by the epidermal growth factor receptor [38].

15.5 Photoacoustic Imaging (PAI)

PAI is also a newer hybrid biomedical imaging modality that combines optical imaging and can be used for both functional and anatomic imaging [39]. The principle here is the generation of ultrasonic waves, which can be detected by ultrasound, producing images. Exogenous contrast agents are needed for MI, as there is a difference in the absorption spectra of various tissues [40]. Different non-plasmonic or plasmonic NPs, such as gold nanoshells, nanorods, or nanocages, can promote light absorption, improving the contrast of PAI signals [41]. The use of perylene-diimide (PDI)-based NIR-absorptive organic NPs (micelle-enveloped PDI) will allow for more efficient contrast agents for PAI of deep brain tumors in living mice. The encapsulation of these in amphiphilic molecules, which are water-soluble PDI NPs, can be easily synthesized and exhibit PAI properties [42]. PAI of targeted SWNTs can also contribute to non-invasive cancer imaging by monitoring nanotherapeutics in living subjects [43–46]. Melanin nanoparticles (MNPs) can also be an important intrinsic biopolymer, with better biocompatibility and biodegradability, superior intrinsic photoacoustic (PA) properties, and the ability to bind various drugs, making them efficient endogenous nanosystems for imaging-guided chemotherapy [47, 48]. The use of Sorafenib (SRF) to create the PEGylated-melanin-drug system, labeled with ^{64}Cu, can be used as an important PET/PAI dual-modality imaging-guided therapy [49]. The use of various ultrasmall MNPs will improve the loading of $^{64}Cu^{2+}$ and Fe^{3+} for trimodal (PET/MRI/PAI) imaging [49, 50].

15.6 Ultrasound Agents

Ultrasound is one of the established modalities with good contrast, better sensitivity, and is easily available at a lower cost, despite the fact that some organs, like the lungs, are not highly sensitive. Various nanoscale US agents are also being used [51–53]. Continuous research has led to better sensitivity of the US agents, using them as markers to increase the sensitivity for tumor evaluation [54] (Fig. 15.3). Anderson et al. [55] showed that the use of modified microbubbles (MBs) in combination with cRGD ligands can increase sensitivity five-fold in the recombinant αvβ3 integrin, compared to control groups. Yan et al. [56] demonstrated the use of iRGD-lipopeptides in the MB membrane, which increased the binding of the iRGD-MBs by improving the static and dynamic conditions of the MBs. Willmann et al. [57] showed that MBs can be used with engineered cystine knot (knottin) peptides, providing better-targeting ligands for αvβ3 integrin, using contrast-enhanced US imaging to assess the vascularity of tumors. This modification of both VEGFR2 and αvβ3 integrin [58] results in a multi-ligand approach, leading to better and higher signals compared to single-targeting counterparts (Fig. 15.4).

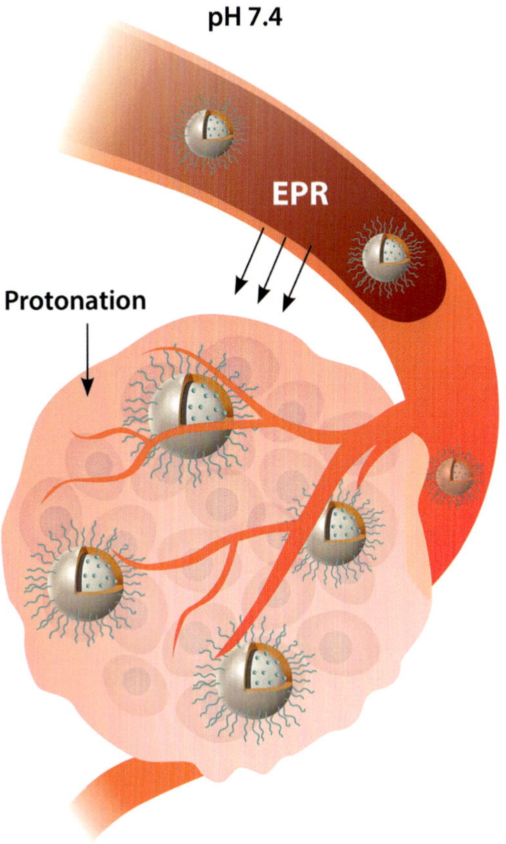

Fig. 15.3 EPR pronation at the normal pH with pronation

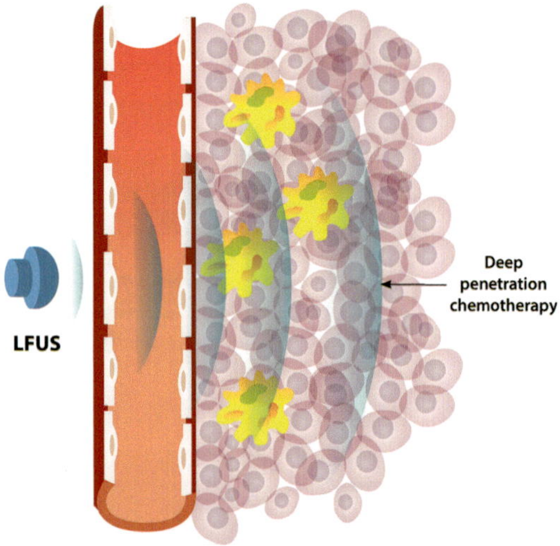

Fig. 15.4 LFUS with focused ultrasound mechanism for deep penetration chemotherapy

15.7 Magnetic Nanoparticles

MRI is now a well-established, non-radiating modality that is non-invasive and has very high resolution for soft tissue contrast, which can delineate anatomical information more effectively. MRI NPs, such as iron oxide NPs (IONPs), have better sensitivity because they shorten T1 and T2 relaxation times. Superparamagnetic iron oxide (SPIO) or ultrasmall superparamagnetic iron oxide (USPIO) NPs have an affinity for reticuloendothelial cells [59], the liver [60–62], spleen [63], and lymph nodes. The basis for this is simple: they accumulate in tumors via the EPR effect [64, 65]. This can be used as a preferential cross-application of MI [66]. These particles are dual T1- and T2-weighted MRI agents, providing better accuracy and biodistribution for in vivo imaging [67]. SPIO NPs are already used as better contrast agents for magnetic resonance (MR) imaging [68], as they have higher magnetization values and smaller sizes, typically less than 100 nm [69]. Magnetic NPs can also be used in laser-induced thermal therapy and can target tissues with less toxicity [70]. The surface coatings of these NPs are nontoxic and biocompatible, allowing for specific delivery to targeted areas. Magnetic NPs can be combined with various drugs, proteins, enzymes, antibodies, or nucleotides, and these can be targeted to areas of interest by applying an external magnetic field. Gd-based agents are designed in such a way that their signal intensity at the target site does not correspond directly to the concentration of the imaging moiety.

15.8 Nanoparticles for Nuclear Imaging

The availability of micro scanners, necessary for small-animal imaging, can be used for various applications in biological processes [71]. PET imaging is a well-established modality, but it has disadvantages due to its very low resolution. This can be improved by combining PET imaging with NPs in the form of radioactive ^{64}Cu, which provides better qualities than traditional radioisotopes [72–74]. Multimodality contrast agents are contrast agents that can be used in two different imaging modalities, such as PET and optical imaging, or PET and MRI. One such combination is PET for management planning and optical imaging for therapeutic intervention. Near-infrared fluorescence (NIRF) imaging, developed by Cai et al. [75], can be used as a PET imaging agent conjugated with ^{64}Cu. These dual or multimodality probes offer better contrast and reduced dosage, making them advantageous for in vivo NIRF imaging with reduced toxicity [76, 77]. Chen et al. [78] showed that the synthesis of ^{64}Cu-DOTA-QD-VEGF nanoprobes can target the vasculature, specifically the VEGF-VEGFR interaction. Nahrendorf et al. [79, 80] demonstrated that dextran-coated SPIO NPs can be synthesized with ^{64}Cu and an NIR fluorophore, allowing fluorescence molecular tomography (FMT) and PET imaging of tumor-associated macrophages. Yang et al. [81] showed that multimodal RGD-targeted SPIO NPs, labeled with ^{64}Cu, can be used for monitoring drug delivery to solid tumors using PET/MR hybrid imaging [81]. Porphyrin, a photosensitizing agent, is used in fluorescent imaging as a dye. Liu et al. [82] developed the ^{64}Cu-porphorysome nanoplatform, which can detect skeletal metastases more precisely with dual-modality imaging, such as optical-PET dual-modal imaging. Chen et al. [83, 84] studied mesoporous silica NPs (HMSNs) to detect tumor angiogenesis, using CD105 as a marker. Liu et al. [44] showed that dual Raman and PET tumor-targeting imaging is more effective by using the Zirconium-89 (^{89}Zr) PET radionuclide [84–88]. Ruggiero et al. [89] successfully used a newer antibody-labeled ^{89}Zr-SWNT, with the monomeric vascular endothelial-cadherin (VE-cad) epitope, employing nanotechnology for tumor angiogenesis. Lee et al. [90] used polymer-coated UCNPs, specifically dimeric RGD-conjugated UCNPs, as various imaging modalities, including optical, MRI, and PET, with higher affinity for targeting the $\alpha v\beta 3$ integrin [90]. Lijowski et al. showed that the T1-based $\alpha v\beta 3$-targeted ^{99m}Tc-Gd NPs are highly sensitive and specific for the localization of tumor neovascularization and can be identified through SPECT-MR imaging [91]. Hu et al. [92] showed that the ^{111}In-perfluorocarbon can be modified by $\alpha v\beta 3$ integrin and identified as a PET-MR agent, functioning as a multimodal agent. Various agents are used for radiolabeling with PET radionuclides, such as ^{64}Cu, and they are also specific target agents for EGFR-targeted PET/optical/MR imaging probes, used for the diagnosis of EGFR-positive tumors [93].

15.9 Challenges and Future Perspectives

Various NPs, as mentioned earlier, are made from different materials and vary in size, shape, composition, crystallinity properties, and magnetism. The reticuloendothelial system (RES), for example, can be used for phagocytizing NPs, causing them to enter the bloodstream and accumulate in the liver, spleen, and bone marrow. The toxicity of NPs is an important consideration, as they may reduce the permeability of various barriers, such as the blood-brain barrier (BBB). Once injected, biodistribution depends on several factors, including size, morphology, charge, and surface chemistry. The ideal size for NPs is typically less than 100 nm, which facilitates better circulation for tumor targeting and drug release.

The phenomenon of NPs entering and accumulating in tumors is based on the enhanced permeability and retention (EPR) effect, caused by aberrant vasculature and lymphatic infiltration [94]. Some agents, such as nano-colloid technetium-99m (^{99m}Tc), can be used for lymphoscintigraphy in conjunction with MRI-enabled cancer detection [95–99]. Certain agents like ^{99m}Tc sulfur nano-colloids are capable of evaluating bone marrow injuries related to bone metastasis [100–103]. Various ^{99m}Tc nano-colloids have been developed for detecting different cancers with MRI [104, 105]. Iron oxide NPs have been used as imaging agents for abdominal imaging [106], but due to toxicity issues, they were later withdrawn from use [106, 107]. Another non-oncological use of NPs includes iron oxide NP ferumoxytol, which is used in the management of iron deficiency anemia. Ferumoxytol is frequently used as an MRI contrast agent in different clinical applications.

15.10 Current Clinical Practice and Limitations

Various imaging modalities are used for tumor imaging [108–113]. Most MI techniques are based on PET and SPECT, which use even smaller antibody-based radiotracers. However, much research is still needed, as they lack the sensitivity required to detect smaller lesions across all the modalities used in nanoimaging [108, 114–121], with depth penetration being a key limitation [122, 123]. The impact of depth penetration on image resolution has been studied by Naumova et al. [124] and Stacy et al. [125]. Other applications of MI include targeting tumor receptors, cancer-related enzymatic activities, and metabolic pathways. Tumor-associated receptors, such as folate receptors, are used in solid tumors [126]. Some trials have demonstrated the use of the folate receptor alpha in breast cancer to visualize the intracellular components [127]. The Raman imaging approach, which uses inelastic photon scattering with various NPs, is another promising modality [128–131]. Human epidermal growth factor receptor 2 (HER2) is present in 20–25% of breast cancers with different phenotypes [132, 133]. Conjugated ultrasmall fluorescent silica NPs can circulate through the blood with better tumor accumulation. Cathepsin-targeted activity-based probes can be used as linking agents to gold NPs (10 nm in diameter). One study showed that NPs at the tumor site exhibited a more than five- to sixfold increase in chemical exchange saturation transfer compared to normal cells. This work also demonstrated a correlation between Furin levels and therapeutic response in tumors. However, MRI field strength has limitations regarding image sensitivity and specificity when compared to CT or MRI [134–136].

Manganese 21 (Mn21) is also used as a pH-sensitive contrast agent for calcium phosphate NPs [137].

15.11 Image-Guided Cancer Interventions

15.11.1 Intraoperative Tumor Resection

In the early stages of cancer, surgery remains the best management option. The development of near-infrared (NIR) fluorophores has enhanced intraoperative tumor visualization. ICG and methylene blue are the two primary agents currently used, although they have limitations such as limited penetration (1 cm), reduced sensitivity for smaller lesions [122], and an inability to differentiate benign from malignant lesions [138]. These agents also have significant drawbacks, including poor photostability, a high plasma protein binding rate, and aggregation in aqueous solutions [139]. Despite these limitations, NPs can be valuable for evaluating lesions, aiding in their staging, and improving surgical efficiency, especially when diagnosed at earlier stages [140, 141]. Nanoparticle carriers can enhance tumor targeting using various amino acids, folic acid, hyaluronic acid, or antibodies against tumor-specific surface receptors [139, 142].

15.11.2 Intraoperative Sentinel Lymph Node Mapping and Resection

In the case of primary tumor resection, sentinel lymph node biopsy plays a crucial role in determining nodal involvement. This process relies on various preoperative sentinel lymph node detection techniques, including CT, MRI, SPECT, PET, or hybrid imaging methods to assess nodal status. However, these techniques have limitations in detecting infectious or inflammatory processes, as well as metastases, due to their lower sensitivity. Additionally, direct analysis during the intraoperative setting is challenging and nonspecific [143].

15.11.3 Image-Guided Radiation Therapy

Radiation therapy is a key treatment modality and a cornerstone of precision radiation oncology. CT-guided radiation therapy is increasingly used to manage various tumors, including those in the prostate, brain, head and neck, lung, breast, and extremities. However, it is less sensitive for abdominal organs due to respiratory motion [144, 145]. MRI-based radiation therapy, utilizing linear accelerators, provides greater clinical benefits [144, 146]. PET/CT-guided radiation therapy is now more effective in identifying the most viable tumor regions, allowing for targeted radiation dosing based on specific indications [147].

15.11.4 The Nanoparticle-Based Image-Guided Radiation Therapy

The concept of radiosensitizers, which use NPs to damage tumor cells and surrounding tissues, can enhance the efficacy of image-guided radiation therapy through a theranostic approach [148]. One such example is silica-based NPs, which, when combined with bismuth and gadolinium, form dual-energy agents that are effective for both CT and MRI imaging. These NPs improve the precision of radiation therapy by targeting tumor cells more effectively while reducing damage to healthy tissues.

15.11.5 Image-Guided Photodynamic and Photothermal Therapy

With the advancements in cancer treatment, the potential use of various photosensitizing agents capable of killing cancer cells is being explored. Since the penetration of these agents is typically shallow in many tissues, photodynamic therapy (PDT) is primarily applied to treat superficial malignancies or lesions. These techniques, however, have not yet been widely applicable to deeper tissues [149]. The use of internal light sources and NPs for photodynamic therapy is emerging as a potential solution.

Kotagiri et al. [150] studied the use of a nano-photosensitizer, titanium dioxide, which has a small diameter of 18 nm. This nanoparticle can absorb Cerenkov radiation, which occurs when positrons or electrons travel faster than the speed of light through biological tissues. Kamkaew et al. [151] showed that certain photosensitizers exhibit better photodynamic therapy efficacy but with limited tissue penetration. Parchur et al. [152] demonstrated the use of hepatic image-guided photothermal therapy, employing CT, MRI, and optical imaging.

15.12 Imaging Applications in Monitoring Responses to Immunotherapy

Current immune checkpoint inhibitor therapies and chimeric antigen receptor T-cell therapies have distinct management protocols for various malignancies [153]. An important concept in monitoring these therapies is pseudo-progression, which refers to the apparent tumor growth that occurs due to inflammation rather than cancer progression [154]. FDG-PET/CT is a valuable imaging modality for monitoring immune checkpoint inhibitors and chimeric antigen receptor T-cell therapies [155]. It helps differentiate between inflammatory cells and tumor cells, and can assess tumor glucose metabolism [156, 157]. PET can also identify inflammation caused by other pathologies, which is important in the context of immunotherapy [158]. Nanoparticle-based imaging plays a crucial role in targeting specific cells in the body [159].

T-cell imaging, in particular, is valuable for evaluating immunotherapies [160]. For example, dendrimer-based gold NPs have been used in fluorescence imaging to monitor immunotherapy responses [161, 162].

Macrophage-targeting NPs are also critical in evaluating macrophage-based theranostic protocols through PET or MRI imaging [163]. Magnetic NPs can be employed in MRI to track macrophage activity by targeting the CD44 receptor. Kamat et al. [164] demonstrated that PET imaging platforms can track tumor-associated macrophages, offering insights into tumor progression and immune responses. Additionally, copper-64 (^{64}Cu)-labeled polyglucose NPs have been utilized in clinical applications to identify macrophage infiltration in both normal and tumor tissues [165].

15.13 Immune-Checkpoint Receptor Imaging

Non-invasive imaging probes are essential for identifying responders and non-responders to treatments, such as those targeting the PD-L1 checkpoint. Gold nanoparticles (NPs) targeting PD-L1 have demonstrated high tumor uptake and can differentiate between tumors and inflammatory processes. Rare-earth NPs, known for their bright luminescent signals, are also valuable in such imaging applications [166, 167].

The use of various NP-based imaging agents enables penetration through different cellular barriers, despite concerns about toxicity and stability.

Different cell components like the fibroblasts, leukocytes, and endothelial cells, can define tumor growth and metastasis.

As a result, the role of nanoimaging in cancer has widened the scope of imaging and therapeutic options leading to the theranostics approach.

15.14 Cancer Therapies

Chemotherapeutic agents offer significant therapeutic benefits, but they also have limitations such as solubility issues, suboptimal pharmacokinetics, insufficient accumulation in target tissues, and lower efficiency with higher toxicity. To address these drawbacks, various formulations have been developed, including liposomes, polymer-drug conjugates, polymeric micelles, lipid NPs, dendrimers, and antibody-drug conjugates. These carriers, typically ranging from 1 to 100 nm in size, provide improved pharmacokinetics and biodistribution by enhancing solubility, preventing accumulation in healthy tissues, and ensuring better deposition at target sites, ultimately increasing the therapeutic index. Due to these factors, nanotherapeutics have been designed for targeted delivery to specific tissues.

Cancer nanomedicine's clinical applications are seen in various probe forms and protocols, enabling better study of tumor accumulation. The tumor vasculature and microenvironment are crucial for evaluating drug delivery systems, with the goal of improving the efficiency of radiotherapy, ultrasound-based therapies, hyperthermia, and phototherapy. The theranostic approach, combining diagnostic imaging and therapy, is key in optimizing cancer treatment strategies by enhancing drug delivery to tumors.

Hybrid imaging techniques, such as CT-FMT (computed tomography-fluorescence molecular tomography), provide non-invasive methods to visualize and assess various organs, including the heart, liver, and kidneys, helping to track the effectiveness of these treatments.

Optical imaging offers excellent multispectral imaging properties, making it ideal for targeting different nanocarriers and assessing tumor accumulation. The area under the curve (AUC) for tumor accumulation is typically higher, reflecting more efficient cellular internalization in various tumors.

15.15 Real-Time Surgery Monitoring

NPs can be utilized in multiple compartments, offering versatile applications in imaging and therapy. The ideal whole-body imaging contrast agent should provide high resolution, superior sensitivity, and the ability to function across multiple modalities. One such approach is multimodal imaging, which can incorporate ultrasmall superparamagnetic iron oxide nanoparticles (SPIONs) combined with fluorochromes. This results in dual-modality imaging agents that can be used for both MRI and fluorescence imaging [168]. MRI-fluorescent NPs are increasingly used in whole-body imaging protocols, including local intraoperative imaging C-dots, for instance, are small (<8 nm) silica NPs that incorporate near-infrared fluorochromes and are coated with a radioactive PET tracer molecule for enhanced imaging. These NPs can be excreted by the kidneys and are being investigated in clinical trials to improve the signal intensity of optical imaging [143]. Polymeric NPs, derived from hyaluronic acids, are another example. These NPs can entrap imaging agents such as ICG and fluorochromes, which improve tumor contrast [169]. Nanotechnology is also revolutionizing intraoperative imaging by using fluorescent dyes; however, limitations like autofluorescence and photobleaching remain challenges [170].

These agents use different autofluorescence or photobleaching, based on the different spectral fingerprints, to identify unequal evaluation in the range [171]. These features are being used for the analysis of SERRS NPs, which can be seen in various cancer imaging agents and can reveal the true microscopic spread beyond the locoregional micrometastases [172]. The core–shell architecture of gold–silica SE(R)RS NPs can be applied with various whole-body imaging modalities like MRI or PET, which offer better deep penetration into deeper tissues, providing three-dimensional imaging in the operating room through photoacoustic tomography [173]. Various nanoparticle designs can be achieved through processes such as nanofabrication and nanoparticle synthesis methodologies, aiming for better modular nanoparticle designs that can be adapted for both therapeutic and imaging theranostic nanoagents [174] in the future.

15.16 Passive Nanotargeting Delivery of Radionuclides for Tumor Nuclear Imaging

The important existing nanotargeted nuclear imaging modalities are gamma-imaging, SPECT, and PET. One important entity is liposomes, which are self-assembling colloidal particles forming a spherical bilayer within small phospholipid vesicles made from a dried lipid mixture [175]. Liposomes are now established as important nanocarriers in the delivery of radionuclides for imaging. Liposomes can be used along with the ^{99m}Tc, ^{111}In, and ^{67}Ga radionuclides for gamma-imaging and also for monitoring drug responses in various preclinical and clinical studies [176–179]. Liposome formulation and radionuclide encapsulation can be evaluated based on various biodistributions, pharmacokinetics, and nuclear imaging for advanced local cancer [180]. The conventional ^{111}In-based liposome (Vescan) used in preclinical and clinical studies can be radiolabeled to form radiolabeled liposomes for imaging cancer and inflammation [179]. Various novel amphiphilic probes can be used for tracking different liposomes and developing new probes [180]. Liposomes can be tagged with ^{18}F and ^{64}Cu for different diagnostic imaging applications and real-time liposomal tracking in vivo [181–183]. Whole-body scintigraphy can also be used for better tumor evaluation of the ^{111}In-liposomes 48 h post-injection [184]. Different metabolites, such as the ^{111}In-liposome, can serve as bifunctional imaging and bimodal therapeutic agents for evaluating the efficacy of radiochemotherapeutics in xenografts [185].

15.17 Passive Nanotargeting Delivery of Radionuclides for Tumor Radiotherapy

Different analytical dosimetry parameters are being used with radiotracers like the ^{131}I, ^{90}Y, ^{188}Re, and ^{67}Cu radionuclide-labeled liposomes, which are optimal liposome systems for radiotherapy and chemotherapy delivery [186]. More efficient solid tumors can be targeted with radiolabeled pegylated liposomes in the management of solid tumors, including head and neck cancers.

Microbubbles (MBs) can be used as probes for US imaging, as they have better echogenicity for ideal contrast-enhanced imaging and function as blood pool agents for evaluating blood perfusion in organs with good blood flow. These images are important for assessing different stages of angiogenesis or angiostenosis in diseased tissues or organs [187, 188]. In principle, MBs are one of the important US-sensitive delivery systems for drug release at various tumor sites. The sonoporation effect on the extracellular matrix with cell membrane penetration increases drug release at tumor sites. The biggest drawback is that they lack theranostic properties due to the leaky tumor neovasculature and their smaller size. Different nanoscale echo-contrast agents are being developed, including perfluorocarbons (PFC)-filled nanoemulsions or nanodroplets, which are in the 250–500 nm size range. These nanoscale echo-contrast agents are composed of phospholipids, protein components like albumin, and copolymers, which enhance emulsification effects. Rapoport et al. showed that different polymeric micelles with various PFCs can form nanodroplets, leading to better accumulation at tumor sites. This ultrasound-triggered drug release uses nanodroplets for both imaging and The use of polymersome-based theranostic ultrasonographic nanoprobes is capable of accumulating in tumors with larger nanodroplets (250–400 nm). Under specific pH conditions and

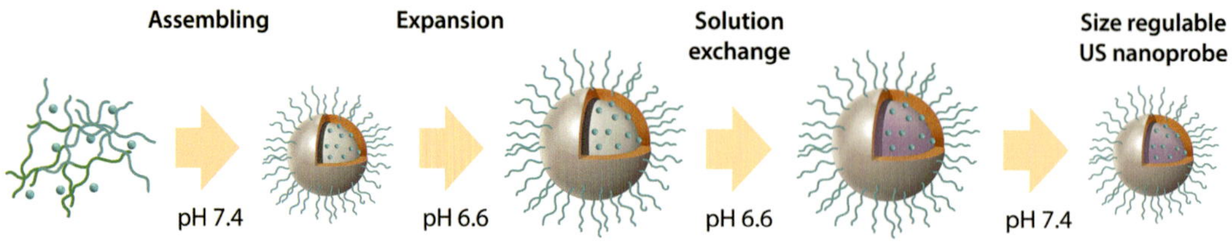

Fig. 15.5 Diagram depicting the role of pH in different settings for assembling, expansion, solution exchange and size based nanoprobe

with the use of either infrared light for surface tumors or ultrasound irradiation for deeper lesions, phase transition of encapsulated PFC occurs to form gas-filled bubbles with diameters larger than 500 nm, which are readily detectable by diagnostic ultrasound (Fig. 15.5). Furthermore, continual low-frequency ultrasound (LFUS) irradiation of the tumor site breaks the bubbles to trigger drug release. We believe such PFC-filled nanoprobes can be used effectively as on-demand ultrasound-based theranostic agents against cancer.

Cancer nanomedicine's clinical applications are seen in different forms of probes and protocols, studying tumor accumulation. Tumor vasculature and the tumor microenvironment are important for evaluating various drug delivery systems. Different parameters are available for assessing tumor vasculature and the tumor microenvironment, enabling more focused and improved drug delivery for radiotherapy, ultrasound-based therapies, hyperthermia, and phototherapy [172, 174, 175, 189]. The theranostic approach is applied in different drug delivery methods as anticancer nanotherapy for patients.

Optical imaging, especially hybrid CT-FMT (computed tomography-fluorescence molecular tomography), is valuable for its non-invasiveness and ability to target organs such as the heart, liver, and kidneys due to its diverse optical properties [175–182].

Optical imaging also offers excellent multispectral imaging properties, which are useful for targeting different nanocarriers, and tumor accumulation (AUC) has a higher range in cellular internalization in various tumors [183–187, 190–192].

15.18 Real-Time Surgery Monitoring

NPs are also capable of being used in cases of chronic renal failure, which is important for evaluating SPIONs in lymph node staging [193–200]. One nanoparticle type can be utilized in different nanoparticle compartments. The ideal whole-body imaging contrast agent offers high resolution, high sensitivity, and multimodality uses [143, 168–171, 200–210]. Multimodal imaging can use ultrasmall SPIONs with fluorochromes, leading to MRI-fluorescent dual-modality imaging agents [211]. MRI-fluorescent NPs are used in whole-body imaging protocols, such as local intraoperative imaging with C-dots. These C-dots are very small (<8 nm) silica NPs containing near-infrared fluorochromes, which can be coated with radioactive PET tracer molecules and are excreted by the kidneys [212, 213]. These PET tracers allow efficient preoperative mapping with different fluorochromes for dynamic evaluation. These NPs are in clinical trials and are also used [214] to improve the signal intensity of optical imaging. Polymeric NPs, derivatives of hyaluronic acid, are used for tumor contrast with entrapped ICG and fluorochromes [215]. Nanotechnology is also applied in intraoperative imaging using various fluorescent dyes, though they have certain limitations despite these properties [216].

These agents use different autofluorescence or photobleaching characteristics, based on various spectral fingerprints, to identify unequal evaluations in the range [217]. These features are applied in the analysis of SERRS NPs, which are used in cancer imaging agents capable of revealing the true microscopic spread beyond locoregional micrometastases [218]. The core–shell architecture of gold–silica SE(R)RS NPs is utilized by various whole-body imaging modalities, such as MRI or PET [219–225], which offer deeper tissue penetration and three-dimensional imaging in the operating room via photoacoustic tomography [226]. Various nanoparticle designs are achieved through nanofabrication and nanoparticle synthesis methodologies, aiming for modular nanoparticle designs suitable for both therapeutic and imaging theranostic nano agents in the future [227].

15.19 Passive Nanotargeting Delivery of Radionuclides for Tumor Nuclear Imaging

The important existing nanotargeted nuclear imaging modalities are gamma-imaging, SPECT, and PET. One important entity is liposomes, which are self-assembling colloidal particles forming a spherical bilayer within small phospholipid vesicles made from a dried lipid mixture [228]. Liposomes are now established as important nanocarriers for delivering radionuclides for imaging. Liposomes can be used along with the ^{99m}Tc, ^{111}In, and ^{67}Ga radionuclides for gamma-imaging and for monitoring drug responses in various preclinical and clinical studies [229–236]. Liposome formulation and radionuclide encapsulation can be evaluated based on various biodistributions, pharmacokinetics, and nuclear imaging for advanced local cancer [237]. The conventional ^{111}In-based liposome (Vescan) used in preclinical and clinical studies can be radiolabeled to form radiolabeled liposomes for imaging cancer and inflammation [238]. Various novel amphiphilic probes can be used to track different liposomes for developing additional probes [239]. Liposomes can be tagged with ^{18}F and ^{64}Cu for various diagnostic imaging applications and real-time liposomal tracking in vivo [240]. Whole-body scintigraphy can also be used for better tumor evaluation of ^{111}In-liposomes after 48 h post-injection [241]. Different metabolites, such as ^{111}In-liposomes, can be used as bifunctional imaging agents, with bimodal therapeutic efficacy in radiochemotherapy being identified in xenografts [242, 243].

15.20 Passive Targeting Delivery of Radionuclides for Tumor Radiotherapy

Different analytical dosimetry parameters are being used with radiotracers like ^{131}I, ^{90}Y, ^{188}Re, and ^{67}Cu radionuclide-labeled liposomes, which serve as optimal liposome systems for delivering radiotherapy and chemotherapy [244]. More efficient targeting of solid tumors is achieved with radiolabeled pegylated liposomes, which are used in managing solid tumors, including head and neck cancers.

15.20.1 Microbubbles (MBs)

Microbubbles can be used as probes for US imaging due to their better echogenicity, making them ideal for contrast-enhanced imaging. They function as blood pool agents for evaluating blood perfusion in organs with good blood flow. These images are important for evaluating different stages of angiogenesis or angiostenosis in diseased tissues or organs [245]. In principle, MBs are important US-sensitive delivery systems used for drug release at various tumor sites. Through the sonoporation effect, extracellular matrix penetration combined with cell membrane penetration leads to increased drug release at tumor sites [246]. The biggest drawback is that they lack theranostic properties due to leaky tumor neovasculature and their smaller sizes [247]. Different nanoscale echo-contrast agents are being developed, including perfluorocarbon (PFC)-filled nanoemulsions or nanodroplets in the 250–500 nm size range [248]. These nanoscale echo-contrast agents are composed of phospholipids, protein components like albumin, and copolymers, which provide better emulsification effects. Rapoport et al. demonstrated that different polymeric micelles with various PFCs can form nanodroplets, leading to better accumulation at the tumor site. This ultrasound-triggered drug release utilizes nanodroplets for both imaging and therapeutic drug delivery agents [249].

The polymersome-based theranostic ultrasonographic nanoprobe can accumulate in tumors, forming larger nanodroplets (250–400 nm). With infrared light for surface tumors or ultrasound irradiation for deeper lesions, the phase transition of encapsulated PFC occurs, forming gas-filled bubbles with diameters larger than 500 nm, which are readily detectable by diagnostic ultrasound. Furthermore, continual LFUS irradiation at the tumor site breaks the bubbles, triggering drug release. We believe such PFC-filled nanoprobes can be effectively used as on-demand ultrasound-based theranostic agents against cancer.

Nanotherapeutics are also important for creating compatible materials like lipids and various imaging agents, which can be designed and synthesized for favorable biodistribution. Various image-guided cancer interventions are being developed for better monitoring of NP-based imaging agents.

15.21 Conclusion

With various technological advancements, more effective diagnostic, therapeutic, and theranostic options are emerging. The use of nanocarriers is seen as an important aspect for targeting molecules such as antibodies, peptides, or ligands on the surface to create efficient nanocarriers. They can also be multifunctional, capable of carrying various radionuclides and chemotherapeutic agents to target tumor cells. Various factors, including nanoparticle composition and structure, specific targeting, and minimized off-target effects, contribute to improved drug delivery systems.

The use of various chemotherapeutic agents presents both advantages and drawbacks, such as solubility issues, suboptimal pharmacokinetics, low accumulation at the target site, and reduced efficiency with higher toxicity. Different formulations are being developed, such as liposomes, polymer-drug conjugates, polymeric micelles, polymeric and lipid NPs, dendrimers, and antibody-drug conjugates. These carriers, typically very small in size (1–100 nm), exhibit improved pharmacokinetics and biodistribution by increasing solubility, reducing accumulation in healthy tissues, and enhancing target tissue deposition, thereby increasing the therapeutic index. Due to these factors, various nanotherapeutics can be designed for specific tissues.

References

1. Adams JY, Johnson M, Sato M, Berger F, Gambhir SS, Carey M, et al. Visualization of advanced human prostate cancer lesions in living mice by a targeted gene transfer vector and optical imaging. Nat Med. 2002;8(8):891–7.
2. Massoud TF, Gambhir SS. Molecular imaging in living subjects: seeing fundamental biological processes in a new light. Genes Dev. 2003;17(5):545–80.
3. James ML, Gambhir SS. A molecular imaging primer: modalities, imaging agents, and applications. Physiol Rev. 2012;92(2):897–965.
4. Koo H, Huh MS, Sun IC, Yuk SH, Choi K, Kim K, et al. In vivo targeted delivery of nanoparticles for theranosis. Acc Chem Res. 2011;44(10):1018–28.
5. Wiwanitkit V. Glomerular pore size corresponding to albumin molecular size, an explanation for underlying structural pathology leading to albuminuria at nanolevel. Ren Fail. 2006;28(1):101.
6. Ullman EF, Schwarzberg M, Rubenstein KE. Fluorescent excitation transfer immunoassay. A general method for determination of antigens. J Biol Chem. 1976;251(14):4172–8.
7. Ferrari M. Cancer nanotechnology: opportunities and challenges. Nat Rev Cancer. 2005;5(3):161–71.
8. Piner RD, Zhu J, Xu F, Hong S, Mirkin CA. "Dip-Pen" nanolithography. Science. 1999;283(5402):661–3.
9. Canelas DA, Herlihy KP, DeSimone JM. Top-down particle fabrication: control of size and shape for diagnostic imaging and drug delivery. Wiley Interdiscip Rev Nanomed Nanobiotechnol. 2009;1(4):391–404.
10. Key J, Leary JF. Nanoparticles for multimodal in vivo imaging in nanomedicine. Int J Nanomedicine. 2014;9:711–26.
11. He X, Wang K, Cheng Z. In vivo near-infrared fluorescence imaging of cancer with nanoparticle-based probes. Wiley Interdiscip Rev Nanomed Nanobiotechnol. 2010;2(4):349–66.
12. He X, Gao J, Gambhir SS, Cheng Z. Near-infrared fluorescent nanoprobes for cancer molecular imaging: status and challenges. Trends Mol Med. 2010;16(12):574–83.

13. Alivisatos P. The use of nanocrystals in biological detection. Nat Biotechnol. 2004;22(1):47–52.
14. Michalet X, Pinaud FF, Bentolila LA, Tsay JM, Doose S, Li JJ, et al. Quantum dots for live cells, in vivo imaging, and diagnostics. Science. 2005;307(5709):538–44.
15. Medintz IL, Uyeda HT, Goldman ER, Mattoussi H. Quantum dot bioconjugates for imaging, labelling and sensing. Nat Mater. 2005;4(6):435–46.
16. Gao J, Chen K, Luong R, Bouley DM, Mao H, Qiao T, et al. A novel clinically translatable fluorescent nanoparticle for targeted molecular imaging of tumors in living subjects. Nano Lett. 2012;12(1):281–6.
17. Smith BR, Cheng Z, De A, Koh AL, Sinclair R, Gambhir SS. Real-time intravital imaging of RGD-quantum dot binding to luminal endothelium in mouse tumor neovasculature. Nano Lett. 2008;8(9):2599–606.
18. Cai W, Hsu AR, Li ZB, Chen X. Are quantum dots ready for in vivo imaging in human subjects? Nanoscale Res Lett. 2007;2(6):265–81.
19. Schipper ML, Cheng Z, Lee SW, Bentolila LA, Iyer G, Rao J, et al. microPET-based biodistribution of quantum dots in living mice. J Nucl Med. 2007;48(9):1511–8.
20. Schipper ML, Iyer G, Koh AL, Cheng Z, Ebenstein Y, Aharoni A, et al. Particle size, surface coating, and PEGylation influence the biodistribution of quantum dots in living mice. Small. 2009;5(1):126–34.
21. Gao J, Chen K, Xie R, Xie J, Lee S, Cheng Z, et al. Ultrasmall near-infrared non-cadmium quantum dots for in vivo tumor imaging. Small. 2010;6(2):256–61.
22. Gao J, Chen K, Miao Z, Ren G, Chen X, Gambhir SS, et al. Affibody-based nanoprobes for HER2-expressing cell and tumor imaging. Biomaterials. 2011;32(8):2141–8.
23. DeCoste SD, Farinelli W, Flotte T, Anderson RR. Dye enhanced laser welding for skin closure. Lasers Surg Med. 1992;12(1):25–32.
24. Gianella A, Jarzyna PA, Mani V, Ramachandran S, Calcagno C, Tang J, et al. Multifunctional nanoemulsion platform for imaging guided therapy evaluated in experimental cancer. ACS Nano. 2011;5(6):4422–33.
25. Santra S, Kaittanis C, Grimm J, Perez JM. Drug/dye-loaded, multifunctional iron oxide nanoparticles for combined targeted cancer therapy and dual optical/magnetic resonance imaging. Small. 2009;5(16):1862–8.
26. Yan X, Niu G, Lin J, Jin AJ, Hu H, Tang Y, et al. Enhanced fluorescence imaging guided photodynamic therapy of sinoporphyrin sodium loaded graphene oxide. Biomaterials. 2015;42:94–102.
27. Yan X, Hu H, Lin J, Jin AJ, Niu G, Zhang S, et al. Optical and photoacoustic dual-modality imaging guided synergistic photodynamic/photothermal therapies. Nanoscale. 2015;7(6):2520–6.
28. Luo S, Zhang E, Su Y, Cheng T, Shi C. A review of NIR dyes in cancer targeting and imaging. Biomaterials. 2011;32(29):7127–38.
29. Liu H, Zhang X, Xing B, Han P, Gambhir SS, Cheng Z. Radiation-luminescence-excited quantum dots for in vivo multiplexed optical imaging. Small. 2010;6(10):1087–91.
30. Yang Y, Shao Q, Deng R, Wang C, Teng X, Cheng K, et al. In vitro and in vivo uncaging and bioluminescence imaging by using photocaged upconversion nanoparticles. Angew Chem Int Ed Engl. 2012;51(13):3125–9.
31. Mitchell GS, Gill RK, Boucher DL, Li C, Cherry SR. In vivo Cerenkov luminescence imaging: a new tool for molecular imaging. Philos Trans A Math Phys Eng Sci. 1955;2011(369):4605–19.
32. Carpenter CM, Sun C, Pratx G, Rao R, Xing L. Hybrid x-ray/optical luminescence imaging: characterization of experimental conditions. Med Phys. 2010;37(8):4011–8.
33. Pratx G, Carpenter CM, Sun C, Rao RP, Xing L. Tomographic molecular imaging of x-ray-excitable nanoparticles. Opt Lett. 2010;35(20):3345–7.
34. Pratx G, Carpenter CM, Sun C, Xing L. X-ray luminescence computed tomography via selective excitation: a feasibility study. IEEE Trans Med Imaging. 2010;29(12):1992–9.
35. Sun C, Pratx G, Carpenter CM, Liu H, Cheng Z, Gambhir SS, et al. Synthesis and radioluminescence of PEGylated Eu(3+)-doped nanophosphors as bioimaging probes. Adv Mater. 2011;23(24):H195–9.
36. Keren S, Zavaleta C, Cheng Z, de la Zerda A, Gheysens O, Gambhir SS. Noninvasive molecular imaging of small living subjects using Raman spectroscopy. Proc Natl Acad Sci USA. 2008;105(15):5844–9.
37. Zavaleta CL, Hartman KB, Miao Z, James ML, Kempen P, Thakor AS, et al. Preclinical evaluation of Raman nanoparticle biodistribution for their potential use in clinical endoscopy imaging. Small. 2011;7(15):2232–40.
38. Jokerst JV, Miao Z, Zavaleta C, Cheng Z, Gambhir SS. Affibody-functionalized gold-silica nanoparticles for Raman molecular imaging of the epidermal growth factor receptor. Small. 2011;7(5):625–33.
39. Grinvald A, Lieke E, Frostig RD, Gilbert CD, Wiesel TN. Functional architecture of cortex revealed by optical imaging of intrinsic signals. Nature. 1986;324(6095):361–4.
40. He Y, Tang Z, Chen Z, Wan W, Li J. A novel photoacoustic tomography based on a time-resolved technique and an acoustic lens imaging system. Phys Med Biol. 2006;51(10):2671–80.
41. Wang LV, Hu S. Photoacoustic tomography: in vivo imaging from organelles to organs. Science. 2012;335(6075):1458–62.
42. Fan Q, Cheng K, Yang Z, Zhang R, Yang M, Hu X, et al. Perylene-diimide based nanoparticles as highly efficient photoacoustic agents for deep

brain tumor imaging in living mice. Adv Mater. 2015;27(5):843–7.
43. Bianco A, Kostarelos K, Prato M. Applications of carbon nanotubes in drug delivery. Curr Opin Chem Biol. 2005;9(6):674–9.
44. Liu Z, Cai W, He L, Nakayama N, Chen K, Sun X, et al. In vivo biodistribution and highly efficient tumour targeting of carbon nanotubes in mice. Nat Nanotechnol. 2007;2(1):47–52.
45. De la Zerda A, Zavaleta C, Keren S, Vaithilingam S, Bodapati S, Liu Z, et al. Carbon nanotubes as photoacoustic molecular imaging agents in living mice. Nat Nanotechnol. 2008;3(9):557–62.
46. De la Zerda A, Liu Z, Bodapati S, Teed R, Vaithilingam S, Khuri-Yakub BT, et al. Ultrahigh sensitivity carbon nanotube agents for photoacoustic molecular imaging in living mice. Nano Lett. 2010;10(6):2168–72.
47. Ren G, Miao Z, Liu H, Jiang L, Limpa-Amara N, Mahmood A, et al. Melanin-targeted preclinical PET imaging of melanoma metastasis. J Nucl Med. 2009;50(10):1692–9.
48. Cheng Z, Mahmood A, Li H, Davison A, Jones AG. [99mTcOAADT]-(CH2)2-NEt2: a potential small-molecule single-photon emission computed tomography probe for imaging metastatic melanoma. Cancer Res. 2005;65(12):4979–86.
49. Zhang R, Fan Q, Yang M, Cheng K, Lu X, Zhang L, et al. Engineering melanin nanoparticles as an efficient drug-delivery system for imaging-guided chemotherapy. Adv Mater. 2015;27(34):5063–9.
50. Yang M, Fan Q, Zhang R, Cheng K, Yan J, Pan D, et al. Dragon fruit-like biocage as an iron trapping nanoplatform for high efficiency targeted cancer multimodality imaging. Biomaterials. 2015;69:30–7.
51. Bloch SH, Dayton PA, Ferrara KW. Targeted imaging using ultrasound contrast agents. Progress and opportunities for clinical and research applications. IEEE Eng Med Biol Mag. 2004;23(5):18–29.
52. Liu R, Tian B, Gearing M, Hunter S, Ye K, Mao Z. Cdk5-mediated regulation of the PIKE-A-Akt pathway and glioblastoma cell invasion. Proc Natl Acad Sci USA. 2008;105(21):7570–5.
53. Zhou J, Patel TR, Sirianni RW, Strohbehn G, Zheng MQ, Duong N, et al. Highly penetrative, drug-loaded nanocarriers improve treatment of glioblastoma. Proc Natl Acad Sci USA. 2013;110(29):11751–6.
54. Kiessling F, Huppert J, Zhang C, Jayapaul J, Zwick S, Woenne EC, et al. RGD-labeled USPIO inhibits adhesion and endocytotic activity of αvβ3-integrin-expressing glioma cells and only accumulates in the vascular tumor compartment. Radiology. 2009;253(2):462–9.
55. Anderson CR, Hu X, Zhang H, Tlaxca J, Declèves AE, Houghtaling R, et al. Ultrasound molecular imaging of tumor angiogenesis with an integrin targeted microbubble contrast agent. Investig Radiol. 2011;46(4):215–24.
56. Yan F, Xu X, Chen Y, Deng Z, Liu H, Xu J, et al. A lipopeptide-based αvβ3 integrin-targeted ultrasound contrast agent for molecular imaging of tumor angiogenesis. Ultrasound Med Biol. 2015;41(10):2765–73.
57. Willmann JK, Kimura RH, Deshpande N, Lutz AM, Cochran JR, Gambhir SS. Targeted contrast-enhanced ultrasound imaging of tumor angiogenesis with contrast microbubbles conjugated to integrin-binding knottin peptides. J Nucl Med. 2010;51(3):433–40.
58. Willmann JK, Lutz AM, Paulmurugan R, Patel MR, Chu P, Rosenberg J, et al. Dual-targeted contrast agent for US assessment of tumor angiogenesis in vivo. Radiology. 2008;248(3):936–44.
59. Chavanpatil MD, Khdair A, Panyam J. Nanoparticles for cellular drug delivery: mechanisms and factors influencing delivery. J Nanosci Nanotechnol. 2006;6(9–10):2651–63.
60. Shamsi K, Balzer T, Saini S, Ros PR, Nelson RC, Carter EC, et al. Superparamagnetic iron oxide particles (SH U 555 A): evaluation of efficacy in three doses for hepatic MR imaging. Radiology. 1998;206(2):365–71.
61. Reimer P, Jähnke N, Fiebich M, Schima W, Deckers F, Marx C, et al. Hepatic lesion detection and characterization: value of nonenhanced MR imaging, superparamagnetic iron oxide-enhanced MR imaging, and spiral CT-ROC analysis. Radiology. 2000;217(1):152–8.
62. Bu L, Xie J, Chen K, Huang J, Aguilar ZP, Wang A, et al. Assessment and comparison of magnetic nanoparticles as MRI contrast agents in a rodent model of human hepatocellular carcinoma. Contrast Media Mol Imaging. 2012;7(4):363–72.
63. de Marco G, Bogdanov A, Marecos E, Moore A, Simonova M, Weissleder R. MR imaging of gene delivery to the central nervous system with an artificial vector. Radiology. 1998;208(1):65–71.
64. Gupta AK, Curtis AS. Surface modified superparamagnetic nanoparticles for drug delivery: interaction studies with human fibroblasts in culture. J Mater Sci Mater Med. 2004;15(4):493–6.
65. Akbarzadeh A, Samiei M, Davaran S. Magnetic nanoparticles: preparation, physical properties, and applications in biomedicine. Nanoscale Res Lett. 2012;7(1):144.
66. Thorek DL, Chen AK, Czupryna J, Tsourkas A. Superparamagnetic iron oxide nanoparticle probes for molecular imaging. Ann Biomed Eng. 2006;34(1):23–38.
67. Cheng K, Yang M, Zhang R, Qin C, Su X, Cheng Z. Hybrid nanotrimers for dual T1 and T2-weighted magnetic resonance imaging. ACS Nano. 2014;8(10):9884–96.
68. Yoffe S, Leshuk T, Everett P, Gu F. Superparamagnetic iron oxide nanoparticles (SPIONs): synthesis and surface modification techniques for use with MRI

and other biomedical applications. Curr Pharm Des. 2013;19(3):493–509.
69. Gupta AK, Gupta M. Synthesis and surface engineering of iron oxide nanoparticles for biomedical applications. Biomaterials. 2005;26(18):3995–4021.
70. Yu MK, Jeong YY, Park J, Park S, Kim JW, Min JJ, et al. Drug-loaded superparamagnetic iron oxide nanoparticles for combined cancer imaging and therapy in vivo. Angew Chem Int Ed Engl. 2008;47(29):5362–5.
71. Phelps ME, Hoffman EJ, Huang SC, Ter-Pogossian MM. Effect of positron range on spatial resolution. J Nucl Med. 1975;16(7):649–52.
72. Shokeen M, Anderson CJ. Molecular imaging of cancer with copper-64 radiopharmaceuticals and positron emission tomography (PET). Acc Chem Res. 2009;42(7):832–41.
73. Cutler CS, Hennkens HM, Sisay N, Huclier-Markai S, Jurisson SS. Radiometals for combined imaging and therapy. Chem Rev. 2013;113(2):858–83.
74. Wadas TJ, Wong EH, Weisman GR, Anderson CJ. Coordinating radiometals of copper, gallium, indium, yttrium, and zirconium for PET and SPECT imaging of disease. Chem Rev. 2010;110(5):2858–902.
75. Cai W, Chen K, Li ZB, Gambhir SS, Chen X. Dual-function probe for PET and near-infrared fluorescence imaging of tumor vasculature. J Nucl Med. 2007;48(11):1862–70.
76. Kirchner C, Liedl T, Kudera S, Pellegrino T, Muñoz Javier A, Gaub HE, et al. Cytotoxicity of colloidal CdSe and CdSe/ZnS nanoparticles. Nano Lett. 2005;5(2):331–8.
77. Cai W, Shin DW, Chen K, Gheysens O, Cao Q, Wang SX, et al. Peptide-labeled near-infrared quantum dots for imaging tumor vasculature in living subjects. Nano Lett. 2006;6(4):669–76.
78. Chen K, Li ZB, Wang H, Cai W, Chen X. Dual-modality optical and positron emission tomography imaging of vascular endothelial growth factor receptor on tumor vasculature using quantum dots. Eur J Nucl Med Mol Imaging. 2008;35(12):2235–44.
79. Nahrendorf M, Zhang H, Hembrador S, Panizzi P, Sosnovik DE, Aikawa E, et al. Nanoparticle PET-CT imaging of macrophages in inflammatory atherosclerosis. Circulation. 2008;117(3):379–87.
80. Nahrendorf M, Keliher E, Marinelli B, Waterman P, Feruglio PF, Fexon L, et al. Hybrid PET-optical imaging using targeted probes. Proc Natl Acad Sci USA. 2010;107(17):7910–5.
81. Yang X, Hong H, Grailer JJ, Rowland IJ, Javadi A, Hurley SA, et al. cRGD-functionalized, DOX-conjugated, and 64Cu-labeled superparamagnetic iron oxide nanoparticles for targeted anticancer drug delivery and PET/MR imaging. Biomaterials. 2011;32(17):4151–60.
82. Liu TW, Macdonald TD, Jin CS, Gold JM, Bristow RG, Wilson BC, et al. Inherently multimodal nanoparticle-driven tracking and real-time delineation of orthotopic prostate tumors and micrometastases. ACS Nano. 2013;7(5):4221–32.
83. Chen F, Hong H, Shi S, Goel S, Valdovinos HF, Hernandez R, et al. Engineering of hollow mesoporous silica nanoparticles for remarkably enhanced tumor active targeting efficacy. Sci Rep. 2014;4:5080.
84. Verel I, Visser GW, Boellaard R, Stigter-van Walsum M, Snow GB, van Dongen GA. 89Zr immuno-PET: comprehensive procedures for the production of 89Zr-labeled monoclonal antibodies. J Nucl Med. 2003;44(8):1271–81.
85. Holland JP, Sheh Y, Lewis JS. Standardized methods for the production of high specific-activity zirconium-89. Nucl Med Biol. 2009;36(7):729–39.
86. Holland JP, Williamson MJ, Lewis JS. Unconventional nuclides for radiopharmaceuticals. Mol Imaging. 2010;9(1):1–20.
87. Holland JP, Caldas-Lopes E, Divilov V, Longo VA, Taldone T, Zatorska D, et al. Measuring the pharmacodynamic effects of a novel Hsp90 inhibitor on HER2/neu expression in mice using 89Zr-DFO-trastuzumab. PLoS One. 2010;5(1):e8859.
88. Holland JP, Divilov V, Bander NH, Smith-Jones PM, Larson SM, Lewis JS. 89ZrDFO-J591 for immunoPET of prostate-specific membrane antigen expression in vivo. J Nucl Med. 2010;51(8):1293–300.
89. Ruggiero A, Villa CH, Holland JP, Sprinkle SR, May C, Lewis JS, et al. Imaging and treating tumor vasculature with targeted radiolabeled carbon nanotubes. Int J Nanomedicine. 2010;5:783–802.
90. Lee J, Lee TS, Ryu J, Hong S, Kang M, Im K, et al. RGD peptide-conjugated multimodal NaGdF4:Yb3+/Er3+ nanophosphors for upconversion luminescence, MR, and PET imaging of tumor angiogenesis. J Nucl Med. 2013;54(1):96–103.
91. Lijowski M, Caruthers S, Hu G, Zhang H, Scott MJ, Williams T, et al. High sensitivity: high-resolution SPECT-CT/MR molecular imaging of angiogenesis in the Vx2 model. Investig Radiol. 2009;44(1):15–22.
92. Hu G, Lijowski M, Zhang H, Partlow KC, Caruthers SD, Kiefer G, et al. Imaging of Vx-2 rabbit tumors with $\alpha\nu\beta3$-integrin-targeted 111In nanoparticles. Int J Cancer. 2007;120(9):1951–7.
93. Yang M, Cheng K, Qi S, Liu H, Jiang Y, Jiang H, et al. Affibody modified and radiolabeled gold-iron oxide hetero-nanostructures for tumor PET, optical and MR imaging. Biomaterials. 2013;34(11):2796–806.
94. Prabhakar U, Maeda H, Jain RK, et al. Challenges and key considerations of the enhanced permeability and retention effect for nanomedicine drug delivery in oncology. Cancer Res. 2013;73(8):2412–7.
95. Thakor AS, Jokerst JV, Ghanouni P, Campbell JL, Mittra E, Gambhir SS. Clinically approved nanoparticle imaging agents. J Nucl Med. 2016;57(12):1833–7.
96. Widding A, Stilbo I, Hansen SW, Hansen HH, Rossing N. Scintigraphy with nanocolloid Tc 99m in patients with small cell lung cancer, with special ref-

erence to bone marrow and hepatic metastasis. Eur J Nucl Med. 1990;16(8–10):717–9.
97. Haddock G, Gray HW, McKillop JH, Bessent RG, Kirk D. 99mTc-nanocolloid bone marrow scintigraphy in prostatic cancer. Br J Urol. 1989;63(5):497–502.
98. Coda C, Cartia GL, Ciambellotti E. Scintigraphy of bone marrow for the evaluation of injuries caused by antiblastic radiotherapy [in Italian]. Minerva Med. 1987;78(21):1611–5.
99. Ciambellotti E, Cartia GL, Coda C. Scintigraphy of the bone marrow for the evaluation of injuries caused by antiblastic agents [in Italian]. Radiol Med (Torino). 1988;75(1–2):78–82.
100. Koops HS, Doting MH, de Vries J, et al. Sentinel node biopsy as a surgical staging method for solid cancers. Radiother Oncol. 1999;51(1):1–7.
101. Kapteijn BA, Nieweg OE, Valdés Olmos RA, et al. Reproducibility of lymphoscintigraphy for lymphatic mapping in cutaneous melanoma. J Nucl Med. 1996;37(6):972–5.
102. Kapteijn BA, Nieweg OE, Liem I, et al. Localizing the sentinel node in cutaneous melanoma: gamma probe detection versus blue dye. Ann Surg Oncol. 1997;4(2):156–60.
103. Borgstein PJ, Pijpers R, Comans EF, van Diest PJ, Boom RP, Meijer S. Sentinel lymph node biopsy in breast cancer: guidelines and pitfalls of lymphoscintigraphy and gamma probe detection. J Am Coll Surg. 1998;186(3):275–83.
104. Weissleder R, Reimer P, Lee AS, Wittenberg J, Brady TJ. MR receptor imaging: ultrasmall iron oxide particles targeted to asialoglycoprotein receptors. AJR Am J Roentgenol. 1990;155(6):1161–7.
105. Reimer P, Weissleder R, Lee AS, Wittenberg J, Brady TJ. Receptor imaging: application to MR imaging of liver cancer. Radiology. 1990;177(3):729–34.
106. Crist RM, Dasa SSK, Liu CH, Clogston JD, Dobrovolskaia MA, Stern ST. Challenges in the development of nanoparticle-based imaging agents: characterization and biology. Wiley Interdiscip Rev Nanomed Nanobiotechnol. 2021;13(1):e1665.
107. Ma YY, Jin KT, Wang SB, et al. Molecular imaging of cancer with nanoparticle-based theranostic probes. Contrast Media Mol Imaging. 2017;2017:1026270.
108. Bermo MS, Koppula B, Kumar M, Leblond A, Matesan MC. The peritoneum: what nuclear radiologists need to know. Semin Nucl Med. 2020;50(5):405–18.
109. Cooney TM, Cohen KJ, Guimaraes CV, et al. Response assessment in diffuse intrinsic pontine glioma: recommendations from the Response Assessment in Pediatric Neuro-Oncology (RAPNO) working group. Lancet Oncol. 2020;21(6):e330–6.
110. Erker C, Tamrazi B, Poussaint TY, et al. Response assessment in paediatric high-grade glioma: recommendations from the Response Assessment in Pediatric Neuro-Oncology (RAPNO) working group. Lancet Oncol. 2020;21(6):e317–29.
111. Oberheim Bush NA, Hervey-Jumper SL, Berger MS. Management of glioblastoma, present and future. World Neurosurg. 2019;131:328–38.
112. Panda A, Goenka AH, Hope TA, Veit-Haibach P. PET/magnetic resonance imaging applications in abdomen and pelvis. Magn Reson Imaging Clin N Am. 2020;28(3):369–80.
113. Patel CM, Sahdev A, Reznek RHCT. CT, MRI and PET imaging in peritoneal malignancy. Cancer Imaging. 2011;11(1):123–39.
114. Hu J, Feng W, Hua J, et al. A high spatial resolution in vivo 1 H magnetic resonance spectroscopic imaging technique for the human breast at 3 T. Med Phys. 2009;36(11):4870–7.
115. Arrivé L, Hodoul M, Arbache A, Slavikova-Boucher L, Menu Y, El Mouhadi S. Magnetic resonance cholangiography: current and future perspectives. Clin Res Hepatol Gastroenterol. 2015;39(6):659–64.
116. Halfdanarson TR, Strosberg JR, Tang L, et al. The North American Neuroendocrine Tumor Society consensus guidelines for surveillance and medical management of pancreatic neuroendocrine tumors. Pancreas. 2020;49(7):863–81.
117. Kichler A, Jang S. Chronic pancreatitis: epidemiology, diagnosis, and management updates. Drugs. 2020;80(12):1155–68.
118. Elmohr MM, Elsayes KM, Pickhardt PJ. Non-neoplastic conditions mimicking peritoneal carcinomatosis at CT imaging. Br J Radiol. 2020;93(1113):20200401.
119. Alcantara D, Leal MP, García-Bocanegra I, García-Martín ML. Molecular imaging of breast cancer: present and future directions. Front Chem. 2014;2:112.
120. Villanueva-Meyer JE, Mabray MC, Cha S. Current clinical brain tumor imaging. Neurosurgery. 2017;81(3):397–415.
121. Akhoundova D, Hiltbrunner S, Mader C, et al. 18F-FET PET for diagnosis of pseudoprogression of brain metastases in patients with non–small cell lung cancer. Clin Nucl Med. 2020;45(2):113–7.
122. van Manen L, Handgraaf HJM, Diana M, et al. A practical guide for the use of indocyanine green and methylene blue in fluorescence-guided abdominal surgery. J Surg Oncol. 2018;118(2):283–300.
123. Haba S, Yamao K, Bhatia V, et al. Diagnostic ability and factors affecting accuracy of endoscopic ultrasound-guided fine needle aspiration for pancreatic solid lesions: Japanese large single center experience. J Gastroenterol. 2013;48(8):973–81.
124. Naumova AV, Modo M, Moore A, Murry CE, Frank JA. Clinical imaging in regenerative medicine. Nat Biotechnol. 2014;32(8):804–18.
125. Stacy MR, Sinusas AJ. Emerging imaging modalities in regenerative medicine. Curr Pathobiol Rep. 2015;3(1):27–36. [Published correction appears in Curr Pathobiol Rep 2018;6(3):199].
126. Scaranti M, Cojocaru E, Banerjee S, Banerji U. Exploiting the folate receptor a in oncology. Nat Rev Clin Oncol. 2020;17(6):349–59.

127. Meier R, Henning TD, Boddington S, et al. Breast cancers: MR imaging of folate-receptor expression with the folate-specific nanoparticle P1133. Radiology. 2010;255(2):527–35.
128. Chen F, Ma K, Madajewski B, et al. Ultrasmall targeted nanoparticles with engineered antibody fragments for imaging detection of HER2-overexpressing breast cancer. Nat Commun. 2018;9(1):4141.
129. Oseledchyk A, Andreou C, Wall MA, Kircher MF. Folate-targeted surface-enhanced resonance Raman scattering nanoprobe ratiometry for detection of microscopic ovarian cancer. ACS Nano. 2017;11(2):1488–97.
130. Glover B, Teare J, Patel N. The status of advanced imaging techniques for optical biopsy of colonic polyps. Clin Transl Gastroenterol. 2020;11(3):e00130.
131. Zavaleta CL, Garai E, Liu JT, et al. A Raman-based endoscopic strategy for multiplexed molecular imaging. Proc Natl Acad Sci USA. 2013;110(25):E2288–97.
132. Park JW, Neve RM, Szollosi J, Benz CC. Unraveling the biologic and clinical complexities of HER2. Clin Breast Cancer. 2008;8(5):392–401.
133. Bensch F, Brouwers AH, Lub-de Hooge MN, et al. 89Zr-trastuzumab PET supports clinical decision making in breast cancer patients, when HER2 status cannot be determined by standard work up. Eur J Nucl Med Mol Imaging. 2018;45(13):2300–6.
134. Tsvirkun D, Ben-Nun Y, Merquiol E, et al. CT imaging of enzymatic activity in cancer using covalent probes reveal a size-dependent pattern. J Am Chem Soc. 2018;140(38):12010–20.
135. Motiei M, Dreifuss T, Betzer O, et al. Differentiating between cancer and inflammation: a metabolic-based method for functional computed tomography imaging. ACS Nano. 2016;10(3):3469–77.
136. Yuan Y, Zhang J, Qi X, et al. Furin-mediated intracellular self-assembly of olsalazine nanoparticles for enhanced magnetic resonance imaging and tumour therapy. Nat Mater. 2019;18(12):1376–83.
137. Mi P, Kokuryo D, Cabral H, et al. A pH-activatable nanoparticle with signal-amplification capabilities for non-invasive imaging of tumour malignancy. Nat Nanotechnol. 2016;11(8):724–30.
138. Manny TB, Krane LS, Hemal AK. Indocyanine green cannot predict malignancy in partial nephrectomy: histopathologic correlation with fluorescence pattern in 100 patients. J Endourol. 2013;27(7):918–21.
139. Egloff-Juras C, Bezdetnaya L, Dolivet G, Lassalle HP. NIR fluorescence-guided tumor surgery: new strategies for the use of indocyanine green. Int J Nanomedicine. 2019;14:7823–38.
140. Wojtynek NE, Mohs AM. Image-guided tumor surgery: the emerging role of nanotechnology. Wiley Interdiscip Rev Nanomed Nanobiotechnol. 2020;12(4):e1624.
141. Thawani JP, Amirshaghaghi A, Yan L, Stein JM, Liu J, Tsourkas A. Photoacoustic-guided surgery with indocyanine green-coated superparamagnetic iron oxide nanoparticle clusters. Small. 2017;13(37):1701300.
142. Wojtynek NE, Olson MT, Bielecki TA, et al. Nanoparticle formulation of indocyanine green improves image-guided surgery in a murine model of breast cancer. Mol Imaging Biol. 2020;22(4):891–903.
143. Bradbury MS, Pauliah M, Zanzonico P, Wiesner U, Patel S. Intraoperative mapping of sentinel lymph node metastases using a clinically translated ultrasmall silica nanoparticle. Wiley Interdiscip Rev Nanomed Nanobiotechnol. 2016;8(4):535–53.
144. Grégoire V, Guckenberger M, Haustermans K, et al. Image guidance in radiation therapy for better cure of cancer. Mol Oncol. 2020;14(7):1470–91.
145. Giaj-Levra N, Borghetti P, Bruni A, et al. Current radiotherapy techniques in NSCLC: challenges and potential solutions. Expert Rev Anticancer Ther. 2020;20(5):387–402.
146. Kurz C, Buizza G, Landry G, et al. Medical physics challenges in clinical MR-guided radiotherapy. Radiat Oncol. 2020;15(1):93.
147. Giraud N, Popinat G, Regaieg H, Tonnelet D, Vera P. Positron-emission tomography-guided radiation therapy: ongoing projects and future hopes. Cancer Radiother. 2020;24(5):437–43.
148. Detappe A, Thomas E, Tibbitt MW, et al. Ultrasmall silica-based bismuth gadolinium nanoparticles for dual magnetic resonance-computed tomography image guided radiation therapy. Nano Lett. 2017;17(3):1733–40.
149. Li X, Lovell JF, Yoon J, Chen X. Clinical development and potential of photothermal and photodynamic therapies for cancer. Nat Rev Clin Oncol. 2020;17(11):657–74.
150. Kotagiri N, Sudlow GP, Akers WJ, Achilefu S. Breaking the depth dependency of phototherapy with Cerenkov radiation and low-radiance-responsive nanophoto-sensitizers. Nat Nanotechnol. 2015;10(4):370–9.
151. Kamkaew A, Cheng L, Goel S, et al. Cerenkov radiation induced photodynamic therapy using chlorin e6-loaded hollow mesoporous silica nanoparticles. ACS Appl Mater Interfaces. 2016;8(40):26630–7.
152. Parchur AK, Sharma G, Jagtap JM, et al. Vascular interventional radiology-guided photothermal therapy of colorectal cancer liver metastasis with theranostic gold nanorods. ACS Nano. 2018;12(7):6597–611.
153. Postow MA, Sidlow R, Hellmann MD. Immune-related adverse events associated with immune checkpoint blockade. N Engl J Med. 2018;378(2):158–68.
154. Liu CH, Abrams ND, Carrick DM, et al. Imaging inflammation and its resolution in health and disease: current status, clinical needs, challenges, and opportunities. FASEB J. 2019;33(12):13085–97.
155. Mirzaei HR, Mirzaei H, Namdar A, Rahmati M, Till BG, Hadjati J. Predictive and therapeutic biomarkers in chimeric antigen receptor T-cell therapy: a clinical perspective. J Cell Physiol. 2019;234(5):5827–41.

156. Evangelista L, Sepulcri M, Pasello G. PET/CT and the response to immunotherapy in lung cancer. Curr Radiopharm. 2020;13(3):177–84.
157. Lang D, Wahl G, Poier N, et al. Impact of PET/CT for assessing response to immunotherapy-a clinical perspective. J Clin Med. 2020;9(11):E3483.
158. Nimmagadda S. Quantifying PD-L1 expression to monitor immune checkpoint therapy: opportunities and challenges. Cancers (Basel). 2020;12(11):E3173.
159. Ahrens ET, Bulte JW. Tracking immune cells in vivo using magnetic resonance imaging. Nat Rev Immunol. 2013;13(10):755–63.
160. Alam IS, Mayer AT, Sagiv-Barfi I, et al. Imaging activated T cells predicts response to cancer vaccines. J Clin Invest. 2018;128(6):2569–80.
161. Chen M, Betzer O, Fan Y, et al. Multifunctional dendrimer-entrapped gold nanoparticles for labeling and tracking T cells via dual-modal computed tomography and fluorescence imaging. Biomacromolecules. 2020;21(4):1587–95.
162. Chen W, Xie Y, Wang M, Li C. Recent advances on rare earth upconversion nanomaterials for combined tumor near-infrared photoimmunotherapy. Front Chem. 2020;8:596658.
163. Mukherjee S, Sonanini D, Maurer A, Daldrup-Link HE. The yin and yang of imaging tumor associated macrophages with PET and MRI. Theranostics. 2019;9(25):7730–48.
164. Kamat M, El-Boubbou K, Zhu DC, et al. Hyaluronic acid immobilized magnetic nanoparticles for active targeting and imaging of macrophages. Bioconjug Chem. 2010;21(11):2128–35.
165. Kim HY, Li R, Ng TSC, et al. Quantitative imaging of tumor-associated macrophages and their response to therapy using 64Cu-labeled macrin. ACS Nano. 2018;12(12):12015–29.
166. Meir R, Shamalov K, Sadan T, et al. Fast image-guided stratification using anti-programmed death ligand 1 gold nanoparticles for cancer immunotherapy. ACS Nano. 2017;11(11):11127–34.
167. Zhong Y, Ma Z, Wang F, et al. In vivo molecular imaging for immunotherapy using ultra-bright near-infrared-IIb rare-earth nanoparticles. Nat Biotechnol. 2019;37(11):1322–31.
168. Kircher MF, Mahmood U, King RS, Weissleder R, Josephson L. A multimodal nanoparticle for preoperative magnetic resonance imaging and intraoperative optical brain tumor delineation. Cancer Res. 2003;63:8122–5.
169. Hill TK, Kelkar SS, Wojtynek NE, Souchek JJ, Payne WM, Stumpf K, Marini FC, Mohs AM. Near infrared fluorescent nanoparticles derived from hyaluronic acid improve tumor contrast for image-guided surgery. Theranostics. 2016;6:2314–28.
170. Keereweer S, Van Driel PB, Snoeks TJ, Kerrebijn JD, Baatenburg de Jong RJ, Vahrmeijer AL, Sterenborg HJ, Löwik CW. Optical image-guided cancer surgery: challenges and limitations. Clin Cancer Res. 2013;19:3745–54.
171. Harmsen S, Bedics MA, Wall MA, Huang R, Detty MR, Kircher MF. Rational design of a chalcogenopyrylium-based surface-enhanced resonance Raman scattering nanoprobe with attomolar sensitivity. Nat Commun. 2015;6:6570.
172. Ojha T, Pathak V, Shi Y, Hennink WE, Moonen CTW, Storm G, et al. Pharmacological and physical vessel modulation strategies to improve EPR-mediated drug targeting to tumors. Adv Drug Deliv Rev. 2017;119:44–60.
173. Dasgupta A, Liu M, Ojha T, Storm G, Kiessling F, Lammers T. Ultrasound-mediated drug delivery to the brain: principles, progress and prospects. Drug Discov Today Technol. 2016;20:41–8.
174. Doleschel D, Rix A, Arns S, Palmowski K, Gremse F, Merkle R, et al. Erythropoietin improves the accumulation and therapeutic effects of carboplatin by enhancing tumor vascularization and perfusion. Theranostics. 2015;5(8):905–18.
175. Chen H, Zhang W, Zhu G, Xie J, Chen X. Rethinking cancer nanotheranostics. Nat Rev Mater. 2017;2:17024.
176. Ojha T, Rizzo L, Storm G, Kiessling FLT. Image-guided drug delivery: preclinical applications and clinical translation. Expert Opin Drug Deliv. 2015;12(8):1203–7.
177. Theek B, Gremse F, Kunjachan S, Fokong S, Pola R, Pechar M, et al. Characterizing EPR-mediated passive drug targeting using contrast-enhanced functional ultrasound imaging. J Control Release. 2014;182:83–9.
178. Kiessling F, Gaetjens J, Palmowski M. Application of molecular ultrasound for imaging integrin expression. Theranostics. 2012;1:127–34.
179. Karageorgis A, Dufort S, Sancey L, Henry M, Hirsjarvi S, Passirani C, et al. An MRI-based classification scheme to predict passive access of 5 to 50-nm large nanoparticles to tumors. Sci Rep. 2016;6:1–10.
180. Lee H-Y, Li Z, Chen K, Hsu AR, Xu C, Xie J, et al. PET/MRI dual-modality tumor imaging using arginine-glycine-aspartic (RGD)-conjugated radiolabeled iron oxide nanoparticles. J Nucl Med. 2008;49(8):1371–9.
181. Pérez-Medina C, Abdel-Atti D, Tang J, Zhao Y, Fayad ZA, Lewis JS, et al. Nanoreporter PET predicts the efficacy of anti-cancer nanotherapy. Nat Commun. 2016;7:11838.
182. Mertens ME, Frese J, Bölükbas DA, Hrdlicka L, Golombek S, Koch S, et al. FMN-coated fluorescent USPIO for cell labeling and non-invasive MR imaging in tissue engineering. Theranostics. 2014;4(10):1002–13.
183. Niu G, Chen X. When radionuclides meet nanoparticles. Nat Nanotechnol. 2018;13(5):359–60.
184. Lee DE, Koo H, Sun IC, Ryu JH, Kim K, Kwon IC. Multifunctional nanoparticles for multimodal imaging and theragnosis. Chem Soc Rev. 2012;41(7):2656–72.

185. Xie J, Lee S, Chen X. Nanoparticle-based theranostic agents. Adv Drug Deliv Rev. 2010;62(11):1064–79.
186. Yang Z, Song J, Tang W, Fan W, Dai Y, Shen Z, et al. Stimuli-responsive nanotheranostics for real-time monitoring drug release by photoacoustic imaging. Theranostics. 2019;9(2):526–36.
187. Lammers T, Kiessling F, Hennink WE, Storm G. Nanotheranostics and image-guided drug delivery: current concepts and future directions. Mol Pharm. 2010;7(6):1899–912.
188. Afsharzadeh M, Hashemi M, Mokhtarzadeh A, Abnous K, Ramezani M. Recent advances in co-delivery systems based on polymeric nanoparticle for cancer treatment. Artif Cells Nanomed Biotechnol. 2018;46:1095–110.
189. Dasgupta A, Liu M, Ojha T, Storm G, Kiessling F, Lammers T. Ultrasound-mediated drug delivery to the brain: principles, progress and prospects. Drug Discov Today Technol. 2016;20(xx):41–8:41.
190. Lee H, Shields AF, Siegel BA, Miller KD, Krop I, Ma CX, et al. 64Cu-MM-302 positron emission tomography quantifies variability of enhanced permeability and retention of nanoparticles in relation to treatment response in patients with metastatic breast cancer. Clin Cancer Res. 2017;23(15):4190–202.
191. Lammers T, Aime S, Hennink WE, Storm G, Kiessling F. Theranostic nanomedicine. Acc Chem Res. 2011;44(10):1029–38.
192. Kunjachan S, Gremse F, Theek B, Koczera P, Pola R, Pechar M, et al. Noninvasive optical imaging of nanomedicine biodistribution. ACS Nano. 2013;7(1):252–62.
193. Lammers T, Kühnlein R, Kissel M, Subr V, Etrych T, Pola R, et al. Effect of physicochemical modification on the biodistribution and tumor accumulation of HPMA copolymers. J Control Release. 2005;110(1):103–18.
194. Banerjee SR, Foss CA, Horhota A, Pullambhatla M, McDonnell K, Zale S, et al. 111In- and IRDye800CW-labeled PLA–PEG nanoparticle for imaging prostate specific membrane antigen-expressing tissues. Biomacromolecules. 2017;18(1):201–9.
195. Gremse F, Theek B, Kunjachan S, Lederle W, Pardo A, Barth S, et al. Absorption reconstruction improves biodistribution assessment of fluorescent nanoprobes using hybrid fluorescence-mediated t tomography. Theranostics. 2014;4(10):960–71.
196. Gremse F, Stärk M, Ehling J, Menzel JR, Lammers T, Kiessling F. Imalytics preclinical: interactive analysis of biomedical volume data. Theranostics. 2016;6(3):328–41.
197. Al Rawashdeh W, Zuo S, Melle A, Appold L, Koletnik S, Tsvetkova Y, et al. Noninvasive assessment of elimination and retention using CT-FMT and kinetic whole-body modeling. Theranostics. 2017;7(6):1499–510.
198. Gerlowski LE, Jain RK. Microvascular permeability of normal and neoplastic tissues. Microvasc Res. 1986;31(3):288–305.
199. Kunjachan S, Pola R, Gremse F, Theek B, Ehling J, Moeckel D, et al. Passive versus active tumor targeting using RGD- and NGR-modified polymeric nanomedicines. Nano Lett. 2014;14(2):972–81.
200. Terreno E, Uggeri F, Aime S. Image guided therapy: the advent of theranostic agents. J Control Release. 2012;161(2):328–37.
201. Huang Q, Wang Y, Chen X, Wang Y, Li Z, Du S, Wang L, Chen S. Nanotheranostics. 2018;2:21–41.
202. Harisinghani MG, Barentsz J, Hahn PF, Deserno WM, Tabatabaei S, van de Kaa CH, de la Rosette J, Weissleder R. Noninvasive detection of clinically occult lymph-node metastases in prostate cancer. N Engl J Med. 2003;348:2491–9.
203. Harmsen S, Huang R, Wall MA, Karabeber H, Samii JM, Spaliviero M, White JR, Monette S, O'Connor R, Pitter KL, Sastra SA, Saborowski M, Holland EC, Singer S, Olive KP, Lowe SW, Blasberg RG, Kircher MF. Surface-enhanced resonance Raman scattering nanostars for high-precision cancer imaging. Sci Transl Med. 2015;7:271ra7.
204. Kircher MF, de la Zerda A, Jokerst JV, Zavaleta CL, Kempen PJ, Mittra E, Pitter K, Huang R, Campos C, Habte F, Sinclair R, Brennan CW, Mellinghoff IK, Holland EC, Gambhir SS. A brain tumor molecular imaging strategy using a new triple-modality MRI-photoacoustic-Raman nanoparticle. Nat Med. 2012;18:829–34.
205. Wall MA, Shaffer TM, Harmsen S, Tschaharganeh DF, Huang CH, Lowe SW, Drain CM, Kircher MF. Chelator-free radiolabeling of SERRS nanoparticles for whole-body PET and intraoperative Raman imaging. Theranostics. 2017;7:3068–77.
206. Li E, Yang Y, Hao G, Yi X, Zhang S, Pan Y, Xing B, Gao M. Multifunctional magnetic mesoporous silica nanoagents for in vivo enzyme-responsive drug delivery and MR imaging. Nanotheranostics. 2018;2:233–42.
207. Chen K, Li Z-B, Wang H, Cai W, Chen X. Dual-modality optical and positron emission tomography imaging of vascular endothelial growth factor receptor on tumor vasculature using quantum dots. Eur J Nucl Med Mol Imaging. 2008;35(12):2235–44.
208. Saad M, Garbuzenko OB, Ber E, et al. Receptor targeted polymers, dendrimeromes: which nanocarrier is the most efficient for tumor-specific treatment and imaging? J Control Release. 2008;130(2):107–14.
209. Liu X, Wang Y, Nakamura K, et al. Auger radiation-induced, antisense-mediated cytotoxicity of tumor cells using a 3-component streptavidin-delivery nanoparticle with ^{111}In. J Nucl Med. 2009;50(4):582–90.
210. Urakami T, Akai S, Katayama Y, Harada N, Tsukada H, Oku N. Novel amphiphilic probes for [18F]-radiolabeling preformed liposomes and determination of liposomal trafficking by positron emission tomography. J Med Chem. 2007;50(26):6454–7.
211. Oku N. Delivery of contrast agents for positron emission tomography imaging by liposomes. Adv Drug Deliv Rev. 1999;37(1–3):53–61.

212. Marik J, Tartis MS, Zhang H, et al. Long-circulating liposomes radiolabeled with [18F]fluorodipalmitin ([18F]FDP). Nucl Med Biol. 2007;34(2):165–71.
213. Seo JW, Zhang H, Kukis DL, Meares CF, Ferrara KW. A novel method to label preformed liposomes with 64Cu for positron emission tomography (PET) imaging. Bioconjug Chem. 2008;19(12):2577–84.
214. Wang H-E, Yu H-M, Lu Y-C, et al. Internal radiotherapy and dosimetric study for 111In/177Lu-pegylated liposomes conjugates in tumor-bearing mice. Nucl Inst Methods Phys Res A. 2006;569(2):533–7.
215. Lee W-C, Hwang J-J, Tseng Y-L, et al. Therapeutic efficacy evaluation of 111In-VNB-liposome on human colorectal adenocarcinoma HT-29/luc mouse xenografts. Nucl Inst Methods Phys Res A. 2006;569(2):497–504.
216. Chang Y-J, Chang C-H, Chang T-J, et al. Biodistribution, pharmacokinetics and micro-SPECT/CT imaging of 188ReBMEDA-liposome in a C26 murine colon carcinoma solid tumor animal model. Anticancer Res. 2007;27(4B):2217–25.
217. Liu P, Chen G, Zhang J. A review of liposomes as a drug delivery system: current status of approved products, regulatory environments, and future perspectives. Molecules. 2022;27:1372.
218. Schneider M. SonoVue, a new ultrasound contrast agent. Eur J Radiol. 1999;9:S347–8.
219. Gorce J-M, Arditi M, Schneider M. Influence of bubble size distribution on the echogenicity of ultrasound contrast agents: a study of SonoVue™. Investig Radiol. 2000;35:661–71.
220. Chung DJ, Cho SH, Lee JM, Hahn ST. Effect of microbubble contrast agent during high intensity focused ultrasound ablation on rabbit liver in vivo. Eur J Radiol. 2012;81:e519–23.
221. Ogihara-Umeda I, Sasaki T, Kojima S, Nishigori H. Optimal radiolabeled liposomes for tumor imaging. J Nucl Med. 1996;37(2):326–32.
222. Harrington KJ, Lewanski CR, Stewart JSW. Liposomes as vehicles for targeted therapy of cancer. Part 2: clinical development. Clin Oncol. 2000;12(1):16–24.
223. Kleiter MM, Yu D, Mohammadian LA, et al. A tracer dose of technetium-99m-labeled liposomes can estimate the effect of hyperthermia on intratumoral doxil extravasation. Clin Cancer Res. 2006;12(22):6800–7.
224. Jensen GM, Bunch TH. Conventional liposome performance and evaluation: lessons from the development of Vescan. J Liposome Res. 2007;17(3–4):121–37.
225. Goins A. Radiolabeled lipid nanoparticles for diagnostic imaging. Expert Opin Med Diagn. 2008;2(7):853–73.
226. Huynh E, Leung BY, Helfield BL, Shakiba M, Gandier J-A, Jin CS, Master ER, Wilson BC, Goertz DE, Zheng G. In situ conversion of porphyrin microbubbles to nanoparticles for multimodality imaging. Nat Nanotechnol. 2015;10:325–32.
227. Lindner JR. Microbubbles in medical imaging: current applications and future directions. Nat Rev Drug Discov. 2001;3:527–33.
228. Unger EC, Hersh E, Vannan M, Matsunaga TO, McCreery T. Local drug and gene delivery through microbubbles. Prog Cardiovasc Dis. 2001;44:45–54.
229. Lee H, Lee E, Kim DK, Jang NK, Jeong YY, Jon S. Antibiofouling polymer-coated superparamagnetic iron oxide nanoparticles as potential magnetic resonance contrast agents for in vivo cancer imaging. J Am Chem Soc. 2006;128:7383–9.
230. Yin TH, Wang P, Li JG, Zheng RQ, Zheng B, Cheng D, Li R, Lai J, Shuai XT. Ultrasound-sensitive siRNA-loaded nanobubbles formed by hetero-assembly of polymeric micelles and liposomes and their therapeutic effect in gliomas. Biomaterials. 2013;34:4532–43.
231. Vivès E, Schmidt J, Pèlegrin A. Cell-penetrating and cell-targeting peptides in drug delivery. Biochim Biophys Acta. 2008;1786:126–38.
232. Gao ZG, Fain HD, Rapoport N. Controlled and targeted tumor chemotherapy by micellar-encapsulated drug and ultrasound. J Control Release. 2005;102:203–22.
233. Mahmood U, Tung C-H, Bogdanov A Jr, Weissleder R. Near-infrared optical imaging of protease activity for tumor detection. Radiology. 1999;213:866–70.
234. Oeffinger BE, Wheatley MA. Development and characterization of a nano-scale contrast agent. Ultrasonics. 2004;42:343–7.
235. Zhou Y, Wang Z, Chen Y, Shen H, Luo Z, Li A, Wang Q, Ran H, Li P, Song W. Microbubbles from gas-generating perfluorohexane nanoemulsions for targeted temperature-sensitive ultrasonography and synergistic HIFU ablation of tumors. Adv Mater. 2013;25:4123–30.
236. Maeda H, Wu J, Sawa T, Matsumura Y, Hori K. Tumor vascular permeability and the EPR effect in macromolecular therapeutics: a review. J Control Release. 2000;65:271–84.
237. Li JG, Zhang L, Lin YJ, Xiao H, Zuo MX, Cheng D, Shuai XT. A pH-sensitive prodrug micelle self-assembled from multi-doxorubicin-tailed polyethylene glycol for cancer therapy. RSC Adv. 2016;6:9160–3.
238. Matsunaga TO, Sheeran PS, Luois S, Streeter JE, Mullin LB, Banerjee B, Dayton PA. Phase-change nanoparticles using highly volatile perfluorocarbons: toward a platform for extravascular ultrasound imaging. Theranostics. 2012;2:1185–98.
239. Sheeran PS, Wong VP, Luois S, McFarland RJ, Ross WD, Feingold S, Matsunaga TO, Dayton PA. Decafluorobutane as a phase-change contrast agent for low-energy extravascular ultrasonic imaging. Ultrasound Med Biol. 2011;37:1518–30.
240. Chen CC, Sheeran PS, Wu SY, Olumolade OO, Dayton PA, Konofagou EE. Targeted drug delivery with focused ultrasound-induced blood-brain barrier opening using acoustically-activated nanodroplets. J Control Release. 2013;172:795–804.

241. Teng Z, Wang R, Zhou Y, Kolios M, Wang Y, Zhang N, Wang Z, Zheng Y, Lu G. A magnetic droplet vaporization approach using perfluorohexane-encapsulated magnetic mesoporous particles for ultrasound imaging and tumor ablation. Biomaterials. 2017;134:43–50.
242. Zhou Y, Wang R, Teng Z, Wang Z, Hu B, Kolios M, Chen H, Zhang N, Wang Y, Li P, Wu X, Lu GM, Chen Y, Zheng YY. Magnetic nanoparticle-promoted droplet vaporization for in vivo stimuli-responsive cancer theranostics. NPG Asia Mater. 2016;8:e313–20.
243. Gupta R, Shea J, Scafe C, Shurlygina A, Rapoport N. Polymeric micelles and nanoemulsions as drug carriers: therapeutic efficacy, toxicity, and drug resistance. J Control Release. 2015;212:70–7.
244. Xu JS, Chen Y, Deng LM, Liu JX, Cao Y, Li P, Ran HT, Zheng YY, Wang ZG. Microwave-activated nanodroplet vaporization for highly efficient tumor ablation with real-time monitoring performance. Biomaterials. 2016;106:264–75.
245. Zhao Y, Song W, Wang D, Ran H, Wang R, Yao Y, Wang Z, Zheng Y, Li P. Phase-shifted PFH@PLGA/Fe3O4 nanocapsules for MRI/US imaging and photothermal therapy with near-infrared irradiation. ACS Appl Mater Int. 2015;7:14231–42.
246. Jian J, Liu CB, Gong YP, Su L, Zhang B, Wang ZG, Wang D, Zhou Y, Xu FF, Li P, Zheng YY, Song L, Zhou XY. India ink incorporated multifunctional phase-transition nanodroplets for photoacoustic/ultrasound dual-modality imaging and photoacoustic effect based tumor therapy. Theranostics. 2014;4:1026–38.
247. Rapoport N, Nam KH, Gupta R, Gao Z, Mohan P, Payne A, Todd N, Liu X, Kim T, Shea J, Scaife C, Parker DL, Jeong EK, Kennedy AM. Ultrasound-mediated tumor imaging and nanotherapy using drug loaded, block copolymer stabilized perfluorocarbon nanoemulsions. J Control Release. 2011;153:4–15.
248. Rapoport N, Gupta R, Kim YS, O'Neill BE. Polymeric micelles and nanoemulsions as tumor-targeted drug carriers: insight through intravital imaging. J Control Release. 2015;206:153–60.
249. Rapoport N. Drug-loaded perfluorocarbon nanodroplets for ultrasound-mediated drug delivery. Adv Exp Med Biol. 2016;880:221–41.

Nanoimaging in Neurology 16

16.1 Gold Nanocages and Photoacoustic Imaging

Gold nanocages are one of the most important and will be the most used particles in the coming years in comparison with previous applications. The physical properties of these nanoparticles will be highly porous, and the hollowness of the cube-shaped structures and their measurements will be compatible, having sizes in the range of 30–200 nm [1]. This is based on the surface plasmon resonance peaks, which can be modified by the extent of the maximal optical absorption in the near-infrared spectrum. Biological tissues will have different spectra and also different attenuations for the different tissues, and their amount is very low [2]. The various particles are the various gold nanoparticles that can be used for optical imaging and can be identified as potential contrast agents for optical and photoacoustic imaging [3–7]. In this photoacoustic imaging, the use of gold nanocages can be used for studying the peaks of the biochemical components in a particular region of interest [8]. As we know, photoacoustic tomography is also an important hybrid imaging technique where the sonographic and optical imaging modalities detect the photons, which are absorbed sonographically, as energy is absorbed due to the photoacoustic effect, where the acoustic signals are generated by the electromagnetic waves [7]. The principle of optical and radio-frequency waves is capable of penetrating deeper tissues with increased depth of penetration [7]. The spatial resolution has more depth in the range of 50 times more than compared to optical imaging alone [2]. The contrast agents like gold nanocages are being tuned to the near-infrared region, which will be closer and can be diagnosed by photoacoustic imaging [5, 9]. The more effective systemic delivery of nanoparticles is based on passive targeting, in which there will be more permeability and retention effect [10]. Due to the derangements of the vascular and lymphatic channels, there will be better targeting by the passive process [10]. The nanoparticles can enter within the cell by attaching surface antibodies like albumin, transferrin, folate, lipoproteins, mannose, etc., causing the endocytosis mechanism [11]. As the size of the gold nanospheres is very small, they are capable of entering the smaller vascular channels and being used as photoacoustic contrast agents [5] (Fig. 16.1). The biodistributions of the nanospheres range from 1.5 to 71 h, which is very comfortable, with very little toxicity, and are also capable of 3D imaging. The biggest drawback is the inability to define the clear tumor margins despite the disruption of the blood-brain barrier in the postoperative scenario. Trimodal imaging is another entity that will be useful to a larger extent, which can be the combination of MR imaging, Raman, and photoacoustic imaging, with an excellent depiction of even the tumor parenchyma margins in

S. Shaikh, *Nanoimaging - Future of Precision Medicine*,
https://doi.org/10.1007/978-981-97-7292-6_16

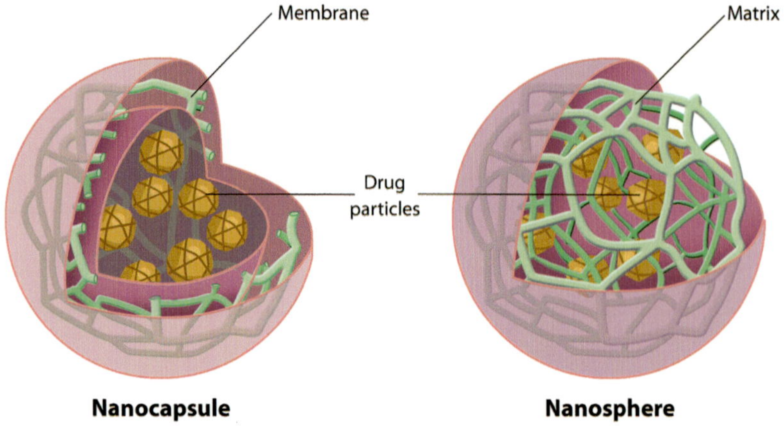

Fig. 16.1 Nano capsule with membrane and nanosphere with a matrix showing the drug particles within

both the microscopic levels in the pre- and post-operative scenarios [6]. The combination of all these three modal nanoparticles is made up of a gold core, silica coating, and maleimide–gadolinium-tetra-azacyclododecanetetra-acetic acid, which is linked into the layer of silica. These trimodal nanoparticles will accumulate in the tumor by the principle of increased permeability with the retention effect, and this will lead to a longer half-life. These have an advantage as they have preferential accumulation in the tumor tissue only.

16.2 Evaluation and Validation of Antiangiogenic Therapy

The role of nanotechnology is evolving to a greater extent due to its property as an angiogenesis inhibitor, which plays an important role in the prevention of the spread of recurrent glioblastoma [12, 13]. Angiogenesis imaging has various applications, such as monitoring the treatment response. MR imaging is currently the best imaging modality for the evaluation, but it has the peculiar property of the "pseudo-response" phenomenon. In this phenomenon, the permeability into the blood vessels is reduced, and the decreased enhancement pattern on different T1-weighted images gives the pseudo-response pattern [14, 15] (Fig. 16.2). Superparamagnetic iron oxide (SPIO) nanoparticles are used in the iron replacement therapy regimens in chronic kidney disease patients [16]. Ferumoxytol is an important "blood pool" agent, which does not extravasate and stays in the vascular compartment. Research has shown that ferumoxytol is better than gadolinium in differentiating between pseudoprogression and true tumor progression in patients with glioblastoma multiforme, which can be confirmed by the cerebral blood volume due to susceptibility-weighted contrast-enhanced MR imaging. Ferumoxytol is more sensitive for detecting tumor progression and can distinguish between pseudoprogression and gadolinium [17]. The various endothelial molecules are categorized for over-expression and vascular proliferation in tumor growth, and these are used as different neovascular proliferation agents that can show the pattern of angiogenesis [18–21].

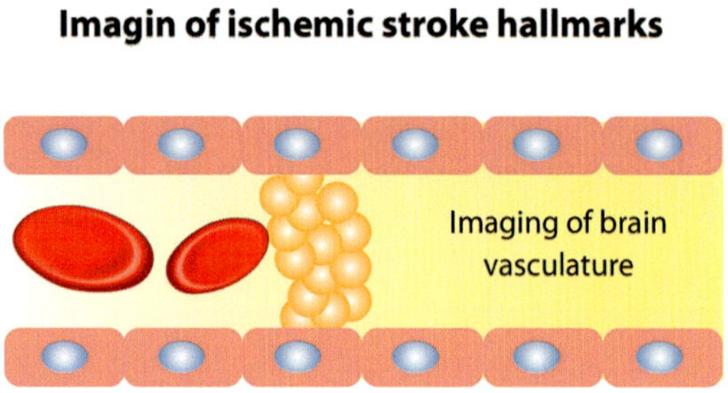

Fig. 16.2 Brain vasculature in ischemic stroke in nanoimaging

Optical imaging is another imaging modality in the near-infrared window (650–900 nm), which has various target agents used for integrin receptors, to better identify tumor margin interfaces [22]. As there are different cycloserine agents, they will play a better role in imaging low-grade gliomas [23]. One of the studies shows the use of the functionalized cyclic fluorophore (IRDye800CW), which is added to determine the amount of uptake in the evaluation of breast tissue, comprising normal healthy tissue, adjacent tumor margins, and also the nodal status [24].

16.3 Detection of Nitric Oxide

The use of nitric oxide as a form of the molecule is derived from L-arginine and nitric oxide (NO) synthase. This NO has various physiological functions that are related to neurotransmission and neovascularization, which can lead to neurotoxicity along with tissue injury, tissue apoptosis, and ischemia [25, 26]. One of the agents is nitrotyrosine, which is used in the evaluation of Alzheimer's disease and Parkinson's disease [27]. The use of NO is very important for research, and this will be able to use other methods to detect nitric oxide, which has various spectroscopic and fluorescent components of the dyes [28, 29]. One of the disadvantages is the amount of NO in 3D, which can't be evaluated separately [30]. Positron-emission tomography imaging is used for the quantification of endotoxin-induced nitric oxide synthase activation in normal healthy volunteers.

16.4 Stem Cell Tracking with MR Imaging

We all know that regenerative therapy is the newest form of management for various pathologies, with therapeutic applications used in various neurodegenerative [31, 32], ischemic [33], and traumatic [34] brain pathologies, which are due to various systemic pathologies like diabetes [35] (Fig. 16.3). The common locations involved are the sub-granular zone of the hippocampal dentate gyrus, subventricular, and periventricular regions [36]. Various newer techniques are used for neuronal regeneration, employing various scaffolds [37], different transcription factors [38], and stem cell implantation [39]. Stem cell imaging can also be done using different stem cell-based therapeu-

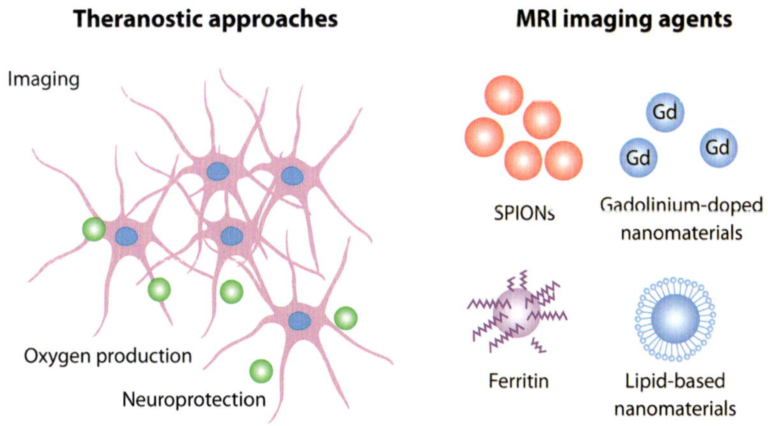

Fig. 16.3 The different theranostic approaches showing the imaging, production, and protection MRI showing the SPIONs, gadolinium particles, ferritin and lipid based nanoparticles

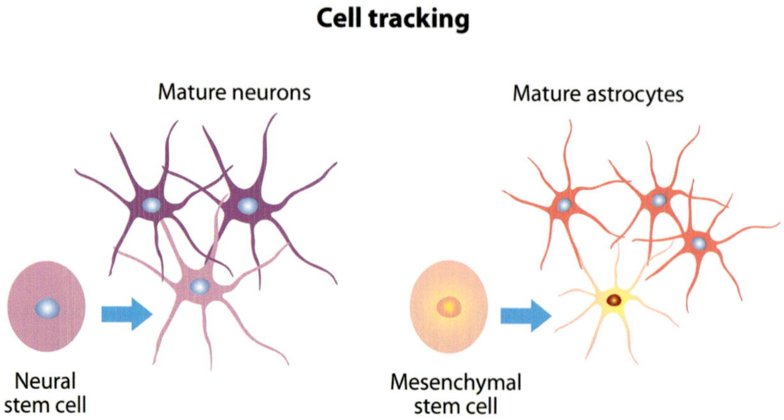

Fig. 16.4 Cell tracking by neural stem cells to mature neurons and mesenchymal stem cells to mature astrocytes

tic options for MR imaging, with better sensitivities through the use of superparamagnetic iron oxide nanoparticles [40] (Fig. 16.4). This SPIO nanoparticle biomechanism is based on normal iron metabolism, which is biocompatible with the human body [41]. Various studies have already shown better identification, labeling, and performance. Here, all neurology-related procedures can be done by first labeling, followed by tracking various pathological cells in conditions like orbit melanoma, neural stem cells in head injury, and marrow stem cells in chronic spinal cord injury [40]. SPIO nanoparticles can also be used for long-term in vivo cell tracking [42]. SPIO-loaded neural stem cells can also be monitored using MR imaging, immunohistochemistry, and bioluminescence as a single nanotechnology. The prolongation of the MR signal was associated with cell death and phagocytic contrast retention. The concept of SPIO materials having shorter half-life for in vivo SPIO-labeled cell imaging is known [43]. The non-use of different parameters for evaluation can lead to misinterpretation of the performance.

16.5 Noninvasive Detection of Neural Progenitor Cells in Living Brains by MR Imaging

Pericytes are one of the most important cells for the manufacturing of various multipotent progenitor cells in adults [44], which are found in the central nervous system at the junction of the inner and outer vascular basement membranes of various capillaries. These cells are regulated by various components, such as neovascularization and the blood-brain barrier [45–47]. The various neural progenitor cells can demonstrate various vascular pathologies in animal imaging using targeted MR imaging contrast agents [48]. The double gene transcript–targeted MR imaging can be used for the identification of various pericytes. The angiogenesis factor is used in various conditions, such as Alzheimer's disease [49] and diabetes mellitus [50], and for the evaluation of pericyte activity in the brain (Fig. 16.5).

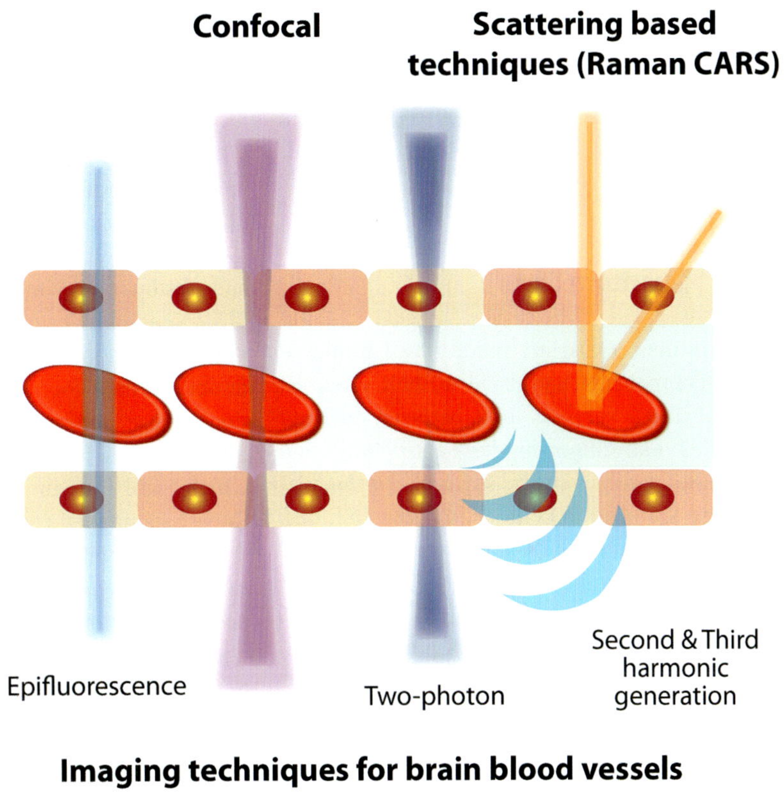

Fig. 16.5 Confocal Raman scattering showing the different imaging techniques in brain blood labelled cells with epifluorescence, two photon and second and third harmonics

16.6 Enhanced Brain Delivery of Blood-Brain Barrier Impermeable Probes for Optical and MR Imaging

Polymeric nanoparticles contain both nanospheres and nanocapsules, which are spherical in shape and made up of biodegradable structures with a central cavity [51] and a matrix [52], respectively. These nanoparticles are made up of surfactants that can bypass the binding and adhesion receptor-based transcytosis seen on the BBB endothelial cells [53, 54]. This allows the potential for in vivo molecular imaging [55] and shows that the delivery of various compounds will not cross the targeted probe for imaging lesions, such as in Alzheimer's disease.

16.7 Nanobodies

Nanobodies are a type of antibody first discovered in the serum of the camel species *Camelus dromedarius*. These are made up of very heavy chains, comprising a very small size, with low immunogenicity and more binding capabilities [56]. Nanobodies are primarily used for cancer therapy [57, 58], antivenoms [59], inflammatory conditions [60], and immunoimaging [60–64]. Nanobodies are capable of identifying various amyloid vascular and parenchymal beta deposits, which help differentiate Alzheimer's disease and cerebral amyloid angiopathy [65]. Nanobodies can distinguish between vascular or parenchymal amyloid deposits, along with the ability to identify their permeability across the blood-brain barrier. There is an active transport mechanism that carries the nanobodies through the blood-brain barrier [66], but there are some limitations to crossing the blood-brain barrier [65]. Different nanobodies are also being designed for BBB permeability.

16.8 Quantum Dots

Quantum dots (QDs) are made up of semiconductor components and have a shell made of polymer. These serve as links between different molecules like oligopeptides, antibodies, and many other particles on the surface. For this reason, they are used in cellular imaging [67, 68], detection of amino acids and proteins [69], and immunological studies in cancer [70]. The molar extinction of QDs is in the range of 10–50 times greater than organic dyes, with contrast that is 20 times brighter [71]. The photobleaching capability is very low, making them useful for biological imaging [72]. QDs are fused to different epidermal growth factors and monoclonal antibodies that target the epidermal growth factor receptor, to identify and differentiate between the normal brain parenchyma and glioma [73, 74]. One advantage is that QD–targeted fluorescent probes are used during surgery to differentiate the tumor from normal brain tissue, especially in low-grade tumors, which are difficult to distinguish. This provides an easy tool to identify and plan for the radical extension of tumor excision.

16.9 Role of Nanoparticles in CNS

Nanoparticles have the biggest advantage for the evaluation of various neurons and glia and are capable of identifying the size, number, development, and various responses, which can be evaluated by fluorescence microscopy using QDss. Some nanoparticles can function in a dual way, having double responses in different cell metabolisms [75, 76]. Cell signaling and cell patterns can be easily identified. Synaptic activity and neurotransmitter activity can be used to identify different neurotransmitter receptors.

16.9.1 Iron Oxide Nanoparticles for MR Imaging of Brain Tumors

Ultrasmall paramagnetic iron oxide-labeled nanoparticles are documented as the first nanoagents to be used in human research. MR imaging agents are now well-established for preoperative and postoperative imaging of human neoplasms [73]. Nanoparticles used in MR imaging can provide better contrast and enhancement in brain tumors, even a couple of hours after administration. These contrast agents can accumulate in different cells like macrophages and astrocytes, which will be in and around the tumors. Superparamagnetic nanoparticles are used for the analysis of targeted tumor therapy in brain tumors. Nanoparticles also function like various neuroprotection devices, developing newer agents that cause fewer side effects or less injury. Some conditions, such as primary or secondary injuries to the cells, can lead to the accumulation of various products, resulting in the derangement of transporter proteins, changes in intracellular calcium concentrations, and apoptosis [77]. One important factor is that different antioxidants can destroy free radicals [78]. Nanoparticles can be useful as therapeutic agents that prevent CNS injury and provide neuroprotection.

Stem cell migration has great potential for the regeneration of damaged brain tissue due to various reasons such as trauma, degeneration, or cerebral infarction caused by different enzymatic changes [79, 80]. Stem cells are clinically used in different forms of enzymatic changes for various pediatric leukodystrophies, one of which is Krabbe disease [81]. In current scenarios, stem cells are observed in CNS tissue. Iron oxide nanoparticles are tracked by MR imaging [82]. Iron oxide-labeled stem cells can be implanted in the CNS injury site under MRI [83, 84]. MR imaging plays a significant role in tracking labeled stem cell migration in the CNS, promoting tissue regeneration and increased reparative mechanisms for repairing CNS cells [85]. Cell signaling is also important as it facilitates cell growth, differentiation, and communication [85]. Neural progenitor cells can also differentiate into various cell types to repair degenerating CNS tissues.

Another important aspect is the use of nanoparticles as sensors in the CNS, which can track different physiological processes in the local environment [86]. Nanoswitches, an established iron oxide particle, are capable of monitoring physiological and abnormal changes in the local nanoenvironment [87]. These agents are composed of various molecules such as glucose, enzymes, and other protein particles. Based on their assembly and disassembly states, they have different magnetic properties, which can be identified by different T2 relaxation times, allowing for the identification of various agents.

The assessment of tumors is also important, with angiogenesis mechanisms being a key criterion [88]. MRI evaluation typically uses gadolinium particles, which have very low signal-intensity contrast, and this is a limitation of MRI. Neovascularization can be identified as nanoparticles express on the luminal surface of angioblasts, acting as a driving vehicle for therapy delivery [89, 90]. Another important capability is the use of photosensitizing molecules, which can be activated by light, making it easier to identify microvascular injury and cell death through phototherapy.

Nanomaterials in the evaluation of brain tissue can lead to less perfusion or more perfusion, and if not intervened in timely, can lead to ischemia and the next infarction. The identification is important as reperfusion can be done immediately with timely diagnosis, preventing permanent damage to the brain parenchyma. Different nanoparticles like various catalase and superoxide dismutase can also be used to prevent different inflammatory changes as well as neural degeneration [91, 92]. Some of the nanoparticles will have different endogenous antioxidant systems [93]. Oxygen-reactive polymer (ORP) is being used in the evaluation of brain injury in cases of brain trauma [94]. The neuronal injury can be prevented by the mechanism of apoptosis and also by the mechanism of the regeneration of the cells. The anti-apoptotic and growth peptides are some of those that are used for protection in various stroke models [95]. Basic fibroblast growth factor (bFGF) can reduce peptide inhibition by reducing the total infarct volume [95]. The nanoparticle is often loaded with different antibodies which will be against different transferrin receptor-1 located on the BBB endothelium for enabling receptor-mediated transcytosis and also facilitating different therapeutics seen to protect neurological injury [96, 97]. Excitotoxicity and neuroinflammation are important mediators to prevent different post-HCA neurological injuries and will act as the preventing severe side effects [97–101]. The basic principle in Alzheimer's disease is that it is a very crippling neurodegenerative disease, and the hallmark is the toxic amyloid beta protein (Aβ) aggregates. Gold nanoparticles have the potential for prevention along with the reduction of the severity of Alzheimer's disease [102]. Gold nanoparticles are being linked to some amino acid chains by synthesizing them so that the selective Aβ aggregates will be able to be identified. The different nanomaterial-based approaches for neural regeneration are being added to various mechanisms like mechanical, catalytic, conductive, optical, magnetic, and cytocompatible properties, having different engineering aspects in various tissues [103, 104]. Along with trauma and various neurodegenerative disorders, like Alzheimer's and Parkinson's disease, research has focused on neural regeneration. The biggest drawback is that the CNS cannot regenerate completely [105–108]. The glial cells and the extracellular components are responsible for the injury. These will be using myelin-associated inhibitors, like myelin-associated glycoprotein (MAG), and pathway activation along with the upregulation of latent matrix proteoglycans [109]. The delaying of Wallerian degeneration can be made possible due to the injury and preventing the glial scar tissue without the functional aspect [110]. The three ways the repair can be done are through the process of cytocompatibility with mechanical properties and also electrical conductivity [111]. Cytocompatibility is based on the principle of neural growth promotion without any inflammatory or infective processes. Electrical conductivity is also the major decisive factor, where electrical stimulation can lead to different neural regenerations, leading to the stabilization of various synaptic connections, leading to synaptogenesis [104]. There are different shortcomings which will impact nerve functioning [112]. Local inflammation can be used to prevent rejection or disease transmission, which will lead to the failure of the biomaterial implants [113, 114]. Various biomaterials, like silicon probes, periprosthetic devices, and various polymers, are excellent sources of nerve conduction and can impact glial scarring [115, 116] by increasing the mechanical and electrical properties for enhancing growth for CNS repair [117]. Carbon nanotubes are cylindrical nanostructures made up of graphene sheets wrapped in themselves. These can have single-walled CNTs (SWCNTs) or multi-walled CNTs (MWCNTs). These CNTs can facilitate neuronal growth in various aspects [118]. One of these studies has demonstrated that different MWCNTs, which are greater in length, as well as the number of various neuronal branches, can enhance growth [119]. Hu et al. have shown that neurite growth patterns, based on different parameters like length, branching pattern, and other factors, are capable of proper functioning [120]. Lovat and colleagues have shown that the electrical signal will be transferred in neural networks [121]. These SWCNT matrices are seen as having very high biocompatibility, causing different neuronal differentiation [122]. The neural-electrical interface between cells and fibers is based on different mechanical and electrical properties [123]. Nanobiocomposites are being designed for better and more specific goals to suppress glial scar formation by adding high-surface-energy carbon nanofibers (CNFs) into composite polymers [124]. This is based on the molecular weight of the polymer and also the surface charge, resulting in better methods for neural tissue engineering [125]. The nanostructured CNFs also show astrocyte proliferation, which also reduces glial scar formation [124]. Gabay and colleagues showed that the neurite-attractive CNT can impact different substrates based on different networks [126]. More recently, Sulejczak et al. have shown that CNTs are important in the restoration of neural pathways damaged by excessive scarring after injury [127]. CNF's neurotoxicity also needs to be evaluated, as it has a lot of potential for neurotoxicity. Another aspect is the combination of CNTs and stem cell technology for various neurological disorders, where the role of multipotent stem cells and nanoscaffolds has a promising future [128]. Better delivery and assistance to regeneration can be useful for better stem cell differentiation, where growth factors need to be more embedded in the stem cells [129, 130]. Jan and colleagues showed that layer-by-layer adding stem cells into neurons will be an important pathway for regeneration [129]. Adding neural stem cells (NSCs) into neurons can add select siRNAs, which induce RNA interference [131–135]. Thus, this nanoparticle delivery platform is made up of the silica NP monolayer, which is coated with siRNA and extracellular matrix proteins. The mechanisms underlying these processes and their activation may open the door to more effective scaffold designs and ultimately pave the way for effective CNS repair. Thus, these will result in better evaluation of the repair mechanisms in the CNS.

The important aspect of various nanomaterials for molecular and ion sensing will be monitoring the stimulation of various ion channels, as these are the internal and external environments of a cell. Throughout the body, more than 300 different types of these ion channels are integrated with cell membranes [136]. There will be more integration in the various channels of the membranes, various synapses with synaptogenesis, and also the cell volume. Ion channels are for receiving, processing, and transmitting signals for any cell processes. This will also be there for different ligands and neurotransmitters, which will change the different voltage based on the requirement of the signaling process [137]. The activation process is equally important, as more or less activation will also aggravate different neurological conditions like stroke, AIDS dementia complex, amyotrophic lateral sclerosis, and Alzheimer's disease. Here, the combination of molecular-based technology and electrophysiology is playing an important role in various changes in genetic mutations, expressing ligand- and voltage-gated ion channels, which are being labeled as channelopathies [138]. This channelization has led to the identification of various hereditary conditions that will cause various gene mutations [138, 139]. Here, the role of molecular biology, nanotechnology, and electrophysiology is also helping to make the physiology for understanding ion channels and also to monitor the flow of ions in the neuronal circuit based on different signaling components [140]. With more advances in various sensors and electrodes in nanoimaging, which will be monitoring the ions that will be flowing in the brain neuronal circuit, this should be biocompatible [141]. Carbon nanofiber-based wireless neurotransmitter concentration systems (WINCS) have greater selectivity and sensitivity for neurotransmitter detection. Even the concentrations of neurotransmitters like dopamine (DA), serotonin (5-HT), and ascorbic acid can be quantified [133]. Wireless carbon nanofiber-based neurotransmitters can be evaluated for earlier and better detection of oxygen and dopamine [132]. The potential for various CNF electrodes for brain stimulation can be achieved by better knowledge of nervous system disorders [142]. Ion flow quantification can be done for various reasons in different ways, as a potential measure of Cl^-, Na^+, and K^+ Sodium flow in neural cells and along the transmembrane sodium is important for the evaluation of various functions [143]. Fluorescence imaging can also be used to measure Na^+ concentrations [144, 145]. One of the studies showed that the half-life of a Na^+ dye can be increased by encapsulating it in the dendrimer and also using the Na^+ nanoprobe to probe physiological processes in various tissues [146]. Different optical sensors are also important for the knowledge of iron flow measurements in the nervous system. These optical sensors can quantify the amount of light in the electrical signal and are capable of identifying the electrical and various neuronal interactions [147]. Genetically encoded fluorescent voltage sensors are also available to detect total neural activity in vivo and in vitro [148, 149]. QDs can be used for voltage sensing and for the electrical activation of various cells [150]. Voltage-based photo-physical properties of QDs can be assessed efficiently [151]. The different modeled parameters can be useful for the QDs in voltage imaging, based on various parameters like QDs' optical properties, neurophysiological, and optical instrumentation parameters [135]. Ion channels in cells are also being managed by magnetically heating NPs [152]. NPs are very sensitive to radio waves and are altered based on different temperature-sensitive ion channels. Superparamagnetic ferrite NPs can also be used for various temperature-sensitive transient receptor potential V1 (TRPV1) [153]. Piezoelectric materials are also important for the analysis of various mechanical distortions due to various electrical responses. Piezoelectric materials are capable of enhancing various neural regeneration in spinal cord and cortical neuron lesions in animal imaging and also for the myelination status [154, 155]. Guo et al. showed that the newer low-cost, small, and long-lasting human-motion-driven triboelectric nanogenerator (TENG) has the property of electroconductivity for the self-powered electrical stimulation system in various neural differentiation [156]. Nanomaterials and piezoelectric components will play an important role in the evaluation of the electrical stimulus and its signaling molecules, having different piezoelectric properties. Some of the agents are zinc oxide and silicon dioxide components, having various applications [157]. Zinc oxide, when combined with various materials, will have much flexibility in shape, leading to a change in the piezoelectric effect properties [158]. The potential of nanotechnology in the neuroscience-based piezoelectric effect has a lot of potential.

16.9.2 Neurotoxicity of Nanomaterials (Nanoneurotoxicity)

Various components like proteins, lipid bilayers, and nucleic acids can be used for the quantification of the toxicity of nanomaterials [159]. The different aspects include the permeability of the cell membrane, which can have different toxic effects based on the various biochemical environments of the cell and the internal physiological system [160, 161]. A couple of the different aspects of toxicity are disruption of the cell, bioaccumulation at various levels, and the freedom to move in intracellular and extracellular areas. Metal oxide NPs will have a significant impact on the toxicity of cells, causing oxidative stress, pathological changes, and different inflammatory responses [104]. Again, they are based on various regulations concerning the manufacture, use, transportation, sale, or disposal of such nanomaterials [162]. Thermomechanical properties can also be used in various applications, including tissue engineering [156]. Zhu et al. have shown a significant amount of genotoxicity due to DNA involvement [153]. Many metal NPs will have very high interactions in different cellular environments, leading to side effects on neurons and brain function [104, 163]. The chemical and physical properties of nanomaterials and their use in biological systems and the BBB are based on various physical properties in vitro and in vivo [164]. Investigations of functional nanomaterials will incorporate the chemical nature, biological interactions, and their impact on living systems, including all side effects [160, 161]. Still, with these various pros and cons, further research on the design of different nanopharmaceuticals can be used for the betterment of life.

16.10 Conclusion

Nanotechnology in neurological applications is seen as having a major impact on stem cell therapy, immunoimaging and immunotherapy, lesion characterization, and therapeutic options for particular diseases. The identification, diagnosis, and therapeutic options in autoimmune and degenerative diseases can explain and be useful for the majority of neuronal pathologies. Thus, we are in the era of research advances for various neurological conditions in the concept of personalized medicine and personalized imaging.

References

1. Sun Y, Xia Y. Shape-controlled synthesis of gold and silver nanoparticles. Science. 2002;298:2176–9.
2. Li W, Brown PK, Wang LV, et al. Gold nanocages as contrast agents for photoacoustic imaging. Contrast Media Mol Imaging. 2011;6:370–7.
3. Weissleder R. A clearer vision for in vivo imaging. Nat Biotechnol. 2001;19:316–7.
4. Chen J, Saeki F, Wiley BJ, et al. Gold nanocages: bioconjugation and their potential use as optical imaging contrast agents. Nano Lett. 2005;5:473–7.
5. Lu W, Huang Q, Ku G, et al. Photoacoustic imaging of living mouse brain vasculature using hollow gold nanospheres. Biomaterials. 2010;31:2617–26.
6. Kircher MF, de la Zerda A, Jokerst JV, et al. A brain tumor molecular imaging strategy using a new triple-modality MRI-photoacoustic Raman nanoparticle. Nat Med. 2012;18:829–34.
7. Zhang Y, Hong H, Cai W. Photoacoustic imaging. Cold Spring Harb Protoc. 2011;2011:pdb.top065508.
8. Hu M, Chen J, Li ZY, et al. Gold nanostructures: engineering their plasmonic properties for biomedical applications. Chem Soc Rev. 2006;35:1084–94.
9. Yang X, Skrabalak SE, Li ZY, et al. Photoacoustic tomography of a rat cerebral cortex in vivo with au nanocages as an optical contrast agent. Nano Lett. 2007;7:3798–802.
10. Maeda H, Fang J, Inutsuka T, et al. Vascular permeability enhancement in solid tumor: various factors, mechanisms involved and its implications. Int Immunopharmacol. 2003;3:319–28.
11. Peer DK, Hong S, Farokhzad OC, et al. Nanocarriers as an emerging platform for cancer therapy. Nat Nanotechnol. 2007;2:751–60.
12. Gerstner ER, Batchelor TT. Antiangiogenic therapy for glioblastoma. Cancer J. 2012;18:45–50.
13. Sorensen AG, Emblem KE, Polaskova P, et al. Increased survival of glioblastoma patients who respond to antiangiogenic therapy with elevated blood perfusion. Cancer Res. 2012;72:402–7.
14. Clarke JL, Chang S. Pseudoprogression and pseudoresponse: challenges in brain tumor imaging. Curr Neurol Neurosci Rep. 2009;9:241–6.
15. Pope WB, Young JR, Ellingson BM. Advances in MRI assessment of gliomas and response to anti-VEGF therapy. Curr Neurol Neurosci Rep. 2011;11:336–44.
16. Lu M, Cohen MH, Rieves D, et al. FDA report: ferumoxytol for intravenous iron therapy in adult patients with chronic kidney disease. Am J Hematol. 2010;85:315–9.
17. Gahramanov S, Raslan AM, Muldoon LL, et al. Potential for differentiation of pseudoprogression from true tumor progression with dynamic susceptibility-weighted contrast-enhanced magnetic resonance imaging using ferumoxytol vs. gadoteridol: a pilot study. Int J Radiat Oncol Biol Phys. 2011;79:514–23.
18. Shattil SJ. Function and regulation of the beta 3 integrins in hemostasis and vascular biology. Thromb Haemost. 1995;74:149–55.
19. De S, Razorenova O, McCabe NP, et al. VEGF-integrin interplay controls tumor growth and vascularization. Proc Natl Acad Sci USA. 2005;102:7589–94.

20. Varner JA, Cheresh DA. Tumor angiogenesis and the role of vascular cell integrin alphavbeta3. Important Adv Oncol. 1996:69–87.
21. Folkman J. Angiogenesis in cancer, vascular, rheumatoid and other diseases. Nat Med. 1995;1:27–31.
22. Huang R, Vider J, Kovar JL, et al. Integrin alphavbeta3-targeted IRDye 800CW near-infrared imaging of glioblastoma. Clin Cancer Res. 2012;18:5731–40.
23. Schnell O, Krebs B, Wagner E, et al. Expression of integrin alphavbeta3in gliomas correlates with tumor grade and is not restricted to tumor vasculature. Brain Pathol. 2008;18:378–86.
24. University Medical Center Groningen. VEGF-targeted fluorescent tracer imaging in breast cancer. http://clinicaltrials.gov/ct2/show/record/NCT01508572. Accessed 8 Nov 2012.
25. Vincent SR. Nitric oxide neurons and neurotransmission. Prog Neurobiol. 2010;90:246–55.
26. Calabrese V, Mancuso C, Calvani M, et al. Nitric oxide in the central nervous system: neuroprotection versus neurotoxicity. Nat Rev Neurosci. 2007;8:766–75.
27. Garthwaite J, Boulton CL. Nitric oxide signaling in the central nervous system. Annu Rev Physiol. 1995;57:683–706.
28. Hong H, Sun J, Cai W. Multimodality imaging of nitric oxide and nitric oxide synthases. Free Radic Biol Med. 2009;47:684–98.
29. Jo A, Do H, Jhon GJ, et al. Electrochemical nanosensor for real-time direct imaging of nitric oxide in living brain. Anal Chem. 2011;83:8314–9.
30. Shim JH, Lee Y. Amperometric nitric oxide microsensor based on nanopore-platinized platinum: the application for imaging NO concentrations. Anal Chem. 2009;81:8571–6.
31. Washington University School of Medicine. PET imaging of endotoxin-induced iNOS activation. http://clinicaltrials.gov/ct2/show/NCT01407796. Accessed 18 Nov 2012.
32. Borlongan CV. Recent preclinical evidence advancing cell therapy for Alzheimer's disease. Exp Neurol. 2012;237:142–6.
33. Song M, Mohamad O, Gu X, et al. Restoration of intracortical and thalamocortical circuits after transplantation of bone marrow mesenchymal stem cells into the ischemic brain of mice. Cell Transplant. 2013;22:2001–15.
34. Ugoya SO, Tu J. Bench to bedside of neural stem cell in traumatic brain injury. Stem Cells Int. 2012;2012:141624.
35. Kuwabara T, Asashima M. Regenerative medicine using adult neural stem cells: the potential for diabetes therapy and other pharmaceutical applications. J Mol Cell Biol. 2012;4:133–9.
36. Gage FH, Ray J, Fisher LJ. Isolation, characterization, and use of stem cells from the CNS. Annu Rev Neurosci. 1995;18:159–92.
37. Migliorini E, Grenci G, Ban J, et al. Acceleration of neuronal precursors differentiation induced by substrate nanotopography. Biotechnol Bioeng. 2011;108:2736–46.
38. Liu H, Zhang SC. Specification of neuronal and glial subtypes from human pluripotent stem cells. Cell Mol Life Sci. 2011;68:3995–4008.
39. Schwarz SC, Schwarz J. Translation of stem cell therapy for neurological diseases. Transl Res. 2010;156:155–60.
40. Bulte JW. In vivo MRI cell tracking: clinical studies. AJR Am J Roentgenol. 2009;193:314–25.
41. Ferrucci JT, Stark DD. Iron oxide-enhanced MR imaging of the liver and spleen: review of the first 5 years. AJR Am J Roentgenol. 1990;155:943–50.
42. Berman SC, Galpoththawela C, Gilad AA, et al. Long-term MR cell tracking of neural stem cells grafted in immunocompetent versus immunodeficient mice reveals distinct differences in contrast between live and dead cells. Magn Reson Med. 2011;65:564–74.
43. University of Edinburgh. Cell tracking using superparamagnetic particles of iron oxide (SPIO) and magnetic resonance imaging (MRI): a pilot study. http://www.clinicaltrials.gov/ct2/show/NCT00972946. Accessed 18 Nov 2012.
44. Dore-Duffy P, Katychev A, Wang X, et al. CNS microvascular pericytes exhibit multipotential stem cell activity. J Cereb Blood Flow Metab. 2006;26:613–24.
45. Rodriguez-Baeza A, Reina-De La Torre F, Ortega-Sanchez M, et al. Perivascular structures in corrosion casts of the human central nervous system: a confocal laser and scanning electron microscope study. Anat Rec. 1998;252:176–84.
46. Itoh Y, Suzuki N. Control of brain capillary blood flow. J Cereb Blood Flow Metab. 2012;32:1167–76.
47. Bergers G, Song S. The role of pericytes in blood-vessel formation and maintenance. Neuro Oncol. 2005;7:452–64.
48. Liu CH, Ren JQ, You Z, et al. Noninvasive detection of neural progenitor cells in living brains by MRI. FASEB J. 2012;26:1652–62.
49. Kalaria RN, Cohen DL, Premkumar DR, et al. Vascular endothelial growth factor in Alzheimer's disease and experimental cerebral ischemia. Brain Res Mol Brain Res. 1998;62:101–5.
50. Zhu M, Bi X, Jia Q, et al. The possible mechanism for impaired angiogenesis after transient focal ischemia in type 2 diabetic GK rats: different expressions of angiostatin and vascular endothelial growth factor. Biomed Pharmacother. 2010;64:208–13.
51. Leary SP, Liu CY, Apuzzo ML. Toward the emergence of nanoneurosurgery. Part III. Nanomedicine: targeted nanotherapy, nanosurgery, and progress toward the realization of nanoneurosurgery. Neurosurgery. 2006;58:1009–26; discussion 1026.
52. Pinto Reis C, Neufeld RJ, Ribeiro AJ, et al. Nanoencapsulation I. Methods for preparation of drug-loaded polymeric nanoparticles. Nanomedicine. 2006;2:8–21.

53. Steiniger SC, Kreuter J, Khalansky AS, et al. Chemotherapy of glioblastomain rats using doxorubicin-loaded nanoparticles. Int J Cancer. 2004;109:759–67.
54. Hasadsri L, Kreuter J, Hattori H, et al. Functional protein delivery into neurons using polymeric nanoparticles. J Biol Chem. 2009;284:6972–81.
55. Koffie RM, Farrar CT, Saidi LJ, et al. Nanoparticles enhance brain delivery of blood-brain barrier-impermeable probes for in vivo optical and magnetic resonance imaging. Proc Natl Acad Sci USA. 2011;108:18837–42.
56. Hamers-Casterman C, Atarhouch T, Muyldermans S, et al. Naturally occurring antibodies devoid of light chains. Nature. 1993;363:446–8.
57. Tijink BM, Laeremans T, Budde M, et al. Improved tumor targeting of anti-epidermal growth factor receptor nanobodies through albumin binding: taking advantage of modular nanobody technology. Mol Cancer Ther. 2008;7:2288–97.
58. Roovers RC, Vosjan MJ, Laeremans T, et al. A biparatopic anti-EGFR nanobody efficiently inhibits solid tumour growth. Int J Cancer. 2011;129:2013–24.
59. Hmila I, Saerens D, Ben Abderrazek R, et al. A bispecific nanobody to provide full protection against lethal scorpion envenoming. FASEB J. 2010;24:3479–89.
60. Coppieters K, Dreier T, Silence K, et al. Formatted anti-tumor necrosis factor alpha VHH proteins derived from camelids show superior potency and targeting to inflamed joints in a murine model of collagen-induced arthritis. Arthritis Rheum. 2006;54:1856–66.
61. Devoogdt N, Xavier C, Hernot S, et al. Molecular imaging using nanobodies: a case study. Methods Mol Biol. 2012;911:559–67.
62. Vaneycken I, D'Huyvetter M, Hernot S, et al. Immuno-imaging using nanobodies. Curr Opin Biotechnol. 2011;22:877–81.
63. Vaneycken I, Devoogdt N, Van Gassen N, et al. Preclinical screening of anti-HER2 nanobodies for molecular imaging of breast cancer. FASEB J. 2011;25:2433–46.
64. Schoonooghe S, Laoui D, Van Ginderachter JA, et al. Novel applications of nanobodies for in vivo bio-imaging of inflamed tissues in inflammatory diseases and cancer. Immunobiology. 2012;217:1266–72.
65. Nabuurs RJ, Rutgers KS, Welling MM, et al. In vivo detection of amyloid-beta deposits using heavy chain antibody fragments in a transgenic mouse model for Alzheimer's disease. PLoS One. 2012;7:e38284.
66. Rutgers KS, Nabuurs RJ, van den Berg SA, et al. Transmigration of beta amyloid specific heavy chain antibody fragments across the in vitro blood-brain barrier. Neuroscience. 2011;190:37–42.
67. Pathak S, Choi SK, Arnheim N, et al. Hydroxylated quantum dots as luminescent probes for in situ hybridization. J Am Chem Soc. 2001;123:4103–4.
68. Bruchez M Jr, Moronne M, Gin P, et al. Semiconductor nanocrystals as fluorescent biological labels. Science. 1998;281:2013–6.
69. Sukhanova A, Devy J, Venteo L, et al. Biocompatible fluorescent nanocrystals for immunolabeling of membrane proteins and cells. Anal Biochem. 2004;324:60–7.
70. Sun W, Subrahmanyam PB, East JE, et al. Connecting the dots: artificial antigen presenting cell-mediated modulation of natural killer T cells. J Interf Cytokine Res. 2012;32:505–16.
71. Smith AM, Dave S, Nie S, et al. Multicolor quantum dots for molecular diagnostics of cancer. Expert Rev Mol Diagn. 2006;6:231–44.
72. Kantelhardt SR, Caarls W, de Vries AH, et al. Specific visualization of glioma cells in living low-grade tumor tissue. PLoS One. 2010;5:e11323.
73. Neuwelt EA, Varallyay P, Bago AG, et al. Imaging of iron oxide nanoparticles by MR and light microscopy in patients with malignant brain tumours. Neuropathol Appl Neurobiol. 2004;30:456–71.
74. Pathak S, Cao E, Davidson MC, et al. Quantum dot applications to neuroscience: new tools for probing neurons and glia. J Neurosci. 2006;26:1893–5.
75. Vu TQ, Maddipati R, Blute TA, et al. Peptide-conjugated quantum dots activate neuronal receptors and initiate downstream signaling of neurite growth. Nano Lett. 2006;5:603–7.
76. Dahan M, Levi S, Luccardini C, et al. Diffusion dynamics of glycine receptors revealed by single-quantum dot tracking. Science. 2003;302:442–5.
77. Yin JJ, Lao F, Fu PP, et al. The scavenging of reactive oxygen species and the potential for cell protection by functionalized fullerene materials. Biomaterials. 2009;30:611–21.
78. Dugan LL, Lovett EG, Quick KL, et al. Fullerene-based antioxidants and neurodegenerative disorders. Parkinsonism Relat Disord. 2001;7:243–6.
79. Jeffery ND, McBain SC, Dobson J, et al. Uptake of systemically administered magnetic nanoparticles (MNPs) in areas of experimental spinal cord injury (SCI). J Tissue Eng Regen Med. 2009;3:153–7.
80. Delcroix GJ, Jacquart M, Lemaire L, et al. Mesenchymal and neural stem cells labeled with HEDP-coated SPIO nanoparticles: in vitro characterization and migration potential in rat brain. Brain Res. 2009;1255:18–31. Epub 2008 Dec 11.
81. Escolar ML, Poe MD, Provenzale JM, et al. Transplantation of umbilical-cord blood in babies with infantile Krabbe's disease. N Engl J Med. 2005;352:2069–81.
82. Rice HE, Hsu EW, Sheng H, et al. Superparamagnetic iron oxide labeling and transplantation of adipose-derived stem cells in middle cerebral artery occlusion-injured mice. AJR Am J Roentgenol. 2007;188:1101–8.
83. Jendelova P, Herynek V, Urdzíkova L, et al. Magnetic resonance tracking of transplanted bone marrow and

embryonic stem cells labeled by iron oxide nanoparticles in rat brain and spinal cord. J Neurosci Res. 2004;76:232–43.
84. Silva GA. Neuroscience nanotechnology: progress, opportunities and challenges. Nat Rev Neurosci. 2006;7:65–74.
85. Silva GA, Czeisler C, Niece KL, et al. Selective differentiation of neural progenitor cells by high-epitope density nanofibers. Science. 2004;303:1352–5.
86. Sun EY, Weissleder R, Josephson L. Continuous analyte sensing with magnetic nanoswitches. Small. 2006;2:1144–7.
87. Sosnovik DE, Weissleder R. Emerging concepts in molecular MRI. Curr Opin Biotechnol. 2007;18:4–10.
88. Winter PM, Caruthers SD, Kassner A, et al. Molecular imaging of angiogenesis in nascent Vx-2 rabbit tumors using a novel 3-targeted nanoparticle and 1.5 Tesla magnetic resonance imaging. Cancer Res. 2003;63:5838–43.
89. Reddy GR, Bhojani MS, McConville P, et al. Vascular targeted nanoparticles for imaging and treatment of brain tumors. Clin Cancer Res. 2006;12:6677–86.
90. Loane DJ, Faden AI. Trends Pharmacol Sci. 2010;31:596.
91. Cornelius C, Crupi R, Calabrese V, Graziano A, Milone P, Pennisi G, Radak Z, Calabrese EJ, Cuzzocrea S. Antioxid Redox Signal. 2013;19:836.
92. Petro M, Jaffer H, Yang J, Kabu S, Morris VB, Labhasetwar V. Biomaterials. 2016;81:169.
93. Bailey ZS, Nilson E, Bates JA, Oyalowo A, Hockey KS, Sajja VS, Thorpe C, Rogers H, Dunn B, Frey AS, Billings MJ, Sholar CA, Hermundstad A, Kumar C, VandeVord PJ, Rzigalinski BA. J Neurotrauma. 2016;37:1452. https://doi.org/10.1089/neu.2016.4644.
94. Xu J, Ypma M, Chiarelli PA, Park J, Ellenbogen RG, Stayton PS, Mourad PD, Lee D, Convertine AJ, Kievit FM. Adv Funct Mater. 2016;26:4124.
95. Yemisci M, Caban S, Gursoy Ozdemir Y, Lule S, Novoa-Carballal R, Riguera R, Fernandez Megia E, Andrieux K, Couvreur P, Capan Y, Dalkara T. Cereb J Blood Flow. 2015;35:469.
96. Selnes OA, Gottesman RF, Grega MA, Baumgartner WA, Zeger SL, McKhann GM, Engl N. J Med. 2012;366:250.
97. Mishra MK, Beaty CA, Lesniak WG, Kambhampati SP, Zhang F, Wilson MA, Blue ME, Troncoso JC, Kannan S, Johnston MV, Baumgartner WA, Kannan RM. ACS Nano. 2014;8:2134.
98. Engels M, Bilgic E, Pinto A, Vasquez E, Wollschläger L, Steinbrenner H, Kellermann K, Akhyari P, Lichtenberg A, Boeken U. J Inflamm. 2014;11:26.
99. Barreiro CJ, Williams JA, Fitton TP, Lange MS, Blue ME, Kratz L, Barker PB, Degaonkar M, Gott VL, Troncoso JC, Johnston MV, Baumgartner WA. Ann Thorac Surg. 2006;81:1593.
100. Williams JA, Barreiro CJ, Nwakanma LU, Lange MS, Kratz LE, Blue ME, Berrong J, Patel ND, Gott VL, Troncoso JC, Johnston MV, Baumgartner WA. Ann Thorac Surg. 2006;81:2235.
101. Sang LYH, Liang YX, Li Y, Wong WM, Tay DKC, So KF, Ellis-Behnke RG, Wu W, Cheung RTF. Nanomedicine. 2015;11:611.
102. Kogan MJ, Bastus NG, Amigo R, Grillo-Bosch D, Araya E, Turiel A, Labarta A, Giralt E, Puntes VF. Nano Lett. 2006;6:110.
103. Shah S, Solanki A, Lee KB. Acc Chem Res. 2016;49:17.
104. Sever M, Mammadov B, Gecer M, Guler MO, Tekinay AB. Therapeutic nanomaterials. John Wiley & Sons, Inc.; 2016. p. 33.
105. Chen MS, Huber AB, van der Haar ME, Frank M, Schnell L, Spillmann AA, Christ F, Schwab ME. Nature. 2000;403:434.
106. Yiu G, He Z. Nat Rev Neurosci. 2006;7:617.
107. Back SA, Tuohy TM, Chen H, Wallingford N, Craig A, Struve J, Luo NL, Banine F, Liu Y, Chang A, Trapp BD, Bebo BF Jr, Rao MS, Sherman LS. Nat Med. 2005;11:966.
108. Yiu G, He Z. Curr Opin Neurobiol. 2003;13:545.
109. Yang P, Yang Z. J Neurol Sci. 2012;312:1.
110. Fawcett JW, Asher RA. Brain Res Bull. 1999;49:377.
111. Zito K, Svoboda K. Neuron. 2002;35:1015.
112. Wolford LM, Stevao ELL. Proc (Bayl Univ Med Cent). 2003;16:152.
113. Barker RA, Widner H. NeuroRx. 2004;1:472.
114. Borlongan CV, Stahl CE, Cameron DF, Saporta S, Freeman TB, Cahill DW, Sanberg PR. Neurol Res. 1996;18:297.
115. Polikov VS, Block ML, Fellous JM, Hong JS, Reichert WM. Biomaterials. 2006;27:5368.
116. Reier PJ, Houle JD. Adv Neurol. 1988;47:87.
117. Fattahi P, Yang G, Kim G, Abidian MR. Adv Mater. 1846;2014:26.
118. Hwang JY, Shin US, Jang WC, Hyun JK, Wall IB, Kim HW. Nanoscale. 2013;5:487.
119. Mattson MP, Haddon RC, Rao AM. J Mol Neurosci. 2000;14:175.
120. Hu H, Ni Y, Montana V, Haddon RC, Parpura V. Nano Lett. 2004;4:507.
121. Lovat V, Pantarotto D, Lagostena L, Cacciari B, Grandolfo M, Righi M, Spalluto G, Prato M, Ballerini L. Nano Lett. 2005;5:1107.
122. Gheith VASMK. Adv Mater. 2005;17:2663.
123. Nguyen-Vu TDB, Chen H, Cassell AM, Andrews RJ, Meyyappan M, Li J. IEEE Trans Biomed Eng. 2007;54:1121.
124. McKenzie JL, Waid MC, Shi R, Webster TJ. Biomaterials. 2004;25:1309.
125. Shah S, Yin PT, Uehara TM, Chueng STD, Yang L, Lee KB. Adv Mater. 2014;26:3673.
126. Gabay T, Jakobs E, Ben-Jacob E, Hanein Y. Phys A Stat Mech Appl. 2005;350:611.
127. Sulejczak D, Andrychowski J, Kowalczyk T, Nakielski P, Frontczak-Baniewicz M, Kowalewski T. Folia Neuropathol. 2014;52:56.
128. Chandrakanthan V, Yeola A, Kwan JC, Oliver RA, Qiao Q, Kang YC, Zarzour P, Beck D, Boelen

L, Unnikrishnan A, Villanueva JE, Nunez AC, Knezevic K, Palu C, Nasrallah R, Carnell M, Macmillan A, Whan R, Yu Y, Hardy P, Grey ST, Gladbach A, Delerue F, Ittner L, Mobbs R, Walkley CR, Purton LE, Ward RL, Wong JWH, Hesson LB, Walsh W, Pimanda JE. Proc Natl Acad Sci USA. 2016;113:E2306.
129. Jan E, Kotov NA. Nano Lett. 2007;7:1123.
130. Solanki A, Shah S, Yin PT, Lee KB. Sci Rep. 2013;3:3.
131. Oller-Salvia B, Sanchez-Navarro M, Giralt E, Teixido M. Chem Soc Rev. 2016;45:707.
132. Chen CH, Lin CT, Hsu WL, Chang YC, Yeh SR, Li LJ, Yao DJ. Nanomedicine. 2013;9:600.
133. Kumar A, Chen F, Mozhi A, Zhang X, Zhao Y, Xue X, Hao Y, Zhang X, Wang PC, Liang XJ. Nanoscale. 2013;5:8307.
134. Rodrigues SF, Fiel LA, Shimada AL, Pereira NR, Guterres SS, Pohlmann AR, Farsky SH. J Biomed Nanotechnol. 2016;12:986.
135. Zhang L, Wang Y, Tang Y, Jiao Z, Xie C, Zhang H, Gu P, Wei X, Yang G-Y, Gu H, Zhang C. Nanoscale. 2013;5:4506.
136. Gabashvili I, Sokolowski BA, Morton C, Giersch AS. JARO. 2007;8:305.
137. Spillane J, Kullmann DM, Hanna MG. J Neurol Neurosurg Psychiatry. 2016;87:37.
138. Kim JB. Korean J Pediatr. 2014;57:1.
139. Rose MR. BMJ. 1998;316:1104.
140. Pancrazio JJ. Nanomedicine. 2008;3:823.
141. Seymour JP, Wu F, Wise KD, Yoon E. Microsyst Nanoeng. 2017;3:16066.
142. Lippincott-Schwartz J. Annu Rev Biochem. 2011;80:327.
143. Minta A, Tsien RY. J Biol Chem. 1989;264:19449.
144. Dubach JM, Das S, Rosenzweig A, Clark HA. Proc Natl Acad Sci USA. 2009;106:16145.
145. Meier SD, Kovalchuk Y, Rose CR. Neurosci J Methods. 2006;155:251.
146. Lamy CM, Sallin O, Loussert C, Chatton JY. ACS Nano. 2012;6:1176.
147. Lehmann-Horn F, Jurkat-Rott K. J Neurol Neurosurg Psychiatry. 2003;74:1466.
148. Akemann W, Mutoh H, Perron A, Rossier J, Knopfel T. Nat Methods. 2010;7:643.
149. Campagnola PJ, Loew LM. Nat Biotechnol. 2003;21:1356.
150. Peterka DS, Takahashi H, Yuste R. Neuron. 2011;69:9.
151. Lugo K, Miao X, Rieke F, Lin LY. Biomed Opt Express. 2012;3:447.
152. Huang H, Delikanli S, Zeng H, Ferkey DM, Pralle A. Nat Nanotechnol. 2010;5:602.
153. Ghasemi-Mobarakeh L, Prabhakaran MP, Morshed M, Nasr-Esfahani MH, Baharvand H, Kiani S, Al-Deyab S, Ramakrishna S. Tissue Eng J Regen Med. 2011;5:E17.
154. Royo Gascon N, Wininger M, Scheinbeim JI, Firestein BL, Craelius W. Ann Biomed Eng. 2013;41:112.
155. Wen J, Liu M. J Mol Neurosci. 2014;52:323.
156. Guo W, Zhang X, Yu X, Wang S, Qiu J, Tang W, Li L, Liu H, Wang ZL. ACS Nano. 2016;10:5086.
157. Theodorou GS, Kontonasaki E, Theocharidou A, Bakopoulou A, Bousnaki M, Hadjichristou C, Papachristou E, Papadopoulou L, Kantiranis NA, Chrissafis K, Paraskevopoulos KM, Koidis PT. Int J Biomater. 2016;2016:11.
158. Xiang HJ, Yang JL, Hou JG, Zhu QS. Appl Phys Lett. 2006;89:223111.
159. Kumar A, Zhang X, Liang XJ. Biotechnol Adv. 2013;31:593.
160. Murphy CJ, Vartanian AM, Geiger FM, Hamers RJ, Pedersen J, Cui Q, Haynes CL, Carlson EE, Hernandez R, Klaper RD, Orr G, Rosenzweig ZE. ACS Cent Sci. 2015;1:117.
161. Klaper R, Arndt D, Bozich J, Dominguez G. Analyst. 2014;139:882.
162. Zhang DA, Rand E, Marsh M, Andrews RJ, Lee KH, Meyyappan M, Koehne JE. Mol Neurobiol. 2013;48:380.
163. Yizhar O, Fenno LE, Davidson TJ, Mogri M, Deisseroth K. Neuron. 2011;71:9.
164. Zhao SY, Liu XJ, Xu Z, Ren HY, Deng B, Tang M, Lu LL, Fu XF, Peng HL, Liu ZF, Duan XJ. Nano Lett. 2016;16:7731.

Nanoimaging in Cardiology 17

17.1 Nanotechnology and Cardiovascular Diseases

The manifold use of nanotechnology is well-established in cancer management [1–6], and now this technology is also being used in biomedical research. The role of cardiovascular imaging is for diagnosing pathologies, drug delivery, stem cell therapy, tissue engineering, stent surgery, and various combinations of these. Nanomaterials have significantly improved clinical imaging avenues for various cardiac causes [7]. Well-established cardiovascular imaging is now one of the important tools for diagnosing these diseases [8]. The basic composition of the research in this area is important for evaluating various components, such as biocompatibility, size distribution (as very small nanosizes are used), high penetration power, and good contrast ability [9, 10]. Paramagnetic nanoparticles, iodinated nanoparticles, and quantum dots provide better contrast abilities [11]. Nanosurface tunability is also an important feature for multimodal evaluation, with specific targets and molecules for targeted delivery, where various imaging probes and therapeutic molecules can be used [12, 13]. Nanoparticles are more stable than chemical imaging probes [14]. Nanoforms will provide better contrast in imaging due to increased conjugation of the nanoparticles. All these factors are important and more viable in cardiovascular imaging, which is non-invasive in nature. The concept of nanoimaging is for better understanding of pathologies so that relevant applications can be tailored for personalized medicine (Fig. 17.1).

Nanocardiovascular-imaging applications fall into four major categories: thrombus imaging, stem cell imaging, graft imaging, and various theranostic approaches (Fig. 17.2). The most common cardiac cause of morbidity is acute coronary syndrome (ACS) [15–19]. Plaque morphology is based on various components, including macrophages, smooth muscle, endothelial cells, mesenchymal cells, and many other elements [20]. Plaque evaluation is crucial for better and more precise management, and it also requires contrast agents for assessment [21, 22]. Any contrast agent will have some level of toxicity and non-uniform distribution in the body, which can lead to systemic toxicity. Due to these limitations, the role of nanoimaging is evolving. One effective method for early detection and characterization of plaques is cardiac magnetic resonance imaging (CMRI), in which gadolinium is used; however, it is not completely safe [23]. To minimize the toxic effects, gadolinium nanoparticles can be used, which reduce toxicity [24]. Gadolinium toxicity is caused by free ions, so the addition of chelating agents can help reduce this risk [25, 26]. However, some leakage of free ions may still occur due to the GD-chelate

S. Shaikh, *Nanoimaging - Future of Precision Medicine*,
https://doi.org/10.1007/978-981-97-7292-6_17

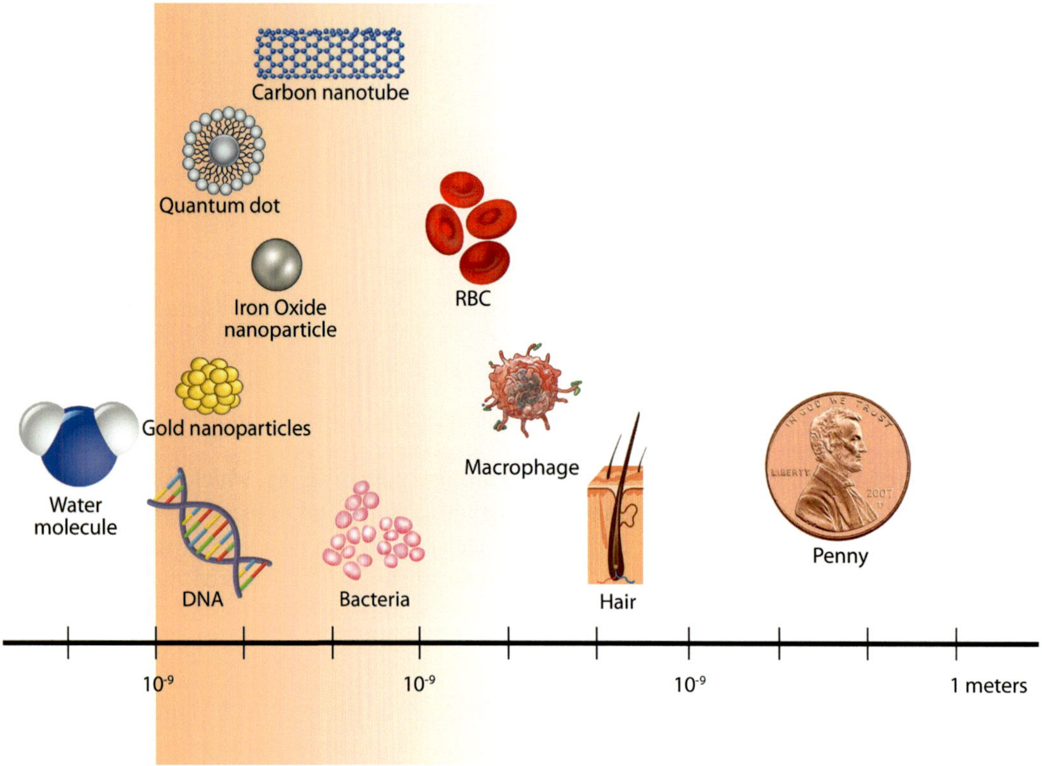

Fig. 17.1 Range of sizes of different nanoparticles

complex [27]. Nanogadolinium agents have larger loading capacity and increased contrast stability [28, 29]. The use of anti-fibrin antibody nanoparticles is highly sensitive for identifying susceptible plaques due to the fibrin present in the plaques [30]. Toxicity can be further reduced by adding Gd to the nanoparticles [31]. Despite the toxicity concerns, new non-toxic imaging agents are being developed for MR imaging. Newer agents, such as ultra-small superparamagnetic iron oxide nanoparticles (SPIONs), provide higher sensitivity with reduced toxicity [32]. SPIONs offer better sensitivity in CMRI and have fewer side effects [33]. Perfluorocarbon nanoparticles, which contain fluorine, offer excellent contrast when used in CMRI [3]. Due to their smaller sizes, SPIONs can combine surface ligands for better targeting and therapy compared to the GD-chelate complex. Lysosomes and free iron particles can alter the interior of cells by depositing an iron pool [34]. Atherosclerotic plaques are categorized into two types: stable fibrous plaques and unstable lipid plaques. The unstable plaques are dynamic, causing thrombosis [35]. These vulnerable plaques are primarily composed of lipids and are covered by a thin fibrous cap. Any injury to this cap can result in the detachment of the plaque [36]. This plaque detachment can block arteries, reducing blood circulation and leading to muscle necrosis. Therefore, identifying these vulnerable plaques is crucial to prevent destabilization and allow for targeted therapy [37]. Therapeutic options focus on the thin fibrous cap of plaques, which secretes proteolytic enzymes that dissolve the cap [38]. Macrophages play a key role in evaluating vulnerable plaques [39]. The phagocytic activity of macrophages is vital for identifying these plaques, and it is believed that macrophages can enhance the contrast of nanomaterials [40]. Thus, nanoform-loaded inflammatory macrophages on plaques can be easily identified using non-invasive imaging techniques [41–45] (Fig. 17.3). The evaluation of nanomaterials can be done by

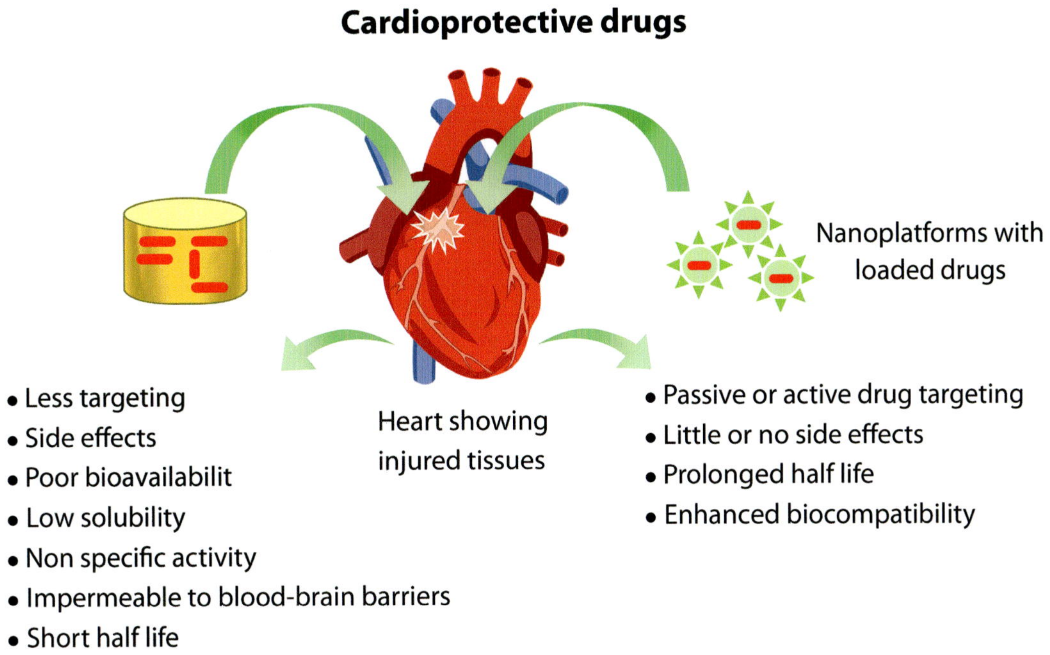

Fig. 17.2 Cardioprotective drugs with nanoplatforms loaded with drugs showing different actions of the drugs

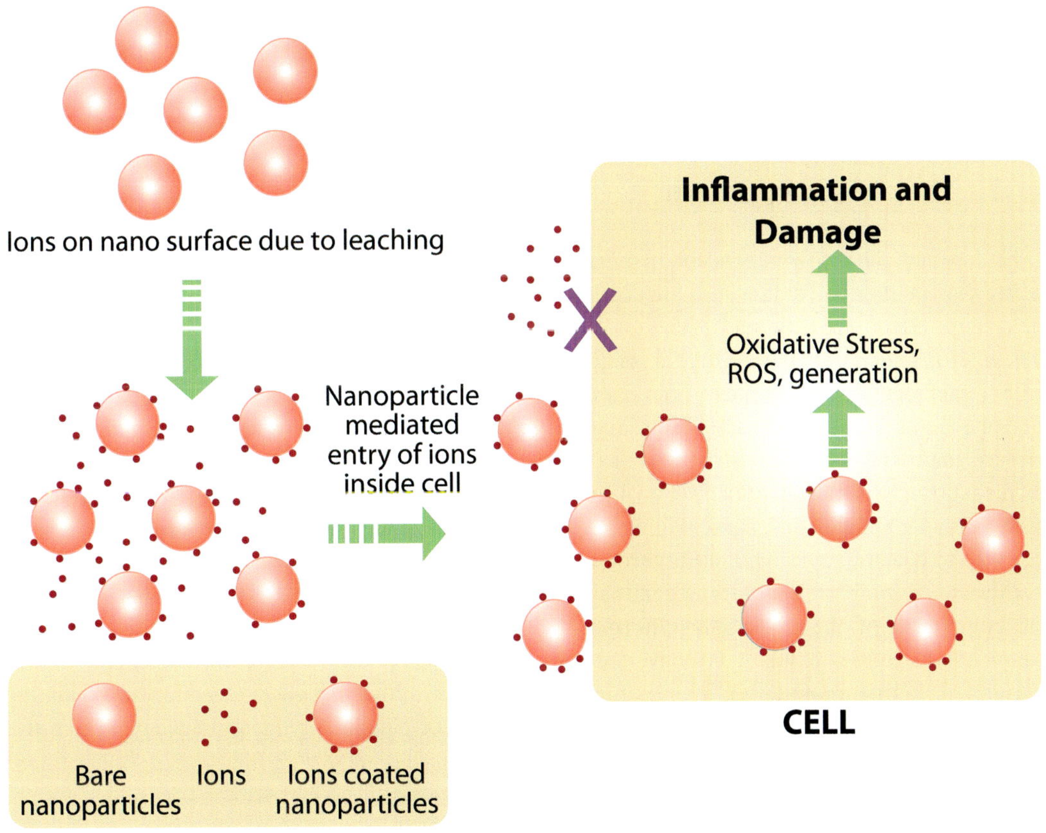

Fig. 17.3 The process of inflammation. The role of the nanoparticles for the entry into the cell with different mechanisms of actions like nanoparticle coating and nanoparticle entry in the cell

combining macrophages loaded with SPIONs and CMRI. CT imaging can be performed when iodinated gold nanoparticles are used as better CT contrast agents. Gold nanoparticles are three times more sensitive than other photon absorption methods [46, 47]. Gold nanoparticles provide superior contrast, are more inert, and the addition of gold particles to the nanoparticles enhances their properties [48]. When gold nanoparticles are coated with other lipid components, they enable multicolor CT imaging [49]. Plaques, in different nanoforms, can be specifically targeted. Amino acids used by plaque-specific proteases can easily identify inflammatory plaques [50]. Factor XIII plays a role in acute thrombus formation by crosslinking fibrin. Fibrin-specific peptides are capable of providing better contrast for evaluating acute thrombus, especially as the thrombus size increases [51]. Atherosclerotic inflammation is caused by reactive oxygen species (ROS) and reactive nitrogen species (RNS). Sometimes, local antioxidants can increase oxidative stress and cause further tissue injury [52]. Imaging of ROS/RNS can be developed using reactive nitrogen species and myeloperoxidase (MPO) [52–57].

17.2 Theranostic Approach

"Theranostics" is a newer concept in imaging, where therapeutics and diagnostics are combined in a single entity [58]. Some specialists define this as a diagnostic procedure that is integrated with therapeutic molecules or devices. Theranostic particles can be added to diagnostic probes, improving contrast and serving as both therapeutic and diagnostic markers (Fig. 17.4). In cardiovascular theranostics, nanoimaging plays a key role in detecting and facilitating thrombolysis/fibrinolysis of thrombi. Plasminogen is a critical component of the thrombolytic/fibrinolytic pathway, where plasminogen is converted into plasmin (a serine proteinase) by plasminogen activators. This process degrades fibrin and other

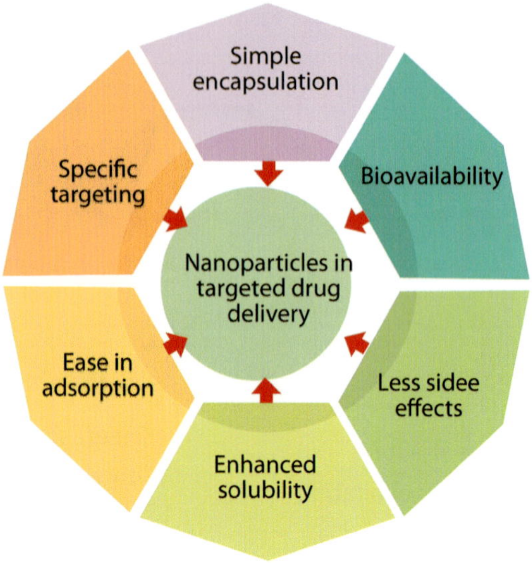

Fig. 17.4 Role of nanoparticles in targeted drug delivery

proteins, leading to a reduction in plaque volume [59]. Currently, the most effective agent for this process is recombinant tissue plasminogen activator (rTPA), which has proven efficient in plaque dissolution [60]. Nano-based theranostics aid in reducing plaque volume by combining with rTPA. Iron oxide nanoparticles combined with rTPA enhance clot dissolution sensitivity [54]. Thrombolytic nanoparticles coated with fluorophores offer the advantage of enabling plaque identification and characterization, as well as providing therapeutic plaque reduction through this technology. Another important aspect of theranostics is the inhibition of angiogenesis in plaques to prevent plaque growth [61]. Angiogenesis within plaques contributes to plaque progression and rupture, which are precursors to myocardial infarction [62]. Integrin $\alpha v\beta 3$ is a key marker of angiogenesis and is found in angiogenic vasculature, making it a critical target for detecting plaque angiogenesis [63]. The use of ultra-small superparamagnetic iron oxide nanoparticles targeting the integrin $\alpha v\beta 3$ receptor allows for early detection and quantification of plaque angiogenesis using MRI as the

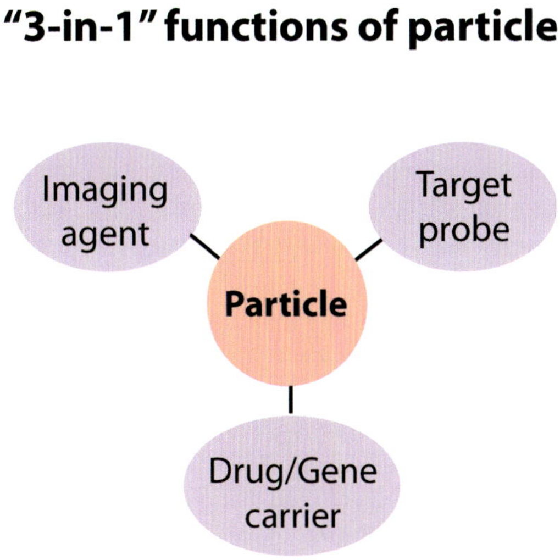

Fig. 17.5 Nanoparticles' role in imaging, target particle and dru and gene carrier

imaging modality, along with integrin αvβ3-targeted perfluorocarbon nanoparticles [64, 65]. Theranostic nanoparticles with paramagnetic properties can detect and inhibit neovascularization [56]. Magnetic nanoparticles, such as infrared fluorophores, can also be activated to aid in evaluating atherosclerotic plaques. When administered, these nanoparticles are taken up by macrophages and are subsequently killed due to the phototoxic effect, an essential feature of theranostics in atherosclerosis [55] (Fig. 17.5).

17.3 Stem Cell Imaging

Stem cell technology is now an important entity for the management of various cardiovascular diseases [66, 67]. One of the best uses is the treatment of the infarcted myocardium [68]. The different mesenchymal stem cells (MSCs) in the bone marrow stromal cells will be capable of identifying the different types of cardiomyocytes [69, 70]. The use of mesenchymal stem cells is capable of making many reversible changes in myocardial infarction [71, 72]. The mesenchymal stem cells will be transported to the different infarcted myocytes depicting the prognosis of the disease. Superparamagnetic iron oxide nanoparticles, when combined with labeled stem cell tracking along with targeting, have shown satisfactory results in chronic myocardial infarction [4, 73].

17.4 Graft Imaging

The end management for end-stage heart failure will be heart transplantation, where the cause can be severe coronary artery disease [74]. After the heart transplantation, the identification of graft rejection cannot be found by routine methods, and one of these is the biopsies, which are a non-viable option [75]. This process is not free from complications and also further leads to fibrosis at the site of the biopsy [76]. One of the recent studies has shown that the use of the fluorophore-tagged iron oxide nanoparticle will be able to identify the underlying cause [54, 55]. With the availability of macrophages and cathepsin (protease), these are important agents used for the identification of graft rejection. Magnetic nanoparticles, when tagged with fluorescent materials, can use macrophages for the identification of the cathepsin activity. Peters and colleagues [77] have shown that the duration of exposure to the traffic and the ambient nanoparticles carbon nanoparticles that are generated show a toxic effect on vascular cells [78]. Similarly, the alveolar wall in the lungs will show [79] cytotoxicity in the lungs, inflammation in the endothelium wall, inhibition of cell growth, and cardiovascular toxicity [80]. The metallic nanoparticles, when causing platelet activation, will result in more cardiac risk [12, 81, 82]. Some of the cosmetic materials with nanoparticles, like titanium oxide and silicon oxide, will lead to plaque progression, dysfunction of the vessels, myocardial ischemic damage, and inhibition of the atrioventricular bundle, etc. [83, 84]. Copper oxide (CuO) nanoparticles are also contributing to oxidative stress, leading to plasminogen activator inhibitor-1 expression, predisposing to myocardial infarction [85]. Nickel nanoparticles are capable of inducing atherosclerosis [86]. Various side effects are also due to the different nanoparticles [87, 88]. Some of the metallic ions are also capable of inducing various oxidative stresses in biological systems [89]. Some of the nanoforms are taken by the cells due to the different nanoparticle applications, as they will have very good penetration power [90]. The modified nanoparticles have better surface stability than the nanoforms [91–94]. The carbon nanoparticle will be amorphous and show a better and super hydrophobic type of toxic effect [95].

17.5 Nano- and Microparticles in Imaging of Cardiovascular Interventions

The imaging of various cardiovascular diseases and the role of the various interventions will be helpful in the evaluation and monitoring of the various primary interventional procedures, which will be localized to various delivery. The nanoimaging will also be more effective for the delivery of Theranostics and also for the evaluation of the therapeutics that are being targeted at various targets [96–99]. With the availability of the various nano- and microparticles, they will lead to the monitoring and also the tracking of the various cardiovascular interventions (Fig. 17.6).

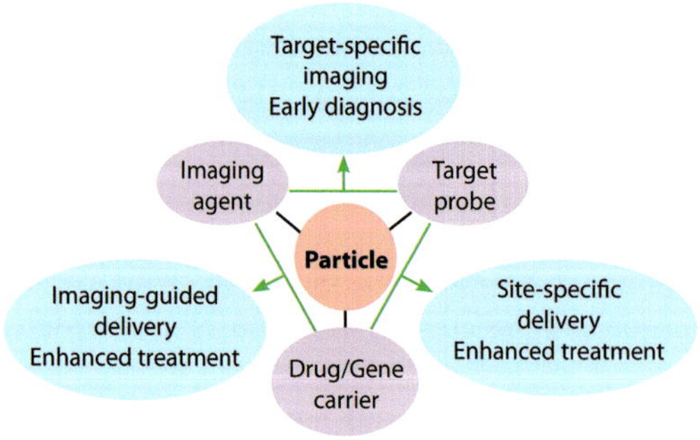

Fig. 17.6 Different actions and roles of the nanoparticles

17.6 Gene and Drug Delivery Imaging

This has shown significant changes in the therapeutics delivery to various targets. The use of ultrasound also plays an important role in the monitoring of various gene delivery [100]. The newer echogenic gene-carrying particles can be seen in various places, which can be attached to the walls of the vessels and also aligned with the target artery walls. Further, this will also lead to better biodistribution and localization at the target sites. Various nanoparticles will result in various diagnostic agents and therapeutic carriers for vascular interventional gene and drug therapy. One of the studies showed that the [101, 102] different drug-carrying nanoparticles will be able to be used for different arterial stent placements, as well as the evaluation of the stent stenosis. The catheter-based gene and drug delivery will be used for primary therapeutic delivery [96]. The molecular targets, like the atherosclerotic plaques along with the ischemic myocardium, will be able to undergo dual loading for both diagnostic and therapeutic applications. Various molecules are being used for the various targets in tissue engineering [96, 98, 101–104]. Fibrin deposition, as already mentioned, will be useful as an important agent and will show the earliest signs of plaque rupture and also intraplaque hemorrhage [105, 106]. These fibrin-specific nanoparticles, which accumulate at the sites of the thrombi, can be detected by US or MR imaging [107, 108] (Fig. 17.7). MR imaging of angiogenesis can also be done when the conjugated paramagnetic nanoparticles are used as markers of angiogenesis [97, 109–111]. With this type of cardiovascular intervention, the transport function of particles will allow for more target-specific imaging, with better distribution of the particles.

Thus, the nanoparticles have various advantages based on the image contrast, electromagnetic properties, and better bio-size compatibility, which can be used for diagnosis and therapy. The technology has shown that the drug-eluting

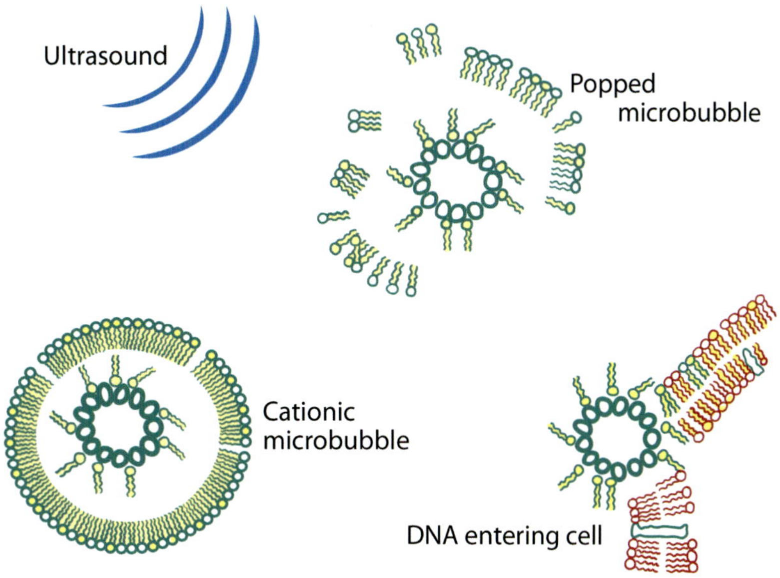

Fig. 17.7 The ultrasound action in the microbubble and DNA entering the cell

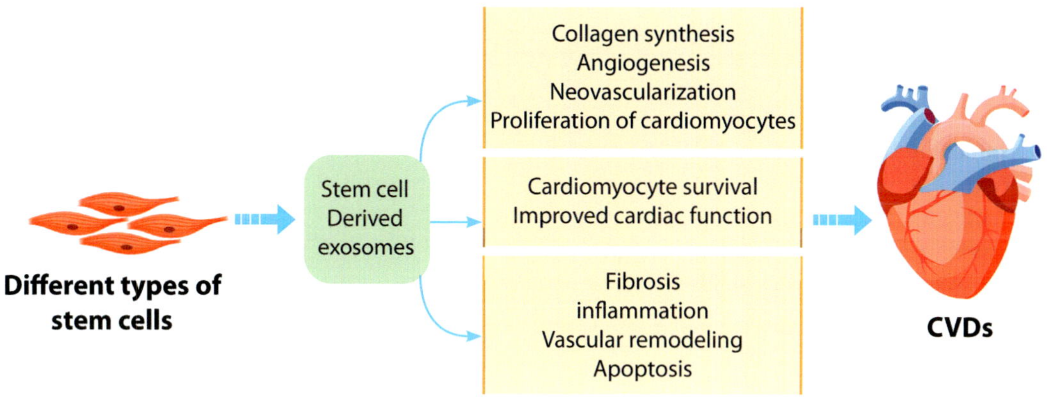

Fig. 17.8 Different types of the stem cells which are derived from exosomes and their actions

stents, which are now nano-mediated drug-eluting stents, will lead to an increase in the half-life of the drug by planned sustained release. These nanoparticle imaging techniques can be monitored for the various drug-eluting stents at the site of stem cell delivery (Fig. 17.8). In conclusion, nanotechnology has made significant advancements in the future imaging of various cardiovascular diseases.

17.7 Cardiovascular Interventions

CVS is an important entity for deciding the specific targets regarding different therapeutics. One type of therapy is nanotherapy, which is non-invasive in nature and can detect the varying amounts of transgenes, different gene-carrying vectors, and also the gene-expressed proteins as well as enzymes. This can be easily identified by various techniques like histology, immunoblotting, autopsy, and biopsy. The different cellular structures can be identified by the various signaling processes, which can define the different processes. Thus, the role of transgene expression with function can be helpful for the different genes in optical imaging and also in the angiogenesis of the ischemic heart, which acts as a therapeutic option [112, 113]. These can be used in specific areas of interest, like the ischemic myocardium, where the VEGF gene expression will be present and is also a component of the monitoring of the ischemic heart. Thus, it is part of the optical-based imaging technique of fluorescent imaging. The different expressions of both the cell surface receptor and the therapeutic protein, where the systemic delivery of the imaging particles can be useful for the better assessment of the function of the therapeutic genes.

17.8 Conclusion

Nanoimaging in the cardiovascular system has evolved in very different ways and has reached the molecular level, where the quantification of angiogenesis can be made, and further, the role of the various particles is also seen based on various receptor algorithms.

References

1. Patra HK, Banerjee S, Chaudhuri U, Lahiri P, Dasgupta AK. Cell selective response to gold nanoparticles. Nanomedicine. 2007;3:111–9.
2. Patra HK, Dasgupta AK. Cancer cell response to nanoparticles: criticality and optimality. Nanomedicine. 2012;8:842–52.
3. Waters EA, Chen J, Yang X, Zhang H, Neumann R, Santeford A, et al. Detection of targeted perfluorocarbon nanoparticle binding using 19F diffusion weighted MR spectroscopy. Magn Reson Med. 2008;60:1232–6.
4. Jasmin, Torres AL, Jelicks L, de Carvalho AC, Spray DC, Mendez-Otero R. Labeling stem cells with superparamagnetic iron oxide nanoparticles: analysis of the labeling efficacy by microscopy and magnetic resonance imaging. Methods Mol Biol. 2012;906:239–52.
5. Zhang Z, Mascheri N, Dharmakumar R, Li D. Cellular magnetic resonance imaging: potential for use in assessing aspects of cardiovascular disease. Cytotherapy. 2008;10:575–86.
6. Von Eschenbach AC. NCI sets goal of eliminating suffering and death due to cancer by 2015. J Natl Med Assoc. 2003;95:637–9.
7. McCarthy JR. Nanomedicine and cardiovascular disease. Curr Cardiovasc Imaging Rep. 2010;3:42–9.
8. Chopard R, Jehl J, Dutheil J, Genon VD, Seronde MF, Kastler B, et al. Evolution of acute coronary syndrome with normal coronary arteries and normal cardiac magnetic resonance imaging. Arch Cardiovasc Dis. 2011;104:509–17.
9. Stelzer R, Hutz RJ. Gold nanoparticles enter rat ovarian granulosa cells and subcellular organelles, and alter in-vitro estrogen accumulation. J Reprod Dev. 2009;55:685–90.
10. Qi L, Wu L, Zheng S, Wang Y, Fu H, Cui D. Cell-penetrating magnetic nanoparticles for highly efficient delivery and intracellular imaging of siRNA. Biomacromolecules. 2012;13:2723–30.
11. Yu SS, et al. Enzymatic- and temperature-sensitive controlled release of ultrasmall superparamagnetic iron oxides (USPIOs). J Nanobiotechnol. 2011;9:7.
12. Deb S, Raja SO, Dasgupta AK, Sarkar R, Chattopadhyay AP, Chaudhuri U, et al. Surface tunability of nanoparticles in modulating platelet functions. Blood Cells Mol Dis. 2012;48:36–44.
13. Karra N, Benita S. The ligand nanoparticle conjugation approach for targeted cancer therapy. Curr Drug Metab. 2012;13:22–41.
14. Park MH, Agasti SS, Creran B, Kim C, Rotello VM. Controlled and sustained release of drugs from dendrimer-nanoparticle composite films. Adv Mater. 2011;23:2839–42, 2843.
15. Deb S, Dasgupta AK. Thrombotic inception at nano-scale. In: Brizzio ME, editor. Acute coronary syndromes. Rijeka: Intech; 2012. http://www.intechopen.com/books/acute-coronary-syndromes/thrombotic-inception-atnano-scale. Accessed 15 Mar 2015.
16. Guha S, Mookerjee S, Guha P, Sardar P, Deb S, Roy PD, et al. Antiplatelet drug resistance in patients with recurrent acute coronary syndrome undergoing conservative management. Indian Heart J. 2009;61:348–52.
17. Guha S, Sardar P, Guha P, Deb S, Karmakar R, Chakraborti P, et al. Dual antiplatelet therapy in ACS: time-dependent variability in platelet aggregation during the first week. Indian Heart J. 2009;61:173–7.
18. Guha S, Sardar P, Guha P, Roy S, Mookerjee S, Chakrabarti P, et al. Dual antiplatelet drug resistance in patients with acute coronary syndrome. Indian Heart J. 2009;61:68–73.
19. Lahiri P, Roy S, Sardar P, Deb S, Chakrabarti P, Guha P, et al. Platelet responsiveness to yohimbine hydrochloride and MRS2179 in the context of the interaction between collagen and epinephrine in

acute coronary syndrome. Blood Cells Mol Dis. 2009;43:105–10.
20. Hyafil F, Cornily JC, Rudd JH, Machac J, Feldman LJ, Fayad ZA. Quantification of inflammation within rabbit atherosclerotic plaques using the macrophage-specific CT contrast agent N1177: a comparison with 18F-FDG PET/CT and histology. J Nucl Med. 2009;50:959–65.
21. Kerwin WS, Naumova A, Storb R, Tapscott SJ, Wang Z. Mapping contrast agent uptake and retention in MRI studies of myocardial perfusion: case control study of dogs with Duchenne muscular dystrophy. Int J Cardiovasc Imaging. 2013;29:819–26.
22. Arifin DR, Kedziorek DA, Fu Y, Chan KW, McMahon MT, Weiss CR, et al. Microencapsulated cell tracking. NMR Biomed. 2013;26:850–9.
23. Perazella MA. Gadolinium-contrast toxicity in patients with kidney disease: nephrotoxicity and nephrogenic systemic fibrosis. Curr Drug Saf. 2008;3:67–75.
24. Cao CY, Shen YY, Wang JD, Li L, Liang GL. Controlled intracellular self-assembly of gadolinium nanoparticles as smart molecular MR contrast agents. Sci Rep. 2013;3:1024.
25. Shellock FG, Kanal E. Safety of magnetic resonance imaging contrast agents. J Magn Reson Imaging. 1999;10:477–84.
26. Pellico J, et al. HAP-multitag, a PET and positive MRI contrast nanotracer for the longitudinal characterization of vascular calcifications in atherosclerosis. ACS Appl Mater Interfaces. 2021;13:45279–90.
27. Grobner T, Prischl FC. Gadolinium and nephrogenic systemic fibrosis. Kidney Int. 2007;72:260–4.
28. Ghaghada KB, Ravoori M, Sabapathy D, Bankson J, Kundra V, Annapragada A. New dual mode gadolinium nanoparticle contrast agent for magnetic resonance imaging. PLoS One. 2009;4:e7628.
29. Korkusuz H, Ulbrich K, Welzel K, Koeberle V, Watcharin W, Bahr U, et al. Transferrin-coated gadolinium nanoparticles as MRI contrast agent. Mol Imaging Biol. 2013;15:148–54.
30. Iafisco M, Alogna A, Miragoli M, Catalucci D. Cardiovascular nanomedicine: the route ahead. Nanomedicine. 2019;14:2391–4.
31. Park JA, Kim HK, Kim JH, Jeong SW, Jung JC, Lee GH, et al. Gold nanoparticles functionalized by gadolinium-DTPA conjugate of cysteine as a multimodal bioimaging agent. Bioorg Med Chem Lett. 2010;20:2287–91.
32. Wang YX. Superparamagnetic iron oxide based MRI contrast agents: current status of clinical application. Quant Imaging Med Surg. 2011;1:35–40.
33. Yilmaz A, Rösch S, Yildiz H, Klumpp S, Sechtem U. First multiparametric cardiovascular magnetic resonance study using ultrasmall superparamagnetic iron oxide nanoparticles in a patient with acute myocardial infarction: new vistas for the clinical application of ultrasmall superparamagnetic iron oxide. Circulation. 2012;126:1932–4.
34. Gu J, Xu H, Han Y, Dai W, Hao W, Wang C, et al. The internalization pathway, metabolic fate and biological effect of superparamagnetic iron oxide nanoparticles in the macrophage-like RAW264.7 cell. Sci China Life Sci. 2011;54:793–805.
35. Finn AV, Nakano M, Narula J, Kolodgie FD, Virmani R. Concept of vulnerable/unstable plaque. Arterioscler Thromb Vasc Biol. 2010;30:1282–92.
36. Waxman S, Ishibashi F, Muller JE. Detection and treatment of vulnerable plaques and vulnerable patients: novel approaches to prevention of coronary events. Circulation. 2006;114:2390–411.
37. Hu C, Luo R, Wang Y. Heart valves cross-linked with erythrocyte membrane drug-loaded nanoparticles as a biomimetic strategy for anti-coagulation, anti-inflammation, anti-calcification, and endothelialization. ACS Appl Mater Interfaces. 2020;12:41113–26.
38. Martinet W, De Meyer GRY. Selective depletion of macrophages in atherosclerotic plaques: myth, hype, or reality? Circ Res. 2007;100:751–3.
39. Jaffer FA, Libby P, Weissleder R. Molecular imaging of cardiovascular disease. Circulation. 2007;116:1052–61.
40. Hirota K, Terada H. Endocytosis of particle formulations by macrophages and its application to clinical treatment, molecular regulation of endocytosis. In: Ceresa B, editor. Biochemistry, genetics and molecular biology. CCBY; 2012. http://www.intechopen.com/books/molecular-regulation-of-endocytosis/endocytosis-of-particleformulations-by-macrophages-and-its-application-to-clinicaltreatment. Accessed 25 Mar 2015.
41. Nahrendorf M, Zhang H, Hembrador S, Panizzi P, Aikawa E, et al. Nanoparticle PET-CT imaging of macrophages in inflammatory atherosclerosis. Circulation. 2008;117:379–87.
42. Kang HW, Kim J, Yu Y, Oh J. Photoacoustic response of magnetic nanoparticles to pulsed laser irradiation. J Korean Phys Soc. 2009;55:2224–8.
43. Cormode DP, Roessl E, Thran A, Skajaa T, Gordon RE, Schlomka JP, et al. Atherosclerotic plaque composition: analysis with multicolor CT and targeted gold nanoparticles. Radiology. 2010;256:774–82.
44. Van Herck JL, De Meyer GR, Martinet W, Salgado RA, Shivalkar B, De Mondt R, et al. Multi-slice computed tomography with N1177 identifies ruptured atherosclerotic plaques in rabbits. Basic Res Cardiol. 2010;105:51–9.
45. Majumudar MD, Yoo J, Keliher EJ, Truelove JJ, Iwamoto Y, Sena B, et al. Polymeric nanoparticle PET/MR imaging allows macrophage detection in atherosclerotic plaques. Circ Res. 2013;112:755–61.
46. Au JT, Craig G, Longo V, Zanzonico P, Mason M, Fong Y, et al. Gold nanoparticles provide bright long-lasting vascular contrast for CT imaging. AJR Am J Roentgenol. 2013;200:1347–51.
47. Kim D, Park S, Lee JH, Jeong YY, Jon S. Antibiofouling polymer-coated gold nanopar-

ticles as a contrast agent for in vivo X-ray computed tomography imaging. J Am Chem Soc. 2007;129:7661–5.
48. Danila D, Johnson E, Kee P. CT imaging of myocardial scars with collagen-targeting gold nanoparticles. Nanomedicine. 2013;9:1067–76.
49. Cormode DP, Naha PC, Fayad ZA. Nanoparticle contrast agents for computed tomography: a focus on micelles. Contrast Media Mol Imaging. 2014;9:37–52.
50. Nahrendorf M, Waterman P, Thurber G, Groves K, Rajopadhye M, Panizzi P, et al. Hybrid in vivo FMT-CT imaging of protease activity in atherosclerosis with customized nanosensors. Arterioscler Thromb Vasc Biol. 2009;29:1444–51.
51. McCarthy JR, Patel P, Botnaru I, Haghayeghi P, Weissleder R, Jaffer FA. Multimodal nanoagents for the detection of intravascular thrombi. Bioconjug Chem. 2009;20:1251–5.
52. Podrez EA, Abu-Soud HM, Hazen SL. Myeloperoxidase generated oxidants and atherosclerosis. Free Radic Biol Med. 2000;28:1717–25.
53. Panizzi P, Nahrendorf M, Wildgruber M, Waterman P, Figueiredo JL, Aikawa E, et al. Oxazine conjugated nanoparticle detects in vivo hypochlorous acid and peroxynitrite generation. J Am Chem Soc. 2009;131:15739–44.
54. McCarthy JR, Sazonova IY, Erdem SS, Hara T, Thompson BD, Patel P, et al. Multifunctional nanoagent for thrombus targeted fibrinolytic therapy. Nanomedicine (Lond). 2012;7:1017–28.
55. Winter PM, Neubauer AM, Caruthers SD, Harris TD, Robertson JD, Williams TA, et al. Endothelial alpha(v)beta3 integrin-targeted fumagillin nanoparticles inhibit angiogenesis in atherosclerosis. Arterioscler Thromb Vasc Biol. 2006;26:2103–9.
56. McCarthy JR, Korngold E, Weissleder R, Jaffer FA. A lightactivated theranostic nanoagent for targeted macrophage ablation in inflammatory atherosclerosis. Small. 2010;6:2041–9.
57. Pacher P, Beckman JS, Liaudet L. Nitric oxide and peroxynitrite in health and disease. Physiol Rev. 2007;87:315–424.
58. Xie J, Lee S, Chen X. Nanoparticle-based theranostic agents. Adv Drug Deliv Rev. 2010;62:1064–79.
59. Meltzer ME, Doggen CJ, de Groot PG, Rosendaal FR, Lisman T. The impact of the fibrinolytic system on the risk of venous and arterial thrombosis. Semin Thromb Hemost. 2009;35:468–77.
60. Tabish TA, Hayat H, Abbas A, Narayan RJ. Graphene quantum dots-based electrochemical biosensing platform for early detection of acute myocardial infarction. Biosensors. 2022;12:77.
61. Hu S, et al. Exosome-eluting stents for vascular healing after ischaemic injury. Nat Biomed Eng. 2021;5:1174–88.
62. Moreno PR, Purushothaman KR, Fuster V, Echeverri D, Truszczynska H, Sharma SK, et al. Plaque neovascularization is increased in ruptured atherosclerotic lesions of human aorta: implications for plaque vulnerability. Circulation. 2004;110:2032–8.
63. Sajid M, Stouffer GA. The role of alpha(v)beta3 integrins in vascular healing. Thromb Haemost. 2002;87:187–93.
64. Zhang C, Jugold M, Woenne EC, Lammers T, Morgenstern B, Mueller MM, et al. Specific targeting of tumor angiogenesis by RGD-conjugated ultrasmall superparamagnetic iron oxide particles using a clinical 1.5-T magnetic resonance scanner. Cancer Res. 2007;67:1555–62.
65. Waters EA, Chen J, Allen JS, Zhang H, Lanza GM, Wickline SA. Detection and quantification of angiogenesis in experimental valve disease with integrin-targeted nanoparticles and 19-fluorine MRI/MRS. J Cardiovasc Magn Reson. 2008;10:43.
66. Mohyeddin-Bonab M, Mohamad-Hassani MR, Alimoghaddam K, Sanatkar M, Gasemi M, Mirkhani H, et al. Autologous in vitro expanded mesenchymal stem cell therapy for human old myocardial infarction. Arch Iran Med. 2007;10:467–73.
67. Poh KK, Sperry E, Young RG, Freyman T, Barringhaus KG, Thompson CA. Repeated direct endomyocardial transplantation of allogeneic mesenchymal stem cells: safety of a high dose, "off-the-shelf", cellular cardiomyoplasty strategy. Int J Cardiol. 2007;117:360–4.
68. Zhang Y, Li TS, Lee ST, Wawrowsky KA, Cheng K, Galang G, et al. Dedifferentiation and proliferation of mammalian cardiomyocytes. PLoS One. 2010;5:e12559.
69. Hasan A, et al. Micro and nanotechnologies in heart valve tissue engineering. Biomaterials. 2016;103:278–92.
70. Toma C, Pittenger MF, Cahill KS, Byrne BJ, Kessler PD. Human mesenchymal stem cells differentiate to a cardiomyocyte phenotype in the adult murine heart. Circulation. 2002;105:93–8.
71. Chen SL, Fang WW, Qian J, Ye F, Liu YH, Shan SJ, et al. Improvement of cardiac function after transplantation of autologous bone marrow mesenchymal stem cells in patients with acute myocardial infarction. Chin Med J. 2004;117:1443–8.
72. Lu M, Zhao S, Liu Q, Jiang S, Song P, Qian H, et al. Transplantation with autologous mesenchymal stem cells after acute myocardial infarction evaluated by magnetic resonance imaging: an experimental study. J Thorac Imaging. 2012;27:125–35.
73. Kedziorek DA, Kraitchman DL. Superparamagnetic iron oxide labeling of stem cells for MRI tracking and delivery in cardiovascular disease. Methods Mol Biol. 2010;660:171–83.
74. Boilson BA, Raichlin E, Park SJ, Kushwaha SS. Device therapy and cardiac transplantation for end-stage heart failure. Curr Probl Cardiol. 2010;35:8–64.
75. Oduk Y, et al. VEGF nanoparticles repair the heart after myocardial infarction. Am J Physiol Heart Circ Physiol. 2018;314:H278–84.
76. Calé R, Almeida M, Gonçalves P, Rebocho MJ, Raposo L, Teles R, et al. Complications of endomyocardial biopsy after heart transplantation: a lesser evil. Rev Port Cardiol. 2012;31:159–62.

77. Peters A, von Klot S, Heier M, Trentinaglia I, Hörmann A, Wichmann HE, et al., Cooperative Health Research in the Region of Augsburg Study Group. Exposure to traffic and the onset of myocardial infarction. N Engl J Med. 2004;351:1721–30.
78. Mills NL, Miller MR, Lucking AJ, Beveridge J, Flint L, Boere AJ, et al. Combustion-derived nanoparticulate induces the adverse vascular effects of diesel exhaust inhalation. Eur Heart J. 2011;32:2660–71.
79. Shimada A, Kawamura N, Okajima M, Kaewamatawong T, Inoue H, Morita T. Translocation pathway of the intratracheally instilled ultrafine particles from the lung into the blood circulation in the mouse. Toxicol Pathol. 2006;34:949–57.
80. Vong LB, et al. Novel angiogenesis therapeutics by redox injectable hydrogel—regulation of local nitric oxide generation for effective cardiovascular therapy. Biomaterials. 2018;167:143–52.
81. Deb S, Patra HK, Lahiri P, Dasgupta AK, Chakrabarti K, Chaudhuri U. Multistability in platelets and their response to gold nanoparticles. Nanomedicine. 2011;7:376–84.
82. Deb S, Chatterjee M, Bhattacharya J, Lahiri P, Chaudhuri U, Pal Choudhuri S, et al. Role of purinergic receptors in platelet-nanoparticle interactions. Nanotoxicology. 2007;1:93–103.
83. Mikkelsen L, Sheykhzade M, Jensen KA, Saber AT, Jacobsen NR, Vogel U, et al. Modest effect on plaque progression and vasodilatory function in atherosclerosis-prone mice exposed to nanosized TiO(2). Part Fibre Toxicol. 2011;8:32.
84. Chen Z, Meng H, Xing G, Yuan H, Zhao F, Liu R. Age-related differences in pulmonary and cardiovascular responses to SiO2 nanoparticle inhalation: nanotoxicity has susceptible population. Environ Sci Technol. 2008;42:8985–92.
85. Yu M, Mo Y, Wan R, Chien S, Zhang X, Zhang Q. Regulation of plasminogen activator inhibitor-1 expression in endothelial cells with exposure to metal nanoparticles. Toxicol Lett. 2010;195:82–9.
86. Kang GS, Gillespie PA, Gunnison A, Moreira AL, Tchou Wong KM, Chen LC. Long-term inhalation exposure to nickel nanoparticles exacerbated atherosclerosis in a susceptible mouse model. Environ Health Perspect. 2011;119:176–81.
87. Jun EA, Lim KM, Kim K, Bae ON, Noh JY, Chung KH, Chung JH. Silver nanoparticles enhance thrombus formation through increased platelet aggregation and procoagulant activity. Nanotoxicology. 2011;5:157–67.
88. Reijnders L. The release of TiO2 and SiO2 nanoparticles from nanocomposites. Polym Degrad Stab. 2009;94:873–6.
89. Jomova K, Valko M. Advances in metal-induced oxidative stress and human disease. Toxicology. 2011;283:65–87.
90. Baroli B, Ennas MG, Loffredo F, Isola M, Pinna R, LópezQuintela MA. Penetration of metallic nanoparticles in human full-thickness skin. J Invest Dermatol. 2007;127:1701–12.
91. Mano SS, Kanehira K, Sonezaki S, Taniguchi A. Effect of polyethylene glycol modification of TiO2 nanoparticles on cytotoxicity and gene expressions in human cell lines. Int J Mol Sci. 2012;13:3703–17.
92. Phenrat T, Long TC, Lowry GV, Veronesi B. Partial oxidation ("aging") and surface modification decrease the toxicity of nanosized zerovalent iron. Environ Sci Technol. 2009;43:195–200.
93. Takahashi H, Niidome Y, Niidome T, Kaneko K, Kawasaki H, Yamada S. Modification of gold nanorods using phosphatidylcholine to reduce cytotoxicity. Langmuir. 2006;22:2–5.
94. Iijima M, Kamiya H. Surface modification for improving the stability of nanoparticles in liquid media. KONA Powder Part J. 2009;27:119–29.
95. Yan L, Zhao F, Li S, Hu Z, Zhao Y. Low-toxic and safe nanomaterials by surface-chemical design, carbon nanotubes, fullerenes, metallofullerenes, and graphenes. Nanoscale. 2011;3:362–82.
96. Kim H, et al. Affinity-driven design of cargo-switching nanoparticles to leverage a cholesterol-rich microenvironment for atherosclerosis therapy. ACS Nano. 2020;14:6519–31.
97. Beldman TJ, et al. Nanoparticle-aided characterization of arterial endothelial architecture during atherosclerosis progression and metabolic therapy. ACS Nano. 2019;13:13759–74.
98. Yin RX, Yang DZ, Wu JZ. Nanoparticle drug- and gene-eluting stents for the prevention and treatment of coronary restenosis. Theranostics. 2014;4:175–200.
99. Huang Y, et al. Platelet-derived nanomotor coated balloon for atherosclerosis combination therapy. J Mater Chem B. 2020;8:5765–75.
100. Du X, Yang Y, Le Visage C, et al. In vivo US monitoring of catheter-based vascular delivery of gene microspheres in pigs: feasibility. Radiology. 2003;228:555–9.
101. Lockhart JH, et al. Self-assembled miRNA-switch nanoparticles target denuded regions and prevent restenosis. Mol Ther. 2021;29:1744–57.
102. Banai S, Chorny M, Gertz SD, et al. Locally delivered nanoencapsulated tyrphostin (AGL2043) reduces neointima formation in balloon-injured rat carotid and stented porcine coronary arteries. Biomaterials. 2005;26:451–61.
103. Fayad ZA, Amirbekian V, Toussaint JF, Fuster V. Identification of interleukin-2 for imaging atherosclerotic inflammation. Eur J Nucl Med Mol Imaging. 2006;33:111–6.
104. Chen JW, Pham W, Weissleder R, Bogdanov A Jr. Human myeloperoxidase: a potential target for molecular MR imaging in atherosclerosis. Magn Reson Med. 2004;52:1021–8.
105. Lameijer M, et al. Efficacy and safety assessment of a TRAF6-targeted nanoimmunotherapy in atherosclerotic mice and non-human primates. Nat Biomed Eng. 2018;2:279–92.
106. Sirol M, Fuster V, Badimon JJ, et al. Chronic thrombus detection with in vivo magnetic resonance imag-

ing and a fibrin-targeted contrast agent. Circulation. 2005;112:1594–600.
107. Morawski AM, Winter PM, Crowder KC, et al. Targeted nanoparticles for quantitative imaging of sparse molecular epitopes with MRI. Magn Reson Med. 2004;51:480–6.
108. Fredman G, et al. Targeted nanoparticles containing the proresolving peptide Ac2-26 protect against advanced atherosclerosis in hypercholesterolemic mice. Sci Transl Med. 2015;7:275ra20.
109. Hood JD, Bednarski M, Frausto R, et al. Tumor regression by targeted gene delivery to the neovasculature. Science. 2002;296:2404–7.
110. Kerr JS, Mousa SA, Slee AM. Alpha(v-)beta(3) integrin in angiogenesis and restenosis. Drug News Perspect. 2001;14:143–50.
111. Winter PM, Morawski AM, Caruthers SD, et al. Molecular imaging of angiogenesis in early-stage atherosclerosis with alpha(v)beta3-integrin-targeted nanoparticles. Circulation. 2003;108:2270–4.
112. Ta HT, et al. Activatable magnetic resonance nanosensor as a potential imaging agent for detecting and discriminating thrombosis. Nanoscale. 2018;10:15103–15.
113. Gao F, Qiu B, Kar S, Zhan X, Hofmann LV, Yang X. Intravascular magnetic resonance/ radiofrequency may enhance gene therapy for prevention of in-stent neointimal hyperplasia. Acad Radiol. 2006;13:526–30.

18 Nanoimaging in Infection and Inflammation

18.1 Introduction

Inflammation imaging in vivo will be the better answer for the evaluation of various disease pathologies, like cancers, atherosclerosis, and neurodegeneration. This will also be very important in detecting inflammation. The advances in technology and the use of nanoparticles (NPs) will be related to the various biomarkers of inflammation, which can diagnose various diseases and prognoses for various preventative and therapeutic strategies (Fig. 18.1).

The newer concepts are related to the various imaging parameters of the macrophages and monocytes. This is also important for different pathogens, along with the different macrophages and monocytes, which have an important role in the evaluation of the different lymphocytes, cytokine, and chemokine secretion-related processes like phagocytosis [1]. Monocytes and macrophages are being targeted for the evaluation of various nanomaterials, like small and ultrasmall paramagnetic iron oxide NPs [2]. The NP uptake is based on different mechanisms, like the phagocytosis related to the monocyte–macrophage system [3]. Again, the entire action of the various NPs is seen by various parameters, like the amount of uptake, which can also be regulated by different populations [4]. Various modalities are capable of imaging the macrophages, like PET, SPECT, optical imaging, and CT.

The imaging of the vascular pathologies as well as the various disease parameters, like plaque detection, quantification, and rupture of the vulnerable plaques [3]. ^{18}F-FDG PET imaging will identify the different infiltrations of the macrophages and can study the various infective and inflammatory vascular plaques. ^{18}F-FDG can identify the macrophage density in a better way due to its sensitivity [5, 6]. TSPO ligands [7] and ^{11}C-PK11195 [8], DOTATATE (^{68}Ga and ^{64}Cu) are also important for the evaluation of various plaques. The macrophage infiltration into the circulation will be able to identify the plaques in the vessel wall, and this, along with the DOTATATE, will have different cellular components as well [9]. The use of various nanobodies [10] and SPIO are being used due to the T2 signal intensity reduction on the high-resolution magnetic resonance imaging (MRI) for quantifying the plaque burden, its quantification, and also the post-therapy changes [11–14]. The ^{18}F-FDG PET and USPIO are being used for the inflammation imaging, and in this comparison, the USPIO is better for plaque inflammation by various parameters [15–19].

S. Shaikh, *Nanoimaging - Future of Precision Medicine*,
https://doi.org/10.1007/978-981-97-7292-6_18

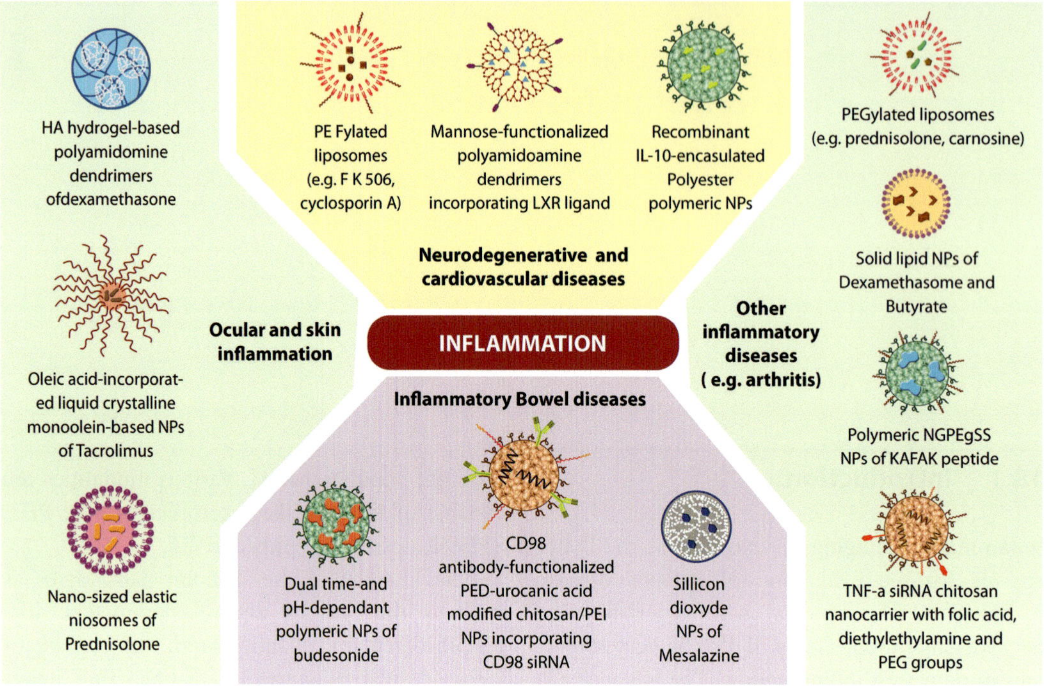

Fig. 18.1 Role of the nanoparticles in the inflammation in different conditions. Ocular and skin infection, neurodegenerative and CVS disorders, inflammatory bowel disease, arthritis and all these showing different mechanisms

18.2 Nanoimaging in Arthritis

Nanoimaging is being used in the evaluation of various types of arthritis at different stages [20], and this will result in a better analysis of various arthritis types [21, 22]. The concept of biosensing will be used for the evaluation of various inflammatory arthritis (Fig. 18.2). SPIONs are being used for the analysis of various inflammations and infections [23]. SPIONs can be used for the evaluation of rheumatoid arthritis [24, 25] by using the principle of targeting SPIONs, which can be attached to the monoclonal anti-CD3 antibody. Gold NPs are also an important entity for the evaluation of rheumatoid arthritis [26]. Gold NPs are being used in various modalities like CT, MRI, Raman spectroscopy, fluorescence, and photoacoustic imaging [27]. Gold nanomaterials make it very easy to evaluate various inflammation and arthritis imaging [28]. Marc et al. [29] showed that gold nanorods can target the necrosis factor. Optoacoustic molecular imaging is also an important entity for the evaluation by the optoacoustic signals, which have different magnitudes for the various arthritic joints in animal imaging. Other factors are also based on various factors like bioactivity, adsorption capacities, and polymer nanomaterials, which will have better protocols [30, 31]. One of the medicinal aspects is the use of Tocilizumab-loaded NPs, which will show RA joint tissue by non-invasive tracking [32]. The inflamed synovial membrane/fluid with the pannus of inflammatory vascular tissue can show the various signals from the inflamed tissues. In another study, Shuyi et al. [33] showed that different arthritis types will develop by using multimodal imaging agents [34] (Fig. 18.3). Nanoceria is also a compatible NP used for various types of arthritis in multimodal imaging [35, 36]. Clinically, arthritis is characterized by various decreased functions leading to deformity [37]. Veigas et al. [38] have shown a more cost-effective approach that can detect different RF markers. The nanoconjugates can accumulate,

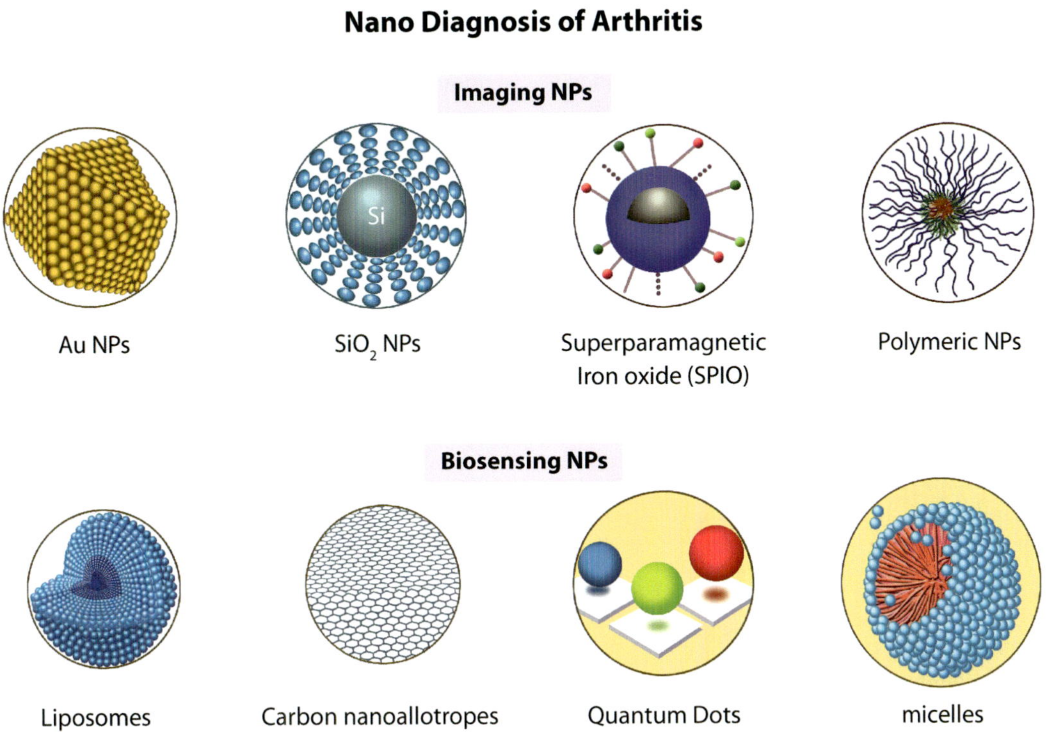

Fig. 18.2 Role of nano diagnosis in arthritis by using the imaging of nanoparticles and biosensing of the nanoparticles

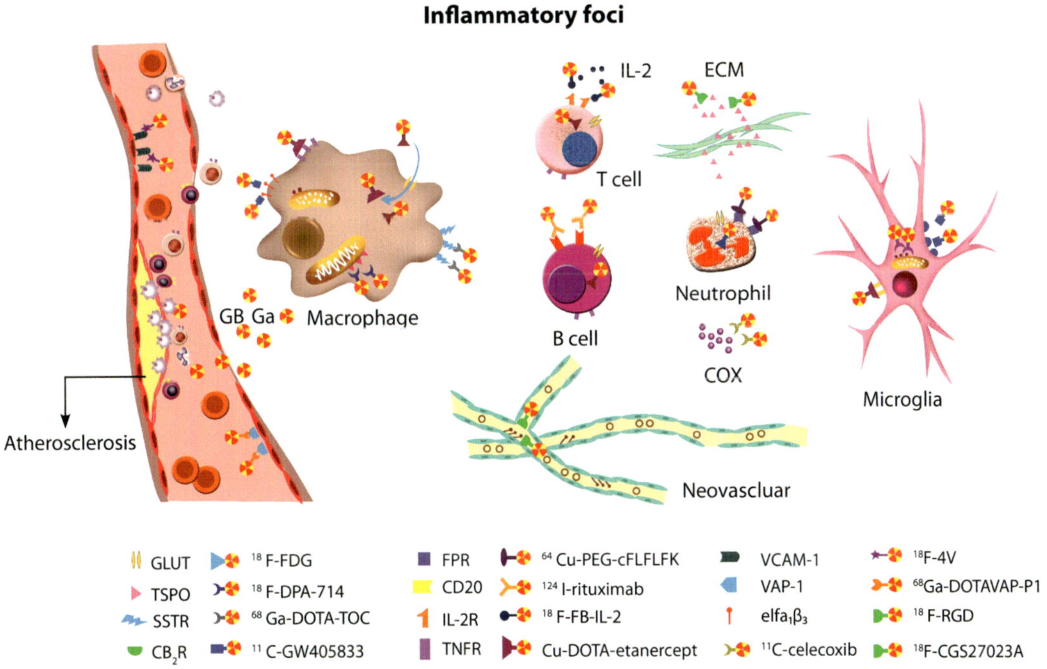

Fig. 18.3 Inflammatory foci showing the different mechanisms of the Inflammatory responses based on various reactions, different reactions and role of various biomarkers and all these having different roles in the evaluation of inflammation

and this can be identified by the change in the photonics-based biosensor applications, which are hybrid NPs having different metallic nanomaterials [39]. Hwang et al. [39] used different nanohybrids, which will use various surface-enhanced Raman scattering (SERS). The use of various mycoplasma pneumonia (MP) and RA shows that RA will show an increased risk [40]. Jia et al. [41] have shown that different MP in various immunoassays can be evaluated.

18.3 Rheumatoid Arthritis

Rheumatoid arthritis is associated with many other diseases, like the human immunodeficiency virus (HIV) infection, which is a global disease [42]. Islam et al. [43] have shown that different biomarkers can be used by activation for the management of various types of arthritis, particularly inflammatory arthritis. The inflammation is based on the amount and extent of the synovial joint, cartilage injury, along with various malformations [44]. The prevalence of this inflammatory disorder is higher in women and is three times more common compared to men [45]. Other organs that can be involved are the heart, lungs, kidneys, and pleura [44]. NSAIDs and disease-modifying antirheumatic drugs (DMARDs) are important therapeutic agents used in RA [45, 46]. The drugs that are being circulated are distributed throughout the entire body, and due to this, they may have more side effects because of greater and longer accumulation in the body. Sometimes, other side effects occur because the distribution is not reaching the disease and inflamed joints. Due to this, there will be more progression of the disease, and it will spread to vital organs, leading to further disease spread [46]. Therefore, newer NPs are being used, and many NPs from all categories are available (Fig. 18.4). These newer agents will have different biocompatibility, with

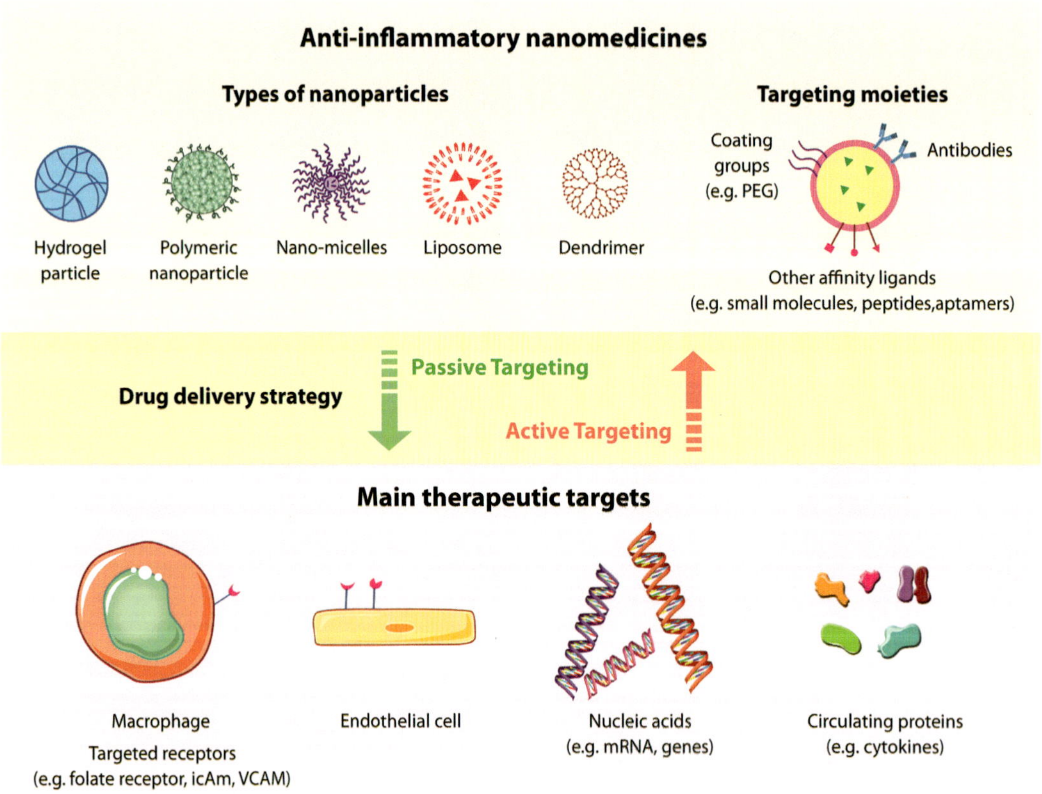

Fig. 18.4 Anti-inflammatory nanomedicines showing the role of the different targeting of nanoparticles, different moieties, drug delivery systems and therapeutic targets

a lack of toxicity, along with drug release, focused, and targeted drug delivery to the damaged and inflamed tissues in RA [44, 47, 48].

18.4 Liposomes

Liposomes are one of the established nanovehicles, made up of various lipid bilayers, which are surrounded by an aqueous core. These agents have lower immunogenicity with very high biocompatibility, with different conjugations of various hydrophilic and lipophilic agents, and similar side effects with better potential due to a better half-life. Another disadvantage is that the circulation time is very short, and they are removed from the circulation faster. Stability is also an issue, as they do not have long stability in physiological and pathological situations. The various liposomes will have different parameters, like osmotic pressure, lipid hydrolysis, and surfactant decomposition. Due to this, these NPs need to increase the integrity of their walls through the polymerization process [14]. In one of the studies by Guimarães et al. [49], one of the drugs, MTX, will be added to various liposomes, which will have various interactions and result in better bonding. Another aspect is that when the liposomes circulate in the body for a longer time, it will result in passive targeting, comprising various parameters like hydrophobicity, surface charge, and particle size. These liposomes have an affinity towards the synovium and leaky vessels at the damaged joints. The endothelial gap is variable in different patients, depending on various parameters [48, 49]. One study shows that the different mechanisms of the liposome's actions and targeting will have different impacts on various biophysical and biochemical reactions. The mechanism of action is that liposomes have significant anti-arthritic effects due to inflammation. Drug delivery can result in leaky vasculature, leading to different cell-mediated sequestration (ELVIS), which will accumulate in the inflamed joints, having better and more focused side effects [48] (Fig. 18.4). One study showed that different thermosensitive liposomes can be loaded into liposomal delivery systems by the principle of microwave hyperthermia. These agents show better-controlled release at the targeted sites with less toxicity. Polymeric NPs also have better biodegradability, biocompatibility, more flexibility, and better mechanical properties [50–52]. The chances of modification of various NPs in the conjugation process with other smaller molecules can improve the efficacy of this anti-RA drug [53]. Synthetic cationic polymers can also be used for the delivery of nucleic acids like DNA and RNA [54]. PEI is commonly used for better cationic charge density in the physiological pH settings for the attachment of nucleic acids [54]. Espinosa-Cano et al. [55] have also shown that different independent regulation mechanisms exist for the different NPs, which can accumulate and also deliver to the tissues. This will be helpful for the prevention of extravasation in the leaky vasculature into the surrounding tissues [56]. This is not possible due to the shorter half-life. Bashir et al. [57] showed that PLGA is better biocompatible and biodegradable among various polymers in drug delivery systems. The TFC NPs, through the process of nanoprecipitation, will show better delivery for depositing at the inflamed synovium [57]. A study by Howard et al. [58] has shown that RNA NPs can suppress different inflammatory macrophages. This will be shown in the form of cartilage, which will be damaged by various approaches in local and systemic inflammation. One of the studies by Ha et al. [59] has shown the significant potential for drug delivery using various multifunctional NPs (MNPs), as there is improved efficiency while simultaneously reducing side effects. The different interactions will show the different MNPs, which can show that fibroblast-like RA synoviocytes can be used [59]. NIR resonance is also an important entity, as the Au shell will accelerate the MTX release from the NPs. The MTX loaded in the MNP therapy can be used for managing RA [59]. One of the advantages of the resonance of the Au shell, combined with the MNPs, is that this is sometimes more efficient than chemotherapy regimens.

18.5 Niosomes

Niosomes are vesicles that are lamellar in shape, microscopic in size, and made up of various surfactants as well as cholesterol. Their advantage is that they can have controlled release of the drugs in the region of interest and also offer different biocompatibility for delayed clearance [60]. These niosomes are prepared in non-ionic surfactant vesicles, which can also impact niosome formation in the range of better hydration temperatures, along with the preparation method [61]. These primary surfactants, like the vesicles, are made up of both ionic and non-ionic surfactants due to their lower toxicity. Various surfactants will have different properties based on the various niosomes [62, 63]. In one of the studies [64], the vesicular carrier as a drug delivery system can also show better drug solubility and retention time, which will be increased at the absorption site. One of the studies by Mujeeb and Sailaja [65] showed the increased action of the drug with fewer side effects.

18.6 Metal NPs

These various metallic NPs have better properties and also thermal applications, and due to this, they are being more widely used [18]. Another advantage of these metallic NPs is that they are very small in size, usually below 100 nm [66].

18.6.1 Silver NPs

Silver NPs (Ag NPs) have anti-inflammatory activities due to their actions based on the action of the inflammatory precursors [67]. The mechanism of action of silver NPs is to reduce the vasculature endothelium [61–63]. One of the studies shows [64] the delivery of the encapsulated piroxicam in the vesicular carrier of non-ionic surfactant having transdermal patches [45]. Liposomal gel is being prepared such that the retention time topically will result in pain, which is due to RA. Mujeeb and Sailaja [65] have shown that RA management will increase the drug action, which has fewer side effects.

Metal NPs are nowadays more often used in various parameters in medicine, especially in the imaging field [18]. As already mentioned, these particles are very small and in the range below 100 nm [66, 67]. The silver NPs will leak into the extracellular space, where the airway wall will be thickened by increasing T helper type-2 (TH2) cell-mediated inflammation (IL-9, IL-4, IL-5, and IL-13) cells. The silver NPs will result in the reduction of cell production [68]. The inflammatory cells will infiltrate into the different cytokines and will be able to increase the amount of synovial inflammation, progressing as different anti-inflammatory macrophage phenotypes [69]. The concept here is that the NPs will dissolve when they enter the intracellular silver ions, and this mechanism is an important aspect of the anti-inflammatory process to remove reactive oxygen species (ROS), where the polarization can occur. Gold NPs [70] are being prepared from the aqueous leaf extracts of M., and they can be identified by X-ray diffraction (XRD) and have a very small diameter of 20–30 nm. In one of the studies, there were reactions between the synoviocytes and macrophages causing inflammatory changes, which can increase the anti-inflammatory reactions. The amount of absorption by the superparamagnetic iron oxide nanoparticles (SPIONs) on the synovial membrane is increased by the polyvinyl alcohol coating. These NPs are seen in the synovium, which will be the different platform for intra-articular drug delivery in both acute and chronic joint diseases [44]. The SPIONs have better drug delivery, which can be better targeted with an external magnetic field. Carneiro et al. [71, 72] have shown that when there is better magnetic targeting withSPIONs this will result in various side effects. Gold SPIONs are also administered intra-peritoneally in collagen-induced arthritis (CIA) via the synovial fluid of CIA [16] (Fig. 18.5). Different nanoformulations will offer more RA therapy and diagnostics, as they are not toxic and are focused on targeting the delivery of drugs [73] (Fig. 18.6).

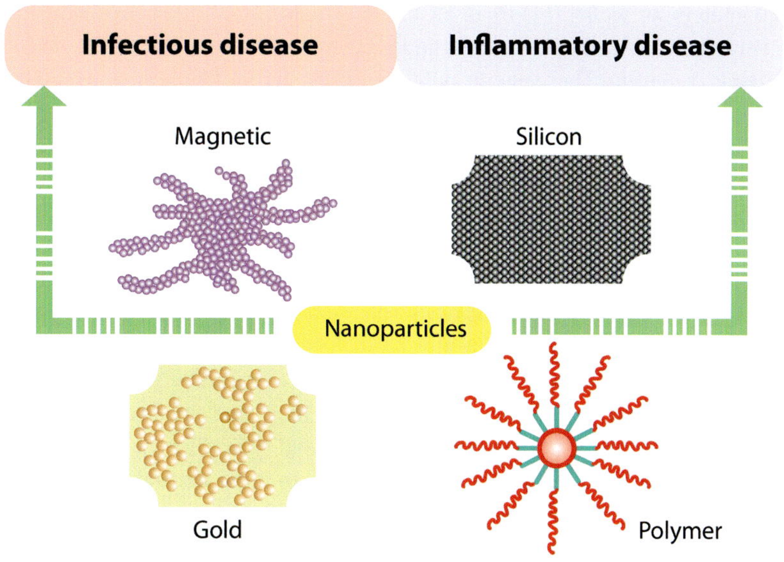

Fig. 18.5 The process of infection and inflammation and the use of the various nanoparticles like magnetic, silicon, gold and polymer

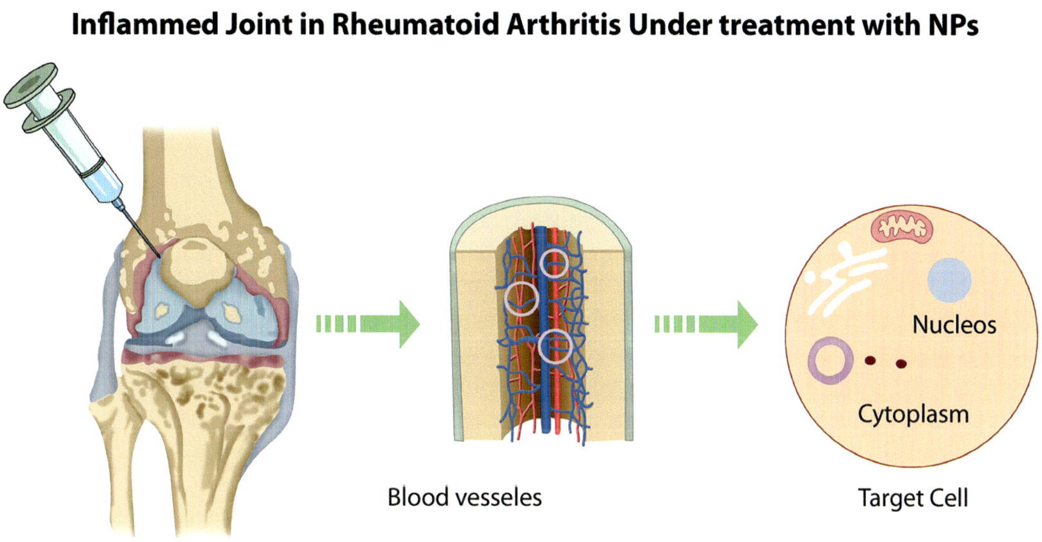

Fig. 18.6 Nanoparticle use in the rheumatoid arthritis inflammation and targeted to joint involved

18.7 Quantum Dots

Quantum dots are inorganic nanomaterials or semiconductor nanocrystals in the size range of 1–10 nm, made up of a central conductor and peripheral shells made of CdS and ZnS [17]. QDs are widely used for precise drug delivery, abnormal cell imaging, and RA diagnosis and treatment [74]. tested mercaptopropionic acid (MPA) nanoconjugates, which can also be useful for better imaging of inflammation.

18.8 Solid Lipid NPs

SLNs have a diameter of 120–200 nm and are used for drug delivery, offering all the advantages in inflammation [67]. Anti-inflammatory properties include precise targeting, less absorption, faster metabolism, and rapid elimination. Arora et al. [75] studied and tried to manage joint pathology. TNF-α, CRP, and oxidative stress in CFA also cause inflammation, which is again dose-dependent on the immunomodulatory parameters as well [76]. Pi is an alkaloid substance derived from various fruits and roots of the Piperaceae group. These agents will have a better response to oral chloroquine suspension [77].

Polymeric micelles are also being used as they have better biocompatibility and biodegradable properties due to the various drug delivery systems [78]. The micelle-based system exists for the management of RA [79]. Glucocorticoids (GLCs) are also present in various rheumatoid arthritis treatments in very high doses. Wang et al. [80–87] have shown a targeted, very low-dose drug delivery system. The advantage of the micelles is that they have a longer half-life and circulate in the blood with better accumulation in the region of interest. Abdollahi et al. [87–90] studied the practical use of dextran stearate polymeric micelles.

18.9 Targeted CT IBD Imaging

CT is also one of the best modalities for imaging and has many advantages, but compared to nuclear medicine, it has lower soft tissue sensitivity. CT is also capable of detecting smaller molecular and cellular targets easily. Nano-based CT contrast agents are available for use in different CT applications in inflammatory bowel disease [91, 92]. These gold NPs will be tagged with NPs based on the surface of isothiocyanate as the surface coating [93], which will be associated with the NP platform for better intracellular uptake as well as NP degradation [94]. The newer concept of Spectral CT is better and more precise compared to routine CT applications [95]. Dual-energy and K-edge subtraction methods will also be a better option due to the deconvolution of the CT signal.

18.10 Targeted MRI Imaging

This is due to the availability of various agents that are being targeted for various cancers and inflammation. These are paramagnetic NPs with chelated (Gd) or manganese (Mn) attached to the surface of various NPs, or superparamagnetic (SPIO) or ferromagnetic iron oxide NPs [86–88]. Gd-targeted NPs, which are being used in various inflammation studies, have bioretention, biotransformation, and various other parameters [86]. Mn is also an endogenous metal ion, and the NPs will be combined with the Mn, which is chelated. These Mn NPs will be targeted in the macrophages of the plaques and show better safety with no toxicity [86, 96, 97]. The iron oxide-based NPs will also show lower toxicity and high biocompatibility. Iron oxide particles have an affinity for the reticuloendothelial system cells and are taken into these cells [88, 98, 99]. The Kupffer cells (macrophages) will degrade into the normal iron pool due to the bonding of transferrin and ferritin. The biodistribution, pharmacokinetics, and targeting of the iron oxide NPs will result in surface modifications. Different iron-based NPs will be added to dextran and other NPs [97, 100, 101]. SPIO NPs, which can be targeted or not targeted, can be added to the MR contrast agent, which is seen in various diseases like atherosclerosis, malignancy, multiple sclerosis, stroke, infection, transplant rejection, and arthritis in all the clinical and preclinical studies [102–104]. MRI, which is targeted, can be more challenging in the imaging of the GIT. A few of the clinical and preclinical studies using SPIO-based inflammation can be done at various cellular levels, which are nonspecific in various inflammatory pathologies [105–110]. Frericks et al. showed that the ultrasmall iron oxide particles can be used quantitatively and qualitatively for IBD, showing more signal loss in T1-, T2-, and T2*-weighted images in inflammation. Oral MRI contrast agents are given for better diagnosis and evaluation of IBD, and also for follow-up [82]. Oral administration of Mn is better as this reduces artifacts due to peristalsis, and the air/lumen interface, allowing better identification of the wall [83, 84]. As Mn is paramagnetic, the uptake of Mn in the cell wall cells will show a better signal with enhancement. Even the iron-laden macrophages can be seen better when administered orally [85]. Various studies in IBD show a bright intestinal wall due to the amount of Mn uptake, all seen in the walls of the inflammation. Iron deposition seen as iron-laden cells can be confirmed by the In111 oxine-labeled macrophages using SPECT. The conclusion is that oral Mn, as well as various SPIO-labeled macrophages, will show MR imaging of inflammation in vivo. Even though MRI is an important molecular imaging modality, it has many limitations. In this scenario, the dual imaging paramagnetic/superparamagnetic approach will be efficiently used in motion blur scenarios.

18.11 Hepatitis

Viral hepatitis is a chronic liver disease that is seen globally. NPs are also being used for the diagnosis of hepatitis due to the various hepatitis B and C viruses. The existing techniques have low sensitivity and efficiency. The newer nanotechnology detection methods use various electrochemical techniques, such as gold-enhanced NPs with magnetic properties, which have shown that hepatitis is a very common condition [111]. Gold NP immunosensors are more compatible with these antibodies [111]. Nano-gold protein chips are used for the detection and antibody analysis of various types of hepatitis B and C simultaneously [112]. Gold NPs are more uniform and stable, making detection and analysis easier [112]. The nano-gold protein chip can be used for the detection of the hepatitis virus in the immune system. Silver staining of the gold NPs is more effective in the use of hepatitis B and C detection [113]. When NPs are added, they mimic the virus and also act as an anti-viral agent [114]. The nasal mucosa is more effective for hepatitis B vaccines, as it stimulates various antibody responses [115]. IgA was found to be useful as various bacteria and viruses adhere to the mucosal wall [116]. In particular, chitosan NPs possess desirable qualities for nasal vaccination, such as biodegradability, low toxicity, and close interaction with the mucosa [115]. In another study, DNA vaccine-loaded SiO_2-conjugated layered double hydroxide (LDH) NPs induced high serum antibody responses in vivo. These SiO_2@LDH NPs significantly promoted T-cell proliferation and skewed T-helper cells toward Th1 polarization [117].

18.12 Tuberculosis

Tuberculosis (TB) is the most common infectious disease caused by *Mycobacterium tuberculosis* (MTB), primarily affecting the respiratory system [118]. According to World Health Organization records, one-third of the world's population is infected with TB, making it the second most deadly infectious disease [119]. Multiple strains of MTB exist, which are difficult to manage despite DNA and RNA screening. The role of gold NPs in TB evaluation shows promising potential, as various gold NP probe assays can detect MTB strains within a few hours [120]. NPs can detect MTB by analyzing color change patterns. The amplification of MTB aids in identifying, analyzing, and detecting these strains [121]. Electrochemical biosensors can capture DNA fragments from MTB, allowing them to be labeled with gold NPs [121]. Aerosol techniques are widely used in the management of TB, and nebulized NPs are being developed for the administration of anti-TB drugs. These nebulized NPs can reduce the daily doses of anti-TB drugs effectively [122, 123]. Nebulized NPs also increase the bioavailability of various drugs, improving their effectiveness [122]. Polymeric NPs facilitate drug absorption in the gastrointestinal tract by adhering to the mucosa. The combination of anti-TB drugs with NPs enhances antimicrobial activity [123]. Iron-based NPs are used with chip sensor systems to detect *Mycobacterium tuberculosis* in less than an hour [124]. Multivalent surface-decorated nanocarriers can be used for efficient drug delivery in antituberculous treatments [125].

18.13 Bone Inflammation

Nanotechnology is increasingly being applied to evaluate and treat inflammation. Metal nanoparticles have an affinity for osteoblasts, enabling them to attach to bone tissue, where osteoblast growth is accelerated [126]. In the osteogenesis process, nanocomposite materials can stimulate various growth patterns in bones. Hydroxyapatite, for example, promotes osteoblast production

[127], and hydroxyapatite nanotubes further enhance osteoblast formation. Carbon nanotubes, with a diameter of around 60 nm, play a role in osseointegration by supporting the growth of other cells like fibroblasts [127]. Titanium (Ti) is a preferred material in nanotechnology due to its ability to prevent corrosion and its excellent biocompatibility. Titanium nanomaterials are in direct contact with bone and prevent inflammation at the surgical implant site [128]. Heparin, known for its anti-inflammatory and anticoagulant properties, enhances osteoblast production and osteogenic activity [129]. Magnetic nanoparticles are also useful in the management of bone diseases and infections. These nanoparticles, guided by a magnetic field, act on osteoblasts, with calcium phosphate helping facilitate bone repair and treatment of bone diseases [130]. Polyamidoamine (PAMAM) dendrimers are employed as anti-inflammatory agents for bones, providing better compatibility due to their primary amino groups and uniformity [131]. SPIONs combined with PLGA particles are used in the treatment of joint inflammation [130].

Nanotechnology is commonly used in the management of various skin infections [132, 133]. These nanomaterials can be designed for controlled drug release over a specific time duration, without causing harm to the skin before or after drug release [134, 135]. Other formulations can be developed in a similar manner [136, 137]. Nitric oxide-embedded nanoparticles are often used for injectable treatments for skin infections [138]. The stratum corneum, the outermost layer of the skin, acts as an important barrier to prevent UV radiation. Sunscreens that contain nanoparticles are essential for consumers, as they help protect the skin from UV damage. Zinc oxide (ZnO) and titanium dioxide (TiO_2) nanoparticles protect the skin by absorbing harmful UVA and UVB radiation, which are then reflected back into the atmosphere. The blocking of UV photons occurs through absorption, reflection, and scattering processes [139, 140]. One study has shown that iron oxide nanoparticles are effective for delivering thrombin [141], which has direct effects on inflammatory cells, fibroblasts, and endothelial cells, playing a key role in inflammation [142]. Antibiotics used for skin inflammation and wounds can be efficiently delivered via nanoparticles, ensuring controlled release and reducing the need for high antibiotic doses [143]. Psoriasis, a condition causing chronic inflammation of the skin and joints, accelerates skin cell multiplication [144]. Nanoparticles facilitate the easy penetration of anti-inflammatory drugs through the skin's various barriers [145, 146]. Nanovesicular formulations enhance both drug accumulation on the skin surface and its permeation below the skin [147].

18.14 Internal Inflammation

Nanotechnology has made significant advancements in improving drug release within internal organs [148, 149]. Peptide conjugates, for example, enhance cellular interactions, providing better protein release and signaling [150]. Peptide nanoparticles in the body can specifically target inflamed cells, ensuring more effective drug concentration at the site of inflammation [150]. Nanoparticles can also trigger internal inflammation due to their small particle size and surface coatings. The ideal nanoparticle size for such applications is between 120 and 200 nm [151]. Nanoparticles composed of PLA/PLGA with PEG grafts produce fewer immune responses, making them suitable for therapeutic applications [151]. Mesoporous silica nanoparticles reduce inflammatory responses in cells by interacting with pro-inflammatory cytokines [152]. Silica particles can track the movement of particles in cells or organs, useful for monitoring in both normal and cancerous tissues. Nanotechnology also plays a critical role in regulating neuroinflammation, especially in conditions like multiple sclerosis. Nanoparticles can deliver therapeutic medication to the brain, addressing neuroinflammation and potentially preventing disease progression [153]. A study showed that gadolinium diethylenetriaminepentaacetic fatty acid (Gd-DTPA-FA) nanoparticles, combining a DTPA-FA ligand with gadolinium acetate, are a more efficient MRI contrast agent for diagnosing early acute pancreatitis [154].

18.15 Conclusion

The use of various nanoparticles in therapeutic applications for infective and inflammatory lesions represents a significant advancement. Nanotechnology is expected to play a key role in future diagnostic and therapeutic approaches, offering innovative solutions for treating and managing these conditions.

References

1. Rua R, McGavern DB. Elucidation of monocyte/macrophage dynamics and function by intravital imaging. J Leukoc Biol. 2015;98:319–32.
2. Neuwelt A, Sidhu N, Hu CA, Mlady G, Eberhardt SC, Sillerud LO. Iron-based superparamagnetic nanoparticle contrast agents for MRI of infection and inflammation. AJR. 2015;204:W302–13.
3. Moore KJ, Tabas I. Macrophages in the pathogenesis of atherosclerosis. Cell. 2011;145:341–55.
4. Weissleder R, Nahrendorf M, Pittet MJ. Imaging macrophages with nanoparticles. Nat Mater. 2014;13:125–38.
5. Ali A, Tawakol A. FDG PET/CT imaging of carotid atherosclerosis. Neuroimaging Clin N Am. 2016;26:45–54.
6. Menezes LJ, Kotze CW, Agu O, et al. Investigating vulnerable atheroma using combined 18F-FDG PET/CT angiography of carotid plaque with immunohistochemical validation. J Nucl Med. 2011;52:1698–703.
7. Foss CA, Bedja D, Mease RC, et al. Molecular imaging of inflammation in the ApoE 2/2 mouse model of atherosclerosis with IodoDPA. Biochem Biophys Res Commun. 2015;461:70 5.
8. Gaemperli O, Shalhoub J, Owen DR, et al. Imaging intraplaque inflammation in carotid atherosclerosis with 11C-PK11195 positron emission tomography/computed tomography. Eur Heart J. 2012;33:1902–10.
9. Malmberg C, Ripa RS, Johnbeck CB, et al. 64Cu-DOTATATE for noninvasive assessment of atherosclerosis in large arteries and its correlation with risk factors: head-to-head comparison with 68Ga-DOTATOC in 60 patients. J Nucl Med. 2015;56:1895–900.
10. Bala G, Blykers A, Xavier C, et al. Targeting of vascular cell adhesion molecule-1 by 18F-labelled nanobodies for PET/CT imaging of inflamed atherosclerotic plaque. Eur Heart J Cardiovasc Imaging. 2016;17:1001. Epub ahead of print.
11. Morishige K, Kacher DF, Libby P, et al. High-resolution magnetic resonance imaging enhanced with superparamagnetic nanoparticles measures macrophage burden in atherosclerosis. Circulation. 2010;122:1707–15.
12. Millon A, Dickson SD, Klink A, et al. Monitoring plaque inflammation in atherosclerotic rabbits with an iron oxide (P904) and 18F-FDG using a combined PET/MR scanner. Atherosclerosis. 2013;228:339–45.
13. Stendahl JC, Sinusas AJ. Nanoparticles for cardiovascular imaging and therapeutic delivery, part 2: radiolabeled probes. J Nucl Med. 2015;56:1637–41.
14. Purohit BS, Ailianou A, Dulguerov N, Becker CD, Ratib O, Becker M. FDG-PET/CT pitfalls in oncological head and neck imaging. Insights Imaging. 2014;5(5):585–602.
15. Das SS, Bharadwaj P, Bilal M, et al. Stimuli-responsive polymeric nanocarriers for drug delivery, imaging, and theragnosis. Polymers (Basel). 2020;12(6):1397.
16. Dadfar SM, Camozzi D, Darguzyte M, et al. Size-isolation of superparamagnetic iron oxide nanoparticles improves MRI, MPI and hyperthermia performance. J Nanobiotechnol. 2020;18(1):22.
17. Malviya G, Galli F, Sonni I, Signore A. Imaging T-lymphocytes in inflammatory diseases: a nuclear medicine approach. Q J Nucl Med Mol Imaging. 2014;58(3):237–57.
18. van de Wiele C, Sathekge M, Maes A. Targeting monocytes and macrophages by means of SPECT and PET. Q J Nucl Med Mol Imaging. 2014;58(3):269–75.
19. Iodice V, Laganà B, Lauri C, et al. Imaging B lymphocytes in autoimmune inflammatory diseases. Q J Nucl Med Mol Imaging. 2014;58(3):258–68.
20. Jo J, Tian C, Xu G, Sarazin J, Schiopu E, Gandikota G, Wang X. Photoacoustic tomography for human musculoskeletal imaging and inflammatory arthritis detection. Photoacoustics. 2018;12:82–9.
21. Glaudemans AW, Quintero AM, Signore A. PET/MRI in infectious and inflammatory diseases: will it be a useful improvement? Berlin/Heidelberg: Springer; 2012.
22. Cormode DP, Skajaa T, Fayad ZA, Mulder WJ. Nanotechnology in medical imaging: probe design and applications. Arterioscler Thromb Vasc Biol. 2009;29:992–1000.
23. Dardzinski BJ, Schmithorst VJ, Holland SK, Boivin GP, Imagawa T, Watanabe S, Lewis JM, Hirsch R. MR imaging of murine arthritis using ultrasmall superparamagnetic iron oxide particles. Magn Reson Imaging. 2001;19:1209–16.
24. Lutz AM, Seemayer C, Corot C, Gay RE, Goepfert K, Michel BA, Marincek B, Gay S, Weishaupt D. Detection of synovial macrophages in an experimental rabbit model of antigen-induced arthritis: ultrasmall superparamagnetic iron oxide–enhanced MR imaging. Radiology. 2004;233:149–57.
25. Chen C-L, Siow TY, Chou C-H, Lin C-H, Lin M-H, Chen Y-C, Hsieh W-Y, Wang S-J, Chang C. Targeted superparamagnetic iron oxide nanoparticles for in vivo magnetic resonance imaging of

T-cells in rheumatoid arthritis. Mol Imaging Biol. 2017;19:233–44.
26. Finkelstein A, Walz D, Batista V, Mizraji M, Roisman F, Misher A. Auranofin. New oral gold compound for treatment of rheumatoid arthritis. Ann Rheum Dis. 1976;35:251–7.
27. Ma J, Liu X, Wang R, Zhang J, Jiang P, Wang Y, Tu G. Bimetallic core-shell 3nanostars with tunable surface plasmon resonance for surface-enhanced Raman scattering. ACS Appl Nano Mater. 2020;3:10885–94.
28. Zhao Y, Liu Y, Li X, Wang H, Zhang Y, Ma H, Wei Q. Label-free ECL immunosensor for the early diagnosis of rheumatoid arthritis based on asymmetric heterogeneous polyaniline-gold nanomaterial. Sens Actuators B Chem. 2018;257:354–61.
29. Fournelle M, Bost W, Tarner IH, Lehmberg T, Weiß E, Lemor R, Dinser R. Antitumor necrosis factor-α antibody-coupled gold nanorods as nanoprobes for molecular optoacoustic imaging in arthritis. Nanomed Nanotechnol Biol Med. 2012;8:346–54.
30. Sarkar S, Levi-Polyachenko N. Conjugated polymer nano-systems for hyperthermia, imaging and drug delivery. Adv Drug Deliv Rev. 2020;163:40–64.
31. Chen J, Qi J, Chen C, Chen J, Liu L, Gao R, Zhang T, Song L, Ding D, Zhang P. Tocilizumab-conjugated polymer nanoparticles for NIR-II photoacoustic-imaging-guided therapy of rheumatoid arthritis. Adv Mater. 2020;32:2003399.
32. Vu-Quang H, Vinding MS, Jakobsen M, Song P, Dagnaes-Hansen F, Nielsen NC, Kjems J. Imaging rheumatoid arthritis in mice using combined near infrared and 19F magnetic resonance modalities. Sci Rep. 2019;9:14314.
33. Xiao S, Tang Y, Lin Y, Lv Z, Chen L. Tracking osteoarthritis progress through cationic nanoprobe-enhanced photoacoustic imaging of cartilage. Acta Biomater. 2020;109:153–62.
34. Rieter WJ, Kim JS, Taylor KM, An H, Lin W, Tarrant T, Lin W. Hybrid silica nanoparticles for multimodal imaging. Angew Chem Int Ed. 2007;46:3680–2.
35. Celardo I, Pedersen JZ, Traversa E, Ghibelli L. Pharmacological potential of cerium oxide nanoparticles. Nanoscale. 2011;3:1411–20.
36. Gavrilă B, Ciofu C, Stoica V. Biomarkers in rheumatoid arthritis, what is new? J Med Life. 2016;9:144.
37. Humby FC, Al Balushi F, Lliso G, Cauli A, Pitzalis C. Can synovial pathobiology integrate with current clinical and imaging prediction models to achieve personalized health care in rheumatoid arthritis? Front Med. 2017;4:41.
38. Veigas B, Matias A, Calmeiro T, Fortunato E, Fernandes AR, Baptista PV. Antibody modified gold nanoparticles for fast colorimetric screening of rheumatoid arthritis. Analyst. 2019;144:3613–9.
39. Hwang EY, Lee JH, Lim DW. Compartmentalized bimetal cluster-poly (aniline) hybrid nanostructures for multiplexed detection of autoantibodies in early diagnosis of rheumatoid arthritis. Sens Actuators B Chem. 2020;321:128482.
40. Chu K-A, Chen W, Hsu CY, Hung Y-M, Wei JC-C. Increased risk of rheumatoid arthritis among patients with Mycoplasma pneumonia: a nationwide population-based cohort study in Taiwan. PLoS One. 2019;14:e0210750.
41. Jia X, Wang C, Rong Z, Li J, Wang K, Qie Z, Xiao R, Wang S. Dual dye-loaded Au@ Ag coupled to a lateral flow immunoassay for the accurate and sensitive detection of Mycoplasma pneumoniae infection. RSC Adv. 2018;8:21243–51.
42. Carroll MB, Fields JH, Clerc PG. Rheumatoid arthritis in patients with HIV: management challenges. Open Access Rheumatol Res Rev. 2016;8:51.
43. Islam S, Shukla S, Bajpai VK, Han Y-K, Huh YS, Kumar A, Ghosh A, Gandhi S. A smart nanosensor for the detection of human immunodeficiency virus and associated cardiovascular and arthritis diseases using functionalized graphene-based transistors. Biosens Bioelectron. 2019;126:792–9.
44. Oliveira IM, Gonçalves C, Reis RL, Oliveira JM. Engineering nanoparticles for targeting rheumatoid arthritis: past, present, and future trends. Nano Res. 2018;11:4489–506.
45. Paradkar M, Vaghela S. Thiocolchicoside niosomal gel formulation for the pain management of rheumatoid arthritis through topical drug delivery. Drug Deliv Lett. 2018;8:159–68.
46. Meka RR, Venkatesha SH, Acharya B, Moudgil KD. Peptide-targeted liposomal delivery of dexamethasone for arthritis therapy. Nanomedicine. 2019;14:1455–69.
47. Chabib L, Ikawati Z, Martien R, Ismail H, Wahyudi MDP, Arimurni DA, Muhtadi WK, Hidayat A. Rheumatoid arthritis and the challenge of using nanoparticles for its treatment. MATEC Web Conf. 2018;154:4005.
48. Wang Q, He L, Fan D, Liang W, Fang J. Improving the anti-inflammatory efficacy of dexamethasone in the treatment of rheumatoid arthritis with polymerized stealth liposomes as a delivery vehicle. J Mater Chem B. 2020;8:1841–51.
49. Guimarães D, Noro J, Loureiro A, Lager F, Renault G, Cavaco-Paulo A, Nogueira E. Increased encapsulation efficiency of methotrexate in liposomes for rheumatoid arthritis therapy. Biomedicines. 2020;8:630.
50. Ren H, He Y, Liang J, Cheng Z, Zhang M, Zhu Y, Hong C, Qin J, Xu X, Wang J. Role of liposome size, surface charge, and PEGylation on rheumatoid arthritis targeting therapy. ACS Appl Mater Interfaces. 2019;11:20304–15.
51. Shen Q, Zhang X, Qi J, Shu G, Du Y, Ying X. Sinomenine hydrochloride loaded thermosensitive liposomes combined with microwave hyperthermia for the treatment of rheumatoid arthritis. Int J Pharm. 2020;576:119001.
52. Xie Y, Tuguntaev RG, Mao C, Chen H, Tao Y, Wang S, Yang B, Guo W. Stimuli-responsive polymeric nanomaterials for rheumatoid arthritis therapy. Biophys Rep. 2020;6:193–210.

53. Jeong M, Park J-H. Nanomedicine for the treatment of rheumatoid arthritis. Mol Pharm. 2020;18:539–49.
54. Yu Z, Reynaud F, Lorscheider M, Tsapis N, Fattal E. Nanomedicines for the delivery of glucocorticoids and nucleic acids as potential alternatives in the treatment of rheumatoid arthritis. Wiley Interdiscip Rev Nanomed Nanobiotechnol. 2020;12:e1630.
55. Espinosa-Cano E, Aguilar MR, Portilla Y, Barber DF, San Román J. Anti-inflammatory polymeric nanoparticles based on ketoprofen and dexamethasone. Pharmaceutics. 2020;12:723.
56. Vyas D, O'Dell KM, Bandy JL, Boyce EG. Tofacitinib: the first Janus kinase (JAK) inhibitor for the treatment of rheumatoid arthritis. Ann Pharmacother. 2013;47:1524–31.
57. Bashir S, Aamir M, Sarfaraz RM, Hussain Z, Sarwer MU, Mahmood A, Akram MR, Qaisar MN. Fabrication, characterization and in vitro release kinetics of tofacitinib-encapsulated polymeric nanoparticles: a promising implication in the treatment of rheumatoid arthritis. Int J Polym Mater Polym Biomater. 2020;70:449–58.
58. Howard KA, Paludan SR, Behlke MA, Besenbacher F, Deleuran B, Kjems J. Chitosan/siRNA nanoparticle–mediated TNF-α knockdown in peritoneal macrophages for anti-inflammatory treatment in a murine arthritis model. Mol Ther. 2009;17:162–8.
59. Md S, Alhakamy NA, Aldawsari HM, Husain M, Kotta S, Abdullah ST, Fahmy UA, Alfaleh MA, Asfour HZ. Formulation design, statistical optimization, and in vitro evaluation of a naringenin nanoemulsion to enhance apoptotic activity in A549 lung cancer cells. Pharmaceuticals. 2020;13:152.
60. Bhardwaj S, Bhatia S. Development and characterization of niosomal gel system using Lallementia royaleana Benth. mucilage for the treatment of rheumatoid arthritis. Iran J Pharm Res. 2020;19:465–82.
61. Bhardwaj P, Tripathi P, Gupta R, Pandey S. Niosomes: a review on niosomal research in the last decade. J Drug Deliv Sci Technol. 2020;56:101581.
62. Lu B, Huang Y, Chen Z, Ye J, Xu H, Chen W, Long X. Niosomal nanocarriers for enhanced skin delivery of quercetin with functions of anti-tyrosinase and antioxidant. Molecules. 2019;24:2322.
63. Ag Seleci D, Seleci M, Walter J-G, Stahl F, Scheper T. Niosomes as nanoparticular drug carriers: fundamentals and recent applications. J Nanomater. 2016;2016:7372306.
64. Subashini Rajaram AS, Dharmalingam SR, Chidambaram K. Fabrication of non-ionic surfactant vesicular gel for effective treatment of rheumatoid arthritis. JEMDS. 2020;9:2289–96.
65. Mujeeb SA, Sailaja AK. Formulation of ibuprofen loaded niosomal gel by different techniques for treating rheumatoid arthritis. J Bionanosci. 2017;11:169–76.
66. Kaur M. Impact of response surface methodology–optimized synthesis parameters on in vitro anti-inflammatory activity of iron nanoparticles synthesized using Ocimum tenuiflorum Linn. BioNanoScience. 2020;10:1–10.
67. Janakiraman K, Krishnaswami V, Rajendran V, Natesan S, Kandasamy R. Novel nano therapeutic materials for the effective treatment of rheumatoid arthritis-recent insights. Mater Today Commun. 2018;17:200–13.
68. Agarwal H, Nakara A, Shanmugam VK. Anti-inflammatory mechanism of various metal and metal oxide nanoparticles synthesized using plant extracts: a review. Biomed Pharmacother. 2019;109:2561–72.
69. Yang Y, Guo L, Wang Z, Liu P, Liu X, Ding J, Zhou W. Targeted silver nanoparticles for rheumatoid arthritis therapy via macrophage apoptosis and re-polarization. Biomaterials. 2020;264:120390.
70. An H, Song Z, Li P, Wang G, Ma B, Wang X. Development of biofabricated gold nanoparticles for the treatment of alleviated arthritis pain. Appl Nanosci. 2020;10:617–22.
71. Park J-Y, Kwon S, Kim S-H, Kang YJ, Khang D. Triamcinolone-gold nanoparticles repolarize synoviocytes and macrophages in an inflamed synovium. ACS Appl Mater Interfaces. 2020;12:38936–49.
72. Carneiro MFH, Machado ART, Antunes LM, Souza TE, Freitas VA, Oliveira LC, Rodrigues JL, Pereira MC, Barbosa F. Gold-coated superparamagnetic iron oxide nanoparticles attenuate collagen-induced arthritis after magnetic targeting. Biol Trace Elem Res. 2020;194:502–13.
73. Albuquerque J, Moura CC, Sarmento B, Reis S. Solid lipid nanoparticles: a potential multifunctional approach towards rheumatoid arthritis theranostics. Molecules. 2015;20:11103–18.
74. Kalangi SK, Swarnakar NK, Sathyavathi R, Narayana Rao D, Jain S, Reddanna P. Synthesis, characterization, and biodistribution of quantum dot-celecoxib conjugate in mouse paw edema model. Oxidative Med Cell Longev. 2018;2018:3090517.
75. Arora R, Kuhad A, Kaur I, Chopra K. Curcumin loaded solid lipid nanoparticles ameliorate adjuvant-induced arthritis in rats. Eur J Pain. 2015;19:940–52.
76. Bhalekar MR, Madgulkar AR, Desale PS, Marium G. Formulation of piperine solid lipid nanoparticles (SLN) for treatment of rheumatoid arthritis. Drug Dev Ind Pharm. 2017;43:1003–10.
77. Alarifi S, Massadeh S, Al-Agamy M, Aamery MA, Al Bekairy A, Yassin AE. Enhancement of ciprofloxacin activity by incorporating it in solid lipid nanoparticles. Trop J Pharm Res. 2020;19:909–18.
78. Hwang D, Ramsey JD, Kabanov AV. Polymeric micelles for the delivery of poorly soluble drugs: from nanoformulation to clinical approval. Adv Drug Deliv Rev. 2020;156:80–118.
79. Yun L, Shang H, Gu H, Zhang N. Polymeric micelles for the treatment of rheumatoid arthritis. Crit Rev Ther Drug Carrier Syst. 2019;36:219–38.
80. Wang Q, Jiang J, Chen W, Jiang H, Zhang Z, Sun X. Targeted delivery of low-dose dexamethasone using PCL–PEG micelles for effective treatment of rheumatoid arthritis. J Control Release. 2016;230:64–72.
81. Abdollahi AR, Firouzian F, Haddadi R, Nourian A. Indomethacin loaded dextran stearate poly-

meric micelles improve adjuvant-induced arthritis in rats: design and in vivo evaluation. Inflammopharmacology. 2020;29:107–21.
82. Hubbell JA, Chilkoti A. Nanomaterials for drug delivery. Science. 2012;337(6092):303–5.
83. Bourzac K. Nanotechnology: carrying drugs. Nature. 2012;491(7425):58–60.
84. Alvarez-Lorenzo C, et al. Light-sensitive intelligent drug delivery systems. Photochem Photobiol. 2009;85(4):848–60.
85. Cholkar K, et al. Optimization of dexamethasone mixed nanomicellar formulation. AAPS PharmSciTech. 2014;15(6):1454–67.
86. Briley-Saebo K, Yeang C, Witztum JL, Tsimikas S. Imaging of oxidation-specific epitopes with targeted nanoparticles to detect high-risk atherosclerotic lesions: progress and future directions. J Cardiovasc Transl Res. 2014;7:719–36.
87. Briley-Saebo KC, Mulder WJ, Mani V, Hyafil F, Amirbekian V, Aguinaldo JG, Fisher EA, Fayad ZA. Magnetic resonance imaging of vulnerable atherosclerotic plaques: current imaging strategies and molecular imaging probes. J Magn Reson Imaging. 2007;26:460–79.
88. Briley-Saebo KC, Cho YS, Shaw PX, Ryu SK, Mani V, Dickson S, Izadmehr E, Green S, Fayad ZA, Tsimikas S. Targeted iron oxide particles for in vivo magnetic resonance detection of atherosclerotic lesions with antibodies directed to oxidation-specific epitopes. J Am Coll Cardiol. 2011;57:337–47.
89. Popovtzer R, Agrawal A, Kotov NA, Popovtzer A, Balter J, Carey TE, Kopelman R. Targeted gold nanoparticles enable molecular CT imaging of cancer. Nano Lett. 2008;8:4593–6.
90. Cai QY, Kim SH, Choi KS, Kim SY, Byun SJ, Kim KW, Park SH, Juhng SK, Yoon KH. Colloidal gold nanoparticles as a blood-pool contrast agent for X-ray computed tomography in mice. Investig Radiol. 2007;42:797–806.
91. Meir R, Motiei M, Popovtzer R. Gold nanoparticles for in vivo cell tracking. Nanomedicine (Lond). 2014;9:2059–69.
92. Betzer O, Shwartz A, Motiei M, Kazimirsky G, Gispan I, Damti E, Brodie C, Yadid G, Popovtzer R. Nanoparticle-based CT imaging technique for longitudinal and quantitative stem cell tracking within the brain: application in neuropsychiatric disorders. ACS Nano. 2014;8:9274–85.
93. Qin J, Peng C, Zhao B, Ye K, Yuan F, Peng Z, Yang X, Huang L, Jiang M, Zhao Q, et al. Noninvasive detection of macrophages in atherosclerotic lesions by computed tomography enhanced with PEGylated gold nanoparticles. Int J Nanomedicine. 2014;9:5575–90.
94. Fede C, Fortunati I, Weber V, Rossetto N, Bertasi F, Petrelli L, Guidolin D, Signorini R, De Caro R, Albertin G, et al. Evaluation of gold nanoparticles toxicity towards human endothelial cells under static and flow conditions. Microvasc Res. 2015;97:147–55.
95. Shikhaliev PM, Fritz SG. Photon counting spectral CT versus conventional CT: comparative evaluation for breast imaging application. Phys Med Biol. 1905;2011:56.
96. Xue S, Qiao J, Jiang J, Hubbard K, White N, Wei L, Li S, Liu ZR, Yang JJ. Design of ProCAs (protein-based Gd(3+) MRI contrast agents) with high dose efficiency and capability for molecular imaging of cancer biomarkers. Med Res Rev. 2014;34:1070–99.
97. Briley-Saebo KC, Nguyen TH, Saeboe AM, Cho YS, Ryu SK, Volkova ER, Dickson S, Leibundgut G, Wiesner P, Green S, et al. In vivo detection of oxidation-specific epitopes in atherosclerotic lesions using biocompatible manganese molecular magnetic imaging probes. J Am Coll Cardiol. 2012;59:616–26.
98. Petersein J, Saini S, Weissleder R. Liver II: iron oxide-based reticuloendothelial contrast agents for MR imaging. Clinical review. Magn Reson Imaging Clin N Am. 1996;4:53–60.
99. Lee SH, Kim BH, Na HB, Hyeon T. Paramagnetic inorganic nanoparticles as T1 MRI contrast agents. Wiley Interdiscip Rev Nanomed Nanobiotechol. 2014;6:196–209.
100. Xiao Y, Lin ZT, Chen Y, Wang H, Deng YL, Le DE, Bin J, Li M, Liao Y, Liu Y, et al. High molecular weight chitosan derivative polymeric micelles encapsulating superparamagnetic iron oxide for tumor-targeted magnetic resonance imaging. Int J Nanomedicine. 2015;10:1155–72.
101. Park J, An K, Hwang Y, Park JG, Noh HJ, Kim JY, Park JH, Hwang NM, Hyeon T. Ultra-large-scale syntheses of monodisperse nanocrystals. Nat Mater. 2004;3:891–5.
102. Tourdias T, Roggerone S, Filippi M, Kanagaki M, Rovaris M, Miller DH, Petry KG, Brochet B, Pruvo JP, Radüe EW, et al. Assessment of disease activity in multiple sclerosis phenotypes with combined gadoliniumand superparamagnetic iron oxide-enhanced MR imaging. Radiology. 2012;264:225–33.
103. Heesakkers RA, Jager GJ, Hovels AM, de Hoop B, van den Bosch HC, Raat F, Witjes JA, Mulders PF, van der Kaa CH, Barentsz JO, et al. Prostate cancer: detection of lymph node metastases outside the routine surgical area with ferumoxtran-10-enhanced MR imaging. Radiology. 2009;251:408–14.
104. Tang TY, Howarth SP, Miller SR, Graves MJ, Patterson AJ, U-King-Im JM, Li ZY, Walsh SR, Brown AP, Kirkpatrick PJ, et al. The ATHEROMA (atorvastatin therapy: effects on reduction of macrophage activity) study. Evaluation using ultrasmall superparamagnetic iron oxide-enhanced magnetic resonance imaging in carotid disease. J Am Coll Cardiol. 2009;53:2039–50.
105. Denis MC, Mahmood U, Benoist C, Mathis D, Weissleder R. Imaging inflammation of the pancreatic islets in type 1 diabetes. Proc Natl Acad Sci USA. 2004;101:12634–9.
106. Gaglia JL, Guimarães AR, Harisinghani M, Turvey SE, Jackson R, Benoist C, Mathis D, Weissleder

R. Noninvasive imaging of pancreatic islet inflammation in type 1A diabetes patients. J Clin Invest. 2011;121:442–5.
107. Beckmann N, Falk R, Zurbrugg S, Dawson J, Engelhardt P. Macrophage infiltration into the rat knee detected by MRI in a model of antigen-induced arthritis. Magn Reson Med. 2003;49:1047–55.
108. Jander S, Schroeter M, Saleh A. Imaging inflammation in acute brain ischemia. Stroke. 2007;38: 642–5.
109. Saleh A, Schroeter M, Jonkmanns C, Hartung HP, Mödder U, Jander S. In vivo MRI of brain inflammation in human ischaemic stroke. Brain. 2004;127:1670–7.
110. Glaudemans AW, Quintero AM, Signore A. PET/MRI in infectious and inflammatory diseases: will it be a useful improvement? Eur J Nucl Med Mol Imaging. 2012;39:745–9.
111. Tang DP, Yuan R, Chai YQ, Zhong X, Liu Y, Dai JY, Zhang LY. Anal Biochem. 2004;333:345–50.
112. Duan L, Wang Y, Li SS, Wan Z, Zhai J. BMC Infect Dis. 2005;5:53.
113. Abraham A, Kannangai R, Sirdharan G. Indian J Med Microbiol. 2008;26:297–301.
114. Klippstein R, Pozo D. Nanomed Nanotechnol Biol Med. 2010;6:523–9.
115. Borges O, Cordeiro-da-Silva A, Tavares J, Santarem N, de Sousa A, Borchard G, Junginger HE. Eur J Pharm Biopharm. 2008;69:405–16.
116. Khatri K, Goyal AK, Gupta PN, Mishra N, Vyas SP. Int J Pharm. 2008;354:235–41.
117. Wang J, Zhu R, Gao B, Wu B, Li K, Sun X, Liu H, Wang S. Biomaterials. 2014;35:466–78.
118. Shingadia D, Novelli V. Lancet Infect Dis. 2003;3:624–32.
119. World Health Organization. Global tuberculosis control: WHO report 2010. World Health Organization; 2010.
120. Veigas B, Machado D, Perdigao J, Portugal I, Couto I, Viveiros M, Baptista PV. Nanotechnology. 2010;21:415101.
121. Wang S, Inci F, De Libero G, Singhal A, Demirci U. Biotechnol Adv. 2013;31:438–49.
122. Sosnik A, Carcaboso AM, Glisoni RJ, Moretton MA, Chiappetta DA. Adv Drug Deliv Rev. 2010;62:547–59.
123. Gelperina S, Kisich K, Iseman MD, Heifets L. Am J Respir Crit Care Med. 2005;172:1487–90.
124. Chun AL. Nat Nanotechnol. 2009;4:1–2.
125. D'Addio SM, Baldassano S, Shi L, Cheung L, Adamson DH, Bruzek M, Anthony JE, Laskin DL, Sinko PJ, Prud'homme RK. J Control Release. 2013;168:41–9.
126. Tran N, Webster TJ. Wiley Interdiscip Rev Nanomed Nanobiotechnol. 2009;1:336–51.
127. Zhang L, Webster TJ. Nano Today. 2009;4:66–80.
128. Carmody EE, Schwarz EM, Puzas JE, Rosier RN, O'Keefe RJ. Arthritis Rheum. 2002;46:1298–308.
129. Kim SE, Song SH, Yun YP, Choi BJ, Kwon IK, Bae MS, Moon HJ, Kwon YD. Biomaterials. 2011;32:366–73.
130. Gu W, Wu C, Chen J, Xiao Y. Int J Nanomedicine. 2013;8:2305–17.
131. Thomas TP, Goonewardena SN, Majoros IJ, Kotlyar A, Cao Z, Leroueil PR, Baker JR Jr. Arthritis Rheum. 2011;63:2671–80.
132. Li W, Li X, Wang Q, Pan Y, Wang T, Wang H, Song R, Deng H. Carbohydr Polym. 2014;99:218–25.
133. Xin S, Li X, Wang Q, Huang R, Xu X, Lei Z, Deng H. J Biomed Nanotechnol. 2014;10:803–10.
134. Cevc G, Vierl U. J Control Release. 2010;141:277–99.
135. Wang Q, Zhang N, Hu X, Yang J, Du Y. J Biomed Mater Res A. 2007;82:122–8.
136. Wang Q, Du Y-M, Fan L-H, Liu H, Wang X-H. Wuhan Univ J Nat Sci. 2003;6:13.
137. Wang Q, Zhang N, Hu X, Yang J, Du Y. J Biomed Mater Res A. 2008;803(85):881–7.
138. Englander L, Friedman A. Clin Aesthet Dermatol. 2010;3:45–50.
139. Popov AP, Zvyagin AV, Lademann J, Roberts MS, Sanchez W, Priezzhev AV, Myllylä R. J Biomed Nanotechnol. 2010;6:432–51.
140. Kim SY, Jeong SH, Lee EY, Park Y-H, Bae HC, Jang YS, Maeng EH, Kim M-K, Son SW. Toxicol Environ Health Sci. 2012;3:258–61.
141. Ziv-Polat O, Topaz M, Brosh T, Margel S. Biomaterials. 2010;31:741–7.
142. Carney D, Mann R, Redin W, Pernia S, Berry D, Heggers J, Hayward P, Robson M, Christie J, Annable C. J Clin Invest. 1992;89:1469.
143. Tocco I, Zavan B, Bassetto F, Vindigni V. J Nanomaterials. 2012;2012:1–11.
144. dos Santos FK, Oyafuso MH, Kiill CP, Daflon-Germiao M, Chorilli M. Curr Nanosci. 2013;9:159–67.
145. Shah PP, Desai PR, Singh M. J Control Release. 2012;158:336–45.
146. Shah PP, Desai PR, Patel AR, Singh MS. Biomaterials. 2012;33:1607–17.
147. Caddeo C, Sales OD, Valenti D, Saurí AR, Fadda AM, Manconi M. Int J Pharm. 2013;443:128–36.
148. Du D, Chang N, Sun S, Li M, Yu H, Liu M, Liu X, Wang G, Li H, Liu X, Wang Q, Peng H. J Control Release. 2014;182:99–110.
149. Liu M, Li M, Sun S, Li B, Du D, Sun J, Cao F, Li H, Jia F, Wang T, Chang N, Yu H, Wang Q, Peng H. Biomaterials. 2014;35:3697–707.
150. Webber MJ, Kessler JA, Stupp SI. J Intern Med. 2010;267:71–88.
151. Higaki M. Medical Online; 2009. p. 112–7.
152. Lee S, Yun HS, Kim SH. Biomaterials. 2011;32:9434–43.
153. Singh AV, Khare M, Gade WN, Zamboni P. Autoimmune Dis. 2012;2012:160830.
154. Zhang H-W, Wang L-Q, Xiang Q-F, Zhong Q, Chen L-M, Xu C-X, Xiang X-H, Xu B, Meng F, Wan Y-Q. Biomaterials. 2014;35:356–67.

19 Preclinical Imaging

19.1 Introduction

Nanotechnology has evolved to a greater extent due to the manipulation of atomic and molecular scale matter in the preparation of new structures, devices, and materials. We all know that nanotechnology uses materials in sizes within the range of 1 and 100 nm [1]. The history of nanotechnology dates back 2000 years; the first use of gold nanoparticles was in the use of sulfide nanocrystals in hair dye for coloring glass in the ancient Greek era and also in the Roman era [2, 3]. The word "nanos" in Greek means dwarf [2, 3]. Feynman was the first person who initiated the concept of nanotechnology in 1959 [2]. Later, Smalley was awarded the Nobel Prize in Chemistry for his discovery of a new carbon nanotube, and he defined nanotechnology "as the art and science of building stuff that does stuff at the nanometer scale" [3]. Now, with a lot of advancements, nanotechnology has become interdisciplinary, involving different fields, and the medical field is one of the most developed. The exact definition of nanomedicine is the knowledge and skill of manipulating and exploiting the unique chemical, physical, electrical, optical, and biological attributes of natural or synthesized material at the nanoscale for various medical purposes, such as timely prevention, early detection, and targeted treatment of disease [4–9]. The radionuclides that are nanotargeted are used for diagnosis as well as for treatment, and these are based on various aspects like half-life, decay times, penetration range, and emission properties.

19.2 Preclinical Imaging

With the advent of research, nanomaterials are becoming more and more sensitive, so these techniques are capable of detecting single cells even before symptoms appear [7, 10, 11]. This fusion of nanomaterials with molecular imaging devices allows for more and more precise diagnoses at the molecular level [7]. This is possible due to single nanoparticle tagging with imaging agents, with different targeting ligands, bypassing biological barriers and resulting in increased targeting efficacy [12]. With extensive experimental research, these radio-labeled nanoparticles are now becoming stable under many physiological conditions [13]. Radionuclide molecular imaging modalities and molecular magnetic resonance imaging (mMRI) are more sensitive due to quantitative imaging with very low resolution [14, 15].

19.3 Multimodality Imaging

The use of multimodality imaging is based on different resolution patterns, like intrinsic and extrinsic resolution [1, 15], which provide much

S. Shaikh, *Nanoimaging - Future of Precision Medicine*,
https://doi.org/10.1007/978-981-97-7292-6_19

higher accuracy [12]. Multiple imaging modalities can be used as multi-modal imaging when used simultaneously [16]. Nanoparticles have a larger surface area, which is more useful in multifunctional imaging [17] (Fig. 19.1). In tumor angiogenesis, various components include superparamagnetic iron oxide (SPIO) nanoparticles, quantum dots (QDs), and microbubbles [12, 18–21]. The concept of multi-labeling nanoparticles also has some disadvantages [22]. High-density lipoprotein (HDL) cores, along with synthetic nanomaterials like QDs, gold, and iron oxide, can be used for the identification of atherosclerotic plaques [23]. The vascular endothelial growth factor (VEGF), VEGF receptor (VEGFR), and integrin αv β3 are used in the diagnosis and management of diseases [24–26]. Indium-111 (^{111}In) labeled perfluorocarbon nanoparticles in single photon emission tomography (SPET) imaging have shown a greater number of ^{111}In atoms per nanoparticle in carcinoma lung, which shows a significant increase in the

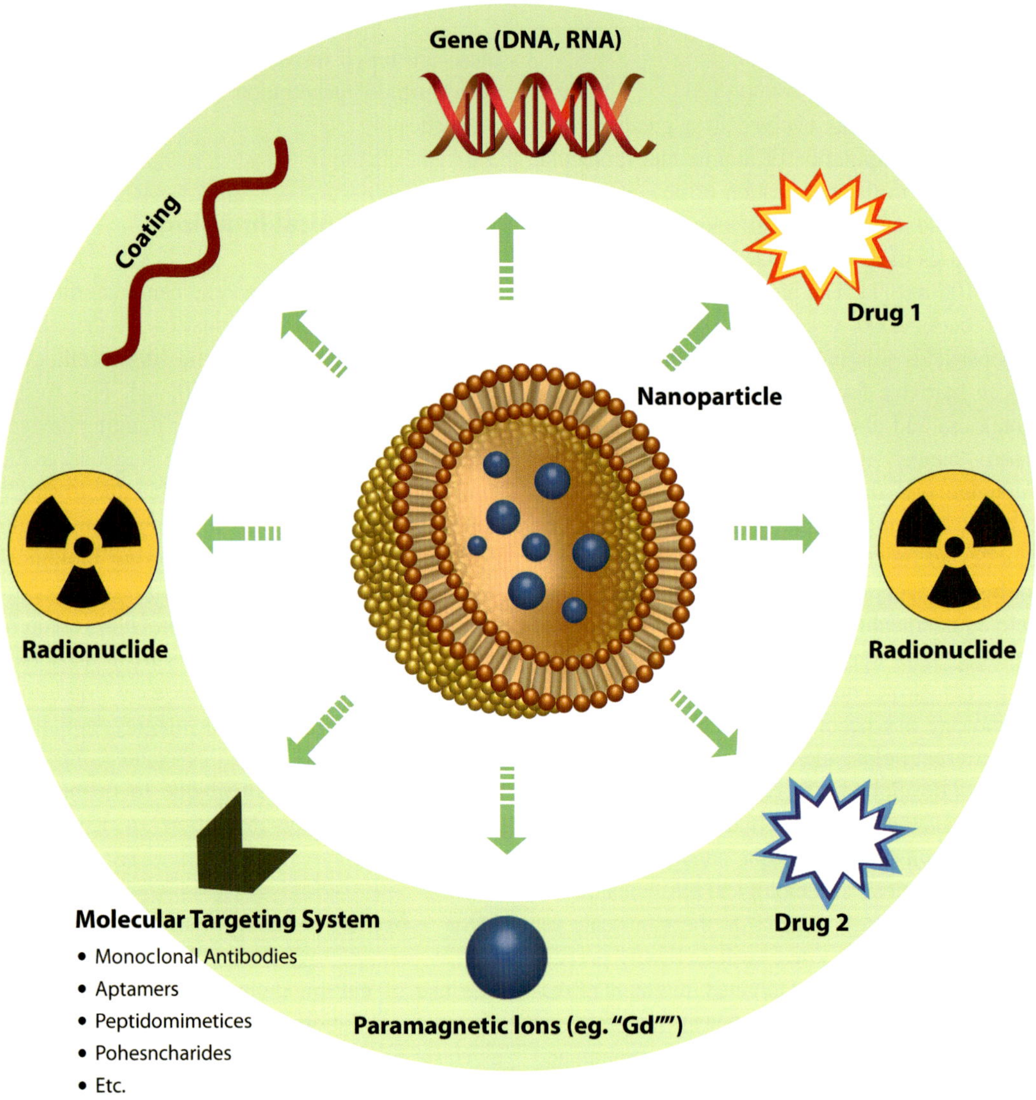

Fig. 19.1 Molecular targeting system showing the role in various applications by use of different biomarkers used for targeting

tumor-to-muscle ratio as well as in mean tumor activity [27]. As a result, newer tumors can be easily identified along with other techniques [27]. The chemical and biological advantages of nanomaterials will be useful for improved transport of contrast agents in ^{99m}Tc, gadolinium, and fluorophores [7].

19.4 Quantum Dots

The various QD-based probes can be used for the evaluation of fluorescence imaging (NIRF) and positron emission tomography (PET) by using integrin αv β3, which is supposed to have better and more precise targeting efficacy for tumors. Due to this, NIRF/PET imaging is capable of detecting more tumors at earlier stages [14, 18, 28]. QDs are being used in a variety of applications like cell tracking, reticuloendothelial system mapping, and tumor targeting [29]. The QD-based dual-functional NIRF and PET imaging has shown better tumor targeting efficacy [29]. The QD-based combination of NIRF and PET imaging has made a better impact on the diagnosis for NIRF-targeted imaging [19, 30–32]. PET/MRI with various integrin expressions using polyaspartic acid (PASP)-coated iron oxide (IO) (PASPIO) nanoparticles linked to DOTA and cyclic arginine-glycine-aspartic (RGD) peptides has made it possible to detect tumors at early stages [16]. Various peptides, viruses, and monoclonal antibodies used against tumor molecules are becoming more specific due to advanced research in tumor targeting and effective drug distribution [11, 33, 34]. The use of various nanoscale liposomes, water-soluble polymers, nanomicelles, and nanodendrimers is becoming more widespread, leading to improved diagnostics, therapeutics, and theranostic components [35, 36] (Fig. 19.2). Different genes [37], peptides [38], proteins [39], and antibodies are being added and tagged to various carbon nanotube (CNT) vehicles in tumor cells [40]. Carbon nanotube drug delivery has become a more precise method of delivering various radionuclides or drugs [41]. Lipid nanocapsules (LNC) are being used as carriers for internal radiotherapy in cases of malignant glioma in Fischer rats, and it was concluded that the tendency to remain in brain tissue is higher, increasing the median survival time. The lipophilic nature of LNC provides

Fig. 19.2 Role of nano nuclear medicine in different aspects of nanoimaging

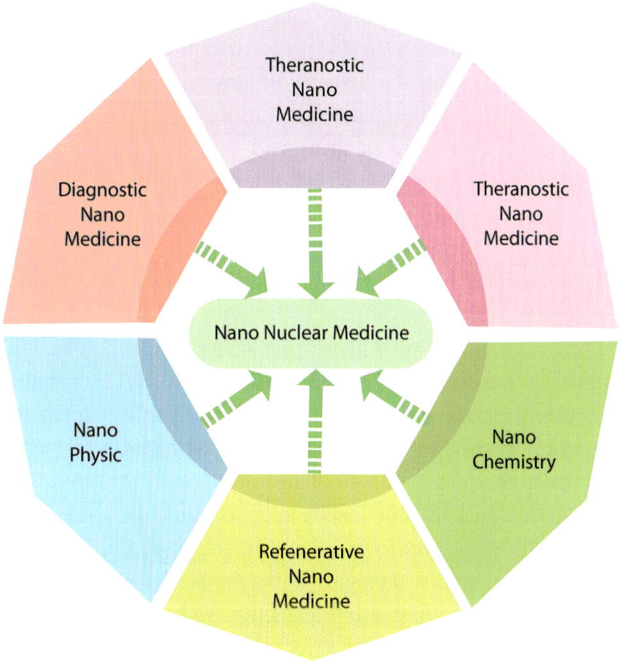

more effective irradiation of tumors [42]. Epidermal growth factor (EGF) with diethylenepentaacetic acid (DTPA) can be added to ^{111}In for the management of breast cancer, reducing the growth rate and improving survival in hormone-resistant breast cancer cells [41]. Monoclonal antibodies are being tagged to ^{111}In for the bifunctional nanostructure leading to Burkitt's lymphoma cells [43]. Theranostics [32] can be used for various multifunctional nanoparticles in diagnostic and therapeutic applications through the use of different nanoparticles [44]. Nanomaterials can be delivered to specific sites and targeted cells for drug delivery in the region of interest [45–47]. Various multifunctional nanoplatforms, like a combination of cuprum-64 (^{64}Cu)-labeled folate-conjugated shells with cross-linked nanoparticles, are used in PET imaging and radiotherapy of tumor cells [48]. Superparamagnetic iron oxide nanoparticles (SPION) with ^{18}F-fluoride in hemagglutinating virus for gene transfer improve distribution in HVJ-Es in injected rodents when magnets are used for PET planning. The magnetic force significantly alters biodistribution. The antibody (mAb)-linked iron oxide nanoparticles (bio probes), when heated, produce an alternating magnetic field (AMF) and are used for thermal therapy of tumors and metastatic disease [49]. Due to the flexibility in the biodistribution of therapeutic vehicles, the delivery of genes, drugs, and tracers has become easier and more feasible [50], and the molecular imaging and treatment of cardiovascular diseases have become easier [51]. The combination of cell membranes with nanoparticles has enhanced gene delivery [52]. The M13 virus has become highly sought-after for cancer evaluation [53]. M13 binding to various cancer cells, along with various nanodevices, improves sensitivity for cancer diagnosis [54]. The use of regenerative medicine with stem cells and the induction of different endogenous repair mechanisms, such as transplantation for restoring damaged and inefficient tissues in conditions like cancer, diabetes, cardiovascular disease, central or peripheral nerve damage, and osteoarthritic diseases, has become more effective [55, 56]. The use of QD in various IO nanoparticles and nanopolymers will aid in various pathologies in regenerative nanomedicine. These can be added and fused with radioisotopes for nuclear medical procedures.

19.5 Quantum Dots Properties

The QDs are seen in the heavy-metal core cadmium selenide (CdSe) and cadmium telluride (CdTe), which have a zinc sulfide (ZnS) shell [56]. The shell thickness can be modified based on different reaction times. Various semiconductor nanocrystals are also being used for cell labeling, cell tracking, imaging, and monitoring of cellular events [57]. QDs are being used and are a mark of multiple cell synergy in multiplex imaging [58, 59], making them better for the long-term labeling of stem cell progenitors [58]. Iron oxide nanoparticles have advantages as they can attach to external cell surfaces, penetrate cells, and penetrate cell tissues, making them more viable [60]. The dextran-coated superparamagnetic iron oxide (SPION) is a better contrast agent for labeling human mesenchymal stem cells (MSC) and human embryonic stem cells (ESC) [61]. The damaged brain cells in brain trauma settings, using MSC labeled with SPION, have a better impact [62]. SPION-labeled stem cells in the cardiovascular system are now more precise [63]. Tissue engineering can also mimic the artificial extracellular matrix, providing better outcomes in tissue engineering [64]. Nanofibrous scaffolds are capable of increasing the stability of implanted stem cells and are important for stem cells, as they can be monitored for better proliferation, differentiation, and viability [65, 66]. The sodium iodide symporter (NIS) has more research articles compared to ^{99m}Tc-pertechnetate imaging [65]. MSC expressing can be designed in the human NIS gene and requires more viability of transplanted human stem cells. Scintigraphy has also shown radionuclide uptake higher than in MSC-NIS not implanted in a scaffold (Fig. 19.3). When using the MSC-NIS/PLLA scaffold mixture, there are better survival rates

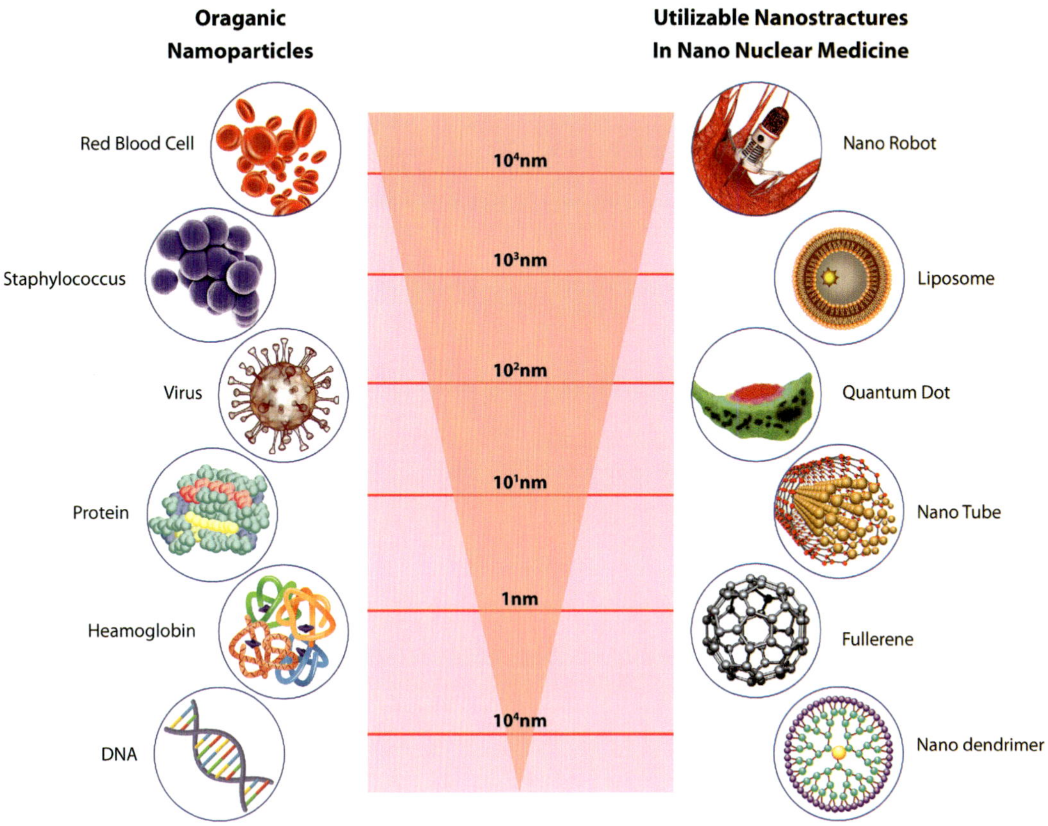

Fig. 19.3 The left column showing the role of the organic nanoparticles and the right row showing the different commonly used nanoparticles in nano nuclear medicine

compared to non-scaffold MSC-NIS. The use of NSC-flu/polymer and MSC-NIS/polymer scaffold complexes is more capable for longitudinal monitoring of implanted stem cells [65]. Nanoparticle-based agents are used in theranostics, which are permitted [67, 68]. With this, nanotechnology is becoming a broad field for medicine and nuclear medicine theranostics, expected to yield practical applications in the future.

19.6 Conclusion

Thus, the use of various nanoimaging techniques in preclinical imaging will provide more valuable information in preclinical imaging aspects and guide the increasing use of nanoimaging for molecular imaging and cell life regulation.

References

1. Ryvolova M, Chomoucka J, Drbohlavova J, Kopel P, Babula P, Hynek D, Adam V, Eckschlager T, Hubalek J, Stiborova M, Kaiser J, Kizek R. Modern micro and nanoparticle-based imaging techniques. Sensors (Basel). 2012;12:14792–820.
2. Torres Martin de Rosales R, Tavare R, Glaria A, Varma G, Protti A, Blower PJ. Applications of nanoparticles in biomedical imaging. Bioconjug Chem. 2011;22:455–65.
3. Hoshyar N, Gray S, Han H, Bao G. The effect of nanoparticle size on in vivo pharmacokinetics and cellular interaction. Nanomedicine (Lond). 2016;11:673–92.
4. Li Y, Xiao Y, Lin HP, Reichel D, Bae Y, Lee EY, Jiang Y, Huang X, Yang C, Wang Z. Applications of nanoparticles in biomedical imaging biomaterials. 2018;188:160–72.
5. Abdollah MRA, Carter TJ, Jones C, Kalber TL, Rajkumar V, Tolner B, Gruettner C, Zaw-Thin M, Baguna Torres J, Ellis M, Robson M, Pedley RB, Mulholland P, de Rosales RTM, Chester KA. Bridging

smart nanosystems with clinically relevant models and advanced imaging for precision drug delivery. ACS Nano. 2018;12:1156–69.
6. Marangoni VS, Neumann O, Henderson L, Kaffes CC, Zhang H, Zhang R, Bishnoi S, Ayala-Orozco C, Zucolotto V, Bankson JA, Nordlander P, Halas NJ. Enhancing T1 magnetic resonance imaging contrast with internalized gadolinium(III) in a multilayer nanoparticle. Proc Natl Acad Sci USA. 2017;114:6960–5.
7. Pan D, Lanza GM, Wickline SA, et al. Nanomedicine: perspective and promises with ligand-directed molecular imaging. Eur J Radiol. 2009;70:274–85.
8. Ting G, Chang CH, Wang HE, et al. Nanotargeted radionuclides for cancer nuclear imaging and internal radiotherapy. J Biomed Biotechnol. 2010;9(5):35–7.
9. Assadi M, Afrasiabi K. An establishment of nano-nuclear medicine fellowship: is it necessary? Iran J Nucl Med. 2010;18(Suppl 1):67.
10. Assadi M, Nabipour I. The role of nanotechnology in nuclear medicine as a limb of molecular medicine with emphasis on medical roadmap of Iran. In: International congress of nuclear medicine and molecular imaging, Iran; 2009.
11. Nanotechnology. European technology platform on nanomedicine nanotechnology for health. 2005 [cited 2010 Jun 10].
12. Cai W, Chen X. Multimodality molecular imaging of tumor angiogenesis. J Nucl Med. 2008;49(Suppl 2):113S–28S.
13. Singh R, Pantarotto D, Lacerda L, et al. Tissue biodistribution and blood clearance rates of intravenously administered carbon nanotube radiotracers. Proc Natl Acad Sci U SA. 2006;103:3357–62.
14. Massoud TF, Gambhir SS. Molecular imaging in living subjects: seeing fundamental biological processes in a new light. Genes Dev. 2003;17:545–80.
15. Gambhir SS. Molecular imaging of cancer with positron emission tomography. Nat Rev Cancer. 2002;2:683–93.
16. Lee HY, Li Z, Chen K, et al. PET/MRI dual-modality tumor imaging using arginine-glycine-aspartic (RGD)-conjugated radiolabeled iron oxide nanoparticles. J Nucl Med. 2008;49:1371–9.
17. Cai W, Chen X. Nanoplatforms for targeted molecular imaging in living subjects. Small. 2007;3:1840–54.
18. Michalet X, Pinaud FF, Bentolila LA, et al. Quantum dots for live cells, in vivo imaging, and diagnostics. Science. 2005;307:538–44.
19. Medintz IL, Uyeda HT, Goldman ER, et al. Quantum dot bioconjugates for imaging, labelling and sensing. Nat Mater. 2005;4:435–46.
20. Larson DR, Zipfel WR, Williams RM, et al. Water-soluble quantum dots for multiphoton fluorescence imaging in vivo. Science. 2003;300:1434–6.
21. Schipper ML, Cheng Z, Lee SW, et al. MicroPET-based biodistribution of quantum dots in living mice. J Nucl Med. 2007;48:1511–8.
22. Wagenaar DJ, Chen JA. Nuclear imaging of vulnerable plaque: contrast improvements through multi-labeling of nanoparticles. In: 2005 IEEE nuclear science symposium conference record, Fajardo; 2005. 5pp.
23. Cormode DP, Skajaa T, van Schooneveld MM, et al. Nanocrystal core high-density lipoproteins: a multimodality contrast agent platform. Nano Lett. 2008;8:3715–23.
24. Cai W, Chen K, Mohamedali KA, et al. PET of vascular endothelial growth factor receptor expression. J Nucl Med. 2006;47:2048–56.
25. Ferrara N, Hillan KJ, Novotny W. Bevacizumab (Avastin), a humanized anti-VEGF monoclonal antibody for cancer therapy. Biochem Biophys Res Commun. 2005;333:328–35.
26. Hood JD, Cheresh DA. Role of integrins in cell invasion and migration. Nat Rev Cancer. 2002;2:91–100.
27. Hu G, Lijowski M, Zhang H, et al. Imaging of Vx-2 rabbit tumors with alpha(nu)beta3-integrin-targeted 111In nanoparticles. Int J Cancer. 2007;120:1951–7.
28. Cai W, Chen K, Li ZB, et al. Dual-function probe for PET and near-infrared fluorescence imaging of tumor vasculature. J Nucl Med. 2007;48:1862–70.
29. Chen K, Li ZB, Wang H, et al. Dual-modality optical and positron emission tomography imaging of vascular endothelial growth factor receptor on tumor vasculature using quantum dots. Eur J Nucl Med Mol Imaging. 2008;35:2235–44.
30. Gao X, Cui Y, Levenson RM, et al. In vivo cancer targeting and imaging with semiconductor quantum dots. Nat Biotechnol. 2004;22:969–76.
31. Lim YT, Kim S, Nakayama A, et al. Selection of quantum dot wavelengths for biomedical assays and imaging. Mol Imaging. 2003;2:50–64.
32. Jackson H, Muhammad O, Daneshvar H, et al. Quantum dots are phagocytized by macrophages and colocalize with experimental gliomas. Neurosurgery. 2007;60:524–9; discussion 529–30.
33. Debbage P, Jaschke W. Molecular imaging with nanoparticles: giant roles for dwarf actors. Histochem Cell Biol. 2008;130:845–75.
34. Reilly RM. Biomolecules as targeting vehicles for in situ radiotherapy of malignancies. In: Knaeblein J, editor. Modern biopharmaceuticals: design, development and optimization, vol. 2. Weinheim: Wiley-VCH; 2005.
35. Mitra A, Nan A, Line BR, et al. Nanocarriers for nuclear imaging and radiotherapy of cancer. Curr Pharm Des. 2006;12:4729–49.
36. Hamoudeh M, Kamleh MA, Diab R, et al. Radionuclides delivery systems for nuclear imaging and radiotherapy of cancer. Adv Drug Deliv Rev. 2008;60:1329–46.
37. Pantarotto D, Singh R, McCarthy D, et al. Functionalized carbon nanotubes for plasmid DNA gene delivery. Angew Chem Int Ed Engl. 2004;43:5242–6.
38. Pantarotto D, Briand JP, Prato M, et al. Translocation of bioactive peptides across cell membranes by carbon nanotubes. Chem Commun (Camb). 2004;1:16–7.

39. Kam NW, O'Connell M, Wisdom JA, et al. Carbon nanotubes as multifunctional biological transporters and near-infrared agents for selective cancer cell destruction. Proc Natl Acad Sci U S A. 2005;102(11):600–5.
40. Reilly RM. Carbon nanotubes: potential benefits and risks of nanotechnology in nuclear medicine. J Nucl Med. 2007;48:1039–42.
41. Reilly R. The immunoreactivity of radiolabeled antibodies—its impact on tumor targeting and strategies for preservation. Cancer Biother Radiopharm. 2004;19:669–72.
42. Allard E, Hindre F, Passirani C, et al. 188Re-loaded lipid nanocapsules as a promising radiopharmaceutical carrier for internal radiotherapy of malignant gliomas. Eur J Nucl Med Mol Imaging. 2008;35:1838–46.
43. McDevitt MR, Chattopadhyay D, Kappel BJ, et al. Tumor targeting with antibody-functionalized, radiolabeled carbon nanotubes. J Nucl Med. 2007;48:1180–9.
44. Lucignani G. Nanoparticles for concurrent multimodality imaging and therapy: the dawn of new theragnostic synergies. Eur J Nucl Med Mol Imaging. 2009;36:869–74.
45. Pastorin G, Wu W, Wieckowski S, et al. Double functionalization of carbon nanotubes for multimodal drug delivery. Chem Commun (Camb). 2006;11:1182–4.
46. Kim G, O'Regan R, Nie S. Biomedical nanotechnology for molecular imaging, profiling, and drug targeting. Conf Proc IEEE Eng Med Biol Soc. 2005;1:714–6.
47. Longmire M, Choyke PL, Kobayashi H. Clearance properties of nano-sized particles and molecules as imaging agents: considerations and caveats. Nanomedicine (Lond). 2008;3:703–17.
48. Rossin R, Pan D, Qi K, et al. 64Cu-labeled folate-conjugated shell cross-linked nanoparticles for tumor imaging and radiotherapy: synthesis, radiolabeling, and biologic evaluation. J Nucl Med. 2005;46:1210–8.
49. DeNardo SJ, DeNardo GL, Natarajan A, et al. Thermal dosimetry predictive of efficacy of 111In-ChL6 nanoparticle AMF-induced thermoablative therapy for human breast cancer in mice. J Nucl Med. 2007;48:437–44.
50. Flexman JA, Cross DJ, Lewellen BL, et al. Magnetically targeted viral envelopes: a PET investigation of initial biodistribution. IEEE Trans Nanobioscience. 2008;7:223–32.
51. Wickline SA, Neubauer AM, Winter PM, et al. Molecular imaging and therapy of atherosclerosis with targeted nanoparticles. J Magn Reson Imaging. 2007;25:667–80.
52. Galanis E, Vile R, Russell SJ. Delivery systems intended for in vivo gene therapy of cancer: targeting and replication competent viral vectors. Crit Rev Oncol Hematol. 2001;38:177–92.
53. Ross PE. Viral nano electronics. Sci Am. 2006;295:52–5.
54. Lee SW, Mao C, Flynn CE, et al. Ordering of quantum dots using genetically engineered viruses. Science. 2002;296:892–5.
55. Muneoka K, Allan CH, Yang X, et al. Mammalian regeneration and regenerative medicine. Birth Defects Res C Embryo Today. 2008;84:265–80.
56. Solanki A, Kim JD, Lee KB. Nanotechnology for regenerative medicine: nanomaterials for stem cell imaging. Nanomedicine (Lond). 2008;3:567–78.
57. Rosen AB, Kelly DJ, Schuldt AJ, et al. Finding fluorescent needles in the cardiac haystack: tracking human mesenchymal stem cells labeled with quantum dots for quantitative in vivo three-dimensional fluorescence analysis. Stem Cells. 2007;25:2128–38.
58. Jaiswal JK, Mattoussi H, Mauro JM, et al. Long-term multiple color imaging of live cells using quantum dot bioconjugates. Nat Biotechnol. 2003;21:47–51.
59. Wu X, Liu H, Liu J, et al. Immunofluorescent labeling of cancer marker Her2 and other cellular targets with semiconductor quantum dots. Nat Biotechnol. 2003;21:41–6.
60. Bulte JW, Kraitchman DL. Monitoring cell therapy using iron oxide MR contrast agents. Curr Pharm Biotechnol. 2004;5:567–84.
61. Corot C, Robert P, Idee JM, et al. Recent advances in iron oxide nanocrystal technology for medical imaging. Adv Drug Deliv Rev. 2006;58:1471–504.
62. Zhu J, Zhou L, XingWu F. Tracking neural stem cells in patients with brain trauma. N Engl J Med. 2006;355:2376–8.
63. Kraitchman DL, Heldman AW, Atalar E, et al. In vivo magnetic resonance imaging of mesenchymal stem cells in myocardial infarction. Circulation. 2003;107:2290–3.
64. Li X, Liu X, Dong W, et al. In vitro evaluation of porous poly(lactic acid) scaffold reinforced by chitin fibers. J Biomed Mater Res B Appl Biomater. 2009;90:503–9.
65. Hwang DW, Jang SJ, Kim YH, et al. Real-time in vivo monitoring of viable stem cells implanted on biocompatible scaffolds. Eur J Nucl Med Mol Imaging. 2008;35:1887–98.
66. Park KI, Teng YD, Snyder EY. The injured brain interacts reciprocally with neural stem cells supported by scaffolds to reconstitute lost tissue. Nat Biotechnol. 2002;20:1111–7.
67. Del Vecchio S, Zannetti A, Fonti R, et al. Nuclear imaging in cancer theranostics. Q J Nucl Med Mol Imaging. 2007;51:152–63.
68. Madanayake NH, Rienzie R, Adassooriya NM. Nanoparticles in nanotheranostics applications. In: Rai M, Jamil B, editors. Nanotheranostics: applications and limitations. Cham: Springer International Publishing; 2019. p. 19–40.

20 In Vivo Nanoimaging

20.1 Nanomaterial for Various Biomedical Applications

In recent times, nanomaterials are used for both diagnostic and therapeutic applications. Various nanoparticles can be magnetic or non-magnetic, including metallic elements like iron, cobalt, nickel, platinum, or combinations of two or more metals [1–3]. Iron is an important metal used in diagnostics and therapeutics, such as iron oxide or bimetallic SPIONs. SPIONs are essential for magnetic resonance imaging (MRI) diagnostics. The magnetic properties and nanoscale size of these magnetic NPs make them suitable for many applications [4, 5]. The advantage of these NPs is their small size and shape, allowing easy interaction with various cell components [6]. The sizes of these nanoparticles are similar to viruses (20–450 nm), proteins (5–50 nm), and cells (10–100 μm), leading to better cellular uptake [6–8]. Magnetic NPs are also used in various applications, including dental [9], neurological [10], and cardiovascular diseases [5]. Various review articles [2, 4, 6, 11–13] show that magnetic nanomaterials can be applied for different magnetic cell separation processes like protein separation [14], used as MRI contrast agents in non-contraindicated magnetic fields [15], hyperthermia cancer therapy [16], enzyme-like SPIONs with catalytic roles, synthesis of SPIONs at high temperatures without thermal decomposition [17], and targeted drug or gene release [18] (Fig. 20.1). Disease management can be site-specific with NP-based biomarkers [19], offering fewer side effects and greater precision [5, 16]. Magnetic ferrous NPs, mostly magnetite and maghemite, are seen as promising agents [20]. These ferrous metallic NPs are used in vivo with advantages for the immune system due to their organic biomolecule composition, providing a better immune response. Therapeutic magnetic NPs have a longer plasma half-life and better biodistribution, making them efficient and safe [15]. Their size allows deeper tissue or capillary penetration with minimal extravasation after intravenous injection. "Theranostic" or "Nanotheranostics" is a planned approach for bioactive compounds with controlled release at targeted sites. However, extravasation in tumor tissues may reduce theranostic potential [2, 21–23]. Understanding pharmacokinetics, including distribution–metabolization–excretion and renal clearance, provides better information [5]. In the reticuloendothelial system, including the liver, spleen, bone marrow, and lymph nodes, mechanisms like opsonization, phagocytosis, and endocytosis facilitate biodistribution [15], based on NP size [23]. Various factors influence NP distribution in different media and proper clearance from the body, as discussed. Magnetic NPs are advantageous because they are moldable to the immune system, more effective, and have fewer

S. Shaikh, *Nanoimaging - Future of Precision Medicine*,
https://doi.org/10.1007/978-981-97-7292-6_20

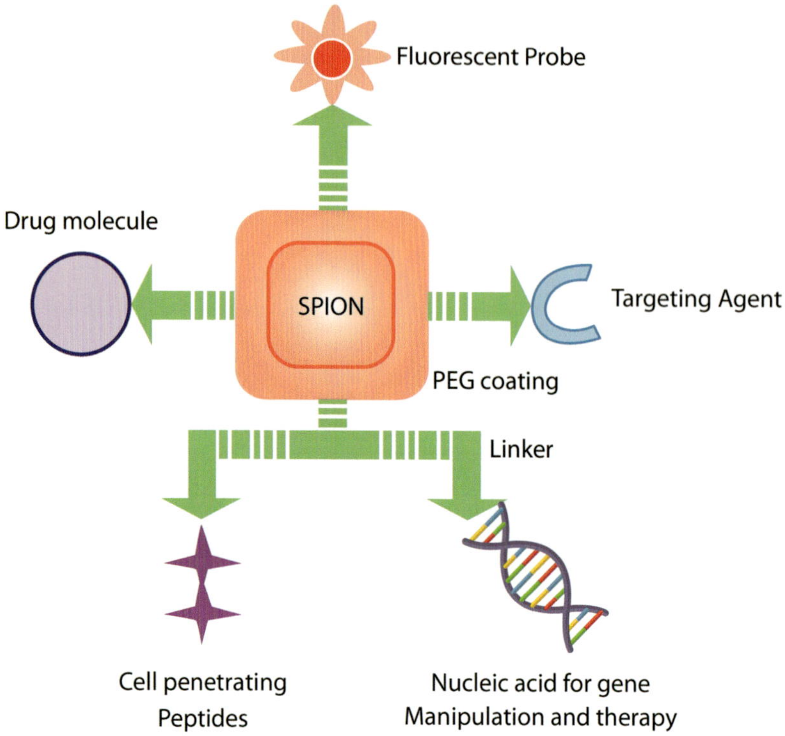

Fig. 20.1 Role of SPION in nanomedicine

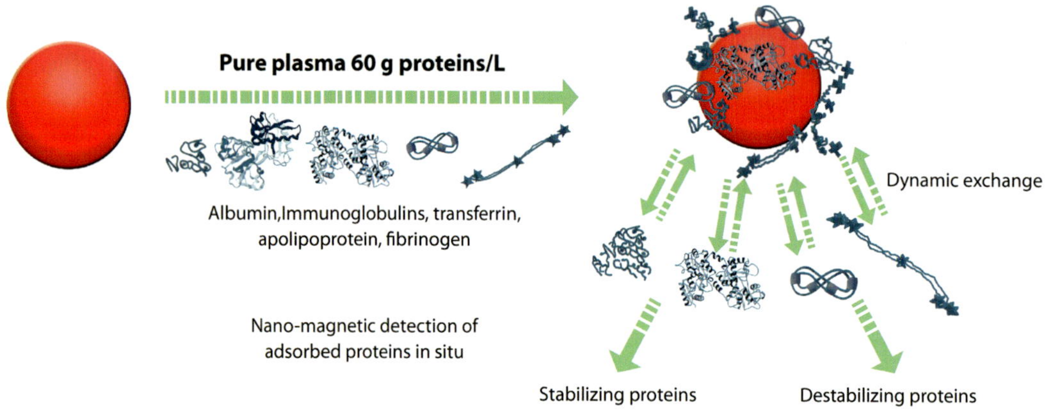

Fig. 20.2 Plasma proteins showing the role in the various nanomedicine detection in situ

side effects [5]. Nanomagnets used for various functions are stable in media (Fig. 20.2). NPs are used for chemical surface modifications to achieve biocompatibility with molecules like polymers, hydrophilic molecules, or silica layers [6, 23]. Magnetic NPs interact with or adhere to cell membranes based on surface nanoprobes and charge [8]. Therefore, various nanoparticles should be discussed concerning metabolism in vivo, especially in blood circulation, for all theranostic applications. Applications can vary depending on nanoparticle properties [24]. Nanoparticles are small with excellent biocompatibility for various applications [15, 23], and magnetic NPs exhibit efficient targeting without injury to nanoparticles in vivo [25, 26]. Metallic nanoparticles are also widely used in therapeutic, diagnostic, and real-time imaging applications.

Ferrous metals are composed of hematite (α-Fe_2O_3), maghemite (γ-Fe_2O_3), and magnetite (Fe_3O_4 or $FeO{\cdot}Fe_2O_3$), commonly found in soil and rocks due to volcanic eruptions and air pollution [3, 27, 28]. Based on size, they are classified into micrometer-sized (300 nm–3.5 μm), standard (10–150 nm), and ultrasmall iron oxide crystals. NP biodistribution depends on size, which should have an optimal average diameter to prevent sequestration [5, 7, 19, 28]. They should also circulate in the bloodstream, accumulate in the region of interest, and prevent vessel embolization. Therefore, ultrasmall SPIONs are used as contrast agents for liver diseases as they can passively target liver cells (Kupffer cells), enhancing blood pool and lymph node imaging [19, 23]. Hematite (α-Fe_2O_3), the oldest iron oxide, has better stability. Maghemite (γ-Fe_2O_3), however, is only kinetically stable. A combination of mixed-valent magnetite allows better distribution of both iron(II) and (III) in crystallographic sublattice sites [3, 27, 29–31].

20.2 Iron Oxide Stability

Iron oxide stability is enhanced by combining various iron oxide nanocrystals with corona or ligand layers, which improve biocompatibility and selective actions [1]. These corona layers help maintain the magnetite phase for better stability. Liu et al. [32] demonstrated that modifying the NP surface enhances stability, improves biodistribution, and increases blood circulation time. Combining gold or silica outer layers provides pH-sensitive properties, creating a suitable surface for grafting bioactive materials for targeted delivery [3, 7, 23, 30]. According to DLVO theory, colloidal stability results from (1) electrostatic repulsion, (2) Van Der Waals attraction, (3) magnetic dipolar forces, and (4) steric repulsion [22, 30]. Surface charge plays a crucial role in stabilizing iron oxides. Mixing positively and negatively charged nanoparticles can neutralize electrolytes in blood, creating a neutral surface that enhances both stability and binding. Steric repulsion is also essential for stability, as it facilitates the attachment of various molecules/ligands, resulting in NP repulsion. The application of polymeric coatings, either in situ or post-synthesis, establishes a steric barrier, further improving stabilization [30, 33]. However, even with these techniques, achieving optimal SPION stability requires additional strategies for biocompatibility under different internal and external conditions [5, 22, 30, 34, 35]. The core-shell structure of SPIONs offers unique physical and chemical properties that enable functionalization through corona layers, creating effective nanoparticles or nano-biomarkers [19] with improved surface functionalization [36] for targeted cancer cell accumulation. This is beneficial in hyperthermia cancer therapy [16, 35], where active targeting helps prevent damage to healthy tissue. Active targeting is especially effective due to the enhanced permeability and retention (EPR) effect in cancerous tissues, where abnormal vascularity and extravasation allow passive targeting through NP accumulation. Despite its advantages, passive targeting may cause cytotoxic side effects. Therefore, active targeting is preferred to reduce potential damage to surrounding healthy tissue [5].

20.3 Nanoparticle Iron Oxide Nanocrystals

Iron oxide nanocrystals with surface coatings, such as corona or ligand layers, are more biocompatible and offer specific biological functionality, which aids in resistance to oxidation and erosion [1, 31]. For magnetite to retain its phase, corona protection is essential. This can be achieved by using core-shell or ligand layers as stabilizing agents. Liu et al. [32] demonstrated that surface modification improves nanoparticle stability, biodistribution, and blood circulation time, thanks to the hydrophilic or lipophilic properties introduced. Adding gold or silica layers further enhances pH-sensitive properties, allowing for more effective targeting of bioactive materials [3, 7, 23, 30]. The Derjaguin-Landau-Verwey-Overbeek (DLVO) theory explains colloidal stability in terms of electrostatic repulsion and Van der Waals attraction. Iron oxide nanoparticle stability relies on four key forces: (1) electrostatic repulsion, (2) Van der Waals attraction, (3) magnetic dipolar forces, and (4) steric repulsion [22, 30]. Van der Waals forces can cause NP aggregation by lowering interfacial energy, thereby reducing the overall surface area. Surface charge, through its impact on zeta potential, is critical for maintaining colloidal stability. A neutral surface minimizes nonspecific binding and preserves stability, which is especially useful in physiological environments. Zeta potential, associated with the Debye-Hückel length, serves as a measure of NP stability in biological conditions. Steric repulsion also plays a vital role in stability by preventing particles from coming too close, particularly when molecules or ligands are attached to NP surfaces. In situ or post-synthesis polymeric coatings offer a robust steric barrier by affecting polymer conformation and chain length, which can further enhance stability [30, 33].

20.4 Coprecipitation

Coprecipitation is a common and effective method for producing nanoparticles, often involving partial oxidation or a stoichiometric mixture of Fe(3+) and Fe(2+) salts [37, 38]. In this process, magnetite stability is influenced by factors like particle size, colloidal stability, and resistance to oxidation, all of which are managed through the coprecipitation technique. Surface composition and charge are critical for controlling the nucleation and growth steps. Typically, coprecipitation is conducted in water, with various electrolytes present, as factors such as pH, ionic strength, and the counterions of metallic salts play essential roles in nanoparticle stabilization.

20.5 Conclusion

In vivo imaging has advanced significantly, enabling more precise diagnosis and innovative applications through the use of various agents.

References

1. Lenzi E, Henriksen-Lacey M, Molina B, Langer J, de Albuquerque CDL, Jimenez D, de Aberasturi LM, Liz-Marzán. Combination of live cell surface-enhanced Raman scattering imaging with chemometrics to study intracellular nanoparticle dynamics. ACS Sensors. 2022;7(6):1747–56.
2. Belić D, Fragueiro O, Salah D, Beckett A, Volk M, Brust M. Imaging of nanoscale gold in "intact" biological cells by environmental electron microscopy. J Phys Chem C. 2021;125(50):27865–75.
3. Teja AS, Koh P. Synthesis, properties, and applications of magnetic iron oxide nanoparticles. Prog Cryst Growth Charact Mater. 2009;55:22–45.
4. Goesmann H, Feldmann C. Nanoparticulate functional materials. Angew Chem Int Ed. 2010;49:1362–95.
5. Sun C, Lee JSH, Zhang M. Magnetic nanoparticles in MR imaging and drug delivery. Adv Drug Deliv Rev. 2008;60:1252–65.
6. Pankhurst QA, Connolly J, Jones SK, Dobson J. Applications of magnetic nanoparticles in biomedicine. J Phys D Appl Phys. 2003;36:R167.
7. Na HB, Song IC, Hyeon T. Inorganic nanoparticles for MRI contrast agents. Adv Mater. 2009;21:2133–48.
8. Pimpha N, Chaleawlert-Umpon S, Sunintaboon P. Core/shell polymethyl methacrylate/polyethyleneimine particles incorporating large amounts of iron oxide nanoparticles prepared by emulsifier-free emulsion polymerization. Polymer. 2012;53:2015–22.
9. Zhu X, Xiong H, Zhou Q, Zhao Z, Zhang Y, Li Y, Wang S, Shi S. A pH-activatable MnCO3 nanoparticle for improved magnetic resonance imaging of tumor malignancy and metastasis. ACS Appl Mater Interfaces. 2021;13(16):18462–71.
10. Wadghiri YZ, Li J, Wang J, Hoang DM, Sun Y, Xu H, Tsui W, Li Y, Boutajangout A, Wang A, et al. Detection of amyloid plaques targeted by bifunctional USPIO in Alzheimer's disease transgenic mice using magnetic resonance microimaging. PLoS One. 2013;8:1–10.
11. Ansari SAMK, Ficiarà E, Ruffinatti FA, Stura I, Argenziano M, Abollino O, Cavalli R, Guiot C, D'Agata F. Magnetic iron oxide nanoparticles: synthesis, characterization and functionalization for biomedical applications in the central nervous system. Materials. 2019;12:465.
12. Wu W, Wu Z, Yu T, Jiang C, Kim W-S. Recent progress on magnetic iron oxide nanoparticles: synthesis, surface functional strategies, and biomedical applications. Sci Technol Adv Mater. 2015;16:23501.
13. Kharisov BI, Rasika Dias HV, Kharissova OV, Manuel Jiménez-Pérez V, Olvera Pérez B, Muñoz Flores B. Iron-containing nanomaterials: synthesis, properties, and environmental applications. RSC Adv. 2012;2:9325–58.
14. Scarabelli L, Sun M, Zhuo X, Yoo S, Millstone JE, Jones MR, Liz-Marzán LM. Plate-like colloidal metal nanoparticles. Chem Rev. 2023;123(7):3493–542.

15. Grasset F, Duguet E, Mornet S, Vasseur S, Grasset F, Duguet E. Magnetic nanoparticle design for medical diagnosis and therapy. J Mater Chem. 2004;14:2161–75.
16. Bañobre-López M, Teijeiro A, Rivas J. Magnetic nanoparticle-based hyperthermia for cancer treatment. Rep Pract Oncol Radiother. 2013;18:397–400.
17. Gao L, Fan K, Yan X. Iron oxide nanozyme: a multifunctional enzyme mimetic for biomedical applications. Theranostics. 2017;7:3207–27.
18. Sahoo B, Devi KSP, Dutta S, Maiti TK, Pramanik P, Dhara D. Biocompatible mesoporous silica-coated superparamagnetic manganese ferrite nanoparticles for targeted drug delivery and MR imaging applications. J Colloid Interface Sci. 2014;431:31–41.
19. Hahn MA, Singh AK, Sharma P, Brown SC, Moudgil BM. Nanoparticles as contrast agents for in vivo bioimaging: current status and future perspectives. Anal Bioanal Chem. 2011;399:3–27.
20. Gupta AK, Gupta M. Synthesis and surface engineering of iron oxide nanoparticles for biomedical applications. Biomaterials. 2005;26:3995–4021.
21. Mouli SK, Tyler P, McDevitt JL, Eifler AC, Guo Y, Nicolai J, Lewandowski RJ, Li W, Procissi D, Ryu RK, et al. Image-guided local delivery strategies enhance therapeutic nanoparticle uptake in solid tumors. ACS Nano. 2013;7:7724–33.
22. Mahmoudi M, Sant S, Wang B, Laurent S, Sen T. Superparamagnetic iron oxide nanoparticles (SPIONs): development, surface modification and applications in chemotherapy. Adv Drug Deliv Rev. 2011;63:24–46.
23. Cormode DP, Skajaa T, Fayad ZA, Mulder WJM. Nanotechnology in medical imaging: probe design and applications. Arterioscler Thromb Vasc Biol. 2009;29:992–1000.
24. Kolhatkar AG, Jamison AC, Litvinov D, Willson RC, Lee TR. Tuning the magnetic properties of nanoparticles. Int J Mol Sci. 2013;14:15977–6009.
25. Bardhan R, Chen W, Bartels M, Perez-Torres C, Botero MF, McAninch RW, Contreras A, Schiff R, Pautler RG, Halas NJ, et al. Tracking of multimodal therapeutic nanocomplexes targeting breast cancer in vivo. Nano Lett. 2010;10:4920–8.
26. Zheng W, Gao F, Gu H. Magnetic polymer nanospheres with high and uniform magnetite content. J Magn Magn Mater. 2005;288:403–10.
27. Bruce IJ, Taylor J, Todd M, Davies MJ, Borioni E, Sangregorio C, Sen T. Synthesis, characterisation and application of silica-magnetite nanocomposites. J Magn Magn Mater. 2004;284:145–60.
28. Singh N, Jenkins GJS, Asadi R, Doak SH. Potential toxicity of superparamagnetic iron oxide nanoparticles (SPION). Nano Rev. 2010;1:5358.
29. Leonov I, Yaresko AN. On the Verwey charge ordering in magnetite. J Phys Condens Matter. 2007;19:021001.
30. Laurent S, Forge D, Port M, Roch A, Robic C, Vander Elst L, Muller RN. Magnetic iron oxide nanoparticles: synthesis, stabilization, vectorization, physicochemical characterizations and biological applications. Chem Rev. 2008;108:2064–110.
31. Arias SL, Pessan PJ, Vieira PA, Lima MT, Delbem CA, Monteiro RD. Iron oxide nanoparticles for biomedical applications: a perspective on synthesis, drugs, antimicrobial activity, and toxicity. Antibiotics. 2018;7:46.
32. Lim J, Yeap SP, Che HX, Low SC. Characterization of magnetic nanoparticle by dynamic light scattering. Nanoscale Res Lett. 2013;8:381–95.
33. Lodhia J, Mandarano G, Ferris NJ, Eu P, Cowell SF. Development and use of iron oxide nanoparticles (Part 1): Synthesis of iron oxide nanoparticles for MRI. Biomed Imaging Interv J. 2010;6:e12.
34. Dulińska-Litewka J, Łazarczyk A, Hałubiec P, Szafrański O, Karnas K, Karewicz A. Superparamagnetic iron oxide nanoparticles—current and prospective medical applications. Materials. 2019;12:617.
35. Wu W, He Q, Jiang C. Magnetic iron oxide nanoparticles: synthesis and surface functionalization strategies. Nanoscale Res Lett. 2008;3:397–415.
36. Santhosh PB, Ulrih NP. Multifunctional superparamagnetic iron oxide nanoparticles: promising tools in cancer theranostics. Cancer Lett. 2013;336:8–17.
37. Li F, Liang Z, Liu J, Sun J, Xi H, Zhao M, Liu J, Bai R, Kim D, Sun X, Hyeon T, Ling D. Dynamically reversible iron oxide nanoparticle assemblies for targeted amplification of T1-weighted magnetic resonance imaging of tumors. Nano Lett. 2019;19(7):4213–20.
38. Ferrer-Ugalde A, Sandoval S, Pulagam KR, Muñoz-Juan A, Laromaine A, Llop J, Tobias G, Núñez R. Radiolabeled cobaltabis(dicarbollide) anion–graphene oxide nanocomposites for in vivo bioimaging and boron delivery. ACS Appl Nano Mater. 2021;4(2):1613–25.

21 Development of Nanotheronostics

21.1 Introduction

'Theranostics'Theranostics' has two different definitions. Theranostic nanoparticles (TNPs) are being developed that can be used for the monitoring and management of disease. One definition is the use of an appropriate diagnostic methodology for the personalization of a separate therapeutic intervention [1, 2]. Another shorter and concise definition is TNPs, which offer various opportunities for passive and active targeting, with the use of environmentally responsive drug release, molecular imaging, and other therapeutic functions on a single platform (Fig. 21.1). One of the major hurdles is that TNPs will be used for targeting new biomarkers, with less toxicity, better stability, reasonable costs, and, most importantly, patents to control all these applications [3, 4]. Due to newer agents, the safety aspects of nanoparticles need to be carefully considered [5]. Diagnostic nanoagents will be chosen based on better tissue resolution, optimum wash-in/wash-out kinetics, and clearance times of these nanoparticles [6, 155, p. 1060]. Therapeutic nanoparticles require a longer half-life, slow clearance, greater sensitivity in imaging, better accuracy for targeting lesions, and drug release at the target sites. Optical imaging, magnetic resonance imaging (MRI), nuclear imaging, computed tomography (CT), and ultrasound (US) [7] are the modalities, and each modality has its advantages and disadvantages.

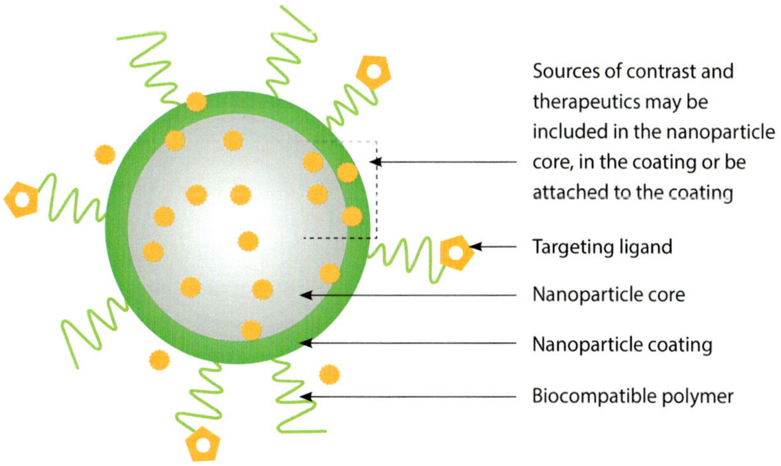

Fig. 21.1 The structure of the Nanoparticle sphere, coating as well as attachment

S. Shaikh, *Nanoimaging - Future of Precision Medicine*,
https://doi.org/10.1007/978-981-97-7292-6_21

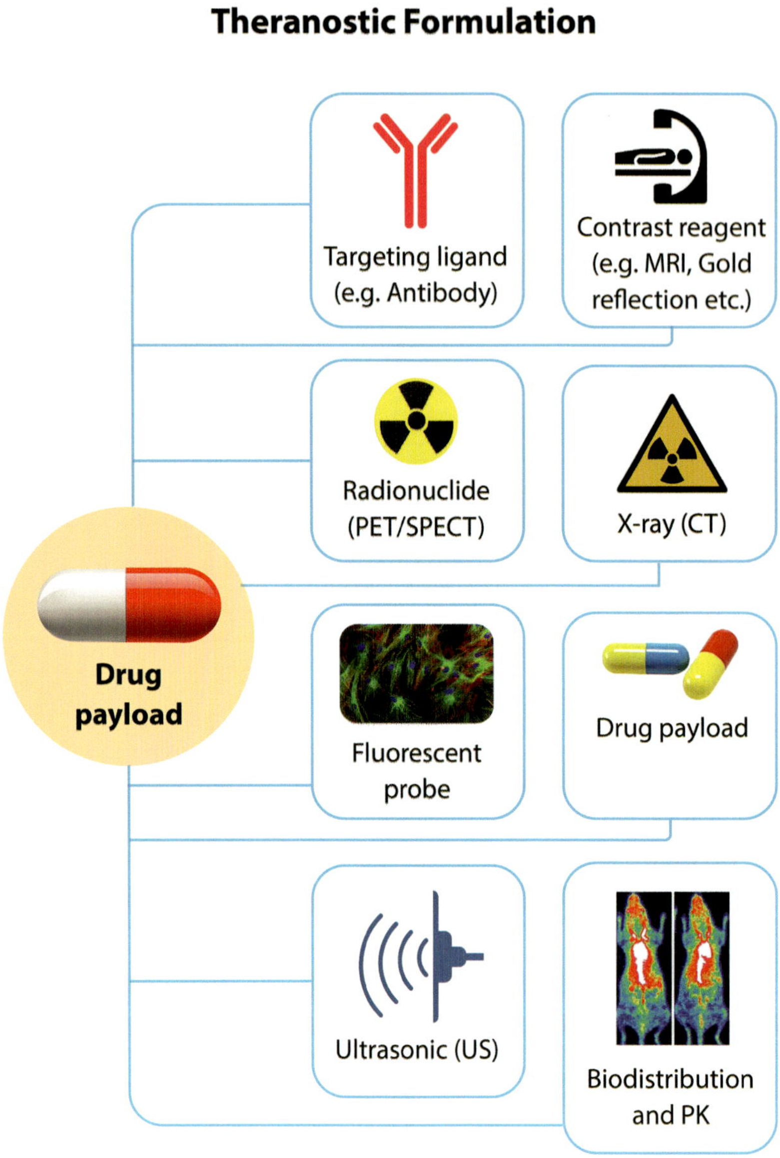

Fig. 21.2 Theronostics formulations showing the different imaging modalities also including the modes for drug payload

As we all know, molecular imaging plays a role in better evaluating various biological processes at the cellular and subcellular levels by using different specific molecular probes or contrast agents, with existing molecular imaging modalities including MRI, CT, US, optical imaging (bioluminescence and fluorescence), single-photon emission computed tomography (SPECT), and positron emission tomography (PET) [8] (Fig. 21.2).

21.2 Optical Imaging

Optical imaging is an important and futuristic molecular imaging modality, and a lot of clinical work is being conducted in this modality. The basic principle is the use of photons emitted from various bioluminescent or fluorescent probes. The biggest advantage is the detection of low-energy photons visible in the spectrum of visible to near-infrared (NIR) light; this visualization helps in identifying various pathologies, which can increase spatial resolution without any ionizing radiation. On the contrary, the biggest disadvantage is the very poor tissue depth penetration (0–2 cm). Another disadvantage is the high susceptibility to noise due to tissue scattering of photons in the light spectrum (395–600 nm) [7]. The issue of autofluorescence and light absorption by proteins (257–280 nm), heme groups (maximum absorbance at 560 nm), and even water (above 900 nm) creates problems for optical imaging [9]. Still, in the NIR range (700–900 nm), there is a significant advantage due to reduced autofluorescence, reduced tissue scattering, and greater depth of penetration, which is needed for in vivo imaging.

21.3 Magnetic Resonance Imaging

MRI images are produced by applying a magnetic field to hydrogen atoms. Relaxation occurs when the radiofrequency pulses are applied, causing the nuclei to return to their original position, which forms the image [10]. The use of contrast agents is based on the relaxation of T1 and T2 parameters of water, allowing for easier identification and quantification of different pathologies. Gadolinium (Gd) and manganese (Mn) are paramagnetic molecules, and when administered along with nanomolecules, macromolecules, or nanoparticles, their interaction can produce the image. Tagging magnetic iron-oxide nanoparticles, which are superparamagnetic substances, on the surface can be useful as a better contrast agent. The use of different contrast agents may also cause issues related to the accumulation of particles, leading to toxicity, which needs to be addressed. For tumor and vascular imaging, gadolinium has some toxicity and slow excretion, leading to greater accumulation of the contrast agent.

21.4 Radionuclide-Based Imaging

The γ-ray emissions and electromagnetic radiation are also an important basis for SPECT and PET. The radiopharmaceuticals administered in the body can be easily identified by SPECT and PET image acquisition, which is based on the low background signal with very little amplification, allowing gamma rays to be detected due to the megavolt range. SPECT, as compared to PET, has lower sensitivity. Another advantage of SPECT is that the preparation of radionuclides is much simpler than PET radionuclides. SPECT and PET are quantitative techniques and have an advantage over anatomical imaging modalities [11]. PET has poor spatial resolution, and to minimize this, the use of CT in hybrid imaging significantly improves the sensitivity of hybrid imaging. The concept of theranostics will also contribute to better sensitivity [12].

21.5 Computed Tomography (CT)

The role of CT is to provide complementary structural information along with the physiological information from PET. When the X-ray beam passes through the body, there will be different X-ray attenuation, which is based on the attenuation coefficient, itself dependent on the atomic number and electron density of the tissues. Due to differential absorption, the identification of various tissues, like bone, fat, and air, produces better contrast of anatomical structures [13]. CT contrast agents are lower molecular weight, having prolonged circulation in the blood pool, making them useful in vascular pathologies. Nanoparticles with higher atomic numbers, such as iodine, bismuth, or gold, can be used as CT contrast agents [14]. These CT contrasts need to combine solid nanoparticles or liposomes with iodinated molecules [13]. Gold nanoparticles are the most common CT contrast agents [14, 15].

21.6 Ultrasound

US is a common clinical imaging modality due to factors like low cost, speed, simplicity, and safety, and is based on the principle of sound reflection from organs [16]. US contrast agents are based on different acoustic properties, such as gas [14]. Chemical conjugation with various drugs in complex forms can be used for various optical imaging applications, providing more quantitative information and better contrast based on the dynamics of the contrast agents across imaging modalities [17–21]. Various imaging modalities are based on different principles. The use of fluorescent dyes and quantum dots (QDs) with bioluminescence principles (10–15 to 10–17) and fluorescence (10–9 to 10–12) with 2–5 mm depth will have better sensitivity, capable of providing more functional information without radiation exposure, though with very low resolution and limited tissue penetration. Computed tomography uses higher contrast agents in the range of 50–200 μm, with very high spatial resolution, making it capable of differentiating between tissues. MRI uses various para- or superparamagnetic metals (e.g., Gd, Mn) in the range of 10–3 to 10–5, with 25–100 μm depth, and offers higher resolution with no ionizing radiation, capable of providing more physiological and anatomical information. Gamma scintigraphy (PET and SPECT) uses various radionuclides with lower radiation but lower resolution. The polypeptide polymer and many newer drug conjugates are being synthetically developed. HPMA is one drug conjugate that is non-toxic, non-immunogenic, and offers better stability in circulation [22, 23]. These synthetic polymers have longer circulation times, less toxicity, more precise passive tumor targeting, and can also act as active targeting moieties or contrast agents [24–26]. Various contrast agents are also associated with HPMA copolymer conjugates for in vitro and in vivo molecular imaging [27]. An HPMA-Dox conjugate is being evaluated in the first clinical trial [28]. Conjugates of two chemotherapeutic agents (Dox and gemcitabine) and also I-131 can be added to the HPMA copolymer using various methods [29]. The concept of copolymerization allows for the combination of single or multimodal diagnostic agents and chemotherapeutic drugs. The RGD-targeted polymer labeled with In-111 and HPMA along with Gd will result in different copolymerizations [30, 31]. Polymer-drug conjugation is important for enhancing pharmacokinetics and therapeutic index, leading to better drug loading, minimized production costs, and improved reliability. The use of natural and synthetic polymers will be a better combination to overcome these barriers.

21.7 Dendrimers

Polymeric dendrimers are nanostructures with many branches and are able to control various polymerization generations. The planar molecule is converted into a spherical shape due to the polymerization process, resulting in better therapeutics and contrast agents with greater efficiency [32]. The polymerization and synthesis of various nanoparticles, such as dendrimers, yield better molecular weight and chemical composition [33]. The potential of dendrimers as theranostic agents can be evaluated for greater safety and efficacy [34]. One study shows the combination of dendrimer with fluorescein isothiocyanate via recombinant Fibroblast Growth Factor-1 for precise imaging and tumor targeting, achieved by meticulous targeting and cellular internalization [35]. Novel release mechanisms for targeted dendrimers, with controlled photo-sensitive linkers, provide a better drug release mechanism [36]. The in vivo and in vitro capabilities of biodegradable dendrimers show improved biocompatibility, biodistribution, and pharmacokinetics [37–39]. One study demonstrated the viability of dendrimers as MRI contrast agents [40]. The multiple functionalization sites on dendrimers allow precise attachment of targeting ligands like monoclonal antibodies, peptides, and folate to the surface [41]. Dendrimers can be modified for improved targeting of receptors in ovarian cancer [42], enabling better image contrast and drug localization [43]. One study demonstrated biodegradable spaces that allow excretion of the contrast agent [44]. Developing non-toxic, biodegradable dendrimers enhances safety and future applications [33]. These starburst structures of dendrimers enable multivalent attachment of chelators and targeting moieties, allowing selective imaging in angiogenic mouse models [45]. Kobayashi and colleagues suggested better target delivery of In-111-labeled dendrimers in a mouse tumor model [46]. These agents exhibit better biodistribution with faster blood clearance and preferential accumulation at the tumor site; second-generation PAMAM dendrimers with Y-88 showed similar results [47]. Yardanov and colleagues have demonstrated the iodinated dendritic nanoparticle G4-(DMAA-IPA)37 as a better CT imaging agent [48]. Fourth-generation PAMAM dendrimers conjugated with DMAA-IPA exhibit improved iodine capacity for CT contrast. Dendrimers serve as effective imaging agent carriers for studies like tumor regression, gene delivery, and molecular imaging [4, 32, 49].

21.8 Vesicles

Vesicular structures have valuable clinical therapeutic and diagnostic properties. The two major classes of vesicles are liposomes and polymer vesicles, which allow for two types of encapsulation: covalent and non-covalent. These vesicles can be used as diagnostic and therapeutic agents with broad applications [50]. Polymeric vesicles, also known as polymersomes, have dual capabilities, functioning as both drug and contrast agent carriers [51]. Clinically, vesicles are useful for better systemic drug circulation, offering improved safety and a better therapeutic index [4, 17]. Some nano MR agents produce color-mapped images (tumor region) for cancer targeting, functioning as multifunctional magneto-polymeric nanohybrids that demonstrate tumor growth [52]. Various liposomes are prepared by loading cytarabine and daunorubicin into a lipid mixture for enhanced loading efficiency [53]. Some of these have improved pharmacokinetics, selective leukemia uptake, and cytotoxicity, enhancing in vivo efficacy due to prolonged drug exposure in tissues [54, 55]. Fluorescently labeled agents with intrinsic anthracycline fluorescence, like daunorubicin, can be used for in vivo imaging [52, 54], along with liposomal formulations that are used for targeting, therapeutic, and imaging functionalities [56–60]. Liposomes can serve as diagnostic agents labeled with QDs, Mn, and Gd (both aqueous [61] and chelated [62]), and various radionuclides, such as Ga-67, In-111, and Tc-99 m (anchored to the lipid bilayer or within the aqueous core), iodine-based agents [13], or even gas (echogenic liposomes) [63, 64]. These properties make them effective theranostic agents. The role of liposomes is multifunctional without affecting stability, which is why they are commonly used for theranostics. The role of polymer vesicles or polymersomes involves the use of diblock or triblock copolymers, which include polylactic acid (PLA), polyglycolic acid (PGA), polylactic-co-glycolic acid (PLGA), polycaprolactone (PCL), chitosan, and PEG [65–68, 155, p. 1061]. Block copolymers aid in forming the core membrane and can be stabilized in aqueous media by the polymeric block, enhancing encapsulation efficiencies [69–72]. Chemotherapeutic polymersomes demonstrate increased in vitro and in vivo efficiency, coordinated release, improved cellular internalization, and enhanced pharmacokinetics [70, 73]. Specific targeting by polymeric vesicles enables controlled release of hydrophobic dyes [74]. Gene delivery is achievable in vitro and in vivo, facilitating better molecular imaging of polymersomes [72, 75, 76]. Gd-DTPA- and iron-oxide-loaded PLGA nanoparticles can be utilized with US and MRI [77, 78], showing improved efficacy with rapid clearance and reduced toxicity [79]. Superparamagnetic iron-oxide (SPIO) demonstrates enhanced liver accumulation, though it has associated toxicity. Various polymersomes exhibit improved thickness, stability, pharmacodynamics, and systemic circulation, contributing to greater tumor accumulation.

21.9 Micelles

Micellar nanoparticles are among the most effective drug carriers, offering enhanced uniformity in size and structure, along with amphiphilic materials and multifunctional capabilities [80]. Micellar structures, especially polymeric micelles, can also function as drug carriers [81]. The hydrophobic block forms the micellar core, encapsulated within a hydrophilic portion. These micelles demonstrate good biocompatibility and stealth-like properties due to functionalization [82–84]. Various clinical trials evaluate the safety and efficacy of micelles for treating metastatic breast cancer, solid tumors, and non-small-cell lung cancer [85–88]. Micelles effectively target organs, tissues, and disease sites, and various chelation parameters can be applied on the micellar surface. Chelates exhibit improved stability under physiological conditions, addressing toxicity concerns [89–96]. Micelles labeled with In-111 have demonstrated improved stability [97, 98]. These micelles require stability and hydrophobic molecules for therapeutic applications [83, 99–103].

The core-shell structure of micelles can incorporate metals, metal oxides, polymers, and other materials, providing enhanced hydrophobic effects for drug association and release, as well as improved physiological stability [15, 104]. This stability extends to size and shape retention [105, 106]. Metal nanoparticles within micelles allow for precise delivery options [106, 107]. These micelles can carry drugs, genes, and contrast agents for optimized biodistribution, particularly within the reticuloendothelial system (RES) [107]. Nanoparticles like paclitaxel are combined with polymer matrix components for multimodal imaging [108]. FDA-approved SPIO (superparamagnetic iron oxide) nanoparticles are used for intravascular imaging and to address iron deficiency anemia [109–111]. SPIO can be labeled to specific cells for tracking cell behavior [112]. Various fluorescent semiconductor QD nanocrystals have biological applications, often using cadmium selenide cores and zinc sulfide shells [113–115]. QDs can be tuned between 450 nm and 850 nm, with adjustments to size and chemical composition. QDs provide signal brightness 10–100 times greater than that of organic fluorophores [116–120, 155, p. 1062]. They are also developed for single-molecule labeling and to study biomolecular dynamics in live cells [121]. Gold nanoshells are used for photothermal tumor ablation [122]. When administered intravenously, these nanoshells accumulate in tumors and, when exposed, lead to thermal ablation [122–124]. Recent advancements have introduced luminescent gold nanoparticles with emissions ranging from 400 to 1200 nm, suitable for optical imaging [125–127]. Gold nanoparticles are popular for their stability, high X-ray absorption, low toxicity, and prolonged plasma half-life [127]. However, core-shell nanoparticles still face challenges, including toxicity, rapid clearance, poor biodistribution, and limited biodegradation. Various gases are made up of denatured albumin, phospholipids, or various polymers which will have better targeting of the molecules and therapeutics [128, 129]. Gaseous nanoparticles, often made from denatured albumin, phospholipids, or various polymers, can enhance molecule targeting and therapeutics [128, 129]. Microbubbles are used for imaging inflammation, angiogenesis, intravascular thrombus, and tumors, showcasing their potential for diagnostic and therapeutic purposes [130]. In one study, in vivo tumor xenograft models showed that microbubbles accumulate in tumor tissues, enabling controlled drug release [131, 132]. This demonstrates the potential for microbubbles in theranostics by creating the plasmonic nanobubbles using gold, which offers improved therapeutic outcomes [133].

21.10 Carbon Nanotubes

Carbon nanotubes (CNTs) are cylindrical tubes that can be single-walled or multi-walled, providing enhanced stability [134]. These CNTs possess tunable properties with multifunctional potential. Single-walled CNTs can be conjugated with targeted therapeutics to improve formulation [135]. Conjugating single-walled CNTs with cisplatin and epidermal growth factor enhances selective targeting of tumor cells [136]. These CNT formulations offer improved therapeutics with reduced toxicity [137–139]. Near-infrared (NIR) and Raman signals have been detected from subcutaneous tumors in mice, highlighting their diagnostic potential [140]. However, CNTs currently lack optimal MRI contrast capabilities [141–148]. Functionalization efforts are focused on targeted imaging applications, particularly in lymphoid malignancies [149, 150]. Some studies have demonstrated the use of radiolabeled single-walled CNTs that can be visualized through PET scanning [8, 151–154]. Thus, CNTs present sig-

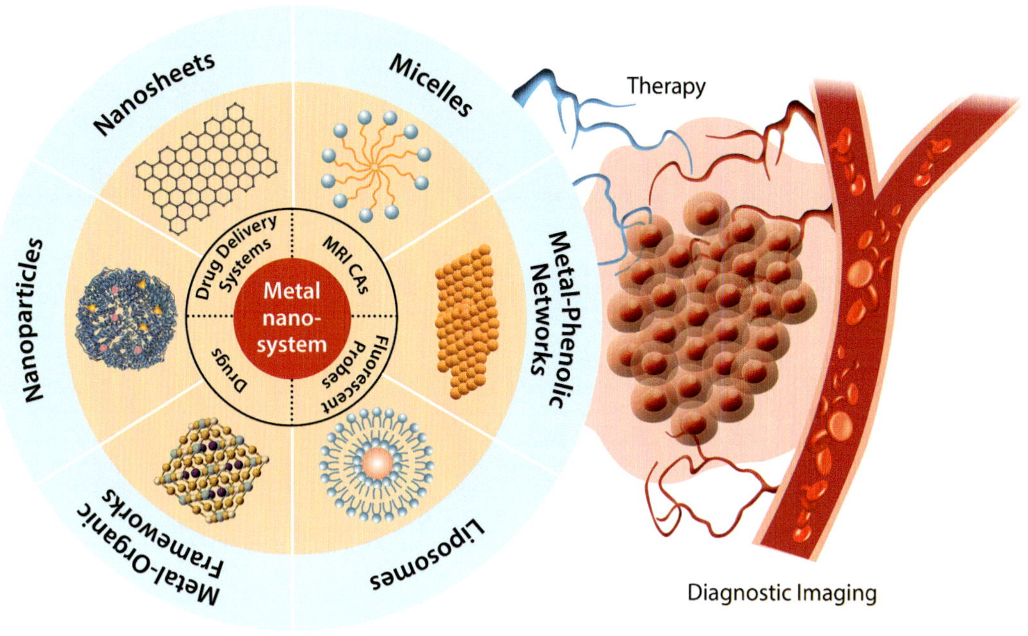

Fig. 21.3 The different nanoparticle uses in the diagnostic and therapeutic nanoimaging

nificant potential for theranostic applications, offering both therapeutic delivery and contrast agent functionality in various nanoparticle formulations.

Thus theranostics is the ideal blend of the diagnosis as well as the therapeutic aspects of imaging (Fig. 21.3).

21.11 Conclusion

The advent of nanoparticles has revolutionized multiple treatment modalities, representing a major milestone in nanoimaging. This innovation enables a deeper understanding of disease pathology through more precise imaging and enhanced therapeutic delivery.

References

1. Miele E, Spinelli GP, Tomao F, Tomao S. Albumin-bound formulation of paclitaxel (Abraxane ABI-007) in the treatment of breast cancer. Int J Nanomedicine. 2009;4:99–105.
2. Pene F, Courtine E, Cariou A, Mira JP. Toward theragnostics. Crit Care Med. 2009;37:S50–8.
3. Medina C, Santos-Martinez MJ, Radomski A, Corrigan OI, Radomski MW. Nanoparticles: pharmacological and toxicological significance. Br J Pharmacol. 2007;150:552–8.
4. Adiseshaiah PP, Hall JB, McNeil SE. Nanomaterial standards for efficacy and toxicity assessment. Wiley Interdiscip Rev Nanomed Nanobiotechnol. 2010;2:99–112.
5. Tinkle SS. Maximizing safe design of engineered nanomaterials: the NIH and NIEHS research perspective. Wiley Interdiscip Rev Nanomed Nanobiotechnol. 2010;2:88–98.
6. Nahrendorf M, Waterman P, Thurber G, et al. Hybrid in vivo FMT-CT imaging of protease activity in atherosclerosis with customized nanosensors. Arterioscler Thromb Vasc Biol. 2009;29:1444–51.
7. Debbage P, Jaschke W. Molecular imaging with nanoparticles: giant roles for dwarf actors. Histochem Cell Biol. 2008;130:845–75.
8. Massoud TF, Gambhir SS. Molecular imaging in living subjects: seeing fundamental biological processes in a new light. Genes Dev. 2003;17:545–80.
9. Park K, Lee S, Kang E, Kim K, Choi K, Kwon IC. New generation of multifunctional nanoparticles for cancer imaging and therapy. Adv Funct Mater. 2009;19:1553–66.
10. Richard C, de Chermont Qle M, Scherman D. Nanoparticles for imaging and tumor gene delivery. Tumori. 2008;94:264–70.
11. Beer AJ, Schwaiger M. Imaging of integrin alphavbeta3 expression. Cancer Metastasis Rev. 2008;27:631–44.

12. Zalutsky MR, Reardon DA, Pozzi OR, Vaidyanathan G, Bigner DD. Targeted alpha-particle radiotherapy with 211at-labeled monoclonal antibodies. Nucl Med Biol. 2007;34:779–85.
13. Mukundan S Jr, Ghaghada KB, Badea CT, et al. A liposomal nanoscale contrast agent for preclinical CT in mice. AJR Am J Roentgenol. 2006;186:300–7.
14. Blomley MJ, Cooke JC, Unger EC, Monaghan MJ, Cosgrove DO. Microbubble contrast agents: a new era in ultrasound. BMJ. 2001;322:1222–5.
15. Haglund E, Seale-Goldsmith MM, Leary JF. Design of multifunctional nanomedical systems. Ann Biomed Eng. 2009;37:2048–63.
16. Hawkins MJ, Soon-Shiong P, Desai N. Protein nanoparticles as drug carriers in clinical medicine. Adv Drug Deliv Rev. 2008;60:876–85.
17. Gaitanis A, Staal S. Liposomal doxorubicin and nab-paclitaxel: nanoparticle cancer chemotherapy in current clinical use. Methods Mol Biol. 2010;624:385–92.
18. Dreher MR, Raucher D, Balu N, Michael Colvin O, Ludeman SM, Chilkoti A. Evaluation of an elastin-like polypeptide-doxorubicin conjugate for cancer therapy. J Control Release. 2003;91:31–43.
19. Furgeson DY, Dreher MR, Chilkoti A. Structural optimization of a "smart" doxorubicin-polypeptide conjugate for thermally targeted delivery to solid tumors. J Control Release. 2006;110:362–9.
20. Bidwell GL III, Davis AN, Fokt I, Priebe W, Raucher D. A thermally targeted elastin-like polypeptide-doxorubicin conjugate overcomes drug resistance. Investig New Drugs. 2007;25:313–26.
21. Bidwell GL III, Fokt I, Priebe W, Raucher D. Development of elastin-like polypeptide for thermally targeted delivery of doxorubicin. Biochem Pharmacol. 2007;73:620–31.
22. Lammers T, Ulbrich K. HPMA copolymers: 30 years of advances. Adv Drug Deliv Rev. 2010;62:119–21.
23. Kopecek J, Kopeckova P. HPMA copolymers: origins, early developments, present, and future. Adv Drug Deliv Rev. 2010;62:122–49.
24. Lammers T. Improving the efficacy of combined modality anticancer therapy using HPMA copolymer-based nanomedicine formulations. Adv Drug Deliv Rev. 2010;62:203–30.
25. Pike DB, Ghandehari H. HPMA copolymer-cyclic RGD conjugates for tumor targeting. Adv Drug Deliv Rev. 2010;62:167–83.
26. Pola R, Studenovsky M, Pechar M, et al. HPMA-copolymer conjugates targeted to tumor endothelium using synthetic oligopeptides. J Drug Target. 2009;17:763–76.
27. Lu ZR. Molecular imaging of HPMA copolymers: visualizing drug delivery in cell, mouse and man. Adv Drug Deliv Rev. 2010;62:246–57.
28. Vasey PA, Kaye SB, Morrison R, et al. Phase I clinical and pharmacokinetic study of PK1 [n-(2-hydroxypropyl) methacrylamide copolymer doxorubicin]: first member of a new class of chemotherapeutic agents–drug-polymer conjugates. Cancer research campaign phase I/II committee. Clin Cancer Res. 1999;5:83–94.
29. Lammers T, Subr V, Ulbrich K, et al. Simultaneous delivery of doxorubicin and gemcitabine to tumors in vivo using prototypic polymeric drug carriers. Biomaterials. 2009;30:3466–75.
30. Borgman MP, Coleman T, Kolhatkar RB, Geyser-Stoops S, Line BR, Ghandehari H. Tumor-targeted HPMA copolymer-(RGDfK)-(CHX-A″-DTPA) conjugates show increased kidney accumulation. J Control Release. 2008;132:193–9.
31. Wang Y, Ye F, Jeong EK, Sun Y, Parker DL, Lu ZR. Noninvasive visualization of pharmacokinetics, biodistribution and tumor targeting of poly[n-(2-hydroxypropyl)methacrylamide] in mice using contrast enhanced mri. Pharm Res. 2007;24:1208–16.
32. Surendiran A, Sandhiya S, Pradhan SC, Adithan C. Novel applications of nanotechnology in medicine. Indian J Med Res. 2009;130:689–701.
33. Lee CC, MacKay JA, Frechet JM, Szoka FC. Designing dendrimers for biological applications. Nat Biotechnol. 2005;23:1517–26.
34. Rupp R, Rosenthal SL, Stanberry LR. Vivagel (spl7013 gel): a candidate dendrimer— microbicide for the prevention of hiv and hsv infection. Int J Nanomedicine. 2007;2:561–6.
35. Thomas TP, Shukla R, Kotlyar A, Kukowska-Latallo J, Baker JR Jr. Dendrimerbased tumor cell targeting of fibroblast growth factor-1. Bioorg Med Chem Lett. 2010;20:700–3.
36. Choi SK, Thomas T, Li MH, Kotlyar A, Desai A, Baker JR Jr. Light-controlled release of caged doxorubicin from folate receptor-targeting pamam dendrimer nanoconjugate. Chem Commun (Camb). 2010;46:2632–4.
37. Padilla De Jesus OL, Ihre HR, Gagne L, Frechet JM, Szoka FC Jr. Polyester dendritic systems for drug delivery applications: in vitro and in vivo evaluation. Bioconjug Chem. 2002;13:453–61.
38. Lee CC, Gillies ER, Fox ME, et al. A single dose of doxorubicin-functionalized bow-tie dendrimer cures mice bearing c-26 colon carcinomas. Proc Natl Acad Sci USA. 2006;103:16649–54.
39. Guillaudeu SJ, Fox ME, Haidar YM, Dy EE, Szoka FC, Frechet JM. Pegylated dendrimers with core functionality for biological applications. Bioconjug Chem. 2008;19:461–9.
40. Wiener EC, Brechbiel MW, Brothers H, et al. Dendrimer-based metal-chelates — a new class of magneticresonance-imaging contrast agents. Magn Reson Med. 1994;31:1–8.
41. Kularatne SA, Low PS. Targeting of nanoparticles: folate receptor. Methods Mol Biol. 2010;624:249–65.
42. Konda SD, Aref M, Wang S, Brechbiel M, Wiener EC. Specific targeting of folatedendrimer MRI contrast agents to the high affinity folate receptor expressed in ovarian tumor xenografts. MAGMA. 2001;12:104–13.
43. Singh P, Gupta U, Asthana A, Jain NK. Folate and folate-PEG-PAMAM dendrimers: synthesis, char-

acterization, and targeted anticancer drug delivery potential in tumor bearing mice. Bioconjug Chem. 2008;19:2239–52.
44. Xu RZ, Wang YL, Wang XL, Jeong EK, Parker DL, Lu ZR. In vivo evaluation of a PAMAM-cystamine-(Gd-DO3A) conjugate as a biodegradable macromolecular mri contrast agent. Exp Biol Med. 2007;232:1081–9.
45. Almutairi A, Rossin R, Shokeen M, et al. Biodegradable dendritic positron-emitting nanoprobes for the noninvasive imaging of angiogenesis. Proc Natl Acad Sci USA. 2009;106:685–90.
46. Kobayashi H, Sato N, Saga T, et al. Monoclonal antibody–dendrimer conjugates enable radiolabeling of antibody with markedly high specific activity with minimal loss of immunoreactivity. Eur J Nucl Med. 2000;27:1334–9.
47. Kobayashi H, Wu C, Kim MK, Paik CH, Carrasquillo JA, Brechbiel MW. Evaluation of the in vivo biodistribution of indium-111 and yttrium-88 labeled dendrimer-1B4M-DTPA and its conjugation with anti-tac monoclonal antibody. Bioconjug Chem. 1999;10:103–11.
48. Yordanov AT, Lodder AL, Woller EK, et al. Novel iodinated dendritic nanoparticles for computed tomography (CT) imaging. Nano Lett. 2002;2:595–9.
49. Mansour HM, Rhee YS, Wu X. Nanomedicine in pulmonary delivery. Int J Nanomedicine. 2009;4:299–319.
50. Torchilin VP. Recent advances with liposomes as pharmaceutical carriers. Nat Rev Drug Discov. 2005;4:145–60.
51. Levine DH, Ghoroghchian PP, Freudenberg J, et al. Polymersomes: a new multi-functional tool for cancer diagnosis and therapy. Methods. 2008;46:25–32.
52. Dicko A, Kwak S, Frazier AA, Mayer LD, Liboiron BD. Biophysical characterization of a liposomal formulation of cytarabine and daunorubicin. Int J Pharm. 2010;391:248–59.
53. Tardi P, Johnstone S, Harasym N, et al. In vivo maintenance of synergistic cytarabine: daunorubicin ratios greatly enhances therapeutic efficacy. Leuk Res. 2009;33:129–39.
54. Lim WS, Tardi PG, Dos Santos N, et al. Leukemia-selective uptake and cytotoxicity of cpx-351, a synergistic fixed-ratio cytarabine: Daunorubicin formulation, in bone marrow xenografts. Leuk Res. 2010;34:1214–23.
55. Bayne WF, Mayer LD, Swenson CE. Pharmacokinetics of CPX-351 (cytarabine/daunorubicin HCl) liposome injection in the mouse. J Pharm Sci. 2009;98:2540–8.
56. Erdogan S, Torchilin VP. Gadolinium-loaded polychelating polymer-containing tumor-targeted liposomes. Methods Mol Biol. 2010;605:321–34.
57. Al-Jamal WT, Al-Jamal KT, Tian B, Cakebread A, Halket JM, Kostarelos K. Tumor targeting of functionalized quantum dot-liposome hybrids by intravenous administration. Mol Pharm. 2009;6:520–30.
58. Strijkers GJ, Kluza E, Van Tilborg GA, et al. Paramagnetic and fluorescent liposomes for target-specific imaging and therapy of tumor angiogenesis. Angiogenesis. 2010;13:161–73.
59. Ponce AM, Vujaskovic Z, Yuan F, Needham D, Dewhirst MW. Hyperthermia mediated liposomal drug delivery. Int J Hyperth. 2006;22:205–13.
60. Chen Q, Tong S, Dewhirst MW, Yuan F. Targeting tumor microvessels using doxorubicin encapsulated in a novel thermosensitive liposome. Mol Cancer Ther. 2004;3:1311–7.
61. Unger EC, Winokur T, MacDougall P, et al. Hepatic metastases: liposomal Gd-DTPA-enhanced MR imaging. Radiology. 1989;171:81–5.
62. Trubetskoy VS, Cannillo JA, Milshtein A, Wolf GL, Torchilin VP. Controlled delivery of Gd-containing liposomes to lymph nodes: surface modification may enhance MRI contrast properties. Magn Reson Imaging. 1995;13:31–7.
63. Alkan-Onyuksel H, Demos SM, Lanza GM, et al. Development of inherently echogenic liposomes as an ultrasonic contrast agent. J Pharm Sci. 1996;85:486–90.
64. Demos SM, Onyuksel H, Gilbert J, et al. In vitro targeting of antibody-conjugated echogenic liposomes for site-specific ultrasonic image enhancement. J Pharm Sci. 1997;86:167–71.
65. Discher BM, Won YY, Ege DS, et al. Polymersomes: tough vesicles made from diblock copolymers. Science. 1999;284:1143–6.
66. Discher DE, Eisenberg A. Polymer vesicles. Science. 2002;297:967–73.
67. Jain JP, Kumar N. Self assembly of amphiphilic (PEG)(3)-PLA copolymer as polymersomes: preparation, characterization, and their evaluation as drug carrier. Biomacromolecules. 2010;11:1027–35.
68. Giacomelli C, Schmidt V, Aissou K, Borsali R. Block copolymer systems: from single chain to self-assembled nanostructures. Langmuir. 2010;26:15734–44.
69. Upadhyay KK, Le Meins JF, Misra A, et al. Biomimetic doxorubicin loaded polymersomes from hyaluronan-block-poly(gamma-benzyl glutamate) copolymers. Biomacromolecules. 2009;10:2802–8.
70. Upadhyay KK, Bhatt AN, Castro E, et al. In vitro and in vivo evaluation of docetaxel loaded biodegradable polymersomes. Macromol Biosci. 2010;10:503–12.
71. Meng F, Engbers GH, Feijen J. Biodegradable polymersomes as a basis for artificial cells: encapsulation, release and targeting. J Control Release. 2005;101:187–98.
72. Christian DA, Cai S, Bowen DM, Kim Y, Pajerowski JD, Discher DE. Polymersome carriers: from self-assembly to siRNA and protein therapeutics. Eur J Pharm Biopharm. 2009;71:463–74.
73. Upadhyay KK, Bhatt AN, Mishra AK, et al. The intracellular drug delivery and anti tumor activity of doxorubicin loaded poly (gamma-benzyl L-glutamate)-B-hyaluronan polymersomes. Biomaterials. 2010;31:2882–92.

74. Mei H, Shi W, Pang Z, et al. EGFP-EGF1 protein-conjugated PEG-PLA nanoparticles for tissue factor targeted drug delivery. Biomaterials. 2010;31:5619–26.
75. Kim Y, Tewari M, Pajeroski DJ, et al. Efficient nuclear delivery and nuclear body localization of antisense oligonucleotides using degradable polymersomes, Conf. Proc IEEE Eng Med Biol Soc. 2006;1:4350–3.
76. Kim Y, Tewari M, Pajerowski JD, et al. Polymersome delivery of siRNA and antisense oligonucleotides. J Control Release. 2009;134:132–40.
77. Ao M, Wang Z, Ran H, et al. GdDTPA-loaded PLGA microbubbles as both ultrasound contrast agent and MRI contrast agent—a feasibility research. J Biomed Mater Res B Appl Biomater. 2010;93:551–6.
78. Lee SJ, Jeong JR, Shin SC, et al. Nanoparticles of magnetic ferric oxides encapsulated with poly(D, L latide-co-glycolide) and their applications to magnetic resonance imaging contrast agent. J Magn Magn Mater. 2004;272–276:2432–3.
79. Prashant C, Dipak M, Yang CT, Chuang KH, Jun D, Feng SS. Superparamagnetic iron oxide–loaded poly(lactic acid)-D-alpha-tocopherol polyethylene glycol 1000 succinate copolymer nanoparticles as mri contrast agent. Biomaterials. 2010;31:5588–97.
80. Liu J, Lee H, Allen C. Formulation of drugs in block copolymer micelles: drug loading and release. Curr Pharm Des. 2006;12:4685–701.
81. Torchilin VP. Micellar nanocarriers: pharmaceutical perspectives. Pharm Res. 2007;24:1–16.
82. Torchilin VP. Targeted polymeric micelles for delivery of poorly soluble drugs. Cell Mol Life Sci. 2004;61:2549–59.
83. Talelli M, Rijcken CJ, van Nostrum CF, Storm G, Hennink WE. Micelles based on HPMA copolymers. Adv Drug Deliv Rev. 2010;62:231–9.
84. Nasongkla N, Bey E, Ren J, et al. Multifunctional polymeric micelles as cancer-targeted, MRI ultrasensitive drug delivery systems. Nano Lett. 2006;6:2427–30.
85. Kim SC, Kim DW, Shim YH, et al. In vivo evaluation of polymeric micellar paclitaxel formulation: toxicity and efficacy. J Control Release. 2001;72:191–202.
86. Lee KS, Chung HC, Im SA, et al. Multicenter phase II trial of Genexol-PM, a Cremophor-free, polymeric micelle formulation of paclitaxel, in patients with metastatic breast cancer. Breast Cancer Res Treat. 2008;108:241–50.
87. Kim DW, Kim SY, Kim HK, et al. Multicenter phase II trial of Genexol-PM, a novel Cremophor-free, polymeric micelle formulation of paclitaxel, with cisplatin in patients with advanced non-small-cell lung cancer. Ann Oncol. 2007;18:2009–14.
88. Lim WT, Tan EH, Toh CK, et al. Phase I pharmacokinetic study of a weekly liposomal paclitaxel formulation (Genexol-PM) in patients with solid tumors. Ann Oncol. 2010;21:382–8.
89. Kabalka GW, Davis MA, Holmberg E, Maruyama K, Huang L. Gadolinium-labeled liposomes containing amphiphilic Gd-DTPA derivatives of varying chain-length — targeted MRI contrast enhancement agents for the liver. Magn Reson Imaging. 1991;9:373–7.
90. Schwendener RA, Wuthrich R, Duewell S, Wehrli E, von Schulthess GK. A pharmacokinetic and MRI study of unilamellar gadolinium-, manganese-, and iron-DTPA-stearate liposomes as organ-specific contrast agents. Investig Radiol. 1990;25:922–32.
91. Kabalka GW, Davis MA, Holmberg E, Maruyama K, Huang L. Gadoliniumlabeled liposomes containing amphiphilic Gd-DTPA derivatives of varying chain length: targeted MRI contrast enhancement agents for the liver. Magn Reson Imaging. 1991;9:373–7.
92. Schwendener RA, Wüthrich R, Duewell S, Westera G, von Schulthess GK. Small unilamellar liposomes as magnetic resonance contrast agents loaded with paramagnetic Mn-, Gd-, and Fe-DTPA – stearate complexes. Int J Pharm. 1989;49:249–59.
93. Jaggi M, Khar RK, Chauhan UPS, Gangal SV. Liposomes as carriers of 99mTc glucoheptonate for liver imaging. Int J Pharm. 1991;69:77–9.
94. Di Bartolo N, Sargeson AM, Smith SV. New 64Cu PET imaging agents for personalised medicine and drug development using the hexa-aza cage, SarAr. Org Biomol Chem. 2006;4:3350–7.
95. Voss SD, Smith SV, DiBartolo N, et al. Positron emission tomography (PET) imaging of neuroblastoma and melanoma with 64Cu-SarAr immunoconjugates. Proc Natl Acad Sci USA. 2007;104:17489–93.
96. Cai H, Li Z, Huang CW, Park R, Shahinian AH, Conti PS. An improved synthesis and biological evaluation of a new cage-like bifunctional chelator, 4-((8-amino-3,6,10,13,16,19-hexaazabicyclo[6.6.6] icosane-1-ylamino)methyl) benzoic acid, for 64Cu radiopharmaceuticals. Nucl Med Biol. 2010;37:57–65.
97. Trubetskoy VS, Frank-Kamenetsky MD, Whiteman KR, Wolf GL, Torchilin VP. Stable polymeric micelles: lymphangiographic contrast media for gamma scintigraphy and magnetic resonance imaging. Acad Radiol. 1996;3:232–8.
98. Weissig V, Whiteman KR, Torchilin VP. Accumulation of protein-loaded long-circulating micelles and liposomes in subcutaneous Lewis lung carcinoma in mice. Pharm Res. 1998;15:1552–6.
99. Nurunnabi M, Cho KJ, Choi JS, Huh KM, Lee YK. Targeted near-IR QDs-loaded micelles for cancer therapy and imaging. Biomaterials. 2010;31:5436–44.
100. Trivedi R, Kompella UB. Nanomicellar formulations for sustained drug delivery: strategies and underlying principles. Nanomedicine (Lond). 2010;5:485–505.
101. Zhang W, Li Y, Liu L, et al. Amphiphilic toothbrush-like copolymers based on poly(ethylene glycol) and poly(epsilon caprolactone) as drug carriers with enhanced properties. Biomacromolecules. 2010;11:1331–8.

102. Akiba I, Terada N, Hashida S, et al. Encapsulation of a hydrophobic drug into a polymer-micelle core explored with synchrotron SAXS. Langmuir. 2010;26:7544–51.
103. Li Y, Xiao K, Luo J, Lee J, Pan S, Lam KS. A novel size-tunable nanocarrier system for targeted anticancer drug delivery. J Control Release. 2010;144:314–23.
104. Liu S, Han MY. Silica-coated metal nanoparticles. Chem Asian J. 2010;5:36–45.
105. Wei A, Leonov AP, Wei Q. Gold nanorods: multifunctional agents for cancer imaging and therapy. Methods Mol Biol. 2010;624:119–30.
106. Sokolov K, Tam J, Travis K, et al. Cancer imaging and therapy with metal nanoparticles. Conf Proc IEEE Eng Med Biol Soc. 2009;2009:2005–7.
107. Shubayev VI, Pisanic TR II, Jin S. Magnetic nanoparticles for theragnostics. Adv Drug Deliv Rev. 2009;61:467–77.
108. Santra S, Kaittanis C, Grimm J, Perez JM. Drug/dye-loaded, multifunctional iron oxide nanoparticles for combined targeted cancer therapy and dual optical/magnetic resonance imaging. Small. 2009;5:1862–8.
109. Schwenk MH. Ferumoxytol: a new intravenous iron preparation for the treatment of iron deficiency anemia in patients with chronic kidney disease. Pharmacotherapy. 2010;30:70–9.
110. Lu M, Cohen MH, Rieves D, Pazdur R. FDA report: ferumoxytol for intravenous iron therapy in adult patients with chronic kidney disease. Am J Hematol. 2010;85:315–9.
111. Senpan A, Caruthers SD, Rhee I, et al. Conquering the dark side: colloidal iron oxide nanoparticles. ACS Nano. 2009;3:3917–26.
112. Pittet MJ, Swirski FK, Reynolds F, Josephson L, Weissleder R. Labeling of immune cells for in vivo imaging using magnetofluorescent nanoparticles. Nat Protoc. 2006;1:73–9.
113. Pan J, Liu Y, Feng SS. Multifunctional nanoparticles of biodegradable copolymer blend for cancer diagnosis and treatment. Nanomedicine (Lond) 2010;5:347–60.
114. Yildiz I, McCaughan B, Cruickshank SF, Callan JF, Raymo FM. Biocompatible CdSe-ZnS core–shell quantum dots coated with hydrophilic polythiols. Langmuir. 2009;25:7090–6.
115. Yildiz I, Deniz E, McCaughan B, Cruickshank SF, Callan JF, Raymo FM. Hydrophilic CdSe-ZnS core–shell quantum dots with reactive functional groups on their surface. Langmuir. 2010;26:11503–11.
116. Ballou B, Ernst LA, Andreko S, Harper T, Fitzpatrick JA, Waggoner AS, Bruchez MP. Sentinel lymph node imaging using quantum dots in mouse tumor models. Bioconjug Chem. 2007;18:389–96.
117. Kim S, Lim YT, Soltesz EG, et al. Near-infrared fluorescent type II quantum dots for sentinel lymph node mapping. Nat Biotechnol. 2004;22:93–7.
118. Gao X, Cui Y, Levenson RM, Chung LW, Nie S. In vivo cancer targeting and imaging with semiconductor quantum dots. Nat Biotechnol. 2004;22:969–76.
119. Wu X, Liu H, Liu J, et al. Immunofluorescent labeling of cancer marker Her2 and other cellular targets with semiconductor quantum dots. Nat Biotechnol. 2003;21:41–6.
120. Bentolila LA, Ebenstein Y, Weiss S. Quantum dots for in vivo small-animal imaging. J Nucl Med. 2009;50:493–6.
121. Chang YP, Pinaud F, Antelman J, Weiss S. Tracking bio-molecules in live cells using quantum dots. J Biophotonics. 2008;1:287–98.
122. O'Neal DP, Hirsch LR, Halas NJ, Payne JD, West JL. Photo-thermal tumor ablation in mice using near infrared-absorbing nanoparticles. Cancer Lett. 2004;209:171–6.
123. Schwartz JA, Shetty AM, Price RE, et al. Feasibility study of particle-assisted laser ablation of brain tumors in orthotopic canine model. Cancer Res. 2009;69:1659–67.
124. Goodrich GP, Bao L, Gill-Sharp K, Sang KL, Wang J, Payne JD. Photothermal therapy in a murine colon cancer model using near-infrared absorbing gold nanorods. J Biomed Opt. 2010;15:018001.
125. Montalti M, Zaccheroni N, Prodi L, O'Reilly N, James SL. Enhanced sensitized NIR luminescence from gold nanoparticles via energy transfer from surface-bound fluorophores. J Am Chem Soc. 2007;129:2418–9.
126. Zheng J, Petty JT, Dickson RM. High quantum yield blue emission from water-soluble Au8 nanodots. J Am Chem Soc. 2003;125:7780–1.
127. Hainfeld JF, Slatkin DN, Focella TM, Smilowitz HM. Gold nanoparticles: a new X-ray contrast agent. Br J Radiol. 2006;79:248–53.
128. Landini L, Santarelli MF, Positano V. Ultrasound techniques for drug delivery in cardiovascular medicine. Curr Drug Discov Technol. 2008;5:328–32.
129. Krupka TM, Solorio L, Wilson RE, Wu H, Azar N, Exner AA. Formulation and characterization of echogenic lipid-pluronic nanobubbles. Mol Pharm. 2010;7:49–59.
130. Ferrara KW, Borden MA, Zhang H. Lipid-shelled vehicles: engineering for ultrasound molecular imaging and drug delivery. Acc Chem Res. 2009;42:881–92.
131. Gao Z, Kennedy AM, Christensen DA, Rapoport NY. Drug-loaded nano/microbubbles for combining ultrasonography and targeted chemotherapy. Ultrasonics. 2008;48:260–70.
132. Rapoport N, Gao Z, Kennedy A. Multifunctional nanoparticles for combining ultrasonic tumor imaging and targeted chemotherapy. J Natl Cancer Inst. 2007;99:1095–106.
133. Lukianova-Hleb EY, Hanna EY, Hafner JH, Lapotko DO. Tunable plasmonic nanobubbles for cell theranostics. Nanotechnology. 2010;21:85102.
134. Dennany L, Sherrell P, Chen J, Innis PC, Wallace GG, Minett AI. Epr characterisation of platinum nanopar-

ticle functionalised carbon nanotube hybrid materials. Phys Chem Chem Phys. 2010;12:4135–41.
135. Dhar S, Liu Z, Thomale J, Dai H, Lippard SJ. Targeted single-wall carbon nanotube-mediated Pt(IV) prodrug delivery using folate as a homing device. J Am Chem Soc. 2008;130:11467–76.
136. Bhirde AA, Patel V, Gavard J, et al. Targeted killing of cancer cells in vivo and in vitro with EGF-directed carbon nanotube-based drug delivery. ACS Nano. 2009;3:307–16.
137. Shvedova AA, Kagan VE. The role of nanotoxicology in realizing the 'helping without harm' paradigm of nanomedicine: lessons from studies of pulmonary effects of single-walled carbon nanotubes. J Intern Med. 2010;267:106–18.
138. Sanchez-Cano C, Hannon MJ. Novel and emerging approaches for the delivery of metallo-drugs. Dalton Trans. 2009;(48):10702–11.
139. Aillon KL, Xie Y, El-Gendy N, Berkland CJ, Forrest ML. Effects of nanomaterial physicochemical properties on in vivo toxicity. Adv Drug Deliv Rev. 2009;61:457–66.
140. Keren S, Zavaleta C, Cheng Z, de la Zerda A, Gheysens O, Gambhir SS. Noninvasive molecular imaging of small living subjects using Raman spectroscopy. Proc Natl Acad Sci USA. 2008;105: 5844–9.
141. Liu Z, Cai W, He L, et al. In vivo biodistribution and highly efficient tumour targeting of carbon nanotubes in mice. Nat Nanotechnol. 2007;2:47–52.
142. Baiker M, Milles J, Dijkstra J, et al. Atlas-based whole-body segmentation of mice from low-contrast micro-CT data. Med Image Anal. 2010;14:723–37.
143. Radu CG, Shu CJ, Nair-Gill E, et al. Molecular imaging of lymphoid organs and immune activation by positron emission tomography with a new [18F]-labeled 2′-deoxycytidine analog. Nat Med. 2008;14:783–8.
144. Le LP, Rivera AA, Glasgow JN, et al. Infectivity enhancement for adenoviral transduction of canine osteosarcoma cells. Gene Ther. 2006;13:389–99.
145. Snyder CS, Kaushal S, Kono Y, Tran Cao HS, Hoffman RM, Bouvet M. Complementarity of ultrasound and fluorescence imaging in an orthotopic mouse model of pancreatic cancer. BMC Cancer. 2009;9:106.
146. Yang J, Lee CH, Ko HJ, et al. Multifunctional magneto-polymeric nanohybrids for targeted detection and synergistic therapeutic effects on breast cancer. Angew Chem Int Ed Engl. 2007;46:8836–9.
147. Medarova Z, Pham W, Farrar C, Petkova V, Moore A. In vivo imaging of siRNA delivery and silencing in tumors. Nat Med. 2007;13:372–7.
148. Cai W, Shin DW, Chen K, et al. Peptide-labeled near-infrared quantum dots for imaging tumor vasculature in living subjects. Nano Lett. 2006;6:669–76.
149. Weissleder R. Scaling down imaging: molecular mapping of cancer in mice. Nat Rev Cancer. 2002;2:11–8.
150. Cormode DP, Skajaa T, Fayad ZA, Mulder WJ. Nanotechnology in medical imaging: probe design and applications. Arterioscler Thromb Vasc Biol. 2009;29:992–1000.
151. Kim D, Park S, Lee JH, Jeong YY, Jon S. Antibiofouling polymer-coated gold nanoparticles as a contrast agent for in vivo X-ray computed tomography imaging. J Am Chem Soc. 2007;129:7661–5.
152. Mody VV, Nounou MI, Bikram M. Novel nanomedicine-based mri contrast agents for gynecological malignancies. Adv Drug Deliv Rev. 2009;61:795–807.
153. Ventura A, Kirsch DG, McLaughlin ME, et al. Restoration of p53 function leads to tumour regression in vivo. Nature. 2007;445:661–5.
154. Fu D-X, Tanhehco Y, Chen J, et al. Bortezomib-induced enzyme-targeted radiation therapy in herpesvirus-associated tumors. Nat Med. 2008;14: 1118–22.
155. Janib SM, Moses AS, MacKay JA. Imaging and drug delivery using theranostic nanoparticles. Adv Drug Deliv Rev. 2010;62:1052–63.

22 Nanoimaging Theronostics: Part I

22.1 Theranostics

Theranostics is a new terminology used for the simultaneous use of diagnosis as well as treatment. This definition of theranostics was coined by Warner [1], which states that it includes both diagnosis and therapy. The different theranostic mechanisms are due to various applications. This term also defines "an integrated nanotherapeutic system which can diagnose, deliver targeted therapy, and monitor the response to therapy." This concept is useful for nanotechnology imaging contrast agents, which can be used simultaneously for therapeutics [2–4]. The combination of these techniques results in better management of particular diseases and reduces risks and costs [5–9]. These diseases cause more morbidity and mortality, addressed by faster diagnosis and treatment (Fig. 22.1). Most research is focused on oncology. This theranostic agent will enable better detection and therapy for various conditions, having more diagnostic strategies and greater therapeutic efficiency, with minimal toxicity resulting in fewer side effects [10–12]. Due to ongoing research, these theranostics are used in different diseases, imparting more contrast and nanoparticle ingredients. The basic component is the use of different contrast agents (Fig. 22.2). In imaging mechanisms, magnetic resonance imaging (MRI) shows dedicated research using various contrast agents. Various particles are integrated to minimize side effects by using different components like gadolinium, iron oxide, gold, silver, and other metals. Nanoparticles are used for both the diagnosis and therapeutics of cancer [13–15]. Some theranostic agents can carry drugs and contrast agents, for example, manganese oxide nanoagents [16] and silica nanoparticles with magnetic and fluorescent tags [17]. The preparation of magnet plasmonic nanoparticles is also an important area, where magnetic components are combined with metal nanoparticles or shells [18–23].

Designing nanoparticles for enhanced margination

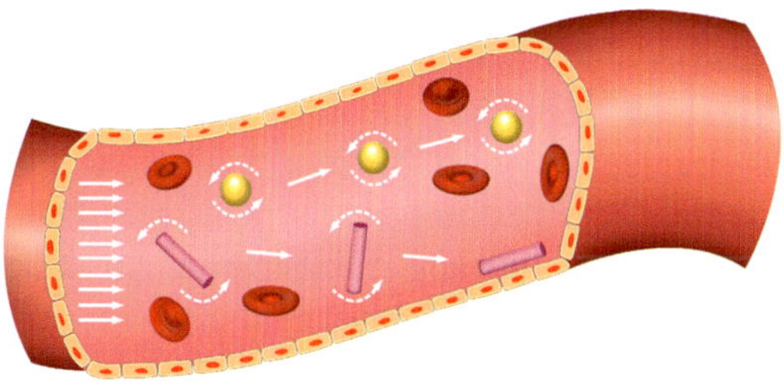

Fig. 22.1 Nanoparticle design for better migration

Iron oxide nanoparticles are used in various cancer theranostics as they are better contrast agents with superparamagnetic activity, better biocompatibility, and very low cost, making them easily accessible. Various IONPs appear in two important forms: magnetite and hematite. These forms have different surface modifications with various inorganic molecules, ligands, and polymeric and non-polymeric stabilizers, allowing

S. Shaikh, *Nanoimaging - Future of Precision Medicine*,
https://doi.org/10.1007/978-981-97-7292-6_22

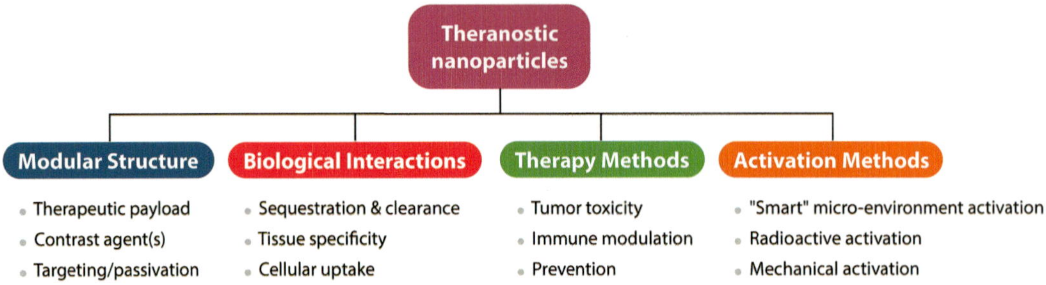

Fig. 22.2 Different Theronostics mechanisms showing modulation, therapy, activation, and Biochemical activations

them to be used in various applications. There can be a combination of different therapies and diagnostic agents that may vary and can be combined. For example, methotrexate (MTX) can be added to superparamagnetic iron oxide nanoparticles (SPIONs), which are used in MRI and also for drug delivery [24]. MTX is a folate antagonist, making it an important chemotherapeutic agent in cancer therapy. Another important drug is trastuzumab, a monoclonal antibody with various cell membrane receptors, which can be coupled with IONPs, as studied by Huh et al. [25] and Lee et al. [26]. It was also shown by Lee et al. that there are different all-in-one target-cell-specific magnetic nanoparticles used for imaging and siRNA delivery [10]. Magnetic nanovectors [27] can serve as better cell transcytosis and intracellular trafficking agents, providing a more efficient route for siRNA delivery. The magnetic properties of various IONPs make them suitable for different cancer gene therapies. Doxorubicin (DOX) can also be combined with modified IONPs for treating lung carcinoma [28]. Piao et al. [29] demonstrated IONPs loaded with DOX for improved therapeutic action. Paclitaxel (PTX) can be used in combination with IONPs for enhanced therapy and diagnosis [30]. Yang et al. [31] demonstrated the use of prepared cRGD-functionalized, DOX-conjugated material and 64Cu-labeled SPIONs in anticancer therapy, used in positron emission tomography (PET)/MRI of tumors as integrin $\alpha v\beta 3$ expression agents. An important aspect of these agents is their stability in various tumor-targeting multifunctional agents, such as polymer vesicles loaded with SPIONs and DOX, providing a better target agent for cancer therapy [31, 32].

22.2 Carbon Nanotubes and Nanocarriers in Cancer Theranostics

Carbon nanotubes (CNTs) are single or multilayered cylindrical structures made up of various allotropes of carbon. They have combined mechanical and electronic properties for better use. One of their uses is thermal ablation, which is commonly used in gene and drug delivery. These CNTs have higher uptake by cells and are effective for the delivery of various active agents. One study shows that core-shell nanotubes, when coated with zirconium oxide, have better loading of various active materials [33]. This optical absorbance can be used in infrared (NIR) photothermal ablation therapy. These agents, along with the nanocarriers, provide better theranostic results, where different CNTs and nanocarriers can be used in drug loading. Liu et al. [34] demonstrated that detection, along with drug delivery, can be done in vitro and in vivo by loading DOX onto CNTs (Fig. 22.3). One study [35] demonstrated that DOX loading on copolymer-coated CNTs can be achieved. The study by Yang et al. [31] showed that magnetic CNTs deliver gemcitabine, an antibiotic used in the treatment of lymph node metastasis. The use of paramagnetic gadolinium nanoparticles coupled with CNTs can lead to better and higher performance in MRI [36]. Loading and delivery of DNA plasmids were shown by Pantarroto et al. [9], Singh et al. [37], and Kam et al. [38], who demonstrated that CNT functionalization using disulfide bonds can be used for siRNA delivery [39]. CNTs can also be used as transporters for siRNA delivery to human T cells and primary cells [40].

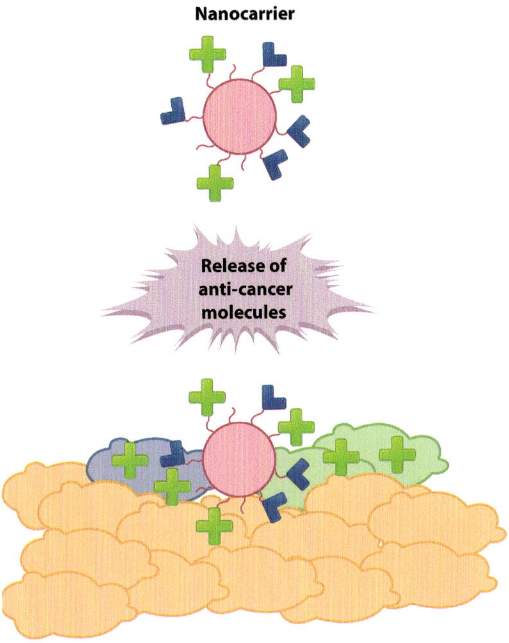

Fig. 22.3 Nanocarrier showing the release of the anticancer molecules

22.3 Quantum Dots in Cancer Theranostics

Quantum dots (QDs) are composed of inorganic semiconductor fluorophores, which can be used in transitions for various groups in the periodic table. They are capable of being used for imaging as well as drug delivery, and drug delivery is improved due to more flexible linking for the variable sizes of the nanoparticle core. This flexibility results in less toxicity [41] (Fig. 22.4). QDs are one of the important entities primarily used for therapeutic aspects compared to diagnostic imaging. A study by Gao et al. [42] showed that QD-based conjugates can accumulate in mice with tumors. The QDs are formed as conjugates [43] based on various novel conjugates of QD-aptamer and DOX for targeted cancer imaging, therapy, and sensing. This will be useful as an important tool for targeting prostate cancer cells. Delivery and imaging can be done with the

Fig. 22.4 Nanoparticle mechanism of actions, different actions seen

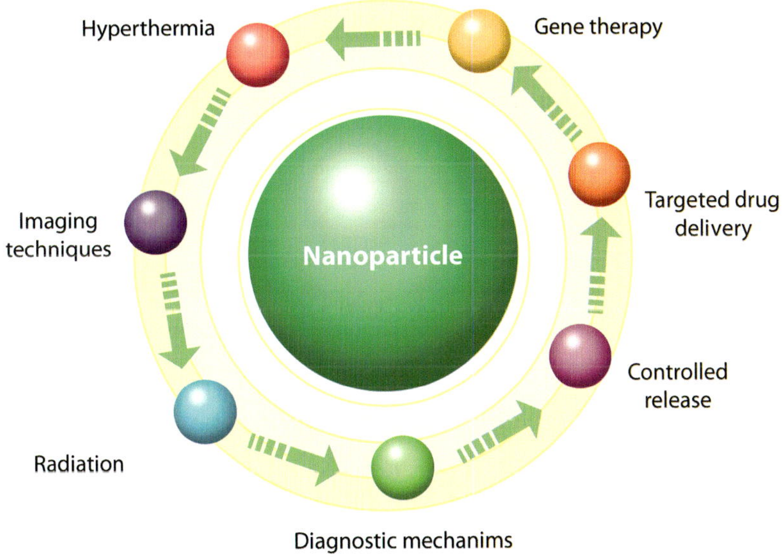

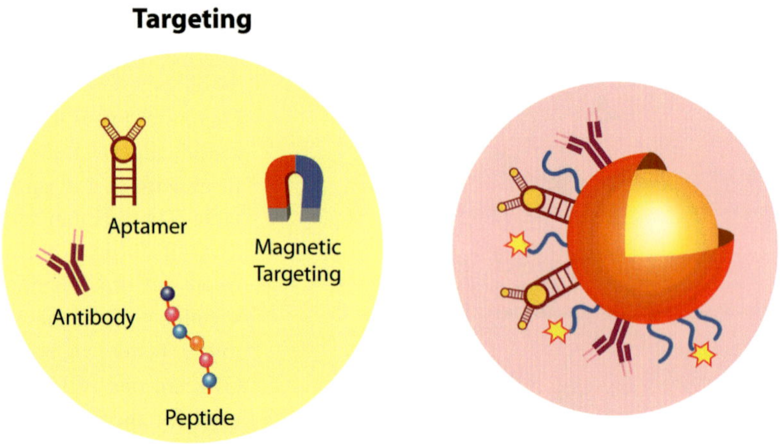

Fig. 22.5 The Left image shows targeting agents and the right one is the cell targeted

help of fluorescence imaging with QDs. A study [44] prepared various conjugates, which can be used in siRNA delivery. The use of gold in cancer management is based on the principles of theranostics. Melancon et al. [45] showed the functional gold-based nanoshells with various optical and magnetic properties, which can be conjugated to various targeting agents, such as superparamagnetic gold-coated iron oxide nanoshells. These gold nanoprobes have different theranostic applications, particularly in prostate cancer [46]. The theranostics of the cells can be seen as components of various plasmonic nanobubbles, which have better selectivity for bubble generation in specific-target cancer cells, with both diagnostic and therapeutic properties. The study demonstrated the use of a nanoplatform as a theranostic agent by coating a gold nanostar with silica [47]. The use of various gold nanorods is important in silica for X-ray/CT imaging with more selective targeting, enhanced radiation therapy (RT), and photothermal therapy (PTT). Gold is also used in drug delivery by using PTX-functionalized gold nanoparticles [48]. The photodynamic therapeutic agent Pc4, when combined with gold nanoparticles, is shown in in vivo photodynamic therapy [49]. Various bifunctional gold nanoshells with an SPIO core can be used for MRI and PTT [50] (Fig. 22.5). Gold-capped mesoporous silica nanospheres can also be loaded with the anticancer drug PTX [51] and are proposed as a mesoporous system capped with magnetic nanoparticles [52]. Various drugs co-encapsulated with water-insoluble drugs and two-photon-absorbing fluorescent dyes can be added to photodynamic therapy [53] for various multifunctional nanoprobes used for imaging and therapy.

22.4 Theranostic Applications

The different theranostic approaches for cancer aim at both diagnosis and drug delivery. The use of various multifunctional nanoparticles is also useful in illustrating applications in cancer diagnosis and therapy [54] as part of theranostic applications. Various studies have shown the use of different theranostic agents that are more useful for in vitro and in vivo evaluation. The formation of the nanocomposite, such as graphene oxide with iron oxide, is used for imaging as well as intratumoral drug delivery. It also includes the biocompatible polymer (PEG), which is capable of reducing side effects. This is also important for photothermal therapy in various tumors and tumor ablation [55, 56] (Fig. 22.6).

Alzheimer's disease is one of the most common neuropathological conditions, caused by the accumulation of amyloid plaques due to Aβ protein aggregation, along with intracellular neurofibrillary tangles (NFTs) formed by the aggregation of hyperphosphorylated tau protein [57–59]. There is a loss of spatial and temporal resolution due to abnormal vocabulary. The characterization of patients is based on early onset, sporadic onset,

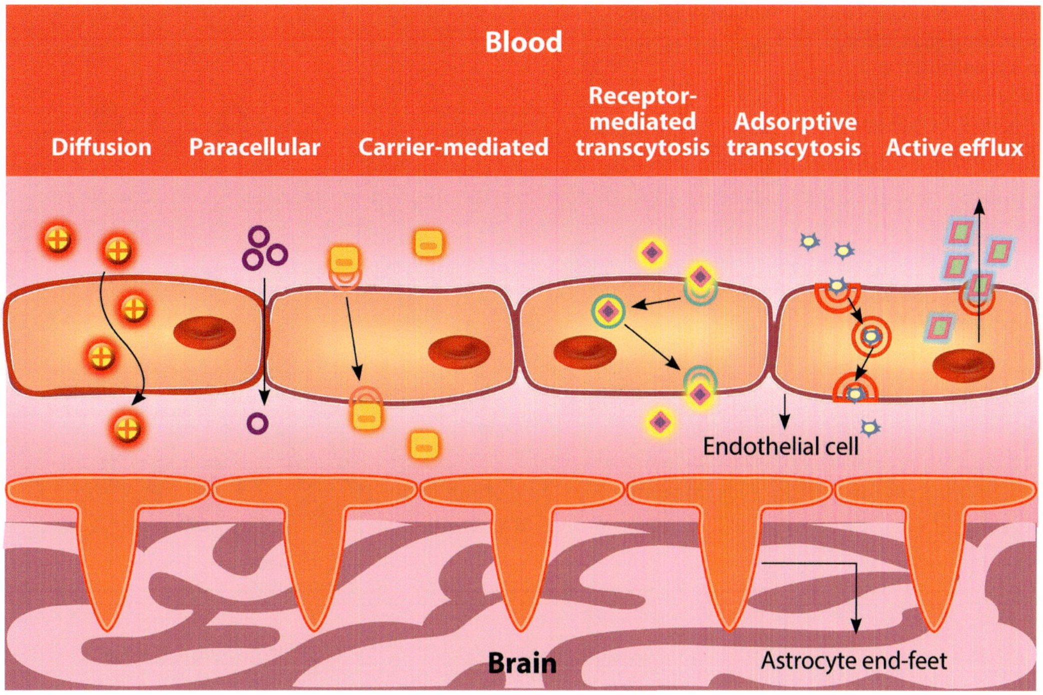

Fig. 22.6 Blood-brain barrier mechanisms in the cells with various processes

or late onset [60, 61]. The early onset of symptoms is around 65 years, often seen with mutations in the presenilin 1 (PS1) and presenilin 2 (PS2) genes. This is due to increased activity in the strongly self-aggregating Aβ42 peptide and is familial in presentation [62–65]. The exact cause is not known, but several predisposing factors include gene mutations and factors affecting lifestyle, smoking, hypertension, heart disease, stroke, and aging [66]. The primary cause is the aggregation and accumulation of Aβ42 in various pathological processes, leading to the disruption of synaptic connections, reduced neurotransmitter release, microglial activation, and astrocyte involvement due to chronic inflammatory responses, ultimately resulting in neuronal loss and dementia. As a result, there is an accumulation of non-fibrillar oligomeric Aβ deposits, which are neurotoxic, rather than fibrillar Aβ deposits in the AD brain [67, 68]. The development of nanoplatforms for imaging has led to very effective drug delivery systems for various CNS diseases [56, 69]. Polymeric NPs have more surface modifications, enabling drug delivery across the blood-brain barrier (BBB) for neurodegenerative diseases. Liposomal NPs can be used in Aβ-targeted lipid conjugates, leading to amyloid plaque deposition due to passage through the BBB [70, 71]. The NPs have been taken up by the brain in vivo or in situ without changes in permeability or integrity. They were also found to be stable at pH 6–8 in aqueous suspended medium and preserved at 4–8 °C, with no aggregation found with serum salts. The release of d-penicillamine from NPs was achieved using dithiothreitol as a reducing agent [72]. The use of various nanovehicles (SNVs) can be seen traveling through the BBB in cerebrovascular amyloid angiopathy (CAA) and AD. The use of gold NPs and weak microwave fields [73–75] has also been noted. Drug delivery to the brain is facilitated by the administration of lectin-modified PEG-poly(lactic-co-glycolic acid) (PEG-PLGA) intranasally.

Theranostic applications of NPs are seen in various neurodegenerative disorders, and this is also demonstrated in the successful targeted delivery of various amyloid plaques in the AD

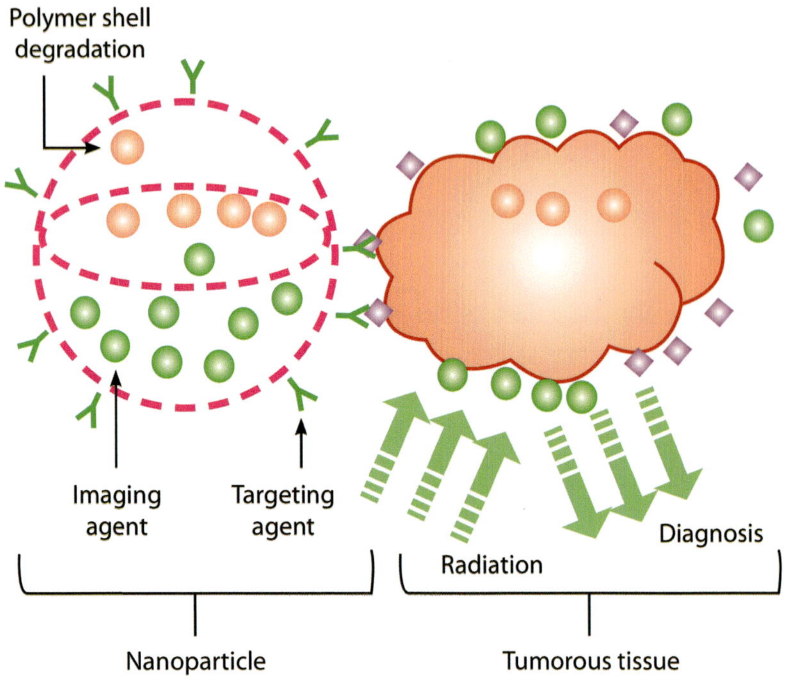

Fig. 22.7 Nanoparticles as Imaging and targeting agents on the left and the radiation and drug actions on right side of the image

mice model. The use of near-infrared (NIR) fluorescence imaging can also be seen in the fluorescence signal before and after IV injection [76, 77]. These NPs will have two targeting peptides, which have more affinity for the ligands that can pass through the BBB, an important component in targeting the various amyloid plaques.

Epilepsy is a CNS disorder in which the nerve cell activity of the brain leads to epileptic seizures, which can be generalized seizures or partial seizures. Both hemispheres are affected in generalized seizures [78, 79]. Various subtypes of seizures are seen, such as absence seizure, grand mal seizure, myoclonic seizure, atonic seizure, clonic seizure, tonic seizure, and partial seizures, each based on the different areas of the brain. Various causes of epilepsy include infections, stroke, head trauma, arteriovenous malformations, cerebrovascular diseases, and perinatal injuries. Changes in ionic concentrations, such as K+ concentrations, and reduced extracellular Ca++, will cause failure [80, 81]. The use of nanochips can release drugs in comparison with other drug delivery methods. The surface charge of NPs will lead to better distribution in the brain due to the fabrication process [82–85]. Different drug delivery systems will use nanogels for MRI applications. Electro-responsive hydrogel NPs (ERHNPs) will be developed, which will facilitate more drug release due to the external electric field [86–89]. One important aspect is the loss of dopamine (DA)-producing neurons, causing clinical conditions such as bradykinesia, rigidity, and resting tremor [90–96]. Various factors increase the risk, which will raise the cytosolic and nuclear fractions. NP applications in therapeutics and diagnostics will result in targeted drug delivery for many neurodegenerative diseases [97–101] (Fig. 22.7).

22.5 Optical Imaging

Optical imaging is a new imaging technique comprising bioluminescent, fluorescent, or even Raman probes. Optical imaging is safe as it uses non-ionizing radiation and is not costly. This imaging technique can be used for theranostic nanoparticle uptake and drug release. Optical imaging has limited tissue penetration (up to the range of 2 cm) and more background noise in the form of autofluorescence [102–105]. The use of fluorescence resonance energy transfer (FRET) is also capable of monitoring apoptosis via the theranostic nanoparticle. The single nanoparticle will result in visualized therapeutic self-reporting. This can be used in cancer therapy by the nanoreporter approach. Nanoparticles are capable of mRNA detection, which can target, bind, and detect the mRNA of survivin, a gene used in cancer diagnosis and treatment [106–108].

22.6 NIR Fluorescence (NIRF)

NIRF has higher spatial resolution and sensitivity, but there is very little absorption and autofluorescence [109–114]. NIRF is based on various cell-based or cell-inspired theranostic particles that have better therapeutic efficacy and self-monitoring. Nanoparticle delivery is increased when they interfere with various mesenchymal stem cells (MSCs). With different photoluminescent properties, QDs have also been used as a promising nanomaterial for theranostics using optical imaging.

22.7 Raman Spectroscopy

Raman imaging is capable of identifying inelastically scattered photons, which will provide more illumination on the metallic surface, leading to a stronger electromagnetic field. This is known as the surface-enhanced Raman scattering (SERS) effect [115–117] (Fig. 22.8). The newer metallic nanoparticle will have more precise ultrasensitive detection due to the unique Raman-active layer. The Raman nanoparticle effect will be due to various factors [118–127]. The theranostic nanoparticles will be capable of photothermal therapy, and this is aided by Raman imaging.

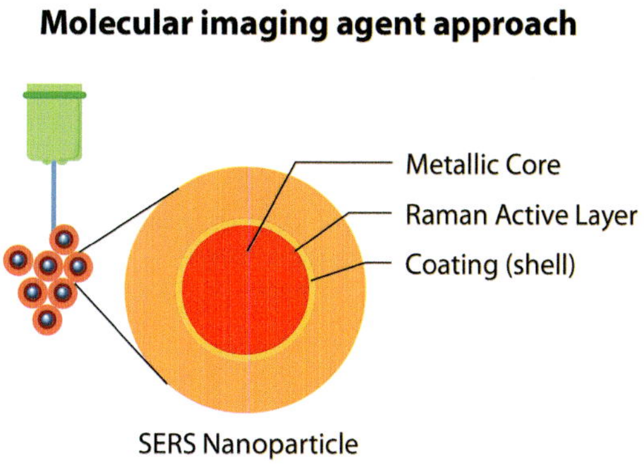

Fig. 22.8 Molecular Imaging agent with nanoparticle actions

22.8 Conclusion

The use of different photoacoustic properties will be more helpful for the better integration of Raman, PET, and photoacoustic imaging.

References

1. Warner S. Diagnostics plus therapy theronostics. Scientist. 2004;18:38–9.
2. Sumer B, Gao J. Theranostic nanomedicine for cancer. Nanomedicine. 2008;3:137–40.
3. Huang P, et al. Folic acid-conjugated silica modified gold nanorods for X-ray/CT imaging-guided dual-mode radiation and photo-thermal therapy. Biomaterials. 2011;32:9796–809.
4. Prabaharan M, et al. Gold nanoparticles with a monolayer of doxorubicin-conjugated amphiphilic block copolymer for tumor-targeted drug delivery. Biomaterials. 2009;30:6065–75.
5. Lee J-H, et al. All-in-one target-cell specific magnetic nanoparticles for simultaneous molecular imaging and siRNA delivery. Angew Chem Int Ed Engl. 2009;48:4174–9.
6. Yang X, et al. Multifunctional SPIO/DOX-loaded wormlike polymer vesicles for cancer therapy and MR imaging. Biomaterials. 2010;31:9065–73.
7. Roy I, et al. Ceramic-based nanoparticles entrapping water-insoluble photosensitizing anticancer drugs: a novel drug-carrier system for photodynamic therapy. J Am Chem Soc. 2003;125:7860–5.
8. Park J-H, et al. Biodegradable luminescent porous silicon nanoparticles for in vivo applications. Nat Mater. 2009;8:331–6.

9. Pantarotto D, et al. Functionalized carbon nanotubes for plasmid DNA gene delivery. Angew Chem Int Ed Engl. 2004;43:5242–6.
10. Liu Z, et al. Drug delivery with carbon nanotubes for in vivo cancer treatment. Cancer Res. 2008;68:6652–60.
11. Savla R, et al. Tumor targeted quantum dotmucin 1 aptamer-doxorubicin conjugate for imaging and treatment of cancer. J Control Release. 2011;153:16–22.
12. Yuan J, et al. Anticancer drug-DNA interactions measured using a photo induced electron-transfer mechanism based on luminescent quantum dots. Anal Chem. 2009;81:362–8.
13. Li Z, et al. One-pot reaction to synthesize water-soluble magnetite nanocrystals. Chem Mater. 2004;16:1391–3.
14. Morel AL, et al. Sonochemical approach to the synthesis of Fe3O4/SiO2 core/shell nanoparticles with tunable properties. ACS Nano. 2008;2:847–56.
15. Souza KC, et al. Mesoporous silica–magnetite nanocomposite synthesized by using a neutral surfactant. Nanotechnology. 2008;19:185603.
16. Liong M, et al. Multifunctional inorganic nanoparticles for imaging, targeting, and drug delivery. ACS Nano. 2008;2:889–96.
17. Insin N, et al. Incorporation of iron oxide nanoparticles and quantum dots into silica microspheres. ACS Nano. 2008;2:197–202.
18. Kang L-J, et al. Stabilization of superparamagnetic iron oxide core–gold shell nanoparticles in high ionic strength media. Langmuir. 2009;25:13384.
19. Moriggi L, et al. Gold nanoparticles functionalized with gadolinium chelates as highrelaxivity MRI contrast agents. J Am Chem Soc. 2009;131:10828–9.
20. Levin CS, et al. Magnetic–plasmonic core–shell nanoparticles. ACS Nano. 2009;3:1379–88.
21. Van Berkel K-Y, et al. A simple route to multimodal composite nanoparticles. Macromolecules. 2009;42:1425–7.
22. Zhao N, Gao MY. Magnetic Janus particles prepared by a flame synthetic approach: synthesis, characterizations and properties. Adv Mater. 2009;21:184–7.
23. Kamei K, et al. Direct cell entry of gold/iron-oxide magnetic nanoparticles in adenovirus mediated gene delivery. Biomaterials. 2009;30:1809–14.
24. Kohler N, et al. Methotrexate-modified superparamagnetic nanoparticles and their intracellular uptake into human cancer cells. Langmuir. 2005;21:8858–64.
25. Huh YM, et al. In vivo magnetic resonance detection of cancer by using multifunctional magnetic nanocrystals. J Am Chem Soc. 2005;127:12387–91.
26. Lee J-H, et al. Artificially engineered magnetic nanoparticles for ultrasensitive molecular imaging. Nat Med. 2007;13:95–9.
27. Veiseh O, et al. Cell transcytosing poly-arginine coated magnetic nanovector for safe and effective siRNA delivery. Biomaterials. 2011;32:5717–25.
28. Yu MK. Drug-loaded superparamagnetic iron oxide nanoparticles for combined cancer imaging and therapy in vivo. Angew Chem Int Ed Engl. 2008;47:5362–5.
29. Piao Y, et al. Wrap-bake-peel process for nanostructural transformation from beta-FeOOH nanorods to biocompatible iron oxide nanocapsules. Nat Mater. 2008;7:242–7.
30. Jain K, et al. Magnetic nanoparticles with dual functional properties: drug delivery and magnetic resonance imaging. Biomaterials. 2008;29:4012–21.
31. Yang X, et al. cRGD-functionalized, DOXconjugated and 64Cu-labeled superparamagnetic iron oxide nanoparticles for targeted anticancer drug delivery and PET/MR imaging. Biomaterials. 2011;32:4151–60.
32. Xu H, et al. Polymer encapsulated upconversion nanoparticle/iron oxide nanocomposites for multimodal imaging and magnetic targeted drug delivery. Biomaterials. 2011;32:9364–73.
33. Pal K, et al. Microstructural investigations of zirconium oxide on core–shell structure of carbon nanotubes. J Nanopart Res. 2011;13:2597–607.
34. Liu Z, et al. Carbon nanotubes in biology and medicine: in vitro and in vivo detection, imaging and drug delivery. Nano Res. 2009;2:85–120.
35. Boucetta HA, et al. Multiwalled carbon nanotube–doxorubicin supramolecular complexes for cancer therapeutics. Chem Commun. 2008;4:459–61.
36. Sitharaman B, et al. Superparamagnetic gadonanotubes are high-performance MRI contrast agents. Chem Commun. 2005;31:3915–7.
37. Singh R, et al. Binding and condensation of plasmid DNA onto functionalized carbon nanotubes: toward the construction of nanotube-based gene delivery vectors. J Am Chem Soc. 2005;127:4388–96.
38. Kam NWS, et al. Functionalization of carbon nanotubes via cleavable disulfide bonds for efficient intracellular delivery of siRNA and potent gene silencing. J Am Chem Soc. 2005;127:12492–3.
39. Liu Z, et al. siRNA delivery into human T cells and primary cells with carbon-nanotube transporters. Angew Chem Int Ed Engl. 2007;46:2023–7.
40. Krishna V, et al. Polyhydroxy fullerenes for non-invasive cancer imaging and therapy. Small. 2010;62:2236–41.
41. Qi L, Gao X. Emerging application of quantum dots for drug delivery and therapy. Expert Opin Drug Deliv. 2008;5:263–7.
42. Gao X, et al. In vivo cancer targeting and imaging with semiconductor quantum dots. Nat Biotechnol. 2004;22:969–76.
43. Bagalkot V, et al. Quantum dot-aptamer conjugates for synchronous cancer imaging, therapy, and sensing of drug delivery based on bi-fluorescence resonance energy transfer. Nano Lett. 2007;7:3065–70.
44. Derfus AM, et al. Targeted quantum dot conjugates for siRNA delivery. Bioconjug Chem. 2007;18:1391–6.

45. Melancon MP, et al. Targeted multifunctional gold-based nanoshells for magnetic resonance-guided laser ablation of head and neck cancer. Biomaterials. 2011;32:7600–8.
46. Ekaterina Y, et al. Selective and self-guided micro-ablation of tissue with plasmonic nanobubbles. J Surg Res. 2011;166:e3–e13.
47. Fales AM, et al. Silica-coated gold nanostars for combined surface-enhanced Raman scattering (SERS) detection and singlet-oxygen generation: a potential nanoplatform for theranostics. Langmuir. 2011;27:12186–90.
48. Gibson JD, et al. Paclitaxel-functionalized gold nanoparticles. J Am Chem Soc. 2007;129:11653–61.
49. Cheng Y, et al. Highly efficient drug delivery with gold nanoparticle vectors for in vivo photodynamic therapy of cancer. J Am Chem Soc. 2008;130:10643–7.
50. Ji X, et al. Bifunctional gold nanoshells with a superparamagnetic iron oxide-silica core suitable for both MR imaging and photothermal therapy. J Phys Chem C. 2007;111:6245–51.
51. Vivero-Escoto JL, et al. Photoinduced intracellular controlled release drug delivery in human cells by gold-capped mesoporous silica nanosphere. J Am Chem Soc. 2009;131:3462–3.
52. Giri S, et al. Stimuli-responsive controlled release delivery system based on mesoporous silica nanorods capped with magnetic nanoparticles. Angew Chem Int Ed Engl. 2005;44:5038–44.
53. Kim S, et al. Organically modified silica nanoparticles co-encapsulating photosensitizing drug and aggregation-enhanced two-photon absorbing fluorescent dye aggregates for two-photon photodynamic therapy. J Am Chem Soc. 2007;129:2669–75.
54. Yang H-M, et al. Multifunctional poly (aspartic acid) nanoparticles containing iron oxide nanocrystals and doxorubicin for simultaneous cancer diagnosis and therapy. Colloids Surf A Physicochem Eng Asp. 2011;391:208–15.
55. Yang K, et al. Multimodal imaging guided photothermal therapy using functionalized graphene nanosheets anchored with magnetic nanoparticles. Adv Mater. 2012;24:1868–72.
56. Gao Q, Zhang J, Gao J, Zhang Z, Zhu H, Wang D. Gold nanoparticles in cancer theranostics. Front Bioeng Biotechnol. 2021;9:647905.
57. Kaur IP, Bhandari R, Bhandari S, Kakkar V. Potential of solid lipid nanoparticles in brain targeting. J Control Release. 2008;127(2):97–109.
58. Faraji AH, Wipf P. Nanoparticles in cellular drug delivery. Bioorg Med Chem. 2009;17(8):2950–62.
59. Kreuter J, Ramge P, Petrov V, et al. Direct evidence that polysorbate-80-coated poly(butylcyanoacrylate) nanoparticles deliver drugs to the CNS via specific mechanisms requiring prior binding of drug to the nanoparticles. Pharm Res. 2003;20(3):409–16.
60. Volicer L. Management of severe Alzheimer's disease and end-of-life issues. Clin Geriatr Med. 2001;17(2):377–91.
61. Forchetti CM. Treating patients with moderate to severe Alzheimer's disease: implications of recent pharmacologic studies. Prim Care Companion J Clin Psychiatry. 2005;7(4):155–61.
62. Selkoe DJ, Schenk D. Alzheimer's disease: molecular understanding predicts amyloid-based therapeutics. Annu Rev Pharmacol Toxicol. 2003;43(1):545–84.
63. Mahmoudi M, Shokrgozar MA. Multifunctional stable fluorescent magnetic nanoparticles. Chem Commun. 2012;48(33):3957–9.
64. Helzner EP, Scarmeas N, Cosentino S, Tang MX, Schupf N, Stern Y. Survival in Alzheimer disease: a multiethnic, population-based study of incident cases. Neurology. 2008;71(19):1489–95.
65. Xie J, Brayne C, Matthews FE, Medical Research Council Cognitive Function and Ageing Study Collaborators. Survival times in people with dementia: analysis from population-based cohort study with 14 year follow-up. BMJ. 2008;336(7638):258–62.
66. Allinson TM, Parkin ET, Turner AJ, Hooper NM. ADAMs family members as amyloid precursor protein alpha-secretases. J Neurosci Res. 2003;74(3):342–52.
67. Dickson DW, Crystal HA, Bevona C, Honer W, Vincent I, Davies P. Correlations of synaptic and pathological markers with cognition of the elderly. Neurobiol Aging. 1995;16(3):285–98.
68. Terry RD, Masliah E, Salmon DP, et al. Physical basis of cognitive alterations in Alzheimer's disease: synapse loss is the major correlate of cognitive impairment. Ann Neurol. 1991;30(4):572–80.
69. Begley DJ. Delivery of therapeutic agents to the central nervous system: the problems and the possibilities. Pharmacol Ther. 2004;104(1):29–45.
70. Roney C, Kulkarni P, Arora V, et al. Targeted nanoparticles for drug delivery through the blood–brain barrier for Alzheimer's disease. J Control Release. 2005;108(2):193–214.
71. Tanifum EA, Dasgupta I, Srivastava M, et al. Intravenous delivery of targeted liposomes to amyloid-β pathology in APP/PSEN1 transgenic mice. PLoS One. 2012;7(10):e48515.
72. Cui Z, Lockman PR, Atwood CS, et al. Novel D-penicillamine carrying nanoparticles for metal chelation therapy in Alzheimer's and other CNS diseases. Eur J Pharm Biopharm. 2005;59(2):263–72.
73. Agyare EK, Curran GL, Ramakrishnan M, Yu CC, Poduslo JF, Kandimalla KK. Development of a smart nanovehicle to target cerebrovascular amyloid deposits and brain parenchymal plaques observed in Alzheimer's disease and cerebral amyloid angiopathy. Pharm Res. 2008;25(11):2674–84.
74. Bastus NG, Kogan MJ, Amigo R, et al. Gold nanoparticles for selective and remote heating of β-amyloid protein aggregates. Mater Sci Eng C. 2007;27(5–8):1236–40.
75. Zhang C, Chen J, Feng C, et al. Intranasal nanoparticles of basic fibroblast growth factor for brain delivery to treat Alzheimer's disease. Int J Pharm. 2014;461(1–2):192–202.

76. Zhang C, Wan X, Zheng X, et al. Dual-functional nanoparticles targeting amyloid plaques in the brains of Alzheimer's disease mice. Biomaterials. 2014;35(1):456–65.
77. Li Y, Xu D, Ho SL, Li HW, Yang R, Wong MS. A theranostic agent for in vivo near-infrared imaging of β-amyloid species and inhibition of β-amyloid aggregation. Biomaterials. 2016;94:84–92.
78. Fisher RS, Ho J. Potential new methods for antiepileptic drug delivery. CNS Drugs. 2002;16(9):579–93.
79. De Tiège X, Laufs H, Boyd SG, et al. EEG-fMRI in children with pharmacoresistant focal epilepsy. Epilepsia. 2007;48(2):385–9.
80. Moody WJ, Futamachi KJ, Prince DA. Extracellular potassium activity during epileptogenesis. Exp Neurol. 1974;42(2):248–63.
81. Rutecki PA, Lebeda FJ, Johnston D. Epileptiform activity induced by changes in extracellular potassium in hippocampus. J Neurophysiol. 1985;54(5):1363–74.
82. Huang WC, Hu SH, Liu KH, Chen SY, Liu DM. A flexible drug delivery chip for the magnetically controlled release of anti-epileptic drugs. J Control Release. 2009;139(3):221–8.
83. Couvreur P, Vauthier C. Nanotechnology: intelligent design to treat complex disease. Pharm Res. 2006;23(7):1417–50.
84. Birnbaum DT, Kosmala JD, Brannon-Peppas L. Optimization of preparation techniques for poly(lactic acid-co-glycolic acid) nanoparticles. J Nanopart Res. 2000;2(2):173–81.
85. Ito F, Fujimori H, Makino K. Incorporation of water-soluble drugs in PLGA microspheres. Colloids Surf B Biointerfaces. 2007;54(2):173–8.
86. Friese A, Seiller E, Quack G, Lorenz B, Kreuter J. Increase of the duration of the anticonvulsive activity of a novel NMDA receptor antagonist using poly (butylcyanoacrylate) nanoparticles as a parenteral controlled release system. Eur J Pharm Biopharm. 2000;49(2):103–9.
87. Ying X, Wang Y, Liang J, et al. Angiopep-conjugated electro-responsive hydrogel nanoparticles: therapeutic potential for epilepsy. Angew Chem Int Ed Engl. 2014;53(46):12436–40.
88. Hsiao MH, Larsson M, Larsson A, et al. Design and characterization of a novel amphiphilic chitosan nanocapsule-based thermo-gelling biogel with sustained in vivo release of the hydrophilic anti-epilepsy drug ethosuximide. J Control Release. 2012;161(3):942–8.
89. Fu T, Kong Q, Sheng H, Gao L. Value of functionalized superparamagnetic iron oxide nanoparticles in the diagnosis and treatment of acute temporal lobe epilepsy on MRI. Neural Plast. 2016;2016:2412958.
90. Singh N, Pillay V, Choonara YE. Advances in the treatment of Parkinson's disease. Prog Neurobiol. 2007;81(1):29–44.
91. Tang WL, Tang WH, Li SD. Cancer theranostic applications of lipid-based nanoparticles. Drug Discov Today. 2018;23:1159–66.
92. Sivasubramanian M, Chuang YC, Chen NT, Lo LW. Seeing better and going deeper in cancer nanotheranostics. Int J Mol Sci. 2019;20(14):3490.
93. Yang HY, Jang MS, Li Y, et al. Hierarchical tumor acidity-responsive self-assembled magnetic nanotheranostics for bimodal bioimaging and photodynamic therapy. J Control Release. 2019;301:157–65.
94. Sarkar SK, Khater Y, Kulkarni A, et al. Feedback mediated cancer therapy: a FRET-based nanoreporter approach. Paper presented at: Proc. SPIE. 2014, 916607–916607-8. https://doi.org/10.1117/12.2061379
95. Prigodich AE, Seferos DS, Massich MD, et al. NanoFlares for mRNA regulation and detection. ACS Nano. 2009;3:2147–52.
96. Choi KY, Jeon EJ, Yoon HY, et al. Theranostic nanoparticles based on PEGylated hyaluronic acid for the diagnosis, therapy and monitoring of colon cancer. Biomaterials. 2012;33:6186–93.
97. Choi KY, Chung H, Min KH, et al. Self-assembled hyaluronic acid nanoparticles for active tumor targeting. Biomaterials. 2010;31:106–14.
98. Huang X, Zhang F, Wang H, et al. Mesenchymal stem cell-based cell engineering with multifunctional mesoporous silica nanoparticles for tumor delivery. Biomaterials. 2013;34:1772–80.
99. Paul A, Jana A, Karthik S, et al. Photoresponsive real time monitoring silicon quantum dots for regulated delivery of anticancer drugs. J Mater Chem B. 2016;4:521–8.
100. Sun C, Gao M, Zhang X. Surface-enhanced Raman scattering (SERS) imaging-guided real-time photothermal ablation of target cancer cells using polydopamine-encapsulated gold nanorods as multifunctional agents. Anal Bioanal Chem. 2017;409:4915–26.
101. Song J, Yang X, Yang Z, et al. Rational design of branched nanoporous gold nanoshells with enhanced physico-optical properties for optical imaging and cancer therapy. ACS Nano. 2017;11:6102–13.
102. Basuki JS, Duong HTT, Macmillan A, et al. Using fluorescence lifetime imaging microscopy to monitor theranostic nanoparticle uptake and intracellular doxorubicin release. ACS Nano. 2013;7:10175–89.
103. Chung EJ, Mlinar LB, Nord K, et al. Monocyte-targeting supramolecular micellar assemblies: a molecular diagnostic tool for atherosclerosis. Adv Healthc Mater. 2015;4:367–76.
104. Chung EJ, Mlinar LB, Sugimoto MJ, et al. In vivo biodistribution and clearance of peptide amphiphile micelles. Nanomedicine. 2015;11:479–87.
105. Santra S, Kaittanis C, Perez JM. Cytochrome C encapsulating theranostic nanoparticles: a novel bifunctional system for targeted delivery of therapeutic membrane-impermeable proteins to tumors and imaging of cancer therapy. Mol Pharm. 2010;7:1209–22.
106. Huang X, Swierczewska M, Choi KY, et al. Multiplex imaging of an intracellular proteolytic

cascade by using a broad-spectrum nanoquencher. Angew Chem. 2012;51:1625–30.
107. Zhang J, Liang Y-C, Lin X, et al. Self-monitoring and self-delivery of photosensitizer-doped nanoparticles for highly effective combination cancer therapy in vitro and in vivo. ACS Nano. 2015;9:9741–56.
108. Ding Q, Zhan Q, Zhou X, et al. Theranostic upconversion nanobeacons for tumor mRNA ratiometric fluorescence detection and imaging-monitored drug delivery. Small. 2016;12:5944–53.
109. He X, Gao J, Gambhir SS, et al. Near-infrared fluorescent nanoprobes for cancer molecular imaging: status and challenges. Trends Mol Med. 2010;16:574–83.
110. Zhang Q, Du Y, Li Y, et al. Gx1-conjugated endostar nanoparticle: a new drug delivery system for anti-colorectal cancer in vivo. Paper presented at: Proc. SPIE. 2014, 895606–895606-7. https://doi.org/10.1117/12.2035910.
111. Bremer C, Tung C-H, Weissleder R. In vivo molecular target assessment of matrix metalloproteinase inhibition. Nat Med. 2001;7:743–8.
112. Yang L, Sajja HK, Cao Z, et al. uPAR-targeted optical imaging contrasts as theranostic agents for tumor margin detection. Theranostics. 2014;4:106–18.
113. Luo Z, Zheng M, Zhao P, et al. Self-monitoring artificial red cells with sufficient oxygen supply for enhanced photodynamic therapy. Sci Rep. 2016;6:23393.
114. Tan WB, Jiang S, Zhang Y. Quantum-dot based nanoparticles for targeted silencing of HER2/neu gene via RNA interference. Biomaterials. 2007;28:1565–71.
115. Fleischmann M, Hendra PJ, McQuillan AJ. Raman spectra of pyridine adsorbed at a silver electrode. Chem Phys Lett. 1974;26:163–6.
116. Keren S, Zavaleta C, Cheng Z, et al. Noninvasive molecular imaging of small living subjects using Raman spectroscopy. Proc Natl Acad Sci USA. 2008;105:5844–9.
117. Qian X, Zhou X, Nie S. Surface-enhanced Raman nanoparticle beacons based on bioconjugated gold nanocrystals and long range plasmonic coupling. J Am Chem Soc. 2008;130:14934–5.
118. Bohndiek SE, Wagadarikar A, Zavaleta CL, et al. A small animal Raman instrument for rapid, wide-area, spectroscopic imaging. Proc Natl Acad Sci USA. 2013;110:12408–13.
119. Garai E, Sensarn S, Zavaleta CL, et al. A real-time clinical endoscopic system for intraluminal, multiplexed imaging of surface-enhanced Raman scattering nanoparticles. PLoS One. 2015;10:e0123185.
120. Jokerst JV, Miao Z, Zavaleta C, et al. Affibody functionalized gold-silica nanoparticles for Raman molecular imaging of the epidermal growth factor receptor. Small. 2011;7:625–33.
121. Kircher MF, de la Zerda A, Jokerst JV, et al. A brain tumor molecular imaging strategy using a new triple modality MRI-photoacoustic-Raman nanoparticle. Nat Med. 2012;18:829–34.
122. McVeigh PZ, Mallia RJ, Veillieux I, et al. Development of widefield SERS imaging endoscope. Paper presented at: Proc. SPIE. 2012, 8217. https://doi.org/10.1117/12.907304
123. Wang Y, Kang S, Khan A, et al. Quantitative molecular phenotyping with topically applied SERS nanoparticles for intraoperative guidance of breast cancer lumpectomy. Sci Rep. 2016;6:21242.
124. Wang YW, Doerksen JD, Kang S, et al. Multiplexed molecular imaging of fresh tissue surfaces enabled by convection-enhanced topical staining with SERS-coded nanoparticles. Small. 2016;12:5612–21.
125. Xie HN, Stevenson R, Stone N, et al. Tracking bisphosphonates through a 20 mm thick porcine tissue by using surface-enhanced spatially offset Raman spectroscopy. Angew Chem. 2012;51:8509–11.
126. Zavaleta CL, Garai E, Liu JT, et al. A Raman-based endoscopic strategy for multiplexed molecular imaging. Proc Natl Acad Sci USA. 2013;110:E2288–97.
127. Zavaleta CL, Smith BR, Walton I, et al. Multiplexed imaging of surface enhanced Raman scattering nanotags in living mice using noninvasive Raman spectroscopy. Proc Natl Acad Sci USA. 2009;106:13511–6.

23 Nanotheronostics: Part II

23.1 Cancer Theronostics

Cancer therapy is an important component of the management of cancer cells in various ways. Different nanocancer therapies are available and are based on various interactions that occur between the cancer cells and the nanoparticles (NPs) under the concept of theranostics (Fig. 23.1).

S. Shaikh, *Nanoimaging - Future of Precision Medicine*,
https://doi.org/10.1007/978-981-97-7292-6_23

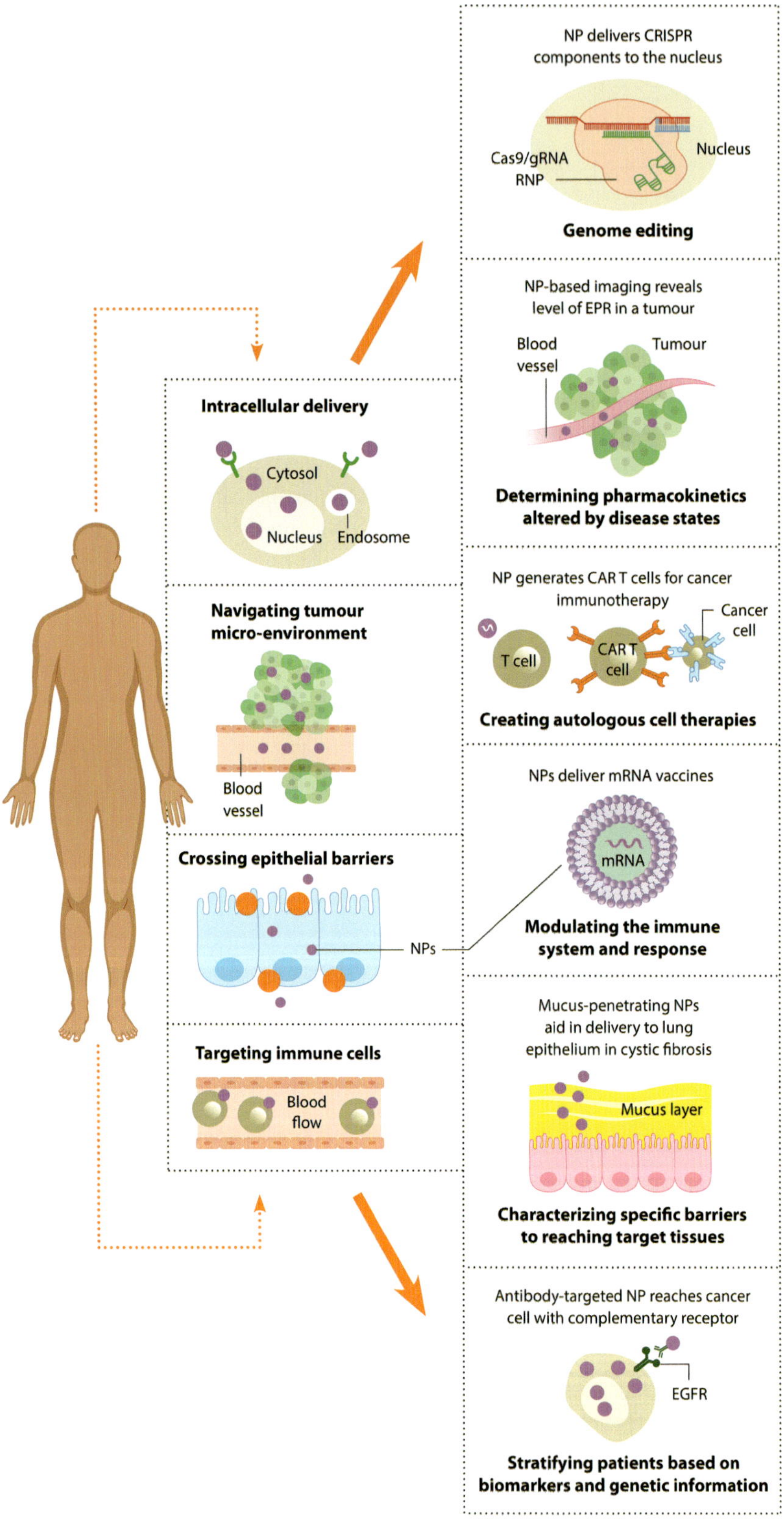

Fig. 23.1 Different theranostic application for various nanoapplication

23.2 Photothermal Therapy

Photothermal therapy is based on the concept of temperature in the cancer cells, and the best criterion is hyperthermia-based cancer therapy. The basic concept here is to kill the cancer cells by focusing on them without perturbing the surrounding normal cells (Fig. 23.2). Our body cells and tissues are not capable of interacting with the NIR-absorbing chromophores. The important landmark will be the use of various laser wavelengths in the tissue, which is known as the optical window in the range of (700–1000 nm), which will minimize the heating of different tissues; thus, the gold NPs (Au-NPs) will have better absorption in the tissues. These Au-NPs and NIR will be used for the selective heating of tumors through the use of various NPs [1]. These different Au-NPs, along with components like thiol and amine groups, will be used to target specific antibodies and drug products. Colloidal gold can absorb the focused wavelengths, resulting in the plasmon surface resonance conducive to altering the hyperthermic properties, which will be helpful in the management of cancer. The Au-NPs localized plasmon surface resonance will lead to a change in the shape and size of the particle, leading to various photothermal and photoacoustic properties by which the wavelengths of various can be utilized and excreted through the excretory system [2]. The most common drawback is that there is no uniform distribution of the heat in the tumor, and this tumor leads to different delivery options, such as the silica-gold nanoshells for the delivery of PTT [3]. The gold-based NPs have more biocompatibility, better conversion of heat, and are also capable of absorbing light, with better tumor penetration. The various NPs with better photothermal properties are nanoshells, nanocages, nanorods, and nanostars. The Au NPs will be more absorbing and can safely penetrate a depth of 2–3 cm in tissues [4]. This nanoplatform can correlate the precise heat generation of plasmonic NPs as the anticancer agent, and this can be done using various human tumor xenografts with the 2-deoxy-2-[F-18]-fluoro-D-glucose (18F-FDG) PET imaging. This shows the capabilities of PET for monitoring the response to treatment [5]. Magnetic NPs, such as iron oxide, are being used as better photothermal agents. These can be used for the focused tumor site in magnetically important regions, having a better absorption coefficient. This can be used for more dedicated drug release when heat is generated along with PTT and chemotherapy [6]. For this, the NPs that can be labeled as the most useful will be based on size, shape, dispersibility, and response to light, along with minimal injury to the surrounding tissues. The most common agents that are available are gold, palladium, silver, germanium, and carbon-based NPs [7]. The red blood cell-coated NPs will show better efficacy, leading to a significant reduction of the tumor volumes, where the NPs can have the impact of photothermal conversion, which will be effective for blood retention from the red blood cell coating. The use of Au NPs, which can be encapsulated by antitumor drugs, such as paclitaxel, will lead to heat generation, causing a stronger anticancer effect [8].

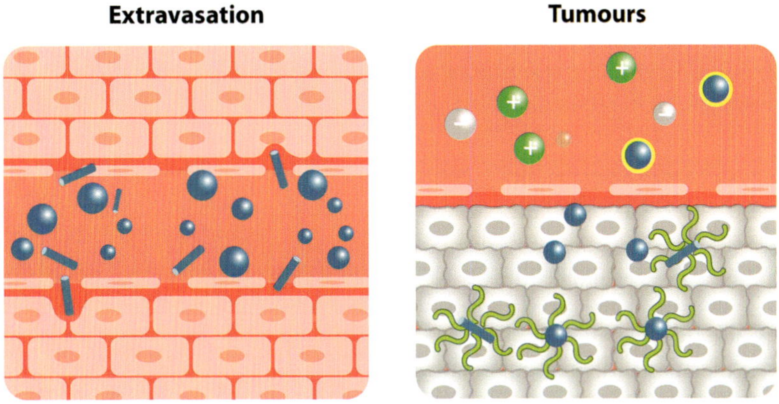

Fig. 23.2 Extravasations for different applications in nanoimaging

23.3 Photodynamic Therapy (PDT)

Photodynamic therapy is a type of therapy where light is used along with oxygen and photosensitizers (PS), which are used to destroy various cancer cells and pathogenic bacteria. Photodynamic therapy is also a type of noninvasive therapy but has a selective affinity for the malignant cells, resulting in damage to the tumor cells by the processes of apoptosis, necrosis, and autophagy. This can be transmitted directly into the various tumor sites as well as the vascular system. There will be the production of molecular oxygen, along with light having more specific wavelengths, which are applied, leading to the damage of the cancer cells. Along with this, there will be hemostasis, more vessel constriction, and also the breakdown of mechanisms causing tumor cell death. Au-NPs and porphyrins are among the agents used for the management of cancer in PDT. The NPs are used as transporters for the porphyrins. The silica NPs (80 nm) will be coated with xylan–TPPOH, which will have significant phototoxic effects, leading to more efficient and higher anticancer efficacy [9, 10]. Different photosensitizers will be conjugated to different antibodies, proteins, peptides, and other ligands with more specific cellular receptors, which will be used in targeted PDT. The process of photocatalysis can result in better light penetration in tissues [11, 12]. The use of NPs will show increased uptake and better phototoxicity. This will result in better drug release, leading to more effective photodynamic therapy for metastatic tumors. The different photosensitizers will result in starvation therapy, as well as better enhancement of PDT [13]. Different photosensitizers will enhance various anticancer therapies, leading to ROS production [14].

23.4 Chemotherapy

The different chemotherapeutic agents will result in better chemotherapeutic drugs for the management of various cancers. The use of paclitaxel, cisplatin, tamoxifen, trastuzumab, docetaxel, and various other drugs will result in better targeting of drug delivery. Different NPs will use different regimes of various chemotherapy agents, including polymer-based NPs, liposomal NPs, metal-based NPs (Au-NPs, SPIONP), carbon-based NPs, mesoporous silica NPs, and protein-based NPs [15]. Various newer-generation NPs will result in different clinical investigations in the form of various anticancer therapies. This will result in active targeting based on various environmental changes. A variety of NP formulations have been approved by the FDA and EMA for the treatment of a wide range of cancers. Some examples are pegylated liposomal doxorubicin and liposomal daunorubicin, which are available in the United States. Nonpegylated liposomal doxorubicin is approved in Europe. Some of the particles will result in better therapy by using different NP albumin-bound particles [16]. Different proteins, as well as various peptides, will be conjugated to the surface of NPs to enhance efficiency for anticancer drugs. The use of serum glycoproteins can be used as ligands in the chemotherapy evaluation of drug delivery [17]. Another important aspect is the toxicity of the drugs due to more accumulation in tumor cells as well as in normal cells. A classic example is the anti-malarial drug chloroquine [18]. Another important entity is the use of Au-NPs, which have better stability, a larger surface area-to-volume ratio, and better plasmon resonance, with multiple functions (Fig. 23.3). The ideal agent should be nontoxic, nonimmunogenic in nature, have high permeability, and retention effects, making it more accessible to tumors, with better penetration and accumulation [19] (Fig. 23.4). The use of cisplatin as a genotoxic agent will have more and better use in combination with radiation or other chemotherapeutic agents, with more precise selectivity for the tumor. The different categories are various organic NPs, which will be used for the transport of cisplatin, such as liposomes, polymeric NPs, polymeric micelles, and dendrimers. Other subsets are inorganic NPs, such as Au NPs, ferromagnetic NPs, and mesoporous silica NPs. A few are hybrid NPs, such as CNTs, nanoscale coordination polymers, and polysilsesquioxane NPs [20]. Of these, various

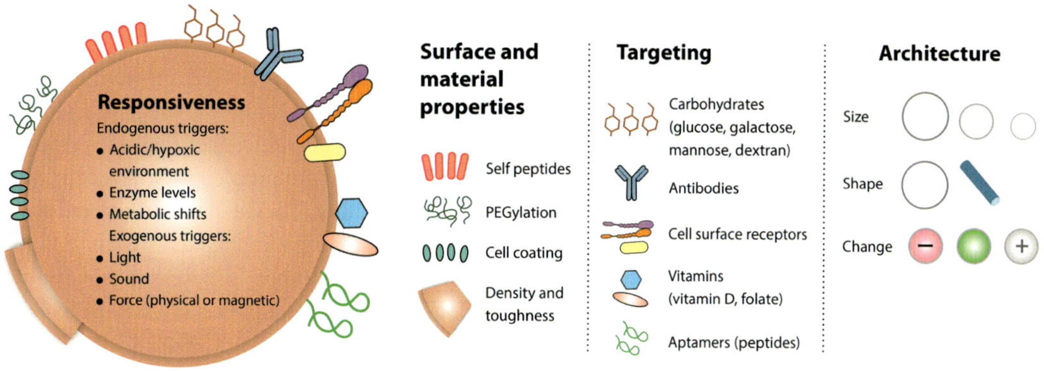

Fig. 23.3 Use of the nanomedicine for the various intracellular actions

Fig. 23.4 Different mechanisms of actions in various applications

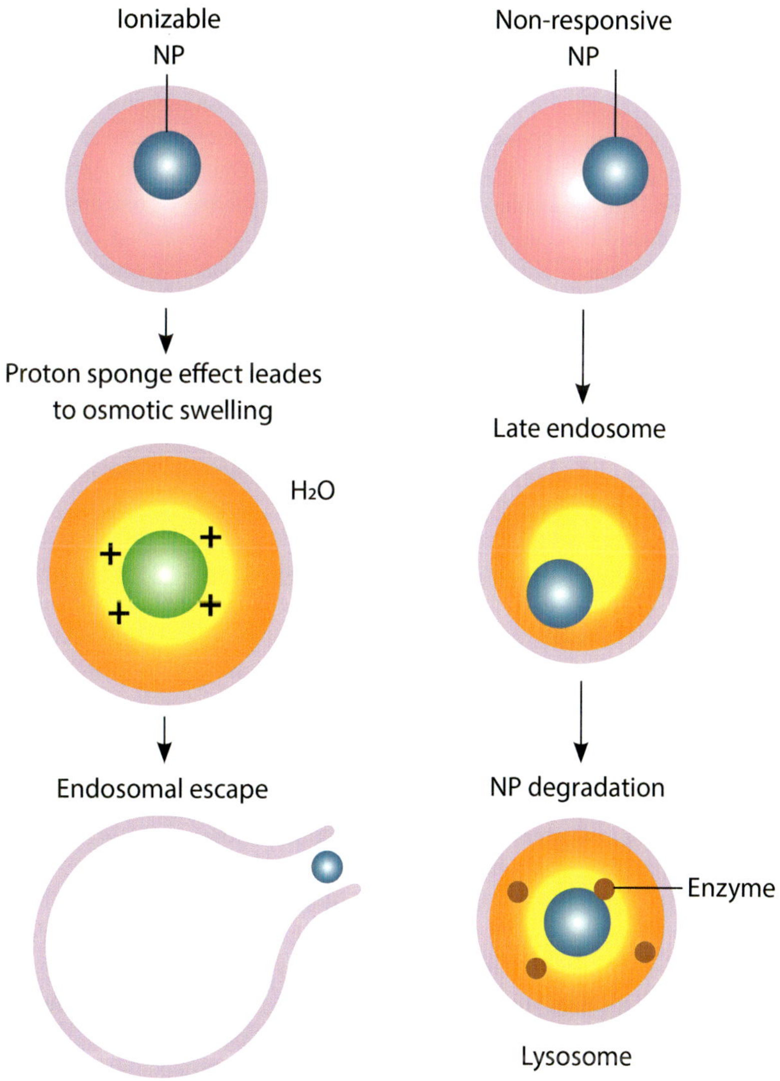

organic NPs are most commonly used due to better and increased circulation, which has a better half-life for tumor accumulation. The use of various NPs can be used for better targeting of drugs to the region of interest, such as in a theranostic platform with multiple functions [21]. Multidrug-loaded NPs are more efficient for multidrug-resistant cells [22]. The principle of hypoxia is to promote better drug efficacy due to the invasiveness of tumor cells and chemoresistance. Tumor-associated macrophages (TAMs) are also capable of generating hypoxia through a selective procedure, which has a simple principle: targeted delivery of MnO_2 to the hypoxic area [23]. Nanomicelles, when used, can be useful for various purposes. Mannose will have better cellular uptake, and different oligomeric hyaluronic acid can be used as the targeting material. Nanodandelions are very small in size and can enter tissue, providing better antitumor activity in tumor mice [24]. Sustained-release NPs can maintain higher drug concentrations in the region of interest. The use of NPs increases the penetration and accumulation of inhaled drugs at the tumor site for better chemotherapy delivery [25].

23.5 Immunotherapy

Immunotherapy is also an important form of therapy and is widely used in various applications with nanotechnology. This immunotherapy will better stimulate or activate the patient's immune system, and it is used on various cancer cells [26]. Immunotherapy has fewer side effects as compared to traditional forms of chemotherapy. Five types of cancer immunotherapy are lymphocyte-promoting cytokines, agonistic antibodies against co-stimulatory receptors, checkpoint inhibitors, engineered T cells such as chimeric antigen receptor (CAR T) and T-cell receptor (TCR) T cells, and cancer vaccines. The use of NPs will better target the T cells in the blood or transport mRNA to the various cancer cells [27]. NP systems are an important tool for more effective antigen delivery, creating a robust and better immune response by stimulating the immune system, in which the natural killer cell response is triggered for better management of various cancers [28].

The use of various antigen-presenting cells will have more tumor antigens, which are capable of triggering a better anticancer immune response [29]. NPs have had multiple roles in recent days in immunotherapy. They are capable of reducing tumor-associated macrophages, which can be seen as various tumor suppressor agents, an important entity for various nano-CRISPR-Cas9 delivery systems, acting as CAR T [30]. The use of immunotherapy will result in the targeting of various tumor environments for better use in cancer treatment. The NPs used in immunotherapy act as therapeutic agents that can be conjugated for targeting, function, and also the tumor model [31]. One of the potent stimulators is the interferon receptor (STING) agonist, which has lesser efficiency at various micromolar concentrations. Due to the micromolar concentrations, the immune response is >100-fold. The use of NP-mediated cytosolic delivery for various STING agonists will result in synergizing effects, leading to better and enhanced immunotherapy [32]. One of the important cancer immunotherapies used for uniformity is related to the simple shape and size compared to market-available agents, as well as the ability to self-assemble into highly repetitive nanostructures. Immunotherapy will have better chemical attachments, higher cargo capacity, and better tolerance to higher temperatures and pH [33]. Metallic NPs are also used more commonly and efficiently in immunotherapy. These metallic NPs, like Au NPs as well as ovalbumin (OVA), are used because they have better antitumor cytotoxic T-cell response. These metallic NPs can also be used for immunotherapy and PTT [34]. The use of immune response rates will also be able to control the immune checkpoint inhibitor anti-PD-1 [35, 36].

23.6 NPs in Precision Medicine

Precision medicine is the newer concept of patient-centric management, which can be improved by better therapeutic responses [37, 38]. Due to the development of various biomarkers, these studies have more significant hallmarks [39]. Different patient-dependent stratifications are being applied, with more precise NP-based roles in clinical trials [40]. The use of various NPs can also result from different precision medicine concepts [38]. Early detection of pancreatic cancer can be achieved, as it is more prone to the absorption of various nanoflakes [41]. The different graphene oxides will combine with various proteins, showing strong adsorption with lower plasma levels [41]. Some studies use other magnetic NPs [42] or Au-NPs [43, 44], which are not complicated to use and have better biomarker detection rates, saving time and leading to better sample evaluation and processing. The NPs are also capable of changing the tumor microenvironment, leading to better accumulation and penetration, which improves drug efficacy in relation to the particular therapy [45–47]. Tumor-associated endothelial cells with NP-delivered microRNA can alter the tumor vasculature and result in better outcomes for cancer therapies [46]. Bio-inspired lipoproteins can be used to remodel tumors and improve NP accessibility to cancer cells by 27-fold [45]. Photothermal NPs are also important for improving the activity of CAR T cells against solid tumors [48]. One of the important roles of NPs is to modulate immune activation or suppression mechanisms, better sensitizing various cancer cells to therapeutics, which will lead to more favorable responses for precision treatments [37, 49–56]. The role of precision medicine is based on varied patient demographics and better delivery with increased efficacy for precision medicines. Many advances can be made in genome sequencing and biomarker detection, which will be more helpful in managing specific diseases.

23.7 NPs for Genome Evaluation

Recent advances in CRISPR, transcription activator-like effector nuclease (TALEN), and zinc-finger nuclease (ZFN) technologies are also advancing biomedical research, drug development, discovery, and gene therapy [57–63]. Genetic analysis is important for Mendelian diseases, as there are approximately 3000 human genes contributing to about 5% of rare diseases [63–69]. Due to genome-editing techniques, there will be more opportunities for the efficient and safe delivery of treatments, with enhanced penetration into cells and reduced toxicity [70]. The delivery of genome-editing systems can also involve different multicomponents, which face barriers in both the extracellular and intracellular regions, targeting the genome of the cells. Lipid-based and polymer-based NPs will carry more nucleic acids in vivo, aiding better clinical development [53, 55, 57, 66]. The LNP siRNA drug is used in the management of amyloidosis [27]. In genomics, NPs play a crucial role in management but also present safety issues [71, 72].

23.8 Intracellular Targeting

The intracellular use of NP-based components is useful for endocytosis and phagocytosis through various nucleic acids, acting as receptor-based intracellular mediators [73]. Different cationic materials are used for the analysis of nucleic acids, facilitating endosomal escape. Various materials, including lipid materials [51, 52, 74] as well as different polymers [75–78]. These can be done by the concept of the passive and active targeting of the different materials, which helps in endocytic uptake [67]. Gene editing requires DNA access [50, 58], which can be achieved by using smaller NPs for endosomal escape [67, 79]. NP properties also influence intracellular trafficking [54, 65]. An important identification parameter is identifying defects in mitochondrial DNA, which is useful for genetic and mitochondrial DNA engineering [67].

Cystic fibrosis is a disease evaluated for genetic defects in the gene, where there is a defect in the transmembrane conductance regulator (CFTR) protein [80]. This life-threatening condition is a monogenic disorder that can be treated with gene therapy. However, many research articles show that gene expression and delivery failure hinder this gene therapy [81, 82]. Several barriers prevent gene expression [83–85]. Various NP formulations are being developed for

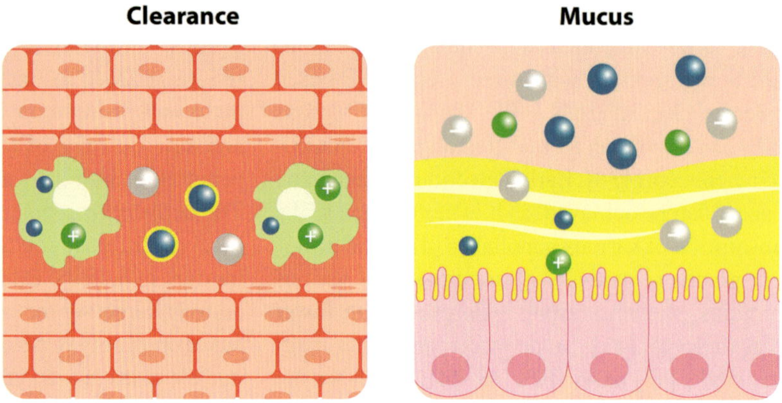

Fig. 23.5 Role of the clearance in the respiratory system as well as in the mucous

the delivery of genetic material, such as mRNA, which is quite effective [68, 86]. Since mucus is an important factor for NPs, smaller NPs will have better penetration [83, 85]. PEGylation can improve the penetration of cystic fibrosis therapy in mucus ex vivo [87]. Protein function can also provide a better depiction of the disease [84]. In the potential applications of NP therapies, these NPs can be injected into the umbilical vessel, amniotic fluid, or specific fetal tissues [88–90]. Early-stage disease identification can be achieved using NP therapy, which will accumulate at the solid tumor site [63].

Nanotechnology is now being used for various theranostic applications. At the molecular scale, nanovaccines can be used for immunization and are capable of both therapeutic and imaging purposes, enabling early identification and more focused, targeted treatment, along with monitoring the treatment response [91–93, p. 123]. Various nanobiosensors offer enhanced stability and biocompatibility, allowing the detection of even very small amounts of bacteria and viruses [94]. Different amplification assays using these nanobiosensors can aid in the diagnosis of viral diseases [94]. The SARS-CoV-2 virus, as well as other viruses, can be identified within 6–10 h when suspected in the respiratory system [92]. Photonic crystals, metallic NPs, graphene, and carbon nanotubes (CNTs) are some of the agents used for virus detection [94, 95] (Fig. 23.3). Nanoplatforms are being employed for analysis based on parameters such as stability, specificity, and sensitivity [96]. The formation of surface plasmon resonance-based biosensors helps in targeting the virus [97]. Identification parameters are crucial for better detection of various protein markers [98–100]. Sensors can be useful for evaluating pathogens, enabling early detection through optical fluorescence or visual virus detection [101–103]. The design of a microfluidic platform with an enhanced surface area for a rolling circle amplification reaction has reduced virus detection time to just 15 min [103] (Fig. 23.5).

23.9 Biosensors

Biosensors are composed of different biomolecules (antibodies, nucleic acids, or antigens) present on the sensor's surface. Target-related probes can induce various physicochemical changes [104–118] by utilizing different electrochemical assays, which are highly sensitive to as low as 100 fM [103, 119, 120]. Au-NPs-based assays are a significant form of nanoplatforms used in the detection of various viruses [119, 120]. One important application is the use of multifunctional Staphylococcus aureus-based nanoprobes for early detection of viral antibodies [121], or dual-functional plasmonic photothermal biosensors [122]. Other components include field-effect transistor-based biosensors [123], and graphene sheet sensors, which can distinguish between non-infected and infected groups [123]. Graphene oxide (GO) and its derivatives exhibit excellent antiviral effects [124], inactivating the virus before it enters cells. Silver-graphene nanocomposites can prevent coronavirus growth [125], making them promising agents for virus inactivation [126]. Silver-coated NPs impact the NP size and virus-NP interactions [127]. Glutathione-capped silver-based nanoclusters have been shown to inhibit virus growth [128]. Quantum dots are highly sensitive and can be used for viral detection in patient care [127, 129]. Curcumin-based cationic carbon dots show promise against coronavirus. Silica-coated NPs are important for preventing virus growth [105, 106], with a broader spectrum and high potency [107]. Nanoclay systems can prevent early infection or electrostatically bind to viruses, offering no protection 24 h post-infection [108]. Polyethylene glycol has strong antiviral activity [109], as does protoporphyrin-modified multiwalled CNTs.

SARS-CoV-2 viruses can be used to edit the RNA virus with replication and fate [110, 111]. The use of nanoplatforms like mesoporous silica or Au-NPs and nanoblades can be used for better viral genome editing [112, 113]. Theranostics, the concept, is to measure the efficiency of the antiviral agents with increased [114–118]. Different chemical combinations will result in more efficient and improved efficiency of drugs [130, 131]. The use of various CNT-based nanosystems has been evolving as an important editing tool for various viral genomes with increased activity [132]. The use of lipid NPs is also important for the evaluation of therapeutic options [133–137], like in vaccination for various viral diseases [138]. The use of various mRNA-lipid NPs will also be able to achieve better and higher efficiency for targeting various genes [139, 140]. The use of siRNA will be the best host for various cells that are capable of replication of various viral diseases like SARS-CoV by disrupting virus RNA, preventing the replication of the viruses, and inhibiting their growth; due to this, it has higher antiviral capabilities, preventing mutation in siRNA [141–143]. The concept here is that the use of various siRNA results in reduced viral load and replication, which prevents the lungs from diffuse alveolar damage. Due to this, it will become a safer agent for various targeted therapies as well as antiviral protocols in antiviral regimes [144]. These two siRNAs will have better-targeted M (membrane) genes, with more and better siRNAs that can inhibit the SARS-CoV membrane gene [145, 146]. The different siRNA-loaded lipid NPs can target different viruses, preventing various lethal symptoms [147, 148]. As we all know, the liver is very prone to viral infections, and this will have different outcomes for COVID-19 infections [149–151]. Due to ongoing research, there will be a better application of various NP vectors, leading to more promising delivery of nucleic acids for different diseases [152, 153]. One of the studies has shown that polyethylene glycol-lipid will be used for the better development of mRNA formulations, resulting in enhanced stability due to more efficient delivery of mRNA to the lungs [93, p. 124, 154], and along with this, other components like cholesterol, phospholipids, polyethylene glycol lipids, and cationic lipids may be optimized for better delivery of the nucleic acids [155, 156]. Lipid NPs can also tune the NP internal charge for the delivery of therapeutics and cellular fate of NPs [157]. The best of the nanomaterials will be used for the delivery of antigens for RNA vaccines [158]. Different photothermal heating will result in the inactivation of the virus, which can be done by NIR laser irradiation. The mRNA vaccines can also be replaced by the higher potential for various production cycles [159]. The mRNA vaccine, which can be entrapped by the lipid NPs, causes better immune responses along with better potential and tolerability of the vaccine [160, 161]. Different components like liposomes, dendrimers, nanoemulsions, and Au-NPs will be used for better mRNA-based vaccine delivery [162–167]. Different novel technologies will result in better cellular functions and more inflammatory reactions, along with improving the immune system [168–172]. The lipid-based NPs are also able to deliver siRNA or mRNA, leading to better immune response for the polymer type [173, 174]. The nanostructures will cause more and better mutations of viruses, leading to more efficiency for the viral vector and better action against SARS-CoV-2 [175–177]. Various macrophages and stimulated inflammatory reactions will cause better adjuvants for better effects of nanomaterials on the immune system [178–180]. For viral infections, there will be more and better nanomaterials, such as CNTs, graphene, polystyrene particles, and nanodiamonds, for immune system activation [181–183]. T helper cells are also important for better infection control in COVID-19 [184, 185]. Various nanoplatforms can be used for various responses, preventing inflammatory cytokines, as well as targeted immunosuppressive agents causing more vague distribution of drug and side effects, which will have an important impact on precise immunosuppressive therapy [186–188]. The nanodiamonds will also be capable of carrying various anti-inflammatory drugs, which will cause significant anti-inflammatory changes due to decreased infiltration of macrophages and pro-inflammatory mediators (TNF-α and iNOS) [189]. The use of various porous carbon-based nanomaterials can help with better cytokine (e.g., TNF-α or IL-6) adsorption due to their pore size, which will be important for the induction of inflammation [190, 191]. The use of photodynamic therapy will help in the inactivation of SARS-CoV-2 by ROS generation as well as cell death [192, 193]. Targeting as well as efficient viral inhibition can be done [194, 195]. Because of the multivalent interactions between a cell and a virus, the application of monovalent medications for inhibiting the entry and spread of viruses has not been a successful approach [194]. In biological systems, multivalent interactions are critically involved in adhesion, recognition, and signaling events [196, 197]. The viral adherence to various receptors leads to different cellular uptake, as well as delivery of genetic materials, resulting in more novel infections [198]. This will be more useful for various fungal, bacterial, or viral infections and can also be associated with drug resistance [199, 200]. Multivalent NPs, such as Au-NPs and Ag-NPs, will have better functions in preventing the entry of the virus [201–205]. Different NPs are being used as broad-spectrum antiviral agents, which are non-toxic [206–208]. The role of various nanocarriers is also important, as they will have better stability for the delivery of various therapeutic agents and better strength for binding [209]. Different ligands are being used for better use as receptors and also for shielding, resulting in better interaction with a virus. The use of various multivalent therapeutic agents will provide better flexibility, and broad-spectrum antiviral effects are also crucial in preventing virus entry [210, 211]. The use of nontoxic nanogels will shield the surface of the virus, making them efficient for viral entry. Another important entity is the use of various multivalent orthopoxvirus particles for better binding and infection inhibition efficiency, acting as important agents for cell entrance monitoring [212]. Different electron microscopes and bioassays are also being used to better understand the interactions between the virus and sulfated polymer-covered graphene sheets, acting as modulators [213]. Another important aspect is the safety issues of nanomaterials, which will monitor the entrance to the cell surface [93, p. 125, 214, 215]. Different inhibitory effects of nanosponges will act against SARS-CoV-2 [216, 217]. Inorganic nanoplatforms can also be used in various physicochemical properties based on their functions in proteins [218]. Another example can be seen in the form of virus-like nanocarriers, which can be used to circulate through various biological barriers, to adapt to different viral adaptations [219–221]. Recent research has shown the use of various pharmacological effects for the creation of different functional systems for nanodelivery [222–224]. Due to changes in various molecular and structural biology along with genomics, there is still a long way to go for better design, which is more effective for various antiviral medications [225]. Different computer simulations will analyze the information, which will be more precise for antiviral therapeutics, especially in the form of nucleotide inhibitors [226]. The multi-faceted advancements in pharmacogenomics, biological mechanisms, translational bioinformatics, and newer ligands designed will facilitate better use for highly mutating viruses, along with vaccines designed for antiviral agents against the virus [227–229]. Mammalian cells need to be accustomed to better targeting and delivery, identifying various pharmacological profiles of NPs [220, 230, 231]. The use of different path mechanisms for evaluating infection nature, better identification, characterization, various defense mechanisms, and finally for management, leads to the concept of Theranostics [232].

23.10 AI-Based Advancements

These will have various techniques that can be used in different mechanical nanoplatforms, which will be used as future potentials for developing technologies that can be used for making different nanoarchitectures, along with different computations that will have a better impact on the nanostructures in biosystems [233, 234]. These highly durable nanomachines enable monitoring of the activities and internal chemistry of organs and access to the malfunctioning regions [235]. Nanorobots can be biologically activated for better sensing, signaling, actuation, and cell entry, which can combat different diseases. Nanoswimmers can enter the targeted therapy areas by navigating the fluid structures, using DNA nanomachines, leading to better targeting for therapeutics [236, 237], and this process can lead to theranostic interventions. With the possibility of different chemical hazards, the use of nanorobots is for cell repair, tissue repair, targeting genes, drug delivery, drug efficiency improvement, and ultimately monitoring the patient. Nanobots can be transmitted wirelessly and can be used for modifying management protocols [235, 238]. These nanorobots can also be used for the destruction of various toxic agents and for different responses to nanorobots in a variety of disorders and viral infections [239].

23.11 MRI-Guided Nanomachines

These nanomachines are also important for tracking as well as actuating NPs, which can be used along with precise magnetic particle localizations [240]. Different nanostructures are being used for better biological functions, which can be observed at the cellular level [241]. Folate can be attached to the nanorobot, which can be used for drug delivery for viral infections in humans [242]. The nanorobots will be released for better-targeted release due to the alteration of pH and temperature [243, 244]. The nucleic acid-based nanorobots are capable of biosensing and identifying better flow rates for the delivery of biological activators, with analysis of intra- or inter-molecular forces, which will be capable of manipulating the NPs, making them more favorable for the nanoplatforms, different chemical reactions, and also for apoptosis [245, 246], making it more conducive to theranostics [247]. The deeper the knowledge of the nanorobots, the more interactions and applications [248, 249]. The α-*N*-acetyl-glycosaminidase (α-NAGA) protein overexpression can be stimulated by the nanorobots by sending different electromagnetic signals that are capable of sensing the concentration in the bloodstream [250]. Radiofrequency devices that can identify the nanorobots will be a complementary asset for the evaluation [251–253]. Hematological parameters are also an important entity for evaluation, prognosis, and progression of the disease [254]. Different tumor markers can identify markers, such as C-reactive protein, and other inflammatory markers like tumor necrosis factor-α or interleukins 4 and 6, which can be monitored as these are tumor markers [254]. The use of mesenchymal stem cells will be more critical for COVID-19 patients with severe conditions [255]. The use of nano-based tools can be used for the inactivation of SARS-CoV-2, leading to the fragmentation of the viral genome, causing irreversible virus inactivation [256].

23.12 Conclusion

Thus, the use of nanotheranostics is important for the evaluation of various cancers and also plays an important role in the management of these cancers. The future is more realistic due to the concept of personalized medicine.

References

1. Hirschberg H, Madsen SJ. Cell mediated photothermal therapy of brain tumors. J Neuroimmune Pharmacol. 2017;12:99–106.
2. Vines JB, Yoon JH, Ryu NE, Lim DJ, Park H. Gold nanoparticles for photothermal cancer therapy. Front Chem. 2019;7:167.
3. Simón M, Norregaard K, Jørgensen JT, Oddershede LB, Kjaer A. Fractionated photothermal therapy in a murine tumor model: comparison with single dose. Int J Nanomedicine. 2019;14:5369–79.
4. Riley RS, Day ES. Gold nanoparticle-mediated photothermal therapy: applications and opportunities for multimodal cancer treatment. Wiley Interdiscip Rev Nanomed Nanobiotechnol. 2017;9(4):e1449.
5. Jørgensen JT, Norregaard K, Tian P, Bendix PM, Kjaer A, Oddershede LB. Single particle and PET-based platform for identifying optimal plasmonic nano-heaters for photothermal cancer therapy. Sci Rep. 2016;6:30076.
6. Estelrich J, Antònia Busquets M. Iron oxide nanoparticles in photothermal therapy. Molecules. 2018;23:1567.
7. Vines JB, Lim DJ, Park H. Contemporary polymer-based nanoparticle systems for photothermal therapy. Polymers. 2018;10:1357.
8. Wu M, Le W, Mei T, et al. Cell membrane camouflaged nanoparticles: a new biomimetic platform for cancer photothermal therapy. Int J Nanomedicine. 2019;14:4431–48.
9. Mokoena DR, George BP, Abrahamse H. Enhancing breast cancer treatment using a combination of cannabidiol and gold nanoparticles for photodynamic therapy. Int J Mol Sci. 2019;20:4771.
10. Bretin L, Pinon A, Bouramtane S, et al. Photodynamic therapy activity of new human colorectal cancer. Cancers. 2019;11:1474.
11. Abrahamse H, Hamblin MR. New photosensitizers for photodynamic therapy. Biochem J. 2016;473:347–64.
12. Honors CN, Kruger CA, Abrahamse H. Photodynamic therapy for metastatic melanoma treatment: a review. Technol Cancer Res Treat. 2018;17:1–15.
13. Yu Z, Zhou P, Pan W, Li N, Tang B. A biomimetic nanoreactor for synergistic chemiexcited photodynamic therapy and starvation therapy against tumor metastasis. Nat Commun. 2018;9:1–9.
14. Zhu T, Shi L, Yu C, et al. Ferroptosis promotes photodynamic therapy: supramolecular photosensitizer-inducer nanodrug for enhanced cancer treatment. Theranostics. 2019;9:3293–307.
15. Liyanage PY, Hettiarachchi SD, Zhou Y, et al. Nanoparticle-mediated targeted drug delivery for breast cancer treatment. Biochim Biophys Acta Rev Cancer. 2019;1871:419–33.
16. Lee MS, Dees EC, Wang AZ. Nanoparticle-delivered chemotherapy: old drugs in new packages. Oncology. 2017;31:198–208.
17. Press D. Application of active targeting nanoparticle delivery system for chemotherapeutic drugs and traditional/herbal medicines in cancer therapy: a systematic review. Int J Nanomedicine. 2018;13:3921.
18. Pelt J, Busatto S, Ferrari M, Thompson EA, Mody K, Wolfram J. Chloroquine and nanoparticle drug delivery: a promising combination. Pharmacol Ther. 2018;191:43–9.
19. Singh P, Pandit S, Mokkapati VRSS, Garg A, Ravikumar V, Mijakovic I. Gold nanoparticles in diagnostics and therapeutics for human cancer. Int J Mol Sci. 2018;19:1979.
20. Duan X, He C, Kron SJ, Lin W. Nanoparticle formulations of cisplatin for cancer therapy. Wiley Interdiscip Rev Nanomed Nanobiotechnol. 2016;8:776–91.
21. Meng F, Han N, Yeo Y. Organic nanoparticle systems for spatiotemporal control of multimodal chemotherapy. Expert Opin Drug Deliv. 2017;14:427–46.
22. Fisusi FA, Akala EO. Drug combinations in breast cancer therapy. Pharm Nanotechnol. 2019;7:3–23.
23. Song M, Liu T, Shi C, Zhang X, Chen X. Bioconjugated manganese dioxide nanoparticles enhance chemotherapy response by priming tumor-associated macrophages toward M1-like phenotype and attenuating tumor hypoxia. ACS Nano. 2016;10:633–47.
24. Wang B, Zhang W, Zhou X, et al. Development of dual-targeted nano-dandelion based on an oligomeric hyaluronic acid polymer targeting tumor-associated macrophages for combination therapy of non-small cell lung cancer. Drug Deliv. 2019;26:1265–79.
25. Mangal S, Gao W, Li T, Zhou QT. Pulmonary delivery of nanoparticle chemotherapy for the treatment of lung cancers: challenges and opportunities. Acta Pharmacol Sin. 2017;38:782–97.
26. Saleh T, Shojaosadati SA. Multifunctional nanoparticles for cancer immunotherapy. Hum Vaccin Immunother. 2016;12:1863–75.
27. Riley RS, June CH, Langer R, Mitchell MJ. Delivery technologies for cancer immunotherapy. Nat Rev Drug Discov. 2019;18:175–96.
28. Wen R, Umeano AC, Kou Y, Xu J, Farooqi AA. Nanoparticle systems for cancer vaccine. Nanomedicine. 2019;8:627–48.
29. Park W, Heo Y-J, Han DK. New opportunities for nanoparticles in cancer immunotherapy. Biomater Res. 2018;22:1–10.
30. Sau S, Alsaab HO, Bhise K, Alzhrani R, Nabil G, Iyer AK. Multifunctional nanoparticles for cancer immunotherapy: a groundbreaking approach for reprogramming malfunctioned tumor environment. J Control Release. 2018;274:24–34.
31. Surendran SP, Moon MJ, Park R, Jeong YY. Bioactive nanoparticles for cancer immunotherapy. Int J Mol Sci. 2018;19:3877.
32. Wilson DR, Sen R, Sunshine JC, Pardoll DM, Green JJ, Kim YJ. Biodegradable STING agonist nanoparticles for enhanced cancer immunotherapy. Nanomedicine. 2018;14:237–46.

33. Shoeb E, Hefferon K. Future of cancer immunotherapy using plant virus-based nanoparticles. Future Sci OA. 2019;5(7):FSO401.
34. Evans ER, Bugga P, Asthana V, Drezek R. Metallic nanoparticles for cancer immunotherapy. Mater Today. 2018;21:673–85.
35. Jia Y, Omri A, Krishnan L, McCluskie MJ. Potential applications of nanoparticles in cancer immunotherapy. Hum Vaccin Immunother. 2017;13:63–74.
36. Rodell CB, Arlauckas SP, Cuccarese MF, et al. TLR7/8-agonist-loaded nanoparticles promote the polarization of tumour-associated macrophages to enhance cancer immunotherapy. Nat Biomed Eng. 2018;2:578–88.
37. Xie J, Lee S, Chen X. Nanoparticle-based theranostic agents. Adv Drug Deliv Rev. 2010;62(11):1064–79.
38. Aguado BA, Grim JC, Rosales AM, Watson-Capps JJ, Anseth KS. Engineering precision biomaterials for personalized medicine. Sci Transl Med. 2018;10:8645.
39. van der Meel R, et al. Smart cancer nanomedicine. Nat Nanotechnol. 2019;14:1007–17.
40. Scheetz L, et al. Engineering patient-specific cancer immunotherapies. Nat Biomed Eng. 2019;3:768–82.
41. Papi M, et al. Converting the personalized biomolecular corona of graphene oxide nanoflakes into a high-throughput diagnostic test for early cancer detection. Nanoscale. 2019;11:15339–46.
42. Shaban M, Hasanzadeh M, Solhi E. An Fe3O4/PEDOT: PSS nanocomposite as an advanced electroconductive material for the biosensing of the prostate-specific antigen in unprocessed human plasma samples. Anal Methods. 2019;11:5661–72.
43. Zhang T, et al. The synergistic effect of Au-COF nanosheets and artificial peroxidase Au@ZIF-8(NiPd) rhombic dodecahedra for signal amplification for biomarker detection. Nanoscale. 2019;11:20221–7.
44. Kaur A, Shimoni O, Wallach M. Novel screening test for celiac disease using peptide functionalised gold nanoparticles. World J Gastroenterol. 2018;24:5379–90.
45. Tan T, et al. Bioinspired lipoproteins-mediated photothermia remodels tumor stroma to improve cancer cell accessibility of second nanoparticles. Nat Commun. 2019;10:3322.
46. Wilson RA, et al. MicroRNA regulation of endothelial TREX1 reprograms the tumour microenvironment. Nat Commun. 2016;7:13597.
47. Papa AL, et al. Ultrasound-sensitive nanoparticle aggregates for targeted drug delivery. Biomaterials. 2017;139:187–94.
48. Chen Q, et al. Photothermal therapy promotes tumor infiltration and antitumor activity of CAR T cells. Adv Mater. 2019;31:e1900192.
49. Patel RB, et al. Development of an in situ cancer vaccine via combinational radiation and bacterial membrane-coated nanoparticles. Adv Mater. 2019;31:e1902626.
50. Adekiya TA, Kondiah PPD, Choonara YE, Kumar P, Pillay V. A review of nanotechnology for targeted anti-schistosomal therapy. Front Bioeng Biotechnol. 2020;8:32. https://doi.org/10.3389/fbioe.2020.00032.
51. Chivere VT, Kondiah PPD, Choonara YE, Pillay V. Nanotechnology-based biopolymeric oral delivery platforms for advanced cancer treatment. Cancers. 2020;12:522. https://doi.org/10.3390/cancers12020522.
52. Deng Y, Zhang X, Shen H, et al. Application of the nano-drug delivery system in treatment of cardiovascular diseases. Front Bioeng Biotechnol. 2020;7:489. https://doi.org/10.3389/fbioe.2019.00489.
53. Gao P, Pan W, Li N, Tang B. Boosting cancer therapy with organelle-targeted nanomaterials. ACS Appl Mater Interfaces. 2019;11:26529–58. https://doi.org/10.1021/acsami.9b01370.
54. Mizrahy S, Raz SR, Hasgaard M, et al. Hyaluronan-coated nanoparticles: the influence of the molecular weight on CD44-hyaluronan interactions and on the immune response. J Control Release. 2011;156:231–8. https://doi.org/10.1016/j.jconrel.2011.06.031.
55. Roma-Rodrigues C, Pombo I, Raposo L, Pedrosa P, Fernandes AR, Baptista PV. Nanotheranostics targeting the tumor microenvironment. Front Bioeng Biotechnol. 2019;7:197. https://doi.org/10.3389/fbioe.2019.00197.
56. Swain S, Sahu PK, Beg S, Babu SM. Nanoparticles for cancer targeting: current and future directions. Curr Drug Deliv. 2016;13:1290–302. https://doi.org/10.2174/1567201813666160713121122.
57. Thorat ND, Bauer J. Functional smart hybrid nanostructures based nanotheranostic approach for advanced cancer treatment. Appl Surf Sci. 2020;527:146809. https://doi.org/10.1016/j.apsusc.2020.146809.
58. Yang Y, Wang L, Wan B, Gu Y, Li X. Optically active nanomaterials for bioimaging and targeted therapy. Front Bioeng Biotechnol. 2019;7:320. https://doi.org/10.3389/fbioe.2019.00320.
59. Yao VJ, D'Angelo S, Butler KS, et al. Ligand-targeted theranostic nanomedicines against cancer. J Control Release. 2016;240:267–86. https://doi.org/10.1016/j.jconrel.2016.01.002.
60. Li D, He S, Wu Y, et al. Excretable lanthanide nanoparticle for biomedical imaging and surgical navigation in the second near-infrared window. Adv Sci. 2019;6:1902042. https://doi.org/10.1002/advs.201902042.
61. Beik J, Jafariyan M, Montazerabadi A, et al. The benefits of folic acid-modified gold nanoparticles in CT-based molecular imaging: radiation dose reduction and image contrast enhancement. Artif Cells Nanomed Biotechnol. 2018;46:1993–2001. https://doi.org/10.1080/21691401.2017.1408019.
62. Astashkina AI, Jones CF, Thiagarajan G, et al. Nanoparticle toxicity assessment using an *in vitro* 3-D kidney organoid culture model. Biomaterials. 2014;35:6323–31. https://doi.org/10.1016/j.biomaterials.2014.04.060.

63. Abdolahinia ED, Nadri S, Rahbarghazi R, Barar J, Aghanejad A, Omidi Y. Enhanced penetration and cytotoxicity of metformin and collagenase conjugated gold nanoparticles in breast cancer spheroids. Life Sci. 2019;231:116545. https://doi.org/10.1016/j.lfs.2019.116545.
64. Chiodo F, Marradi M, Park J, et al. Galactofuranose-coated gold nanoparticles elicit a pro-inflammatory response in human monocyte-derived dendritic cells and are recognized by DC-SIGN. ACS Chem Biol. 2014;9:383–9. https://doi.org/10.1021/cb4008265.
65. Correia AL, Bissell MJ. The tumor microenvironment is a dominant force in multidrug resistance. Drug Resist Updat. 2012;15:39–49. https://doi.org/10.1016/j.drup.2012.01.006.
66. Elzoghby AO, Hemasa AL, Freag MS. Hybrid protein-inorganic nanoparticles: from tumor-targeted drug delivery to cancer imaging. J Control Release. 2016;243:303–22. https://doi.org/10.1016/j.jconrel.2016.10.023.
67. Guo J, Rahme K, He Y, Li LL, Holmes JD, O'Driscoll CM. Gold nanoparticles enlighten the future of cancer theranostics. Int J Nanomedicine. 2017;12:6131–52. https://doi.org/10.2147/IJN.S140772.
68. Huang YJ, Hsu SH. TRAIL-functionalized gold nanoparticles selectively trigger apoptosis in polarized macrophages. Nano. 2017;1:326–37. https://doi.org/10.7150/ntno.2023369.
69. Yin H, Kauffman KJ, Anderson DG. Delivery technologies for genome editing. Nat Rev Drug Discov. 2017;16:387–99.
70. Yin H, et al. Non-viral vectors for gene-based therapy. Nat Rev Genet. 2014;15:541–55.
71. Cavazzana-Calvo M, et al. Gene therapy of human severe combined immunodeficiency (SCID)-X1 disease. Science. 2000;288:669–72.
72. Hacein-Bey-Abina S, et al. LMO2-associated clonal T cell proliferation in two patients after gene therapy for SCID-X1. Science. 2003;302:1181–5.
73. Whitehead KA, Langer R, Anderson DG. Knocking down barriers: advances in siRNA delivery. Nat Rev Drug Discov. 2009;8:129–38.
74. Semple SC, et al. Rational design of cationic lipids for siRNA delivery. Nat Biotechnol. 2010;28:172–6.
75. Davis ME, et al. Evidence of RNAi in humans from systemically administered siRNA via targeted nanoparticles. Nature. 2010;464:1067–70.
76. Mangraviti A, et al. Polymeric nanoparticles for nonviral gene therapy extend brain tumor survival in vivo. ACS Nano. 2015;9:1236–49.
77. Sarett SM, et al. Lipophilic siRNA targets albumin in situ and promotes bioavailability, tumor penetration, and carrier-free gene silencing. Proc Natl Acad Sci USA. 2017;114:E6490–7.
78. He Y, et al. Polyethyleneimine/DNA polyplexes with reduction-sensitive hyaluronic acid derivatives shielding for targeted gene delivery. Biomaterials. 2013;34:1235–45.
79. Xu C-F, Chen G-J, Luo Y-L, et al. Rational designs of in vivo CRISPR–Cas delivery systems. Adv Drug Deliv Rev. 2021;168:3–29. https://doi.org/10.1016/j.addr.2019.11.005.
80. Maule G, et al. Allele specific repair of splicing mutations in cystic fibrosis through AsCas12a genome editing. Nat Commun. 2019;10:3556.
81. Moss RB, et al. Repeated aerosolized AAV-CFTR for treatment of cystic fibrosis: a randomized placebo-controlled phase 2B trial. Hum Gene Ther. 2007;18:726–32.
82. Alton EWFW, et al. Repeated nebulisation of nonviral CFTR gene therapy in patients with cystic fibrosis: a randomised, double-blind, placebo-controlled, phase 2b trial. Lancet Respir Med. 2015;3:684–91.
83. Guan S, et al. Self-assembled peptide–poloxamine nanoparticles enable in vitro and in vivo genome restoration for cystic fibrosis. Nat Nanotechnol. 2019;14:287–97.
84. Robinson E, et al. Lipid nanoparticle-delivered chemically modified mRNA restores chloride secretion in cystic fibrosis. Mol Ther. 2018;26:2034–46.
85. Witten J, Samad T, Ribbeck K. Selective permeability of mucus barriers. Curr Opin Biotechnol. 2018;52:124–33.
86. Patel AK, et al. Inhaled nanoformulated mRNA polyplexes for protein production in lung epithelium. Adv Mater. 2019;31:1805116.
87. Suk JS, et al. Lung gene therapy with highly compacted DNA nanoparticles that overcome the mucus barrier. J Control Release. 2014;178:8–17.
88. Deweerdt S. The fix is in utero. Nature. 2018;564:6–8.
89. Massaro G, et al. Fetal gene therapy for neurodegenerative disease of infants. Nat Med. 2018;24:1317–23.
90. Ricciardi AS, et al. In utero nanoparticle delivery for site-specific genome editing. Nat Commun. 2018;9:1–11.
91. Kubik T, Bogunia-Kubik K, Sugisaka M. Nanotechnology on duty in medical applications. Curr Pharm Biotechnol. 2005;6:17–33.
92. Hassanzadeh P. Nanopharmaceuticals: innovative theranostics for the neurological disorders. Biomed Rev. 2014;25:25–34.
93. Hassanzadeh P. Nanotheranostics against COVID-19: from multivalent to immune-targeted materials. J Control Release. 2020;328:112–26.
94. Hassanzadeh P. New perspectives in biosensor technology. Gastroenterol Hepatol Bed Bench. 2010;3(3):105–7.
95. Palestino G, Garcia-Silva I, Gonzalez-Ortega O, Rosales-Mendoza S. Can nanotechnology help in the fight against COVID-19? Expert Rev Anti Infect Ther. 2020;1:1–17. https://doi.org/10.1080/14787210.2020.1776115.
96. Wang X, Liu L-H, Ramstroem O, Yan M. Engineering nanomaterial surfaces for biomedical applications. Exp Biol Med. 2009;234:1128–39.
97. Shi L, Sun Q, He J, et al. Development of SPR biosensor for simultaneous detection of multiplex respiratory viruses. Biomed Mater Eng. 2015;26:S2207–16.

98. Cho SJ, Woo HM, Kim KS, et al. Novel system for detecting SARS coronavirus nucleocapsid protein using an ssDNA aptamer. J Biosci Bioeng. 2011;112(6):535–40.
99. Park TJ, Lee SK, Yoo SM, et al. Development of reflective biosensor using fabrication of functionalized photonic nanocrystals. J Nanosci Nanotechnol. 2011;11(1):632–7.
100. Ishikawa FN, Curreli M, Olson CA, et al. Importance of controlling nanotube density for highly sensitive and reliable biosensors functional in physiological conditions. ACS Nano. 2010;4(11):6914–22.
101. Wang J, Zhang P, Li CM, Li YF, Huang CZ. A highly selective and colorimetric assay of lysine by molecular-driven gold nanorods assembly. Biosens Bioelectron. 2012;34:197–201.
102. Jung IY, You JB, Choi BR, et al. Highly sensitive molecular detection platform for robust and facile diagnosis of middle east respiratory syndrome (MERS) corona virus. Adv Healthc Mater. 2016;5(17):2168–73.
103. Ting Zhao VX, Wong TI, Zheng XT, Tan YN, Zho X. Colorimetric biosensors for point-of-care virus detections. Mater Sci Energy Technol. 2020;3:237–49.
104. Ting D, Dong N, Fang L, et al. Multisite inhibitors for enteric coronavirus: antiviral cationic carbon dots based on curcumin. ACS Appl Nano Mater. 2018;1(10):5451–9.
105. Botequim D, Maia J, Lino MMF, et al. Nanoparticles and surfaces presenting antifungal, antibacterial and antiviral properties. Langmuir. 2012;28(20):7646–56.
106. Lee JH, Ozorowski G, Ward AB. Cryo-EM structure of a native, fully glycosylated, cleaved HIV-1 envelope trimer. Science. 2016;351(6277):1043–8.
107. Liang JJ, Wei JC, Lee YL, et al. Surfactant-modified nanoclay exhibits an antiviral activity with high potency and broad spectrum. J Virol. 2014;88(8):4218–28.
108. Wang MC, Lin JJ, Tseng HJ, et al. Characterization, antimicrobial activities, and biocompatibility of organically modified clays and their nanocomposites with polyurethane. ACS Appl Mater Interfaces. 2012;4(1):338–50.
109. Miyako E, Nagata H, Hirano K, et al. Photoinduced antiviral carbon nanohorns. Nanotechnology. 2008;19(7):075106.
110. Banerjee I, Douaisi MP, Mondal D, et al. Light-activated nanotube–porphyrin conjugates as effective antiviral agents. Nanotechnology. 2012;23(10):105101.
111. Di Giorgio S, et al. Evidence for host-dependent RNA editing in the transcriptome of SARS-CoV-2. Sci Adv. 2020;6(25):eabb5813.
112. Di Giorgio S, Martignano F, Torcia MG, Mattiuz G, Conticello SG. Evidence for host-dependent RNA editing in the transcriptome of SARS-CoV-2. Sci Adv. 2020;6:eabb5813. https://doi.org/10.1126/sciadv.abb5813.
113. Esvelt KM, Wang HH. Genome-scale engineering for systems and synthetic biology. Mol Syst Biol. 2013;9(1):641. https://doi.org/10.1038/msb.2012.66.
114. Hassanzadeh P, Arbabi E, Rostami F, Atyabi F, Dinarvand R. Carbon nanotubes prolong the regulatory action of nerve growth factor on the endocannabinoid signaling. Physiol Pharmacol. 2015;19:167–76.
115. Hassanzadeh P, Arbabi E, Atyabi F, Dinarvand R. Application of carbon nanotubes as the carriers of the cannabinoid, 2-arachidonoylglycerol: towards a novel treatment strategy in colitis. Life Sci. 2017;179:66–72.
116. Hassanzadeh P, Arbabi E, Atyabi F, Dinarvand R. Nerve growth factor-carbon nanotube complex exerts prolonged protective effects in an in vitro model of ischemic stroke. Life Sci. 2017;179:15–22.
117. Hassanzadeh P, Atyabi F, Dinarvand R. Application of carbon nanotubes for controlled release of growth factors or endocannabinoids: a breakthrough in biomedicine. Biomed Rev. 2017;27:41. https://doi.org/10.14748/bmr.v27.2105.
118. Hassanzadeh P, Arbabi E, Atyabi F, Dinarvand R. Carbon nanotubes provide longer lasting gastroprotective effects for anandamide in stress-induced gastric ulcer in rat. Physiol Pharmacol. 2018;22:38–47.
119. Hoy CF, Kushiro K, Yamaoka Y, et al. Rapid multiplex microfiber-based immunoassay for anti-MERS-CoV antibody detection. Sens Biosensing Res. 2019;26:100304.
120. Layqah LA, Eissa S. An electrochemical immunosensor for the corona virus associated with the Middle East respiratory syndrome using an array of gold nanoparticle-modified carbon electrodes. Mikrochim Acta. 2019;186(4):224.
121. Qiao J, Li Y, Wei C, et al. Rapid detection of viral antibodies based on multifunctional Staphylococcus aureus nanobioprobes. Enzym Microb Technol. 2016;95:94–9.
122. Qiu G, Gai Z, Tao Y, et al. Dual-functional plasmonic photothermal biosensors for highly accurate severe acute respiratory syndrome coronavirus 2 detection. ACS Nano. 2020;14(5):5268–77.
123. Seo G, Lee G, Kim MJ, et al. Rapid detection of COVID-19 causative virus (SARS-CoV-2) in human nasopharyngeal swab specimens using field-effect transistor-based biosensor. ACS Nano. 2020;14(4):5135–42.
124. Ye S, Shao K, Li Z, et al. Antiviral activity of graphene oxide: how sharp edged structure and charge matter. ACS Appl Mater Interfaces. 2015;7(38):21571–9.
125. Chen YN, Hsueh YH, Hsieh CT, et al. Antiviral activity of graphene–silver nanocomposites against non-enveloped and enveloped viruses. Int J Environ Res Public Health. 2016;13(4):430.

126. Park S, Park HH, Kim SY, et al. Antiviral properties of silver nanoparticles on a magnetic hybrid colloid. Appl Environ Microbiol. 2014;80(8):2343–50.
127. Elechiguerra JL, Burt JL, Morones JR, et al. Interaction of silver nanoparticles with HIV-1. J Nanobiotechnology. 2005;3(1):6.
128. Du T, Liang J, Dong N, et al. Glutathione-capped Ag2S nanoclusters inhibit coronavirus proliferation through blockage of viral RNA synthesis and budding. ACS Appl Mater Interfaces. 2018;10(5):4369–78.
129. Loczechin A, Seron K, Barras A, et al. Functional carbon quantum dots as medical countermeasures to human coronavirus. ACS Appl Mater Interfaces. 2019;11(46):42964–74.
130. Zhu S, Li J, Huang A-G, Huang J-Q, Huang Y-Q, Wang G-X. Anti-betanodavirus activity of isoprinosine and improved efficacy using carbon nanotubes based drug delivery system. Aquaculture. 2019;512:734377.
131. Zhu B, Liu G-L, Ling F, Wang G-X. Carbon nanotube-based nanocarrier loaded with ribavirin against grass carp reovirus. Antivir Res. 2015;118:29–38.
132. Demirer GS, et al. High aspect ratio nanomaterials enable delivery of functional genetic material without DNA integration in mature plants. Nat Nanotechnol. 2019;14:456–64.
133. Hassanzadeh P, Atyabi F, Dinarvand R, Dehpour AR, Azhdarzadeh M, Dinarvand M. Application of nanostructured lipid carriers: the prolonged protective effects for sesamol in in vitro and in vivo models of ischemic stroke via activation of PI3K signalling pathway. DARU J Pharm Sci. 2017;25:25. https://doi.org/10.1186/s40199-017-0191-z.
134. Hassanzadeh P, Arbabi E, Atyabi F, Dinarvand R. Ferulic acid-loaded nanostructured lipid carriers: a promising nanoformulation against the ischemic neural injuries. Life Sci. 2018;193:64–76.
135. Hassanzadeh P, Arbabi E, Rostami F, Atyabi F, Dinarvand R. Aerosol delivery of ferulic acid-loaded nanostructured lipid carriers: a promising treatment approach against the respiratory disorders. Physiol Pharmacol. 2017;21:331–42.
136. Hassanzadeh P, Atyabi F, Dinarvand R. Nanoencapsulation: a promising strategy for biomedical applications of ferulic acid. Biomed Rev. 2017;28:22–30.
137. Hassanzadeh P. Nanopharmaceuticals: innovative theranostics for the neurological disorders. Biomed. Rev. 2014;25:25–34.
138. Pardi N, et al. Administration of nucleoside-modified mRNA encoding broadly neutralizing antibody protects humanized mice from HIV-1 challenge. Nat Commun. 2017;8:14630.
139. Conway A, et al. Non-viral delivery of zinc finger nuclease mRNA enables highly efficient in vivo genome editing of multiple therapeutic gene targets. Mol Ther. 2019;27:866–77.
140. Akinc A, Maier MA, Manoharan M, et al. The Onpattro story and the clinical translation of nanomedicines containing nucleic acid-based drugs. Nat Nanotechnol. 2019;14:1084–7.
141. Chang-Jer W, Chan Y-L. Antiviral applications of RNAi for coronavirus. Expert Opin Investig Drugs. 2006;15:89–97.
142. Zheng B-J, Guan Y, Tang Q, et al. Prophylactic and therapeutic effects of small interfering RNA targeting SARS-coronavirus. Antivir Ther. 2004;9:365–74.
143. Leonard JN, Schaffer DV. Antiviral RNAi therapy: emerging approaches for hitting a moving target. Gene Ther. 2006;13:532–40.
144. Li B-J, Tang Q, Cheng D, et al. Using siRNA in prophylactic and therapeutic regimens against SARS coronavirus in Rhesus macaque. Nat Med. 2005;11:944.
145. Frieman M, Baric R. Mechanisms of severe acute respiratory syndrome pathogenesis and innate immunomodulation. Microbiol Mol Biol Rev. 2008;72:672–85.
146. Wang Y, Cao Y-L, Yang F, Zhang Y, Wang S-H, Liu L. Small interfering RNA effectively inhibits the expression of SARS coronavirus membrane gene at two novel targeting sites. Molecules. 2010;15:7197–207.
147. Watanabe T, Umehara T, Yasui F, et al. Liver target delivery of small interfering RNA to the HCV gene by lactosylated cationic liposome. J Hepatol. 2007;47(6):744–50.
148. Torrecilla J, Rodríguez-Gascón A, Solinís MÁ, del PozoRodríguez A. Lipid nanoparticles as carriers for RNAi against viral infections: current status and future perspectives. Biomed Res Int. 2014;2014:161794. https://doi.org/10.1155/2014/161794.
149. Spengler U. Liver disease associated with non-hepatitis viruses. In: Encyclopedia of gastroenterology. Academic Press; 2020. p. 363–76.
150. Cheng Q, Wei T, Jia Y, et al. Dendrimer-based lipid nanoparticles deliver therapeutic FAH mRNA to normalize liver function and extend survival in a mouse model of hepatorenal tyrosinemia type I. Adv Mater. 2018;30(52):e1805308.
151. van der Meel R. Nanotechnology for organ-tunable gene editing. Nat Nanotechnol. 2020;15:253–5.
152. van der Meel R, Sulheim E, Shi Y, Kiessling F, Mulder WJM, Lammers T. Smart cancer nanomedicine, strategic directions to improve translation and exploitation. Nat Nanotechnol. 2019;14(11):1007–17.
153. Kaczmarek JC, Patel AK, Kauffman KJ, et al. Polymer–lipid nanoparticles for systemic delivery of mRNA to the lungs. Angew Chem. 2016;128:1–6.
154. Kowalski PS, Rudra A, Miao L, Anderson DG. Delivering the messenger: advances in technologies for therapeutic mRNA delivery. Mol Ther. 2019;27:710–28.
155. Kulkarni JA, Witzigmann D, Chen S, Cullis PR, van der Meel R. Lipid nanoparticle technology for clinical translation of siRNA therapeutics. Acc Chem Res. 2019;52:2435–44.
156. Cheng Q, Wei T, Farbiak L, Johnson LT, Dilliard SA, Siegwart DJ. Selective organ targeting (SORT)

nanoparticles for tissue-specific mRNA delivery and CRISPR–Cas gene editing. Nat Nanotechnol. 2020;15:313–20.
157. Khan I, Saeed K, Khan I. Nanoparticles: properties, applications and toxicities. Arab J Chem. 2019;12:908–31.
158. Reichmuth A, Oberli M, Jeklenec A, Langer RS. mRNA vaccine delivery using lipid nanoparticles. Ther Deliv. 2016;7(5):319–34.
159. Deering RP, Kommareddy S, Ulmer JB, Brito LA, Geall AJ. Nucleic acid vaccines: prospects for non-viral delivery of mRNA vaccines. Expert Opin Drug Deliv. 2014;11:885–99.
160. John S, Yuzhakov O, Woods A, et al. Multi-antigenic human cytomegalovirus mRNA vaccines that elicit potent humoral and cell-mediated immunity. Vaccine. 2018;36:1689–99.
161. Hassett KJ, Benenato KE, Jacquinet E, et al. Optimization of lipid nanoparticles for intramuscular administration of mRNA vaccines. Mol Ther. 2019;15(1):1–11.
162. Ohno S, Kohyama S, Taneichi M, et al. Synthetic peptides coupled to the surface of liposomes effectively induce SARS coronavirus-specific cytotoxic T lymphocytes and viral clearance in HLA-A* 0201 transgenic mice. Vaccine. 2009;27(29):3912–20.
163. Sekimukai H, Iwata-Yoshikawa N, Fukushi S, et al. Gold nanoparticle-adjuvanted S protein induces a strong antigen-specific IgG response against severe acute respiratory syndrome-related coronavirus infection, but fails to induce protective antibodies and limit eosinophilic infiltration in lungs. Microbiol Immunol. 2020;64(1):33–51.
164. Lu X, Chen Y, Bai B, et al. Immune responses against severe acute respiratory syndrome coronavirus induced by virus-like particles in mice. Immunology. 2007;122(4):496–502.
165. Liu YV, Massare MJ, Barnard DL, et al. Chimeric severe acute respiratory syndrome coronavirus (SARS-CoV) S glycoprotein and influenza matrix 1 efficiently form virus-like particles (VLPs) that protect mice against challenge with SARS-CoV. Vaccine. 2011;29(38):6606–13.
166. Plummer EM, Manchester M. Viral nanoparticles and virus like particles: platforms for contemporary vaccine design. Wiley Interdiscip Rev Nanomed Nanobiotechnol. 2011;3:174–96.
167. Shin MD, Shukla S, Chung YH, et al. COVID-19 vaccine development and a potential nanomaterial path forward. Nat Nanotechnol. 2020;15:646–55.
168. Read SA, Obeid S, Ahlenstiel C, Ahlenstiel G. The role of zinc in antiviral immunity. Adv Nutr. 2019;10:696–710.
169. Dong L, Hu S, Gao J. Discovering drugs to treat coronavirus disease 2019 (COVID-19). Drug Discov Ther. 2020;14:58–60.
170. Wimalawansa SJ. Global epidemic of coronavirus—covid-19: what can we do to minimize risks. Eur J Biomed. 2020;7:432–8.
171. Hirano T, Murakami M. COVID-19: a new virus, but a familiar receptor and cytokine release syndrome. Immunity. 2020;52:731.
172. Yu M, Wu J, Shi J, Farokhzad OC. Nanotechnology for protein delivery: overview and perspectives. J Control Release. 2016;240:24–37.
173. Li B, Zhang X, Dong Y. Nanoscale platforms for messenger RNA delivery. Wiley Interdiscip Rev Nanomed Nanobiotechnol. 2019;11:e1530.
174. van der Zee R, Jiskoot W, van Eden W, Broere F. PLGA, PLGA-TMC and TMCTPP nanoparticles differentially modulate the outcome of nasal vaccination by inducing tolerance or enhancing humoral immunity. PLoS One. 2011;6:e26684.
175. Amanat F, Krammer F. SARS-CoV-2 vaccines: status report. Immunity. 2020;52:583–9.
176. Gao Q, Bao L, Mao H, et al. Development of an inactivated vaccine candidate for SARS-CoV-2. Science. 2020;369:77–81.
177. Knudsen NP, Olsen A, Buonsanti C, et al. Different human vaccine adjuvants promote distinct antigen-independent immunological signatures tailored to different pathogens. Sci Rep. 2016;6:19570.
178. Sun B, Xia T. Nanomaterial-based vaccine adjuvants. J Mater Chem B. 2016;4:5496–509.
179. Ma J, Liu R, Wang X, et al. Crucial role of lateral size for graphene oxide in activating macrophages and stimulating pro-inflammatory responses in cells and animals. ACS Nano. 2015;9:10498–515.
180. Gazzi A, Fusco L, Orecchioni M, et al. Graphene, other carbon nanomaterials and the immune system: toward nanoimmunity-by-design. J Phys Mater. 2020;3:034009.
181. Orecchioni M, Bedognetti D, Sgarrella F, Marincola FM, Bianco A, Delogu LG. Impact of carbon nanotubes and graphene on immune cells. J Transl Med. 2014;12:138.
182. Fuchs AK, Syrovets T, Haas KA, et al. Carboxyl- and amino-functionalized polystyrene nanoparticles differentially affect the polarization profile of M1 and M2 macrophage subsets. Biomaterials. 2016;85:78–87.
183. Fusco L, Avitabile E, Armuzza V, et al. Impact of the surface functionalization on nanodiamond biocompatibility: a comprehensive view on human blood immune cells. Carbon. 2020;160:390–404.
184. Orecchioni M, Bedognetti D, Newman L, et al. Single-cell mass cytometry and transcriptome profiling reveal the impact of graphene on human immune cells. Nat Commun. 2017;8:1109.
185. Prompetchara E, Ketloy C, Palaga T. Immune responses in COVID-19 and potential vaccines: lessons learned from SARS and MERS epidemic. Asian Pac J Allergy Immunol. 2020;(38):1–9.
186. Mehta P, McAuley DF, Brown M, Sanchez E, Tattersall RS, Manson JJ. COVID-19: consider cytokine storm syndromes and immunosuppression. Lancet. 2020;395:1033–4.
187. Al-Lawati H, Aliabadi HM, Makhmalzadeh BS, Lavasanifar A. Nanomedicine for immunosuppressive therapy: achievements in pre-clinical and clinical research. Expert Opin Drug Deliv. 2018;15:397–418.

188. Chen Y, Feng Z, Diao B, et al. The novel severe acute respiratory syndrome coronavirus 2 (SARS-CoV-2) directly decimates human spleens and lymph nodes. medRxiv. 2020. https://doi.org/10.1101/2020.03.27.20045427.
189. Pentecost A, Kim MJ, Jeon S, et al. Immunomodulatory nanodiamond aggregate-based platform for the treatment of rheumatoid arthritis. Regen Biomater. 2019;6:163–74.
190. Presser V, Yeon SH, Vakifahmetoglu C, et al. Hierarchical porous carbide-derived carbons for the removal of cytokines from blood plasma. Adv Healthc Mater. 2012;1:796–800.
191. Seredych M, Haines B, Sokolova V, et al. Graphene-based materials for the fast removal of cytokines from blood plasma. ACS Appl Bio Mater. 2018;1:436–43.
192. Wainwright M. Local treatment of viral disease using photodynamic therapy. Int J Antimicrob Agents. 2003;21:510–20.
193. Wiehe A, O'Brien JM, Senge MO. Trends and targets in antiviral phototherapy. Photochem Photobiol Sci. 2019;18:2565–612.
194. Bhatia S, Camacho LC, Haag R. Pathogen inhibition by multivalent ligand architectures. J Am Chem Soc. 2016;138:8654–66.
195. De Clercq E. Antiviral drugs in current clinical use. J Clin Virol. 2004;30:115–33.
196. Fasting C, Schalley CA, Weber M, et al. Multivalency as a chemical organization and action principle. Angew Chem Int Ed Engl. 2012;51(42):10472–98.
197. Bhatia S, Dimde M, Haag R. Multivalent glycoconjugates as vaccines and potential drug candidates. MedChemComm. 2014;5:862.
198. Bhella D. The role of cellular adhesion molecules in virus attachment and entry. Philos Trans R Soc B. 2015;370:20140035.
199. Annu Moscona A. Medical management of influenza infection. Rev Med. 2008;59:397.
200. Schirmer P, Holodniy M. Oseltamivir for treatment and prophylaxis of influenza infection. Expert Opin Drug Saf. 2009;8:357.
201. Papp I, Sieben C, Sisson AL, et al. Inhibition of influenza virus activity by multivalent glycoarchitectures with matched sizes. Chembiochem. 2011;12:887.
202. Qi Z, Bharate P, Lai CH, et al. Multivalency at interfaces: supramolecular carbohydrate-functionalized graphene derivatives for bacterial capture, release, and disinfection. Nano Lett. 2015;15:6051.
203. Baram-Pinto D, Shukla S, Gedanken A, Sarid R. Inhibition of HSV-1 attachment, entry, and cell-to-cell spread by functionalized multivalent gold nanoparticles. Small. 2010;6:1044–50.
204. Hu RL, Li SR, Kong FJ, Hou RJ, Guan XL, Guo F. Inhibition effect of silver nanoparticles on herpes simplex virus 2. Genet Mol Res. 2014;13:7022–8.
205. Ziem B, Azab W, Gholami MF, Rabe JP, Osterrieder N, Haag R. Size-dependent inhibition of herpesvirus cellular entry by polyvalent nanoarchitectures. Nanoscale. 2017;9:3774–83.
206. Vonnemann J, Sieben C, Wolff C, et al. Virus inhibition induced by polyvalent nanoparticles of different sizes. Nanoscale. 2014;6:2353–60.
207. Papp I, Sieben C, Ludwig K, et al. Inhibition of influenza virus infection by multivalent sialic-acid functionalized gold nanoparticles. Small. 2010;6:2900–6.
208. Cagno V, Andreozzi P, D'Alicarnasso M, et al. Broad-spectrum non-toxic antiviral nanoparticles with a virucidal inhibition mechanism. Nat Mater. 2018;17:195–203.
209. Han Y, Král P. Computational design of ACE2-based peptide inhibitors of SARSCoV-2. ACS Nano. 2020;14:5143–7.
210. Papp I, Sieben C, Sisson AL, et al. Inhibition of influenza virus activity by multivalent glycoarchitectures with matched sizes. Chembiochem. 2011;12:887–95.
211. Dey P, Bergmann T, Cuellar-Camacho JL, et al. Multivalent flexible nanogels exhibit broad-spectrum antiviral activity by blocking virus entry. ACS Nano. 2018;12:6429–42.
212. Ziem B, Thien H, Achazi K, et al. Highly efficient multivalent 2D nanosystems for inhibition of orthopoxvirus particles. Adv Healthc Mater. 2016;5:2922–30.
213. Gholami MF, Lauster D, Ludwig K, et al. Functionalized graphene as extracellular matrix mimics: toward well-defined 2D nanomaterials for multivalent virus interactions. Adv Funct Mater. 2017;27(15):1606477.
214. Jastrzębska AM, Kurtycz P, Olszyna AR. Recent advances in graphene family materials toxicity investigations. J Nanopart Res. 2012;14(12):1320.
215. Steinhilber D, Sisson AL, Mangoldt D, Welker P, Licha K, Haag R. Synthesis, reductive cleavage, and cellular interaction studies of biodegradable, polyglycerol nanogels. Adv Funct Mater. 2010;20:4133–8.
216. Zhang Q, Honko A, Zhou J, et al. Cellular nanosponges inhibit SARS-CoV-2 infectivity. Nano Lett. 2020;20(7):5570–4.
217. Parodi A, Molinaro R, Sushnitha M, et al. Bio-inspired engineering of cell- and virus-like nanoparticles for drug delivery. Biomaterials. 2017;147:155–68.
218. Kotov NA. Inorganic nanoparticles as protein mimics. Science. 2010;330:188.
219. Czapar AE, Zheng Y-R, Riddell IA, et al. Tobacco mosaic virus delivery of phenanthriplatin for cancer therapy. ACS Nano. 2016;10:4119–26.
220. Mullard A. Flooded by the torrent: the COVID-19 drug pipeline. Lancet. 2020;395:1245–6.
221. Senanayake SL. Drug repurposing strategies for COVID-19. Future Drug Discov. 2020;2(2):FDD40.
222. Hassanzadeh P, Atyabi F, Dinarvand R. Ignoring the modeling approaches: towards the shadowy paths in nanomedicine. J Control Release. 2018;280:58–75.
223. Hassanzadeh P, Atyabi F, Dinarvand R. Application of modelling and nanotechnology-based approaches: the emergence of breakthroughs in theranostics

of central nervous system disorders. Life Sci. 2017;182:93–103.
224. Hassanzadeh P. Computational modelling: moonlighting on the neuroscience and medicine. Biomed Rev. 2013;24:25–31.
225. Ferreira LG, Dos Santos RN, Oliva G, Andricopulo AD. Molecular docking and structure-based drug design strategies. Molecules. 2015;20(7):13384–421.
226. Šebera J, Dubankova A, Sychrovsky V, Ruzek D, Boura E, Nencka R. The structural model of Zika virus RNA-dependent RNA polymerase in complex with RNA for rational design of novel nucleotide inhibitors. Sci Rep. 2018;8(1):11132.
227. Shameer K, Johnson KW, Readhead B. Rapid therapeutic recommendations in the context of a global public health crisis using translational bioinformatics approaches: a proof-of-concept study using Nipah virus infection. bioRxiv. https://doi.org/10.1101/333021.
228. Zhang TF, Chen Q. Identification of contaminant sources in enclosed environments by inverse CFD modeling. Indoor Air. 2007;17:167–77.
229. Mazumdar S, Poussou SB, Lin C-H, Isukapalli SS, Plesniak MW, Chen Q. Impact of scaling and body movement on contaminant transport in airliner cabins. Atmos Environ. 2011;45:6019–28.
230. Lunnoo T, Assawakhajornsak J, Puangmali T. In silico study of gold nanoparticle uptake into a mammalian cell: interplay of size, shape, surface charge, and aggregation. J Phys Chem C. 2019;123:3801–10.
231. Ostaszewski M, Mazein A, Gillespie ME, et al. COVID-19 disease map, building a computational repository of SARS-CoV-2 virus-host interaction mechanisms. Sci Data. 2020;7:136. https://doi.org/10.1038/s41597-020-0477-8.
232. Ostaszewski M, Gebel S, Kuperstein I, et al. Community-driven roadmap for integrated disease maps. Brief Bioinform. 2019;20:659–70.
233. Sacha GM, Varona P. Artificial intelligence in nanotechnology. Nanotechnology. 2013;24:1–13. https://doi.org/10.1088/0957-4484/24/45/452002.
234. Cavalcanti A, Shirinzadeh B, Freitas RA Jr, Kretly LC. Medical nanorobot architecture based on nano bioelectronics. Recent Pat Nanotechnol. 2007;1:1 10.
235. Yamaan Saadeh BS, Vyas D. Nanorobotic applications in medicine: current proposals and designs. Am J Robot Surg. 2014;1(1):4–11.
236. Murphy D, Challacombe B, Nedas T, et al. Equipment and technology in robotics. Arch Esp Urol. 2007;60:349–54.
237. Hassanzadeh P, Atyabi F, Dinarvand R. Creation of nanorobots: both state-of-the science and state-of-the-art. Biomed Rev. 2016;27:37–44.
238. Patel GM, Patel GC, Patel RB, Patel JK, Patel M. Nanorobot: a versatile tool in nanomedicine. J Drug Target. 2006;14:63–7.
239. Lee YS, Bae JY, Koo HY, Lee YB, Choi WS. A remote-controlled generation of gold@polydopamine (core@ shell) nanoparticles via physical-chemical stimuli of polydopamine/gold composites. Sci Rep. 2016;6:22650.
240. Martel S, Mathieu JB, Felfoul O, et al. A computer-assisted protocol for endovascular target interventions using a clinical MRI system for controlling untethered microdevices and future nanorobots. Comput Aided Surg. 2008;13:340–52.
241. Balasubramanian S, Kagan D, Jack HCM, et al. Micromachine-enabled capture and isolation of cancer cells in complex media. Angew Chem Int Ed Engl. 2011;50:4161–4.
242. Patel RR. Nanorobotics ideas in nanomedicine. Asian J Pharm Sci Res. 2013;3:15–22.
243. Martel S, Felfoul O, Mohammadi M, Mathieu JB. Interventional procedure based on nanorobots propelled and steered by flagellated magnetotactic bacteria for direct targeting of tumors in the human body. Conf Proc IEEE Eng Med Biol Soc. 2008;2008:2497–500.
244. Somanna MB. Nanobots: the future of medical treatments. Int J Sci Tech Res. 2015;4:276–8.
245. Torelli E, Marini M, Palmano S, et al. A DNA origami nanorobot controlled by nucleic acid hybridization. Small. 2014;10:2918–26.
246. Douglas SM, Bachelet I, Church GM. A logic-gated nanorobot for targeted transport of molecular payloads. Science. 2012;335:831–4.
247. Behkam B, Sitti M. Bacterial flagella-based propulsion and on/off motion control of microscale objects. Appl Phys Lett. 2007;90:1–3.
248. Dubey A, Sharma G, Mavroidis C, Tomassone MS, Nikitczuk K, Yarmushc ML. Computational studies of viral protein nano-actuators. J Comp Theor Nanosci. 2004;1:18–28.
249. Skehel JJ, Wiley DC. Receptor binding and membrane fusion in virus entry: the influenza hemagglutinin. Annu Rev Biochem. 2000;69:531–69.
250. Ahuja SP, Myers JR. A survey on wireless grid computing. J Supercomput. 2006;37(1):3–21.
251. Ricciardi L, Pitz I, Al-Sarawi SF, Varadan V, Abbott D. Investigation into the future of RFID in biomedical applications. Proc SPIE Int Soc Opt Eng. 2003;5119:199–209.
252. de Oliveira Toledo SL, Nogueira LS, das Graças Carvalho M, Alves Rios DR, de Barros Pinheiro M. COVID-19: review and hematologic impact. Clin Chim Acta. 2020;510:170–6.
253. Singh A, Shaikh A, Singh R, Sing AK. COVID-19: from bench to bed side. Diabetes Metab Syndr. 2020;14:277–81.
254. Tanindi A, Cemri M. Troponin elevation in conditions other than acute coronary syndromes. Vasc Health Risk Manag. 2011;7:597–603.
255. Sahu KK, Siddiqui AD, Cerny J. Mesenchymal stem cells in COVID-19: a journey from bench to bedside. Lab Med. 2020;1:1–12. https://doi.org/10.1093/labmed/lmaa049.
256. Warnes SL, Little ZR, Keevil CW. Human coronavirus 229E remains infectious on common touch surface materials. mBio. 2015;6:e01697-15.

Nano-Spectro Imaging 24

24.1 Introduction

The important development of optical imaging techniques has resulted in better resolution with an optical microscope, which has many limitations. One of the common examples is the use of lenses made of glass, which have limited resolution for optical imaging due to the half-wavelength. Optical imaging in visible light can separate two particles with <300 nm distance. The diffraction limit can reduce the wavelength, which can lead to better and more nanoscale resolution (Fig. 24.1). Another myth is that electrons are considered waves, which have very short wavelengths.

The use of atomic force microscopy-infrared spectroscopy (AFM-IR) is also widely used in biological samples for greater efficiency [1]. The criteria for the broader range of mechanical properties also include AFM-IR signals, which have transduction inefficiencies due to stiffer samples [2, 4]. Tapping-mode AFM-IR [3–5] is an important technique for measuring softer [4, 5], stickier [3], rougher [6], and more displaced samples [5].

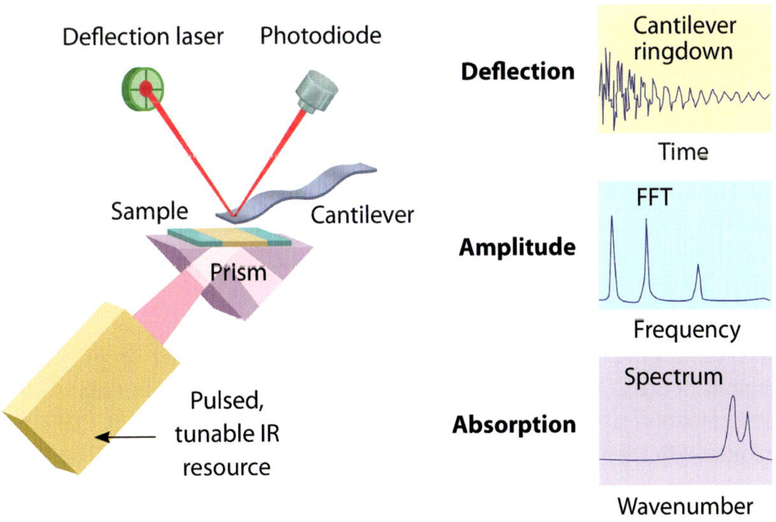

Fig. 24.1 Different mechanisms used in tuneable nanoparticles with different characteristics absorption, amplitude, and deflection

S. Shaikh, *Nanoimaging - Future of Precision Medicine*,
https://doi.org/10.1007/978-981-97-7292-6_24

The study by Dazzi's group, when they were using AFM-IR, focused on biofuels production in bacteria [7, 13], yeasts, and microalgae.

Spectral analysis using AFM-IR can be applied to measure nanoscale absorption. (a) The AFM topography of five representative hexagonal boron-nitride (hBN) frustums shows different aspect ratios (Ar = *d*/*t*). These frustums have a height of 256 nm. Scale bars are 300 nm. (b) The AFM-IR absorption spectra (p-polarization) for better positioning of the AFM tip at the center of representative frustums, as indicated in panel-a. Different polarization methods can be used for measurements of various quantities at selected wavelengths, which are capable of highlighting adjacent fields [11]. These are more precise and useful for analysis in different modes. The Royal Society of Chemistry 2020 has shown the utility of various thicknesses, in excess of 1 mm, and how they can be used for better distribution of biofuel-containing vesicles [7], viruses [11, 12], nanoparticles [9, 14], subcellular structures [15, 26], and proteins [8], all of which can be seen within cells and different organisms. The distribution of drug-delivering nanoparticles, as small as 170 nm, with polylactic acid in macrophages [9], has shown the different leveraging of the particles within intrinsic signatures, using different fluorescent tags that are used of the nanoparticles' biological fate, and this has been shown by the different AFM-IR multivariate analysis applications [10, 13]. Various red blood cells infected by the malarial parasite (Plasmodium falciparum) can be observed in various stages, showing subcellular structures for AFM-IR studies on antimalarial drug interactions with organisms [10]. The discrete wavenumbers correspond to large variances in the spectrum, which can be useful for identifying regions of cells rich in lipids, DNA, hemoglobin, or hemozoin, a byproduct of parasite-induced hemoglobin degradation [10]. The important principal component analysis (PCA) of the AFM-IR spectrum discriminates between Gram-positive and Gram-negative bacteria based on their cell wall spectral characteristics [10]. Gram-negative bacteria have very thin cell walls made up of inner and outer membranes, with a thin peptidoglycan layer between. Gram-positive bacteria have various cell walls composed of an inner membrane and a thick peptidoglycan layer with teichoic acid. There was also a conclusion showing the formation of a cell wall septum, about 45 nm thick, in the bacterium before cell division. Data acquisition can also be slow [10, 16] for the latest developments in spectral acquisition [4]. The use of AFM-IR multivariate analysis is likely to become more commonplace in the future. Different characteristics of various nanostructured drug delivery systems can be seen in lipid–polymer hybrid films [3], liposomes [4], and core–shell polymeric nanoparticles [5]. These all have different contact modes and unfavorable mechanical properties [4]. The different formulations can deliver therapeutic options more effectively than conventional therapies, with better efficiency and fewer side effects. Liposomes, as nanocarriers, can increase therapeutic options for tumor imaging by the EPR effect, with minimal side effects, useful for targeted delivery with better specificity and efficiency (Fig. 24.2). Different AFM-IR techniques allow for label-free identification and mapping the distribution of cytarabine molecules, chemotherapeutic agents, and liposomes [4] in the range of 80 nm diameter. This is useful for better characterization and classification of empty and drug-loaded liposomes, as well as for improving drug encapsulation efficiency for effective dosage. Tapping modes of AFM-IR can also be used for other drugs, such as pipemidic acid, loaded onto the external surface of core–shell polymeric nanoparticles. This formulation helps in better release of various drugs [5].

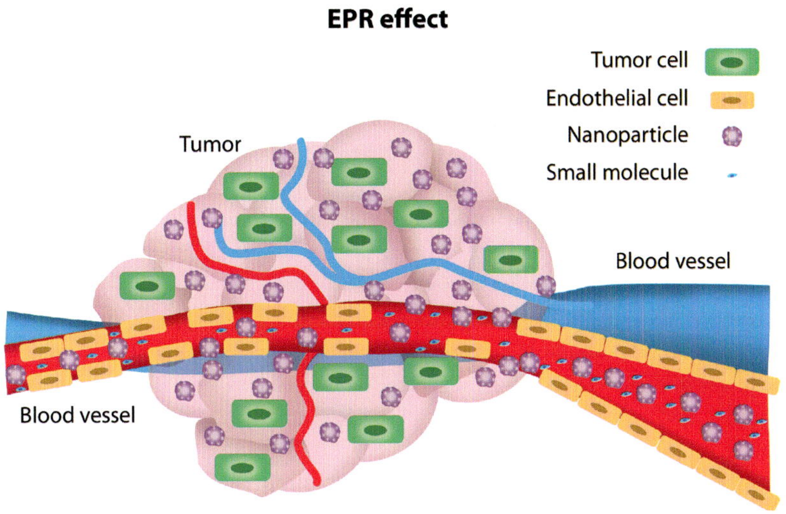

Fig. 24.2 Electroporation effect of the nanoparticles

24.2 Slow Light for the Use of the Nano-Imaging

This will have the ideal imaging system for achieving higher resolution along with the different intrinsic properties of the object. The probing wave will have better energy (or frequency) similar to visible light and have a wavelength that compares to the X-ray or electron wave. The wavelength and frequency of waves can be correlated with the speed of the wave, and this will require visible light, which has the speed of light, and this speed can be altered and reduced. This slow light is capable of providing better and higher spatial resolution when used in optical imaging, resulting in more precise and better intrinsic properties. This process will lead to different optical imaging techniques with a higher refractive index of the material used, like immersion oil, and can be seen in the free space between the object and the lens. The speed of light can be reduced in this space, resulting in better resolution along with an improvement in the resolution. Thus, for high-resolution optical imaging, there is a need for the evanescent wave in the imaging, as these waves cannot propagate. These evanescent waves are closely related to the light source, which is the primary source and has a better-oscillating dipole, while the secondary source, which has better scattering, is the tiny aperture in the form of a tiny particle. It is a bit difficult to create and control the secondary point source, which will be closer to the object. This is the basis of near-field optical imaging, utilizing a tiny aperture or a tiny particle to obtain higher and better resolution, as shown by Synge [17] and experimentally proven in the past [18–20].

The advances in technology have revolutionized the use of infrared and Raman spectroscopic imaging for better resolution on the molecular scale. Various techniques are being used for spectroscopic imaging that can achieve nanometer-scale resolution, including tip-enhanced Raman scattering (TERS), infrared scattering-type scanning near-field optical microscopy (IR s-SNOM), atomic force microscopy infrared (AFM-IR), and photon-induced force microscopy (PiFM).

24.3 TERS Combines Raman Spectroscopy

TERS, Raman Spectroscopy, and s-SNOM are capable of providing the best approach for spectroscopic imaging at the nanoscale, seen beyond the diffraction limit. TERS is based on nanostructured metallic scanning probe microscopes, which are used to enhance the electric field of the incident light at the tip apex, leading to the generation of enhanced Raman scattering from molecules that are in close proximity to the tip. The TERS tip, a Raman image, which provides more precise information, can offer better spatial resolution in the near field. TERS was first introduced in 2000 [21–24] and has made significant advancements in the use of TERS mechanisms for development and application. This has led to various issues. Different theories, technical details, and experimental setups of TERS are found in various sources [25].

The increased use of optical fields in metallic nanostructures will provide unique tools for various applications in different nanoscale devices and spectroscopies. For this, the plasmonic properties of metallic nanostructures, such as the distribution of optical fields and the wavelength, which are more dependent on the surface plasmon resonance at the nanometer scale, will be more precise. The TERS-based technique is used for the more precise evaluation of plasmonic properties at the nanoscale. Due to the frequency-resolved optical information obtained in Raman spectroscopy, along with more research-oriented, innovative molecular-functionalized metallic probes, there will be better field confinement, and the different wavelengths will depend on the localized surface plasmon resonance of gold nanostructures. This will be more successful in achieving spatial resolution better than 11 nm. These results will reveal more inherent plasmonic properties, which will be useful for metallic nanostructures and help in the study of the nature of plasmonic nanostructures and the development of plasmonic applications, such as molecular sensing, energy transfer, or optical storage.

24.4 Conclusion

The use of Nano Spectroscopy is a newer concept and an important one for the analysis and diagnosis of various molecular pathologies. This will be a futuristic modality with many applications, revolutionizing the practice of medicine.

References

1. Schwartz JJ, Jakob DS, Centrone A. A guide to nanoscale IR spectroscopy: resonance enhanced transduction in contact and tapping mode AFM-IR. Chem Soc Rev. 2022;51(13):5248–67. https://doi.org/10.1039/D2CS00095D.
2. Ramer G, Aksyuk VA, Centrone A. Quantitative chemical analysis at the nanoscale using the photothermal induced resonance technique. Anal Chem. 2017;89:13524–31.
3. Tuteja M, Kang M, Leal C, Centrone A. Nanoscale partitioning of paclitaxel in hybrid lipid-polymer membranes. Analyst. 2018;143:3808–13.
4. Wieland K, Ramer G, Weiss VU, Allmaier G, Lendl B, Centrone A. Nanoscale chemical imaging of individual, chemotherapeutic cytarabine-loaded liposomal nanocarriers. Nano Res. 2019;12:197–203.
5. Mathurin J, Pancani E, Deniset-Besseau A, et al. How to unravel the chemical structure and component localization of individual drug-loaded polymeric nanoparticles by using tapping AFM-IR. Analyst. 2018;143:5940–9.
6. Ma X, Beltran V, Ramer G, et al. Revealing the distribution of metal carboxylates in oil paint from the micro- to nanoscale. Angew Chem Int Ed. 2019;58:11652–6.
7. Ramer G, Aksyuk VA, Centrone A. Quantitative chemical analysis at the nanoscale using the PTIR technique. Anal Chem. 2017;89(24):13524–31. https://doi.org/10.1021/acs.analchem.7b03878.38.
8. Lu F, Jin M, Belkin MA. Tip-enhanced infrared nanospectroscopy via molecular expansion force detection. Nat Photonics. 2014;8(4):307–12. https://doi.org/10.1038/nphoton.2013.373.60.
9. Wang C-T, Jiang B, Zhou Y-W, et al. Exploiting the surface-enhanced IR absorption effect in the photothermally induced resonance AFM-IR technique toward nanoscale chemical analysis. Anal Chem. 2019;91(16):10541–8. https://doi.org/10.1021/acs.analchem.9b0155495.
10. Derenne A, Claessens T, Conus C, Goormaghtigh E. Infrared spectroscopy of membrane lipids. In: Roberts GCK, editor. Encyclopedia of biophysics. Berlin: Springer; 2013. p. 1074–81. https://doi.org/10.1007/978-3-642-16712-6_558.97.
11. Li M, Wang H, Li W, Xu XG, Yu Y. Macrophage activation on "phagocytic synapse" arrays: spacing of nanoclustered ligands directs TLR1/2 signaling with an intrinsic limit. Sci Adv. 2020;6(49):eabc8482. https://doi.org/10.1126/sciadv.abc8482.
12. Kenkel S, Mittal S, Bhargava R. Closed-loop atomic force microscopy-infrared spectroscopic imaging for nanoscale molecular characterization. Nat Commun. 2020;11(1):3225. https://doi.org/10.1038/s41467-020-17043-5.
13. Synge EH. A suggested method for extending microscopic resolution into the ultra-microscopic region. Philos Mag. 1928;6:356–62.

14. Pohl DW, Denk W, Lanz M. Optical stethoscopy: image recording with resolution λ/20. Appl Phys Lett. 1984;44:651–3.
15. Kawata S, Inouye Y, Sugiura T. Near-field scanning optical microscope with a laser trapped probe. Jpn J Appl Phys. 1994;33:L1725–7.
16. Mathurin J, Deniset-Besseau A, Dazzi A. Advanced infrared nanospectroscopy using photothermal induced resonance technique, AFMIR: new approach using tapping mode. Acta Phys Polym A. 2020;137(1):29–32. https://doi.org/10.12693/APhysPolA.137.29.
17. Anderson M. Locally enhanced Raman spectroscopy with an atomic force microscope. Appl Phys Lett. 2000;76:3130–2.
18. Hayazawa N, Inouye Y, Sekkat Z, Kawata S. Metallized tip amplification of near-field Raman scattering. Opt Commun. 2000;183:333–6.
19. Pettinger B, Picardi G, Schuster R, Ertl G. Surface enhanced Raman spectroscopy: towards single molecular spectroscopy. Electrochemistry. 2000;68:942–9.
20. Stockle R, Suh Y, Deckert V, Zenobi R. Nanoscale chemical analysis by tip-enhanced Raman spectroscopy. Chem Phys Lett. 2000;318:131–6.
21. Rizevsky S, Zhaliazka K, Dou T, Matveyenka M, Kurouski D. Characterization of substrates and surface-enhancement in atomic force microscopy Infrared analysis of amyloid aggregates. J Phys Chem C. 2022;126(8):4157–62. https://doi.org/10.1021/acs.jpcc.1c09643.
22. Vincent T. Scanning near-field infrared microscopy. Nat Rev Phys. 2021;3(8):537. https://doi.org/10.1038/s42254-021-00337-y.
23. Klementieva O, Sandt C, Martinsson I, Kansiz M, Gouras GK, Borondics F. Super-resolution infrared imaging of polymorphic amyloid aggregates directly in neurons. Adv Sci. 2020;7(6):1903004. https://doi.org/10.1002/advs.201903004.
24. Barth A. Infrared spectroscopy of proteins. Biochim Biophys Acta. 2007;1767(9):1073–101. https://doi.org/10.1016/j.bbabio.2007.06.004.
25. Wieland K, Ramer G, Weiss VU, Allmaier G, Lendl B, Centrone A. Nanoscale chemical imaging of individual chemotherapeutic cytarabine-loaded liposomal nanocarriers. Nano Res. 2019;12(1):197–203.
26. Inouye Y, Kawata S. Near-field scanning optical microscope with a metallic probe tip. Opt Lett. 1994;19:159–61.

Future of Nanoimaging 25

25.1 The Use of the Ultrasound When Combined with NBs in Nucleic Acid Delivery

Ultrasound combined with nanobodies is being used for DNA delivery. This process is based on the mechanism of sonoporation, a phenomenon where different US interactions with the tissues can be identified based on the transmission efficiency [1] (Fig. 25.1). This is important for the site-specific evaluation where there can be genetic materials, which can be used for better tissue non-resonated tissue transfection. These will be able to be used for the transfer of genes by nanobodies, as this will have a significant range of connotations inside nanobubble systems [2–4]. The utilization of cationic lipids is also important in the nanobubbles for the process of electrostatic attachment of DNA, where there can be direct physical assembly of DNA in the surrounding shells when the nanoparticles are made. This shell will be useful for the process of electrostatic interaction of DNA, where there will be covalent binding of DNA-loaded NB vectors along with the utilization of compatible DNA [5]. The different lipid-based NBs in cells and tissues will be able to cause various gene transfections as compared to albumin-based NBs. The major gene transfer along with lipid-based NBs is seen in mice. The use of submicron-sized bubbles along with handheld ultrasonic instruments can act as a gene transfer reagent [6, 7]. The use of submicron-scale bubbles is being developed as a better prospective gene transmission method. This has shown the use of various syntheses of novel bubbles (ranging from 140 to 190 nm in size) that can be used with ultrasound, which will be useful for better efficiency in gene delivery. Gene delivery is usually not accessible by using and increasing the transfection of DNA. One of the methods reported by Takahashi et al. [8] is the use of nanosized PEG-modified liposomes, which have better efficiency for pDNA dissemination (Fig. 25.2). Different cationic lipids will show various pDNA-loaded liposome NBs, which are well-equipped and have better transfection efficiency, better stability, and more efficacy of NBs that will have better transfer of TNF-α genes in tumor cells [9]. The use of the TNF-α encoding plasmid DNA, when loaded into lipid-containing NBs, is also capable of having better transfection of various tumor cells when used under ultrasound irradiation. This property will be useful for the better local production of TNF-a, which can have better induction of the various anticancer properties, more increases in the p53-related apoptosis, and this is leaving causing lesser tumor vessel volume, and lesser tumor size. This will reduce the toxic effects. Watanabe et al. [10] have shown the potency of gene expression that can be achieved together by NBs in the US. The use of gene expression kinetics was accidentally discovered with PET (positron emission tomography) and bioluminescence imaging [10]. Bisazza et al. [4]

S. Shaikh, *Nanoimaging - Future of Precision Medicine*,
https://doi.org/10.1007/978-981-97-7292-6_25

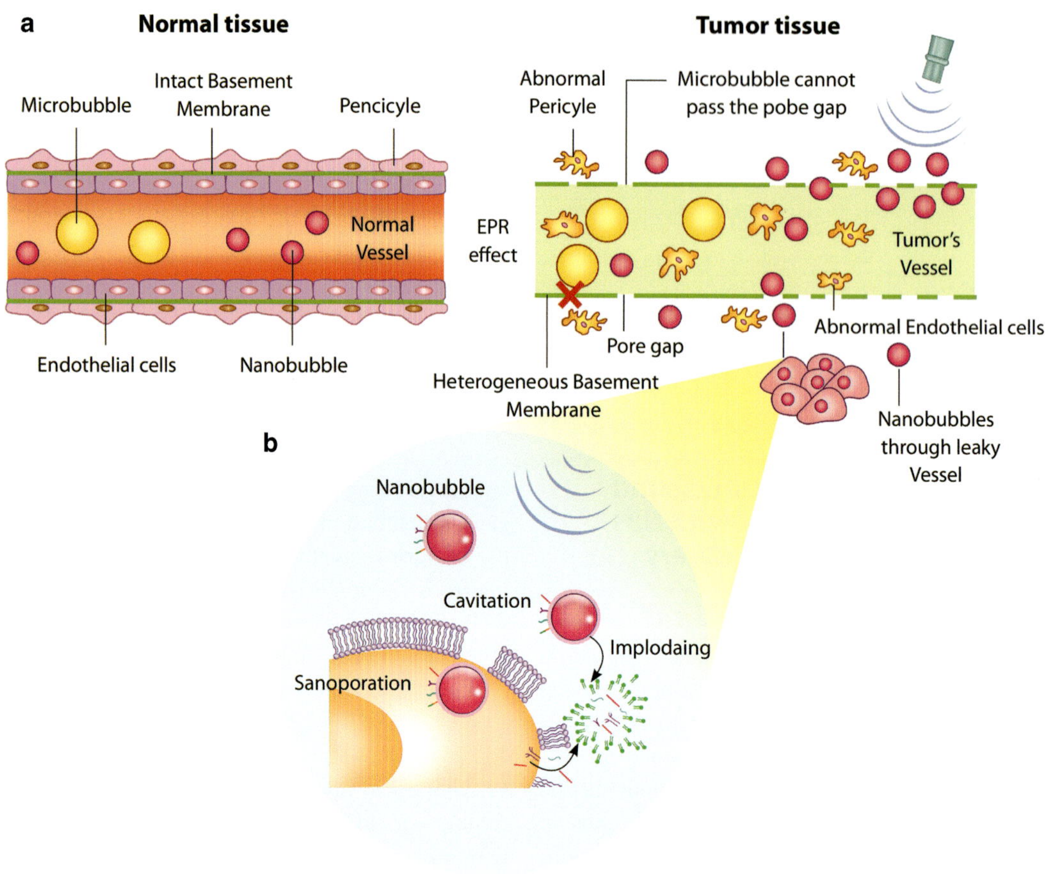

Fig. 25.1 (**a**) Normal tissue the nanobubbles and microbubbles confined in the blood vessel. (**b**) Shows the passage of the nanobubbles and microbubbles in the tumor cells and various mechanisms involved

Fig. 25.2 Perfluoropentane component in the liquid form used as droplet and in gas form used as bubble. The bubble showing different mechanism and uses

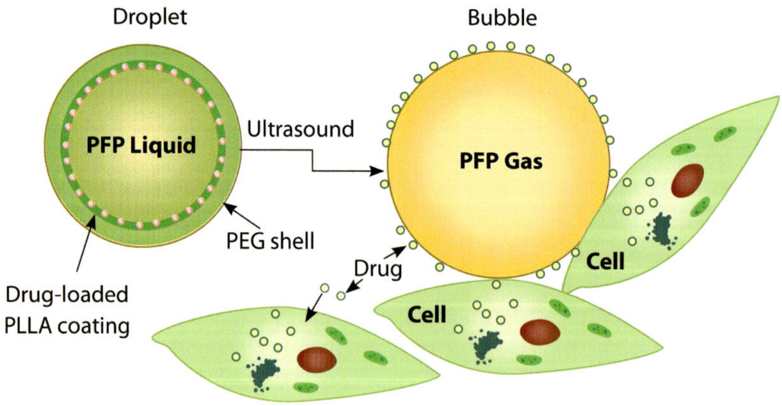

reported different properties of gold-type perunrepentant-core NBs, which have the diethylaminomethyl-dextran shell that encapsulates complex DNA. All these are capable of transfecting plasmid DNA and also defending the cargo [3] (Fig. 25.3). The use of different components like tradecylphosphoric acid, which is capable of increasing surface tension, and different surface modifications [11], along with the polyethyleneimine (PEI) polymer, can be used for enhanced transfection when used along with ultrasound (Fig. 6).

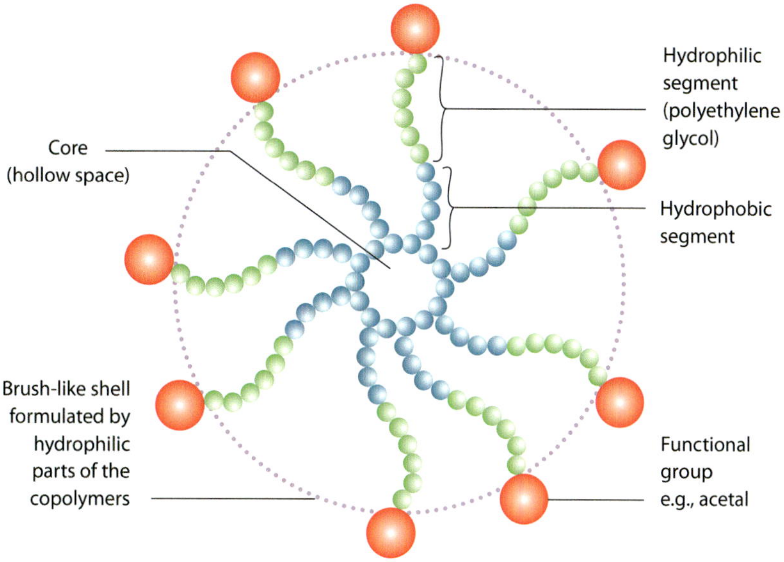

Fig. 25.3 Nanocore as functional group

25.2 Ultrasound Combined with NBs for RNA Delivery

One of the studies by Yin T. et al. [12] shows that NBs can be used along with siRNA for cancer therapy, causing a significant increase in the apoptosis of cancer cells. One example is the better therapeutic role of the glioma variant in nude rodents. This study by Yin T. et al. [13] has shown the use of US-sensitive NBs, which are being used as anticancer agents for siRNA and paclitaxel (PTX), targeting the BCL-2 gene for the management of hepatic tumors. The delivery of siRNA and PTX is achieved by the injection of NBs into the blood circulation in nude mice, with lower-frequency ultrasound enhancing the antitumor activity of these two agents. The use of prostate cancer, by Wang et al. [14], has designed the androgen receptor via the novel electrostatic approach. This will be used in combination with ultrasonic irradiation, with siRNA-loaded NBs repressing cell growth and protein production, and also inhibiting AR mRNA production. The activation of US irradiation can cause the release of CPP-siRNA from the NBs, which will allow better penetration into tumor cells. The role in cancer treatment is via the systemic administration of CPP-siRNA NBs in mice, which decreases c-Myc silencing and impedes cancer proliferation [15–18]. One of the studies by Wu et al. has shown that the [19] novel NB targeting (TNB) can be combined to prevent the interference of RNA (siRNA)–cyanine 5 (Cy5), showing their theranostic ability. The use of the NET-1 siRNA-conjugated TNB control in mice has shown a drastic reduction in tumor growth with increased survival rates. Li et al. [20] have shown that the non-viral shRNA gene delivery method, combining various techniques such as PEI, US, and NBs, is effective for survivors of liver cancer. The PEI-shRNA NBs are also a better method for shRNA transmission, more US-sensitive for siRNA NBs, and more efficient in RNA interference [1]. Different concerns exist regarding the safety of drug and gene delivery systems. These include various safety concerns of NBs, particularly their permeability through ultrasound contrast, as well as the negative impact of ultrasonic irradiation and NB toxicity in clinical applications. The advantage of NBs, as seen in MBs, lies in their different properties for drug targeting and gene delivery. There is a need for very high ultrasonic energy, which will be effective for better drug/gene release by creating cavitations in target tissues or cells [2, 21]. Sonoporation is important for the permeability of various drug deliveries through biological membranes. The use of bubble implosions is capable of inducing more cell permeability to therapeutic agents due to penetration through the cell membranes, and this can also result in biological tissue damage. This tissue damage, induced by ultrasound-mediated transfection, presents safety issues related to the therapeutic applications of this system, which can be mitigated. High-intensity ultrasound is also capable of causing the rapid collapse of bubbles, resulting in the very rapid release of drugs/genes loaded into the foam. This results in non-continuous release of the drugs. The use of ultrasound irradiation also results in ultrasonic heating and damage to biological tissues, where pores on the cell membrane can be created by ultrasonic cavitation, which can be large, impairing the resealing of the pores, and altering different parameters (Fig. 25.4). Ultrasound therapy uses different intensities for different therapeutic purposes [22]. The toxic side effects are also a topic of interest for future research. One study has described that [23] nanobubbles can be used as a secure option for the CCK-8 assay and post-nanobubble injection analysis. Nanobubbles made from phospholipids have very low toxicity and are safe for use in vitro and in vivo [24, 25]. Toxicity is a subject of safety assessment for future research to develop better low-toxic nanomaterials for future translational use [26] (Fig. 25.5).

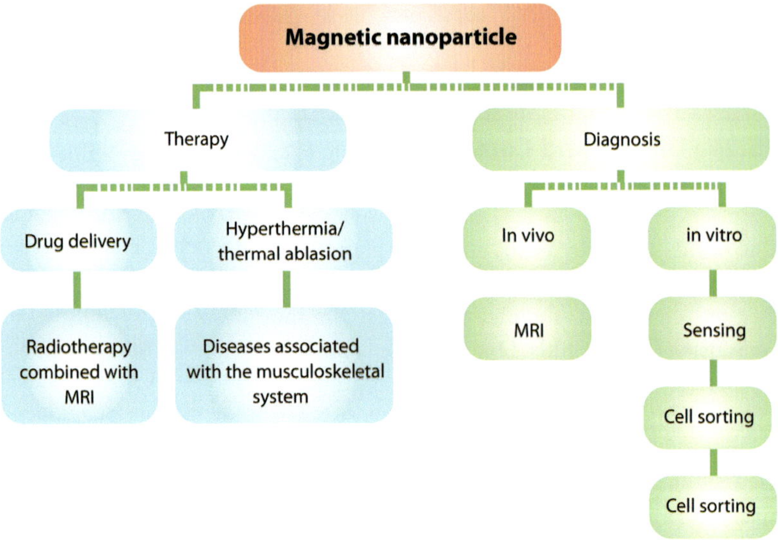

Fig. 25.4 Ultrasound nanoparticles and their different uses

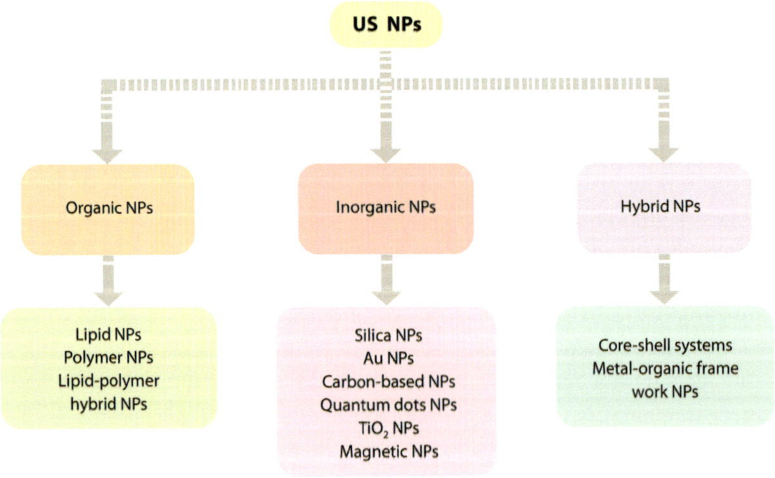

Fig. 25.5 Applications of nanoparticle use in ultrasound in biomedicine

25.3 Stability of NBs

The stability of NBs is an important factor, as most nanoscale bubbles are unstable due to the very high inner pressure at the solid–water interface. The Epstein–Plesset theory [27] states that bubbles smaller than 1000 nm should exist for <0.02 s in water. However, various studies have stated that they can exist for many hours or days [28]. These nanobubbles are capable of expansion and unstable, uncontrolled growth at a particular pressure or temperature. The study by Borkent et al. has shown that at low pressure (6 MPa), there will not be any transmission [29]. However, many studies show that when the temperature increases, causing the boiling of water, these nanoparticles will become stable [30]. Research in the development of ultrasonic probes will result in better ultrasonic imaging as well as improved ultrasonic irradiation, enabling better mediation of tumor-targeted therapy.

The optimization of ultrasonic parameters can result in better acoustic cavitation in the form of vapor bubbles in a liquid, which can collapse due to the acoustic pressure field. These cavitations are classified into two types: stable cavitation and inertial cavitation. Inertial cavitation is capable of generating broadband noise emissions as the bubbles undergo radial oscillations and collapse, which can violently damage tissues [31]. The use of ultrasound in fragment drug-loaded echogenic liposomes (ELIP) will target the more temporal peak in drug or therapeutic effects, leading to a systematic passive release and compromising the ultrasonic parameters. One of the important aspects is the vascular leakage and apoptosis happening in various vascular cells exposed to 1–1.5 MHz pulsed ultrasound, with peak-to-peak pressure amplitudes ranging from 2 to 36 MPa [32]. Another study showed that low-intensity continuous-wave (CW) ultrasound (0.49 MPa peak-to-peak pressure amplitude) resulted in increased delivery of liposome nanobubbles. This resulted in 1 MHz CW ultrasound leading to increased transport of liposome nanobubbles to the endothelium, resulting in stable cavitation [33, 34]. More research and studies need to be explored in the future.

25.4 Nanomaterials in Regenerative Medicine

The term "nanotechnology" refers to an important imaging technique that has resulted in better and smaller sizes in the range of a few hundred nanometers. Until now, we have seen that the use of nanotechnology has led to a wide array of applications for early disease detection, the development of new drugs, delivery devices, therapeutic options, and contrast and diagnostic agents, all of which are required in regenerative medicine. The concept of personalized medicine for individualized therapy and various stem cell applications is tied to the use of different biomaterials [35–39]. Recent advances in different nanocarriers for their applications in regenerative medicine can be seen. Different scaffolds will have a better and more precise design for manufacturing, as they are more porous structures capable of releasing based on different kinetic factors, which will be used for tissue regeneration [40]. By designing better scaffolds, a more biocompatible environment will be created, leading to increased cell infiltration for the reversal and restoration of lost tissues. Additionally, these scaffolds should be able to deliver fully functional tissues in a controlled manner, leading to better integration with the host tissue [41]. Different fibrous and particulate materials are needed for more precise scaffolds, or in combination with various nanocomposites, which will support better replacement. The use of porous scaffolds can be controlled by hierarchical structures, which are customized and also fabricated for core-corona nanoparticles made from ABA-type amphiphilic copolymers. The different carboxylic groups will be forming for better efficiency and can be linked to the core so that the matrix can be added to the matrix [42]. The bioactive, thermo-responsive, substance which can be the injectable scaffold will consist of chitosan–b-glycerophosphate salt formulations with bioactive glass nanoparticles that can be used for various applications in orthopedics [43]. The use of MR nanoparticles and human mesenchymal stem cells (MSCs) for regenerative medicine will show increased intracellular particle uptake when

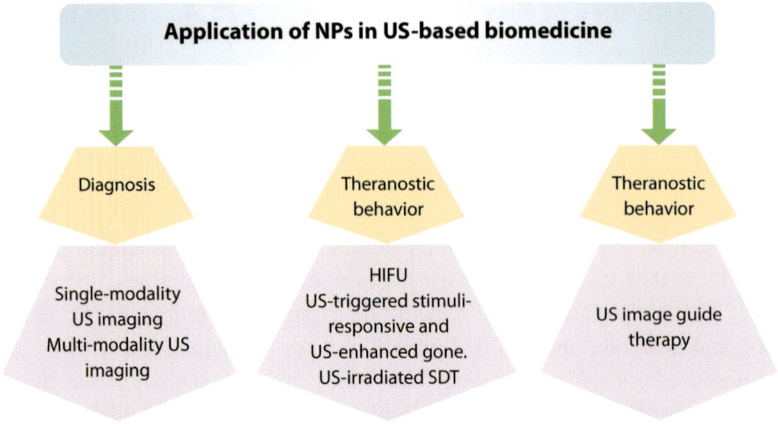

Fig. 25.6 Magnetic nanoparticles in therapy on left side and diagnosis on right side

combined with various cells, allowing for better and more precise examination potential (Fig. 25.6). These particles will have a greater affinity for the outer surface of the particles and toward various metals, allowing for better coatings on implants used for drug delivery or cell labeling [44]. CNTs are capable of selectively differentiating human embryonic stem cells (hESCs) into neuron cells, improving cell viability for neuronal injury compared to single lineage cell transplantation [45]. The smaller-sized MNPs in the range of 1–100 nm will exhibit paramagnetism and support the formation of better nanofibrous scaffolds, resulting in lower shear stresses at the cellular level, which stimulates growth, proliferation, and differentiation of preosteoblast MC3T3-E1 cells [46]. Tissue engineering will be used for cell growth and to identify various pathologies, such as deep calcification, mural thrombus, and tissue necrosis [47]. Various alignments of nanofibers will also impact cellular functions, differentiating between microfibers and individual fiber alignment [48–50]. Different techniques like plasma spraying [51] and plasma-enhanced radiofrequency magnetron sputtering deposition [52] are being explored as fabrication methods. Various nanomaterials for regenerative medicine can be used for highly cross-linked 3D tissue constructs, which impact dendrimer–gelatin nanofiber scaffolds used for delivering multiple functional entities and high payloads of drugs [53–55]. Dendrimers will form various multivalent transient inter-cellular links through the reaction of hydrazide groups for soft tissue engineering [56]. Nanofillers can also enhance cellular adhesion in ceramic-free counterparts [57, 58]. In one study, human osteoprogenitor cells were shown to use nanotopographies to control cell adhesion, cytoskeleton, growth, and production of osteoblastic markers such as osteocalcin and osteopontin in human bone marrow stromal cells [59]. One of the methods is the use of the Langmuir–Blodgett lithography in the nanoimprinting in the case of osteoblasts also for the differentiation in MSCs to the neuronal lineage which is being prepared by the soft lithography. The morphology and motility of smooth muscle cells can be changed on nanopatterned gratings, fabricated using nanoimprinting, which can depict better cell physiology in response to nanotopography by utilizing nanoscale features.

25.5 Conclusion

The future of nanoimaging is very promising, with many applications in diagnostics and therapeutic options, along with various advances in contrast agents.

References

1. Shim MS, Kwon YJ. Stimuli-responsive polymers and nanomaterials for gene delivery and imaging applications. Adv Drug Deliv Rev. 2012;64:1046–59.
2. Movahedi F, Hu RG, Becker DL, Xu C. Stimuli-responsive liposomes for the delivery of nucleic acid therapeutics. Nanomedicine. 2015;11:1575–84.
3. Sirsi SR, Borden MA. Advances in ultrasound mediated gene therapy using microbubble contrast agents. Theranostics. 2012;2:1208–22.

4. Suzuki R, et al. Tumor specific ultrasound enhanced gene transfer in vivo with novel liposomal bubbles. J Control Release. 2008;125(2):137–44.
5. Endo-Takahashi Y, et al. pDNA-loaded bubble liposomes as potential ultrasound imaging and gene delivery agents. Biomaterials. 2013;34(11):2807–13.
6. Horie S, et al. Evaluation of antitumor effects following tumor necrosis factor-a gene delivery using nanobubbles and ultrasound. Cancer Sci. 2011;102(11):2082–9.
7. Watanabe Y, et al. Delivery of Na/I symporter gene into skeletal muscle using nanobubbles and ultrasound: visualization of gene expression by PET. J Nucl Med. 2010;51(6):951–8.
8. Tayier B, et al. Biosynthetic nanobubbles for targeted gene delivery by focused ultrasound. Nanoscale. 2019;11(31):14757–68.
9. Yin T, et al. Ultrasound-sensitive siRNA-loaded nanobubbles formed by hetero-assembly of polymeric micelles and liposomes and their therapeutic effect in gliomas. Biomaterials. 2013;34(18):4532–43.
10. Yin T, et al. Tumor-penetrating codelivery of siRNA and paclitaxel with ultrasound-responsive nanobubbles hetero-assembled from polymeric micelles and liposomes. Biomaterials. 2014;35(22):5932–43.
11. Wang L, et al. Preparation of nanobubbles carrying androgen receptor siRNA and their inhibitory effects on androgen-independent prostate cancer when combined with ultrasonic irradiation. PLoS One. 2014;9(5):e96586.
12. Xie X, et al. Cell-penetrating peptide-siRNA conjugate loaded YSA-modified nanobubbles for ultrasound triggered siRNA delivery. Colloids Surf B Biointerfaces. 2015;136:641–50.
13. Zhou Z, et al. Synergistic effects of ultrasound-targeted microbubble destruction and TAT peptide on gene transfection: an experimental study in vitro and in vivo. J Control Release. 2013;170(3):437–44.
14. Zhu F, et al. Effectiveness of localized ultrasound-targeted microbubble destruction with doxorubicin liposomes in H22 mouse hepatocellular carcinoma model. J Drug Target. 2015;23(4):323–34.
15. Stride E. Physical principles of microbubbles for ultrasound imaging and therapy. Cerebrovasc Dis. 2009;27(2):1 13.
16. Wu B, et al. Preparation of novel targeting nanobubbles conjugated with small interfering RNA for concurrent molecular imaging and gene therapy in vivo. FASEB J. 2019;33(12):14129–36.
17. Li H, et al. Enhanced shRNA delivery by the combination of polyethylenimine, ultrasound, and nanobubbles in liver cancer. Technol Health Care. 2019;27:263–72.
18. Sirsi SR, Borden MA. State-of-the-art materials for ultrasound-triggered drug delivery. Adv Drug Deliv Rev. 2014;72:3–14.
19. Adhikari U, Goliaei A, Berkowitz ML. Nanobubbles, cavitation, shock waves and traumatic brain injury. Phys Chem Chem Phys. 2016;18(48):32638–52.
20. Zhang J, et al. The optimized fabrication of a novel nanobubble for tumor imaging. Front Pharmacol. 2019;31(10):610.
21. Hu C, et al. Ultrasound-mediated nanobubble destruction (UMND) facilitates the delivery of VEGFR2-targeted CD-TK-loaded cationic nanobubbles in the treatment of bladder cancer. J Cancer Res Clin Oncol. 2020;146(6):1415–26.
22. Gao S, et al. Lipid nanobubbles as an ultrasound-triggered artesunate delivery system for imaging-guided, tumortargeted chemotherapy. Onco Targets Ther. 2019;12:1841–50.
23. Cheng B, et al. Influence of nanobubble concentration on blood-brain barrier opening using focused ultrasound under real-time acoustic feedback control. Ultrasound Med Biol. 2019;45(8):2174–87.
24. Cosgrove D, Harvey C. Clinical uses of microbubbles in diagnosis and treatment. Med Biol Eng Comput. 2009;47:813–26.
25. Karpitschka S, et al. Nonintrusive optical visualization of surface nanobubbles. Phys Rev Lett. 2012;109(6):066102.
26. Borkent BM, et al. Superstability of surface nanobubbles. Phys Rev Lett. 2007;98(20):204502.
27. Zhang X, et al. Surface nanobubbles nucleate microdroplets. Phys Rev Lett. 2014;112(14):144503.
28. Holland CK, et al. Direct evidence of cavitation in vivo from diagnostic ultrasound. Ultrasound Med Biol. 1996;22(7):917–25.
29. Hwang JH, et al. Vascular effects induced by combined 1-MHz ultrasound and microbubble contrast agent treatments in vivo. Ultrasound Med Biol. 2005;31(4):553–64.
30. Hitchcock KE, et al. Ultrasound-enhanced delivery of targeted echogenic liposomes in a novel ex vivo mouse aorta model. J Control Release. 2010;144(3):288–95.
31. Liu D, Yang F, Xiong F, Gu N. The smart drug delivery system and its clinical potential. Theranostics. 2016;6:1306–23. https://doi.org/10.7150/thno.14858.
32. O'Neill BE, Vo H, Angstadt M, Li KP, Quinn T, Frenkel V. Pulsed high intensity focused ultrasound mediated nanoparticle delivery: mechanisms and efficacy in murine muscle. Ultrasound Med Biol. 2009;35:416–24.
33. Mura S, Nicolas J, Couvreur P. Stimuli-responsive nanocarriers for drug delivery. Nat Mater. 2013;12:991–1003.
34. El Kadi S, Porter TR, Verouden NJW, van Rossum AC, Kamp O. Contrast ultrasound, sonothrombolysis and sonoperfusion in cardiovascular disease: shifting to theragnostic clinical trials. JACC Cardiovasc Imaging. 2021;15:345–60.
35. Elkhodiry MA, Momah CC, Suwaidi SR, et al. Synergistic nanomedicine: passive, active, and ultrasound-triggered drug delivery in cancer treatment. J Nanosci Nanotechnol. 2016;16:1–18.
36. Canavese G, Ancona A, Racca L, et al. Nanoparticle-assisted ultrasound: a special focus on sonodynamic therapy against cancer. Chem Eng J. 2018;340:155–72.

37. Xing Z, Wang J, Ke H, et al. The fabrication of novel nanobubble ultrasound contrast agent for potential tumor imaging. Nanotechnology. 2010;21:145607.
38. Willerth SM, Sakiyama-Elbert SE. Approaches to neural tissue engineering using scaffolds for drug delivery. Adv Drug Deliv Rev. 2007;59(4–5):325–38.
39. Yao X, Yao H, Li Y, Chen G. Preparation of honeycomb scaffold with hierarchical porous structures by core-crosslinked core-corona nanoparticles. J Colloid Interface Sci. 2009;332(1):165–72.
40. Couto DS, Hong Z, Mano JF. Development of bioactive and biodegradable chitosan-based injectable systems containing bioactive glass nanoparticles. Acta Biomater. 2009;5(1):115–23.
41. Tautzenberger A, Lorenz S, Kreja L, et al. Effect of functionalised fluorescence-labelled nanoparticles on mesenchymal stem cell differentiation. Biomaterials. 2010;31(8):2064–71.
42. Chao T-I, Xiang S, Chen C-S, et al. Carbon nanotubes promote neuron differentiation from human embryonic stem cells. Biochem Biophys Res Commun. 2009;384(4):426–30.
43. Kannarkat JT, Battogtokh J, Philip J, Wilson OC, Mehl PM. Embedding of magnetic nanoparticles in polycaprolactone nanofiber scaffolds to facilitate bone healing and regeneration. J Appl Phys. 2010;107(9, Pt 2):09B307/301–3.
44. Kalfa D, Bel A, Chen-Tournoux A, et al. A polydioxanone electrospun valved patch to replace the right ventricular outflow tract in a growing lamb model. Biomaterials. 2010;31(14):4056–63.
45. Yang F, Murugan R, Wang S, Ramakrishna S. Electrospinning of nano/micro scale poly(L-lactic acid) aligned fibers and their potential in neural tissue engineering. Biomaterials. 2005;26(15): 2603–10.
46. Meng J, Han Z, Kong H, et al. Electrospun aligned nanofibrous composite of MWCNT/polyurethane to enhance vascular endothelium cells proliferation and function. J Biomed Mater Res Part A. 2010;95A(1):312–20.
47. Edwards SL, Church JS, Werkmeister JA, Ramshaw JA. Tubular micro-scale multiwalled carbon nanotube-based scaffolds for tissue engineering. Biomaterials. 2009;30(9):1725–31.
48. Balani K, Anderson R, Laha T, et al. Plasma-sprayed carbon nanotube reinforced hydroxyapatite coatings and their interaction with human osteoblasts in vitro. Biomaterials. 2007;28(4):618–24.
49. Niu L, Kua H, Chua DHC. Bonelike apatite formation utilizing carbon nanotubes as template. Langmuir. 2009;26(6):4069–73.
50. Smith-Freshwater AP, Bowlin GL, Yang H. A novel electrospun dendrimer-gelatin hybrid nanofiber scaffold for tissue regeneration and drug delivery. Mater Res Soc Symp Proc. 2008; 1094E:1094-DD1009–1007.
51. Duan X, Sheardown H. Dendrimer crosslinked collagen as a corneal tissue engineering scaffold: mechanical properties and corneal epithelial cell interactions. Biomaterials. 2006;27(26):4608–17.
52. Duan X, McLaughlin C, Griffith M, Sheardown H. Biofunctionalization of collagen for improved biological response: scaffolds for corneal tissue engineering. Biomaterials. 2007;28(1):78–88.
53. Zhao D, Ong S-M, Yue Z, et al. Dendrimer hydrazides as multivalent transient intercellular linkers. Biomaterials. 2008;29(27):3693–702.
54. Mandoli C, Pagliari F, Pagliari S, et al. Stem cell aligned growth induced by CeO2 nanoparticles in PLGA scaffolds with improved bioactivity for regenerative medicine. Adv Funct Mater. 2010;20(10):1617–24.
55. Verma S, Kumar N. Effect of biomimetic 3D environment of an injectable polymeric scaffold on MG-63 osteoblasticcell response. Mater Sci Eng C. 2010;30:1118–28.
56. Dalby MJ, McCloy D, Robertson M, Wilkinson CD, Oreffo RO. Osteoprogenitor response to defined topographies with nanoscale depths. Biomaterials. 2006;27(8):1306–15.
57. Lenhert S, Meier MB, Meyer U, Chi L, Wiesmann HP. Osteoblast alignment, elongation and migration on grooved polystyrene surfaces patterned by LangmuirBlodgett lithography. Biomaterials. 2005;26(5):563–70.
58. Yim EK, Pang SW, Leong KW. Synthetic nanostructures inducing differentiation of human mesenchymal stem cells into neuronal lineage. Exp Cell Res. 2007;313(9):1820–9.
59. Yim EK, Reano RM, Pang SW, Yee AF, Chen CS, Leong KW. Nanopattern-induced changes in morphology and motility of smooth muscle cells. Biomaterials. 2005;26(26):5405–13.

26 Nanorobots

26.1 Introduction

Nanorobots are defined as controllable machines that are in the size range of the nanometer scale [1] (Fig. 26.1). The use of nanorobotics in various medical applications is based on various functions, and one of the functions is the use of antibodies which are capable of killing or repairing diseased cells. As these nanoparticles (NPs) are compatible with various human cells, they will have a more varied structure and concept which can be used for interacting with the cells [2–4]. Nanorobots can also be programmed to sense different target-specific cells, which can be used for the drug molecules release when the signal is released. These will also work on the concept of the various curative and reconstructive management of the human body at the cellular and subcellular levels for evaluation of the various neoplasms, hepatitis, diabetes, and other diseases. This can also be used for the evaluation of the different materials for the survival of the patients. These are parts of the nanorobotic devices that will be more useful for doing surgeries, as well as the management of the pathologies, and can have very ultra-high accuracy for the localization in drug delivery and further prevent minimal side effects (Fig. 26.2).

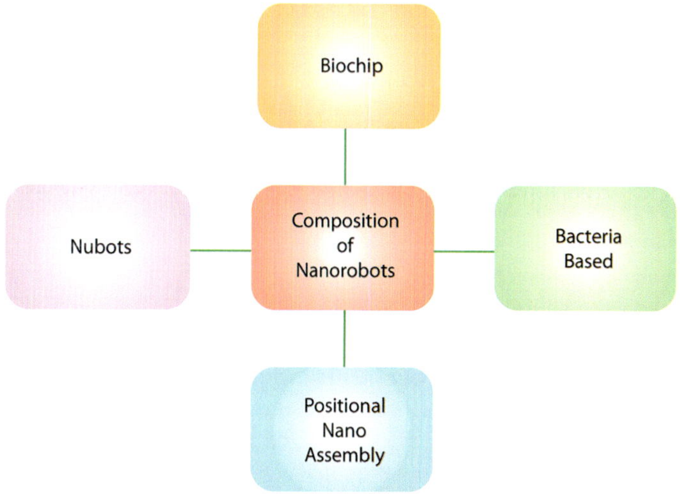

Fig. 26.1 Nanorobots different components and compositions

S. Shaikh, *Nanoimaging - Future of Precision Medicine*,
https://doi.org/10.1007/978-981-97-7292-6_26

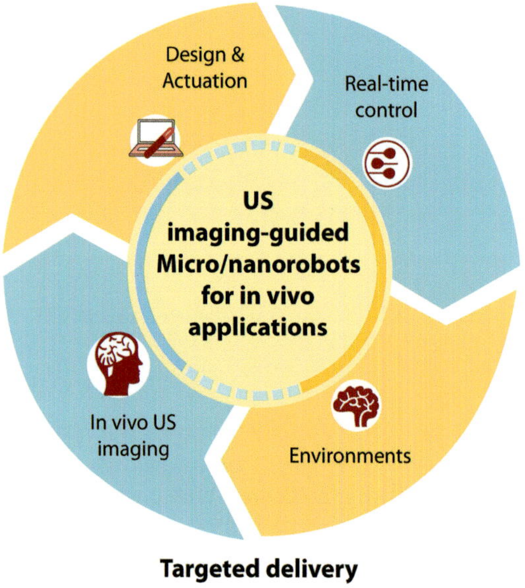

Fig. 26.2 Ultrasound-guided microrobots and nanorobots for targeted delivery

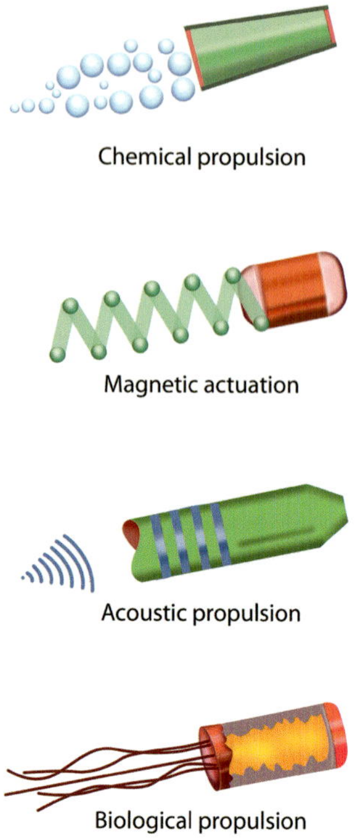

Fig. 26.3 Different mechanisms of the nanorobots: chemical, magnetic, acoustic, and biological applications

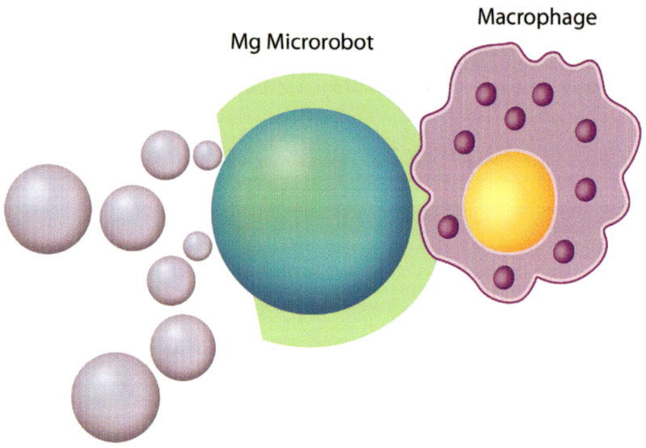

Fig. 26.4 Structure of magnesium robot aligning with macrophages

The use of various nanorobotic systems is very small in size and can be modified into different sizes and act like nanomanipulators [5]. The use of NPs will have a more focused role in biomedicine (Fig. 26.3). The future concept of medicine is personalized precision medicine, and they have a bigger role to play [6]. A couple of the futuristic roles will allow for more accurate and comprehensive diagnosis of diseases. The single level with the pathological landmarks used for the single-molecule level disease, along with the biomolecules, can control the different robotic manipulations. With this, the concept of personalized precision medicine will have a better future due to the development of medical nanorobotics [7–9]. Various nanorobotic devices, including nanorobots and nanomanipulators [10–12], are made up of resin as they have many different biomedical applications and can address various issues in biomedical applications. Nanorobots, as their size is very small, can be used as nanomanipulators and can perform different robotic manipulations based on the nano-sized objects. Cancer is a very important entity that can be used for multifaceted applications like nanomolecules [13–15], which will use various nanorobotic devices like the various probes with multidimensional and multiparametric capabilities, which can be used to diagnose pathological situations of various diseases. These nanorobots and nanomanipulators will be more useful as there will be more interdisciplinary biomedical applications (Fig. 26.4). Various nanorobotic devices [16–22] can be used for various biomedical applications comprising molecular machines, nanomotors,

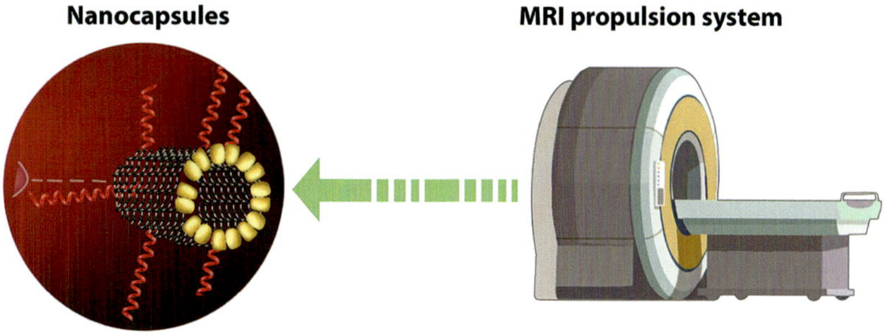

Fig. 26.5 MRI propulsion system targeting the nanocapsule

DNA nanorobotics, and nanomanipulation systems (Fig. 26.5). These will result in the use of various atomic force microscopy (AFM) systems, and also electron microscopy (EM) nanomanipulation systems. This research in nanorobotic systems will have a significant impact on the way the concept of practice in personalized precision medicine evolves.

26.2 DNA Nanorobotics

DNA technology is one of the important aspects in the evaluation of nanorobotic applications (Fig. 26.6). These DNA materials and DNA molecules are capable of making their indigenous predesigned shapes by various processes [23]. These nanorobots can be magneto-robots, which can be associated with their construction by DNA techniques that can be used for various nanorobotic applications. One of the important entities is DNA origami, which was made in 2006 [24]. This is again based on DNA-based technology, which will be better in the use of various complementary "staple" oligonucleotides in the "scaffolds" for the DNA, as well as the 2D and 3D shapes, which will be stabilized in the presence of various thousands of base pairs [25] (Fig. 26.7). This DNA origami can be added along with the self-assembly of various nanometers in the range of micrometer scales, which will be in the giga-Dalton scale [26]. DNA origami is the component that will have many functions for both structural and dynamic nanodevices, which can be important in the formation of nanorobots [27]. In one of the studies by Lund et al. [28], they showed that the robotic behaviors can be based on various substrate molecules of the two-dimensional DNA origami landscape and can also be planned based on DNA technology. In 2018, Kopperger et al. [29] showed a nanoscale robotic arm with electric fields by using DNA origami technology. Various robotic movements can be analyzed and adjusted as per their speed, which can be based on various DNA motor systems. These DNA nanorobots can be added with 2D DNA crystalline particles, which will be seen in different locations [30], and are also capable of better pick-up, transport, and drop-off of various cargos [31]. These will have [28–31] better DNA origami technologies, which can be used for various nanorobotic operations. This DNA origami can also fabricate various DNA devices that can perform various functions like the generation, transmission, and sensing of various types of mechanical forces in the region of the nanoscale [32], like probing various molecular forces along with DNA origami-based components [33]. This can be useful for mapping the 3D orientation in different living cells [34], for guiding various activities of the DNA. These can be used for DNA nanorobotics in various biomedical applications, like various processes such as delivery. These DNA nanorobotic structures can be used for various components of molecular triggers [35]. These DNA nanorobotic systems have been experimentally demonstrated [36, 37] for the use of smart cancer therapy. In 2012, Douglas et al. [38] showed one of the autonomous DNA nanorobots that can be capable of transporting molecular payloads to cells, causing various sensing cells along with surface inputs for various payload deliveries. These nanorobots are also seen in the range of 35 nm × 35 nm × 45 nm on DNA origami technology. These aptamers are being used for various target molecules on the cells,

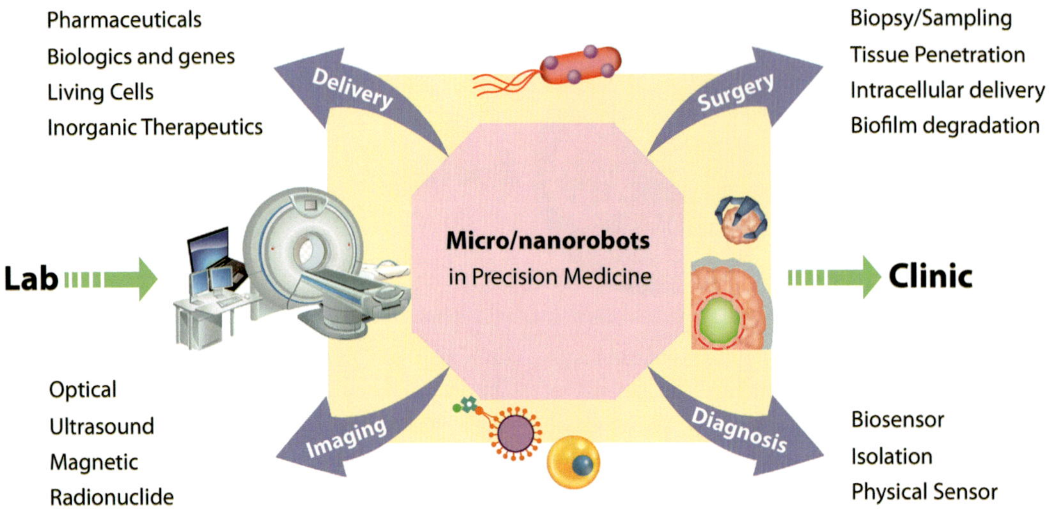

Fig. 26.6 Different applications in microrobots and nanorobots

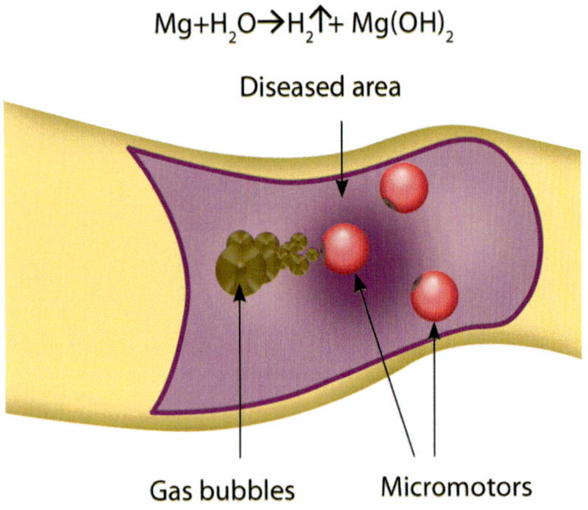

Fig. 26.7 Gas bubble-driven micromotor in the disease area

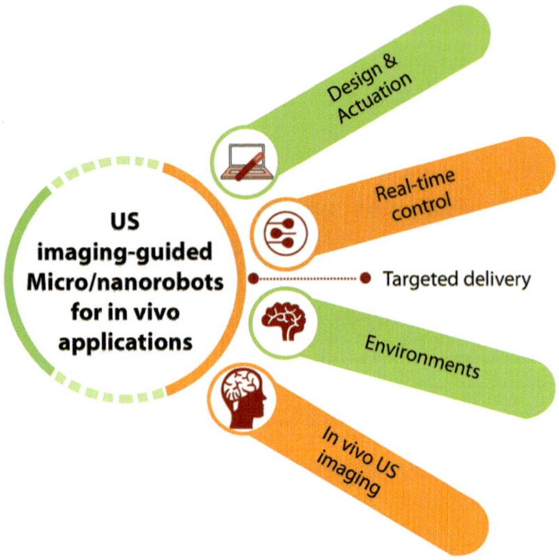

Fig. 26.8 Ultrasound-guided different in vivo imaging

capable of delivering the nanorobots to the target cells. These will show the different prototypes that are effective against cancer cells in vitro. In 2018, Li et al. [39] used in vivo biomedical applications of DNA nanorobots prepared based on different self-assembled DNA origami nanotubes, which have more multiple functional elements. The use of thrombin molecules can be contained in the nanorobots, and this has a significant role as this nanorobot will be saving the thrombin until there is better interaction with the tumor vessel marker nucleolin, which can have small sites that will lead the thrombin to tumor sites. Another component of DNA nanorobots is the capability of delivering tumor-associated blood vessels by ultrasound, which can elicit intravascular thrombosis and cause tumor necrosis and inhibition of tumor growth (Fig. 26.8). Thus, the nanorobots are typically safer and have no reaction in animal models. One of the important researchers has seen that [40], and this will cause the DNA nanorobots. DNA

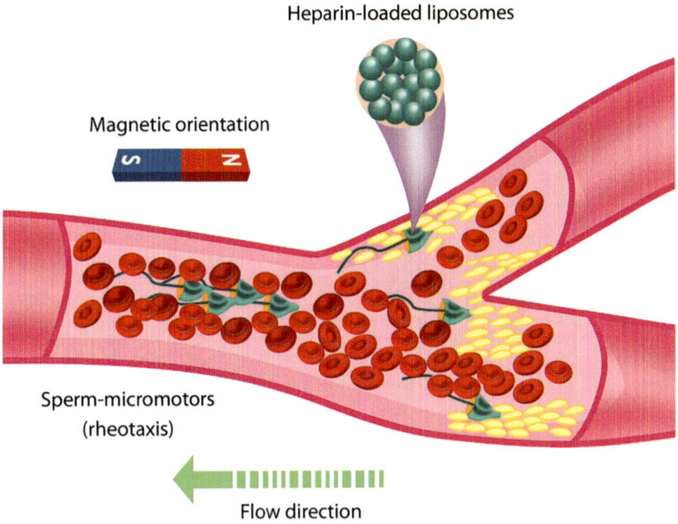

Fig. 26.9 Use of micrometer in the sperm evaluation in a magnetic field

nanorobots are capable of having the best source of therapeutic agents, which are being used for the treatment of renal diseases [38–40] (Fig. 26.9).

26.3 Nanorobotics in Therapeutics

This is becoming one of the important novel therapeutic approaches for various infectious diseases, and the tumor is one of the most commonly used (Fig. 26.10). NPs are also being used for drug delivery to tumor cells. The different biological barriers are commonly used for the [41] NP translation [42]. Different DNA nanorobotics are being used due to their complexity in using three-dimensional robotic tasks, preferably for delivering the drug molecules to the target tumor sites [43], and various animal imaging has shown the different efficacies of these nanorobots in the management of tumors [44]. Due to this, they are the futuristic approach in tumor therapy by nanorobots. Different biomolecules will be used by the nanorobots, which will be the guide and capable of targeting immune cells, leading to the immune response activation [45], thus leading to the development of cancer immunotherapy in a bigger way. Thus, the role of nanorobotics has represented a bigger way for the newer biomedical applications, which will play an important role in the concept of personalized precision medicine. This nanorobotics has come in a very big way, leading to more references [46–52], bigger imaging techniques for nanorobots [46], the navigation platforms for nanorobotic agents [47], in vivo applications for various toxicology topics [48], magnetic resonance imaging (MRI)-based nanorobotics [49], the use of theranostics in cancer by nanorobots [50], nanorobots applications [51], biodistribution of nanorobots in the use of in vivo applications [52]. The newer applications are the better drug carriers, and these are based on specific molecular recognition interactions [44, 53]. The different nanorobots can be active and passive processes; this passive process is also uncontrollable and relies on blood flow and different molecular-specific binding interactions [54]. These can have more specific molecules that can be coated on nanorobots and have the targeting mechanisms as molecules acting on the tumor cells. Presently available nanorobots are being used for in vivo drug delivery. The mechanism is that there will be binding of the nanorobots to the tumors, to which the drugs will be encapsulated into the nanorobots. After this, they are capable of killing the tumor cells, which can be degraded and will be cleared into the immune system [44, 53]. Another way is that the nanorobots will behave like pharmaceutical agents [55].

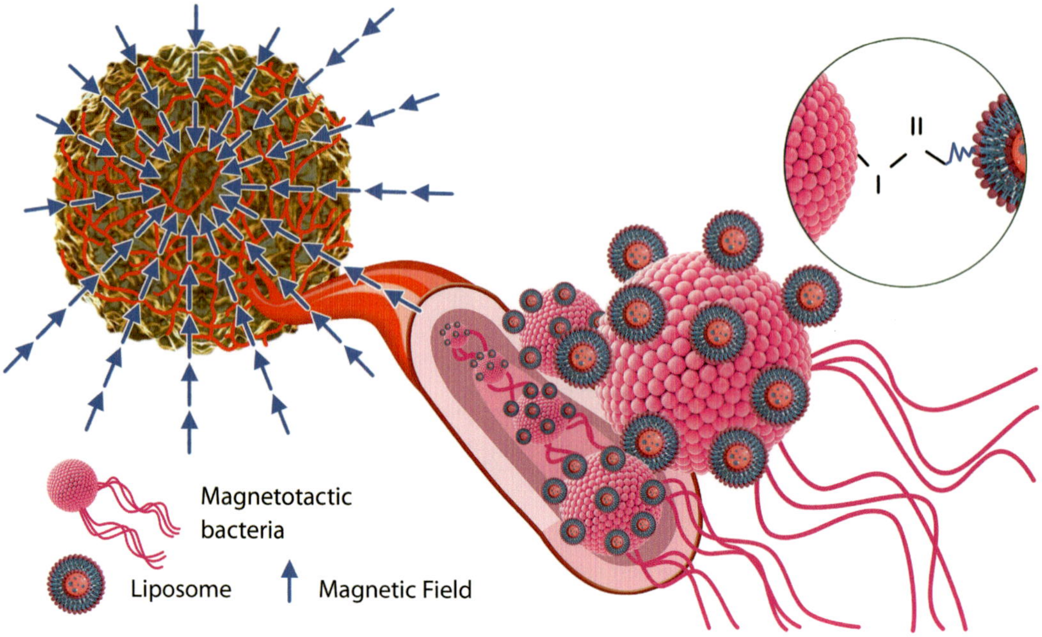

Fig. 26.10 The magnetotactic bacteria use in the liposomes and in a magnetic field

Research is also going on for the reuse of the nanorobots, leading to better efficiency as well as the intelligence of nanorobots, which will be used for various biomedical applications. The various newer developing novel agents are also being used in various physical treatments [56]; these will have different nanorobots used for the degrading of various tumors at the single-cell level. These nanorobots are also used for the management of different diseases, with the use of different nanorobots in various biomedical applications. Various microrobots can also be seen [57, 58], having better and more specific transformations in various biomedical applications. Robotic nanomanipulations based on nanomanipulators are an important entity helpful for the diagnostics and management of various cancers. The various drug responses are again based on the type of involvement of the disease [59, 60]. Still, there are many older methods used for the detection of various cellular biochemical processes [61], along with the pathological changes leading to the changes in the cells [62]. The different aspects are that there will be the physics of the cells, which will be more beneficial for the evaluation of various processes [63]. These nanomanipulators can be used for the better identification of various uses for the better analysis of precision medicine. The biological samples of various patients can be made label-free [64] for the evaluation of all these samples. With the availability of various automated hybrid nanorobotic systems, they will have different nanomanipulators, which will have better advantages for rapidly isolating the patient samples [65] and also possible to have changes in the various biological samples [66]. The AFM nanomanipulators will be used for having better structures and also the properties for various samples in higher spatial resolution [67]. Everything depends on the biological behavior of the various cancers. Different samples can be used for the evaluation of various cancerous cells [68], different immune cells [69], exosomes [70], and different hybrid nanorobotic systems, along with various biomedical applications in the evaluation of biomedicine and robotics, with the use of artificial intelligence in this.

26.4 Applications of Nanorobots

Nanorobots are being used in various applications:

1. Nanorobots in diabetes evaluation
2. Nanorobots in gene therapy
3. Nanorobots in cancer detection and treatment
4. Nanorobots in surgery

26.4.1 Nanorobots in Diabetes

Nanorobots in diabetes are being used for the diagnosis and management of diabetes. Glucose metabolism is the important metabolism for the diagnosis and treatment of diabetes. The use of protein hSGLT3 can be used for the various GI (Gastrointestinal) cholinergic nerve and skeletal muscle functions and can be able to see the concentration inside the cell. This can be an important parameter for the evaluation of diabetes patients. The use of this protein is to identify glucose and glucose metabolism. The nanorobot prototype model is being used in nanobioelectronics. This will be useful for the visibility of various immune reactions in identifying glucose levels in the blood. These nanorobots are being used as chemo sensors in glucose monitoring, which can act as chemical sensors. The image of the blood vessel can be used to see the RBCs (red blood cells) and nanorobots. These nanorobots can be used to track glucose levels. This quantification can detect RF signals and communicate if any gross changes are seen in the glucose levels, which will elicit an alarm on the mobile phone when it is attached.

26.4.2 Nanorobots in Gene Therapy

The use of nanorobots in the evaluation of DNA and proteins will be seen in particular reference for the treatment of various genetic diseases. Chromosomal replacement therapy will be useful for the rectification and repair of genetic maintenance. One of the important aspects will be the coil of DNA in the lower pair of the robot arms. The identification of the upper arms can be used as the source of the regulation of these molecular structures of both DNA and proteins, especially when there is a very large database required for the nanocomputer. The [71] structure of the chain of DNA can also be for 50 nm in the repairing of vessels as compared to various viruses and bacteria. The molecular-level evaluation for various viral infections, arteriosclerosis, and different cancers can be evaluated for the molecular malfunction at the cellular level and cell function for gene expression and protein synthesis. Different genetic materials of the cells can be used in various viral bacteriophage systems, different dendrimers, chemicals, and also different viral carriers. The use of gene clusters and gene copies along partial trisomies and polysomies can be used for the expression of various antisense transcription.

26.4.3 Nanorobots in Cancer

This is used for the detection and management of various cancers and metastasis. This can be useful for targeted drug delivery to reduce the side effects after chemotherapy. These nanorobots are being used for various applications like chemical biosensors, which can identify tumor cells in the very early stages. Another important application is the use of various therapies in cancer, having different hardware architectures as well as for nanobioelectronics in surgery. The surgical nanorobots can be inserted into the body by various applications. These nanorobots can be used for better programming of the various functions, like identifying the pathology and later diagnosing and characterizing the lesions. All this nanomanipulation will be coordinated in the computer, which can use the ultrasound signals that are coded [72].

26.5 Conclusion

The use of robots already has a major impact on various spheres of our lives. But with the use of nanotechnology and imaging, this will further revolutionize the imaging and therapeutic aspects under imaging guidance.

References

1. Ummat A, Dubey A, Mavroidis C. Bio-nanorobotics: a field inspired by nature. In: Bar-Cohen Y, editor. Biomimetics: biologically inspired technologies. Boca Raton London New York: CRC Taylor & Francis; 2006. p. 2.
2. Patel SS, Patel PN. A brief review on nanorobotics applications in medicine and future prospects. Asian J Res Pharm Sci. 2023;13:198. https://doi.org/10.52711/2231-5659.2023.00004.
3. Dang Y, Guan J. Nanoparticle-based drug delivery systems for cancer therapy. Smart Mater Med. 2020;1:10–9. https://doi.org/10.1016/j.smaim.2020.04.001.
4. Gupta RB, Kompella UB. Nanoparticle technology for drug delivery. New York: Taylor & Francis; 2006.
5. Oldham K, et al. Focused issue on micro-/nanorobotics. Int J Intell Robot Appl. 2018;2(4):381–2.
6. Collins FS, Varmus H. A new initiative on precision medicine. N Engl J Med. 2015;372(9):793–5.
7. Copeland RA. The drug-target residence time model: a 10-year retrospective. Nat Rev Drug Discov. 2016;15(2):87–95.
8. Leemans CR, et al. The molecular landscape of head and neck cancer. Nat Rev Cancer. 2018;18(5):269–82.
9. Fernandez-Leiro R, Scheres SHW. Unravelling biological macromolecules with cryo-electron microscopy. Nature. 2016;537(7620):339–46.
10. Dong L, Nelson BJ. Tutorial-robotics in the small part II: nanorobotics. IEEE Robot Autom Mag. 2007;14(3):111–21.
11. Hu C, et al. Soft micro- and nanorobotics. Annu Rev Control Robot Auton Syst. 2018;1:53–75.
12. Halder A, Sun Y. Biocompatible propulsion for biomedical micro/nanorobotics. Biosens Bioelectron. 2019;139:111334.
13. Quail DF, Joyce JA. Microenvironmental regulation of tumor progression and metastasis. Nat Med. 2013;19(11):1423–37.
14. Burrell RA, et al. The causes and consequences of genetic heterogeneity in cancer evolution. Nature. 2013;501(7467):338–45.
15. Anderson RL, et al. A framework for the development of effective anti-metastatic agents. Nat Rev Clin Oncol. 2019;16(3):185–204.
16. Pezzato C, et al. Mastering the non-equilibrium assembly and operation of molecular machines. Chem Soc Rev. 2017;46(18):5491–507.
17. Ellis E, et al. Artificial molecular and nanostructures for advanced nanomachinery. Chem Commun. 2018;54(33):4075–90.
18. Zhang L, et al. Molecular machines with bio-inspired mechanisms. Proc Natl Acad Sci USA. 2018;115(38):9397–404.
19. Xu T, et al. Fuel-free synthetic micro-/nanomachines. Adv Mater. 2017;29(9):1603250.
20. Toebes BJ, et al. Spatial control over catalyst positioning on biodegradable polymeric nanomotors. Nat Commun. 2019;10(1):5308.
21. Li S, et al. Anticancer activities of tumor-killing nanorobots. Trends Biotechnol. 2019;37(6):573–7.
22. Sitti M. Microscale and nanoscale robotics systems. IEEE Robot Autom Mag. 2007;14(1):53–60.
23. Linko V, et al. DNA nanostructures as smart drug-delivery vehicles and molecular devices. Trends Biotechnol. 2015;33(10):586–94.
24. Rothemund PWK. Folding DNA to create nanoscale shapes and patterns. Nature. 2006;440(7082):297–302.
25. Kearney CJ, et al. DNA origami: folded DNA-nanodevices that can direct and interpret cell behavior. Adv Mater. 2016;28(27):5509–24.
26. Schneider F, et al. The sequence of events during folding of a DNA origami. Sci Adv. 2019;5(5):eaaw1412.
27. Zhang F, et al. Structural DNA nanotechnology: state of the art and future perspective. J Am Chem Soc. 2014;136(32):11198–211.
28. Lund K, et al. Molecular robots guided by prescriptive landscapes. Nature. 2010;465(7295):206–10.
29. Kopperger E, et al. A self-assembled nanoscale robotic arm controlled by electric fields. Science. 2018;359(6373):296–301.
30. Ding B, Seeman NC. Operation of a DNA robot arm inserted into a 2D DNA crystalline substrate. Science. 2006;314(5805):1583–5.
31. Thubagere AJ, et al. A cargo-sorting DNA robot. Science. 2017;357(6356):eaan6558.
32. Blanchard AT, Salaita K. Emerging uses of DNA mechanical devices. Science. 2019;365(6458):1080–1.
33. Nickels PC, et al. Molecular force spectroscopy with a DNA origami-based nanoscopic force clamp. Science. 2016;354(6310):305–7.
34. Brockman JM, et al. Mapping the 3D orientation of piconewton integrin traction forces. Nat Methods. 2018;15(2):115–8.
35. Singh HR, et al. A DNA nanorobot uprises against cancer. Trends Mol Med. 2018;24(7):591–3.
36. Tasciotti E. Smart cancer therapy with DNA origami. Nat Biotechnol. 2018;36(3):234–5.
37. Bradley CA. Drug delivery: DNA nanorobots - seek and destroy. Nat Rev Cancer. 2018;18(4):208.
38. Douglas SM, et al. A logic-gated nanorobot for targeted transport of molecular payloads. Science. 2012;335(6070):831–4.
39. Li S, et al. A DNA nanorobot functions as a cancer therapeutic in response to a molecular trigger in vivo. Nat Biotechnol. 2018;36(3):258–64.
40. Jiang D, et al. DNA origami nanostructures can exhibit preferential renal uptake and alleviate acute kidney injury. Nat Biomed Eng. 2018;2(11):865–77.
41. Petros RA, DeSimone JM. Strategies in the design of nanoparticles for therapeutic applications. Nat Rev Drug Discov. 2010;9(8):615–27.
42. Wilhelm S, et al. Analysis of nanoparticle delivery to tumors. Nat Rev Mater. 2016;1:16014.
43. Vartholomeos P, Fruchard M, Ferreira A, Mavroidis C. MRI-guided nanorobotic systems for therapeutic and diagnostic applications. Annu Rev Biomed Eng. 2011;13:157–84.

44. Sharma NN, Mittal RK. Nanorobot movement: challenges and biologically inspired solutions. Int J Smart Sens Intell Syst. 2017;1:87–109.
45. Tapalian SL, et al. Mechanism-driven biomarkers to guide immune checkpoint blockade in cancer therapy. Nat Rev Cancer. 2016;16(5):275–87.
46. Pane S, et al. Imaging technologies for biomedical micro- and nano swimmers. Adv Mater Technol. 2019;4(4):1800575.
47. Martel S. Swimming microorganisms acting as nanorobots versus artificial nanorobotic agents: a perspective view from a historical retrospective on the future of medical nanorobotics in the largest known three-dimensional biomicrofouidic networks. Biomicrofluidics. 2016;10(2):021301.
48. Soto F, Chrostowski R. Frontiers of medical micro/nanorobotics: in vivo applications and commercialization perspectives toward clinical uses. Front Bioeng Biotechnol. 2018;6:170.
49. Dahmen C, et al. Techniques for MRI-based nanorobotics. In: Nanorobotics. Berlin: Springer-Verlag; 2013. p. 301–22.
50. Cerofolini GF, Amato P. Sensing strategies for early diagnosis of cancer by swarm of nanorobots: an evidential paradigm. In: Nanorobotics. Berlin: Springer-Verlag; 2013. p. 331–52.
51. Li J, et al. Enteric micromotor can selectively position and spontaneously propel in the gastrointestinal tract. ACS Nano. 2016;10(10):9536–42.
52. Gao W, et al. Artificial micromotors in the mouse's stomach: a step toward in vivo use of synthetic motors. ACS Nano. 2015;9(1):117–23.
53. Karan S, Majumder DD, Chaudhuri S. Biological response modifier-a nanorobotics control system design for immunotherapy in cancer treatment. In: 1st International Conference on Recent Advances in Information Technology (RAIT). New York, NY: IEEE; 2012. p. 645–50.
54. Kiyoshi M, et al. Affinity improvement of a therapeutic antibody by structure-based computational design: generation of electrostatic interactions in the transition state stabilizes the antibody-antigen complex. PLoS One. 2014;9(1):e87099.
55. Nelson BJ, et al. Microrobots for minimally invasive medicine. Annu Rev Biomed Eng. 2010;12:55–85.
56. Mitchell MJ, et al. Engineering and physical sciences in oncology: challenges and opportunities. Nat Rev Cancer. 2017;17(11):659–75.
57. Li S, et al. Particle robotics based on statistical mechanics of loosely coupled components. Nature. 2019;567(7748):361–5.
58. Xie H, et al. Reconfigurable magnetic microrobot swarm: multimode transformation, locomotion, and manipulation. Sci Robot. 2019;4(28):eaav8006.
59. Pirmohamed M. Personalized pharmacogenomics: predicting efficacy and adverse drug reactions. Annu Rev Genomics Hum Genet. 2014;15:349–70.
60. Sundin I, et al. Improving genomics-based predictions for precision medicine through active elicitation of expert knowledge. Bioinformatics. 2018;34(13):i395–403.
61. Volm M, Efferth T. Prediction of cancer drug resistance and implications for personalized medicine. Front Oncol. 2015;5:282.
62. Suresh S. Biomechanics and biophysics of cancer cells. Acta Biomater. 2007;3(4):413–38.
63. Diz-Munoz A, et al. In pursuit of the mechanics that shape cell surfaces. Nat Phys. 2018;14(7):648–52.
64. Voorde JV, et al. Improving the metabolic fidelity of cancer models with a physiological cell culture medium. Sci Adv. 2019;5(1):eaau7314.
65. da Silva Luz GV, Barros KVG, de Araújo FVC, et al. Nanorobotics in drug delivery systems for treatment of cancer: a review. J Mat Sci Eng A. 2016;6:167–80.
66. Soto F, Wang J, Ahmed R, Demirci U. Medical micro/nanorobots in precision medicine. Adv Sci (Weinh). 2020;7:2002203.
67. Mallouk TE, Sen A. Powering nanorobots. Sci Am. 2009;300:72–7.
68. Ostman A, Augsten M. Cancer-associated fibroblasts and tumor growth-bystanders turning into key players. Curr Opin Gent Dev. 2009;19(1):67–73.
69. Gajewski TF, et al. Innate and adaptive immune cells in the tumor microenvironment. Nat Immunol. 2013;14(10):1014–22.
70. Rajagopal C, Harikumar KB. The origin and functions of exosomes in cancer. Front Oncol. 2018;8:66.
71. Vaughn JR. Over the horizon: potential impact of emerging trends in information and communication technology on disability policy and practice. Washington DC: National Council on Disability; 2006.
72. Abhilash M. Nanorobots. Int J Pharma Bio Sci. 2010;V1(1):1–10.

Nanosafety

27

27.1 Reduction of Drug Toxicity Through Nanomedicine

The nanotherapeutics in the earlier stages were approved as they had far fewer side effects compared to existing drugs. The first nanomedicine is Doxil, which is made up of a liposomal combination of doxorubicin. This Doxil was first approved by the US Food and Drug Administration (FDA) in 1995 for AIDS-related Kaposi's sarcoma and is now used for various applications like multiple myeloma [1–4]. Doxil has less cardiotoxicity compared to doxorubicin [4]. Fewer side effects of nanoimaging can be seen due to the better accumulation of these drugs in the region of interest, ultimately resulting in better therapeutic efficacy. After injection, it will have <0.01% of the injected dose, which is still capable of accumulating in the region of interest in the body [5], and this applies to the same value, which is approximately 1–5% for nanoparticles (NPs) [6] (Fig. 27.1). The basic mechanism is the increased permeability and retention (EPR) effect of NPs in tumor sites. Due to their smaller size, these NPs can easily pass into the vasculature of any tissue. The EPR effect is also due to differences seen in the vasculature of tumors and normal tissues [7, 8]. These cancer blood vessels have fenestrations, permitting better movement of NPs in various tumor tissues. The concept of the EPR effect is not seen in human tumors, and there will be more and better heterogeneity in existing patients and different cancer types [9]. The use of various hydrophobic therapeutics will cause less toxicity in various NPs. These NPs are also capable of better use than various toxic solubilizing agents [10–12]. They can be used in various in vitro and in vivo for non-water-soluble drugs. Dimethyl sulfoxide (DMSO) can also be the agent of choice in cell culture [13, 14]. However, this DMSO agent also has toxicity levels, as seen in various studies [15–18]. Important clinically approved hydrophobic drugs can be formulated for toxic agents. Drugs, namely paclitaxel and docetaxel, are given by adding polyethoxylated castor oil (Cremophor EL) and the dehydrated ethanol mixture [19] and polysorbate 80 [20], respectively. The side effects due to solubilizing agents can be managed by using antihistamines and corticosteroids to prevent side effects [21]. Different NPs will have better results in comparison with various water-soluble drugs, which have less solubility. Abraxane is one of the types of drugs that is made up of various protein-bound paclitaxel components and is now used in metastatic breast cancer [22]. According to the literature, higher doses of Abraxane, in comparison to Cremophor EL-based paclitaxel, are better tolerated by patients [23].

S. Shaikh, *Nanoimaging - Future of Precision Medicine*,
https://doi.org/10.1007/978-981-97-7292-6_27

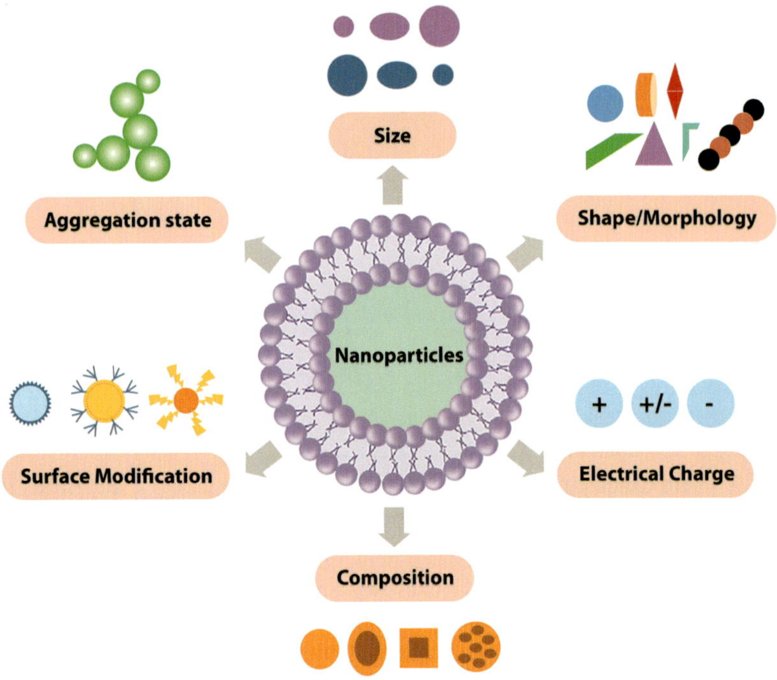

Fig. 27.1 Nanoparticle role and different application related to safety

27.2 Toxicity of Nanoparticles in Medicine

NPs also have some side effects and may lead to cytotoxic effects. Carbon-based NPs have more toxicity in both in vitro and in vivo assays [24–26]. These carbon nanotubes have malignant potential and can cause mesothelioma [27, 28]. Carbon nanotubes come in various shapes [29], and despite their biocompatibility advantages, they can still have toxic effects [30]. These gold NPs can have more metallic nanoparticles with minimal side effects for example the cytotoxic effects of the silver nanoparticles shown by Wolfram et al. [31, 32]. Iron oxide NPs can be used in both in vitro and in vivo studies [33–35], particularly in the area of reactive oxygen species (ROS) [36].

The selective targeting of NPs can prevent damage. The use of pristine NPs causes different surface modifications [37], which can improve compatibility with various hydroxyl groups in gadolinium fullerene, resulting in lower toxicity [3, 38, 39]. Cell viability can be improved by using a polymer coating agent on the surface of iron oxide NPs, leading to better cell viability [40].

27.3 Mechanisms of Nanoparticle Toxicity

The mechanism of toxicity of certain NPs occurs at multiple levels, such as molecular, cellular, and tissue levels. An important concept here is that NPs move through different parts of the body, encountering diverse biological microenvironments, including the blood, extracellular components, cytoplasm, and various cellular organelles. The nano-bio interface involves interactions with biomolecules, cellular components, and various tissues. NP components also have side effects and toxicity.

27.4 Molecular Toxicity of Nanomedicines

Various NPs, when entering the biological environment, interact with various cells and components. Different biomolecules are capable of forming various biological fluids on the surface of NPs [41]. These coatings, labeled as protein corona, are made up of soft and hard layers [42]. The hard layer is in close relation to the NP, while the soft layer is more dynamic. The different characteristics of NPs can impact the hard corona [43, 44], which is made up of approximately 100 proteins [45, 46]. The protein corona can be affected by various NP properties, such as shape, size, and charge [41]. These interactions can change the size of the NPs, which can either increase or decrease in size [46–50]. These size variations can also cause different zeta potentials, making it more anionic [46]. The different endogenous biomolecules undergo some form of transformation when they come in contact with NPs. This impact can pose safety issues for NPs [41, 51, 52]. Some NPs cause changes in various proteins, such as albumin [53], cytochrome c [54], and ribonuclease A [55]. NP interactions can lead to changes in various parameters, including particle size [56–58]. The larger the surface area of NPs, the more interaction with the surface, leading to changes in the structure of the protein. Different conformational changes will affect protein function. These changes will have a greater impact on the binding of transferrin to iron oxide NPs, which may cause irreversible changes in protein structure and function [59]. Intravenous injection of iron oxide NPs will result in alteration and damage to the iron transport mechanisms. NPs can also cause more unfolding of fibrinogen, leading to changes in inflammatory signaling pathways [60]. An important safety parameter is NP-induced protein aggregation when proteins are bound to the NP surface. This can lead to changes in different pathological conditions, such as amyloidosis, which is linked to various diseases [61, 62]. Multiple changes can occur on the surface of the NP, causing abnormal structural changes in protein molecules, NP toxicity, and replenishment of the protein layer [63, 64]. Changes in the densities of the protein coat can result in reduced toxicity [63]. Further changes depend on the affinity of cell membrane integrity [63, 65, 66]. This will also result in the protein corona preventing NP-induced platelet activation and hemolysis [46].

27.5 Cellular Toxicity of Nanomedicines

NPs are also capable of eliciting changes in various toxicity levels, resulting in changes to the cell membrane. This disrupts the membrane integrity, causing leakage of materials into the cell and affecting organelle function. This includes disruption to the lysosome membranes, which are compromised by various components, such as titanium dioxide [67], zinc oxide [68], polystyrene [69], and polycation particles [1]. There will also be significant damage to the lysosomal preventing the release of protons, iron, and hydrolytic enzymes, and causing different stress aggregates [70]. There is also a major mechanism in which polycations are released [71]. The T-cell membrane can maintain different polycation chains, which affect the components of the lipid bilayer, allowing holes to form in the membrane. This process results in the formation of ROS and may cause changes in membrane permeabilization [72, 73]. Different mechanisms are involved in the process of endocytosis [74]. There are also various cellular locations based on the perfect balance between toxicity and movement. Gold NPs also result in the accumulation of autophagosomes [75]. Polyethyleneimine (PEI), a polymeric NP, can damage the mitochondrial membrane, leading to proton leakage through various mechanisms [76]. This disrupts the electron transport system and the synthesis of adenosine triphosphate (ATP), affecting the cell interior [77]. This is an important aspect of the NP mechanism and its exposure.

NPs can cause various forms of radical reactions or indirectly lead to changes in cellular homeostasis [77, 78], leading to different forms of stress [79]. This may also result in cell damage, cell death, apoptosis, necrosis, and autophagic cell death [80]. These outcomes depend on the size [30, 31], dose [81], incubation period [81], and charge of the NPs [82]. The cell death pathways also depend on the various cell lines, resulting in NP-induced cell death [31, 83] (Fig. 27.2).

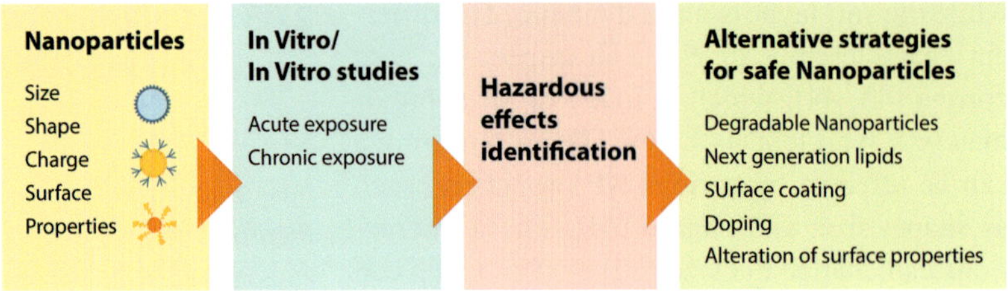

Fig. 27.2 Nanoparticle different features and different structures, reactions and safety protocols used

27.6 Tissue Toxicity of Nanomedicines

NP-induced damage can affect the functioning of various organs, resulting in molecular and cellular defects and damage. One example is the presence of carbon nanotubes in the lungs, which causes lung inflammation [84, 85]. Hepatotoxicity can be caused by NP accumulation, leading to increased levels of liver enzymes in the blood [86]. Cytotoxicity can also activate the immune system, which can sometimes result in differences in how NPs are recognized. Intravenous injection can affect different organs, such as the spleen and liver [87, 88]. Various macrophages can also engulf NPs when they circulate in the body. These phagocytic cells are part of the reticuloendothelial system, and their primary role is to destroy foreign bodies. Immunoglobulins can also bind to various NPs, activating the complement system [89]. The role of the complement system is to identify foreign microparticles and NPs. Another way the system identifies these is by recognizing patterns from bacteria and viruses [90, 91], based on surface morphology, size, composition, charge, and the NP shape, which enhances identification [92]. Different charge activations, such as cationic or anionic charges, and NP hydrophobicity, also contribute to immunological activation [93], based on the damage associated with hydrophobic molecules [94]. The interaction of hydrophobic cellular components can be alarming, as they can cause significant damage to both foreign materials and endogenous plasma proteins, leading to complement system activation [41]. Various proteins can bind to the surface of NPs, which are recognized as foreign bodies. These polymeric NPs are capable of strongly activating the complement system [95]. Different mechanisms are involved, either through direct interactions or binding to Fc receptors [89, 96]. These nanomaterials interact with various systems [97, 98]. Toll-like receptors (TLRs) are also involved in immune reactions, which can increase the impact of the toxicity [99]. NP-induced complement activation can cause significant and gross inflammatory reactions, which may be critical in nature [100, 101]. One example is the nanotherapeutic Doxil, which can cause hypersensitivity reactions, including skin reactions, hypotension, hypertension, respiratory issues, or pain in humans [102] and pigs [103]. These immunological hypersensitivity reactions are mediated through various pathways [104]. Various in vitro assays can also result in NP-induced immunotoxicity through mechanisms that lead to better exposure to serum or plasma [105–108]. One study showed that the impact of the protein corona inhibited natural blood coagulation processes during the use of ethylenediaminetetraacetic acid (EDTA) [108], by inhibiting complement activation [109]. Immune activation in enzyme-linked immunosorbent assays (ELISA) can assess complement activation using NPs with serum or heparin-treated plasma samples. The use of pegylated liposomes [103], carbon nanotubes [110], and polystyrene particles [111] leads to more activation in vitro. Different aspects of cytokine production by peripheral blood mononuclear cells (PBMCs) have been observed [112]. Cytokines like interferon-gamma (IFN-γ), tumor necrosis factor (TNF-α), and interleukin-12 (IL-12) are markers of inflammation. Severe adverse immunological reactions can be identified at various genome levels [113]. The use of NPs can cause adverse reactions that may lead to the use of antifouling agents to prevent protein binding. NPs with PEG can offer protection against protein binding [114–116]. Increased PEG density can result in the accelerated blood clearance (ABC) phenomenon, causing rapid removal of particles from circulation [117–119]. Different lipoproteins can utilize cholesterol-rich liposomes, which may cause severe reactions when NPs are injected in vivo [120, 121]. Some pathogenic organisms cannot be detected due to interference with various mechanisms [122–131].

27.7 Mechanisms of Nanoparticles

NPs also have toxicity, like any other drugs, which depends on the various routes of administration and their exposure. The exposure to NPs can occur via various routes, such as ingestion, injection, inhalation, and skin contact. The organ-specific toxicity of the various NPs is based on the different routes of administration. Pulmonary inhalation of NPs causes various inflammatory reactions, fibrosis, and necrosis of different tissues [132–134]. Application of skin products, or the use of drugs via ingestion or injection, also impacts tissues. These NPs are used in different organs, including the brain, which provides information about tissue permeability [135, 136]. These tissues are affected in various tumors, where the vasculature of the NPs will exhibit the enhanced permeability and retention (EPR) effect [137, 138]. The accumulation of NPs is based on different physical and chemical properties, and NPs can accumulate in different cellular processes, such as oxidative stress [139, 140]. NP circulation depends on the surface area-to-size ratio, which also influences the accumulation in various tissues and organs [141, 142]. The toxicity of various NPs can be seen in different studies involving both short-term and chronic exposures. The dynamics of NPs as drug delivery agents can be studied further [138, 143]. Various studies have shown that the pharmacokinetics of encapsulated drugs are also influenced by the carriers of the drug. NPs, as carriers, exhibit more toxicity in the blood and different organs [132]. Different toxicity levels are observed for various NPs in different target organs due to the generation of reactive oxygen species (ROS), DNA damage, and modifications of protein structures and functions, which impair membrane integrity. Different NPs have properties that facilitate large surface areas for molecular interactions at target areas. Acute systemic exposure to solid NPs is more precise than exposure to metal-containing or metal oxide NPs, which cause oxidative stress, inflammation, and DNA damage [144, 145]. Oxidative stress is induced in the liver, spleen, and kidneys, inhibiting the effects of antioxidants [146–148]. The ROS results from NP-induced stress can cause dysfunction and activate stress-related cell signaling pathways, leading to DNA damage, cell cycle arrest, and apoptosis [147–155]. Various NPs also cause epigenetic modifications, such as changes in DNA methylation status and post-translational modifications of histones [156–159].

Different NPs have different toxicity levels. NP toxicity is based on various biophysical properties, such as size, surface area, surface charge, and aggregation state. This results in different NP characteristics for various organ systems, where more molecular interactions are observed with proteins and macromolecules [160]. NPs also play an important role in cellular uptake and toxicity. These effects are more precise depending on the NP properties, along with the production of ROS in the kidneys [161, 162]. Size-dependent toxicity is also observed in the liver. NPs with a size of 10 nm exhibit higher tissue distribution, leading to dangerous toxic effects in the liver compared to larger NPs (40 and 100 nm) [163, 164]. The size and shape of various fibrous NPs, such as single-walled nanotubes, make it difficult for the body to prevent organ deposition [165]. Chemical properties also influence the pharmacokinetics of NPs; charged NPs show higher accumulation in target organs compared to uncharged NPs. The use of zinc leads to more accumulation in the liver, lungs, and kidneys compared to uncharged NPs [166]. Various polysaccharide conjugations result in greater accumulation of NPs in the brain, liver, and spleen, correlating with their toxicity in these organs [167]. Surface modifications of NPs, with various coating materials, can reduce toxicity. Polyethylene glycol (PEG) is an important entity that reduces the toxicity of NPs, showing better interactions with proteins. Long fibrous particles, such as single-walled nanotubes, lead to clearer and better organ deposition [165]. The surface chemistry of NPs also influences pharmacokinetics. Charged particles exhibit better and higher accumulation in target organs, such as the spleen, compared to uncharged NPs [166]. Different polysaccharides also cause more accumulation in the brain, liver, and spleen, leading to increased toxicity in these organs [167–169]. NP surface properties can also affect potential toxicity. These NPs and their interactions with cells determine NP efficacy and toxicity, which depend on the biodistribution of these particles [170]. NPs show more cellular uptake, even in smaller molecules, which could reduce toxicity in cells [171]. Various NP–cell interactions depend on different properties, and their toxicity is influenced by multiple factors. Efforts are needed to minimize the toxic effects and develop safer and more efficient NPs [172–175].

Endocytosis of NPs is also an important mechanism, with various processes involved, including phagocytosis, pinocytosis, clathrin-dependent endocytosis, caveolae-dependent endocytosis, and clathrin/caveolae-independent endocytosis [173–179]. The different shapes and circulation times affect the amount of cellular uptake, with macrophage phagocytosis showing a longer circulation time [180]. NPs undergo processes like membrane translocation and simple diffusion, influenced by NP size and lipophilicity [181]. Different types of cellular interactions can affect various NP modifications. Gold NPs, for instance, which are based on the various amino acids like aspartate, glycine, leucine, lysine, and serine, attach to the cell surface, impacting intracellular uptake due to different biocompatibility processes compared to unmodified gold NPs along with the different conjugated amino acids [182]. Cells containing various NPs can be seen in the cytoplasm, enclosed by the membrane [183, 184], with hydrophobic NPs interacting with adjacent cells [185–190]. Different nanoparticles will have various polymeric NPs or lipid-based NPs, in the various inorganic NPs comprising of the zinc, iron, and silver oxide NPs, this is seen based on the various free ions along with the different NPs showing the different measurable toxic responses in the concentrations of the 30 ppm (μg/mL). Various NPs of iron oxide and zinc oxide are soluble and toxic, leading to harmful effects at various toxic levels [190–192]. An acidic pH of 5.5 also contributes to the degradation of these NPs, releasing toxic heavy metal ions. These ions can damage cells by inactivating enzymes [28, 193–198].

27.8 Nanoparticle Toxicity

Acute toxicity can be assessed by evaluating various parameters for different cosmetics. The dissolution or degradation of NPs can take a significant amount of time, which leads to increased biodistribution and accumulation, allowing for the assessment of bioaccumulation in the environment. Chronic exposure conditions can also result in long-term effects, as seen in the evaluation of carbon nanotubes in breast cancer metastasis [199–204]. Some NPs, like lipids, proteins, polysaccharides, and surfactants, are approved by the FDA [205]. However, toxicological assays are not capable of predicting long-term side effects of NPs. Some NPs can be labeled as dangerous or "potentially hazardous," and must be handled with special care. The pharmacokinetic properties of NPs, such as their distribution and elimination, should be carefully considered [206–209] (Fig. 27.3). Surface coating of NPs can improve their biocompatibility [209–214]. Certain multi-walled carbon nanotubes (CNTs) are capable of acting as drug nanocarriers, supporting tissue engineering and water purification, but they can also induce inflammation, fibrosis, and cancer progression [215–222].

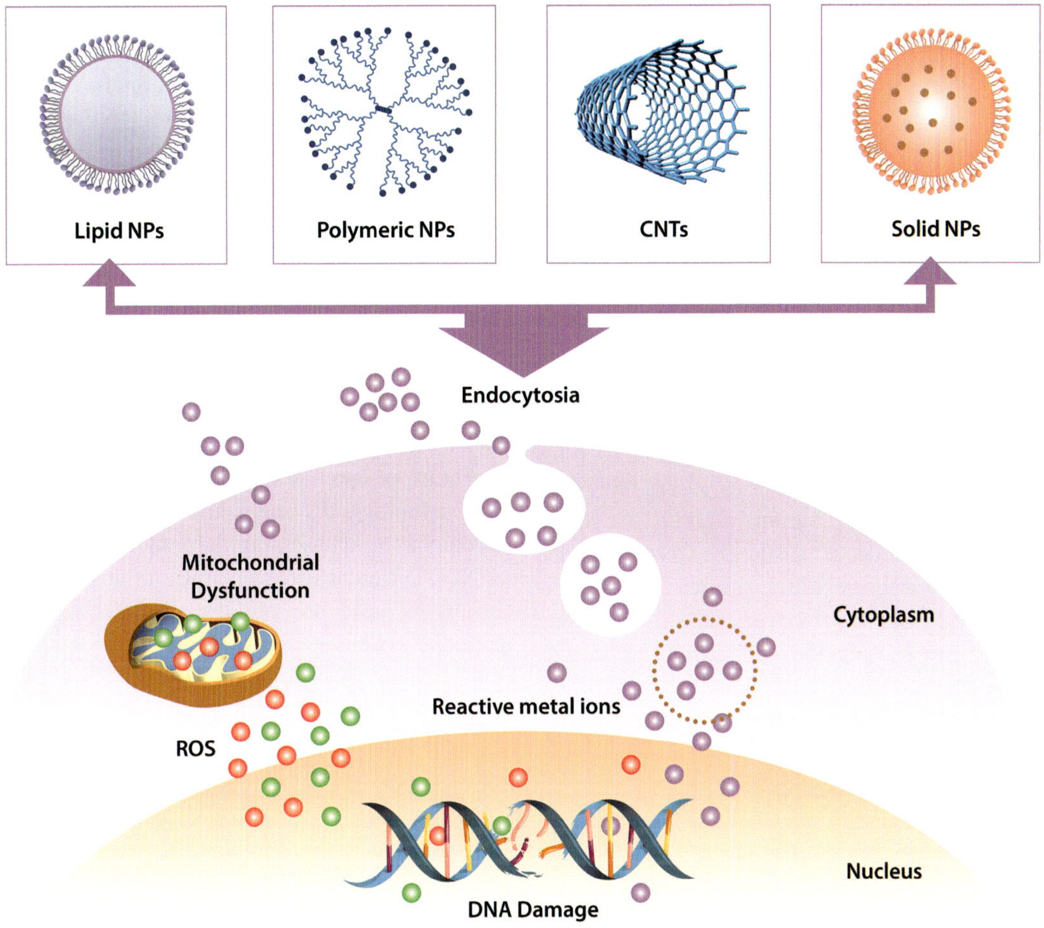

Fig. 27.3 Mitochondrial dysfunction and DNA damage in the toxic effects of the nanoparticles

Dopants such as aluminum, titanium, and iron can reduce the binding energy of various metal ions to oxygen [223–228]. Doping NPs can reduce cytotoxicity through mechanisms like altering charge density and hydrophobicity, which can increase the application of NPs in biomedical uses and targeted drug delivery [44, 229–245]. Surface functionalization of iron oxide NPs with organic compounds like aldehyde, carboxyl, and amino groups stabilizes the high chemical reactivity of these NPs, which decreases their toxicity while enhancing their biological compatibility [246–253]. Another approach involves using antibodies developed through various immune responses.

27.9 Potential Nanorobot Hazards

Nanorobot designs may present potential toxicity issues due to micro/nanoscale motors used in the construction of various nanorobots, raising concerns about the safety of DNA nanorobots in cells. Nanorobots can induce immunological or cytotoxic responses, leading to the formation of thrombi or blood coagulation [254–258]. The use of silver in nanorobots does not appear to cause significant side effects [259–262]. UV light, however, has the potential to damage the skin and increase the risk of skin cancer. Different NPs may have a wide range of potential applications.

27.10 Nanoinformatics and Immunosafety by Design

Computational modeling and informatics of NPs are crucial for the discovery, development, and application of these materials. These nanomaterials exhibit structural heterogeneity and often lack standardized methodologies, adequate studies, and quality characterization [263–270]. Mechanisms like molecular dynamics and density functional theory can enhance our understanding of nano-bio interactions, leading to more impactful knowledge [271–275]. Various nanomaterials can be used for different structural properties as well as biological activities [276–280]. Various algorithms need to be explained in detail causing the QNAR models, like vector machine [281], artificial neural network [282], and decision trees [283–292], among others, and causing the applications in various nanomaterials immune reactions by using different processes like the genomics, transcriptomics, and metabolomics which can be used as the specific immune responses for the use of the dif. This is an important entity that can be used for the various immune characterization of carbon-based materials, integrating data-driven methodologies that are extendable to other 2D materials.

The cytotoxicity of NPs can be identified by various processes and for various applications by the use of the different assays that are suitable for such applications [293–313]. Most available safety tests have been developed for conventional therapeutic agents. Various nanoparticles can be used for different carbon nanotubes which will have toxicity based on various functions like a 60% reduction in viability. Several parameters need to be evaluated to identify carbon nanotubes with carcinogenic effects [314, 315]. Dynamic flow models are useful for studying specific NP interactions with endothelial cells [316–320].

27.11 Regulatory Aspects of Nanomedicine

Various guidelines are applied to different nanotherapeutics to address safety issues. This will be able to show the different physicochemical properties, like the surface area, shape, and drug release, which can be able to prevent the different potential hazards related to the various NP safety as well as the different phases of NP design. Neurological effects and immunological activation can lead to adverse immune reactions.

27.12 Conclusions

Nanomedicine is a novel and diverse field due to its smaller size and wide range of applications. Different drugs are being developed to reduce toxicity and minimize cytotoxic properties.

Future research should focus on understanding the biological systems affected by NPs, with an emphasis on assessing the long-term toxicity of NPs. Different strategies must be developed to ensure the safe use of NPs, considering their potential toxicity.

References

1. Alsina M, Trudel S, Furman RR, et al. A phase I single-agent study of twice-weekly consecutive-day dosing of the proteasome inhibitor carfilzomib in patients with relapsed or refractory multiple myeloma or lymphoma. Clin Cancer Res. 2012;18(17):4830–40.
2. Shley JD, Quinlan CJ, Schroeder VA, et al. Dual carfilzomib and doxorubicin-loaded liposomal nanoparticles for synergistic efficacy in multiple myeloma. Mol Cancer Ther. 2016;15(7):1452–9.
3. Cosco D, Cilurzo F, Maiuolo J, et al. Delivery of miR-34a by chitosan/PLGA nanoplexes for the anticancer treatment of multiple myeloma. Sci Rep. 2015;5:17579.
4. Barenholz Y. Doxil®--the first FDA-approved nano-drug: lessons learned. J Control Release. 2012;160(2):117–34.
5. Li KC, Pandit SD, Guccione S, Bednarski MD. Molecular imaging applications in nanomedicine. Biomed Microdevices. 2004;6(2):113–6.
6. Bae YH, Park K. Targeted drug delivery to tumors: myths, reality and possibility. J Control Release. 2011;153(3):198–205.
7. Matsumura Y, Maeda H. A new concept for macromolecular therapeutics in cancer chemotherapy: mechanism of tumoritropic accumulation of proteins and the antitumor agent smancs. Cancer Res. 1986;46(12 Pt 1):6387–92.
8. Maeda H, Wu J, Sawa T, Matsumura Y, Hori K. Tumor vascular permeability and the EPR effect in macromolecular therapeutics: a review. J Control Release. 2000;65(1–2):271–84.

9. Bogart LK, Pourroy G, Murphy CJ, et al. Nanoparticles for imaging, sensing, and therapeutic intervention. ACS Nano. 2014;8(4):3107–22.
10. Wolfram J, Suri K, Huang Y, et al. Evaluation of anticancer activity of celastrol liposomes in prostate cancer cells. J Microencapsul. 2014;31(5):501–7.
11. Celia C, Trapasso E, Locatelli M, et al. Anticancer activity of liposomal bergamot essential oil (BEO) on human neuroblastoma cells. Colloids Surf B Biointerfaces. 2013;112:548–53.
12. Norvaisas P, Ziemys A. The role of payload hydrophobicity in nanotherapeutic pharmacokinetics. J Pharm Sci. 2014;103(7):2147–56.
13. Yang Y, Wolfram J, Boom K, Fang X, Shen H, Ferrari M. Hesperetin impairs glucose uptake and inhibits proliferation of breast cancer cells. Cell Biochem Funct. 2013;31(5):374–9.
14. Yang Y, Wolfram J, Shen H, Fang X, Ferrari M. Hesperetin: an inhibitor of the transforming growth factor-beta (TGF-beta) signaling pathway. Eur J Med Chem. 2012;58:390–5.
15. Ruiz-Delgado GJ, Mancias-Guerra C, Tamez-Gomez EL, et al. Dimethyl sulfoxide-induced toxicity in cord blood stem cell transplantation: report of three cases and review of the literature. Acta Haematol. 2009;122(1):1–5.
16. Zenhausern R, Tobler A, Leoncini L, Hess OM, Ferrari P. Fatal cardiac arrhythmia after infusion of dimethyl sulfoxide-cryopreserved hematopoietic stem cells in a patient with severe primary cardiac amyloidosis and end-stage renal failure. Ann Hematol. 2000;79(9):523–6.
17. Montaguti P, Melloni E, Cavalletti E. Acute intravenous toxicity of dimethyl sulfoxide, polyethylene glycol 400, dimethylformamide, absolute ethanol, and benzyl alcohol in inbred mouse strains. Arzneimittelforschung. 1994;44(4):566–70.
18. Rubin LF. Toxicity of dimethyl sulfoxide, alone and in combination. Ann N Y Acad Sci. 1975;243:98–103.
19. Gelderblom H, Verweij J, Nooter K, Sparreboom A, Cremophor EL. The drawbacks and advantages of vehicle selection for drug formulation. Eur J Cancer. 2001;37(13):1590–8.
20. Cho HJ, Park JW, Yoon IS, Kim DD. Surface-modified solid lipid nanoparticles for oral delivery of docetaxel: enhanced intestinal absorption and lymphatic uptake. Int J Nanomedicine. 2014;9:495–504.
21. Francis P, Schneider J, Hann L, et al. Phase II trial of docetaxel in patients with platinum-refractory advanced ovarian cancer. J Clin Oncol. 1994;12(11):2301–8.
22. Miele E, Spinelli GP, Miele E, Tomao F, Tomao S. Albumin-bound formulation of paclitaxel (Abraxane ABI-007) in the treatment of breast cancer. Int J Nanomedicine. 2009;4:99–105.
23. Ibrahim NK, Desai N, Legha S, et al. Phase I and pharmacokinetic study of ABI-007, a Cremophor-free, protein-stabilized, nanoparticle formulation of paclitaxel. Clin Cancer Res. 2002;8(5):1038–44.
24. Li Y, Boraschi D. Endotoxin contamination: a key element in the interpretation of nanosafety studies. Nanomedicine. 2016;11:269–87. https://doi.org/10.2217/nnm.15.196.
25. Liu Y, Zhao Y, Sun B, Chen C. Understanding the toxicity of carbon nanotubes. Acc Chem Res. 2013;46(3):702–13.
26. Jia G, Wang H, Yan L, et al. Cytotoxicity of carbon nanomaterials: single-wall nanotube, multi-wall nanotube, and fullerene. Environ Sci Technol. 2005;39(5):1378–83.
27. Donaldson K, Murphy FA, Duffin R, Poland CA. Asbestos, carbon nanotubes and the pleural mesothelium: a review of the hypothesis regarding the role of long fibre retention in the parietal pleura, inflammation and mesothelioma. Part Fibre Toxicol. 2010;7:5.
28. Poland CA, Duffin R, Kinloch I, et al. Carbon nanotubes introduced into the abdominal cavity of mice show asbestos-like pathogenicity in a pilot study. Nat Nanotechnol. 2008;3(7):423–8.
29. Johnston HJ, Hutchison GR, Christensen FM, et al. A critical review of the biological mechanisms underlying the in vivo and in vitro toxicity of carbon nanotubes: the contribution of physicochemical characteristics. Nanotoxicology. 2010;4(2):207–46.
30. Pan Y, Neuss S, Leifert A, et al. Size-dependent cytotoxicity of gold nanoparticles. Small. 2007;3(11):1941–9.
31. Kim TH, Kim M, Park HS, Shin US, Gong MS, Kim HW. Size-dependent cellular toxicity of silver nanoparticles. J Biomed Mater Res A. 2012;100(4):1033–43.
32. Pratsinis A, Hervella P, Leroux JC, Pratsinis SE, Sotiriou GA. Toxicity of silver nanoparticles in macrophages. Small. 2013;9(15):2576–84.
33. Zhu MT, Wang Y, Feng WY, et al. Oxidative stress and apoptosis induced by iron oxide nanoparticles in cultured human umbilical endothelial cells. J Nanosci Nanotechnol. 2010;10(12):8584–90.
34. Wang B, Feng W, Zhu M, et al. Neurotoxicity of low-dose repeatedly intranasal instillation of nano- and submicron-sized ferric oxide particles in mice. J Nanopart Res. 2009;11(1):41–53.
35. Zhu MT, Wang B, Wang Y, et al. Endothelial dysfunction and inflammation induced by iron oxide nanoparticle exposure: risk factors for early athero sclerosis. Toxicol Lett. 2011;203(2):162–71.
36. Wang B, Yin J-J, Zhou X, et al. Physicochemical origin for free radical generation of iron oxide nanoparticles in biomicroenvironment: catalytic activities mediated by surface chemical states. J Phys Chem C. 2012;117(1):383–92.
37. Yan L, Zhao F, Li S, Hu Z, Zhao Y. Low-toxic and safe nanomaterials by surface-chemical design, carbon nanotubes, fullerenes, metallofullerenes, and graphenes. Nanoscale. 2011;3(2):362–82.
38. Meng H, Xing G, Sun B, et al. Potent angiogenesis inhibition by the particulate form of fullerene derivatives. ACS Nano. 2010;4(5):2773–83.

39. Zhao F, Zhao Y, Liu Y, Chang X, Chen C, Zhao Y. Cellular uptake, intracellular trafficking, and cytotoxicity of nanomaterials. Small. 2011;7(10):1322–37.
40. Gupta AK, Wells S. Surface-modified superparamagnetic nanoparticles for drug delivery: preparation, characterization, and cytotoxicity studies. IEEE Trans Nanobioscience. 2004;3(1):66–73.
41. Wolfram J, Yang Y, Shen J, et al. The nano-plasma interface: implications of the protein corona. Colloids Surf B Biointerfaces. 2014;124:17–24.
42. Monopoli MP, Aberg C, Salvati A, Dawson KA. Biomolecular coronas provide the biological identity of nanosized materials. Nat Nanotechnol. 2012;7(12):779–86.
43. Zhu M, Perrett S, Nie G. Understanding the particokinetics of engineered nanomaterials for safe and effective therapeutic applications. Small. 2013;9(9–10):1619–34.
44. Zhu M, Nie G, Meng H, Xia T, Nel A, Zhao Y. Physicochemical properties determine nanomaterial cellular uptake, transport, and fate. Acc Chem Res. 2013;46(3):622–31.
45. Tenzer S, Docter D, Rosfa S, et al. Nanoparticle size is a critical physicochemical determinant of the human blood plasma corona: a comprehensive quantitative proteomic analysis. ACS Nano. 2011;5(9):7155–67.
46. Tenzer S, Docter D, Kuharev J, et al. Rapid formation of plasma protein corona critically affects nanoparticle pathophysiology. Nat Nanotechnol. 2013;8(10):772–81.
47. Mortensen NP, Hurst GB, Wang W, Foster CM, Nallathamby PD, Retterer ST. Dynamic development of the protein corona on silica nanoparticles: composition and role in toxicity. Nanoscale. 2013;5(14):6372–80.
48. Goy-Lopez S, Juarez J, Alatorre-Meda M, et al. Physicochemical characteristics of protein-NP bioconjugates: the role of particle curvature and solution conditions on human serum albumin conformation and fibrillogenesis inhibition. Langmuir. 2012;28(24):9113–26.
49. Monopoli MP, Walczyk D, Campbell A, et al. Physical-chemical aspects of protein corona: relevance to in vitro and in vivo biological impacts of nanoparticles. J Am Chem Soc. 2011;133(8):2525–34.
50. Wolfram J, Suri K, Yang Y, et al. Shrinkage of pegylated and non-pegylated liposomes in serum. Colloids Surf B Biointerfaces. 2014;114C:294–300.
51. Pelaz B, Charron G, Pfeiffer C, et al. Interfacing engineered nanoparticles with biological systems: anticipating adverse nano-bio interactions. Small. 2013;9(9–10):1573–84.
52. Zuo G, Kang SG, Xiu P, Zhao Y, Zhou R. Interactions between proteins and carbon-based nanoparticles: exploring the origin of nanotoxicity at the molecular level. Small. 2013;9(9–10):1546–56.
53. Shang L, Wang Y, Jiang J, Dong S. pH-dependent protein conformational changes in albumin:gold nanoparticle bioconjugates: a spectroscopic study. Langmuir. 2007;23(5):2714–21.
54. Shang W, Nuffer JH, Muniz-Papandrea VA, Colon W, Siegel RW, Dordick JS. Cytochrome C on silica nanoparticles: influence of nanoparticle size on protein structure, stability, and activity. Small. 2009;5(4):470–6.
55. Shang W, Nuffer JH, Dordick JS, Siegel RW. Unfolding of ribonuclease a on silica nanoparticle surfaces. Nano Lett. 2007;7(7):1991–5.
56. Teichroeb JH, Forrest JA, Jones LW. Size-dependent denaturing kinetics of bovine serum albumin adsorbed onto gold nanospheres. Eur Phys J E Soft Matter. 2008;26(4):411–5.
57. Teichroeb JH, Forrest JA, Ngai V, Jones LW. Anomalous thermal denaturing of proteins adsorbed to nanoparticles. Eur Phys J E Soft Matter. 2006;21(1):19–24.
58. Lundqvist M, Sethson I, Jonsson BH. Protein adsorption onto silica nanoparticles: conformational changes depend on the particles' curvature and the protein stability. Langmuir. 2004;20(24):10639–47.
59. Mahmoudi M, Shokrgozar MA, Sardari S, et al. Irreversible changes in protein conformation due to interaction with superparamagnetic iron oxide nanoparticles. Nanoscale. 2011;3(3):1127–38.
60. Deng ZJ, Liang M, Monteiro M, Toth I, Minchin RF. Nanoparticle-induced unfolding of fibrinogen promotes Mac-1 receptor activation and inflammation. Nat Nanotechnol. 2011;6(1):39–44.
61. Linse S, Cabaleiro-Lago C, Xue WF, et al. Nucleation of protein fibrillation by nanoparticles. Proc Natl Acad Sci USA. 2007;104(21):8691–6.
62. Chiti F, Dobson CM. Protein misfolding, functional amyloid, and human disease. Annu Rev Biochem. 2006;75:333–66.
63. Ge C, Du J, Zhao L, et al. Binding of blood proteins to carbon nanotubes reduces cytotoxicity. Proc Natl Acad Sci USA. 2011;108(41):16968–73.
64. Wang L, Li J, Pan J, et al. Revealing the binding structure of the protein corona on gold nanorods using synchrotron radiation-based techniques: understanding the reduced damage in cell membranes. J Am Chem Soc. 2013;135(46):17359–68.
65. Hu W, Peng C, Lv M, et al. Protein corona-mediated mitigation of cytotoxicity of graphene oxide. ACS Nano. 2011;5(5):3693–700.
66. Lesniak A, Fenaroli F, Monopoli MP, Aberg C, Dawson KA, Salvati A. Effects of the presence or absence of a protein corona on silica nanoparticle uptake and impact on cells. ACS Nano. 2012;6(7):5845–57.
67. Hamilton RF, Wu N, Porter D, Buford M, Wolfarth M, Holian A. Particle length-dependent titanium dioxide nanomaterials toxicity and bioactivity. Part Fibre Toxicol. 2009;6:35.
68. Cho WS, Duffin R, Howie SE, et al. Progressive severe lung injury by zinc oxide nanoparticles; the

role of Zn2+ dissolution inside lysosomes. Part Fibre Toxicol. 2011;8:27.
69. Lunov O, Syrovets T, Loos C, et al. Amino-functionalized polystyrene nanoparticles activate the NLRP3 inflammasome in human macrophages. ACS Nano. 2011;5(12):9648–57.
70. Stern ST, Adiseshaiah PP, Crist RM. Autophagy and lysosomal dysfunction as emerging mechanisms of nanomaterial toxicity. Part Fibre Toxicol. 2012;9:20.
71. Boussif O, Lezoualc'h F, Zanta MA, et al. A versatile vector for gene and oligonucleotide transfer into cells in culture and in vivo: polyethylenimine. Proc Natl Acad Sci USA. 1995;92(16):7297–301.
72. Won YY, Sharma R, Konieczny SF. Missing pieces in understanding the intracellular trafficking of polycation/DNA complexes. J Control Release. 2009;139(2):88–93.
73. Berndt C, Kurz T, Selenius M, Fernandes AP, Edgren MR, Brunk UT. Chelation of lysosomal iron protects against ionizing radiation. Biochem J. 2010;432(2):295–301.
74. Iversen TG, Skotland T, Sandvig K. Endocytosis and intracellular transport of nanoparticles: present knowledge and need for future studies. Nano Today. 2011;6(2):176–85.
75. Ma X, Wu Y, Jin S, et al. Gold nanoparticles induce autophagosome accumulation through sizedependent nanoparticle uptake and lysosome impairment. ACS Nano. 2011;5(11):8629–39.
76. Hall A, Larsen AK, Parhamifar L, Meyle KD, Wu LP, Moghimi SM. High resolution respirometry analysis of polyethylenimine-mediated mitochondrial energy crisis and cellular stress: mitochondrial proton leak and inhibition of the electron transport system. Biochim Biophys Acta. 2013;1827(10):1213–25.
77. Yan L, Gu Z, Zhao Y. Chemical mechanisms of the toxicological properties of nanomaterials: generation of intracellular reactive oxygen species. Chem Asian J. 2013;8(10):2342–53.
78. Chen R, Huo L, Shi X, et al. Endoplasmic reticulum stress induced by zinc oxide nanoparticles is an earlier biomarker for nanotoxicological evaluation. ACS Nano. 2014;8(3):2562–74.
79. Zhang W, Wang C, Li Z, et al. Unraveling stress-induced toxicity properties of graphene oxide and the underlying mechanism. Adv Mater. 2012;24(39):5391–7.
80. Andon FT, Fadeel B. Programmed cell death: molecular mechanisms and implications for safety assessment of nanomaterials. Acc Chem Res. 2013;46(3):733–42.
81. Foldbjerg R, Olesen P, Hougaard M, Dang DA, Hoffmann HJ, Autrup H. PVP-coated silver nanoparticles and silver ions induce reactive oxygen species, apoptosis and necrosis in THP-1 monocytes. Toxicol Lett. 2009;190(2):156–62.
82. Schaeublin NM, Braydich-Stolle LK, Schrand AM, et al. Surface charge of gold nanoparticles mediates mechanism of toxicity. Nanoscale. 2011;3(2):410–20.
83. Xia T, Kovochich M, Liong M, Zink JI, Nel AE. Cationic polystyrene nanosphere toxicity depends on cell-specific endocytic and mitochondrial injury pathways. ACS Nano. 2008;2(1):85–96.
84. Warheit DB, Laurence BR, Reed KL, Roach DH, Reynolds GA, Webb TR. Comparative pulmonary toxicity assessment of single-wall carbon nanotubes in rats. Toxicol Sci. 2004;77(1):117–25.
85. Muller J, Huaux F, Moreau N, et al. Respiratory toxicity of multi-wall carbon nanotubes. Toxicol Appl Pharmacol. 2005;207(3):221–31.
86. Kedmi R, Ben-Arie N, Peer D. The systemic toxicity of positively charged lipid nanoparticles and the role of toll-like receptor 4 in immune activation. Biomaterials. 2010;31(26):6867–675.
87. Kumar R, Roy I, Ohulchanskky TY, et al. In vivo biodistribution and clearance studies using multimodal organically modified silica nanoparticles. ACS Nano. 2010;4(2):699–708.
88. Lee MJ, Veiseh O, Bhattarai N, et al. Rapid pharmacokinetic and biodistribution studies using cholorotoxin-conjugated iron oxide nanoparticles: a novel non-radioactive method. PLoS One. 2010;5(3):e9536.
89. Goodridge HS, Underhill DM, Touret N. Mechanisms of Fc receptor and dectin-1 activation for phagocytosis. Traffic. 2012;13(8):1062–71.
90. Moghimi SM, Andersen AJ, Ahmadvand D, Wibroe PP, Andresen TL, Hunter AC. Material properties in complement activation. Adv Drug Deliv Rev. 2011;63(12):1000–7.
91. Degn SE, Thiel S. Humoral pattern recognition and the complement system. Scand J Immunol. 2013;78(2):181–93.
92. Szebeni J. Complement activation-related pseudoallergy: a new class of drug-induced acute immune toxicity. Toxicology. 2005;216(2–3):106–21.
93. Moyano DF, Goldsmith M, Solfiell DJ, et al. Nanoparticle hydrophobicity dictates immune response. J Am Chem Soc. 2012;134(9):3965–7.
94. Seong SY, Matzinger P. Hydrophobicity: an ancient damage-associated molecular pattern that initiates innate immune responses. Nat Rev Immunol. 2004;4(6):469–78.
95. Vauthier C, Persson B, Lindner P, Cabane B. Protein adsorption and complement activation for di-block copolymer nanoparticles. Biomaterials. 2011;32(6):1646–56.
96. Owens DE 3rd, Peppas NA. Opsonization, biodistribution, and pharmacokinetics of polymeric nanoparticles. Int J Pharm. 2006;307(1):93–102.
97. Yang Y, Wolfram J, Fang X, Shen H, Ferrari M. Polyarginine induces an antitumor immune response through binding to toll-like receptor 4. Small. 2014;10(7):1250–4.
98. Tamayo I, Irache JM, Mansilla C, Ochoa-Reparaz J, Lasarte JJ, Gamazo C. Poly(anhydride) nanoparticles act as active Th1 adjuvants through toll-like receptor exploitation. Clin Vaccine Immunol. 2010;17(9):1356–62.

99. Holst B, Raby AC, Hall JE, Labeta MO. Complement takes its toll: an inflammatory crosstalk between toll-like receptors and the receptors for the complement anaphylatoxin C5a. Anaesthesia. 2012;67(1):60–4.
100. Ricklin D, Hajishengallis G, Yang K, Lambris JD. Complement: a key system for immune surveillance and homeostasis. Nat Immunol. 2010;11(9):785–97.
101. Moghimi SM, Farhangrazi ZS. Nanomedicine and the complement paradigm. Nanomedicine. 2013;9(4):458–60.
102. Chanan-Khan A, Szebeni J, Savay S, et al. Complement activation following first exposure to pegylated liposomal doxorubicin (Doxil): possible role in hypersensitivity reactions. Ann Oncol. 2003;14(9):1430–7.
103. Szebeni J, Baranyi L, Savay S, et al. Complement activation-related cardiac anaphylaxis in pigs: role of C5a anaphylatoxin and adenosine in liposome-induced abnormalities in ECG and heart function. Am J Physiol Heart Circ Physiol. 2006;290(3):H1050–8.
104. Pichler WJ, Adam J, Daubner B, Gentinetta T, Keller M, Yerly D. Drug hypersensitivity reactions: pathomechanism and clinical symptoms. Med Clin North Am. 2010;94(4):645–64. xv
105. Wen H, Lei Y, Eun SY, Ting JP. Plexin-A4-semaphorin 3A signaling is required for toll-like receptor- and sepsis-induced cytokine storm. J Exp Med. 2010;207(13):2943–57.
106. Dobrovolskaia MA, Patri AK, Zheng J, et al. Interaction of colloidal gold nanoparticles with human blood: effects on particle size and analysis of plasma protein binding profiles. Nanomedicine. 2009;5(2):106–17.
107. Martel J, Young D, Young A, et al. Comprehensive proteomic analysis of mineral nanoparticles derived from human body fluids and analyzed by liquid chromatography-tandem mass spectrometry. Anal Biochem. 2011;418(1):111–25.
108. Lundqvist M, Stigler J, Elia G, Lynch I, Cedervall T, Dawson KA. Nanoparticle size and surface properties determine the protein corona with possible implications for biological impacts. Proc Natl Acad Sci USA. 2008;105(38):14265–70.
109. Zapf S, Loos M. Effect of EDTA and citrate on the functional activity of the first component of complement, C1, and the C1q subcomponent. Immunobiology. 1985;170(3):123–32.
110. Andersen AJ, Robinson JT, Dai H, Hunter AC, Andresen TL, Moghimi SM. Single-walled carbon nanotube surface control of complement recognition and activation. ACS Nano. 2013;7(2):1108–19.
111. Hamad I, Al-Hanbali O, Hunter AC, Rutt KJ, Andresen TL, Moghimi SM. Distinct polymer architecture mediates switching of complement activation pathways at the nanosphere-serum interface: implications for stealth nanoparticle engineering. ACS Nano. 2010;4(11):6629–38.
112. Hanley C, Thurber A, Hanna C, Punnoose A, Zhang J, Wingett DG. The influences of cell type and ZnO nanoparticle size on immune cell cytotoxicity and cytokine induction. Nanoscale Res Lett. 2009;4(12):1409–20.
113. Daly AK. Using genome-wide association studies to identify genes important in serious adverse drug reactions. Annu Rev Pharmacol Toxicol. 2012;52:21–35.
114. Moghimi SM, Wibroe PP, Helvig SY, Farhangrazi ZS, Hunter AC. Genomic perspectives in interindividual adverse responses following nanomedicine administration: the way forward. Adv Drug Deliv Rev. 2012;64(13):1385–93.
115. Schipper ML, Iyer G, Koh AL, et al. Particle size, surface coating, and PEGylation influence the biodistribution of quantum dots in living mice. Small. 2009;5(1):126–34.
116. Kim HR, Andrieux K, Delomenie C, et al. Analysis of plasma protein adsorption onto PEGylated nanoparticles by complementary methods: 2-DE, CE and Protein Lab-on-chip system. Electrophoresis. 2007;28(13):2252–61.
117. Bartucci R, Pantusa M, Marsh D, Sportelli L. Interaction of human serum albumin with membranes containing polymer-grafted lipids: spin-label ESR studies in the mushroom and brush regimes. Biochim Biophys Acta. 2002;1564(1):237–42.
118. Hamad I, Hunter AC, Szebeni J, Moghimi SM. Poly(ethylene glycol)s generate complement activation products in human serum through increased alternative pathway turnover and a MASP-2-dependent process. Mol Immunol. 2008;46(2):225–32.
119. Knop K, Hoogenboom R, Fischer D, Schubert US. Poly(ethylene glycol) in drug delivery: pros and cons as well as potential alternatives. Angew Chem Int Ed Engl. 2010;49(36):6288–308.
120. Hamad I, Hunter AC, Moghimi SM. Complement monitoring of Pluronic 127 gel and micelles: suppression of copolymer-mediated complement activation by elevated serum levels of HDL, LDL, and apolipoproteins AI and B-100. J Control Release. 2013;170(2):167–74.
121. Moein Moghimi S, Hamad I, Bunger R, et al. Activation of the human complement system by cholesterol-rich and PEGylated liposomes-modulation of cholesterol-rich liposome-mediated complement activation by elevated serum LDL and HDL levels. J Liposome Res. 2006;16(3):167–74.
122. Lambris JD, Ricklin D, Geisbrecht BV. Complement evasion by human pathogens. Nat Rev Microbiol. 2008;6(2):132–42.
123. Tian X, Zhu M, Nie G. How can nanotechnology help membrane vesicle-based cancer immunotherapy development? Hum Vaccin Immunother. 2013;9(1):222–5.
124. Zhu M, Tian X, Song X, et al. Nanoparticle-induced exosomes target antigen-presenting cells to initiate Th1-type immune activation. Small. 2012;8(18):2841–8.

125. Liu Y, Jiao F, Qiu Y, et al. Immunostimulatory properties and enhanced TNF-alpha mediated cellular immunity for tumor therapy by C60(OH)20 nanoparticles. Nanotechnology. 2009;20(41):415102.
126. Yang D, Zhao Y, Guo H, et al. [Gd@C(82)(OH)(22)](n) nanoparticles induce dendritic cell maturation and activate Th1 immune responses. ACS Nano. 2010;4(2):1178–86.
127. Cubillos-Ruiz JR, Engle X, Scarlett UK, et al. Polyethylenimine-based siRNA nanocomplexes reprogram tumor-associated dendritic cells via TLR5 to elicit therapeutic antitumor immunity. J Clin Invest. 2009;119(8):2231–44.
128. Fearon DT, Carroll MC. Regulation of B lymphocyte responses to foreign and self-antigens by the CD19/CD21 complex. Annu Rev Immunol. 2000;18:393–422.
129. Reddy ST, van der Vlies AJ, Simeoni E, et al. Exploiting lymphatic transport and complement activation in nanoparticle vaccines. Nat Biotechnol. 2007;25(10):1159–64.
130. Pontes JF, Grenha A. Multifunctional nanocarriers for lung drug delivery. Nano. 2020;10:183.
131. Nabi B, Rehman S, Aggarwal S, Baboota S, Ali J. Nano-based anti-tubercular drug delivery: an emerging paradigm for improved therapeutic intervention. Drug Deliv Transl Res. 2020;10:1111–21.
132. Zhang CY, Lin W, Gao J, et al. pH-responsive nanoparticles targeted to lungs for improved therapy of acute lung inflammation/injury. ACS Appl Mater Interfaces. 2019;11:16380–90.
133. Shi H, Magaye R, Castranova V, Zhao J. Titanium dioxide nanoparticles: a review of current toxicological data. Part Fibre Toxicol. 2013;10:15.
134. Inoue K, Takano H. Aggravating impact of nanoparticles on immune-mediated pulmonary inflammation. Sci World J. 2011;11:382–90.
135. Liao C, Li Y, Tjong SC. Bactericidal and cytotoxic properties of silver nanoparticles. Int J Mol Sci. 2019;20(2):449.
136. Tang W, Fan W, Lau J, Deng L, Shen Z, Chen X. Emerging blood-brain-barrier-crossing nanotechnology for brain cancer theranostics. Chem Soc Rev. 2019;48:2967–3014.
137. He X, Zhang H, Ma Y, et al. Lung deposition and extrapulmonary translocation of nano-ceria after intratracheal instillation. Nanotechnology. 2010;21:285103.
138. Desai N. Challenges in development of nanoparticle-based therapeutics. AAPS J. 2012;14:282–95.
139. Dick CA, Brown DM, Donaldson K, Stone V. The role of free radicals in the toxic and inflammatory effects of four different ultrafine particle types. Inhal Toxicol. 2003;15:39–52.
140. Donaldson K, Stone V, Tran CL, Kreyling W, Borm PJ. Nanotoxicology. Occup Environ Med. 2004;61:727–8.
141. Isama K. In vitro safety evaluation of nanomaterials–cellular response to metal oxide nanoparticles. Yakugaku Zasshi. 2014;134:731–5.
142. Almeida JP, Chen AL, Foster A, Drezek R. In vivo biodistribution of nanoparticles. Nanomedicine (Lond). 2011;6:815–35.
143. Onoue S, Yamada S, Chan HK. Nanodrugs: pharmacokinetics and safety. Int J Nanomedicine. 2014;9:1025–37.
144. Nemmar A, Yuvaraju P, Beegam S, Yasin J, Kazzam EE, Ali BH. Oxidative stress, inflammation, and DNA damage in multiple organs of mice acutely exposed to amorphous silica nanoparticles. Int J Nanomedicine. 2016;11:919–28.
145. Tran CL, Buchanan D, Cullen RT, Searl A, Jones AD, Donaldson K. Inhalation of poorly soluble particles. II. Influence of particle surface area on inflammation and clearance. Inhal Toxicol. 2000;12:1113–26.
146. Couto D, Freitas M, Costa VM, et al. Biodistribution of polyacrylic acid-coated iron oxide nanoparticles is associated with proinflammatory activation and liver toxicity. J Appl Toxicol. 2016;36:1321–31.
147. Teodoro JS, Silva R, Varela AT, et al. Low-dose, subchronic exposure to silver nanoparticles causes mitochondrial alterations in Sprague-Dawley rats. Nanomedicine (Lond). 2016;11:1359–75.
148. Shrivastava R, Raza S, Yadav A, Kushwaha P, Flora SJ. Effects of sub-acute exposure to TiO2, ZnO and Al2O3 nanoparticles on oxidative stress and histological changes in mouse liver and brain. Drug Chem Toxicol. 2014;37:336–47.
149. Khanna P, Ong C, Bay BH, Baeg GH. Nanotoxicity: an interplay of oxidative stress, inflammation and cell death. Nanomaterials (Basel). 2015;5:1163–80.
150. Wang J, Chen C, Liu Y, et al. Potential neurological lesion after nasal instillation of TiO(2) nanoparticles in the anatase and rutile crystal phases. Toxicol Lett. 2008;183:72–80.
151. Ze Y, Zheng L, Zhao X, et al. Molecular mechanism of titanium dioxide nanoparticles-induced oxidative injury in the brain of mice. Chemosphere. 2013;92:1183–9.
152. Manshian BB, Poelmans J, Saini S, et al. Nanoparticle-induced inflammation can increase tumor malignancy. Acta Biomater. 2017;68:99–112.
153. Pandey RK, Prajapati VK. Molecular and immunological toxic effects of nanoparticles. Int J Biol Macromol. 2018;107(Pt A):1278–93.
154. Roy R, Kumar S, Tripathi A, Das M, Dwivedi PD. Interactive threats of nanoparticles to the biological system. Immunol Lett. 2014;158:79–87.
155. Moon C, Park HJ, Choi YH, Park EM, Castranova V, Kang JL. Pulmonary inflammation after intraperitoneal administration of ultrafine titanium dioxide (TiO2) at rest or in lungs primed with lipopolysaccharide. J Toxicol Environ Health Part A. 2010;73:396–409.
156. Sierra MI, Valdes A, Fernandez AF, Torrecillas R, Fraga MF. The effect of exposure to nanoparticles and nanomaterials on the mammalian epigenome. Int J Nanomedicine. 2016;11:6297–306.

157. Shyamasundar S, Ng CT, Yung LY, Dheen ST, Bay BH. Epigenetic mechanisms in nanomaterial-induced toxicity. Epigenomics. 2015;7:395–411.
158. Smolkova B, El Yamani N, Collins AR, Gutleb AC, Dusinska M. Nanoparticles in food. Epigenetic changes induced by nanomaterials and possible impact on health. Food Chem Toxicol. 2015;77:64–73.
159. Stoccoro A, Karlsson HL, Coppede F, Migliore L. Epigenetic effects of nano-sized materials. Toxicology. 2013;313:3–14.
160. Kreyling WG, Semmler-Behnke M, Seitz J, et al. Size dependence of the translocation of inhaled iridium and carbon nanoparticle aggregates from the lung of rats to the blood and secondary target organs. Inhal Toxicol. 2009;21(Suppl. 1):55–60.
161. Elsaesser A, Howard CV. Toxicology of nanoparticles. Adv Drug Deliv Rev. 2012;64:129–37.
162. Cronin JG, Jones N, Thornton CA, Jenkins GJS, Doak SH, Clift MJD. Nanomaterials and innate immunity: a perspective of the current status in nanosafety. Chem Res Toxicol. 2020;33:1061–73. https://doi.org/10.1021/acs.chemrestox.0c00051.
163. Wu T, Tang M. Toxicity of quantum dots on respiratory system. Inhal Toxicol. 2014;26:128–39.
164. Recordati C, De Maglie M, Bianchessi S, et al. Tissue distribution and acute toxicity of silver after single intravenous administration in mice: nano-specific and size-dependent effects. Part Fibre Toxicol. 2016;13:12.
165. Zoroddu MA, Medici S, Ledda A, Nurchi VM, Lachowicz JI, Peana M. Toxicity of nanoparticles. Curr Med Chem. 2014;21:3837–53.
166. Paek HJ, Lee YJ, Chung HE, et al. Modulation of the pharmacokinetics of zinc oxide nanoparticles and their fates in vivo. Nanoscale. 2013;5:11416–27.
167. Shegokar R, Singh KK. Surface modified nevirapine nanosuspensions for viral reservoir targeting: in vitro and in vivo evaluation. Int J Pharm. 2011;421:341–52.
168. Simpson CA, Salleng KJ, Cliffel DE, Feldheim DL. In vivo toxicity, biodistribution, and clearance of glutathione-coated gold nanoparticles. Nanomedicine. 2013;9:257–63.
169. Buzea C, Pacheco II, Robbie K. Nanomaterials and nanoparticles: sources and toxicity. Biointerphases. 2007;2:Mr17–Mr71.
170. Hoshyar N, Gray S, Han H, Bao G. The effect of nanoparticle size on in vivo pharmacokinetics and cellular interaction. Nanomedicine (Lond). 2016;11:673–92.
171. Adjei IM, Sharma B, Labhasetwar V. Nanoparticles: cellular uptake and cytotoxicity. Adv Exp Med Biol. 2014;811:73–91.
172. Manzanares D, Ceña V. Endocytosis: the nanoparticle and submicron nanocompounds gateway into the cell. Pharmaceutics. 2020;12(4):371.
173. Tang H, Feng X, Zhang T, et al. Stability, pharmacokinetics, biodistribution and safety assessment of folate-conjugated pullulan acetate nanoparticles as cervical cancer targeted drug carriers. J Nanosci Nanotechnol. 2015;15:6405–12.
174. Sutunkova MP, Katsnelson BA, Privalova LI, et al. On the contribution of the phagocytosis and the solubilization to the iron oxide nanoparticles retention in and elimination from lungs under long-term inhalation exposure. Toxicology. 2016;363–364:19–28.
175. Platel A, Carpentier R, Becart E, Mordacq G, Betbeder D, Nesslany F. Influence of the surface charge of PLGA nanoparticles on their in vitro genotoxicity, cytotoxicity, ROS production and endocytosis. J Appl Toxicol. 2016;36:434–44.
176. Behzadi S, Serpooshan V, Tao W, et al. Cellular uptake of nanoparticles: journey inside the cell. Chem Soc Rev. 2017;46:4218–44.
177. Papini E, Tavano R, Mancin F. Opsonins and dysopsonins of nanoparticles: facts, concepts, and methodological guidelines. Front Immunol. 2020;11:567365.
178. Nelemans LC, Gurevich L. Drug delivery with polymeric nanocarriers-cellular uptake mechanisms. Materials (Basel). 2020;13(2):366.
179. Patel S, Kim J, Herrera M, Mukherjee A, Kabanov AV, Sahay G. Brief update on endocytosis of nanomedicines. Adv Drug Deliv Rev. 2019;144:90–111.
180. Champion JA, Mitragotri S. Shape induced inhibition of phagocytosis of polymer particles. Pharm Res. 2009;26:244–9.
181. Borbás E, Sinkó B, Tsinman O, et al. Investigation and mathematical description of the real driving force of passive transport of drug molecules from supersaturated solutions. Mol Pharm. 2016;13:3816–26.
182. Cai H, Yao P. Gold nanoparticles with different amino acid surfaces: serum albumin adsorption, intracellular uptake and cytotoxicity. Colloids Surf B Biointerfaces. 2014;123:900–6.
183. Chithrani BD, Ghazani AA, Chan WC. Determining the size and shape dependence of gold nanoparticle uptake into mammalian cells. Nano Lett. 2006;6:662–8.
184. Nativo P, Prior IA, Brust M. Uptake and intracellular fate of surface-modified gold nanoparticles. ACS Nano. 2008;2:1639–44.
185. Verma A, Uzun O, Hu Y, et al. Surface-structure-regulated cell-membrane penetration by monolayer-protected nanoparticles. Nat Mater. 2008;7:588–95.
186. Shang L, Nienhaus K, Nienhaus GU. Engineered nanoparticles interacting with cells: size matters. J Nanobiotechnol. 2014;12:5.
187. Lewinski N, Colvin V, Drezek R. Cytotoxicity of nanoparticles. Small. 2008;4:26–49.
188. Wick P, Manser P, Limbach LK, et al. The degree and kind of agglomeration affect carbon nanotube cytotoxicity. Toxicol Lett. 2007;168:121–31.
189. Limbach LK, Li Y, Grass RN, et al. Oxide nanoparticle uptake in human lung fibroblasts: effects of particle size, agglomeration, and diffusion at low concentrations. Environ Sci Technol. 2005;39:9370–6.
190. Brunner TJ, Wick P, Manser P, et al. In vitro cytotoxicity of oxide nanoparticles: comparison to asbestos,

silica, and the effect of particle solubility. Environ Sci Technol. 2006;40:4374–81.
191. Franklin NM, Rogers NJ, Apte SC, Batley GE, Gadd GE, Casey PS. Comparative toxicity of nanoparticulate ZnO, bulk ZnO, and ZnCl2 to a freshwater microalga (Pseudokirchneriella subcapitata): the importance of particle solubility. Environ Sci Technol. 2007;41:8484–90.
192. Xia T, Kovochich M, Liong M, et al. Comparison of the mechanism of toxicity of zinc oxide and cerium oxide nanoparticles based on dissolution and oxidative stress properties. ACS Nano. 2008;2:2121–34.
193. Karlsson HL, Cronholm P, Gustafsson J, Moller L. Copper oxide nanoparticles are highly toxic: a comparison between metal oxide nanoparticles and carbon nanotubes. Chem Res Toxicol. 2008;21:1726–32.
194. Sayes CM, Gobin AM, Ausman KD, Mendez J, West JL, Colvin VL. Nano-C60 cytotoxicity is due to lipid peroxidation. Biomaterials. 2005;26:7587–95.
195. Long TC, Saleh N, Tilton RD, Lowry GV, Veronesi B. Titanium dioxide (P25) produces reactive oxygen species in immortalized brain microglia (BV2): implications for nanoparticle neurotoxicity. Environ Sci Technol. 2006;40:4346–52.
196. Mohammadpour R, Dobrovolskaia MA, Cheney DL, Greish KF, Ghandehari H. Subchronic and chronic toxicity evaluation of inorganic nanoparticles for delivery applications. Adv Drug Deliv Rev. 2019;144:112–32.
197. Oberdörster G. Safety assessment for nanotechnology and nanomedicine: concepts of nanotoxicology. J Intern Med. 2010;267:89–105.
198. Kane AB, Hurt RH. Nanotoxicology: the asbestos analogy revisited. Nat Nanotechnol. 2008;3:378–9.
199. Lu X, Zhu Y, Bai R, et al. Long-term pulmonary exposure to multi-walled carbon nanotubes promotes breast cancer metastatic cascades. Nat Nanotechnol. 2019;14:719–27.
200. Yousef MI, Mutar TF, Kamel MAE. Hepato-renal toxicity of oral sub-chronic exposure to aluminum oxide and/or zinc oxide nanoparticles in rats. Toxicol Rep. 2019;6:336–46.
201. Pacheco A, Martins A, Guilhermino L. Toxicological interactions induced by chronic exposure to gold nanoparticles and microplastics mixtures in Daphnia magna. Sci Total Environ. 2018;628–629:474–83.
202. Li J, Schiavo S, Xiangli D, et al. Early ecotoxic effects of ZnO nanoparticle chronic exposure in Mytilus galloprovincialis revealed by transcription of apoptosis and antioxidant-related genes. Ecotoxicology. 2018;27:369–84.
203. Oberdörster G, Ferin J, Lehnert BE. Correlation between particle size, in vivo particle persistence, and lung injury. Environ Health Perspect. 1994;102(Suppl. 5):173–9.
204. Arora S, Rajwade JM, Paknikar KM. Nanotoxicology and in vitro studies: the need of the hour. Toxicol Appl Pharmacol. 2012;258:151–65.
205. McClements DJ. Emulsion design to improve the delivery of functional lipophilic components. Annu Rev Food Sci Technol. 2010;1:241–69.
206. Mortezaee K, Najafi M, Samadian H, Barabadi H, Azarnezhad A, Ahmadi A. Redox interactions and genotoxicity of metal-based nanoparticles: a comprehensive review. Chem Biol Interact. 2019;312:108814.
207. Dobrovolskaia MA, Germolec DR, Weaver JL. Evaluation of nanoparticle immunotoxicity. Nat Nanotechnol. 2009;4:411–4.
208. Ravindran S, Suthar JK, Rokade R, et al. Pharmacokinetics, metabolism, distribution and permeability of nanomedicine. Curr Drug Metab. 2018;19:327–34.
209. Maier MA, Jayaraman M, Matsuda S, et al. Biodegradable lipids enabling rapidly eliminated lipid nanoparticles for systemic delivery of RNAi therapeutics. Mol Ther. 2013;21:1570–8.
210. Schubert J, Chanana M. Coating matters: review on colloidal stability of nanoparticles with biocompatible coatings in biological media, living cells and organisms. Curr Med Chem. 2018;25:4553–86.
211. Wang X, Xia T, Ntim SA, et al. Dispersal state of multiwalled carbon nanotubes elicits profibrogenic cellular responses that correlate with fibrogenesis biomarkers and fibrosis in the murine lung. ACS Nano. 2011;5:9772–87.
212. Schlenoff JB. Zwitteration: coating surfaces with zwitterionic functionality to reduce nonspecific adsorption. Langmuir. 2014;30:9625–36.
213. Pombo García K, Zarschler K, Barbaro L, et al. Zwitterionic-coated "stealth" nanoparticles for biomedical applications: recent advances in countering biomolecular corona formation and uptake by the mononuclear phagocyte system. Small. 2014;10:2516–29.
214. Shen CR, Wu ST, Tsai ZT, et al. Characterization of quaternized chitosantem. Ionic-coated "stealth" nanoparticles forel potential magnetic resonance imaging contrast agent for cell tracking. Polym Int. 2011;60:945–50.
215. De Volder MF, Tawfick SH, Baughman RH, Hart AJ. Carbon nanotubes: present and future commercial applications. Science. 2013;339:535–9.
216. Tang S, Tang Y, Zhong L, et al. Short- and long-term toxicities of multi-walled carbon nanotubes in vivo and in vitro. J Appl Toxicol. 2012;32:900–12.
217. Boyles MS, Young L, Brown DM, et al. Multi-walled carbon nanotube induced frustrated phagocytosis, cytotoxicity and pro-inflammatory conditions in macrophages are length dependent and greater than that of asbestos. Toxicol In Vitro. 2015;29:1513–28.
218. Wang X, Xia T, Duch MC, et al. Pluronic F108 coating decreases the lung fibrosis potential of multiwall carbon nanotubes by reducing lysosomal injury. Nano Lett. 2012;12:3050–61.
219. Mutlu GM, Budinger GR, Green AA, et al. Biocompatible nanoscale dispersion of single-

walled carbon nanotubes minimizes in vivo pulmonary toxicity. Nano Lett. 2010;10:1664–70.
220. Rao GT, Babu B, Stella RJ, Manjari VP, Ravikumar RV. Spectral investigations on undoped and Cu(2)(+) doped ZnO-CdS composite nanopowders. Spectrochim Acta A Mol Biomol Spectrosc. 2015;139:86–93.
221. Adeleye AS, Pokhrel S, Madler L, Keller AA. Influence of nanoparticle doping on the colloidal stability and toxicity of copper oxide nanoparticles in synthetic and natural waters. Water Res. 2018;132:12–22.
222. Ahmad J, Siddiqui MA, Akhtar MJ, et al. Copper doping enhanced the oxidative stress-mediated cytotoxicity of TiO2 nanoparticles in A549 cells. Hum Exp Toxicol. 2018;37:496–507.
223. George S, Pokhrel S, Xia T, et al. Use of a rapid cytotoxicity screening approach to engineer a safer zinc oxide nanoparticle through iron doping. ACS Nano. 2010;4:15–29.
224. Sun B, Pokhrel S, Dunphy DR, et al. Reduction of acute inflammatory effects of fumed silica nanoparticles in the lung by adjusting Silanol display through calcination and metal doping. ACS Nano. 2015;9:9357–72.
225. Xia T, Zhao Y, Sager T, et al. Decreased dissolution of ZnO by iron doping yields nanoparticles with reduced toxicity in the rodent lung and zebrafish embryos. ACS Nano. 2011;5:1223–35.
226. Naatz H, Lin S, Li R, et al. Safe-by-design CuO nanoparticles via Fe-doping, Cu-O bond length variation, and biological assessment in cells and zebrafish embryos. ACS Nano. 2017;11:501–15.
227. Yang G, Phua SZF, Bindra AK, Zhao Y. Degradability and clearance of inorganic nanoparticles for biomedical applications. Adv Mater. 2019;31:e1805730.
228. Yuan Y, Ding J, Xu J, Deng J, Guo J. TiO2 nanoparticles co-doped with silver and nitrogen for antibacterial application. J Nanosci Nanotechnol. 2010;10:4868–74.
229. Teoh WY, Amal R, Madler L. Flame spray pyrolysis: an enabling technology for nanoparticles design and fabrication. Nanoscale. 2010;2:1324–47.
230. Vance ME, Kuiken T, Vejerano EP, et al. Nanotechnology in the real world: redeveloping the nanomaterial consumer products inventory. Beilstein J Nanotechnol. 2015;6:1769–80.
231. Liu J, Feng X, Wei L, Chen L, Song B, Shao L. The toxicology of ion-shedding zinc oxide nanoparticles. Crit Rev Toxicol. 2016;46:348–84.
232. Wang D, Lin Z, Wang T, et al. Where does the toxicity of metal oxide nanoparticles come from: the nanoparticles, the ions, or a combination of both? J Hazard Mater. 2016;308:328–34.
233. Hola K, Markova Z, Zoppellaro G, Tucek J, Zboril R. Tailored functionalization of iron oxide nanoparticles for MRI, drug delivery, magnetic separation and immobilization of biosubstances. Biotechnol Adv. 2015;33:1162–76.
234. Mout R, Moyano DF, Rana S, Rotello VM. Surface functionalization of nanoparticles for nanomedicine. Chem Soc Rev. 2012;41:2539–44.
235. Paramasivam G, Kayambu N, Rabel AM, Sundramoorthy AK, Sundaramurthy A. Anisotropic noble metal nanoparticles: synthesis, surface functionalization and applications in biosensing, bioimaging, drug delivery and theranostics. Acta Biomater. 2017;49:45–65.
236. Nicol JR, Dixon D, Coulter JA. Gold nanoparticle surface functionalization: a necessary requirement in the development of novel nanotherapeutics. Nanomedicine (Lond). 2015;10:1315–26.
237. Blanco E, Shen H, Ferrari M. Principles of nanoparticle design for overcoming biological barriers to drug delivery. Nat Biotechnol. 2015;33:941–51.
238. Li R, Wang X, Ji Z, et al. Surface charge and cellular processing of covalently functionalized multiwall carbon nanotubes determine pulmonary toxicity. ACS Nano. 2013;7:2352–68.
239. Nel AE, Madler L, Velegol D, et al. Understanding biophysicochemical interactions at the nano-bio interface. Nat Mater. 2009;8:543–57.
240. Atale SS, Dyawanapelly S, Jagtap DD, Jain R, Dandekar P. Understanding the nano-bio interactions using real-time surface plasmon resonance tool. Int J Biol Macromol. 2019;123:97–107.
241. Cai P, Zhang X, Wang M, Wu YL, Chen X. Combinatorial nano-bio interfaces. ACS Nano. 2018;12:5078–84.
242. Kettiger H, Schipanski A, Wick P, Huwyler J. Engineered nanomaterial uptake and tissue distribution: from cell to organism. Int J Nanomedicine. 2013;8:3255–69.
243. Zarska M, Novotny F, Havel F, et al. TwoStep mechanism of cellular uptake of cationic gold nanoparticles modified by (16-mercaptohexadecyl) trimethylammonium bromide. Bioconjug Chem. 2016;27:2558–74.
244. Goodman CM, McCusker CD, Yilmaz T, Rotello VM. Toxicity of gold nanoparticles functionalized with cationic and anionic side chains. Bioconjug Chem. 2004;15:897–900.
245. Voinov MA, Sosa Pagan JO, Morrison E, Smirnova TI, Smirnov AI. Surface-mediated production of hydroxyl radicals as a mechanism of iron oxide nanoparticle biotoxicity. J Am Chem Soc. 2011;133:35–41.
246. Thomas R, Park IK, Jeong YY. Magnetic iron oxide nanoparticles for multimodal imaging and therapy of cancer. Int J Mol Sci. 2013;14:15910–30.
247. Wu W, He Q, Jiang C. Magnetic iron oxide nanoparticles: synthesis and surface functionalization strategies. Nanoscale Res Lett. 2008;3:397–415.
248. Cardoso VF, Francesko A, Ribeiro C, Banobre-Lopez M, Martins P, Lanceros-Mendez S. Advances in magnetic nanoparticles for biomedical applications. Adv Healthc Mater. 2018;7(5) https://doi.org/10.1002/adhm.201700845.

249. Schütz CA, Juillerat-Jeanneret L, Mueller H, Lynch I, Riediker M. Therapeutic nanoparticles in clinics and under clinical evaluation. Nanomedicine (Lond). 2013;8:449–67.
250. Anselmo AC, Mitragotri S. Nanoparticles in the clinic: an update. Bioeng Transl Med. 2019;4:e10143.
251. Pease DF, Peterson BA, Gilles S, Hordinsky MK, Bohjanen KA, Skubitz KM. Development of cutaneous squamous cell carcinoma after prolonged exposure to pegylated liposomal doxorubicin and hand-foot syndrome: a newly recognized toxicity. Cancer Chemother Pharmacol. 2019;84:217–21.
252. Kozma GT, Shimizu T, Ishida T, Szebeni J. Anti-PEG antibodies: properties, formation, testing and role in adverse immune reactions to PEGylated nano-biopharmaceuticals. Adv Drug Deliv Rev. 2020;154–155:163–75.
253. Hwang R, Mirshafiee V, Zhu Y, Xia T. Current approaches for safer design of engineered nanomaterials. Ecotoxicol Environ Saf. 2018;166:294–300.
254. Nakamura H, Jun F, Maeda H. Development of next-generation macromolecular drugs based on the EPR effect: challenges and pitfalls. Expert Opin Drug Deliv. 2015;12:53–64.
255. Fang J, Liao L, Yin H, et al. Photodynamic therapy and imaging based on tumor-targeted nanoprobe, polymer-conjugated zinc protoporphyrin. Future Sci OA. 2015;1(3):FSO4.
256. Stendahl JC, Sinusas AJ. Nanoparticles for cardiovascular imaging and therapeutic delivery, part 1: compositions and features. J Nucl Med. 2015;56:1469–75.
257. Surana S, Shenoy AR, Krishnan Y. Designing DNA nanodevices for compatibility with the immune system of higher organisms. Nat Nanotechnol. 2015;10:741.
258. Denkhaus E, Salnikow K. Nickel essentiality, toxicity, and carcinogenicity. Crit Rev Oncol Hematol. 2002;42(1):35–56.
259. Hansen SF. The emergence of nano risk-immunity. Front Nanosci Nanotechnol. 2016;2(3):131–4.
260. Shatkin JA. The future in nanosafety. Nano Lett. 2020;20:1479–80. https://doi.org/10.1021/acs.nanolett.0c00432.
261. Lubick N. Nanosilver toxicity: ions, nanoparticles or both? Environ Sci Technol. 2008;42(23):8617.
262. Wiesner MR, Lowry GV, Jones KL, et al. Decreasing uncertainties in assessing environmental exposure, risk, and ecological implications of nanomaterials. Environ Sci Technol. 2009;43(17):6458–62.
263. De Pablo JJ, Jones B, Kovacs CL, Ozolins V, Ramirez AP. The materials genome initiative, the interplay of experiment, theory and computation. Curr Opin Solid State Mater Sci. 2014;18:99–117. https://doi.org/10.1016/j.cossms.2014.02.003.
264. Schmidt J, Marques MRG, Botti S, Marques MAL. Recent advances and applications of machine learning in solid-state materials science. NPJ Comput Mater. 2019;5:83. https://doi.org/10.1038/s41524-019-0221-0.
265. Schleder GR, Padilha ACM, Acosta CM, Costa M, Fazzio A. From DFT to machine learning: recent approaches to materials science–a review. J Phys Mater. 2019;2:032001. https://doi.org/10.1088/2515-7639/ab084b.
266. Liu C, Chen H, Wang S, et al. Two-dimensional materials for next-generation computing technologies. Nat Nanotechnol. 2020;15:545–57. https://doi.org/10.1038/s41565-020-0724-3.
267. Tawfik SA, Isayev O, Stampfl C, Shapter J, Winkler DA, Ford MJ. Efficient prediction of structural and electronic properties of hybrid 2d materials using complementary DFT and machine learning approaches. Adv Theory Simul. 2019;2:1800128. https://doi.org/10.1002/adts.201800128.
268. Giusti A, Atluri R, Tsekovska R, et al. Nanomaterial grouping: existing approaches and future recommendations. NanoImpact. 2019;16:100182. https://doi.org/10.1016/j.impact.2019.100182.
269. Basei G, Hristozov D, Lamon L, et al. Making use of available and emerging data to predict the hazards of engineered nanomaterials by means of in silico tools: a critical review. NanoImpact. 2019;13:76–99. https://doi.org/10.1016/j.impact.2019.01.003.
270. Karcher S, Willighagen EL, Rumble J, et al. Integration among databases and data sets to support productive nanotechnology: challenges and recommendations. NanoImpact. 2018;9:85–101. https://doi.org/10.1016/j.impact.2017.11.002.
271. Lynch I, Afantitis A, Leonis G, Melagraki G, Valsami-Jones E. Strategy for identification of nanomaterials' critical properties linked to biological impacts: interlinking of experimental and computational approaches. In: Advances in QSAR modeling. Cham: Springer; 2017. p. 385–424.
272. Singh AV, Rosenkranz D, Ansari MHD, et al. Artificial intelligence and machine learning empower advanced biomedical material design to toxicity prediction. Adv Intell Syst. 2020;2:2000084. https://doi.org/10.1002/aisy.202000084.
273. Afantitis A, Melagraki G, Isigonis P, et al. Nanosolveit project: driving nanoinformatics research to develop innovative and integrated tools for in silico nanosafety assessment. Comput Struct Biotechnol J. 2020;18:583–602. https://doi.org/10.1016/j.csbj.2020.02.023.
274. Haase A, Klaessig F. Eu US roadmap nanoinfor matics 2030. EU Nanosafety Cluster; 2018. 127p. https://doi.org/10.5281/zenodo.1486012.
275. Cui Q, Hernandez R, Mason SE, Frauenheim T, Pedersen JA, Geiger F. Sustainable nanotechnology: opportunities and challenges for theoretical/ computational studies. J Phys Chem B. 2016;120:7297–306. https://doi.org/10.1021/acs.jpcb.6b03976.
276. Winkler DA. Role of artificial intelligence and machine learning in nanosafety. Small. 2020;16:2001883. https://doi.org/10.1002/smll.202001883.
277. Lynch I, Afantitis A, Exner T, et al. Can an InChi for nano address the need for a simplified representa-

tion of complex nanomaterials across experimental and nanoinformatics studies? Nano. 2020;10:1–44. https://doi.org/10.3390/nano10122493.
278. Rajan K. Nanoinformatics: data-driven materials design for health and environmental needs. In: Nanotechnology environmental health and safety. William Andrew Publishing; 2014. p. 173–98.
279. Muratov EN, Bajorath J, Sheridan RP, et al. QSAR without borders. Chem Soc Rev. 2020;49:3525–64. https://doi.org/10.1039/d0cs00098a.
280. Winkler DA. Recent advances, and unresolved issues, in the application of computational modelling to the prediction of the biological effects of nanomaterials. Toxicol Appl Pharmacol. 2016;299:96–100. https://doi.org/10.1016/j.taap.2015.12.016.
281. Trinh TX, Ha MK, Choi JS, Byun HG, Yoon TH. Curation of datasets, assessment of their quality and completeness, and nanoSAR classification model development for metallic nanoparticles. Environ Sci Nano. 2018;5:1902–10. https://doi.org/10.1039/c8en00061a.
282. Choi J-S, Ha MK, Trinh TX, Yoon TH, Byun H-G. Towards a generalized toxicity prediction model for oxide nanomaterials using integrated data from different sources. Sci Rep. 2018;8:6110. https://doi.org/10.1038/s41598-018-24483-z.
283. Gajewicz A, Puzyn T, Odziomek K, et al. Decision tree models to classify nanomaterials according to the DF4nanoGrouping scheme. Nanotoxicology. 2018;12:1–17. https://doi.org/10.1080/17435390.2017.1415388.
284. Feng R, Yu F, Xu J, Hu X. Knowledge gaps in immune response and immunotherapy involving nanomaterials: databases and artificial intelligence for material design. Biomaterials. 2021;266:120469. https://doi.org/10.1016/j.biomaterials.2020.120469.
285. Burello E. A mechanistic model for predicting lung inflammogenicity of oxide nanoparticles. Toxicol Sci. 2017;159:339–53. https://doi.org/10.1093/toxsci/kfx136.
286. Ban Z, Yuan P, Yu F, Peng T, Zhou Q, Hu X. Machine learning predicts the functional composition of the protein Corona and the cellular recognition of nanoparticles. Proc Natl Acad Sci USA. 2020;117:10492–9. https://doi.org/10.1073/pnas.1919755117.
287. Quan X, Liu J, Zhou J. Multiscale modeling and simulations of protein adsorption: progresses and perspectives. Curr Opin Colloid Interface Sci. 2019;41:74–85. https://doi.org/10.1016/j.cocis.2018.12.004.
288. Duan Y, Coreas R, Liu Y, et al. Prediction of protein Corona on nanomaterials by machine learning using novel descriptors. NanoImpact. 2020;17:100207. https://doi.org/10.1016/j.impact.2020.100207.
289. Alsharif SA, Power D, Rouse I, Lobaskin V. In silico prediction of protein adsorption energy on titanium dioxide and gold nanoparticles. Nano. 2020;10:1–21. https://doi.org/10.3390/nano10101967.
290. Findlay MR, Freitas DN, Mobed-Miremadi M, Wheeler KE. Machine learning provides predictive analysis into silver nanoparticle protein Corona formation from physicochemical properties. Environ Sci Nano. 2018;5:64–71. https://doi.org/10.1039/C7EN00466D.
291. Le TC, Yin H, Chen R, et al. An experimental and computational approach to the development of ZnO nanoparticles that are safe by design. Small. 2016;12:3568–77. https://doi.org/10.1002/smll.201600597.
292. Mikolajczyk A, Gajewicz A, Mulkiewicz E, et al. Nano-QSAR modeling for ecosafe design of heterogeneous TiO_2-based nano-photocatalysts. Environ Sci Nano. 2018;5:1150–60. https://doi.org/10.1039/C8EN00085A.
293. Puzyn T, Rasulev B, Gajewicz A, et al. Using nano-QSAR to predict the cytotoxicity of metal oxide nanoparticles. Nat Nanotechnol. 2011;6:175–8. https://doi.org/10.1038/nnano.2011.10.
294. Wang W, Sedykh A, Sun H, et al. Predicting nano-bio interactions by integrating nanoparticle libraries and quantitative nanostructure activity relationship modeling. ACS Nano. 2017;11:12641–9. https://doi.org/10.1021/acsnano.7b07093.
295. Martins C, Dreij K, Costa PM. The state-of-the art of environmental toxicogenomics: challenges and perspectives of "omics" approaches directed to toxicant mixtures. Int J Environ Res Public Health. 2019;16:1–16. https://doi.org/10.3390/ijerph16234718.
296. Peng T, Wei C, Yu F, et al. Predicting nanotoxicity by an integrated machine learning and metabolomics approach. Environ Pollut. 2020;267:115434. https://doi.org/10.1016/j.envpol.2020.115434.
297. Serra A, Fratello M, Cattelani L, et al. Transcriptomics in toxicogenomics, part III: data modelling for risk assessment. Nano. 2020;10:708. https://doi.org/10.3390/nano10040708.
298. Ahmad F, Mahmood A, Muhmood T. Machine learning-integrated omics for the risk and safety assessment of nanomaterials. Biomater Sci. 2021;9:1598–608. https://doi.org/10.1039/D0BM01672A.
299. Kinaret PAS, Ndika J, Ilves M, et al. Toxicogenomic profiling of 28 nanomaterials in mouse airways. Adv Sci. 2021;8(10):2004588. https://doi.org/10.1002/advs.202004588.
300. Singh AV, Ansari MHD, Rosenkranz D, et al. Artificial intelligence and machine learning in computational nanotoxicology: unlocking and empowering nanomedicine. Adv Healthc Mater. 2020;9:1901862. https://doi.org/10.1002/adhm.201901862.
301. Murugadoss S, Das N, Godderis L, Mast J, Hoet PH, Ghosh M. Identifying nanodescriptors to predict the toxicity of nanomaterials: a case study on titanium dioxide. Environ Sci Nano. 2021;8(2):580–90. https://doi.org/10.1039/D0EN01031F.

302. Papadiamantis AG, Jänes J, Voyiatzis E, et al. Predicting cytotoxicity of metal oxide nanoparticles using Isalos analytics platform. Nano. 2020;10:1–19. https://doi.org/10.3390/nano10102017.
303. Milosevic A, Romeo D, Wick P. Understanding nanomaterial biotransformation: an unmet challenge to achieving predictive nanotoxicology. Small. 2020;16:e1907650. https://doi.org/10.1002/smll.201907650.
304. Papadiamantis AG, Klaessig FC, Exner TE, et al. Metadata stewardship in nanosafety research: community-driven organisation of metadata schemas to support fair nanoscience data. Nano. 2020;10:2033. https://doi.org/10.3390/nano10102033.
305. Martinez DST, Da Silva GH, de Medeiros AMZ, Khan LU, Papadiamantis AG, Lynch I. Effect of the albumin Corona on the toxicity of combined graphene oxide and cadmium to daphnia magna and integration of the datasets into the NanoCommons Knowledge Base. Nanomaterials (Basel). 2020;10:1936. https://doi.org/10.3390/nano10101936.
306. Worth A, Aschberger K, Asturiol D, et al. Evaluation of the availability and applicability of computational approaches in the safety assessment of nanomaterials. Luxembourg: Publications Office of the European Union; 2017. https://doi.org/10.2760/248139.
307. Varsou DD, Afantitis A, Tsoumanis A, et al. A safe-by-design tool for functionalized nanomaterials through the Enalos nanoinformatics cloud platform. Nanoscale Adv. 2019;1:706–18. https://doi.org/10.1039/c8na00142a.
308. Gazzi A, Fusco L, Orecchioni M, et al. Graphene, other carbon nanomaterials and the immune system: toward nanoimmunity-by-design. J Phys Mater. 2020;3:034009. https://doi.org/10.1088/2515-7639/ab9317.
309. Clift MJ, Raemy DO, Endes C, et al. Can the Ames test provide an insight into nano-object mutagenicity? Investigating the interaction between nano-objects and bacteria. Nanotoxicology. 2013;7(8):1373–85.
310. Ng CT, Li JJ, Bay BH, Yung LY. Current studies into the genotoxic effects of nanomaterials. J Nucleic Acids. 2010;2010:ID947859.
311. Worle-Knirsch JM, Pulskamp K, Krug HF. Oops they did it again! Carbon nanotubes hoax scientists in viability assays. Nano Lett. 2006;6(6):1261–8.
312. Oberdorster G. Safety assessment for nanotechnology and nanomedicine: concepts of nanotoxicology. J Intern Med. 2010;267(1):89–105.
313. Long TC, Tajuba J, Sama P, et al. Nanosize titanium dioxide stimulates reactive oxygen species in brain microglia and damages neurons in vitro. Environ Health Perspect. 2007;115(11):1631–7.
314. Oberdörster G, Yu CP. The carcinogenic potential of inhaled diesel exhaust: a particle effect? J Aerosol Sci. 1990;21(Supplement 1(0)):S397–401.
315. Takagi A, Hirose A, Nishimura T, et al. Induction of mesothelioma in p53+/− mouse by intraperitoneal application of multi-wall carbon nanotube. J Toxicol Sci. 2008;33(1):105–16.
316. Corathers SD, Falciglia M. The role of hyperglycemia in acute illness: supporting evidence and its limitations. Nutrition. 2011;27(3):276–81.
317. Lam CW, James JT, McCluskey R, Hunter RL. Pulmonary toxicity of single-wall carbon nanotubes in mice 7 and 90 days after intratracheal instillation. Toxicol Sci. 2004;77(1):126–34.
318. Ge C, Li Y, Yin J-J, et al. The contributions of metal impurities and tube structure to the toxicity of carbon nanotube materials. NPG Asia Mater. 2012;4:e32.
319. Ge C, Meng L, Xu L, et al. Acute pulmonary and moderate cardiovascular responses of spontaneously hypertensive rats after exposure to single-wall carbon nanotubes. Nanotoxicology. 2012;6(5):526–42.
320. Joris F, Manshian BB, Peynshaert K, De Smedt SC, Braeckmans K, Soenen SJ. Assessing nanoparticle toxicity in cell-based assays: influence of cell culture parameters and optimized models for bridging the in vitro-in vivo gap. Chem Soc Rev. 2013;42(21):8339–59.

GPSR Compliance

The European Union's (EU) General Product Safety Regulation (GPSR) is a set of rules that requires consumer products to be safe and our obligations to ensure this.

If you have any concerns about our products, you can contact us on ProductSafety@springernature.com

In case Publisher is established outside the EU, the EU authorized representative is:

Springer Nature Customer Service Center GmbH
Europaplatz 3
69115 Heidelberg, Germany

Batch number: 10756384

Printed by Printforce, the Netherlands